D1708836

Introduction to Livestock and Companion Animals

Fifth Edition

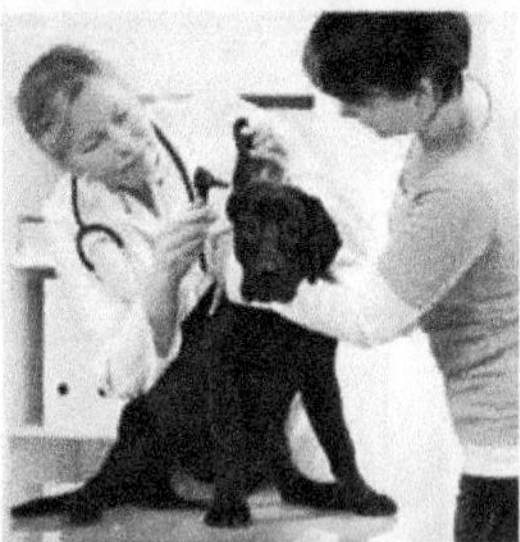

JASPER S. LEE

Agricultural Educator
Clarkesville, Georgia

JIM HUTTER

Missouri State University
Springfield

RICK RUDD

Virginia Tech
Blacksburg

LYLE WESTROM

University of Minnesota
Crookston

AMANDA PATRICK-HEFNER

Oconee County High School
Georgia

Introduction to

LIVESTOCK & COMPANION ANIMALS

Fifth Edition

Jasper S. Lee — Series Editor

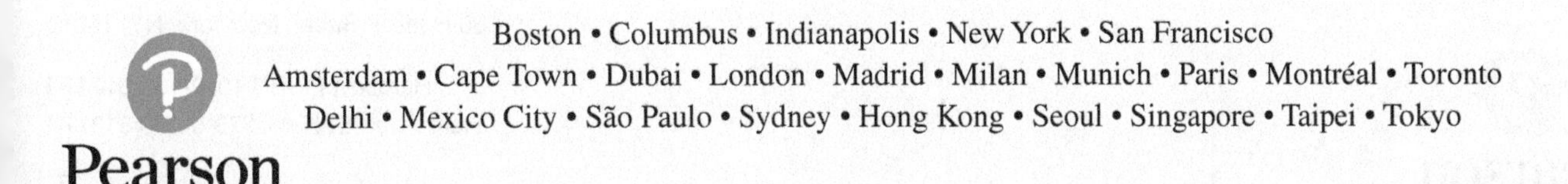

330 Hudson Street, New York, NY 10013

Hardcover ISBN 10: 0-13-451317-7

Hardcover ISBN 13: 978-0-13-451317-1

2 17

Preface

Animals! Animals are all about us. Animals help us in so many essential ways. Our animals, and the products they provide, have changed so much in recent years. Likewise, the animal industry is changing. These changes are having radical impacts on education in animal care and production. A traditional book is no longer appropriate. ***Introduction to Livestock and Companion Animals,* Fifth Edition**, represents the innovative approach used by the best teachers and students in the United States. The focus is on serving the needs of students and teachers. New approaches in the animal industry are undergirded with science principles, including veterinary technology. This book puts the science together with the human social situations and issues faced in animal production and care.

Standards in the Agriculture, Food and Natural Resources (AFNR) Career Cluster are increasingly important to teachers and students. This book embraces the standards set forth for the Animal Systems pathway within the AFNR cluster. The performance elements, performance indicators, and measurement statements of the national standards are integrated into this book. Some, of course, are more advanced than an introductory text. Regardless, the foundation provided will serve students well as they move forward with their study of the Animal Systems pathway.

***Introduction to Livestock and Companion Animals*, Fifth Edition**, has been revised to serve as an introductory textbook in animal science and related industry, including veterinary technology. It is about far more than cattle, swine, and chickens, though these are included in more detail than before. It integrates a multitude of perspectives on animal education. Basics of biology are included. Veterinary technology practices are introduced. Practical connections are addressed in each chapter. Biotechnology has been expanded. Content to meet state standards in biological science and animal courses has been enhanced.

The content on issues in the animal industry is a strong contribution to the book. Never before have so many issues faced animal producers and keepers. This book introduces and expands on these issues. You will want to view this book as your key to future success in the animal industry. Think about it: What is an issue? What are the major emerging issues in animal agriculture?

Traditional areas of animal agriculture are continued in this edition. Chapters present beef, swine, sheep, dairy, horse, and poultry production. Other chapters cover aquatic animals; draft animals; dogs; cats; birds, rodents, and reptiles; and service, laboratory, and exotic animals. Of course, content on career and personal development is included. The content is designed to ensure success for students with limited backgrounds in agriculture. Throughout, a student-friendly approach has been used. Chapter design and content are based on research with students and teachers. Line art and photographs clarify and enhance important concepts. The writing level is appropriate for the intended audience. Type face and size have been selected to enhance interest and readability. The content is presented in a systematic manner.

Introduction to Livestock and Companion Animals, **Fifth Edition**, has much to offer! The major features are full-color design, hundreds of modern and youthful photographs and drawings, science-based content, modern animal subjects, high appeal to users, and carefully written copy.

Whether you take a quick look or do a thorough analysis, you will readily see that it is the introductory animal education book for you!

Jasper S. Lee
Series Editor

Acknowledgments

The authors of ***Introduction to Livestock and Companion Animals,* Fifth Edition,** are grateful to many individuals for their assistance. These persons made it possible for the authors to realize the goal of a revised and redirected book on animals.

Particularly acknowledged for assistance with this edition of the book are several individuals who focus on large and small animal care: Rachel Eddleman, DVM, Georgia; Nancy S. Jackson, DVM, Mississippi; Karen Knarr, DVM, Tennessee; Suzanne Sheldon, DVM, Georgia; Larry Anthony, DVM, Mississippi; Rob Knarr, DVM, Tennessee; Linda Brook, DVM, Georgia; Glenn Farrar, DVM, Georgia; and Melissa Farrar, DVM, Georgia.

Associations and businesses acknowledged here are:

- Red Wagon Enterprises, Farmington, Illinois
- Northeast Veterinary Hospital, Cornelia, Georgia
- Allflex USA, Inc., Dallas / Fort Worth Airport, Texas
- The Vet at Blue Ridge, Watkinsville, Georgia
- Elite Genetics, Waukon, Iowa
- Harris Farms, Coalinga, California
- National Hereford Hog Record Association, Flandreau, South Dakota
- Embrex Incorporated, Research Triangle Park, North Carolina
- Washington Department of Fisheries Hatchery, Wilapa, Washington
- National Swine Registry, West Lafayette, Indiana
- East Ridge Animal Hospital, Chattanooga, Tennessee
- Paso Fino Horse Association, Inc., Plant City, Florida

- Maddox Dairy, Fresno, California
- BarLink Ranch, Madras, Oregon
- College of Veterinary Medicine, Texas A&M University, College Station, Texas
- School of Veterinary Medicine, Purdue University
- American Breeders Service, DeForest, Wisconsin
- Texas Department of Agriculture, Austin, Texas
- American Rambouillet Sheep Breeders Association, San Angelo, Texas
- The American Donkey and Mule Society, Denton, Texas
- Red Bluff Ranch, Thermopolis, Wyoming
- Luck-E-G Nubians, Blanchard, Idaho
- Lucky Three Ranch, Inc., Loveland, Colorado

High schools recognized for their assistance are:

- Switzerland County High School, Vevay, Indiana
- North Decatur High School, Greensburg, Indiana
- Edmond High School, Edmond, Oklahoma
- Elk Grove High School, Elk Grove, California
- Franklin Community School, Franklin, Indiana
- Millsaps Career Center, Starkville, Mississippi
- Yukon High School, Yukon, Oklahoma
- Florin High School, Florin, California
- Oconee County High School, Watkinsville, Georgia
- Pelion High School, Pelion, South Carolina
- Ponderosa High School, Shingle Springs, California
- Thief River Falls High School, Thief River Falls, Minnesota
- Eastern Randolph High School, Ramseur, North Carolina
- Florin High School, Sacramento, California
- Morgan County High School, Madison, Georgia
- Sandra D. O'Connor High School, Helotes, Texas
- James Madison High School, San Antonio, Texas
- Lemoore High School, Lemoore, California

- Central West High School, Fresno, California
- Monticello High School, Monticello, Illinois
- Snyder High School, Snyder, Texas
- Tri-Valley High School, Downs, Illinois
- Chase High School, Forest City, North Carolina
- Park High School, Livingston, Montana
- Alvirne High School, Hudson, New Hampshire
- Franklin County High School, Carnesville, Georgia
- Paxton-Buckley-Loda High School, Paxton, Illinois
- Clovis High School, Clovis, California
- Fleming County High School, Flemingsburg, Kentucky
- Coral Reef High School, Miami, Florida
- Turner Technical High School, Miami, Florida
- East Mississippi Community College, Scooba, Mississippi

Special acknowledgment goes to manuscript reviewers Robert C. Albin, Texas Tech University; Richard Joerger, University of Minnesota; and Tim Chamblee, Mississippi State University. Also acknowledged here are Chris Embry Mohr, Illinois; Austin Bull, North Carolina; and Jody Pollok, Michigan, for their assistance with previous editions.

The assistance of students, staff, and friends of Piedmont College, Demorest, Georgia, is acknowledged. Their help with illustrations and design enhanced the appeal of the book.

Individuals due special acknowledgment in this revision of *Introduction to Livestock and Companion Animals* are Peter McCarthy, President of Emergent Learning, and Jennifer Frew, Editorial Director of Emergent Learning. Their leadership and support throughout the revision process are greatly appreciated.

Contents

Unit One—Animals and People

Unit Two—Animal Biology and Technology

Unit Three—Food Animal Technology

Unit Four—Pleasure and Draft Animal Technology

UNIT ONE

(Courtesy, auremar/Shutterstock)

Animals and People

CHAPTER 1

Animals in Our Lives

OBJECTIVES

This chapter provides basic information on the animal industry. It has the following objectives:

1. Explain the meaning of the animal industry.
2. Identify ways animals help people.
3. Identify trends in the animal industry.
4. Trace the domestication of animals.
5. Describe the role of science in animal production.
6. List the names of common animals based on sex classification and age.
7. Explain the meaning of animal well-being.
8. Identify environmental concerns associated with animal production.
9. Describe the small animal industry.

TERMS

animal domestication
animal industry
animal marketing
animal processing
animal production
animal selection
animal services
animal shelter
animal supplies
animal well-being
botany
by-product
castration
companion animal
dairy cattle
environment
euthanasia
health
industrial farm
life science
livestock
meat animal
mohair
neutering
poultry
reproduction
science
spaying
veterinary medicine
zoology

1–1. Dogs are the favorite companion animals in the United States. (Courtesy, Andresr/Shutterstock)

ANIMALS are important to people. We benefit from them in so many ways. Animal care and production form a large, exciting industry. This industry has many species and varies widely in practices. A few quick examples are: Cattle are raised on Western rangelands and other places. Pets are about everywhere. Poultry are raised in modern houses. Horses are kept on green, grassy pastures and in fine barns. Some of the animals we enjoy are wildlife. Animals are everywhere!

Animal products have so many important uses. People often think of food and clothing as uses of animal products. Uses also include power, companionship, research, medicine, safety, and service. All of these and many other uses are important to people. Think of yourself: How is your life influenced by animals? You may have a favorite dog, cat, or bird that brings pleasure to your life. We always treat animals to promote their well-being.

The animal industry has many good career opportunities. There are jobs in a wide range of animal areas. Some jobs focus on animal care and production. Others deal with gaining important products from animals. Whatever your interests, you will need knowledge and skill to be successful. You can learn a lot about animals from this book.

THE ANIMAL INDUSTRY

The ***animal industry*** is all of the activities in raising animals and meeting the needs people have for animals and animal products. It includes many steps that make modern production possible. The animal industry assures people of quality products. It uses practices that provide proper care of animals. The animal industry is made of three major areas: production, supplies and services, and marketing and processing.

ANIMAL PRODUCTION

Animal production is raising animals for food, companionship, and other uses. How animals are produced has changed. Livestock are raised on farms or ranches, often on a large scale. Other animals may be raised in homes, aquaria, pens, and other small-scale facilities. A good example is the use of kennel facilities to raise dogs.

1–2. Cows, as ruminants, convert pasture forages into meat and milk for human food. (Courtesy, Viorel Sima/ Shutterstock)

Livestock

Livestock are mammals produced on farms and ranches for food and other purposes. They are often given special care to assure good health and growth. Common livestock include cattle, swine (hogs), sheep, horses, and goats. The number of goats has increased markedly in the United States over the last decade. Animals now being produced more widely include llamas, bison, and elk.

Table 1-1. Number and Value of Selected Species on U.S. Farms and Ranches*

Species	Number of Animals	Number of Producers	Value ($)
Cattle (beef and dairy)	92.6 million	935,000	78.2 billion
Hogs and Pigs	64.6 million	69,000	6.8 billion
Sheep and Lambs	5.35 million	na	761 million
Goats (meat, milk, and Angora)	2.86 million**	152,000	na

*Statistical Abstract of the United States reflecting January 1, 2012, data. (accessed July 2, 2012, at www.nass.usda.gov/Publicatons/Todays_Reports/reports.shep0112.pdf.)

**Of these, 172,000 were Angora goats and 360,000 were milk goats. The trend has been a decrease in Angora goats and an increase in milk and meat goats.

Companion Animals

1–3. People of all ages and cultures enjoy animals, such as this boy riding a goat. (Courtesy, Lucian Coman/Shutterstock)

A ***companion animal*** is an animal that is used to provide humans with fun and friendship through close association. Companion animals are often called pets. Common companion animals include dogs, cats, fish, rabbits, hamsters, gerbils, ferrets, and snakes. Some companion animals also provide service such as dogs for people who are visually impaired.

Poultry

Poultry is a group of fowl (birds) that are raised for use as food and other products such as feathers. Both meat and eggs are produced by poultry. Special kinds of feed, facilities, and care are used to keep poultry healthy and growing. Common kinds of poultry are chickens, turkeys, ducks, and geese. New species that are being farmed include quail, ostrich, and emu.

Other Animals

Many kinds of animals are produced; some grow naturally, as is the case with wildlife animals. Food fish, laboratory animals, and exotic animals are included. These

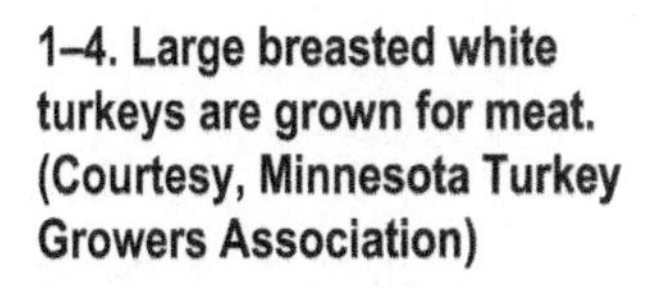

1–4. Large breasted white turkeys are grown for meat. (Courtesy, Minnesota Turkey Growers Association)

1–5. Elk are popular wildlife found in many parts of the United States, particularly the Rocky Mountain areas. (Courtesy, Wesley Aston/Shutterstock)

animals help people in a variety of ways. Wildlife animals, such as bear, moose, bison, fox, elk, and turtles, enrich the lives of humans but are not typically "farmed." Many of the animals are covered in other chapters of this book.

SUPPLIES AND SERVICES

Keeping and producing animals requires supplies and services. With livestock, some supplies are produced on the farm or ranch, such as hay for feed. Most supplies come from off the farm.

Animal suppliles are the inputs needed in animal production. They are typically obtained from an agribusiness that provides animal supplies. These include feed mills and factories that make the equipment needed to raise animals. Dealers make them available in the local area. Common supplies include manufactured feed, cages and feeding equipment, animal medicines, fencing material, and hauling equipment.

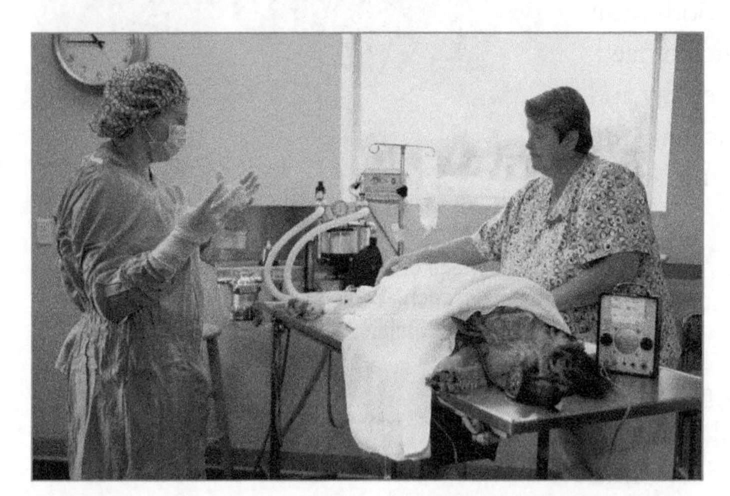

1–6. Animal owners often seek the services of a veterinarian. (This dog will have surgery to remove a tumor.) (Courtesy, Education Images)

Animal services include the help that producers need to raise animals efficiently. Veterinarians and veterinary technicians help assure good animal health. Farriers are people skilled in put-ting shoes on horses. People skilled in shearing sheep (clipping off the wool) provide the service to wool producers. People in animal services may go from one ranch or kennel to another providing help.

MARKETING AND PROCESSING

When animals or their products are ready for the consumer, they are marketed. ***Animal marketing*** is the process of moving animals and their products from the producer to the consumer. It includes many functions. With companion animals, it may include getting animals into stores for customers to buy. With food animals, it involves preparing the animal products for consumption.

Animal producers often have choices about marketing. With food animals, several marketing alternatives exist. Each approach is intended to help the producer make a profit. Four marketing examples are included here:

- Cash marketing–Cash marketing is getting paid for animals at the time they are sold. This may involve delivery to a livestock auction or sold at private sale to another producer, feedlot, processing facility, or other outlet. Producers are typically paid the prices that are prevalent at the time of sale or as agreed upon by buyer and seller. In some cases, the animals or products (eggs, milk, etc.) are sold directly to consumers. The sale of one bull by a cattle producer to another cattle producer is a private sale usually for cash. Cash marketing has somewhat declined in recent years.
- Forward contracting–A marketing arrangement in which the producer and the buyer enter into an agreement before the animal product is produced is known as forward contracting. This approach helps guarantee the producer a market outlet and helps guarantee the buyer (often a processing facility) a supply of animals and/or products. Both the producer and the buyer know the standards that the animals must meet and the price that the buyer will pay the producer (quality assurance standards are often involved to assure a high-grade animal).
- Commodity marketing–A commodity is a basic product that is in original form or has had only primary processing. Beef, pork, and other animal products are traded in commodity markets. An agricultural commodity market is a market that trades in primary goods that are intended for additional manufacture, such as live cattle and hogs. These markets have historically been more prominent with corn, wheat, soybeans, cotton, sugar, coffee, and cocoa. Some people use futures contracts to invest in commodities. The procedures are often more detailed than can be negotiated by an untrained individual; local brokers may be used. The largest commodity exchange is the Chicago Mercantile Exchange (CME). Futures contracts may be traded: A futures contract is an agreement to take or make delivery of a commodity in the future at a specified price. Futures trading is done by a commodity exchange, such as the CME. Internationally, the U.S. is a major trading nation in animal products. Beef, pork, poultry, and other animal products are traded as commodities. Beef, for example, has been exported to Japan, Canada, South Korea, and other nations. The U.S. has imported beef from Australia, New Zealand, Nicaragua, and other nations. Lamb, pork, and other animal products are also imported, with New Zealand lamb being particularly popular in some restaurants.
- Vertical integration–In economics terms, vertical integration is an arrangement in which a company owns the supply chain of production. A good example in the animal industry is broiler production. The processing company will own feed mills, hatcheries, and processing equipment. Agreements will be signed with growers to provide a growing house and the labor to yield birds that are of desired size and quality for harvest. The processor will

likely have field supervisors who visit the poultry farms and specify how the birds are raised to assure the desired product. Trucks owned by the integrator will deliver feed with appropriate nutrients for the stage of growth of the birds. At time of harvest, the integrator (processor) will send a harvest team to load the birds onto trucks owned by the processor. Following processing, the integrator will market products to supermarkets, restaurants, and other outlets, including international markets. With vertical integration, the processor assumes most of the risk associated with an enterprise.

Preparing animal products usually involves processing. ***Animal processing*** is preparing animals or their products for the consumer. What is done varies with the product. Pork is one example. The swine are processed into the desired pieces of meat. These cuts are delivered to a local grocery store. The store may place the meat in attractive packages for displaying in a refrigerated case. Milk and egg products go through a similar marketing process.

All animal products are carefully packaged and stored to assure good quality. Meat, eggs, and milk are inspected by trained government officials and industry personnel. Products that are not wholesome are rejected and are not used for food.

WAYS ANIMALS HELP PEOPLE

Most people are well aware of common animal products. The hamburger, turkey sandwich, and fish filet are foods that we enjoy eating. Animals are used for more than food. They fill important roles as pets, in sports, and to do work.

In processing, very little of an animal is wasted. Many parts of an animal's carcass that were formerly wasted are now used. Some medicines are made from certain animal organs and tissues.

These materials are known as by-products. A ***by-product*** is a product made from the parts of animals that are not used for food, such as bone meal for fertilizer.

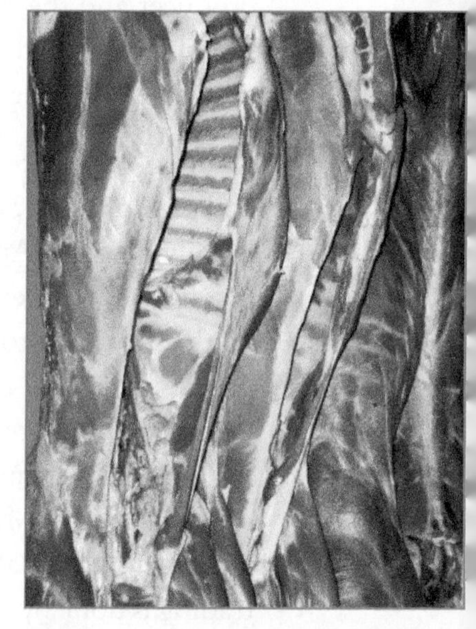

1–7. Hanging lamb carcasses are being chilled in a processing plant. (Courtesy, Steve Lovegrove/Shutterstock)

FOOD

Animals provide many foods that people enjoy. These foods are high in nutrients and help us live healthy lives. Without good food, people are not healthy and do not grow as they should.

Foods from animals primarily include meat, milk, and eggs. Some animals give us more than one kind of food product, such as chickens that provide both meat and eggs.

1–8. Meat is a popular food in the United States. (Courtesy, Monkey Business Images/Shutterstock)

Table 1-2. Common Food Products From Animals

Animal	Food Item Name
Meat	
Cattle	
Younger than 3 months	Veal
Older animals	Beef
Swine (hogs)	Pork
Sheep	
Young (less than 1 year)	Lamb
Older (older than 1 year)	Mutton
Goat	Goat mutton
Chicken	
Young (less than 12 weeks)	Broiler
Neutered young male	Capon
Old hen	Hen
Turkey	Turkey
Fish	Fish
Milk	
Dairy cattle	Milk
Goats	Goat's milk
Eggs	
Chickens	Eggs
Fish	Caviar

We get meat from animals we raise and from wild animals. A ***meat animal*** is an animal raised for its meat. The animals that we raise for food include cattle, fish, turkeys, chickens, swine, and sheep. Horses are used as meat animals in some countries.

Wild animals used for food include deer, rabbit, quail, and fish. Wild animals are not included as livestock. They are known as game or wildlife. Game is wildlife hunted for food or other uses. The meat from animals is known by different names, such as beef from cattle and pork from swine.

Milk is primarily from cattle. Cattle specially grown to produce milk are known as ***dairy cattle***. Goats and a few other animals are sometimes milked.

Eggs are primarily from chickens. A few other species may produce eggs for human food, including guineas and ducks. In some instances, people enjoy fish eggs, known as caviar.

CLOTHING

1–9. Shirts, coats, ties, shoes, belts, jackets, and other clothing may be made from animal fibers, skins, or hides. (Courtesy, Dmitry Kalinovsky/Shutterstock)

Clothing is made from many different animal by-products. In addition, some animals are raised specifically for products to make clothing. These include mink for their fur and certain breeds of sheep raised primarily for wool.

Clothing may be made from animal skin (hide) or hair that grows on the skin. Bones, antlers, and other animal parts may be used. The skin of animals is sometimes known as leather. Cattle, sheep, kangaroo, and snakes are a few animals whose skin is used in making belts, shoes, jackets, and other items. The hair or fleece from several different animals is used to make clothing, such as wool from sheep and fur from mink. A special quality clothing is made from **mohair**, a product of angora goats. Feathers from geese may be used in stuffing pillows. These feathers are known as down.

The most valuable animal product used in clothing is the pearl. Often used in jewelry, the pearl is formed by an oyster when a grain of sand or another foreign body gets inside its shell. Some oysters are grown to produce pearls, known as cultured pearls.

COMPANIONSHIP AND PLEASURE

Companionship and pleasure are important to people. A companion animal is an animal that provides benefits that help people enjoy life. We sometimes call them "pets." Examples include dogs, cats, and ornamental fish. Horses and ponies are examples of animals kept as pets that are too big to stay inside a home. Some animals are used for sporting events, such as horse racing and goat roping. Many different kinds of pleasure animals are found in North America. Dog and horse racing, rodeoing, and similar events are recreational events that involve animals.

1–10. People of all ages enjoy pets. (Courtesy, Ilike/Shutterstock)

SERVICE

Service animals are animals that assist people in living and work. They are used in many ways and may be given special training. Dogs may be used to lead visually impaired people, herd sheep, or guard property. Some animals, such as rats and monkeys, are carefully used in laboratories for finding new medicines to cure human and animal diseases. Dogs in police canine units are helpful against crime. Horses and mules are used as draft animals. They are strong and have the ability to pull heavy loads.

TRENDS IN THE ANIMAL INDUSTRY

Having the animals and animal products we want involves a number of different activities. How we produce animals has changed greatly. Animal producers use supplies and services to help animals grow. Once food animals and their products have been produced, they must be made available to the consumer.

LARGER PRODUCERS

The nature of animal production has become larger and more specialized. There are fewer producers today in North America. A half century ago, many people raised animals for food

CONNECTION

EASY-TO-USE PRODUCTS

Many animal products are prepared into easy-to-use forms. Just go to a local supermarket to see for yourself. Shoppers have many choices to make...beef, pork, chicken, turkey, and fish. All of these choices are due to animal quality and product development.

Easy-to-use products are sometimes known as convenience foods. This means that they are ready-to-eat or can be quickly prepared. In short, they are convenient to use.

The shopper shown here is selecting a chicken breast sandwich product. This product has been cooked, seasoned with lemon pepper, hickory smoked, and thinly-sliced, making it just right for sandwiches. A product in this form is far easier to use than trying to cook and slice the chicken breast yourself! (Courtesy, Education Images)

purposes. Many homes had a few chickens and a cow or so for milk. Nearly every farm was diversified and kept a few animals in addition to crops. Times have changed!

Today, livestock are raised on specialized farms or ranches that are often quite large. New approaches are being used to grow them. Only one kind of animal may be grown. It is becoming unusual to find a large-scale farm that has more than one kind of animal.

INDUSTRIAL FARMS

An ***industrial farm*** is a large-scale corporate-type operation that efficiently produces animals in carefully controlled environments. The animals are often confined in houses specially designed to provide for their well-being. Industrial farming is sometimes known as factory farming but that word has grown in disfavor. Factory farming conjures up the notion that animals are packed into abusive environments. Though the animals could be in such situations, most producers do not abusive their animals.

Hog (swine) production has rapidly moved to an industrial farm approach. One producer may have a large operation that produces far more hogs than several smaller farms. The hogs are managed for maximum growth and efficient use of production inputs.

Many corporations grow for a specific market. You will likely see some of these products in your local grocery store, such as a brand label on poultry.

SOME SMALL PRODUCERS REMAIN

Many small producers are still involved with animals. Service animals, such as draft horses, are often raised on small farms. Companion animals are raised by pet growers or individuals who raise them for a hobby. Animals used in laboratories are often carefully raised under close supervision so that quality standards are met. Many people who have a few acres of land are involved with small-scale livestock production.

1–11. This modern hog facility at the University of Tennessee–Martin bears little resemblance to pig pens of the past. (Courtesy, Education Images)

DEMAND FOR QUALITY

Consumers are increasingly concerned about the quality of animal products. Greater emphasis is on providing products that are wholesome. Consumers are also concerned about the conditions under which animals are raised. Some consumers are willing to pay more for animals that have not been produced in factory-type situations.

DOMESTICATION OF ANIMALS

1–12. Exotic animals may be on display in zoological parks. Here a koala lives in a naturalized habitat at the San Diego Zoo. (The koala [*Phascolarctos cinereus*] is a herbivorous marsupial that is a native of Australia.) (Courtesy, Education Images)

Animals have not always been tame. Some are still wild today. Raising animals began hundreds of years ago. People found that it was easier to tend them than to hunt them in the wild. People also found that they could do a better job of raising the animals if they understood their needs.

Animal domestication is taking animals from nature and keeping them in a controlled environment. It involves taming wild animals and helping them adapt to being raised by humans.

Domestication is more than taming just one or two animals. Domestication involves a gradual process with an entire population or all members of a species. A mutually beneficial relationship develops between humans and the animals. An individual wild animal can be tamed though this is not the same as domestication. In some cases, domestication appears to result in animals becoming accustomed to human presence and bonding with humans—maybe even with changes in genetic make-up. Occasionally, a domesticated animal will revert to living in wild conditions. Swine, horses, cats, and dogs are examples. Domesticated animals that revert to the wild are known as feral animals. Such animals are big problems in some places.

Many years ago, humans were hunters and gatherers. They hunted for meat and eggs and gathered food from wild plants. Humans then began to have gardens near their home and raise animals for convenience.

Domestication is important in the history of how human societies developed. Domestication began thousands of years ago with goats, donkeys, and a few other animals. People are striving to domesticate new species today, such as catfish, alligators, and elk. The dog was one of the first companion animals to be domesticated.

Animal and plant domestication occurred pretty much at the same time. To some extent, the domestication of animals depended on plant domestication (as with corn, which is a major animal feed).

Some of the earliest animals domesticated were: dogs (10,000 or so years BC in an undetermined location), sheep (8,500 years BC in western Asia), cats (8,500 years BC western Asia and nearby area), chickens (6,000 BC in Asia), guinea pigs (5,000 BC in the Andes Mountains), and horses (3,600 BC in Kazakhstan).

Some of the earliest plants domesticated include: fig trees (9,000 BC in Near East), barley (8,500 BC in Near East), rice (8,000 BC in Asia), and maize (corn) (7,000 BC in Central America).

SCIENCE AND TECHNOLOGY IN ANIMAL PRODUCTION

Successfully producing animals requires education. The animal industry depends on people educated in many areas to produce the animals we need. Science and technology have important roles in education about animal production.

SCIENCE AREAS

Science is knowledge about the world we live in. It deals with facts and relationships among the facts. People who have an education in science are better animal producers. Animal science is the special application of basic science in the production of animals.

Areas of life science are very important in successful animal production. ***Life science*** is about living things, and includes ***zoology***, which is the study of animals. The study includes the structure and functions of animals and their parts. It also includes studying how to keep them healthy by meeting their needs.

Another important life science that is indirectly related to animals is botany. ***Botany*** is the study of plants. Since most animals eat plants or plant products, knowing how to produce plants is essential.

Technology is the use of science in the work and lives of people. Producers of animals use technology about animals in their work. Every time a disease is treated or an animal fed, some form of technology is used. The technology could be in the design of equipment used, the way the treatment is given, or in other ways.

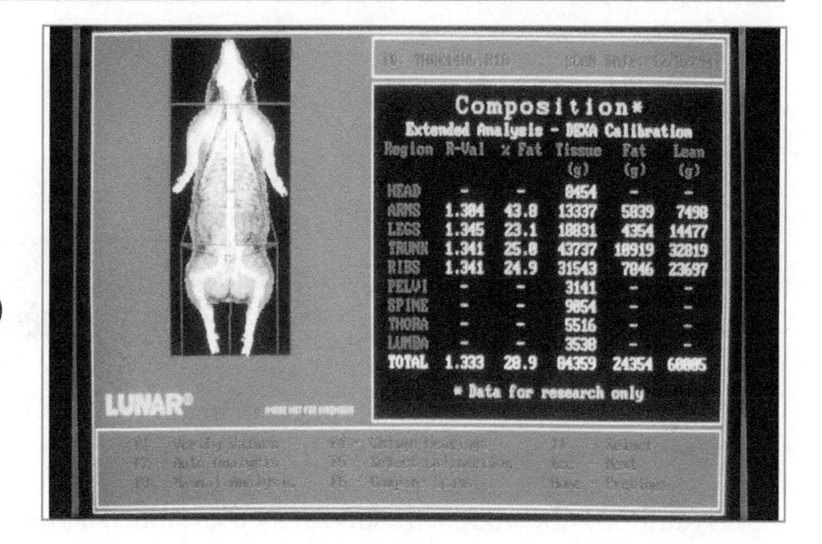

1–13. A technology known as DEXA scan has been used of a live pig to disclose fat and lean tissue measurements. (DEXA stands for "Dual Energy X-ray Absorptiometry.") (Courtesy, Agricultural Research Service, USDA)

AGRISCIENCE

Agriscience is the use of science in producing animals and plants. The plants include flowers, grain crops, and forests. Some of the plants are used as food for animals. Many areas of science are involved. Producers need education in basic areas of science. Thousands of high schools have classes in agriscience to help people learn about animal production.

ANIMAL SCIENCE AND TECHNOLOGY

Science and technology are important in the production of animals. Information from many areas is used in the animal industry. The most widely used information areas are in animal selection, nutrition, animal health, reproduction, and environment.

Animal Selection

Animal selection is choosing animals to achieve desired goals. Animal production begins with selection. Animals and humans share many similarities. Animals are individuals, each with its own personality, body type, and rate of growth. Carefully select the animals to raise. Choose those that fill a need in some unique way. After all, you want a demand to exist for the animal you raise!

Breeding animals are carefully selected for their traits. This is because scientists know that the offspring of animals resemble their parents. If parents have a certain trait, their offspring are highly like to have the trait.

Animal Nutrition

A major role of food animal production is to convert feed into human food. Some animals, such as cattle and sheep, can use bulky feed materials for growth and maintenance of their bodies. Humans have no use for grass and hay as a raw material, but can consume the "hay" when it is converted to beef by a steer.

1–14. Hogs have been improved so that pork is sometimes referred to as "white meat." (Courtesy, Education Images)

Nutrition is the study of the kind of food an animal needs and how it is used by the animal. Much research has been done to learn the nutritional needs of animals. Several areas of science are involved in helping animal producers provide for the needs of animals.

There are many considerations that must be taken into account when preparing a ration for an animal. A ration is an animal's diet. It is what the animal consumes each day. The importance of having a proper ration is illustrated here with a lamb. Lambs require certain nutrients in order to grow. Their digestive systems can only handle certain kinds of feed. A lamb will not grow well on a diet of only hay.

Animal Health

Health is the condition in which an animal is free of disease and all parts of the animal are functioning properly. Many different diseases can attack animals if not prevented. Animal producers can take steps to keep animals healthy.

1–15. Dorset rams being conditioned for breeding season. This requires good nutrition, positive health care, and attention to reproductive potential. (Courtesy, Education Images)

Veterinary medicine is the area that deals with the health and well-being of animals. Veterinarians are people trained in methods of preventing and treating disease and other animal health problems. They often have assistants and technicians who can provide valuable help.

Producers know that it is important to prevent disease. Once an animal gets a disease, it is often very difficult to successfully provide a treatment. The best approach is to keep animals healthy and avoid disease problems. Regular vaccination can be used to prevent some diseases. Having a good place for animals to live helps to keep them healthy. Owners of companion animals should also keep their animals healthy through preventative medicine and care.

Animal Reproduction

Animals must reproduce themselves. ***Reproduction*** is the process by which offspring are produced. People who raise animals need to understand the reproduction process. Large numbers of offspring are often desired.

Efficiently having new animals is essential for most animal producers. For example, a hog producer wants piglets to be born at a certain time and in good health condition. The producer also wants the females to bare as many pigs as practical. This requires careful attention to the reproductive processes in hogs.

People who have companion animals and other animals may want to limit reproduction, such as a family with a cat. They may want to keep animals from producing offspring. This can be achieved with knowledge of the reproduction process.

Animal Environment

A good environment is essential for an animal to survive and grow properly. ***Environment*** is the surroundings of an animal. Some animals require shelter; others can survive in outside conditions without shelter. Producers need to know the environment required and manage animals within the environment.

A good example of the importance of environment is with chicken production. Small, newly-hatched chicks must be in a protected house. The house must have controlled temperature and humidity. Equipment in the house must provide food and water. The chicks must also be protected from predators and other hazards.

ANIMAL CLASSIFICATION

Animals are classified and named by species, age, and sexual condition. This approach is important in correctly identifying animals. The common name of an animal is the most popular way of identification. For example, we refer to canines as "dogs."

The age of an animal gives another distinction. Mature animals and immature animals have different names. For example, a puppy is a young dog.

The sexual condition of an animal is important. Animals are either male or female. Some animals are altered sexually, known as ***neutering***. Neutering is done to keep an animal from reproducing or to get other desired traits. For example, neutering may increase the rate of growth and quality of meat a beef animal produces. In males, neutering involves removing the testicles, which is known as ***castration***. With females, neutering is known as spaying. ***Spaying*** is removing the ovaries or cutting the fallopian tubes so eggs cannot enter the uterus where they may be fertilized and develop as an embryo and fetus.

Tables 1-3 and 1-4 summarize important terms for animals based on their species, age, and sexual condition.

Table 1-3. Classification of Livestock Based on Age and Sexual Condition

Category	Cattle	Goat	Sheep	Hog	Chicken	Turkey
Name	Bovine	Caprine	Ovine	Porcine	Galine	Meleagris
Mature Male	Bull	Buck	Ram	Boar	Rooster	Tom
Mature Female	Cow	Doe	Ewe	Sow	Hen	Hen
Young Male	Bull	Buck Kid	Ram Lamb	Shoat	Cockerel	Tom Poult
Young Female	Heifer	Doeling	Ewe Lamb	Gilt	Pullet	Hen Poult
Altered Male	Steer	Wether	Wether	Barrow	Capon	—
Altered Female	Spayed	Spayed	Spayed	Spayed	—	—
Newborn	Calf	Kid	Lamb	Pig	Chick	Poult
Group	Herd	Herd	Flock	Drove	Flock	Flock

Table 1-4. Classification of Selected Companion Animals Based on Age and Sexual Condition

Category	Dog	Cat	Horse	Guinea Pig	Rabbit
Name	Canine	Feline	Equine	Cavy	Lagomorph
Mature Male	Stud	Tom	Stallion	Male Cavy	Buck
Mature Female	Bitch	Queen	Mare	Female Cavy	Doe
Young Male	Intact	—	Colt	—	—
Young Female	Bitch	—	Filly	—	—
Altered Male	Neuter	Gib	Gelding	—	—
Altered Female	Spay	Spay	Spayed	—	—
Newborn	Puppy	Kitten	Foal	Kit	Kit
Group	Pack	Bevy	Herd	Group	Group

ANIMAL WELL-BEING

1–16. Providing for the needs of animals is essential in their well-being. (Courtesy, Education Images)

1–17. These dairy cows receive excellent care on this Minnesota farm. (Courtesy, Education Images)

Animal well-being is caring for an animal so that its needs are met and it does not suffer. The needs of animals vary depending on their species. Producers need to study the natural conditions under which animals grow best and use that information in caring for their needs.

Animals should not be confined in places or ways that threaten their well-being. Some people refer to this as humane treatment of animals. This applies to producers of animals for food as well as to people who have companion and other animals. For example, leaving a dog in a parked car on a hot day is cruel and may lead to disease or death.

Most producers take steps that assure that an animal has a far better environment than would be present if living in the wild. Using vaccinations to prevent disease is definitely a positive step in animal well-being. Providing proper feed and water is also another positive practice in animal well-being. Facilities that provide protection from a harsh environment are also important. Steps are taken to prevent suffering such as using anesthetic prior to surgery. Animals should never be whipped and beaten. Never leave an animal tied up longer than necessary. Protect animals from bad weather.

Animals are sometimes injured or diseased. They may be suffering from pain or disability associated with the injury or disease. Sometimes animals get struck by cars when they get in a road. People face a dilemma about what to do. They know that there is no hope for the badly injured animal. The animal should likely be euthanatized. ***Euthanasia*** is killing an animal that is suffering. Euthanasia provides an easy and painless death for an animal. Euthanasia is sometimes controversial. Only trained, qualified individuals should practice euthanasia.

Mention of animal well-being usually includes two other areas: animal welfare and animal rights. These two areas often involve human emotion. Laws have been enacted related to these two areas. The laws often have difficulty distinguishing between animal welfare and animal rights. Each deals with how animals are treated. Each also sometimes gets tangled with legal questions about animal justice. Therefore, the distinction is sometimes a matter of interpretation.

ANIMAL WELFARE

Animal welfare is using animals for human purposes that sacrifice the notion that animals have interests. The interests of animals can be traded away. Good justification must exist for humans to sacrifice animals but the criteria used in arriving at a justification vary. All uses of animals should be made humanely and without subjecting the animal to unnecessary pain.

The Federal government, states, and local governments have regulations about animal welfare. Title 7, Chapter 54, of the United States Code contains legal regulations on the transportation, sale, and handling of certain animals. The regulations include companion animals, lab animals, performing animals (such as in a circus), fighting animals, and others. (Use this Web site for more information on Federal laws: **http://www.animal-law.org**)

In 2007, the U.S. Congress passed the Animal Fighting Prohibition Enforcement Act of 2005. This law amended the Federal criminal code relating to birds, dogs, and other animals in interstate commerce. It prohibits selling, buying, transporting, delivering, or receiving animals for the purpose of a fighting venture. It was signed by the President on May 3, 2007. Birds, particularly trained roosters, have been involved in cockfighting. Dogs have been involved in highly-publicized and cruel fighting activities. Felony penalties are provided for those convicted under provisions of the Law.

ANIMAL RIGHTS

Animal rights is the notion that animals have rights like humans. These rights cannot be traded away or sacrificed simply because good consequences may result. People who are animal rights proponents feel that when humans exploit animals, they should consider their rights. Federal and state laws on animal cruelty often address animal rights. In some cases, issues associated with animal rights may wind up in courts of law.

People sometimes don't understand the nature of gaining animal productivity. Animal producers must perform certain management practices if their animals are to efficiently provide the products preferred by consumers. Some organizations speak out about the ways animals are raised. A campaign against some common animal production practices refers to these as "animal mutilation." The Compassion in World Farming organization opposes practices such as tail docking of baby pigs, debeaking of chickens, and castration of calves and lambs.

Many practices are highly useful to today's producers. Every producer should respect and properly care for animals. There is no place in animal production for cruelty. Proper care is the way producers gain productivity. Notions of animal rights may sometimes conflict with acceptable production practices. There are no easy solutions.

ANIMAL ETHICS

Animal ethics is caring for animals to assure their well-being and practicing appropriate standards in relating to other animal owners. Ethics applies to animals used in lab research as well as production, showing, and other uses.

The International Association of Fairs and Expositions has established a National Code of Show Ring Ethics. These guide livestock shows throughout the United States. Some of the ethics relate to animal well-being; others cover honest relationships among people. The Code prohibits the use of certain drugs and surgical procedures, devices that strike or shock animals, and other dishonest and inhumane activities.

ENVIRONMENTAL CONCERNS WITH ANIMAL PRODUCTION

We must be aware of the potential problems animals can cause in our environment. This is true if raised as companions, or for meat, fur, fun, or other reasons.

All animals create waste. Some occasionally die or become diseased. This creates a need to dispose of the dead animal. Animals produce body wastes that require disposal. Large farms create much waste. The wastes can contaminate water and cause health problems. Animals can cause soil erosion, make an area aesthetically displeasing, and create unpleasant odors.

Permits are needed for some types of animal production. Other areas have zoning laws that restrict animal production. Landlords may restrict keeping companion animals in rental property. Leash laws and permit regulations should always be followed.

DISPOSING OF ANIMAL WASTE

When animals defecate and urinate, the waste products must be disposed of properly. Wastes allowed to remain in an area can attract insects and become a source of disease. Animal housing must be cleaned regularly to prevent disease. Animal waste that is allowed to drain into the ground or into a water source can contaminate surface and/or ground water and cause health problems for wildlife and eventually people.

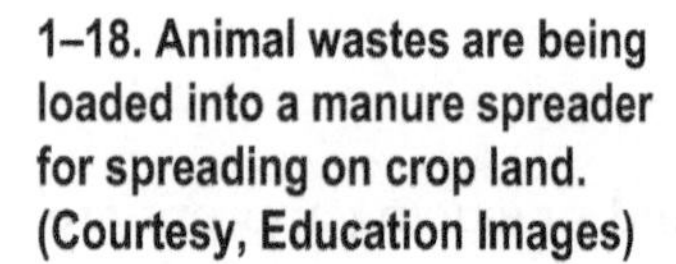

1–18. Animal wastes are being loaded into a manure spreader for spreading on crop land. (Courtesy, Education Images)

Even the waste produced by small animals must be disposed of properly. Although the animal is small, the waste produced is proportionate with their size. Some small animals can be trained to use litter boxes so that waste management is easier. Small animals need to have their cages cleaned regularly to keep them healthy!

1–19. A compost bin is used on this poultry farm to dispose of dead birds and other wastes. (Courtesy, Education Images)

Waste from animal facilities should be disposed of without contaminating soil or water. Many animal producers use animal waste as fertilizer for crops. Small animal waste can be composted. However you dispose of your animals' waste you must have a good plan to do it properly!

OTHER ENVIRONMENTAL CONCERNS

Large animals can create erosion and cause an area to become unsightly. For example, hogs on pasture may root their snouts in the ground looking for food. They can tear up an area of turf with little effort. When left unattended, these areas often become mud pits and are not very pleasing to look at. This also causes soil erosion that can pollute surface water and cause the loss of topsoil. Large animals that wade in water or cross creeks loosen soil and cause it to erode. This degrades water.

1–20. A lagoon waste management system is used at this hog production facility. (The lagoon prevents contamination of nearby streams and other resources.) (Courtesy, Natural Resources Conservation Service, USDA)

SMALL ANIMAL INDUSTRY IN SOCIETY

The unique roles of small animals in society create a special section in the animal industry for separate study. Many people think of the small animal industry as dealing with pets. But, it deals with more than pets and the term companion animal is more appropriate than pet. These animals develop a special bond with their owners and are often kept in their owner's home.

Many times people think that the small animal industry deals with dogs and cats but it is much more: birds, snakes, iguanas, guinea pigs, hamsters, gerbils, and rabbits are a few additional examples. No doubt, you can think of other examples of small animals kept as companion animals in your local community.

As covered earlier in this chapter, small animals are used in so many ways in society: service such as guide dogs, research such as new medical products, produce products such as fur, and, most importantly, companionship. Small animals kept as companions help people lead healthy lifestyles. It is widely stated that people with such animals are less likely to have

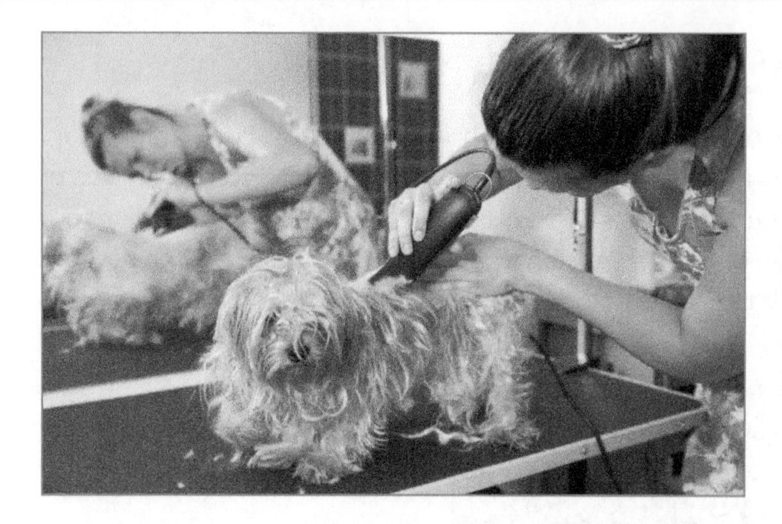

1–21. Grooming is an important part of caring for some small animals. This shows a Yorkshire Terrier being clipped to manage hair and skin disease. (Courtesy, Anton Gvozdikov/Shutterstock)

depression, high blood pressure, and other health problems. Just watching fish in an aquarium is said to reduce muscle tension and pulse rate! Taking a walk with a dog provides needed human exercise as well as promotes health of the dog.

Here are a few facts:

	Dogs	Cats	Birds
Percent of homes in USA with animals	36.5	30.4	3.1
Number of animals in USA (millions)	77.8	85.8	14.3
Veterinary visits per year (household average number)	2.6	1.6	0.3

INDUSTRY SEGMENTS

Small animals have certain requirements, which take financial outlay by the people who own them. Owners spent an estimated $60.59 billion on the upkeep of their animals (aka pets) in 2015. Industry segments have developed to provide these products, such as feed manufacturers, veterinary clinics, and grooming and boarding services. The industry requires people with skills and knowledge to work in all of the areas. Some purchases are made in local pet stores, online, or at veterinary facilities. These contribute to segments in the small animal industry, as follows:

- Feed—Small animals have nutrition needs. Feed is needed based on the requirements of the animal species. Some is canned, while other feed is dry. Feed is manufactured for a specific species, age, and growth stage of an animal. Feed should always be matched to the animal and its stage of life. Small animal owners spent an estimated $23.04 billion for food in 2015.

- Shelter, Housing, and Supplies—Some small animals stay in homes with people; others are housed in kennels, cateries, aviaries, hutches, and the like. Even those kept in homes may be placed in crates or cages some or all of the time. Exercise equipment, feeding and watering bowls, beds, collars and leashes, and the like are used. Manufacturers produce the needed products. In 2015, the estimate is that animal owners spent $14.39 billion for equipment and supplies.
- Health Care—Small animals require certain practices to promote their health and well-being. Most need vaccinations, parasite control, and other care provided by veterinary clinics and hospitals. Occasionally, an animal may need surgery to remove a cyst or to repair a broken bone. Cleaning teeth is a practice needed by some small animals. In 2015, small animal owners spent an estimated $15.73 billion for veterinary care.

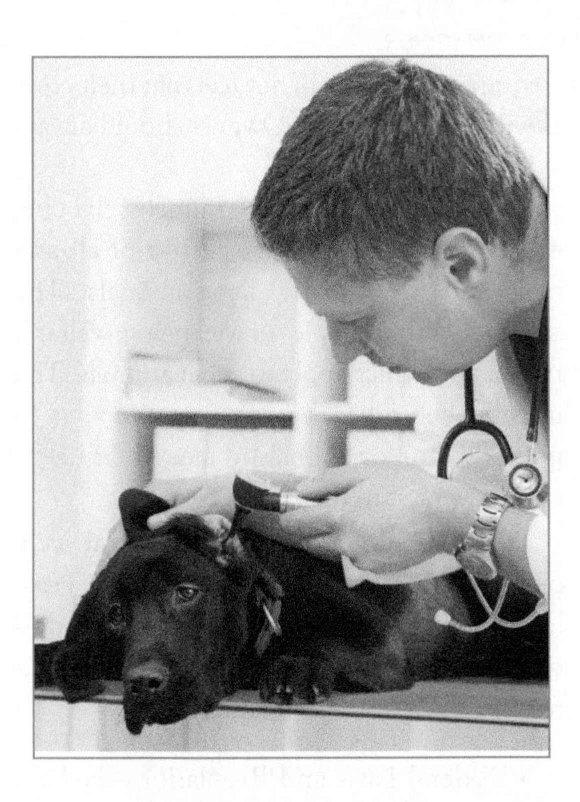

1–22. Veterinary medical care of dogs and other small animals is essential to assure their well-being. (Courtesy, Alexander Raths/Shutterstock)

- Grooming and Boarding—Grooming serves several useful roles, including the hygienic care and cleaning of an animal and enhancing its physical appearance. Brushing is often a daily activity, depending on species and breed. Periodically some animals are clipped to shorten the hair or fur with clippers and scissors. Nails/claws may also be trimmed. Skin health treatments may also be applied. Veterinary hospitals, animal grooming shops, and kennels may offer grooming services. Kennels may also provide boarding services if the owner must be away and unable to care for the animal. Some kennels offer animal training services. The estimated expenditures for grooming and boarding in 2015 were $5.24 billion. Since most animals live several years, these expenditures are typically repeated each year for an animal.
- Small Animal Production—Most small animal owners buy their animals. In 2015, $2.19 billion was spent buying small animals. Some animals are bought from pet stores and kennels while others are bought from individual animal owners that have an extra puppy, kitten, or other species. Individuals may have a couple or so of breeding animals that produce two or more litters a year. The off-

spring are often sold to new owners when the young have reached sufficient maturity. Such owners provide for the well-being of the young, growing animals. Feed, housing, bedding, and veterinary care (including vaccinations) are provided.

Some young animals are produced in places referred to as "mills." A mill (typically a puppy mill) is a place that keeps animals for breeding and production of young to sell. Sometimes, the animals are kept in cramped, dirty quarters that do not properly care for the well-being of the animals. Animal shelters also have animals available. These animals are those that have been rescued from bad situations or abandoned by their owners.

SMALL ANIMAL WELL-BEING AND RIGHTS

Most animal owners know that their animals require proper care. They feel that their companions/pets are well worth the costs associated with keeping them. Owners should always provide for the well-being of their animals.

1–23. Guinea pigs are often used as companion animals and in research. (Courtesy, Monica Butnaru/Shutterstock)

Earlier in this chapter, animal well-being was covered from the perspective of all animals and, particularly, large animals. The same general principles of well-being, welfare, rights, and ethics apply to small animals. The impetus behind each of these concepts is to properly care for animals to meet their needs and prevent suffering.

Some of the standards for animal well-being are based on Federal and state laws, local city or county regulations, and industries associated with small animals. Here are a few examples:

- Federal Laws and Regulations—Federal Acts and regulations intended to promote animal well-being are (along with date of enactment):
  - 1966—The Laboratory Animal Welfare Act was passed by Congress. A major provision was to protect small animals used for

laboratory research and testing. It also offered regulations on the humane care and handling of dogs, cats, and other laboratory animals. As amended, this has remained the only Act on animal welfare. It is administered by the Animal and Plant Health Inspection Service in the U.S. Department of Agriculture.

- 1970–An amendment to the Laboratory Animal Welfare Act of 1966 passed Congress that changed the name of the Act to The Animal Welfare Act of 1966 (referred to as AWA).
- 1976–An amendment to the AWA was passed making it illegal to promote animal fighting, such as with dogs.
- 2007–The Animal Fighting Prohibition Enforcement Act of 2007 was enacted to strengthen provisions of previous laws on animal fighting.
- 2008–Additional amendments to the AWA were enacted to tighten regulations on animal fighting and otherwise protect animals. One provision banned the importation of puppies under six month of age for resale.
- Years Since–Several pieces of legislation have been introduced that would further protect small animals but little action has been taken by Congress. Farm animals are excluded from the AWA.

The AWA is the only law in the United States that regulates small animals used in research and exhibition as well as those that may be used as companions. The species include cats, dogs, hamsters, rabbits, rats, mice, guinea pigs, and other warm-blooded species as may be determined to be covered by the U.S. Secretary of Agriculture (USDA). Examples of species excluded from the Act are birds, certain rats and mice, reptiles, and farm animals. Some of the provisions of AWA as it now stands include licensing animal facilities, animal identification, animal record keeping, facilities and operating standards, health and husbandry standards, transportation standards, and rules of practice. The USDA operates the Animal Welfare Information Center and has a wealth of online information, including the Animal Welfare Act and Regulations "Blue Book." (For more information, go to: https://awic.nal.usda.gov/.)

- State Laws and Regulations–Each state has different laws but all must be within provisions of the Federal AWA. States have laws that focus on the cruel treatment of small animals. Here are a few examples:
  - Failure to provide food, care, and shelter
  - Torturing an animal
  - Abandoning an animal

- Confining or transporting an animal in a cruel manner
- Killing, poisoning, or seriously injuring an animal
- Causing animals to fight each other

A nationally-recognized source of information about animal laws is the Animal Legal and Historical Center at Michigan State University. A Web site operated by the Center can be easily searched for information about any state's laws on animal well-being. For example, Texas residents readily locate laws for their state. The site is: www.animallaw.info. Another national source of animal law information is the Animal Law Resource Center. The Center provides information on legislation and legal matters pertaining to animals. This includes animal cruelty, animal control, and laboratory animal welfare. For information, go to: www.animallaw.com.

- Animal Species and Breed Clubs—Dogs, cats, and other species are supported by associations that seek to promote the welfare, knowledge, and betterment of a species or breed. These associations often provide educational programs and materials. Competition between animal owners and breeds may be set up as dog shows, cat shows, and the like. The American Cat Fanciers Association (ACFA) and the American Kennel Club (AKC) are two examples. (These associations are covered in more detail later in the book.)

1–24. Homeless cats often wonder in cities without food and care. (They are starved and diseased. Such cats could benefit from placement in an animal shelter. Note the unthrifty and malnourished appearance of these cats.) (Courtesy, Photopixel/ Shutterstock)

- Animal Shelters—An ***animal shelter*** is a place that receives and provides care for abandoned, lost, and stray animals, and for those that have been given up by their owners. (Animal shelters are sometimes known as pounds.) Animals in situations where they are being abused may be seized by local authorities and taken to a shelter for care. Animal control officers are found in most local cities and communities. Their work is promote animal well-being and may include capturing viscous or dangerous animals as well as those that are stray. They work in cooperation with shelters. Shelters are in local cities and towns throughout the United States. Some receive partial or full financial support from local government; others are supported by charitable organizations. The animals are given appropriate care to promote their well-being, including feed, health, and grooming. Facilities properly house and keep animals safe. Volunteer individuals often assist hired people at shelters. The animals are typically available for adoption by an individual who can provide proper care. Some local shelters are active with spay and neuter programs.

- Animal welfare associations—A number of organizations promote the care of animals. Some focus on well-being while others focus on production of quality animals. Here are three examples of organizations that focus on well-being: (Note that these often have local associations that help carry out work of the national association.)

  - The Human Society of the United States (HSUS)—Founded in 1954, the HSUS works on a full-range of animal well-being issues. A particular focus in the area of small animal is on puppy mills. The organization is headquartered in Washington, DC, and is affiliated with Humane Society International, a worldwide organization promoting animal well-being. The HSUS has five animal sanctuaries in the United States and a number of local affiliates. The HSUS Web site: www.humanesociety.org.

  - People for the Ethical Treatment of Animals (PETA)—Headquartered in Norfolk, Virginia, PETA was founded by Ingrid Newkirk in 1980. In short, PETA states that animals are not to be abused in any way. PETA is pursuing four major areas: factory farming, fur farming, animals in entertainment, and animals in research and testing. Some of the animals targeted are small animals. Major initiatives against chaining dogs in yards and animal fighting have received national attention. Opposition to the human consumption of animal meat and a few other positions have been controversial. The PETA Web site is: www.peta.org.

- American Society for the Prevention of Cruelty to Animals (ASPCA)—The ASPCA was founded in 1866 by Henry Abergh in New York to help prevent cruelty to animals in the USA. The organization was patterned after the Royal Society for the Prevention of Cruelty to Animals in the United Kingdom, founded in 1824. The ASPCA was the first organization for the humane treatment of animals in the Western Hemisphere. The ASPCA promotes education in the proper care of small animals such as gerbils, hamsters, and guinea pigs through online videos at www.aspca.org/pet-care/small-pet-care. Local ASPCA chapters help promote animal care. The ASPCA also has a program of pet insurance to cover veterinary costs in case of an injury, disease, or other condition. The Web site is: www.aspca.org/.
- Commercial animal care regulations—A wide range of commercial ventures promote animal well-being. Federal, state, and local laws and regulations often apply. Pet stores that sell small animals and the supplies and equipment promote animal well-being. These stores may also offer some veterinary-related care including grooming and vaccinations. Veterinary clinics and hospitals sometimes have reduced-cost programs to spay and neuter animals as well as provide vaccinations. The pet stores often have education programs in caring for and training pets. Other commercial ventures also promote animal well-being, such as airlines have regulations on transporting animals, hotels have regulations on the presence of animals, and parks regulate animal access.

1–25. Animal shelters should have clean facilities and provide for the needs of animals. (Courtesy, Alaettin Yldirim/Shutterstock)

REVIEWING

MAIN IDEAS

The animal industry is a large and diverse field. It includes the production of animals as well as supplies and services and product marketing and processing. In addition, companion animals, service animals, and other animal species are important. Animal domestication has had a big impact on how animals affect the lives of people. We depend on animal products and services every day.

Science and technology are an integral part of modern animal production. Selecting and producing animals that will meet consumer needs involves using science. Providing animals with a balanced diet and adequate care to ensure animal health uses science and technology.

Animal production is the foundation of the livestock and poultry industry. Farm animal production is moving away from the small operator to commercial operations capable of raising large numbers of livestock. Others involved in animal production include pet shops, laboratory technicians, and private individuals.

Animals are classified by their specie, common name, sexual condition, and age.

Animals are used in many ways. Some are used primarily for their meat and food products. Others are used for pleasure, companionship, service, and power. Overall, animals make a huge contribution to the quality of human life.

Animals do pose concerns for the environment. The waste they produce must be properly disposed of. Animals should be managed so they do not cause soil erosion or cause aesthetic damage to property.

QUESTIONS

Answer the following questions using correct spelling and complete sentences.

1. What is the animal industry?
2. What are the major areas of the animal industry?
3. What are four ways small animals are used? Large animals? Explain each.
4. What is animal domestication? Give three examples each of small animals and of large animals that have been domesticated.
5. What are the trends in the animal industry? Briefly explain each.
6. How is science used in the animal industry?
7. What are the five most important areas of the animal industry that involve science and technology?
8. What is a factory farm?
9. What is a neutered animal? How does neutering vary for males and females?
10. What is animal well-being?

11. What animal species are typically part of the small animal industry? Name any two species and indicate why these animals are kept.

12. What are the segments of the small animal industry in society? List four, and briefly describe the importance of each segment in one sentence.

13. What organizations promote the well-being of small animals? Name and briefly discuss any three organizations involved in animal rights.

14. What is an animal shelter? What roles do animal shelters serve?

EVALUATING

Match the term with the correct definition. Write the letter by the term in the blank provided.

a. poultry
b. livestock
c. euthanasia
d. by-product
e. dairy cattle
f. castration
g. companion animal
h. zoology
i. botany
j. health

_______1. Mammals produced on farms and ranches for food and other purposes.

_______2. The condition of an animal in being free of disease.

_______3. Fowl raised for food and other products.

_______4. Parts of animals not used for food that are used to make products.

_______5. Providing painless and easy death of a diseased animal.

_______6. Cattle kept for milk production.

_______7. The study of plants.

_______8. The study of animals.

_______9. Removing the testicles of a male.

______10. Animals used to provide humans with fun and friendship through close association.

EXPLORING

1. Take a field trip to a farm, kennel, or other facility that produces animals. Determine the kinds of animals that are produced. Learn the major practices followed to produce the animals. What is involved in raising the animals? Prepare a written report on your findings. Give an oral report in class.

2. Investigate how animal production has changed in the last 20 or so years. Interview an animal producer who has been in business for a long time. Ask how the production practices have

changed. Discuss probable reasons for farmers raising fewer livestock species and the shrinking numbers of animal producers. Prepare a written report on your findings. Give an oral report in class.

3. Investigate how animals and animal products are marketed in your local community. Tour a livestock auction, egg processing plant, pet store, or other facility. Interview the manager about the nature of the activities involved in marketing the animals. Summarize your findings in a written or oral report.

4. Assess your interests in animals. Indicate how you feel about the items below. Check the response that best represents your feelings about an activity with animals. Discuss your responses with your agriculture teacher. This should help you better understand your interest in animals. Areas that you are unsure about and don't like should be explored. You may find that you really like the activity!

	like	unsure	don't like
feed an animal	____	____	____
groom an animal	____	____	____
care for a sick animal	____	____	____
give a shot to prevent disease	____	____	____
haul an animal	____	____	____
care for a baby animal	____	____	____
train an animal	____	____	____
care for where animals live	____	____	____
exhibit an animal at a show	____	____	____
use science to study animals	____	____	____

5. Investigate livestock commodity markets in your local area. Include all species grown, such as cattle, swine, and poultry. Identify the markets used by producers. Determine the current market prices with newspaper listings, Web site reports, calls to local markets, and with interviews of producers. Another source of daily information is the Web site of the CME Group: www.cmegroups.com/trading/agricultural/. Prepare a report on your findings.

6. Select a domesticated small animal species and investigate when and where it was domesticated. Investigate events associated with its domestication and the importance of the species in today's human life. Use books, online sites, and other sources of information. Prepare a half page report on your findings and give a short oral report in class. You may also do this for a large animal species.

7. Assume you are considering acquiring a new pet, such as a dog, cat, or guinea pig. Select the species and breed. Study the animal's needs, its growth, and mature size. Develop a budget for one year of caring for the animal. Include the initial cost of the animal (if bought) and the feed, equipment/housing, veterinary care, grooming, boarding, training, and other costs. Prepare a summary of the costs. (You can get information by visiting a pet store or veterinary clinic, from online sources, and interviewing friends and relatives.) Be sure to investigate the well-being needs of your pet. Make a report of your findings to your class.

8. Research an organization that is involved with animal rights. (Note that several examples are included in the chapter.) Use online resources, interview local animal welfare authorities, and visit a local animal shelter. Determine the purpose or mission of the organization, sources of funding, available facilities and education programs, and relevant laws. Prepare a report on your findings. Give an oral report in class.

9. Investigate an association that promotes small animal betterment, including animal rights. Prepare a report on your findings. Include the name of the association, species or breed included, purpose or mission of the association, educational and other services provided, and general description of the work of the association. Several examples of associations are: cats-American Cat Fanciers Association (www.acfacat.com/), dogs-American Kennel Club (www.akc.org/), guinea pigs (aka cavies)-American Cavy Breeders Association (ACBA) (www.acbaonline.com/), hamsters-American Fancy Rat and Mouse Association (AFRMA) (www.afrma.org/hamster.htm), and gerbils-The American Gerbil Society (http://agsgerbils.org).

10. Investigate the benefits of small animal ownership. Use books and online sources of information. Interview a neighbor or family member who has a small animal. (You may also find some disadvantages.) Prepare a short oral report for class that summarizes your findings.

11. Investigate important events and dates related to animal welfare and rights. Be sure to include legislative actions as well as other details. Prepare a poster that summarizes your findings. Choose one event that you think is most important and orally discuss that event in class. Be sure to tell why it is important and how it relates to animal welfare and rights.

12. Evaluate the relationship between animal commodities that are produced locally. Investigate beef, pork, lamb, poultry and eggs, and fresh milk. Compare and contrast needs for land, equipment and structures, labor, and feed other inputs. Investigate marketing approaches and assess profitability. Give an oral report in class on your findings.

UNIT TWO

(Courtesy, Alexander Raths/Shutterstock)

Animal Biology and Technology

CHAPTER 2

Animals as Organisms

OBJECTIVES

This chapter covers the basic anatomy and physiology of common animals. It has the following objectives:

1. Explain taxonomy and scientific names.
2. List and describe the major animal groups.
3. Identify the life processes of animals.
4. Explain the structural basis of animals.
5. Describe the anatomy of common animals.
6. Explain the structure and parts of bones.
7. Identify the major organ systems of animals and explain the physiology of each.

TERMS

anatomy
bone
cartilage
cell division
cell specialization
circulatory system
connective tissue
digestive system
epithelial tissue
excretory system
growth plate
homeostasis
integumentary system
invertebrate
Kingdom Animalia
lymphatic system
mammal
marrow
meiosis
mitosis
muscular system
muscular tissue
nervous system
nervous tissue
organ
organ system
physiology
reproductive system
respiratory system
skeletal system
tissue
vertebrae
vertebrate

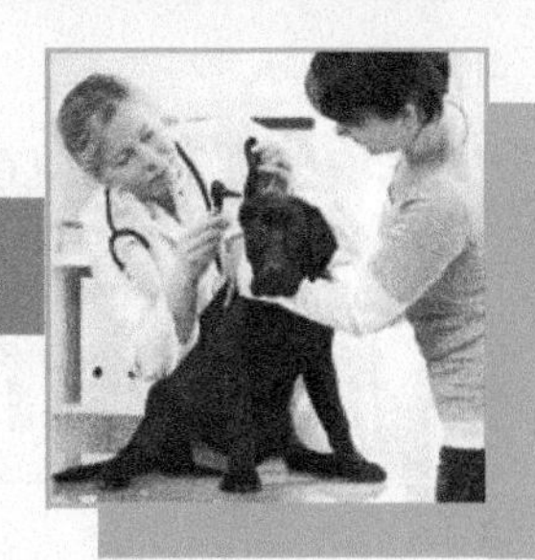

2–1. The rabbit is a species with distinct biological characteristics. Though distinct, the species shares many characteristics with other animals, particularly mammals. (Courtesy, Eric Isselée/Shutterstock)

LIVING animals—in fact, all living organisms—must carry out life processes. As complex creatures, most animals have the same basic systems and needs for life to occur. Cells, tissues, and organs of an animal serve important roles. The organ systems they form carry out life processes. While much has been learned about animals, there is a lot more to learn.

Just as animals have many similarities there are differences. The differences are greater between species that are not so closely related. For example, birds have greater differences with livestock than with other birds. A fish species is more like other fish than it is like dogs or guinea pigs. Unique life processes must occur if members of a species are to thrive and reproduce assuring future generations.

People are better producers, care-givers, and friends of animals if they know animal biology. Knowledge helps people understand why animals respond as they do and how to respond to their. Learning the major internal and external parts and functions of animals is a good place to begin. This chapter provides a good starting point.

TAXONOMY AND NAMES

Animals are organisms. An ***organism*** is any living thing. Organisms carry out processes needed to remain alive and continue their species. Every known organism has a scientific name; most have common names.

TAXONOMY

Taxonomy is the science of classifying organisms. Most classifications of animals are due to physical characteristics. A ***classification system*** is used to distinguish different animals from each other. Classification systems have changed over the years as scientists learned more about animals. Modern DNA and RNA analysis helps in the refinement of classification. Until recent years, Kingdom was the broadest group in the hierarchy but that has changed with the Domain being introduced. The three Domains are:

- Archea—microscopic single-celled organisms lacking a cell nucleus; often inhabit harsh environments.
- Bacteria—microscopic organisms whose cells are not compartmentalized; serve harmful and beneficial roles, with some bacteria causing animal diseases.
- Eukaryota—organisms whose cells have a distinct membrane-bound nucleus; maybe single-celled or multiple-celled organisms; includes animals, plants, fungi, and others.

The hierarchy of classification is used to show the relationships and differences among animals. Note that breeds are not species or subspecies but rather are members of a species that may have a number of breeds. Table 2–1 shows the hierarchy of classification.

Table 2–1. Hierarchy of Classification

Category	Description
Species	Reproductively isolated populations with physical similarities (With some species, further division may be made into subspecies to reflect greater similarities and differences.)
Genus	Contains related species
Family	Contains related genera
Order	Contains related family
Class	Contains related orders
Phylum	Contains related classes
Kingdom	Contains related phyla
Domain	Contains related kingdoms

Moving from bottom to top—more characteristics in common.
Moving from top to bottom—common characteristics are more distinctive.

SCIENTIFIC NAMES

Animals have common and scientific names. A **common name** is the name used in everyday conversation. A problem with common names is that they tend to vary from one place to another.

A **scientific name** is based on the taxonomy of an animal species. Every animal has a two- or three-part scientific name. The name is made of its genus and species (and subspecies in the case of cattle). Scientific names are written with the words in italics or underlined. Only the first letter of the genus is capitalized. Table 2–2 shows the common and scientific names of selected animals.

Table 2–2. Common and Scientific Names of Selected Animals

Common Name	Scientific Name
Hog (domestic)	*Sus scrofa*
Cattle	*Bos primigenius* Subspecies: *B. p. taurus* (cattle of European origin; no hump) Subspecies: *B. p. indicus* (cattle of Indian origin; hump on neck at shoulders and floppy ears)
Sheep	*Ovis aries*
Spanish goat	*Capra pyrenaica*
Horse	*Equus caballus*
Cat (domestic)	*Felis domesticus*
Common hamster	*Cricetus cricetus*
Golden hamster	*Cricetus auratus*
Chicken	*Gallus domesticus*
Turkey	*Meleagris gallopavo*
Channel catfish	*Ictalurus punctatus*
Rainbow trout	*Oncorhynchus mykiss*
Llama	*Lama glama*
Dog (domestic)	*Canis familiaris*
Rabbit (domestic)	*Oryctolagus cuniculus*

ANIMAL GROUPS

Animals are in the ***Kingdom Animalia***. Though all animals are in this Kingdom, they also have unique differences. Some have fur, others have scales or feathers. Some fly, while others walk or swim.

Organisms in the Kingdom Animalia share three traits:

1. Animals are made up of cells. (Cells are the "building blocks" of living organisms.)
2. Animals can move about on their own. (Moving about is known as locomotion. A few animals show very little locomotion, such as the oyster.)
3. Animals get their food from other sources. (Animals take in food, transform and store it, and release it as energy.)

2–2. A model of a horse is being used to learn the basics of horse anatomy. (Courtesy, Education Images)

STRUCTURES AND FUNCTIONS

Within the Kingdom Animalia, animals are grouped many ways. This grouping is based on the structures and functions of the body.

Animals are classified based on the presence of vertebrae. ***Vertebrae*** are bones that form a segmented spinal column. This segmented column is commonly known as a backbone. The presence of vertebrae is associated with the kind of skeletal system of an animal.

Animals are classified into two groups based on vertebrae: vertebrates and invertebrates. A ***vertebrate*** is an animal with a backbone; an ***invertebrate*** is an animal without a backbone. Examples of invertebrates include shrimp, crawfish, honeybees, spiders, mites, earthworms, and snails.

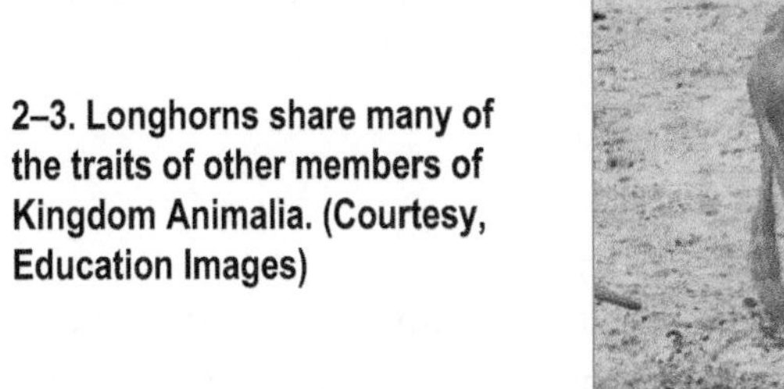

2–3. Longhorns share many of the traits of other members of Kingdom Animalia. (Courtesy, Education Images)

Vertebrates belong to the Phylum Chordata and to the Subphylum Verbrata. They have many common characteristics. They have vertebrae (bones and cartilage) that surround the nerve cord. Bones make up the internal skeleton that provides the body's framework. Vertebrates have a skull that protects the brain. They also have an axial skeleton made of the backbone and skull. Paired limbs (arms and legs) are attached to the axial skeleton. Finally, muscles provide movement by being attached to the skeleton.

CLASSES

Animals are in several classes. The three main classes are birds (Aves), bony fish (Osteichthyes), and mammals (Mammalia). Snakes, shellfish, turtles, and alligators are examples in other classes.

Birds

Some 9,000 species of birds have been identified. Birds belong to the class Aves. Birds can live in the air, on land, or in the water. Most birds are wild; a few have been domesticated. Chickens, turkeys, and ducks are often raised. People are eating more chicken and turkey because these meats are often said to be lower in fat and cholesterol. Wild ducks, geese, pheasant, and quail are popular game birds among sport hunters.

2–4. Birds, such as this Tucan, belong to the class Aves. The Tucan is more like other birds than it is like mammals and fish. (Courtesy, Nici Kuehl/Shutterstock)

Birds are covered with feathers. They are lightweight because of their hollow bones. Their wings help to lift them and make them better flyers. Birds have two legs and internal organ systems similar to most other animals.

The chicken (*Gallus domesticus*) is the most widely raised bird. The turkey (*Meleagris gallopavo*) is the second most commonly produced bird in the United States.

Fish

Fish are aquatic animals. They live in water. There are three classes of fish, but the bony fish (*Osteichthyes*) are the ones commonly raised for food. Scientists have identified about 25,000 species of fish.

Many fish are caught from rivers and oceans, but aquafarming is increasing in popularity. The channel catfish (*Ictalurus punctatus*) is the most commonly raised fish. Trout, tilapia, salmon, and baitfish are also

CAREER PROFILE

ANIMAL TRAINER

An animal trainer works with animals to help them gain the desired skills to meet human needs. Training varies with the kind of animal and the behavior that is desired. Horses are trained for riding. Mules may be trained to pull a heavy load. Cattle are trained to lead in a show halter or to enter a stanchion.

Animal trainers need a good understanding of the biology of the animals they are working with. They need to understand the nature of the animal and how it responds to human demands. Animal trainers need to be patient and always consider the well-being of an animal. Practical experience with animals is essential. Many animal trainers learn the trade under the direction of an experienced trainer.

Jobs for animal trainers are found on farms or ranches where animals are trained. This photo shows a trainer working with a mule to teach it a desired behavior. (Courtesy, Lucky Three Ranch, Inc.)

raised on aquafarms. Many people raise several species of ornamental fish at home or at their business for pets or decorations.

Fish are covered with skin and/or scales. The head, tail, and trunk form the three major parts of the body. The mouth, eyes, and gills are found on the head, while the trunk contains the internal organs. Strong muscles in the tail help the fish to move about.

Fish are physiologically similar to other animals. However, fish filter oxygen from the water as it passes over their gills. Some lay eggs that are fertilized outside the body; others give birth to live young. The body temperature is regulated by the temperature of the water.

2–5. Fish, such as these Nile tilapia (*Oreochromis niloticus*), are quite different from Aves and mammals. (Courtesy, Agricultural Research Service, USDA)

Mammals

Mammals belong to the class Mammalia. A ***mammal*** is a species of animal that has mammary glands. Female mammals have mammary glands that produce milk for the newborn. Male and female mammals mate to reproduce. The female will carry the developing embryo in her uterus.

Mammals have hair, a well-developed brain, a lower jawbone with teeth, and a heart with four chambers. Unlike fish, mammals internally regulate their body temperature. Examples of mammals include cattle, goats, hogs, horses, and sheep.

Male mammals have rudimentary or undeveloped mammary glands. Female mammary glands develop into mature organ systems during puberty.

Many of the animals used for food, companionship, and in other ways are mammals. Examples are (scientific name in parentheses) cattle (*Bos primigenius*), hogs (*Sus scrofa*), sheep (*Ovis aries*), Spanish goat (*Capra pyrenaica*), dog (*Canis familiaris*), and horses (*Equus caballus*).

2–6. As a mammal, a pig is quite a bit different from Aves and bony fish. (Courtesy, Kingsburg (California) High School)

2–7. A cat is a mammal that preys on Aves, fish, and other small mammals. (Courtesy, Kachalkina Veronika/ Shutterstock)

LIFE PROCESSES

Animals carry out certain processes in order to remain alive. When these processes stop, the animal is no longer living. It has died. You will learn about some of these life processes in this chapter. The content is applied specifically to animals.

CHEMICAL PROCESSES

Life is based on chemical processes that occur in the protoplasm. Protoplasm is the mixture of water and proteins that form the living contents of a cell. The nature of the chemical processes determines the kind of animal and life functions that it is performing. As complex organisms, animals carry out a number of life processes that support the living condition in protoplasm.

Since cells are specialized and vary within an animal, the chemical processes also vary. The processes are varied so that the animal can carry out life processes as intended for the particular species. You will learn more about these processes in the section of the chapter that deals with physiology.

ANIMAL LIFE NEEDS

The life processes are:

- Getting and using food—Food provides the nutrients for an animal to carry out life processes. Without proper food, an animal will not grow well and may become diseased and die.
- Movement—Movement includes processes that occur internally as well as locomotion. The ability to move is important for most animals to get their food. It is also important in escaping danger and raising young.
- Circulation—Circulation is moving blood, nutrients, oxygen, and wastes throughout the body.
- Respiration—Respiration is the process of providing oxygen and nutrients to cells. This releases the energy in the nutrients and produces some waste materials.
- Growth and repair—These two processes are important in young and mature animals. Animals grow by increasing the number and size of cells. Repair is replacing cells that have been damaged or no longer function.
- Secretion—Secretion is the production of liquid substances containing hormones and other materials. It is important in eating and digesting food as well as other life processes.
- Sensation—Sensation is the ability for an animal to respond to its environment. It receives information, such as pain, and responds to provide comfort.
- Reproduction—Reproduction is the process by which new members of a species are created. It is not essential for the life of an organism but is essential for a species to perpetuate itself. Animal producers are concerned about promoting and controlling reproductive processes.

THE STRUCTURAL BASIS OF ANIMALS

The structural basis of animals begins with cells. Cells are the basic structures of an animal. They have a definite internal organization and relationship. Most body cells can be seen only with a microscope. Some cells are large such as a chicken egg.

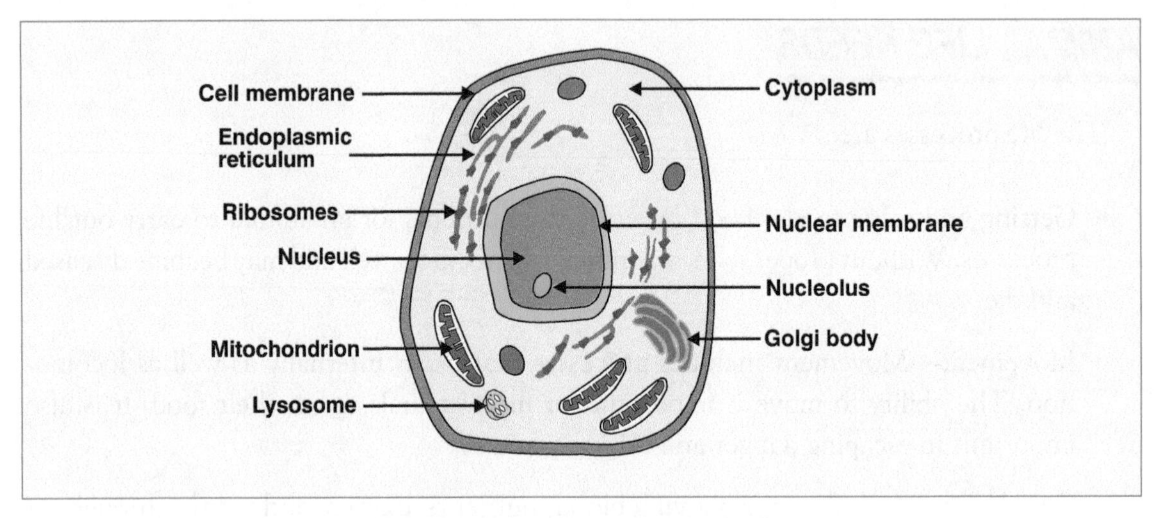

2–8. Major parts of a typical animal cell.

Cells vary in size and shape. Some are quite long, such as a cell in a muscle in the leg of an animal. Other cells are nearly square. All animal cells have membranes that surround the internal parts and create a definite structure. The cell membrane regulates what enters and leaves a cell. Cells have a nucleus that controls cell activity. The nucleus contains chromosomes and genetic material. The nucleus is surrounded by cytoplasm within the cell membrane. The protoplasm is all of the liquid substance inside the membrane.

CELL DIVISION

Cell division is the process of a cell splitting into two cells. The process involves several steps. The new cells contain identical genetic material.

Two types of cell division occur: mitosis and meiosis. **Mitosis** is cell division for growth and repair. The process involves several phases. Division by one cell results in two cells. Animal growth depends on rapid cell division. Good nutrition is needed to promote cell division.

Meiosis is cell division for sexual reproduction. With meiosis, division results in four cells known as daughter cells. Each daughter cell has half the number of chromosomes of the parent cell. The process of fertilization provides the other half of the chromosomes.

CELL SPECIALIZATION

Cell specialization is the difference in cells that allows them to perform unique activities. Organisms have several different kinds of cells. These different cells form groups that work together to perform functions. For example, the cells of eyes are specialized to allow vision to occur.

Tissue

2–9. A peacock has many cells, tissues, organs, and organ systems. All work together to support life processes. (Courtesy, Education Images)

Groups of cells form tissues. A ***tissue*** is a cluster of cells that are alike in structure and activity. Tissues perform specific functions, such as those that form skin or bones. Tissues may group together to form organs.

There are four types of animal tissue:

- Epithelial tissue—***Epithelial tissue*** covers body surfaces and lines body cavities. Its purpose as skin is to protect the body. Epithelial tissue contains cilia that are hairlike extensions. These cilia move materials. For example, cilia in the lungs move dirt and other impurities out of the body. The main organ system of the epithelial tissue is skin.
- Connective tissue—***Connective tissue*** holds and supports body parts. It binds body parts together providing support and protection. It also fills spaces, stores fat, and forms blood cells. Connective tissue helps to form muscular and skeletal cells.
- Muscular tissue—***Muscular tissue*** creates movement of body parts. It is also known as contractile tissue. These tissues not only aid in movement of the entire animal, but also allow the respiratory and digestive systems to function.
- Nervous tissue—***Nervous tissue*** responds to stimuli and transmits nerve impulses. Nervous tissue contains neurons that conduct impulses away from the cell. Environmental stimuli cause reactions to occur.

Organs

An ***organ*** is a group of tissues with a similar function. The tissues work together so that the organ can carry out its purpose. Livers and lungs are examples of organs. Some organs work with other organs to make life possible.

Organ Systems

An ***organ system*** is a group of organs working together to carry out a specific activity. These are the major systems of the body. An example is the digestive system. The mouth, stomach, and intestines form the digestive system. (Each system is covered in more detail later in the chapter.)

ANATOMY

Anatomy and physiology are important in knowing how organisms live and go about life processes. **Anatomy** is the study of the form, shape, and appearance of animals. **Physiology** is the study of the functions of the cells, tissues, organs, and systems of an organism. Knowing anatomy is essential to understanding physiology.

All animals have organ systems that maintain homeostasis. **Homeostasis** is the relative stability of the internal environment. An example of homeostasis is the work of the circulatory system. The circulatory system carries nutrients from the digestive system. It also carries oxygen from the lungs to the cells from the respiratory system via the blood. The circulatory system rids the body of metabolic waste from the cells into the excretory system. The nervous system coordinates the functions of the other systems.

Understanding life processes is important. Studying the anatomy and physiology of livestock helps in understanding the inner workings of the human body. The internal and external features are closely related.

Animal selection and evaluation often involve referring to external features. Knowing the names of the parts is essential in communicating with others. Figures 2–10 through 2–16 illustrate the external parts of several animals.

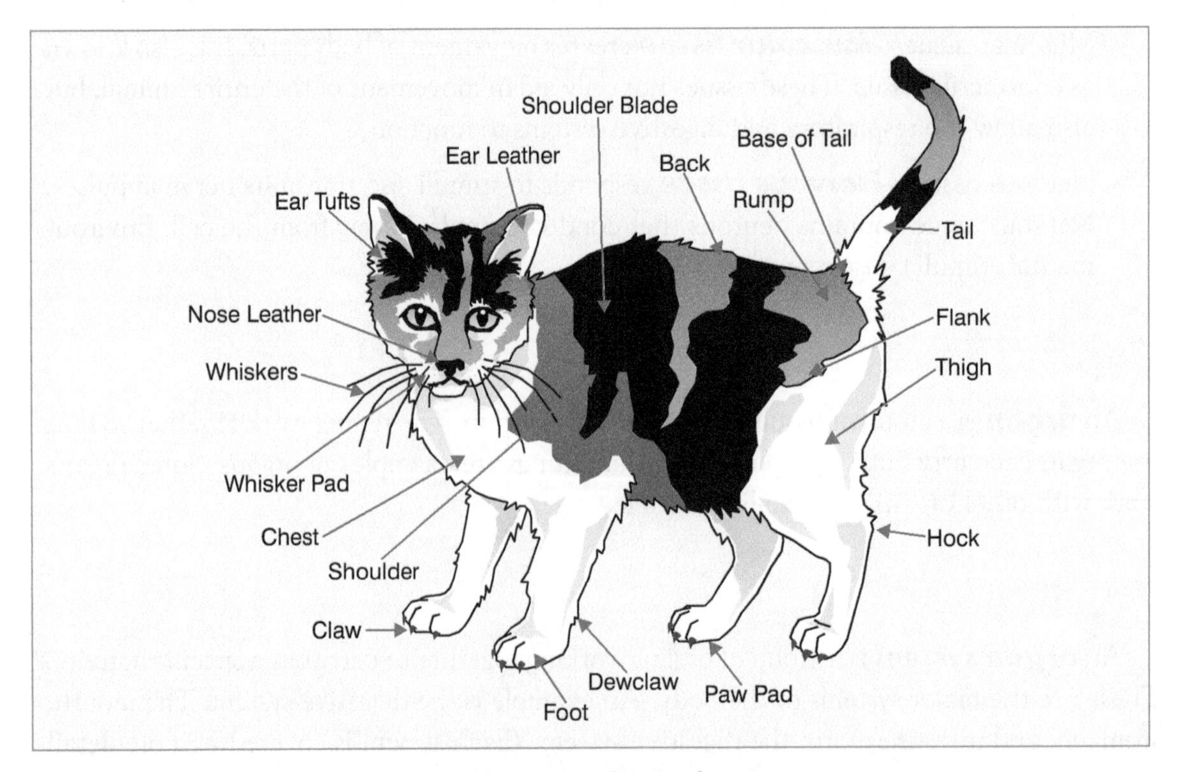

2–10. Major external parts of a cat.

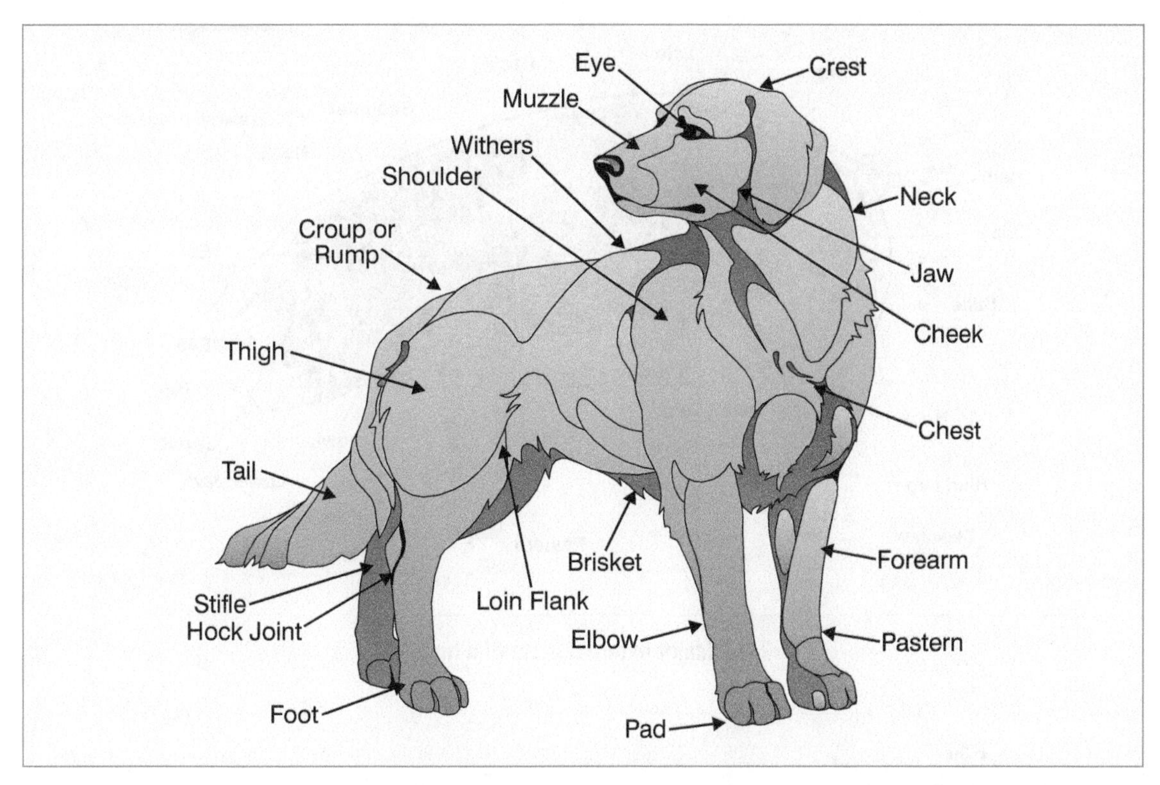

2–11. Major external parts of a dog.

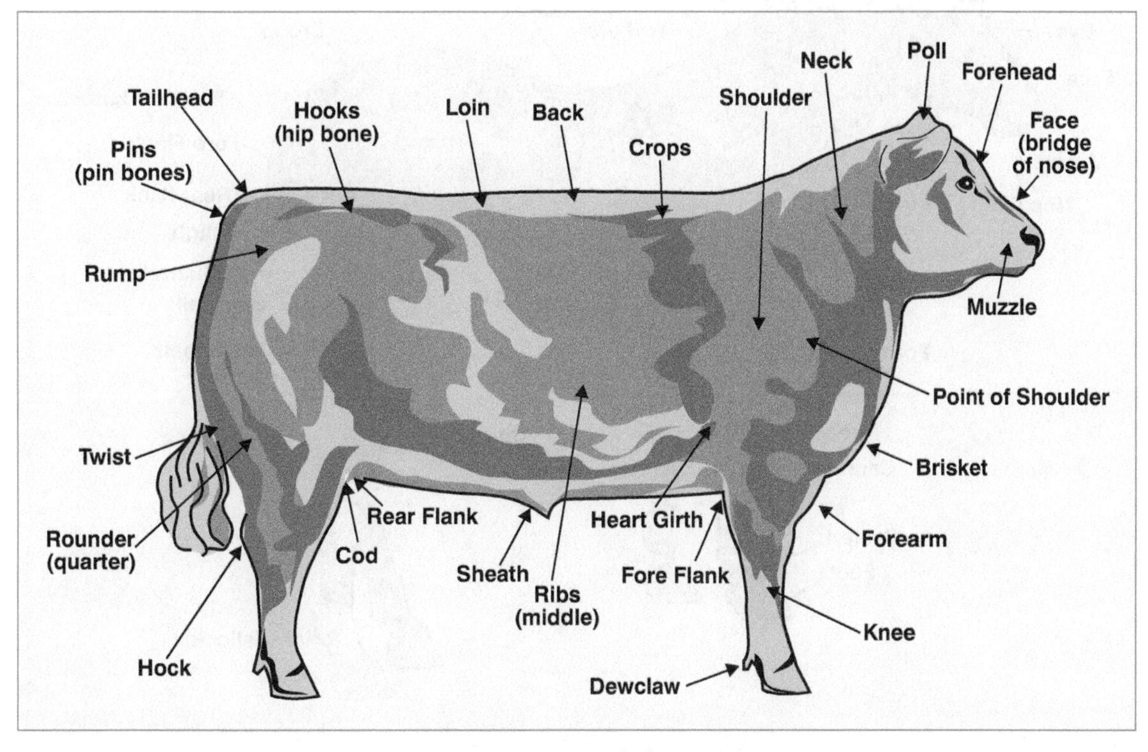

2–12. Major external parts of a steer.

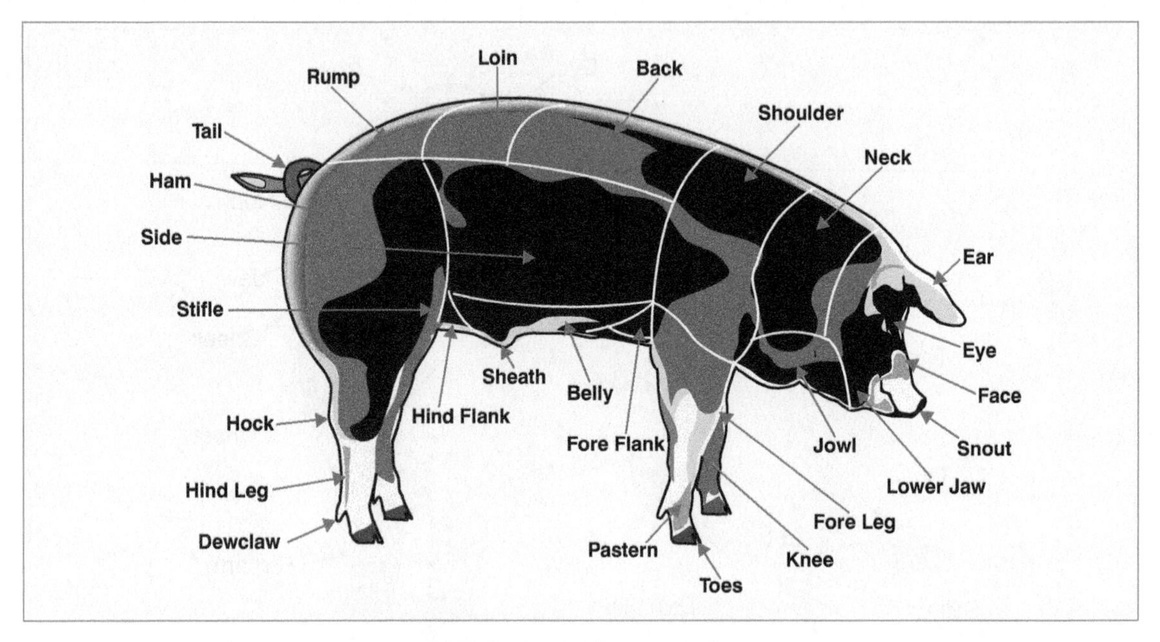

2–13. Major external parts of a hog.

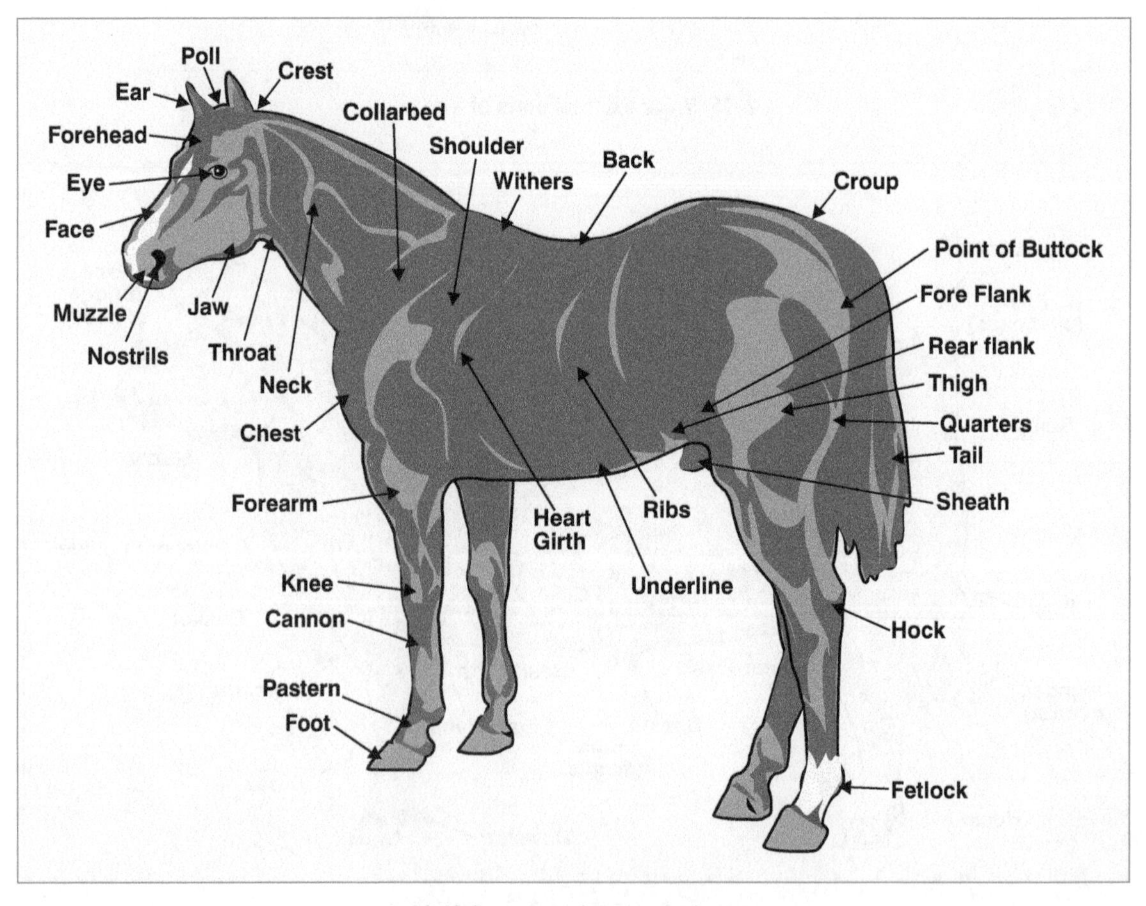

2–14. Major external parts of a horse.

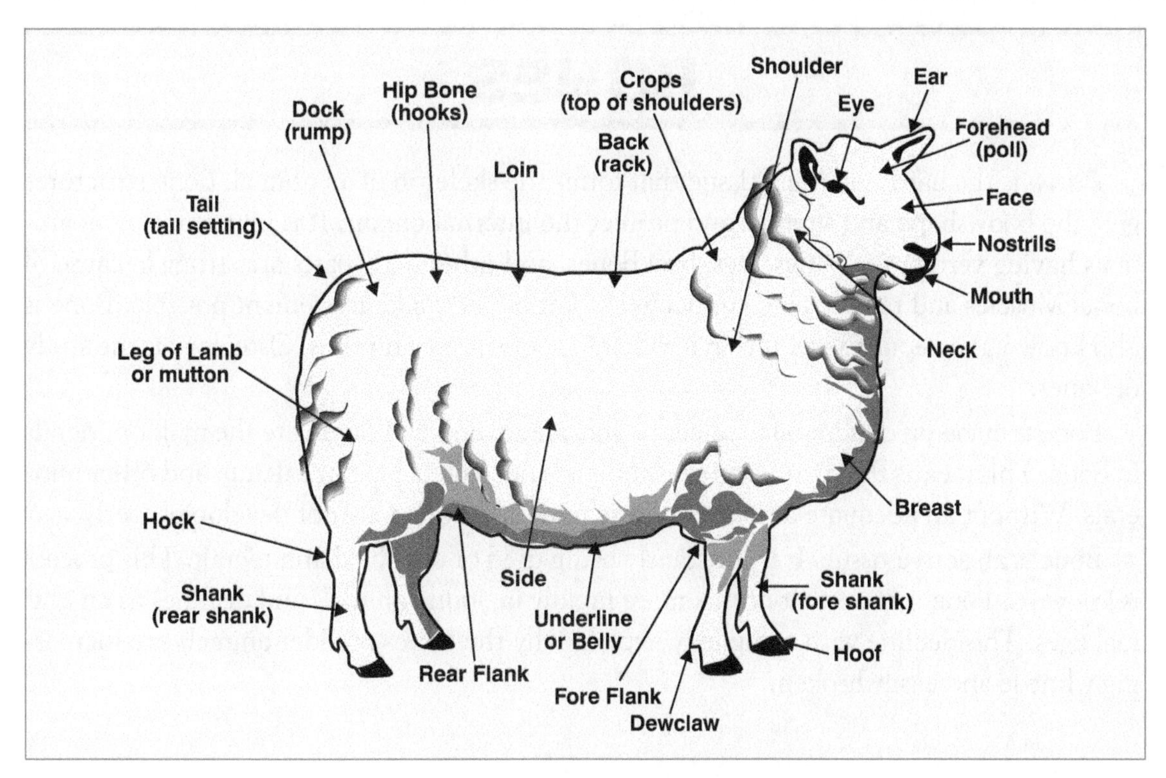

2–15. Major external parts of sheep.

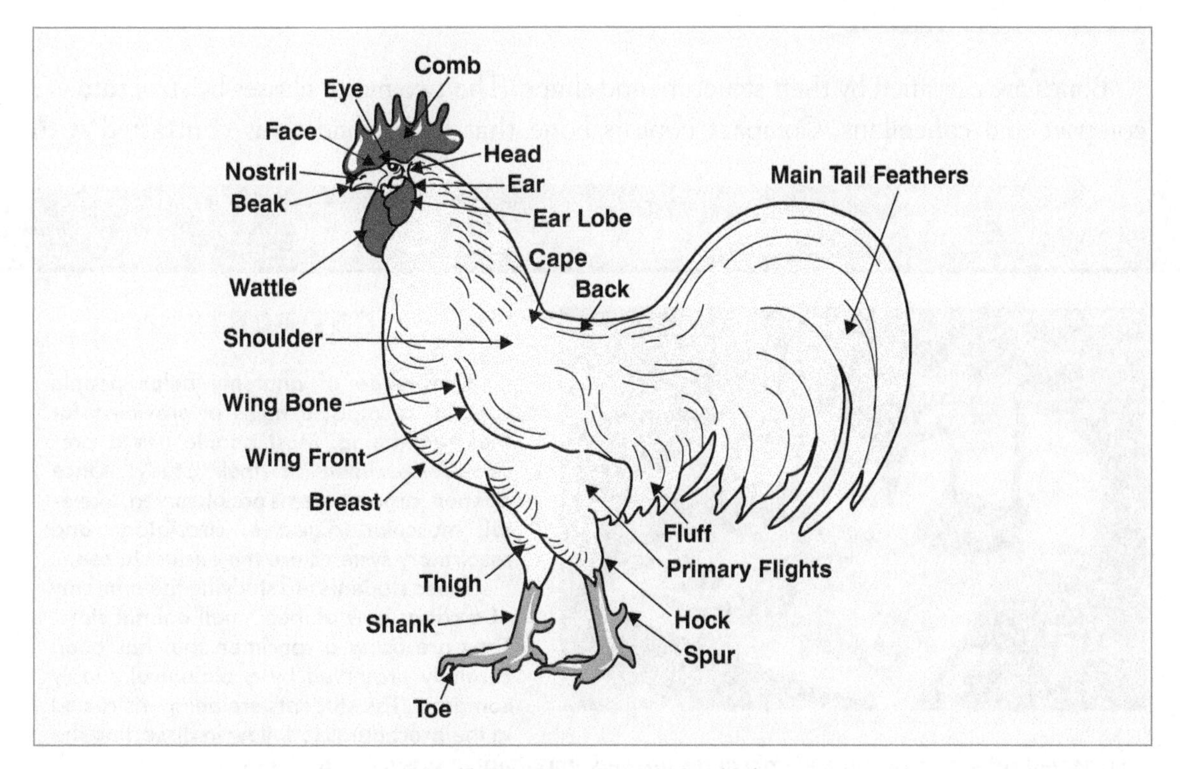

2–16. Major external parts of a chicken.

BONES

Bone is the hard, semirigid tissue that forms the skeleton of an animal. Bone structures give the body shape and support and protect the internal organs. It is important in all animals having vertebrae. Wings, legs, backbones, and other body parts are strong because of bone. Muscles and tendons are connected to bones and make movement possible. Bone is also known as osseous tissue. It is in a wide range of shapes and sizes. Osteology is the study of bones.

Bone is made up of minerals. Calcium, phosphate, and carbonate are the major minerals in bone. This means that young, growing animals need diets high in calcium and other minerals. Without an adequate amount of these minerals, bone may not develop properly.

Bone is an active tissue. It is constantly being used or dissolved and rebuilt. This process is known as bone turnover. It occurs more rapidly in young animals and declines as an animal ages. This decline with aging may explain why the bones of older animals are increasingly brittle and easily broken.

KINDS OF BONES

Bones are classified by their structure and shape. The two major classes by structure are compact and cancellous. Compact bone is bone that looks solid. It is contrasted with

CONNECTION

CAT PARTS

Knowledge of anatomy helps people care for animals. It helps in providing for their well-being. Most people use a preserved specimen in their study. Once opened, organ systems are observed. Skeletal, muscular, digestive, circulatory, and respiratory systems are the easiest to see.

These students are studying the anatomy of a cat as part of their small animal class. They are using a specimen that has been carefully preserved by a biological supply company. The students are being instructed in the procedures to follow in dissecting the preserved cat specimen. Safety procedures are very important in dissection activities.

Students will apply what they have learned in the school lab, on the job, and in caring for their own animals. (Courtesy, Education Images)

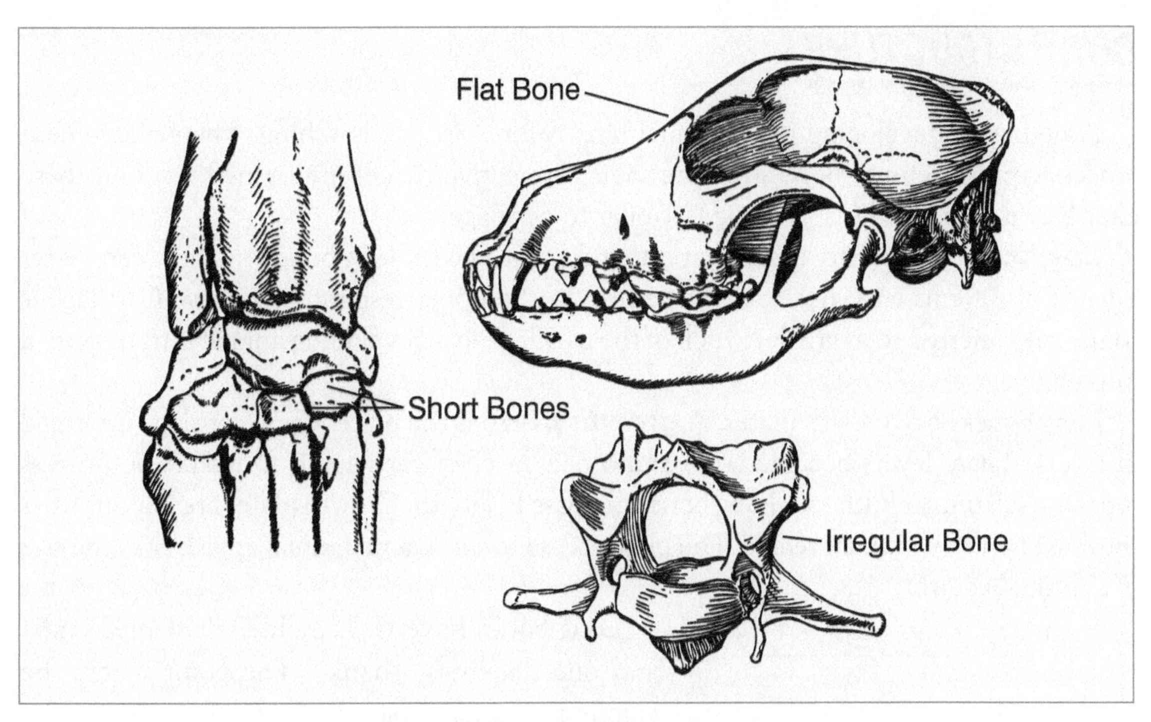

2–17. Shape of flat, irregular, and short bones.

cancellous bone, which has the appearance of visible spaces within the bone. Cancellous bone is located at the ends of marrow.

Two other kinds of bone are woven bone and haversian bone. Woven bone is immature or developing bone that may form compact or cancellous bone. Haversian bone is the bone in older animals that has been "remodeled" to help it retain its strength and usefulness. Haversian bone may be either compact or cancellous bones.

The five shapes of bones are:

- Short—A short bone is cube-shaped. They are the small carpal and tarsal bones.
- Flat—A flat bone is two plates of compact bone with a layer of cancellous bone between. Flat bone forms much of the skull.
- Irregular—An irregular bone is a complex bone with varying shapes. Examples are vertebrae and facial bones.
- Sesamoid—Sesamoid bone is small and embedded in tendons. It is about the size of a sesame seed.
- Long—A long bone is a bone that is longer than wide. These are the bones in legs and other structures. Specific bones include the tibia, femur, humerus, metacarpal, and metatarsal bones. A long bone is divided into the diaphysis and epiphysis.

BONE STRUCTURE

Bone begins developing in an animal long before birth or hatching. The development process creates definite bone structure. Soft connective tissues are formed first and these later become bone. This soft tissue is similar to cartilage.

Cartilage is a rubbery tissue that is found at the ends of long bones and between vertebrae. It also forms ears, nose, and structures in the upper respiratory system. Cartilage is sometimes referred to as gristle. Much of the cartilage in a developing animal is transformed into bone.

Long bones have growth plates. A **growth plate** is the place where cartilage is formed in layers. These layers become bone and a new layer of cartilage is formed. The process repeats itself until adulthood. Long bones increase in length. Growth plates are important as the size of an organism increases. The plates cease to form as an animal approaches the size of a normal adult.

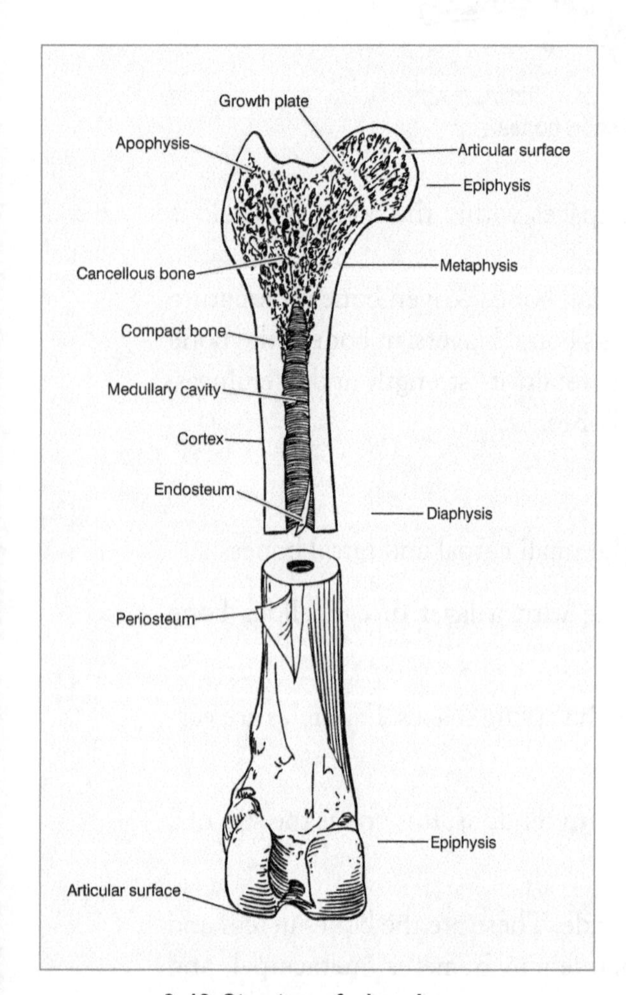

2–18. Structure of a long bone.

Long bones have two epiphysis (enlarged ends) and one diaphysis (shank). The point where the epiphysis and diaphysis join is the metaphysis. The periosteum is a fibrous covering around those bones that are not encased in cartilage. This layer is the attachment site for ligaments and tendons. A ligament is strong connective tissue that connects bones and holds organs in place. A tendon is strong fibrous tissue that connects muscle with bone. Figure 2–18 shows the structure of a long bone.

Most bones have hollow areas. The hollow area is known as the medullary cavity. The lining of the medullary cavity is the endosteum, which is a fibrous-type of tissue. **Marrow** is the material found in the medullary cavity. Marrow may be yellow or red. Yellow marrow is mostly fat tissue. Red marrow is a network of blood vessels and blood-forming cells. All bones have blood vessels and nerves. The compact bone surrounding the medullary cavity is cortex.

Most bones join with other bones. Think of the leg of an animal. For example, bones join at the knee. The articular surface is where two bones join, and a thin layer of cartilage separates the two bones.

MAJOR ANIMAL SYSTEMS AND PHYSIOLOGY

As organisms, animals have several systems that make the living condition possible. The physiology of these organ systems involves essential functions for an animal. The major organ systems and the chief structures in the systems are shown in Table 2–3.

Table 2–3. Organ Systems and Structures

System	Chief Structures
Circulatory	Heart
Digestive	Stomach and intestines
Integumentary	Skin
Endocrine	Ductless glands
Excretory	Kidneys and bladder
Muscular	Muscles
Nervous	Brain, spinal cord, nerves
Reproductive	Ovaries and testes
Respiratory	Lungs
Skeletal	Bones, joints

- Muscular system—The ***muscular system*** acquires materials and energy. It creates body movements, maintains posture, supports the body, and produces heat. In meat animals, good muscling is needed because of its value as food. Figure 2–19 outlines parts of the hog's muscular system.
- Skeletal system—The ***skeletal system*** provides the framework for the body. The major

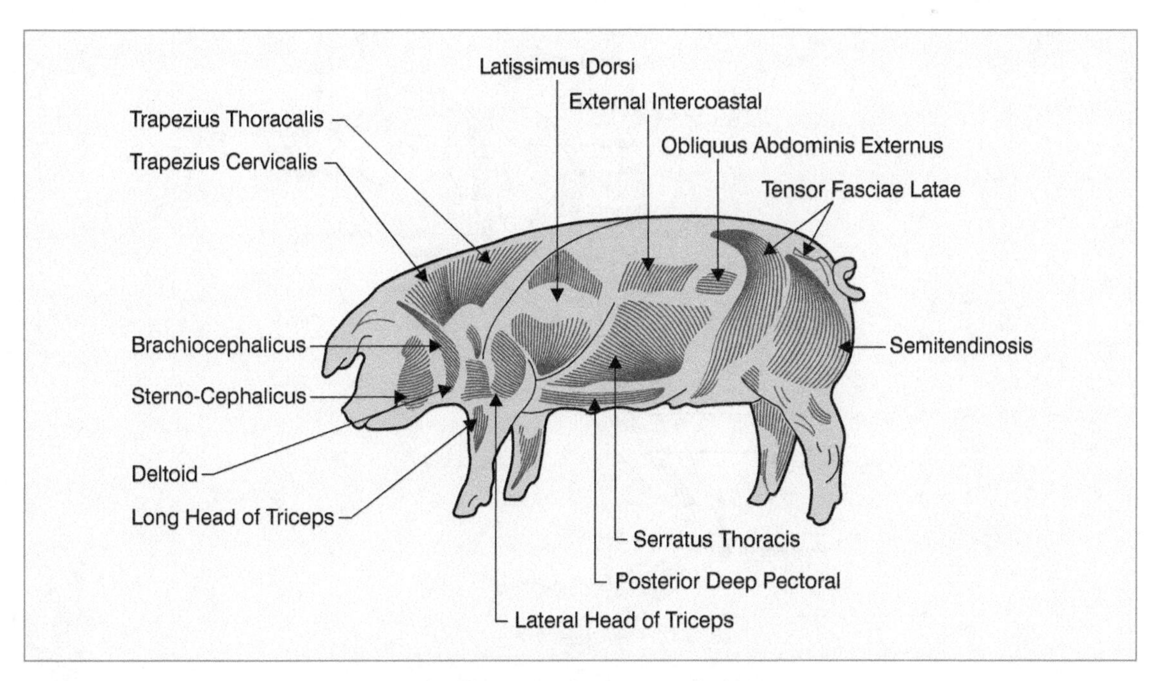

2–19. The muscular system of a hog.

components are bones and cartilage. The skeletal system supports and protects internal organs, stores minerals, and produces blood cells. Skeletal systems should be strong and capable of fulfilling a desired purpose, such as helping a horse run fast. The skeletal system of a chicken and a hog are shown in Figure 2–20.

- Digestive system—The **digestive system** breaks food into smaller parts that are used by the body. Nutrients are made available from these food materials. Digestive

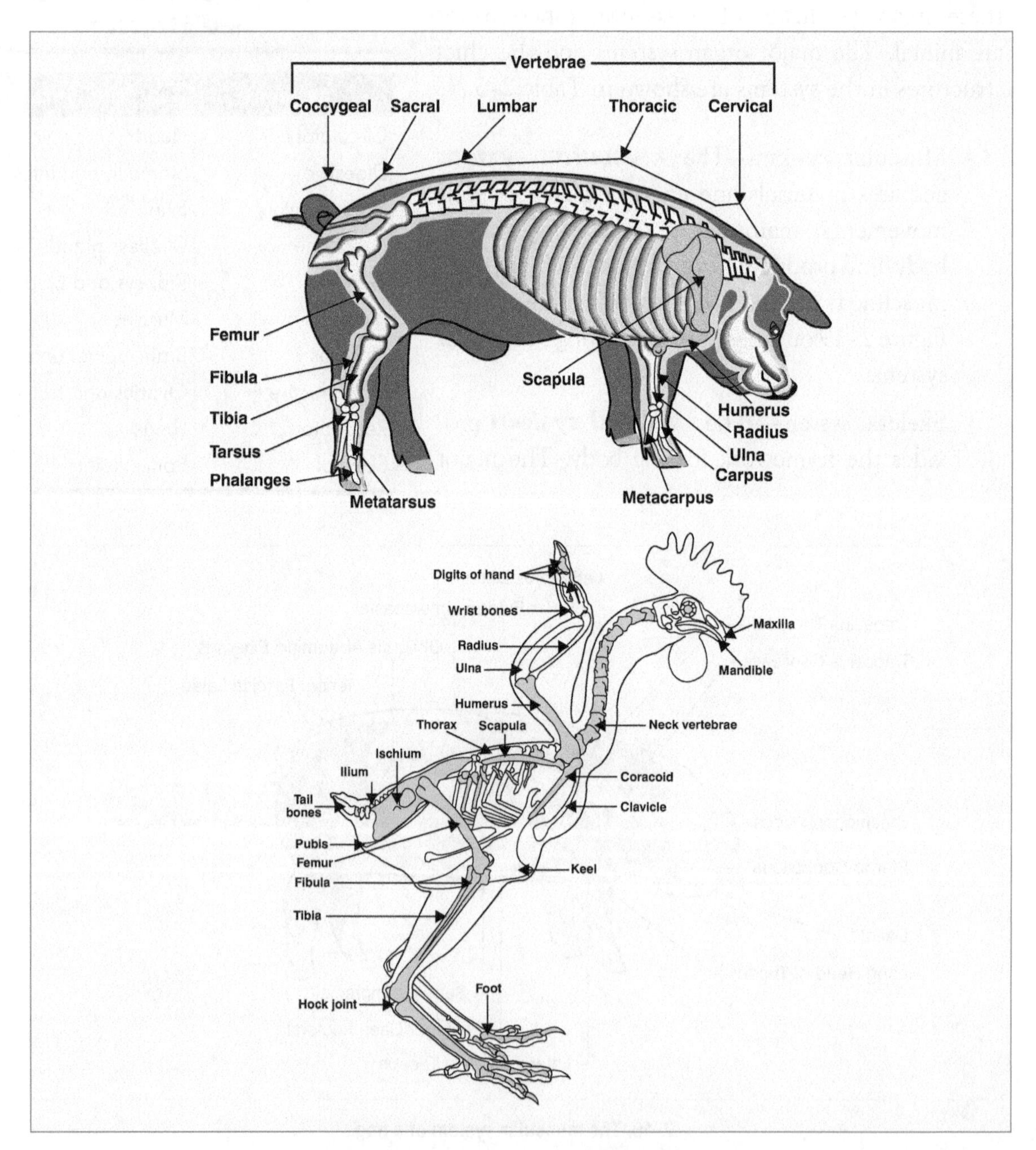

2–20. The skeletal systems of a chicken and a hog.

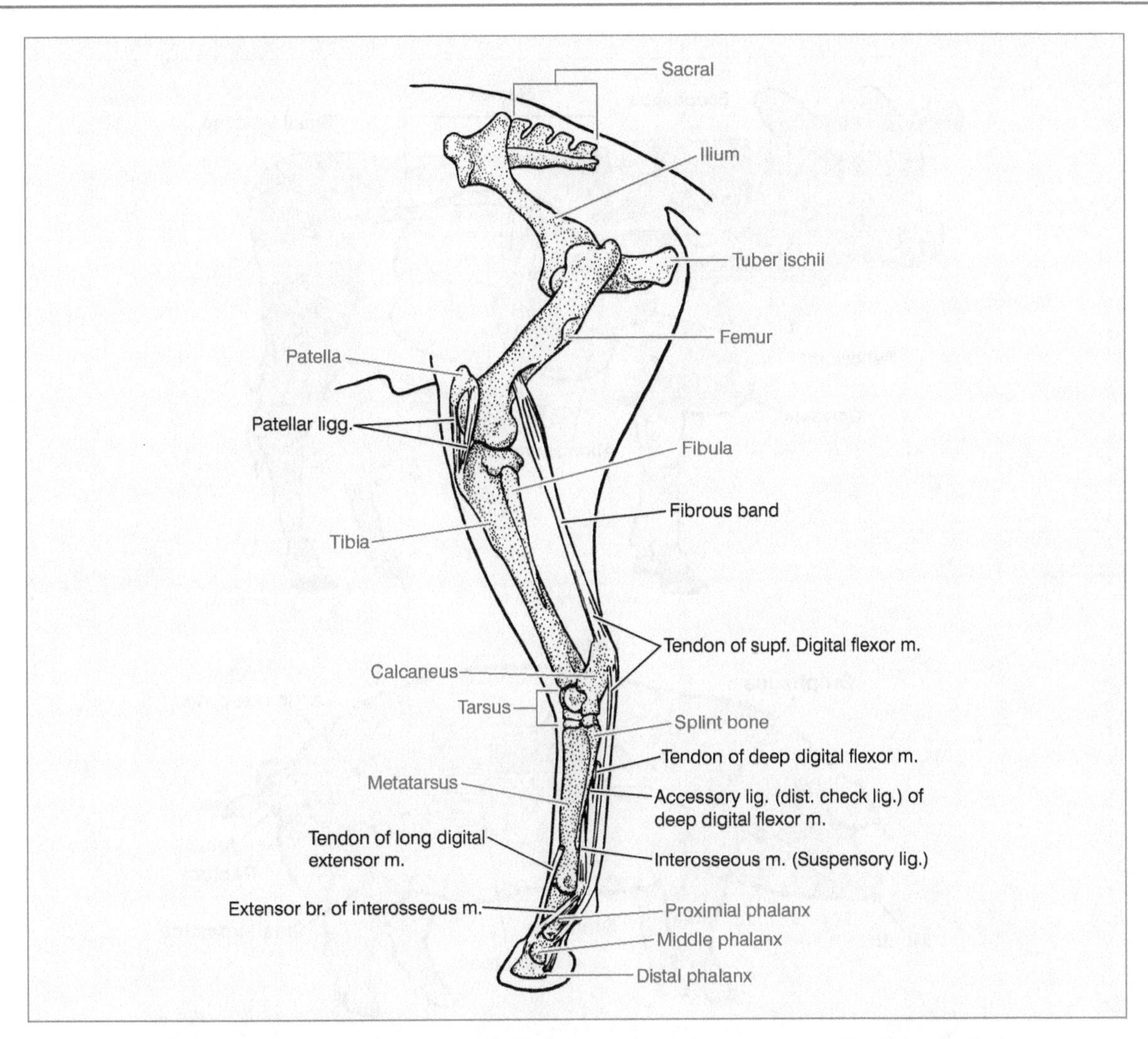

2–21. The structure of bones (red print) and tendons (black print) in the hind leg of a horse.

systems vary. These differences allow animals to consume various food materials. For example, cattle are ruminants. The polygastric system allows the animals to eat grass and other forages. But pigs are different. Pigs have monogastric or nonruminant digestive systems. They must be fed grain and other concentrated food materials. Figure 2–22 outlines the digestive systems of a hog (monogastric or nonruminant), a cow (polygastric or ruminant), and a chicken.

- Respiratory system—The ***respiratory system*** governs gas exchange. Oxygen is taken in through the nostrils. It goes to the lungs where it passes into the blood. This system also maintains blood pH by expelling carbon dioxide. The respiratory systems of fish and some other animals are different. For example, fish have gills for removing oxygen from the water. The oxygen is exchanged in the gill filaments for carbon dioxide in the blood. Figure 2–23 shows the respiratory system of a hog.

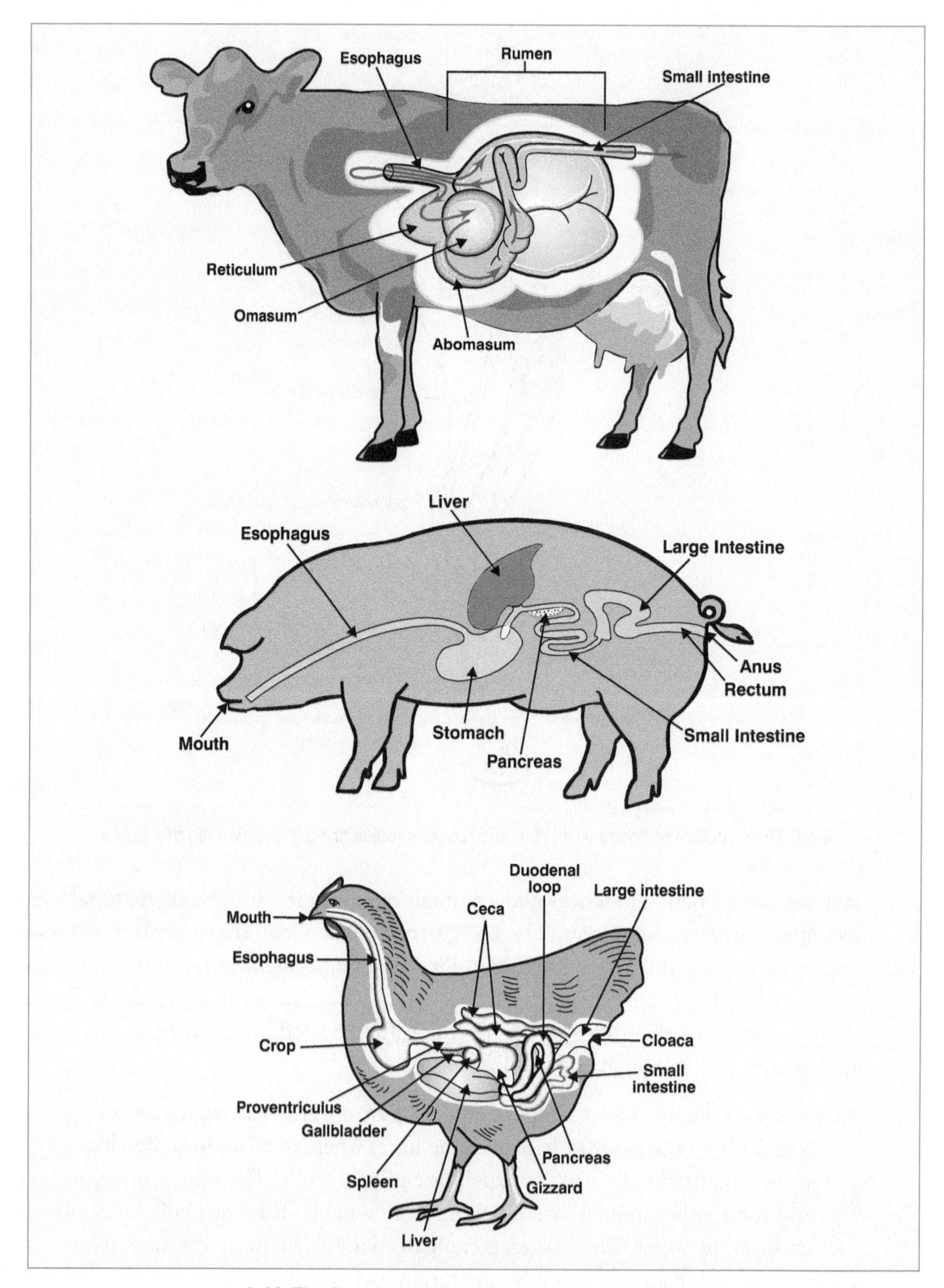

2–22. The digestive systems of a cow, hog, and chicken.

- Circulatory system—Nutrients, oxygen, and metabolic wastes are moved by the ***circulatory system***. Oxygen (from the lungs) and nutrients (absorbed in the small intestine) are transported to the cells, while wastes are removed. The circulatory system also moves hormones and protects against injury and microbes. The system consists of blood, heart, arteries, capillaries, and veins. The circulatory system of a horse is shown in Figure 2–24.

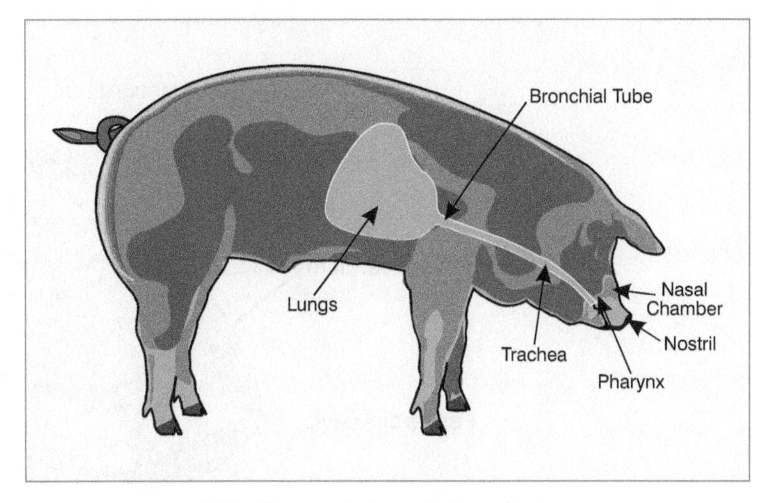

2–23. The respiratory system of a hog.

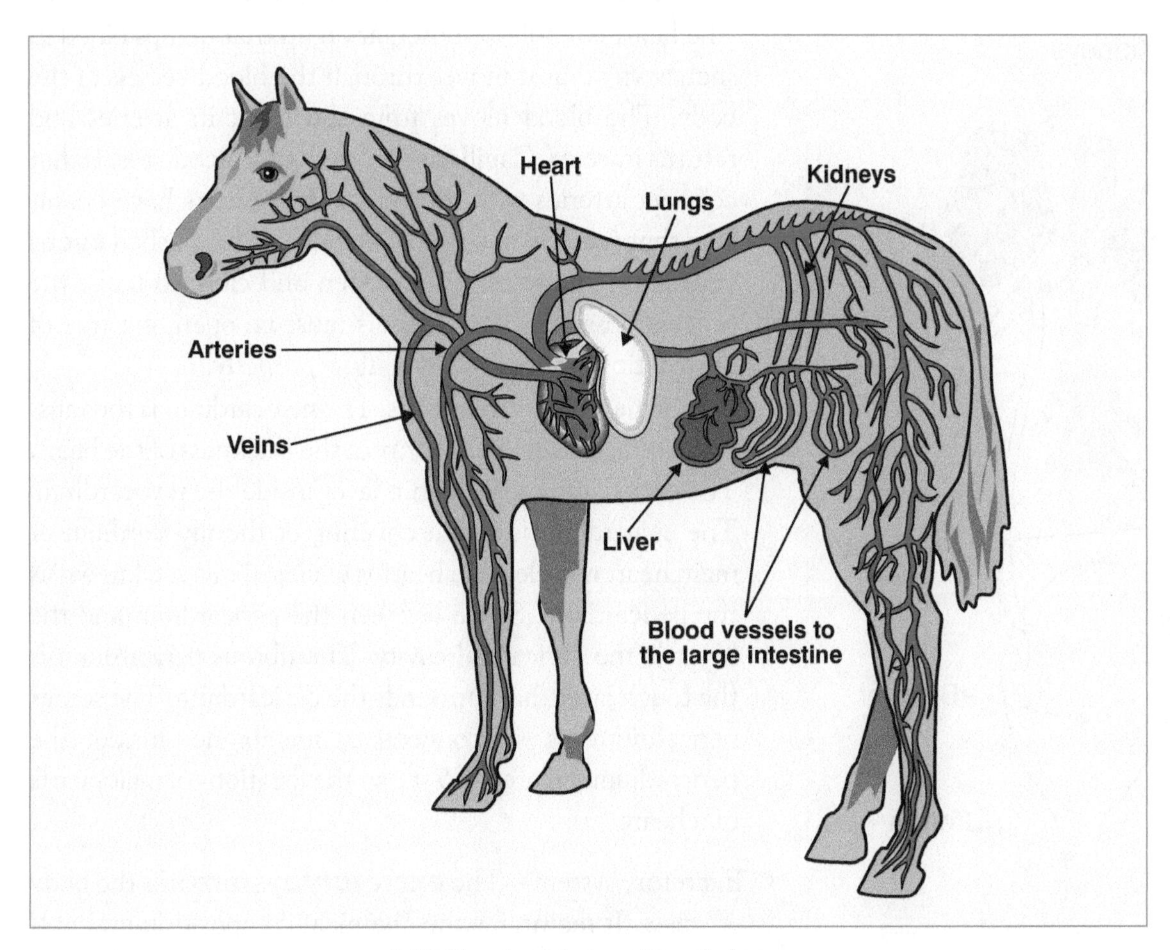

2–24. The circulatory system of a horse.

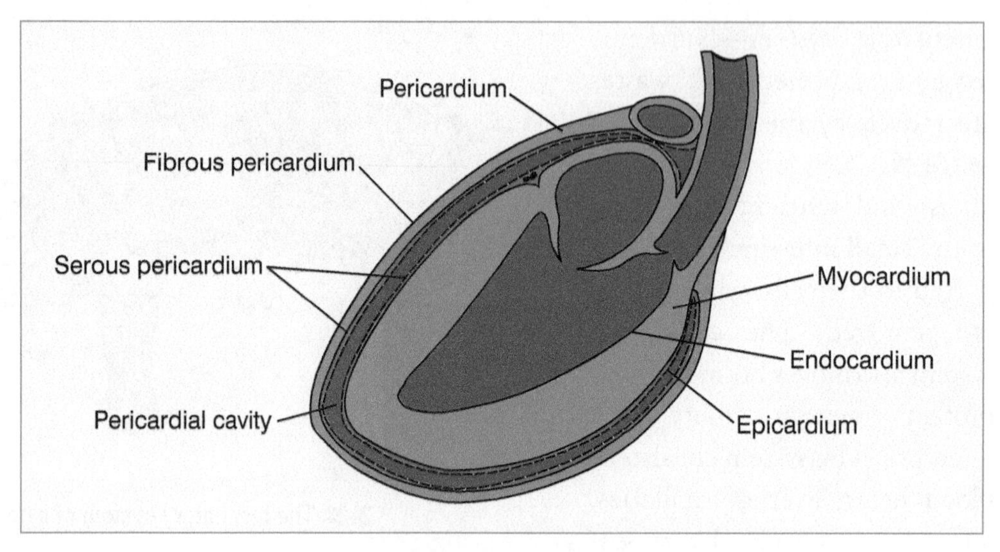

2–25. Major parts of a heart (longitudinal section).

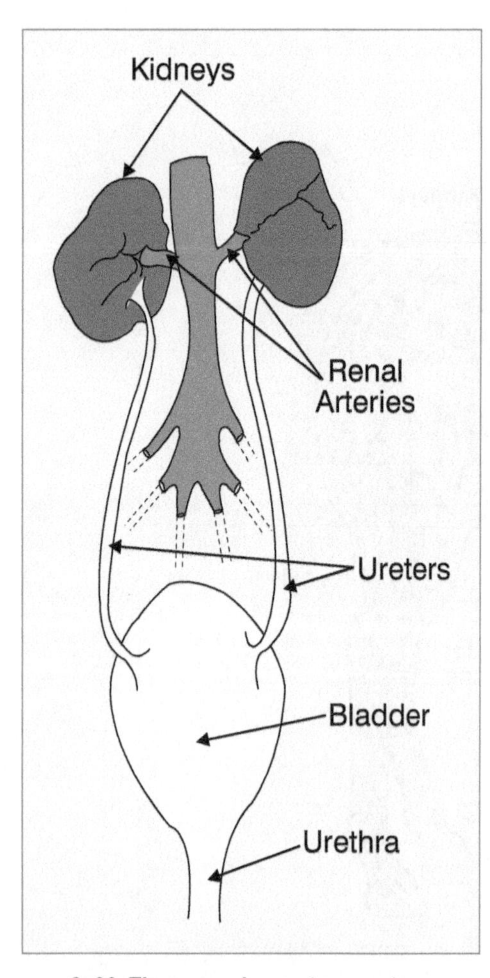

2–26. The general excretory system.

The major organ in the circulatory system is the heart. The heart is a hollow muscular organ that pumps blood in such a way that it moves through the blood vessels of the body. The blood moves from the heart in arteries and returns in veins. Capillaries are very tiny blood vessels that connect arteries with the veins. Most hearts have chambers that receive amounts of blood that are pushed out by contracting muscles. Valves open and close to make the process possible. Blood vessels must be open and free of obstructions for the blood to flow properly.

The heart has three layers. The myocardium is the muscle that makes up the majority of the thickness of the heart. The endocardium is the thin layer inside the myocardium. The epicardium is a thin covering of the myocardium or main heart muscle. The heart is enclosed in a sac known as the pericardium. Space between the pericardium and the heart is the pericardial cavity. The fibrous pericardium is the tough layer that surrounds the pericardium. The serous pericardium is a connective membrane inside the pericardium. Figure 2–25 shows the locations of major parts of a heart.

- Excretory system—The ***excretory system*** rids the body of waste. It maintains the chemical composition and volume of blood and tissue fluid. Wastes are expelled through the lungs, skin, bladder, and anus as the body rids itself of undigested food. Figure 2–26 shows the general excretory system.

- Lymphatic system—The ***lymphatic system*** circulates a clear fluid known as lymph. One role of lymph is to protect the body from disease. The lymphatic system transports excess tissue fluid to the blood, moves fat to the blood, and helps provide immunity against diseases.
- Nervous system—The ***nervous system*** coordinates body activities. In coordination with the endocrine system, nerve impulses react to stimuli. The system regulates other systems and controls learning and memory. The endocrine system secretes hormones that regulate the body's metabolism, growth, and reproductive systems. The nervous

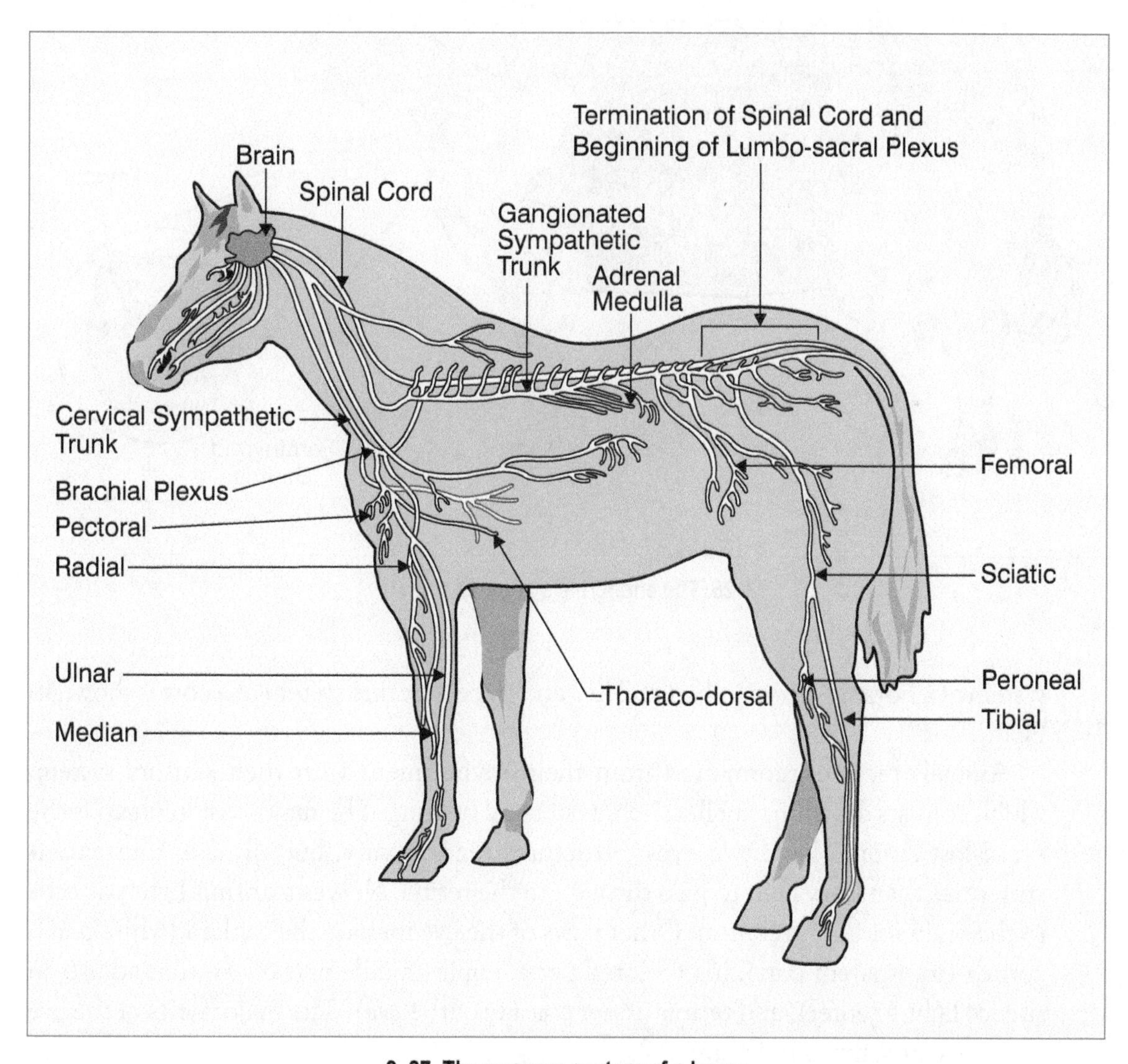

2–27. The nervous system of a horse.

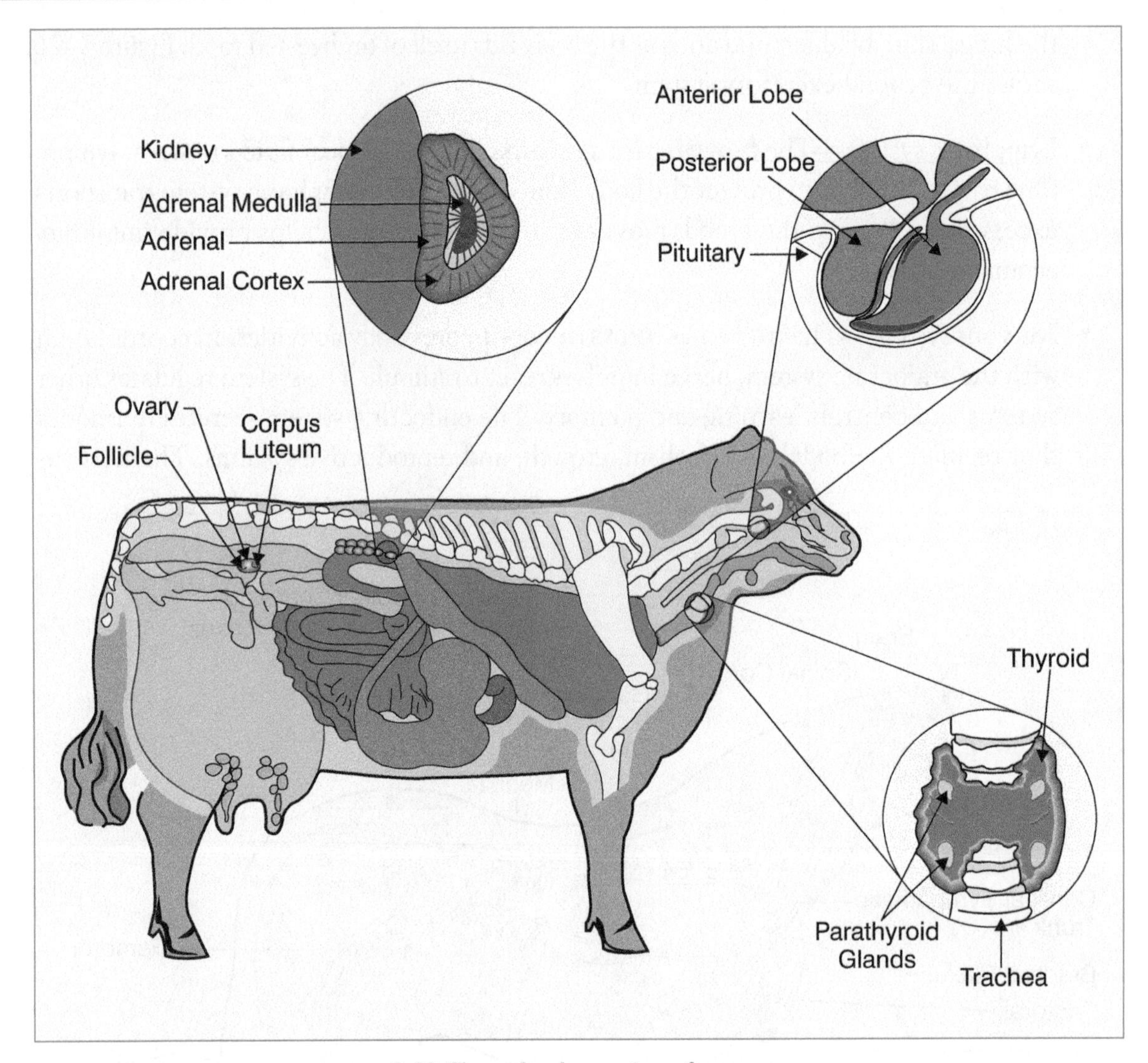

2–28. The endocrine system of a cow.

system of a horse is shown in Figure 2–27 and the endocrine system of a cow is shown in Figure 2–28.

Animals receive information from their environment with their sensory system, which consists of vision, smell, touch, taste, and hearing. The major vision organ is the eye. Most animals have two eyes. Structures tend to vary, but all have transparent structures that allow light to pass through to the retina. Nerves transmit light patterns to the brain for interpretation. Other parts of the eye include the schlera (white part), cornea (transparent part), iris (colored part), pupil (middle part of eye that adjusts in size for light to enter), and retina (inner coating of the eye). The major parts of the eye of a horse are shown in Figure 2–29.

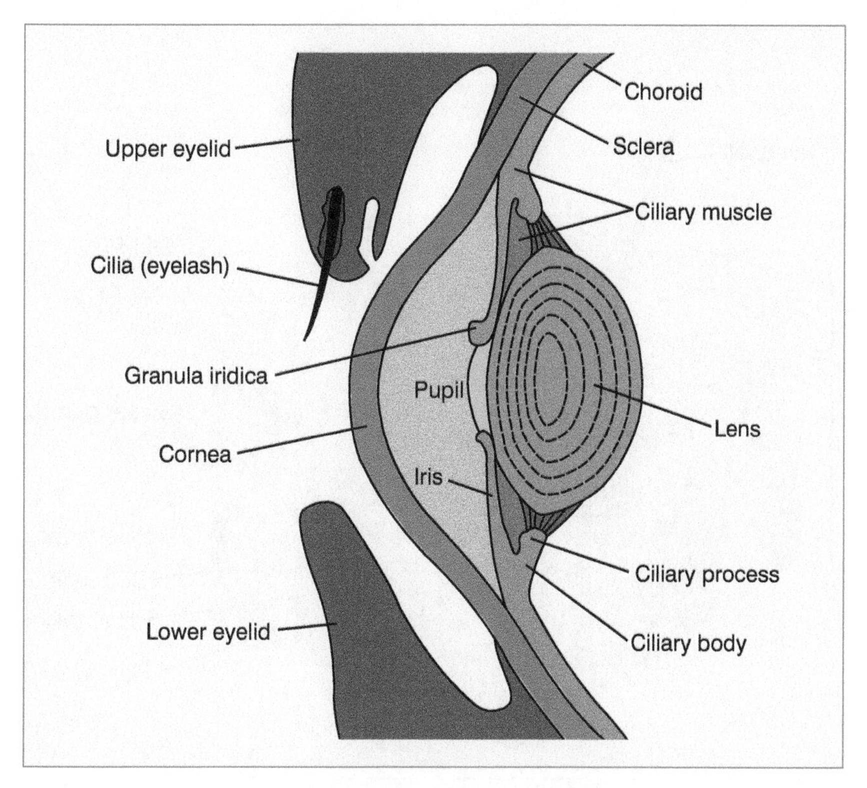

2–29. Major parts of the eye of a horse.

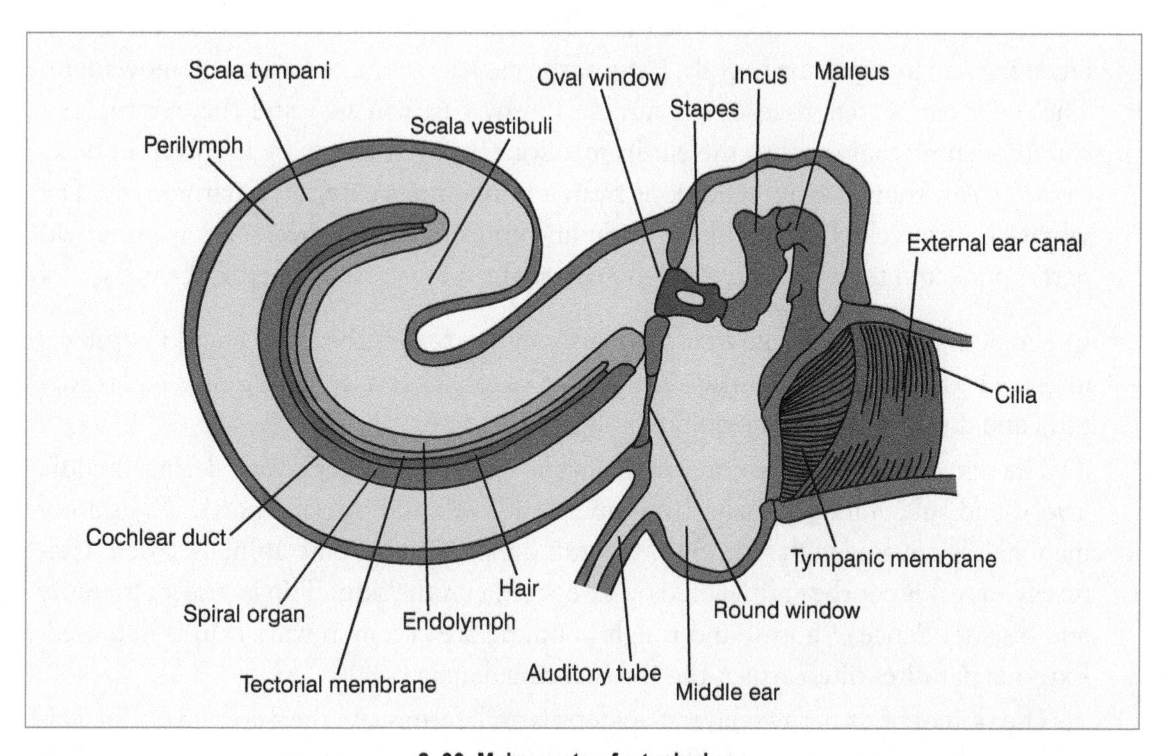

2–30. Major parts of a typical ear.

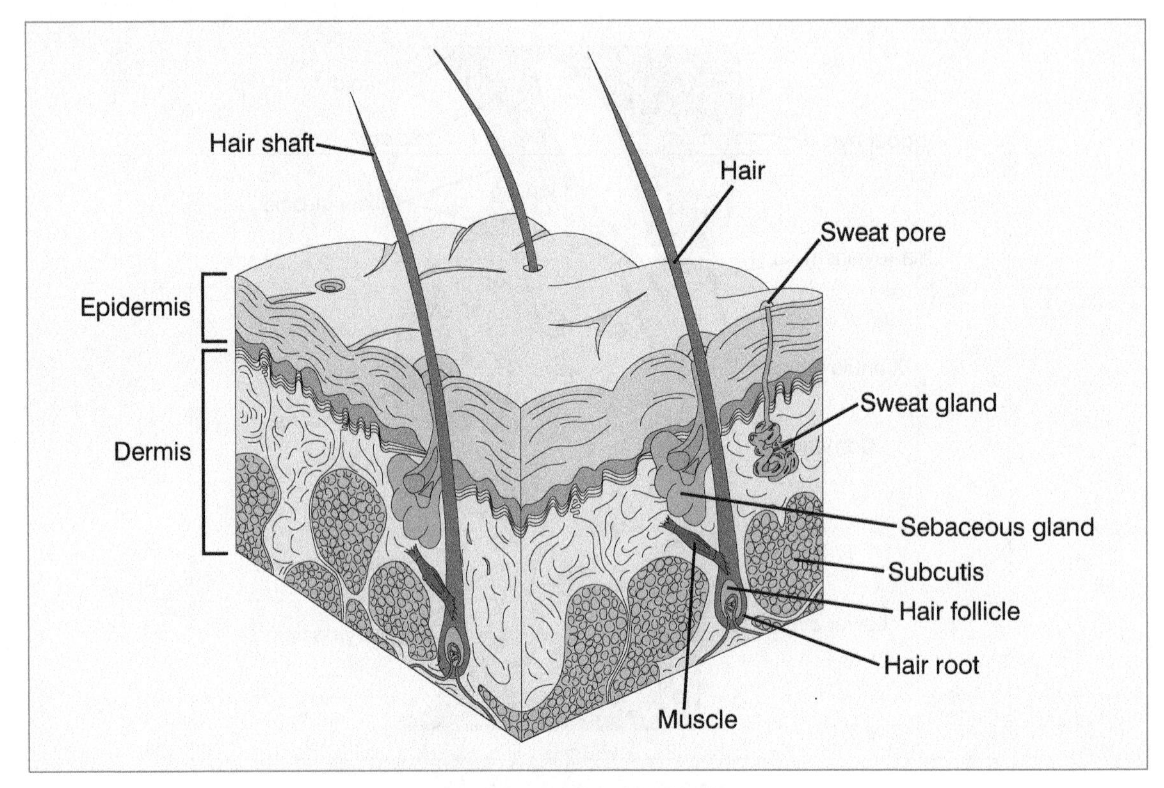

2–31. Sample of skin structure.

The ear is the organ used in hearing. The ear is part of a complicated process of receiving and interpreting sounds. Parts of the ear also promote balance in movement. The outer ear is comprised of the auricle (fleshy part you see) and the external ear canal (opening that leads to the eardrum). Bone and soft tissue are involved in hearing. The eardrum picks up vibrations from an organism's external environment. The vibrations are picked up by the eardrum and transferred as movements to other ear parts connected to nerves. Figure 2–30 shows the major parts of a typical ear.

- Integumentary system—The ***integumentary system*** is the skin. The skin protects internal body tissues from outside dangers. It keeps out foreign materials, such as bacteria and dust. It also helps regulate body temperature.

  The skin of mammals has three main layers: epidermis (outer layer), dermis (middle layer), and subcutaneous tissue (the innermost layer and thickest part). The skin of mammals is covered with hair. Sweat glands are present in some animals. Sweat (primarily water) is odorless until acted on by bacteria on the skin. Skin is subject to injury and disease. Fences, thorns, and rough handling are common ways skin is damaged. External parasites often attack the skin causing damage.

  The skin of birds has two layers: epidermis and dermis. As the outer layer, the epidermis is thin and comprised of flat-type cells that produce keratin. Keratin is a tough

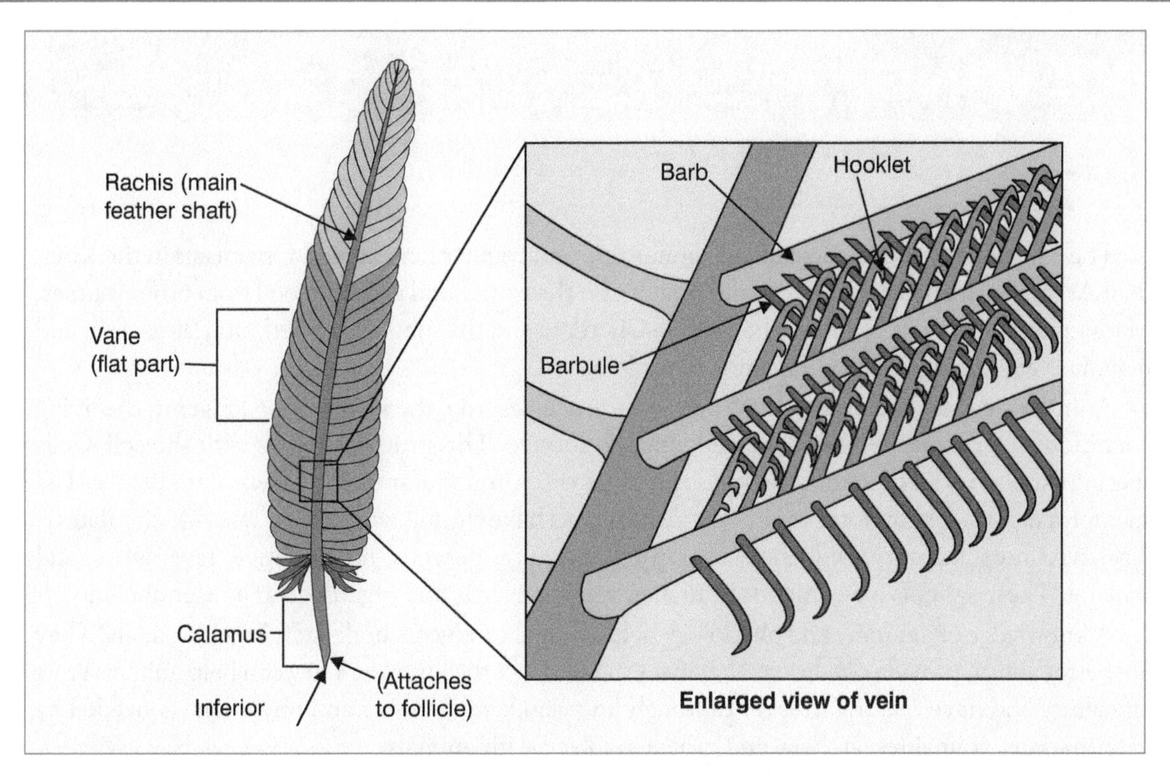

2–32. Major parts of a contour feather.

protein that is used in producing feathers and the outer sheath of beaks and claws. The dermis is thicker and stores fat to provide heat and other nutrients for the bird. The dermis contains blood vessels, nerves, and smooth muscles which regulate body heat by relaxing or "fluffing" the feathers. A cold bird "fluffs" its feathers to hold more heat next to the body.

Feathers are outgrowths of skin. Formed from protein, feathers are nonliving once they are fully developed. Feathers only have sensation at the follicle where they originate. Most birds have several kinds of feathers. Those most prominent that cover a bird's body are known as contour feathers. Small, soft, fluffy feathers on some birds are known as down. Feathers grow from follicles much like hair grows from the skin of mammals.

The condition of feathers reflects the health and status of the bird. Worn feathers are a sign of disease and damage caused by flight or other activity. Mature birds shed feathers and grow new ones in a process of molting (covered in Chapter 12).

- Reproductive system—The ***reproductive system*** produces offspring. New individuals of the same species are produced. The system goes through various processes and functions to assure successful reproduction. Some reproductive system structures are also included with the urinary system. The reproductive system is covered in detail in Chapter 5.

REVIEWING

MAIN IDEAS

The Kingdom Animalia includes all agricultural and companion animals. Organisms in the Kingdom Animalia are made of cells, can move about on their own, and get their food from other sources. Many common animals belong to the phylum Chordata and the subphylum Verbrata. Birds, fish, and mammals are among the most common Chordata.

Animals carry out life processes. When these processes stop, the animal is no longer in the living condition. A structure is needed for an animal to survive. The structure begins with the cell. Cells specialize to carry out unique activities for an organism. Groups of specialized cells form tissues. Tissues form organs. Organs form organ systems. Animals have the following organ systems: circulatory, digestive, integumentary, endocrine, excretory, muscular, nervous, reproductive, respiratory, and skeletal. These systems have important tissues, such as bones, and organs, such as eyes and ears.

A knowledge of anatomy and physiology helps animal producers understand their animals. They are better able to provide conditions that meet the needs of the animals. They can help animals grow efficiently and have healthy lives. A thorough knowledge of anatomy and physiology is needed by veterinarians, animal researchers, and others involved with animals.

QUESTIONS

Answer the following questions using complete sentences and correct spelling.

1. What are the three traits of organisms in the Kingdom Animalia?
2. What are the common characteristics of vertebrates?
3. Compare and contrast birds, bony fishes, and mammals.
4. What are the life processes of animals?
5. What is a cell? Cell specialization?
6. What is an organ system?
7. Distinguish between anatomy and physiology.
8. What are the major systems of animals? What is the function of each?
9. Compare and contrast the digestive systems of a hog, a cow, and a chicken.
10. Why is good muscle development important in selecting animals for meat production?

EVALUATING

Match the term with the correct definition. Write the letter of the term on the line provided.

a. vertebrate
b. anatomy
c. nervous tissue
d. invertebrate
e. physiology
f. homeostasis
g. taxonomy
h. mammal
i. growth plate
j. organ system

_____1. Responds to stimuli and transmits nerve impulses.
_____2. Animal with backbone.
_____3. Constancy of the internal environment of an organism.
_____4. Function of cells, tissues, organs, and systems.
_____5. Form, shape, and appearance of animals.
_____6. Females give birth to live young and secrete milk.
_____7. Animal that does not have a backbone.
_____8. Science of classifying organisms.
_____9. Location in bone development where cartilage is formed.
_____10. A group of organs that works together to carry out a specific function.

EXPLORING

1. Dissect a pig or other farm animal. Compare the various types of tissue and organ systems and their functions. Use proper procedures in your work. Follow all safety practices.

2. Visit a local farm or ranch that raises various agricultural animals or attend a livestock show. Compare and contrast the external parts of the animals. Use a camera to make photographs of the different animals. Prepare a poster or bulletin board that summarizes your observations.

3. Make a sketch on a separate sheet of paper of an animal that is raised on local farms or kept as a companion. Label the major external parts.

4. Visit a local pet shop. Observe the animals that are for sale. Compare the traits of animals of different species. Investigate the care needs of each species for its well-being.

5. Describe the digestive systems of non-ruminant animals such as swine and chickens. Prepare a one-page summary based on at least three references. Include a sketch of the digestive system of one species (refer to Figure 2-22 and make an online search for images of non-ruminant digestive systems).

CHAPTER 3

Animal Nutrition and Feeding

OBJECTIVES

This chapter covers the nutrition and feeding of animals. It has the following objectives:

1. List the major nutrient needs of animals and describe the purpose of each.
2. Relate feed requirements to the structure of digestive systems.
3. Describe the ways animals use nutrients.
4. Describe the types of feedstuffs.
5. Explain how animals are fed.
6. Describe how rations are formulated.

TERMS

abomasum
amino acids
balanced ration
calorie
carbohydrate
carnivore
concentrate
diet
fat
feed
feed analysis
feedstuff
fiber
forage
free access
growth
herbivore
lactation ration
lipid
maintenance
mineral
nutrient
nutrition
omasum
omnivore
palatability
Pearson Square Method
permanent pasture
protein
ration
reproduction ration
reticulum
roughage
rumen
scheduled feeding
stomach
supplement
temporary pasture
vitamin

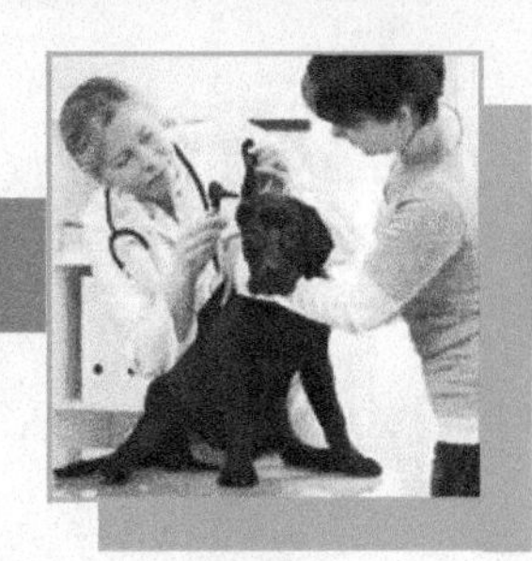

3–1. Japanese quail are being given an experimental feed. (The feed has been supplemented with a yeast extract. This is a test of the feed's efficacy against bacteria that create food poisoning in humans—*Salmonella* and *Campylobacter*.) (Courtesy, Agricultural Research Service, USDA.)

ANIMALS have specific nutrient needs. As an animal producer or keeper, you want to provide feed that meets the needs of your animals. Proper feed provides the nutrients that animals need to grow, be productive, and have good health. Without needed nutrients, animals fail to perform as they should.

Nutrient needs vary by species, age, activity, and other factors. Fortunately, there are ways of determining what your animals need to be fed. Most of the time, it isn't that difficult to meet an animal's needs. Researchers have determined the nutritional needs of common animal species. They have also determined the nutrient content of selected feedstuffs. By knowing the needs of animals and contents of feedstuffs, feed can be provided in the right amounts.

Interestingly, most animals tend to balance their diets if given the opportunity. Animals that graze in good pastures choose which plant materials to consume. Proper attention to feed nutrients is more important when animals are confined. This chapter will introduce you to animal nutrition and feeding.

3–2. Active animals need more energy than those that are not active. (Courtesy, Abramova Kseniya/Shutterstock)

3–3. Commercial feeds may be made with grain, soybean meal, minerals, vitamins, and other substances that promote animal health and well-being. (Courtesy, Education Images)

NUTRITION AND NUTRIENTS

Nutrition is the process by which animals eat and use food. Food is needed by animals to live, grow, lactate, reproduce, and work. All animals must have the food they need for a balanced diet.

Proper animal nutrition increases feed efficiency and rate of gain. It decreases the number of days required to grow a meat animal to market size. Producers want animals to grow fast. They also want them to make efficient use of the feed. Keepers of companion and exotic animals want them to be healthy and content. Feeding too much or the wrong feed is a big waste!

Animals must be fed a diet that meets their needs. If their needs are not met, they will not grow, can get sick, and may die. What animals are fed depends on many factors. Considering differences in digestive systems is important. For example, ruminants eat more grasses because their stomachs allow them to make use of these foods. The stages of life and the activities of the animal also affect feed needs.

NUTRIENT NEEDS OF ANIMALS

Feedstuffs contain nutrients. A ***nutrient*** is a substance that is necessary for an organism to live and grow. Nutrients make it possible for animals to carry out life processes. Nutritional requirements vary with age, stage of development, environmental conditions, activity, and genetic makeup.

A ***ration*** is the total amount of feed an animal has in a 24-hour period. The ration may be fed in certain amounts throughout the day or all at once. Feeding some animals a ration at one time may not be best. A ***balanced ration*** is one that contains all the nutrients that the animal needs in the correct proportions. Too much of a nutrient is wasteful and, in some cases, can harm an animal. A nutrient deficiency results in stunted growth and low production.

The nutrients required by all animals include water, carbohydrates, fats, proteins, minerals, and vitamins.

Water

Water is necessary for an animal to live. Animals can live longer without food than without water. Water is found in all the cells of the body. It makes up about 75 percent of the weight of an animal's body. Water content may be as high as 90 percent in newborn animals. Dairy cows may consume up to 50 gallons of water on a warm day. Adult beef cows may drink 10-20 gallons a day. The amount needed is also related to activity, gestation, and lactation.

3–4. Waterers suspended in a hog feeding facility provide readily-available, high-quality water. (Courtesy, Education Images)

Water has two basic functions:

- Regulating body temperature—Water helps control body temperature because it accumulates, transfers, and loses heat through evaporation.
- Promoting biochemical processes—All biochemical reactions require water. Water is the major component of cells, blood, body tissues, and milk. Without adequate water consumption, disease may develop, such as kidney disease in dogs.

Animals get water in several ways. Most of it is by drinking. Water is also in feed and may be produced through biochemical reactions. Water is lost through urine, feces, sweat, lactation, and vapor from the lungs.

Water should be clean. It should be free of chemical pollutants, mud, and disease-causing organisms. Some producers are now using portable watering systems to make water readily available wherever cattle are located.

Special attention to water supplies is needed in weather extremes. In the winter, cold temperatures cause water to freeze. Animals cannot drink frozen water! Water consumption increases in warm weather. Plenty of water should be available all of the time. Animals should have free access to water. Companion animals, for example, should be able to drink at will from a bowl or other container.

Carbohydrates

Carbohydrates provide energy. They are a major component of plant tissues. Carbohydrates should make up about 75 percent of an animal's diet. Besides energy, carbohy-

3–5. Cereal grains that may be used to provide carbohydrates in livestock feed include oats, barley, and corn. (Courtesy, Education Images)

drates aid in the use of proteins and fats. Carbohydrates provide energy for growth, maintenance, work, reproduction, and lactation.

Three types of carbohydrates are

- Sugars—Sugars are of two kinds: simple sugars (monosaccharides) and double sugars (disaccharides). Glucose and fructose are simple sugars. Sucrose is a double sugar and is used to make table sugar. Glucose is an excellent source of energy for most cells. Sugars are found in many foods, with fruit and milk being two examples. Sugars provide efficient energy.
- Starch—Starch is an important source of energy. It is found in grain, root crops, and other plant materials. Starch may be converted to glucose in the digestive process.
- Fiber—***Fiber*** is the material left after the food has been digested. It is made of plant cell walls and cellulose. Fiber helps the digestive system function smoothly. Fiber absorbs water and provides bulk. It also helps to increase the bacterial population in the rumen of cattle.

The major sources of carbohydrates are cereal grains such as corn, wheat, barley, oats, hay, and rye. Abnormal carbohydrate metabolism can lead to ketosis and diabetes.

Carbohydrates are not stored in the body. They must be eaten every day. A small amount of glucose is always in the blood. Unused glucose and carbohydrates are converted to fats.

Lipids (Fats)

A ***lipid*** is a nutrient that can be dissolved with ether. Ether is a colorless liquid solvent used in nutrition research. Using ether helps determine the nutrients in food materials.

Most lipids are fats or oils. A ***fat*** is a good source of energy. Fats contain the highest amounts of energy, with 2.25 times more energy than carbohydrates. Fats help supply energy for normal body maintenance. Fats also provide a healthy skin, keep the nervous system healthy, give food a good flavor, and carry fat-soluble vitamins A, D, E, and K. Animals do not develop a fat deficiency for energy. However, they can develop a vitamin deficiency.

Sources of fats in animal diets include meat, tallow, vegetable oil, tankage (processed meat and bones), cottonseed, and fish meal. No tankage should be fed ruminants.

Obesity in animals is related to fat intake. Research has shown a genetic link between fat accumulation in an animal and fat intake. This indicates the importance of genetic selection in animal breeding programs. Keepers of companion and exotic animals should regulate fat intake and provide exercise for animals to reduce the likelihood of obesity.

3–6. Feedstuffs may also promote animal well-being such as the Birdsfoot trefoil this cow is grazing. The forage contains tannin which is a natural antibloating compound. (Courtesy, Agricultural Research Service, USDA)

Protein

Protein is a nutrient needed to grow new tissue and repair old tissue. Protein is especially important for weight gain, growth, and gestation. It is used only for energy when carbohydrates and fats are deficient.

3–7. Broilers are fed specially formulated diets to assure rapid growth. (Courtesy, Education Images)

Protein serves many functions. It is found in wool, feathers, horns, claws, beaks, DNA, RNA, skin, hair, hooves, blood plasma, enzymes, hormones, and immune antibodies. Protein is needed every day for basic functions. It is also needed to build and repair cells. Three to five percent of the body's proteins are rebuilt every day. The highest amounts of protein can be found in the muscles of animals.

Protein contains amino acids. ***Amino acids*** are the building blocks of proteins. Organisms use amino acids to build proteins. Twenty-three amino acids have been found in protein. Ten of the amino acids are essential. The essential amino acids are:

- arginine
- histidine
- isoleucine
- leucine
- lysine
- methionine
- phenylalanine
- threonine
- tryptophan
- valine

The ten essential amino acids are used to make the other amino acids that the body needs. Animals with simple stomachs must eat a variety of food. Each food contributes only part of the essential amino acids. They can synthesize all but 10 or 11 of the amino acids.

Ruminants have microbes in their rumen. The microbes synthesize the essential amino acids from nonessential amino acids. The microbes are then digested in the digestive system. This provides the ruminant with essential amino acids. Because of this, much research has been done to increase the efficiency of ruminants.

CONNECTION

ENERGY TO WORK

Working animals need the energy to do their work. For example, a horse that works cattle and carries a rider much of the day uses a lot of energy. Proper nutrition is needed to assure that the horse maintains itself and keeps healthy.

Knowing what to feed requires information about the species of animal, its condition, and the nature of its work. An inactive animal doesn't need as much energy as an active animal. Animal keepers should refer to nutrition sources in selecting the ration to be fed. In some cases, feed is manufactured especially for the condition and use of the animal.

Failure to feed properly results in the animal losing weight and becoming inefficient. In some cases, disease can develop and the animal can become unable to do strenuous activity. This photo shows a healthy horse quite capable of strenuous activity while carrying the rider. (Courtesy, American Quarter Horse Association)

3–8. Baby pigs receive important nutrients from the milk of their mother. These rapidly growing pigs will soon be weaned and provided a manufactured, high-protein feed. (Courtesy, U.S. Department of Agriculture)

Sources of protein include soybean meal, cottonseed meal, fish meal, tankage, skim milk, and alfalfa hay. Protein is the most common nutrient deficiency. Most feedstuffs are low in protein. Supplements may be needed. For example, corn is deficient in lysine and tryptophan (amino acids). The deficiency results in slow growth. By supplementing corn with soybean meal, this problem can be solved.

Symptoms of a protein deficiency include anorexia, slow growth rate, decreased feed efficiency, anemia, edema, low birth weight of young, and lower milk production. Protein consumption is like a chain. When one amino acid is deficient, the whole body suffers.

Young animals need diets with higher amounts of protein than older animals. This is because protein is essential for growth.

Minerals

A ***mineral*** is an inorganic element found in small amounts in the body. Minerals are essential in skeleton growth and necessary for body systems to function properly. Feedstuffs must provide needed minerals.

Soil deficient in a particular mineral may produce crops that are low in the mineral. This can lead to deficiencies in feeds made from the crops. Some of these include selenium, copper, manganese, cobalt, and iodine. Mineral supplements are often used.

Minerals are necessary for a strong skeletal system. Minerals also make up organic compounds needed by muscles, organs, blood cells, and other soft tissue. They regulate many body functions. Examples of sources of minerals are alfalfa hay, cereal grains, bone meal, molasses, and salt.

Macrominerals are minerals required in large amounts. Microminerals, or trace minerals, are minerals required in smaller amounts. They are just as important as macrominerals.

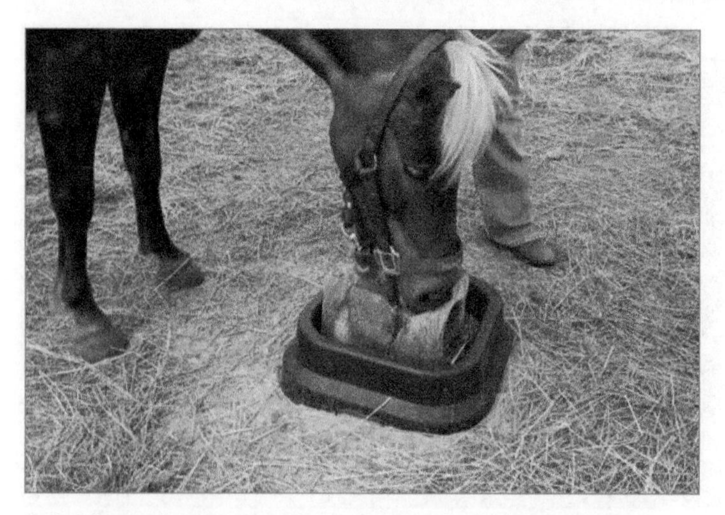

3–9. A horse licks a mineral supplement block to gain needed nutrients in small amounts. (The block is in a holder to keep it clean.) (Courtesy, Education Images)

Here is a summary of several minerals:

- Calcium—Calcium is a macromineral required in the highest amounts. Lack of calcium causes bones and teeth to form improperly. When calcium is not consumed on a daily basis, the body removes it from the bones for use in other body processes.
- Phosphorus—Phosphorus is another macromineral. Phosphorus is a key ingredient in the body's use of protein.
- Iron—Iron deficiency leads to anemia and a tired feeling. Iron is needed to make hemoglobin for red blood cells.
- Iodine—Lack of iodine will lead to fatigue, increased appetite, and a rapid pulse.
- Magnesium—Magnesium deficiency causes muscle tremors and shaking.
- Sodium and potassium—Sodium and potassium are needed to maintain water balance. Sodium and potassium transfer nutrients and waste through the cell membrane.

Deaths due to mineral deficiencies are rare, but inadequate amounts can cause economic losses to the producer. Deficiencies may cause a poor rate of gain, feed inefficiency, decreased reproduction, and a decrease in production of meat, milk, eggs, and wool.

Trace mineralized salt should be made available to animals. This will give most animals all the minerals they need.

Appendix A presents the mineral needs of animals. The functions and deficiency symptoms are included.

Vitamins

A ***vitamin*** is an organic substance needed in small quantities to perform specific functions. Several vitamins are needed. They do not provide energy but are necessary in using energy. Vitamins help to regulate body functions, keep the body healthy, and develop resistance to diseases. The deficiency of a vitamin can lead to disease and death.

Vitamins are either fat soluble or water soluble. Fat-soluble vitamins are stored in the fat and are released as they are needed by the body. These include vitamins A, D, E, and K.

3–10. Alfalfa hay provides needed vitamins for young dairy heifers. (Courtesy, Education Images)

Water-soluble vitamins are dissolved by water and need to be consumed every day. They include vitamin C and the B vitamins.

Ruminants do not have any problems getting the necessary nutrients. Vitamin A may be the only one that is not readily available in most feeds. Vitamin D can easily be obtained when animals are exposed to the sun.

Appendix B lists the functions and deficiency symptoms of vitamins.

FOODS AND DIGESTIVE SYSTEMS

Animal digestive systems vary. Differences in the systems affect the foods they eat. Feedstuffs should be selected based on the digestive system. Using the wrong feedstuffs results in poor growth and can lead to malnutrition. Cattle may perform well on good quality hay but hogs, hamsters, and chickens would not!

3–11. Sheep are herbivores. They eat grasses, legumes, and other succulent plants. (Courtesy, PHB.cz (Richard Semik) /Shutterstock)

FEEDING GROUPS

Animal species eat foods from different sources. The foods suit their digestive systems and natural feeding habits.

3–12. A badger is a carnivore. In the wild, badgers prey on small animals, such as squirrels. In captivity, they are often fed horse meat. (Courtesy, Sue Robinson/Shutterstock)

3–13. Hogs are omnivores. Their diets often include considerable grain, such as corn, and may contain processed sources of animal protein. (Courtesy, Education Images)

The major differences are the sources of the food: plant sources, animal sources, and combinations of plant and animal sources.

- Herbivores—A **herbivore** is an animal that eats foods from plant sources. The food materials may be leaves, stems, twigs, flowers, roots, and seeds. Cattle are herbivores that eat leaves and stems. Other herbivores, such as horses and swine, eat seeds, like grain. Herbivores are the most efficient users of food materials. Most agricultural animals are herbivores.
- Carnivores—A **carnivore** is an animal that eats foods from animal sources. The food is typically the flesh of other animals, particularly herbivores but they may also eat omnivores and, occasionally, other carnivores. Larger animals may eat smaller animals. Carnivores are sometimes divided into two groups:
    - Obligate carnivore—These are animals that depend solely on animal flesh for their food. Wolves, lions, bobcats, and falcons are examples.
    - Facultative carnivore—These are carnivores that consume some non-animal foods. Examples include domesticated dogs and cats that are often fed a manufactured food containing ingredients of both animal and plant origin. In some cases, carnivores are fed meat foods that have been manufactured specifically for them. Carnivores need a meat-based diet. Many exotic animals in zoological parks are carnivores, such as badgers and lions.
- Omnivores—An **omnivore** is an animal that eats both plant and animal food materials. Omnivores eat grain, leaves, meat, and similar food materials. Fish are often omnivores. Hogs will eat meat as well as grain food materials. Hogs fed grain along with a small amount of tankage (processed animal scraps) will gain weight more rapidly than those that are strictly fed a ration of plant-based materials.

DIGESTIVE SYSTEMS

Most animals have one of three types of digestive systems: monogastric, ruminant, and nonruminant.

Monogastric Systems

A monogastric digestive system has a simple stomach. The ***stomach*** is a muscular organ that stores ingested food and moves it into the small intestine. Some preparation for digestion occurs in the stomach.

Acid is secreted by the stomach. The acid results in a low pH of 1.5 to 2.5. The low pH destroys most bacteria and begins breaking down the food materials. Muscular movement of the stomach churns the contents and softens food materials.

Dogs, cats, chickens, hogs, and rats have monogastric stomachs. These animals are fed feedstuffs that are high in nutrients and digestibility. They do not do well on feedstuffs that are high in roughage and fiber.

Ruminant Systems

A ruminant digestive system has a large stomach divided into compartments. The largest section of the stomach is the ***rumen***. The rumen receives ingested food. It contains bacte-

CAREER PROFILE

ANIMAL NUTRITIONIST

An animal nutritionist investigates feeds and feedstuffs to assure that the nutrient needs of animals are met. Duties vary with the nature and size of the facility in which the research is done. It may include designing experiments, carrying out experiments, observing findings of experiments, and writing reports.

A good knowledge of feed chemistry and nutrition is needed. Most nutritionists have advanced degrees in nutrition or an area of animal science. Computer and statistical analysis skills are needed to select and use appropriate data analysis methods. Jobs for animal nutritionists are with universities, animal research facilities, and feed manufacturers.

This photo shows a nutritionist investigating the use of flax oil in feed pellets for rainbow trout. (Courtesy, Agricultural Research Service, USDA)

ria and other microbes that promote fermentation. Ruminant systems are designed for the food materials to be ingested, eructated (belched up), chewed, and swallowed again. The ***reticulum*** is the second segment of the stomach. It has many layer-like projections. The reticulum traps materials that are not food, such as nails or stones that have been ingested. The ***omasum*** follows the reticulum. The omasum is a small compartment that acts as a filter of materials for the next compartment—the ***abomasum***. The abomasum secretes gastric juices that kill the microbes that have passed with the food materials from the rumen. The microbes are digested by the juices (acid) and other substances in the abomasum. The intestines serve in a similar manner to those of the monogastric animal in the absorption of nutrients. (Note that ruminant systems are sometimes known as polygastric.)

Ruminant digestive systems use food materials high in roughage. This is why cattle are known as efficient animals on the grassland areas of the United States. The microbes that grow on the ingested forage in the rumen are sources of nutrients. Sheep, deer, cattle, and goats are examples of ruminants.

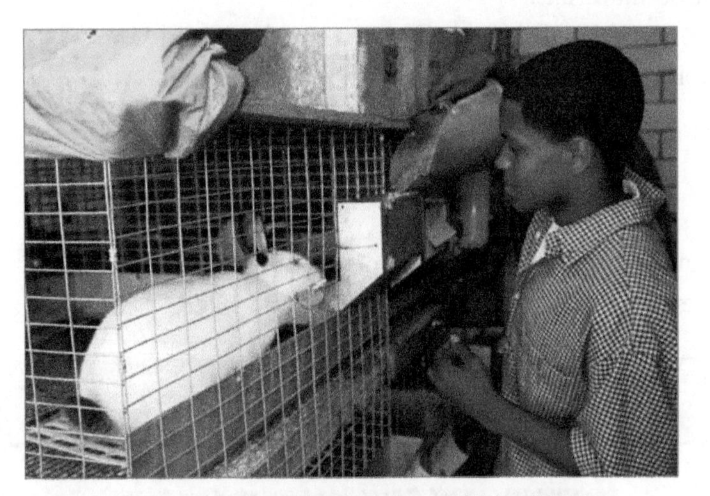

3–14. Rabbits are nonruminant herbivores. (This rabbit is receiving a small-size pellet form of feed manufactured especially for rabbits.) (Courtesy, Education Images)

Nonruminant Herbivores

A nonruminant herbivore is an animal that eats large amounts of roughage but does not have a stomach with several compartments. Horses, rabbits, and guinea pigs are examples of nonruminant herbivores. The digestive systems of these animals do some of the same functions as those of ruminants. These animals often eat forage as well as grains and other concentrated feeds.

NUTRIENT USES AND REQUIREMENTS

Nutrient needs vary with the species, age, size, condition, and environment of an animal. Nutrients are used to maintain bodies and for growth, reproduction, lactation, and work.

MAINTENANCE

Maintenance is keeping the body at a constant state. There is no loss or gain of weight. No matter what an animal is doing, nutrients are required. Animals require energy for internal workings of the body, heat to maintain body temperature, and small quantities of vitamins, minerals, and proteins.

A maintenance diet is usually high in carbohydrates and fats. It contains small amounts of protein, minerals, and vitamins. About one-half of the feed consumed is used for maintenance requirements. Animals on a maintenance diet are usually not growing, lactating, reproducing, or working.

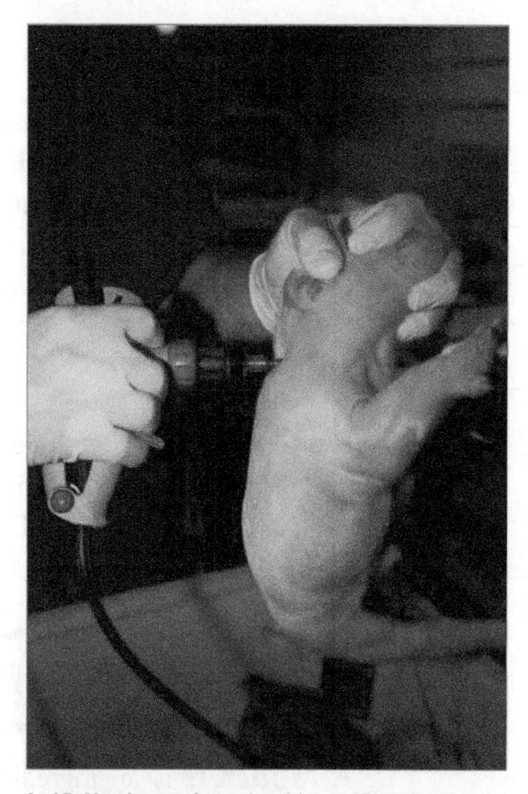

3–15. Newborn pigs raised in confinement are routinely given iron injections. (These injections are to ensure that all nutrient needs are met. When outside, pigs are able to root in the soil and gain iron. Iron is not available in that manner when raised on a concrete floor.) (Courtesy, Education Images)

GROWTH

In addition to maintenance, nutrients are needed for growth. ***Growth*** is an increase in the size of muscle, bones, and organs of the body. Meat production depends on the growth of the animal. Growth requires mostly energy and smaller amounts of the other nutrients. A growth ration is high in protein, vitamins, and minerals. For example, chickens are fed a carefully prepared ration so they grow rapidly.

Young animals and breeding stock should receive a ration that provides plenty of protein. Young animals are expected to grow rapidly and efficiently.

REPRODUCTION

Proper nutrition is essential for efficient reproduction. Poor nutrition is a major reason for reproductive failures. During the first trimester of pregnancy, the fetus requires large amounts of protein, minerals, and vitamins. A ***reproduction ration*** is high in nutrients needed by breeding animals. Females that do not receive a proper ration will give birth to underweight babies or abort the fetus before birth.

Males also need additional nutrients for reproduction. Lack of nutrients can lower sperm production and fertility rates. Males with inadequate diets may be unable to breed the females.

3–16. A lactation diet is needed by a nursing dog. (Courtesy, PHB.cz (Richard Semik) /Shutterstock)

LACTATION

Lactation is the production of milk. Females must have nutrients to produce large quantities of milk. A ***lactation ration*** is given to females who are producing milk. Females that are pregnant and lactating require even higher levels of nutrients.

Since milk is high in protein, calcium, and phosphorus, milking animals need diets high in these nutrients. Dairy cows need a ration that has been specially prepared for them.

WORK

Work and activity require energy. Examples are draft animals, such as mules, that pull heavy loads, race horses, hunting dogs, and guard llamas. Increased carbohydrates and fats supply the needed nutrients. Draft horses should receive additional grain in their diets to assure energy to do the work.

TYPES OF FEEDS

A ***feedstuff*** is an ingredient used in making feed for animals. Common feedstuffs include corn, soybean meal, and barley. Most feedstuffs have been analyzed to determine nutrient content. Some are high in certain nutrients and low in others.

Feed is what animals eat to get nutrients. Some feeds have few nutrients; others are high in nutrients. Most feeds provide more than one nutrient.

Feedstuffs can be fed to provide flavor, color, or texture to increase palatability. ***Palatability*** is how well an animal likes a feed. Some feedstuffs are made more palatable by adding molasses. Many animals like the sweet taste of molasses. A feed high in nutrients is of no benefit if animals refuse to eat it!

3–17. Holstein dairy cows are feeding on quality hay. (Courtesy, Agricultural Research Service, USDA)

FEED CLASSIFICATION

Feeds are classified as forages, concentrates, or supplements.

3–18. Crimson clover (left) and white clover (right) are used in some permanent pastures. (Courtesy, Education Images)

Roughages (Forages)

A ***roughage*** is a feed that is mostly leaves and tender stems of plants. They are also known as forages. A forage is a bulky feedstuff that is high in fiber–roughage. They are also low in protein and energy. Forages are fed because they are relatively inexpensive. Young, tender forage plants are more nutritious than older, mature plants.

3–19. Windrows of wheat plants are sun curing until the proper stage for chopping in making haylage. The wheat was cut as it was approaching maturity or milk stage of kernels. Haylage is chopped at 40-60 percent moisture—silage is 70 percent or more moisture. (Courtesy, Education Images)

Grasses and legumes are grown in pastures as forage. Pastures may be permanent or temporary. A ***permanent pasture*** has plants that live from one year to the next. Crimson clover and white clover are examples of two legumes used in permanent pastures. Most clovers are planted in mixtures with grass. Examples of common grasses are Bermudagrass, Kentucky blue grass, and bromegrass. Native grasses are predominant on rangelands. A ***temporary pasture*** is planted to forages for grazing during one season or year. Most temporary pastures are planted to high-yielding forages, such as millet or milo.

3–20. Round bales of hay are covered for protection from rain. Moisture causes hay to deteriorate and lose nutritional value. (Courtesy, Education Images)

Some forages are harvested for feeding at a later day. Most important of these are hay, silage, and haylage. Hay is typically grasses and legumes that have been cut and sun cured to lower moisture content to 40 percent or less—sometimes as low as 15 percent. Hay is often baled into rectangular bales or large round rolls and stored until needed. Silage is fermented high moisture fodder (such as corn) that is cut, chopped, and stored in air-tight upright or ground silos. Moisture content is 70 percent or more. Haylage is a roughage that has lower moisture than silage but more than hay. It is cut, chopped, and stored in conditions that prevent air contact and often involve fermentation. In some cases, crop residue is also used as forage. They contain high levels of fiber. Cattle often only eat forages. Some cattle in high-production uses are provided concentrates and supplements in addition to forage.

Forages are harvested at the vegetative stage. As the plant matures, it loses energy because the energy is being put into the seed for reproduction. The younger the plant, the more energy and protein it will contain.

Concentrates

A ***concentrate*** is a feed that is high in energy or protein. Concentrates have more energy per pound than forages but are usually lower in protein. Examples of high-energy concentrates are corn, wheat, sorghum, barley, rye, and oats. Examples of high-protein concentrates are soybean oil meal, cottonseed oil meal, and sunflower meals. Higher producing animals need more nutrients from concentrates.

Cattle are usually not fed concentrates in a maintenance ration. They are fed if the forage is of very poor quality. Sheep may also receive concentrates to supplement forage.

Many small animals need concentrates as a regular part of their diet. Cats, dogs, hogs, and similar animals are fed diets appropriate to their needs.

3–21. Three forms of feed concentrates. Finely ground meal is on the left and is for baby animals, such as chicks or fish. The small pellets in the center are for calves, dogs, and growing fish. The large range cubes are for cattle on pasture that need supplemental feed. (Courtesy, Education Images)

Supplements

High-producing animals are given supplements so they get the protein, minerals, and vitamins that they require. A ***supplement*** is a feed material high in a specific nutrient. Some contain several nutrients. Examples of protein supplements include meals made from soybeans, cottonseed, corn gluten, sunflower meal, canola, and coconut.

Protein quality is usually less important to ruminants than nonruminants. Microbes present in the rumen produce protein. Nonprotein sources of nitrogen, such as urea or ammoniated molasses, are available for ruminants to eat.

Salt and mineral blocks can be placed in a feedlot or pasture for free-choice feeding.

GOOD FEEDSTUFFS

All animals require different amounts of nutrients. All feedstuffs provide differing amounts of nutrients.

3–22. Weighing feed allows exact feeding of an individual animal. (Weighing is not very practical in a facility that produces more than a few animals unless an automated computer-based system is used.) (Courtesy, Education Images)

Rations and Diets

A ration should provide the right amount and proportion of nutrients. A ***diet*** is the type and amount of feed an animal receives in its ration. Diet also includes water. Diets should be based on the needs of the animal being fed. Diets also vary according to the nutrient content of the feed.

A good ration should be balanced, have variety, be succulent, be palatable, bulky, economical, and suitable. A balanced ration will increase feed gain, decrease expense, and increase profits. A variety of feeds will make ration-balancing easier and increase palatability. A succulent ration that is juicy and fresh will increase production. Bulky rations aid in digestibility because of the fiber. Economical rations should provide needed nutrients and maximize profits. A ration should be suited to the type of animal.

Some animals have diets that require live food. Snakes, lobsters, hedgehogs, and spiders are examples of such animals. A supply of "live" food must be available and provided to the animal, such as a mouse for a snake.

Feed Analysis

Feed analysis is the nutrient content of a feedstuff or prepared mixed feed. Chemical laboratory methods are used in analyzing feed. The information is important in selecting the diets of animals to assure a balanced ration. Labels on feed containers often report some of the information.

An analysis of a feed sample usually provides information in several areas. These are

- Dry matter—Dry matter is the weight of feed materials after moisture has been removed. Feed quality is based on the proportion of water in feed. Silage has much higher water than No. 2 Yellow Corn.

- Crude protein—Crude protein is the nitrogen content of feed multiplied by 6.25 (a constant factor). Feedstuffs with higher crude protein are typically more nutritious.
- Fat—Fat content is determined by using an ether extract process. The fat is dissolved by the ether. The remaining feed material is weighed and a percentage of fat is calculated.
- Ash—Ash is the residue that remains after feed has been burned at a temperature of 600°C.
- Crude fiber—Crude fiber is determined by boiling the feed material in an acid and using laboratory procedures to dry the feed. The weight before and after drying is determined and used to calculate percent.
- Nitrogen-free extract (NFE)—The percentage of NFE is determined by subtracting the percentage of water, crude protein, ether extract, ash, and crude fiber from 100.
- Nutrients—Feed materials can be analyzed for nutrient content. Examples include calcium, phosphorus, and copper.

HOW ANIMALS ARE FED

How and when animals are fed is important in gaining production and growth. Animals need to consume the recommended amounts of each nutrient without overeating. They need feed for maintenance, growth, reproduction, lactation, and work. Feed can either be fed at scheduled times or be fed free access.

3–23. Feeding quality hay on a free-access basis to cattle provides the energy (and heat) animals need to withstand very cold weather on a range. (Courtesy, Natural Resources Conservation Service, USDA)

FEED QUALITY

Regardless of how animals are fed, the feed quality should be high. This means the feeds are high in needed nutrients, free of toxins and spoilage, and properly stored. Feeds should be made with palatable ingredients appropriate to the digestive system of the species being fed. Some issues with quality include feed that contains bacteria, chemical residues, animal dung, or other impurities. Manufactured feeds should be properly labeled, stored, and of consistent quality. Costs of feeds should be kept low to promote profit but, at the same time, meet the needs of the animals being fed.

FREE ACCESS

Free access or free choice is allowing animals to eat feed whenever they want. The feed is provided in feeders and is always available. Hay is commonly fed free access to cattle. Hogs can be fed concentrates free access because they will not overeat. Cattle do not receive concentrates free access because they will overeat.

Livestock on pasture are commonly provided with salt or mineral blocks free access. Of course, the pasture is free access. Livestock can graze as they wish. Water is also usually provided free access.

3–24. Guinea pig pellets should be nutritious and low in fat. (Pellets should be supplemented with high-quality grass hay or dark, leafy greens such as turnip or dandelion leaves. A high-fat diet may result in a guinea pig becoming obese.) (Courtesy, Education Images)

SCHEDULED FEEDING

Many animals are fed at the same time each day. Feeding times are based on the needs of the animal or management practices.

Scheduled feeding is providing feed at certain times of the day. It is commonly used with cattle because they will overeat. Dairy cows are usually fed at milking time. While they are in the barn, they will each be given an individual ration.

Baby fish, known as fry, are fed once every hour. Their digestive systems will not hold much food because they are small. They are unable to store much energy and thus need to be fed frequently.

Computer feeding systems are becoming more popular with cattle and hogs. Computerized chips can be placed in ear tags or under the skin in the ear. A computer will read the personal identification number and place an amount of feed into a feeding trough.

FORMULATING RATIONS

3–25. A cat is provided a balanced ration. (Courtesy, Education

Providing the feed animals need involves two major areas: knowing the nutrient needs of the animals and knowing the nutrient contents of feedstuffs. Scientists have studied the nutrient needs of animals. Major nutrient needs in most diets are protein and energy. Minerals and vitamins are also important.

Ration formulation is combining various feedstuffs to make a ration (feed) for animals of the same species and condition in life. The ration must be palatable, suited to digestive system, and eaten in the needed amount to meet daily nutrient requirements. It should be made of ingredients (feedstuffs) that keep costs of the feed low so that the animal enterprise will more likely be profitable. It is essential that the feed provide the needed nutrients for maintenance, growth, work, milk production, and reproduction.

USING NUTRITION INFORMATION

Protein and energy are the nutrients in greatest amounts in rations.

Protein is stated as crude protein and is given as a percentage or in grams. Protein needs are higher for young, lactating, and pregnant animals. The needs of an animal must be matched with its diet. Some feedstuffs are higher in protein than others. Most concentrates have higher protein than roughages.

Energy is from carbohydrates, fats, and some proteins in feed. Energy is stated as total digestible nutrients (TDN). The calorie is the modern way of stating energy.

3–26. Read nutrient information on the label of canned food before buying it to be sure the needs of your animal will be met. (Courtesy, Education Images)

7 16374 20070 9

CO-OP

FRY FISH FEED

—Manufactured By—

ALABAMA FARMERS COOPERATIVE, INC.
DECATUR, ALABAMA 35609

GUARANTEED ANALYSIS

Crude Protein, not less than 49.00%
Crude Fat, not less than 9.00%
Crude Fiber, not more than 2.00%

INGREDIENTS: Fish Meal, Meat Meal, Flash Dried Blood Meal, Dried Whey, Ground Corn, Fish Oil, Vitamin A Supplement, Vitamin D-3 Supplement, Vitamin E Supplement, Riboflavin Supplement, Calcium Pantothenate, Niacin Supplement, Vitamin B-12 Supplement, Choline Chloride, Menadione Sodium Bisulfite, Thiamine Mononitrate, Ascorbic Acid, Pyridoxine Hydrochloride, Folic Acid, Ethoxyquin (A preservative), Ground Limestone, and traces of Manganous Oxide, Calcium Iodate, Copper Oxide, Cobalt Carbonate, Zinc Oxide, Iron Carbonate and Sodium Selenite.

Tag Code: 021092

3–27. Sample feed label showing ingredients, analysis, and manufacturer.

3–28. A mill that manufactures feed for a feedlot. (Courtesy, Education Images)

Energy is measured in calories. A ***calorie*** is the amount of heat needed to raise the temperature of 1 gram of water 1 degree C. Most animals require a large number of calories each day. Calories in feed or as requirements are stated as kilocalorie (kcal) or megacalorie (Mcal). A kcal is 1,000 calories. An Mcal is 1 million calories.

Table 3–1. Partial Daily Nutrient Needs of Selected Animals

Species/Condition	Weight (kg)	Selected Nutrient Needs		
		Digestible Energy	Calcium	Crude Protein
Mature working horse (moderate work, such as roping cattle)	900	36.2 (Mcal)	44 (g)	10.4%
Colt (weanling, 6 months)	335	23.4 (Mcal)	44 (g)	14.5%
Swine (weaned pig)	20	3,230 (kcal)	—	18.0%
Swine (pregnant sow)	140	3,340 (kcal)	—	12.0%
Cow (dairy, lactating, and pregnant)	450	19.9 (Mcal)	30 (g)	973 (g)
Rabbit (maintenance, adult)	2	4,200 (kcal)	—	12.0%
Chicken (4 week-old broiler)	0.5	—	0.9%	20.0%

Mcal = 1,000,000 calories

kcal = 1,000 calories

Note: Convert kg to pounds by multiplying kg by 2.2046.

Source: Information in this table was adapted and summarized from multiple sources.

Table 3–2. Examples of Nutrient Content of Selected Feedstuffs

Feed	Dry Matter	Crude Protein	Fat	Ash	NFE*	TDN**	DE***
	(%)						(Mcal/kg)
Roughages							
Alfalfa hay (mid-bloom, sun cured)	91	17.1	3.3	7.8	37.4	52	2.46
Bermudagrass (fresh)	29	4.2	0.6	3.3	13.0	17	0.77
Clover (fresh Ladino)	18	4.4	0.9	1.9	8.1	13	0.60
Millet (foxtail, fresh)	29	2.8	0.9	2.5	13.4	18	0.77
Sorghum fodder (with heads, sun cured)	90	6.2	2.0	8.9	47.4	51	2.24
Wheat straw	90	3.2	1.8	6.9	40.4	40	1.90
Concentrates							
Barley grain(all analysis)	88	11.7	1.7	2.4	67.7	75	3.42
Corn (#2 grain)	87	8.9	4.0	1.2	71.3	80	3.47
Cotton Seed Meal (solvent)	93	41.2	4.7	6.1	28.9	70	3.27
Oats (grain, all analysis)	89	11.9	4.7	3.1	58.9	69	3.00
Soybean meal (solvent)	89	44.4	1.5	6.4	30.6	76	1.45
Other							
Molasses (black strap)	74	4.3	0.2	9.8	59.7	60	2.68

*NFE=nitrogen free extract.

**Based on ruminant digestion.

***DE = digestible energy.

Source: Information in this table adapted from feed composition tables in M. E. Ensminger and R. C. Perry, *Beef Cattle Science*, 7th Edition. Danville, Illinois: Interstate Publishers, Inc., 1997.

Scientists have determined the energy in feedstuffs as well as the daily energy requirements of many animals.

The National Research Council (NRC) has prepared tables that provide the nutrient composition of feedstuffs. All feedstuffs have been given a number. That number is an international feed number (IFN). The number standardizes the identification of feedstuffs worldwide. For example, the IFN for #2 shelled corn is 4-02-931.

Nutritional information about feeds is used to formulate rations. The amount of each nutrient is figured into the ration. This is based on the nutrient requirements of the animal. The information tells how much roughage, concentrate, and supplement is needed. This is done with a calculator or with a computer program. Feed specialists formulate rations for specific needs.

3–29. Lambs are being provided a growth diet. (Courtesy, Education Images)

3–30. Manufactured feed may be sold in bags. (Courtesy, Education Images)

In ration formulation, the feedstuffs are selected to give the least cost ration. Computer programs can also compute feed rations that are economical as well as nutrient balanced. For example, more corn may be in a ration if it costs less than barley.

Manufactured feeds are made to contain specific nutrients. Nutrients are often stated as percentages. Labels attached to feed containers provide nutrient information. Labels give the amount of protein, the ingredients, and the name and address of the manufacturer. Labels may also have a bar code. Bar codes are electronically scanned and used in tracking a feed. The newer QR Code (Quick Response Code) is now used on some food labels.

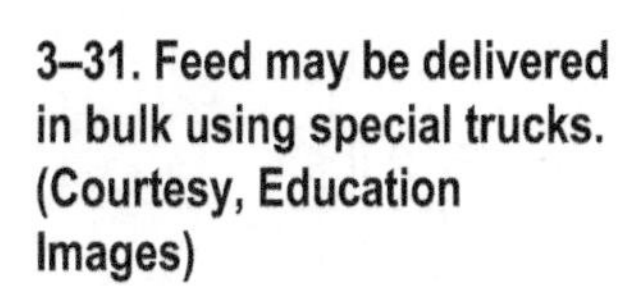

3–31. Feed may be delivered in bulk using special trucks. (Courtesy, Education Images)

STEPS	EXAMPLE
Step 1. Draw a 1- to 2-inch square. Place diagonal lines across the square. **Step 2.** Write the percentage of crude protein needed by the animal in the center of the square where the diagonal lines cross. **Step 3.** Write the feeds to be used at each left corner. Place the percent of crude protein in the feeds after the name of feed.	**Weaned pig weighing 20 kg (44 lbs.)** Corn (#2 grain) 8.9 18.0 Soybean meal 44.4
Step 4. Subtract the smaller of the numbers from the larger numbers. (This involves crude protein needed by the animal and that provided by the feed.) Write the difference at opposite corners.	8.9 — 26.4 parts corn 18.0 44.4 — 9.1 parts soybean meal
Step 5. The numbers at the two right corners are parts of the two feed ingredients that are needed. (Parts can be measured as weight or volume just so the proportion remains as was calculated.)	26.4 parts corn + 9.1 parts soybean meal **35.5 Total Parts**
Step 6. The percentage of each feed needed in the ration can be found by dividing the number of parts by the total parts.	26.4 ÷ 35.5 = 74.4% corn 9.1 ÷ 35.5 = 25.6% soybean meal
Step 7. The amount of each feed ingredient for a large batch of feed is determined by multiplying the percentage of each by the total amount of feed desired.	If batch = 1 ton (2,000 lbs.): shelled corn = 1,488 lbs. soybean meal = 512 lbs. 2,000 lbs. Note: Vitamin and mineral supplements may be needed in the feed.

3–32. How to use the Pearson Square method.

USING THE PEARSON SQUARE METHOD

The ***Pearson Square method*** is a simple way to calculate a ration for a specific animal. It can also be used to calculate ingredients for batches of feed. The method is satisfactory with some animal production. The calculated ration may be adequate in terms of protein but deficient in minerals and vitamins.

Using the Pearson Square method requires information on (1) the nutrient needs of the animal, and (2) the nutrient content of feed. Information on the nutrient content of feed is provided in feed composition tables. The nutrient needs of animals are available in feeding standards tables.

The feedstuffs chosen must be appropriate and practical. It is possible to calculate a ration that is impractical, such as one based on an animal eating a greater quantity of low-quality roughage than it can possibly hold.

Use the example in Figure 3–32 as a guide for calculating other rations. Refer to nutrition books for additional information on feedstuffs and feeding standards.

REVIEWING

MAIN IDEAS

Nutrition is the process by which animals eat and use food. The food is used for maintenance, growth, reproduction, lactation, and activity. Nutrients are substances necessary for an organism to live, grow, and function. The nutrients that all animals require include water, carbohydrates, lipids (fats), proteins, minerals, and vitamins.

A maintenance diet is one in which the animal consumes enough to prevent loss or gain of the body. It is high in carbohydrates and fats. Growing animals need more protein, vitamins, and minerals. A reproduction ration includes extra protein, minerals, and vitamins for the fetus. Lactating animals require additional protein and minerals. Work utilizes high energy feed.

Feedstuffs must be appropriate to the feeding groups and digestive systems. The feeding groups are herbivores (eat plant foods), carnivores (eat animal flesh), and omnivores (eat both plant and flesh foods). Digestive systems are monogastric, ruminant (polygastric), and nonruminant herbivore.

Forages, concentrates, and supplements are the three different types of feed. Forages are feeds made up of leaves and stalks of plants. Concentrates are feeds high in energy and low in protein. Supplements are high in protein, minerals, and vitamins.

Animals can be fed free access at all times or they can be fed at scheduled times.

Feed analysis is used in balancing rations. Nutrition information from the National Research Council is needed. The nutrient content of feed and the needs of the animal are considered in balancing a ration. Feed manufacturing companies use computer programs to formulate rations. The Pearson Square method is a simple way to balance a ration.

QUESTIONS

Answer the following questions using complete sentences and correct spelling.

1. What is nutrition? Why is it necessary to understand the importance of nutrition?
2. What is a ration? Balanced ration?
3. What are nutrients? What are the six nutrients that all animals require?
4. Describe the functions of water.
5. What are carbohydrates? Describe the functions of carbohydrates.
6. What are lipids (fats)? What is the main function of fats?
7. What is the primary function of proteins?
8. What are the feeding groups? Briefly explain each.
9. What are the three types of digestive systems? Briefly explain each.
10. What are minerals? Why are minerals important?
11. Define vitamins. Why are vitamins required by animals?
12. Compare and contrast the five nutrient requirements of animals.
13. Compare and contrast the three types of feeds.
14. Describe the characteristics of good feedstuffs.
15. How are animals fed?
16. What is the Pearson Square method?
17. What is ration formulation? Why are least-cost factors important?

EVALUATING

Match the term with the correct definition. Write the letter of the term on the line provided.

a. nutrition
b. ration
c. fiber
d. protein
e. lactation ration
f. feedstuff
g. concentrate
h. supplement
i. free access
j. scheduled feeding

______1. Providing feed at a certain time.

______2. A nutrient used to build new tissue and repair old tissue.

______3. The process by which animals eat and use food.

______4. The material left after food has been digested.

______5. The amount of feed an animal has in a 24-hour period.

______6. A ration for females producing milk.

______7. Ingredient used to make feed.

______8. Material fed animals for nutrients.

______9. Feed material high in a specific nutrient.

_____10. When feed is available for animals when they want to eat.

EXPLORING

1. Visit a local feed mill. Determine the kinds of feeds sold and used by local animal farms. Look at how the feed is packaged. Observe where the ingredients are stored. Prepare a report on your observations.

2. Interview a local veterinarian concerning animal nutrition and the common problems that they deal with.

3. Interview a nutritionist at a feed company about animal feeding and rations and common problems with which they deal.

4. Tour a local farm. Compare the various rations and feedstuffs that are fed to different animals depending on their ages and uses.

5. Use the Pearson Square method and calculate the ration for an animal.

6. Get a label from an animal feed container, such as horse, cat, dog, or fish feed. Study the label and list the following information: kind of animal to be fed, manufacturer, major ingredients, percent protein, percent fat, calories, and percent fiber. What is your assessment of the quality of the food: good, poor, or unsure? Why?

7. Calculate the costs associated with owning a dog. If the dog is given one can of food each day at a cost of $1.10 per can, what is the cost to feed the dog for a year? If the dog has a check-up and medications twice a year at a cost $60 each time and is groomed four times a year at a cost of $35 per grooming, what is the total annual cost of owning the dog including food?

8. Investigate the use of mammalian protein in the manufacture of animal feeds for ruminant animals, such as goats, sheep, and cows. Particularly focus your investigation on a disease of cattle known as BSE (Bovine Spongiform Encephalopathy or "mad cow"). What is the link between the feeding of meat scraps and BSE? Prepare a report on your findings.

9. Calculate the cost of a ration for animals. Choose an animal listed in Table 3-1 and the feed ingredients in Table 3-2. (Of course, other detailed resources can be used in this activity, such as other animal species and other feedstuffs.) Get feedstuff cost and availability information from local sources. Use the Pearson Square Method. Calculate the cost to make a ton of feed.

CHAPTER 4

Animal Health

OBJECTIVES

This chapter provides guidelines to help producers in areas of animal health. It has the following objectives:

1. Explain animal health, normal behavior, and abnormal behavior.
2. Explain the impact of environment on animal health.
3. List and explain the economic losses caused by poor animal health.
4. Describe common diseases and parasites of animals.
5. Explain how good health is maintained.
6. Select practices for treating disease.

TERMS

anaplasmosis
anthrax
antibiotic
antibody
bacteria
balling gun
behavior
biological
blackleg
brucellosis
coccidiosis
colostrum
contagious disease
disease
disinfectant
dose syringe
ectotherm
endotherm
external parasite
foot and mouth disease
fungi
grubs
health
hog cholera
immunity
implant
injection
internal parasite
isolation
leptospirosis
lice
mastitis
medication
noncontagious disease
oral medication
parasite
pharmaceutical
preconditioning
protozoa
rabies
sanitation
shipping fever
sleeping sickness
topical medication
virus
vital signs

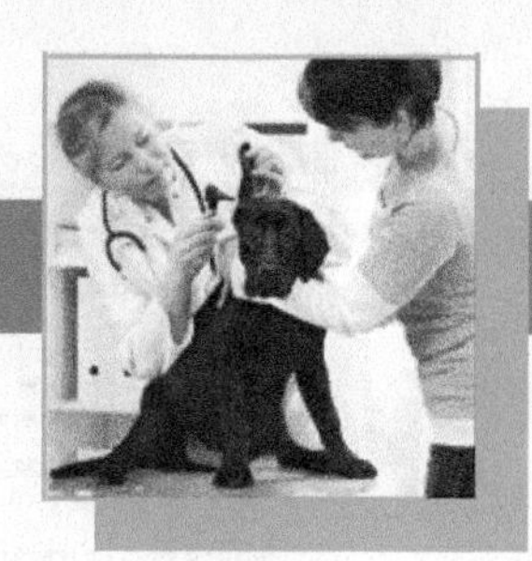

4–1. Healthy animals are fun to raise and have as companions. (Courtesy, Ljupco Smokovski/Shutterstock)

ANIMALS need good health in order to grow and be productive. Life processes are carried out efficiently. Animals that do not have good health fail to grow and produce. Promoting good health among animals is an important part of animal care and well-being.

You can usually readily observe an animal for its health status. Healthy animals have bright eyes, shiny hair coat, clear nostrils, and appropriate weight. They move about alertly and appear to enjoy life. Body processes are performed as they should. Vital signs of an animal may be assessed to determine the presence of a diseased condition. In addition, skin lesions, loss of hair, and dull coat are signs of health problems.

Using appropriate management practices promotes good health. Each species has unique requirements. For example, the environment that promotes health of a fish would not do so for a pig or bird. Some animal species are similar; others are quite different. In short, know your animals!

HEALTH AND MANAGEMENT

Most all practices in efficiently producing animals are related to animal management. Maintaining a good environment and providing proper feed are essential. Following practices to assure reproductive efficiency is also a part of management.

Animal management may also include identifying, docking, castrating, and dehorning some animal species. Identifying may involve using tags, tattoos, brands, DNA profiling, or microchips so that the ownership or heredity of an animal may be identified. Docking is typically used to shorten or remove the tails of lambs as an aid in keeping manure from soiling the wool. Castrating is used to prevent males from unwanted breeding. Dehorning is used to remove or prevent the growth of horns on breeds of animals, particularly cattle, that develop horns. Other practices are used depending on the species of animal, such as floating teeth and shoeing horses. Skills are needed to assure that good techniques are used with these practices. (Most of these are covered in detail in other chapters of the book.)

HEALTH

Health is the condition of an animal in terms of how the functions of life are being performed. An animal is in good health when its life processes are normal. Behavior is related to health. **Behavior** is the reaction of an animal to certain stimuli. It is the manner in which an animal reacts to its environment.

Disease is a disturbance in the functions or structure of an animal. Sometimes, only one organ or body part may be affected. Other times, the entire animal may show disease. In short, disease is the lack of good health.

Some animals that appear healthy may be suffering from pain or disease. Knowing each individual animal will help to easily detect and correct any ill health. An animal keeper should learn the signs of good and ill health.

4–2. Bright eyes, alert behavior, neat appearance, and no discharge from nostrils or mouth indicate good health. (Courtesy, neelsky/Shutterstock)

GOOD HEALTH AND VITAL SIGNS

Good health is the absence of disease. Some signs of good health are: good appetite, alert and content behavior, bright eyes, shiny coat, normal feces and urine, normal vital signs, and normal reproduction.

Health is assessed with vital signs. ***Vital signs*** are the signs of life. Pulse, respiration, and temperature are the vital signs of many animals.

Healthy animals show normal behavior. They will eagerly eat their food. A lingering or lazy animal may be in poor health. Healthy animals will stretch and watch the world around them. Eyes that are clear and alive are very important signs. A shiny coat is evidence of a proper diet. The presence of blood or mucus in the feces is a sure sign of a problem. Table 4–1 lists the normal vital signs of selected animals.

Table 4–1. Normal Signs of Selected Adult Animals

Species	Average Normal Temperature	Normal Pulse Rate	Normal Respiration Rate
	(rectal °F)	(rate/min.)	(rate/min.)
Cattle	101.5	60–70	10–30
Swine	102.6	60–80	8–15
Sheep	102.3	70–80	12–20
Goat	103.8	70–80	12–20
Horse	100.5	32–44	8–16
Chicken	106.0	200–400	15–30

ILL HEALTH

Animals with disease often have behavior that is not normal. If the signs of a healthy animal are not observed, then the animal is in ill health.

Signs of ill health may include one or more of the following: lack of appetite, sunken eyes or discharge from the eyes, discharge from the mouth or nostrils, inactivity, rapid breathing, rapid pulse rate, high temperature, full hair coat, lumps or protrusions on the body, open sores, seclusion, bloody urine or feces, and loss in production levels or weight.

4–3. A swollen eyelid and discharge around the eye are signs of an eye disease. (Courtesy, Education Images)

IMPACT OF ENVIRONMENT ON ANIMAL HEALTH

Undue stress can be caused by an animal's environment. A change in the environment or an unsuitable environment will create stress. This causes the animal to become more susceptible to diseases and parasites.

Important factors affecting the health of an animal include temperature, light, moisture, movement, and pollution.

TEMPERATURE

Specific temperature ranges are best for different animal species. If the temperature is too high or too low, health problems may occur.

Animals are endothermic or ectothermic. An ***endotherm*** is an animal that strives to maintain a constant body temperature. Disease and extreme temperature conditions may make it impossible to do so. Dogs, cats, and horses are examples of endothermic animals.

An ***ectotherm*** is an animal that adjusts its body temperature to that of its environment. Fish, reptiles, and frogs are examples of ectotherms.

4–4. As ectothermic animals, these koi fish require water that provides a good environment. (Courtesy, discpicture/Shutterstock)

4–5. These newborn piglets have an environment designed just for them with a heat lamp and facility to protect them from being mashed. (Courtesy, Education Images)

Endothermic animals adjust to temperature extremes as well as possible. During hot weather they may sweat, pant, or wade in cool water. Confined animals need proper ventilation and fans to keep cool. Animals on pasture need plenty of fresh water and shade trees.

During cold weather, animals will burn more energy to produce body heat. They must be fed more carbohydrates and fats to produce this heat. The muscles of the animal will involuntarily contract or shiver to burn energy and keep the body warm.

Younger animals have a more difficult time maintaining body temperature than older animals. They require protection from weather extremes.

LIGHT

Light affects most animals. Chickens are likely to be affected by light more than other animals. However, animals with light-colored skin, such as hogs, may become burned by too much sunshine. In laying hens, light causes the pituitary gland to release hormones into the bloodstream. This causes them to grow faster and lay more eggs. Electric lights are often on in broiler and laying houses to take advantage of this.

Color of the animal affects its reaction with light. This is because darker colors absorb light rays and lighter colors reflect light rays. Cattle with light colors around their eyes are more likely to have eye disease. Animals with light-colored skin may be more susceptible to disease. For example, white cats are more likely to get skin disease than those of darker colors.

4–6. These young chicks are in a lighted environment with nutritious feed, good water, clean litter, and temperature management. (Courtesy, Education Images)

MOISTURE

Moisture is due to humidity and precipitation. Air with high humidity (moisture in the air) will not cool animals as well as dry air. The perspiration on the skin of animals will not evaporate as fast in a humid location.

Diseases live and reproduce in moisture. Animals raised in confinement are more likely to have disease problems. Ventilation allows air to move about and is needed to remove the moisture from a confined barn. Proper ventilation lowers the inside temperature and decreases the chance of a disease outbreak.

MOVING

Lifting and moving animals creates stress. Stress lowers resistance to disease. When animals are corralled, hauled, or chased onto trucks, their stress level increases. Loading and hauling animals exposes them to environmental conditions, such as rain, snow, wind, and hot or cold temperatures. Trailers and loading chutes should be free of protruding nails, broken boards, wire, and debris. Animals can be injured by sharp objects. Animals shipped on airplanes have stress. Also, water, temperature, and food needs must be met.

Animals in good physical condition are better able to cope with stress. Producers may precondition animals. This is done by gradually acclimating them to their new environment.

POLLUTION

Pollution of water sources, feed, and the environment may cause animals to develop a disease. Water and feed sources may be poisoned by pesticides or industrial waste. Storing or hauling feeds in areas that have contained chemicals may contaminate feedstuffs. Pastures sprayed with chemicals may pose a pollution problem.

LOSSES DUE TO POOR HEALTH

It is easy to see the economic loss to producers if animals die. The greatest losses due to poor animal health are less obvious. Diseases lower production and threaten humans.

DEATH

Death is the most obvious loss due to poor animal health. Dead animals are of no value. The producer has lost all that was invested. Costs go up because of the expense of disposing of the animals.

Dead animals should be buried, burned, or, with chickens, composted. Producers may need help with the proper methods of disposal. A veterinarian or health official is a good

CONNECTION

UNRAVELING HOLLOW TAIL

Hollow tail was once a problem that some cattle had in the late winter. Producers were concerned. The cattle were quite thin, sluggish, and had rough hair coats. Some got down and died. Examination revealed that the tail was hollow (hence the name hollow tail).

Producers began a treatment for hollow tail. They would cut into the tail with a knife and fill the hollow with salt. The animals would jump, run, and do most anything to try to escape the pain. No longer sluggish, the cattle were said to be recovering.

Hollow tail was unraveled by researchers. They found that the hollow in the tail was due to severe malnutrition. The animals were starving and had used fat and other tissue in the tail to support life processes. The animals were often wormy. What was being done to treat hollow tail was wrong. It did not reflect the real problem nor animal well-being. Producers now know that hollow tail results from not having enough food.

This photo shows a cow with the signs of hollow tail. With feed and care, she will overcome the condition. (Courtesy, Education Images)

source of information. Improperly disposing of animals may create bad odors and attract coyotes or wild dogs, who may, in turn, spread disease to other animals and people.

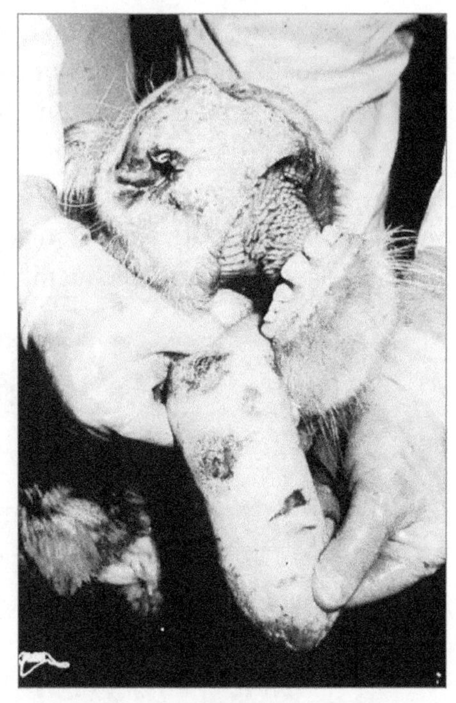

4–7. Lesions on the tongue of this beef animal are signs of foot and mouth disease. (Courtesy, U.S. Department of Agriculture)

LOWER PRODUCTION

Healthy animals mean increased profits for owners. Poor health may result in lower production levels. Some easy-to-see results of poor health are failure to reproduce or breed, slow growth rates, decreased milk production, decreased meat production and quality, and increased costs of production. All of these lead to decreased profits for producers.

HUMAN DISEASE

Diseased animals may spread diseases to humans. Examples are anthrax, brucellosis, leptospirosis, rabies, trichinosis, and tuberculosis.

Brucellosis, or Bang's disease, is found in cattle. Humans may get undulant fever from cattle infected with brucellosis. Drinking raw milk from an infected cow is one way humans may contract the disease. Another source of contamination is getting milk or other fluids from an infected animal on open

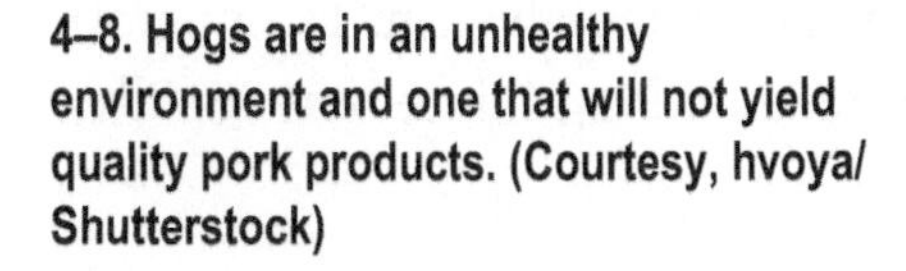

4–8. Hogs are in an unhealthy environment and one that will not yield quality pork products. (Courtesy, hvoya/Shutterstock)

wounds. Dairy cattle are regularly vaccinated and checked for this disease. Pasteurization makes milk safe for humans to drink.

Trichinosis is caused by a small worm in the flesh of infected hogs. When pork is not cooked thoroughly, the disease may be transferred to humans. Very few hogs have this disease, since people have stopped feeding raw garbage with pork scraps in it to hogs. Properly cooked pork and clean cooking areas have also helped eliminate this disease. Always wash your hands after touching raw meat.

COMMON DISEASES AND PARASITES

Disease is any disorder that keeps animals from carrying out life processes in normal ways. Symptoms are not the disease itself but a sign of a problem.

CLASSIFICATION OF DISEASES

Knowing how a disease is spread will help to better control and prevent it. Diseases are either contagious or noncontagious.

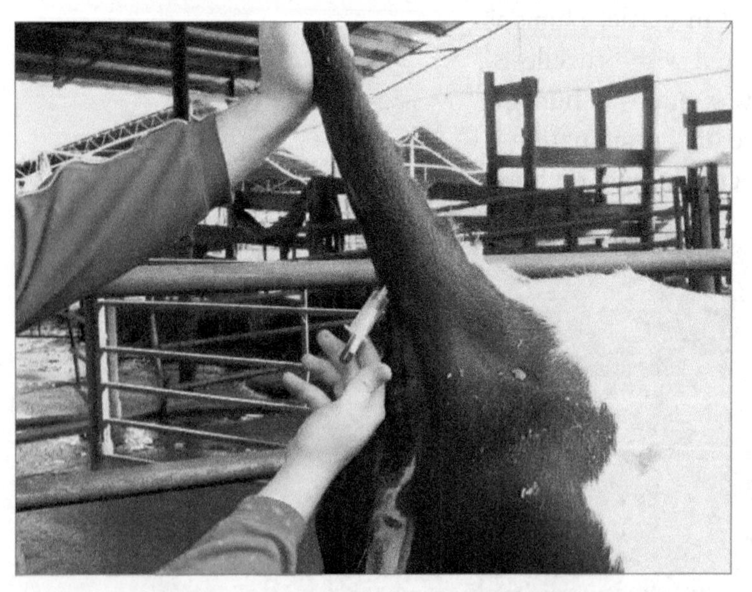

4–9. A blood sample is being taken from a blood vessel under the tail of a cow for analysis. The sample will undergo laboratory analysis for disease organisms or abnormalities. (Courtesy, Nancy S. Jackson, DVM, Mississippi)

Contagious Diseases

A ***contagious disease*** is a disease spread by direct or indirect contact with other animals. Pathogens, such as viruses, bacteria, fungi, protozoa, and parasites, cause contagious diseases. Pathogens are organisms that can produce disease. They may produce a toxin or poison that upsets the normal body metabolism. The microbes are small and reproduce very rapidly.

Viral diseases. A ***virus*** is a tiny disease-producing particle too small to be seen with an ordinary microscope. Viruses closely resemble the DNA of animal cells. This makes them hard to control. It is difficult to find a chemical that will kill the virus without killing the animal's DNA. Vaccinating animals is the most effective way of controlling viruses. Examples of viral diseases include rabies, hog cholera, smallpox, and distemper.

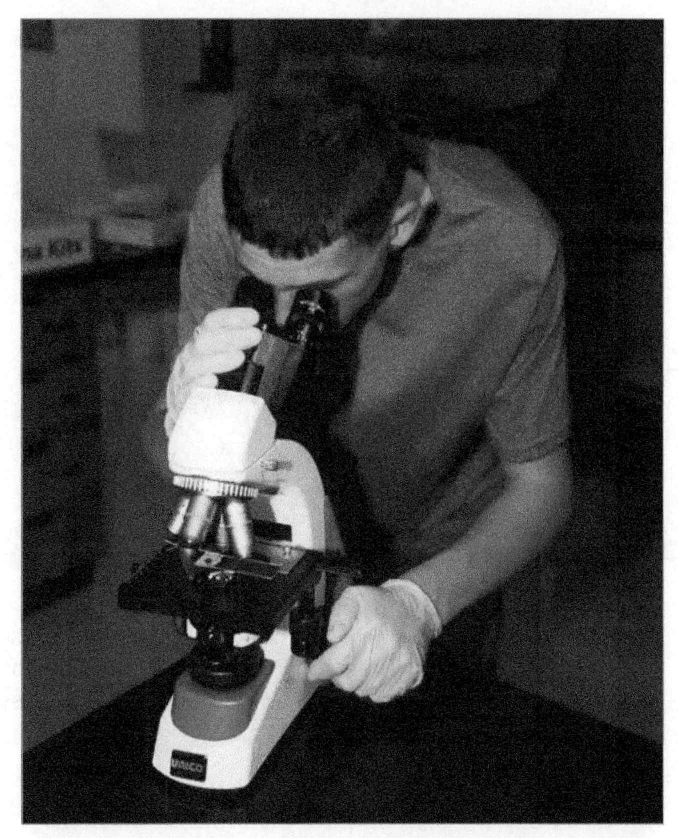

4–10. A microscope is used to check a swab sample collected from the ear of a dog for disease. (Courtesy, Education Images)

Bacteria. ***Bacteria*** are one-celled organisms. They are microbes that are sometimes called germs. Many pathogenic bacteria form spores that become resistant to control measures. Examples of diseases caused by bacteria include tuberculosis, brucellosis, leptospirosis, mastitis, and tetanus.

Fungal diseases. ***Fungi*** are unicellular organisms that generally cause diseases on the outside of the body. Conditions caused by fungal diseases may result in secondary infections. Bacteria may be responsible for secondary infections. Examples of fungal diseases are ringworm and coccidiosis.

Protozoa. ***Protozoa*** are unicellular organisms. They are the simplest forms of animal life. Examples of diseases caused by protozoa include malaria and anaplasmosis.

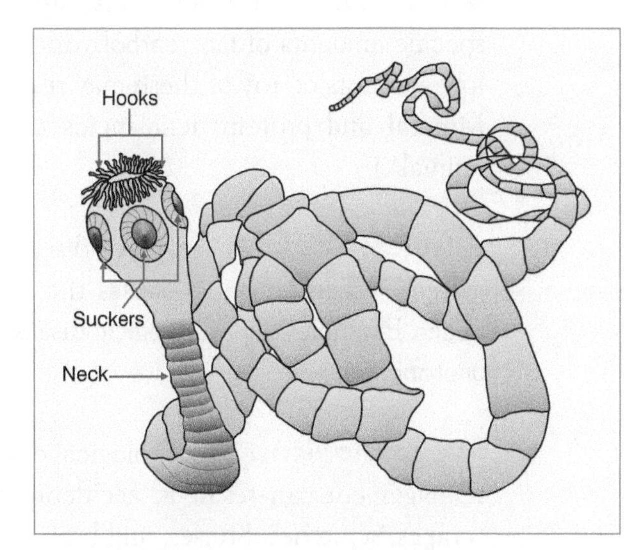

4–11. The tapeworm is an internal parasite. It is a flat, segmented worm with suckers and hooks for attaching to the intestines. Tapeworms can grow to several feet long in large animals.

Parasites. A ***parasite*** is a multicellular organism that lives in or on another animal. Parasites receive their nutrients from their host animal.

An ***internal parasite*** lives inside the host. These parasites are found in the digestive system, muscles, or other tissues. Internal parasites enter the host when it is eating contaminated feed or when eating feed on a dirty floor. Most internal parasites are obtained by feces being in food.

4–12. An adult female blacklegged tick (*Ixodes scapularis*) engorged on blood is shown here. The tick may transmit disease when it bites a host. They live on low vegetation in brushy, woody areas where mice, deer, and other mammals frequent. (Courtesy, Agricultural Research Service, USDA)

Pastures with short grass increase the likelihood of ingesting parasites. Examples of internal parasites are tapeworms, trichina, roundworms, cattle grubs, and hookworms.

An ***external parasite*** lives on the external parts of the animal. They obtain their food by biting or sucking the blood and tissue of the animal. External parasites can transfer contagious diseases or parasites to other animals. Examples of external parasites include ticks, fleas, lice, mites, and leeches.

Noncontagious Diseases

A ***noncontagious disease*** is a disease that is not spread by casual contact. It is caused by a nutritional, physiological, or morphological problem or is genetics related.

Nutritional. Most nutritional diseases result from an unbalanced diet. Animals require specific amounts of fats, carbohydrates, proteins, minerals, and vitamins. Too high or too low amounts of any of these may result in a nutritional disease and decrease profitability. Mineral and protein deficiencies are common. (Chapter 3 discusses the nutrition of animals.)

Physiological. This type of disease results from a defect of a tissue, organ, or organ system. When an organ causes the body's metabolism to be off balance, other problems occur. Examples of physiological diseases include milk fever, heart failure, birth defects, and acetonemia.

Morphological. Morphological diseases are related to physical injuries to animals. Poor management can result in accident or injury to animals. Possible injuries include cuts, scrapes, scratches, bruises, and broken bones. These decrease the efficiency of the animal. Good management practices can decrease the likelihood of these problems. Pick up loose wire, old lumber, and equipment. Remove protruding nails and broken boards. Cattle may have magnets placed in their digestive systems to collect nails and wire that they eat. Feed mills often run feed over magnets to remove any metal particles from the ingredients.

Genetics. Some diseases are genetics-related. This means that the heredity of the animal predisposes it toward certain diseases. An example is pigmentation around eyes. Animals that because of heredity have light skin coloring around the eyes are more likely to have eye disease. Other genetics-related diseases are connected to anatomy and physiology.

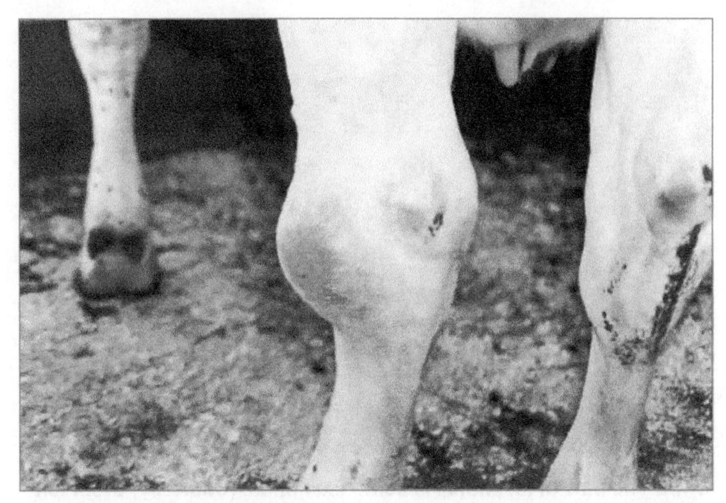

4–13. A hock problem on a dairy cow likely resulting from an injury. (Courtesy, Education Images)

SELECTED DISEASES OF ANIMALS

Some diseases affect only certain species while others affect several species. Roundworms may affect dogs, cats, hogs, cattle, sheep, horses, poultry, and fish. Blackleg disease, however, primarily affects cattle.

Several common animal diseases are briefly described here. Proper diagnosis and treatment of a disease may require a veterinarian.

- Anaplasmosis—***Anaplasmosis*** is a parasitic disease caused by a protozoan that attacks red blood corpuscles. Cattle are primarily affected. In chronic anaplasmosis, animals become anemic and may overcome the disease on their own. In acute forms, animals have rapid heartbeat, muscular tremors, and loss of appetite. Sick animals may become aggressive and want to fight. Death may occur in a few days. The disease is prevented by immunizing cattle. Some treatments are available if the disease is caught early.
- Anthrax—***Anthrax*** is an acute infectious disease that affects most endothermic animals. It most frequently affects cattle during the summer when they are on pasture. Affected animals have a fever, rapid respiration, and swelling on the neck. Animals will die suddenly. Prevention of anthrax includes vaccinating animals, controlling flies, and sanitizing their environment. Penicillin may be effective if given in large doses.

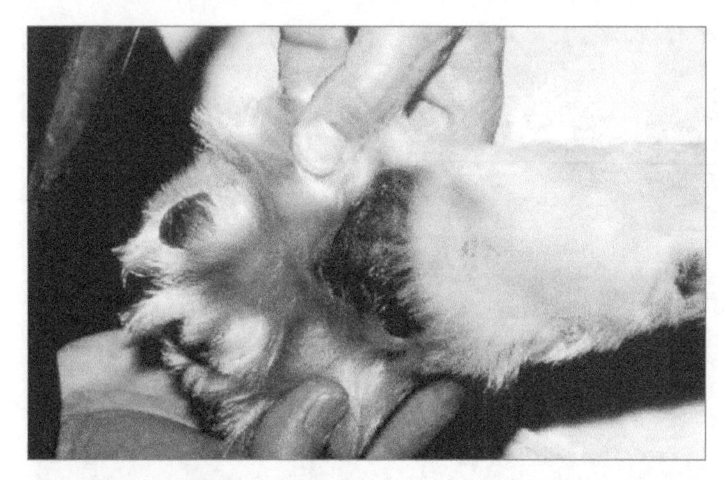

4–14. The paw of an animal is being examined. (Courtesy, U.S. Fish and Wildlife Service)

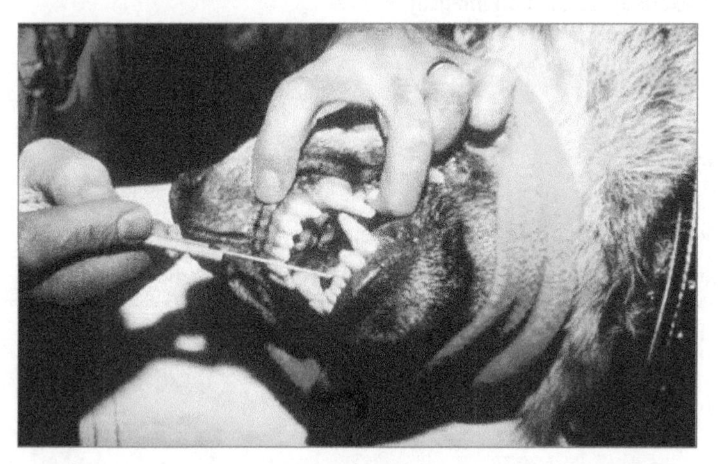

4–15. The teeth of an animal are being examined. (Courtesy, U.S. Fish and Wildlife Service)

- Bang's (Brucellosis)—Bang's or ***brucellosis*** affects cattle, sheep, goats, and hogs. The reproductive tract of the female is infected. Developing fetuses may be aborted. Cows may have to be bred several times before they become pregnant. When they do become pregnant, abortion may end pregnancy in only a few weeks. Bang's is prevented by vaccinating heifers. Sanitation, testing cattle, and bringing only Bang's-free cattle into a herd help to control this disease. Humans can get this disease from infected animals or their milk. In humans, this is known as undulant fever. If humans survive undulant fever, they may suffer permanent crippling and disability.
- Blackleg—***Blackleg*** is an acute, highly infectious disease that usually results in death. It primarily affects cattle, although sheep and goats can also get it. Symptoms include high fever; swelling in the neck and shoulder; and muscles in the neck, shoulder, and thighs crackle when mashed. Eventually animals lose their appetite and die. Few animals recover from blackleg. Blackleg is prevented by vaccinating calves with a bacterin, which is a vaccine made from killed or inactive bacteria.
- Porcine parvovirus—***Parvovirus*** is a common disease among hogs, with similar strains affecting dogs, some birds, and rodents. In hogs, the reproductive tract of the female is affected so that fetal pigs are aborted or they may be born dead. Boars may transfer the disease in semen during breeding. A vaccine is available. Using artificial insemination helps prevent spread of parvovirus.
- Coccidiosis—***Coccidiosis*** is a parasitic disease affecting chickens, turkeys, ducks, geese, and game birds. It is caused by a protozoan. Symptoms include bloody feces, ruffled feathers, unthrifty or sick appearance, and pale coloring. Medications known as anticoccidials are put in feed and water to prevent and treat this disease. Since it is

transmitted by infected birds, wild birds should be kept away from poultry flocks and infected areas should be sanitized.

- Equine sleeping sickness—Horses are affected by ***sleeping sickness***. It is caused by a virus transmitted by insects that bite horses, mules, and wild rodents. Affected animals walk aimlessly crashing into things; later they appear sleepy, cannot swallow, grind their teeth, and possibly go blind. Some horses recover, but most will die within two to four days. No effective treatment has been developed, but horses can be vaccinated to develop immunity.
- Foot and mouth disease—***Foot and mouth disease*** affects animals that have cloven or divided feet. It is a highly contagious disease caused by a virus. There is no known treatment or vaccine. Quarantine is the best control at the present time. No animals or meat can be imported into the United States with this disease. Animals with the disease get watery blisters in the mouth and on the skin around the hooves. Teats of females may also blister. Animals will have a high fever. The United States does not currently have a foot and mouth disease problem, but many other countries do.
- Grubs (warbles)—Cattle ***grubs*** are internal parasites caused by heel flies. The fly lays its eggs in the summer around the heels of cattle. When they hatch as larva, they go

CONNECTION

MAKING A DIAGNOSIS

The calf shown here is obviously abnormal. Why? What is the disease? The picture gives some information to help in making a diagnosis. Other information may be needed about the calf. Even the best veterinary medical experts may be baffled by some animal health conditions.

The condition of this calf may be one or more of the following:

- Congenital malformation known as arthrogryposis.
- Caused by maternal (mother) ingestion of toxins during gestation.
- Caused by maternal deficiencies or excesses of certain trace elements.
- Due to maternal infection with a virus such as bovine virus diarrhea.
- Due to physical agents such as embryo manipulation after conception.
- Have no clearly established cause.

Such occurrences as shown here are rare. No absolute diagnosis is available. More information is needed, including careful laboratory tests. When given proper care, most animals will exhibit normal development for their species. (Courtesy, University of Minnesota-Crookston Dairy Club, and Nancy S. Jackson, DVM, Mississippi)

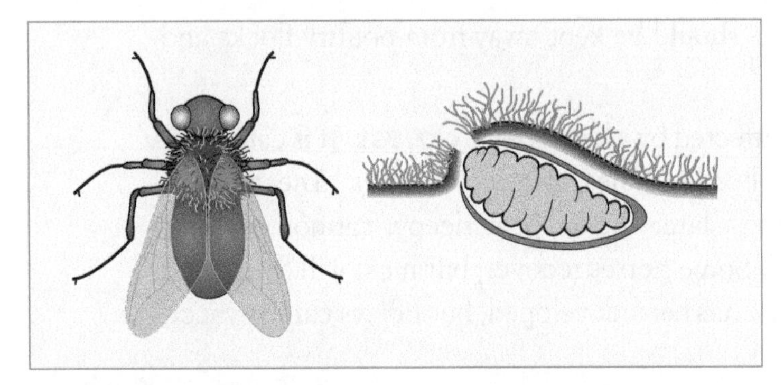

4–16. Drawing of a cattle grub fly and larva under the skin. Larva damage the hide and meat.

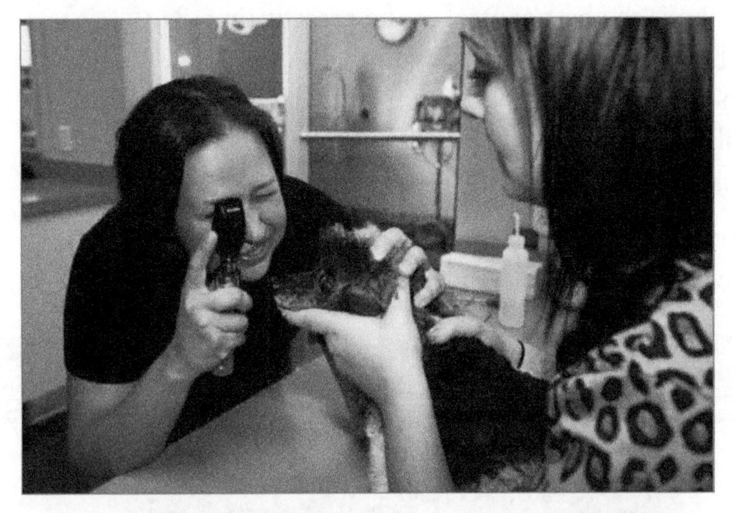

4–17. Routine health examinations are used to promote well-being (This shows an examination of the eyes of a visually-impaired dog using an ophthalmoscope.) (Courtesy, Education Images)

into the skin and move through the body until they reach the back in a few months. In the back, the grubs cause bumps to swell in late winter and spring. The larva (about $^3/_4$ inch or 2 cm long) will come out of the bumps and grow into an adult fly. A good fly control program will prevent grubs. Several treatments are available and should be followed according to their directions. Grubs damage the hide by making holes in it as well as the flesh around it. Cattle grubs cause more injury to cattle than any other insect.

- Leptospirosis—***Leptospirosis*** is a bacterial disease that affects cattle, dogs, sheep, and most other farm and companion animals. Symptoms include high fever, poor appetite, and bloody urine. Females will abort their fetuses. Antibiotics are sometimes used to treat leptospirosis. Animals may be vaccinated against this disease.
- Lice—***Lice*** are external parasites that attack cattle, hogs, and other species. Lice are small insects that suck their host's blood. Animals may become anemic. Lice also cause animals to itch; they will often be seen scratching or rubbing against trees or posts. Lice usually appear in the winter. Back rubbers and other means can be used to administer insecticides to control lice.
- Mastitis—***Mastitis*** is the leading cause of profit loss in dairy cattle. Mastitis is a bacterial disease that affects female cattle, sheep, goats, and swine. It is an inflammation in the udder that interferes with milk production. Chronic mastitis causes the milk to be thick or lumpy. Acute mastitis is indicated by a fever and a hard, warm udder. Laboratory tests are used to detect most mastitis. Chronic mastitis that is not treated can cause the death of the animal.

- Poisonous plants—Poisonous plants are sometimes problems. Examples of poisonous plants include low larkspur, oak, tall larkspur, timber milk vetch, and water hemlock. Most poisonous plants cannot be totally eliminated. Possible loss is reduced with good pasture management. Routine weed control will kill most poisonous plants. Knowing the common poisonous plants in the area prevents losses. Study the symptoms of poisoned animals. Provide good pasture and feed for range animals. Promptly treat animals that may have been poisoned.
- Rabies—***Rabies*** is a disease that occurs in nearly all endothermic (warm-blooded) animals. It can be transferred to humans. Caused by a virus, rabies results in death. Signs include irritable and aggressive behavior, avoiding light, fever, vomiting, and diarrhea. Some animals become frenzied, vicious, and attack everything that moves. Lips may be drawn to expose the teeth. Once the disease develops, it is fatal. Incubation is three to eight weeks. Autopsy of the brain is a sure method of diagnosis. Vaccination can prevent rabies. Dogs, cats, bats, foxes, skunks, and raccoons are examples of animals where rabies may be a problem. In recent years, skunks have had 40 percent of the diagnosed cases. Never touch or pet an animal that you do not know. Humans bitten or scratched by an infected animal should receive immediate medical attention.
- Roundworms and tapeworms—Roundworms, tapeworms, and other worms are parasites that can infect all animals. They are found in the intestines and the stomach. Tapeworms can grow several feet long. When they become this large, they use much of the feed that the animal has eaten. Animals with low infestations of small-sized worms may not show symptoms. As the number and size of worms in the digestive tract increases, the animal loses weight, becomes anemic, and may have diarrhea. Sanitation is a good control measure.
- Shipping fever—Cattle and sheep of all ages can get ***shipping fever***, though it is more of a problem for younger animals. It is an environmental disease caused by conditions animals encounter when hauled or sold. Animals will develop a high temperature, discharge from the eyes and nostrils, and may cough. Most will have difficulty breathing. It is more likely to affect thin, underfed animals that are hauled long distances. Cattle should be vaccinated three to four weeks before being moved.

4–18. Feed should be free of poisonous substances and provide nutrients. (Courtesy, Education Images)

- Poisonous feed—Animals sometimes eat poisonous food. Feed may contain substances that poison animals. Molded feedstuffs should be avoided. Poisonous plants may grow in pastures and create problems with animals that graze. Feed contamination is sometimes a source of poisoning. Some animals are allergic to specific foods. Companion animals typically do not do well on human foods. For example, dogs should not be fed chocolate.

MAINTAINING GOOD HEALTH

A key to success with animals is keeping them healthy. Using good management will increase production and decrease losses due to poor health. An understanding of the animal's bodily defenses will lead to healthy management practices.

BODY DEFENSES

The first line of defense of an animal is the epithelial tissue or skin and mucous membranes. Once a foreign substance has gotten past the epithelial tissue the body uses other mechanisms.

The digestive tract is very high in acidity. This depresses bacterial growth. Tissue fluids (lymph) contain leukocytes that neutralize or engulf pathogens. The liver and other lymph organs will trap pathogens until they can be destroyed.

When an inflammatory reaction occurs, redness will be present. Redness is due to blood flow to the infected area. The area will feel hot, and swelling or edema will occur. Increased sensitivity or pain is felt by the animal. All of these are the body's attempt at destroying

4–19. Many animal owners have routine health exams of their animals. (Courtesy, College of Veterinary Medicine, Texas A&M University)

pathogens. In addition, metabolic activity and immune reaction (development of antibodies) will increase.

HEALTH MANAGEMENT PRACTICES: DISEASE CONTROL, TREATMENT, AND PREVENTION

Animal keepers should use appropriate practices (methods) to maintain good animal health. Several of these are described here. These may be useful in controlling, treating, and preventing disease.

Environmental Conditions

The environmental conditions that the animal lives in must be checked regularly. Space requirements should be met. Allow enough room for animals; do not overcrowd. Ventilation is needed to have a continuous supply of fresh and clean air. Providing a clean and dry place to lay helps to decrease the disease potential.

Sanitation

Keeping areas where animals are raised and fed clean is ***sanitation***. Filth carries disease. Sanitation decreases the chances of animals contracting disease.

CAREER PROFILE

VETERINARIAN

A veterinarian is a highly trained professional who is involved in promoting animal health. The work may involve treating diseased animals and following practices to prevent disease. Lab tests and other activities may be used to assure proper diagnosis.

Veterinarians need degrees in veterinary medicine. These are available from universities with veterinary medicine programs. Practical experience with animals is highly desired. A good background in the biological sciences is helpful.

Veterinarians work in clinics or with companies and agencies that need the services of a skilled animal health provider. Some work with agencies that deal with animals used for human foods. This photo shows a veterinarian using ultrasound to assess the condition of a horse's leg. (Courtesy, College of Veterinary Medicine, Texas A&M University)

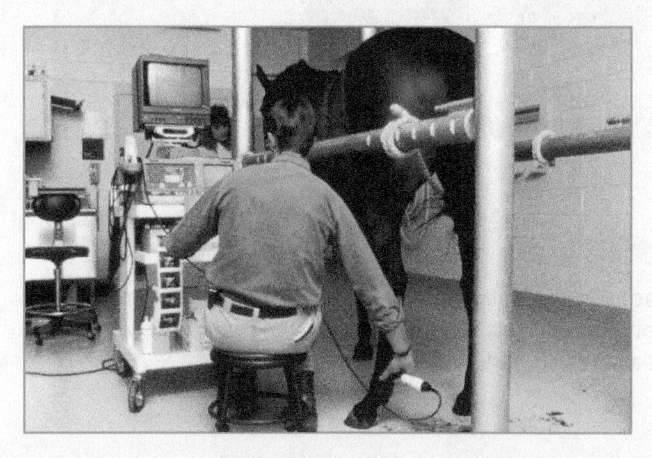

4–20. Two practices in sanitation are using a sanitizing solution on the tires of vehicles that come onto a farm and using a dip for shoe soles. (Courtesy, Education Images)

Dead animals should be removed and disposed of properly. Barns and facilities should be cleaned regularly. Waste and manure should be disposed of properly and regularly. Many farms have fenced lagoons or pits to temporarily store animal waste.

A ***disinfectant*** is a substance that destroys the causes of disease. Facilities, especially where young are raised, should be cleaned with a disinfectant regularly. Disinfectants include alcohol, iodine, lime, chlorine bleach, and soap. People should always follow safety precautions in using disinfectants.

Clean facilities also increase the natural beauty of the surroundings and lessen susceptibility to disease.

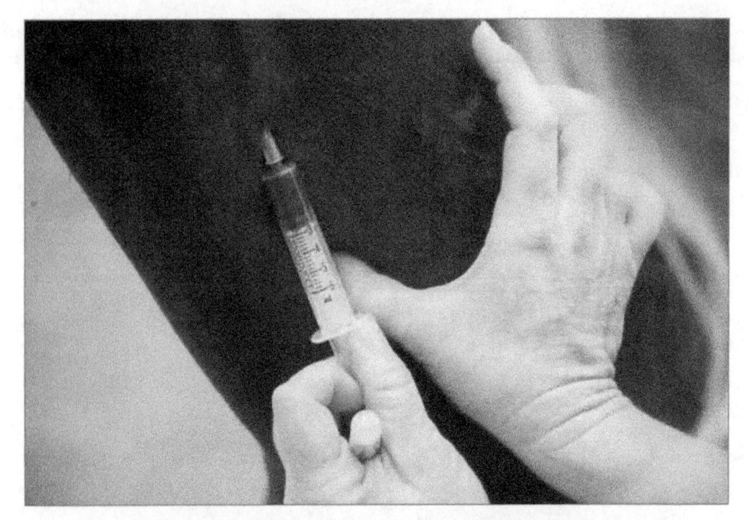

4–21. Disease diagnosis may require collection of a blood sample for laboratory testing. (This sample is being taken from a vein in the neck of a horse.) (Courtesy, Education Images)

Proper Nutrition

Animals that have proper nutrition are more resistant to disease and stress. Providing the proper ration is important. Poorly fed, weak animals do not have the resistance needed to fend off diseases.

Isolation

Isolation is separating diseased and non-diseased animals. Diseases are often spread by animal contact. Isolating diseased animals decreases the risk of spreading a disease.

New animals brought to a farm should be isolated for three to four weeks. If the animals have a disease, the signs should be visible in that time. If no diseases are seen, the animals may be turned in with the herd or flock.

Restrict Truck and Equipment Traffic

Diseases can be brought to a farm by trucks, farm equipment, and other vehicles. Trucks and equipment should be disinfected between trips to other farms or feedlots. Mud, manure, sawdust, and straw can harbor disease.

Restrict Human Access

People may bring diseases with them as they travel from farm to farm. Diseases can be transferred on shoes or boots. Some farms require everyone to walk through a boot tub of disinfectant solution. Other places restrict access to facilities to only a few individuals. Breeding farms may require visitors to wear plastic slip-on boots or to view the facility with video cameras.

Preconditioning

Preparing animals for stress is ***preconditioning***. Stocker calves and other animals that are to be hauled are often preconditioned. Preconditioning is important with any activity that might cause stress. Castration and dehorning should be done by the time cattle are two months old, so as not to coincide with transfer to another lot or weaning. Calves should be weaned and started on feed thirty days before being sold. The amount of handling and moving time should be kept to a minimum.

4–22. Newly hatched chicks are being hand vaccinated. (Courtesy, Education Images)

Immunization

Immunity means that an animal resists disease. An animal often develops immunity when it has a disease. Artificial immunization (as acquired through vaccination) is used to protect animals from disease.

4–23. Young calves need colostrum from their mother's milk (Courtesy, U.S. Department of Agriculture)

Antibodies are immune substances produced in the body. Immunity can be developed to many different diseases, thus allowing animals to withstand exposure to disease even when coming in contact with it.

Vaccines, serums, bacterins, and toxoids are used to help animals develop artificial immunity to diseases. Animals should be vaccinated to prevent common diseases.

Mothers often pass immunity to newborn young. Young mammals get immunity by antibody transfer from the mother through the placenta and colostrum. The placenta is the organ that unites a fetus with the uterus in which it is developing. ***Colostrum*** is a substance found in milk shortly after giving birth. The presence of colostrum quickly drops a few hours after birth.

Animals may be fed vaccines in their feed or given shots. Most vaccines contain dead or living organisms that cause the disease.

TREATING DISEASE

Animals that get a disease need medication to help them in trying to overcome the disease. A ***medication*** is a substance that is used to prevent and control disease. Animal producers are well aware that prevention is preferred to treatment. Treatments for some diseases may not be effective.

The assistance of a qualified person may be needed to identify and treat a disease. A disease must be properly identified before a treatment is selected.

KINDS OF MEDICATIONS

Several types of medications are available. Some can only be used by a licensed veterinarian. Proper use is essential. The animal being treated can be injured or killed by improper use.

Two general kinds of medications are used: biologicals and pharmaceuticals. A ***biological*** is a medicine primarily used to prevent disease. The purpose is often to develop immunity in an animal. A ***pharmaceutical*** is a medicine used to treat a diseased animal. Pharmaceuticals reduce or prevent the continued growth of microbes that cause disease.

- Antibiotics—An ***antibiotic*** is a substance produced by an organism that will inhibit or kill another organism. Antibiotics are used in diluted form to treat diseases caused by microbes. Penicillin, terramycin, and streptomycin are the best known antibiotics. These should be used only when needed. Overusing antibiotics may result in the disease pathogens developing resistance to them. The antibiotic would lose effectiveness.
- Pesticides—Pesticides are used on parasites. These should be carefully used. Both internal and external parasites are controlled with pesticides. Wormers, insecticides, and miticides are examples. Wormers are also known as anthemintics or dewormers.
- Dietary supplements—A dietary supplement is a substance that provides vitamins, minerals, or other nutrients that help animals overcome disease.
- Sulfonamides—The sulfonamides are commonly known as sulfa drugs. Several sulfa drugs are used. These agents stop the growth of infecting germs without poisoning the animal. The sulfas have been used for many years.

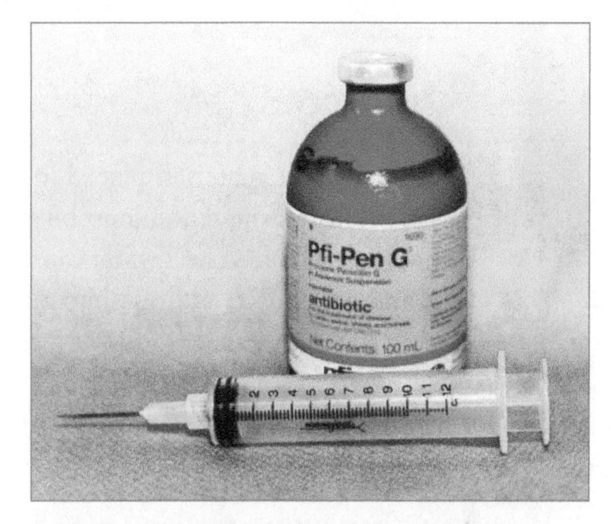

4–24. An antibiotic that is given with a syringe and hypodermic needle. (Courtesy, Education Images)

4–25. Four examples of medicines are (from left): liquid for eyes, powder for superficial wounds, liquid for skin fungal infections, and spray-on wound dressing for use on horses and ponies. (Courtesy, Education Images)

- Others—Many other medications are available. These often use destroyed or altered germs. Vaccines, serums, bacterins, and toxoids are most common. In addition to treatment, most of these are primarily to prevent disease by developing immunity.

ADMINISTERING MEDICATIONS

Medications must reach sources of infection or need. This requires proper administration. Several methods are used. Select the method based on the type of medication and kind of disease. Be sure the medication is approved for the intended use.

Additives

Additives are materials placed in food and water. A variety of medications are suited to this method. The materials must be added at the proper rate. As the animal eats, medication is ingested. The feed and water must be used only for the animals as approved. The use of feed and water with additives must usually be stopped for a period before animals or their products are consumed, known as the withdrawal time.

Injections

An ***injection*** is a medication given directly into the bloodstream, tissues, muscle, or body cavity. Injections are sometimes known as shots. A hypodermic needle and syringe is typically used. Sanitation around injection sites is important. This may involve washing the

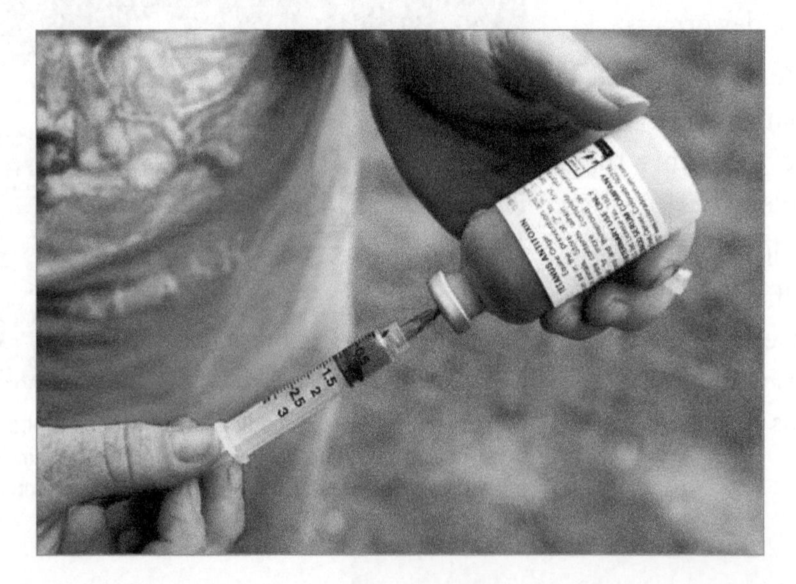

4–26. Filling a syringe with tetanus antitoxin before giving an injection. (Courtesy, Education Images)

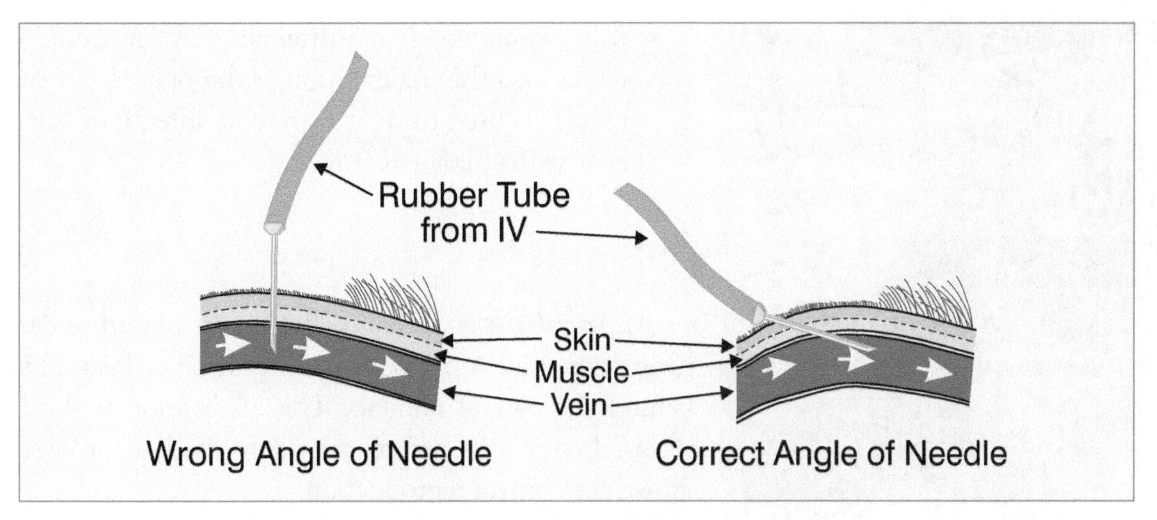

4–27. Wrong and correct way of placing an IV needle in a vein.

area and cleaning it with a cotton ball containing isopropyl alcohol. Careful attention to dose size is needed. Doses are measured in cubic centimeters (cc).

Several kinds or sites of injections are used. Medications vary in where they are given. This information is given on the label that accompanies injectable materials. The kinds are

- Intradermal—An intradermal injection is made into the skin. Since the skin is often thin, care is needed. Few drugs are given this way. Tests for disease (such as TB) may involve skin injections.
- Subcutaneous—A subcutaneous injection is made just beneath the skin. These injections are given in easy-to-get-to loose skin. With cattle, the site of the injection is often on the neck or behind the shoulder.
- Intramuscular—An intramuscular injection is made into a muscle. The most frequent sites are in the hindquarters and shoulders but never into the ham of a hog. Penicillin, for example, is given with an intramuscular injection.
- Intravenous—An intravenous (often known as IV) injection is made into the veins of an animal. These injections are used when a quick response is needed, such as to save the life of an animal.
- Intranasal—An intranasal injection is given through the nose. It involves spraying the medication into the nostrils. Intranasal injections are often used with innoculants and do not involve a hypodermic needle and syringe.
- Intraperitoneal—An intraperitoneal injection is made into the body cavity. It is most commonly used with cattle. The site is in the hollow of the flank. These are used when a quick response is needed.

4–28. Cattle moving through a dipping vat to treat for external parasites. (Courtesy, Agricultural Research Service, USDA)

- Intramammary—An intramammary injection is made into the udder through the opening in the teat. It is used to treat mammary infections, such as mastitis in dairy cattle.

Implants

An ***implant*** is a small pellet that is placed underneath the skin. Most implants are in fairly loose skin behind the ears of animals. The medication is slowly released. Implants are more widely used to promote growth or control reproduction.

Topicals

A ***topical medication*** is one that is placed on the skin or surface area of an animal. These medications may be sprayed, poured, or dusted on animals. In some cases, the animals are dipped into tubs or vats of solution. Backrubbers or other devices that animals rub against can be used as topical applicators. Topical medications are typically for abrasions (cuts or wounds) or to control external parasites.

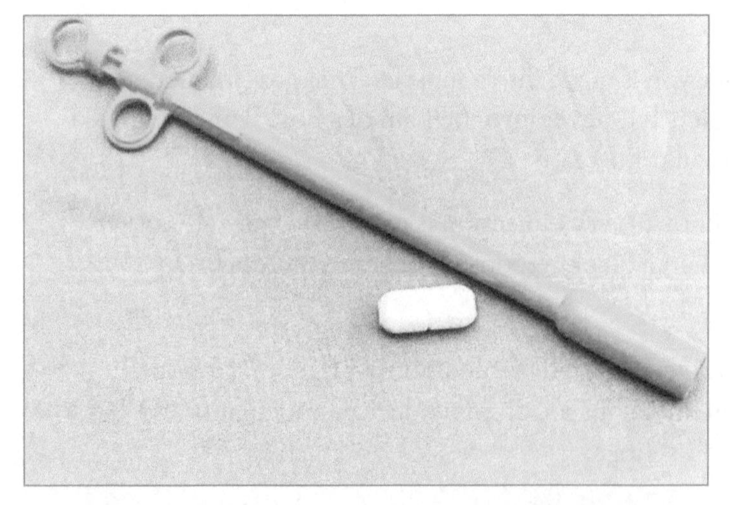

4–29. A balling gun with bolus. (The bolus is placed in the end of the gun and released into the animal's throat.) (Courtesy, Education Images)

Oral

An ***oral medication*** is given through the mouth. The forms of medication may be liquid, pill, bolus, or as an additive to the feed or water. These medications are often for digestive problems or to control internal parasites. The sulfa drugs are also given orally. Liquids are given with a ***dose syringe***. A pill or bolus may be given with a ***balling gun***.

Some animals are particularly choosy about what they eat. For example, a small pill can be concealed in a piece of hotdog wiener for a dog. With a cat, one person restrains the animal and another opens its mouth and quickly places the pill as far back as possible. The throat is stroked to promote swallowing.

As with all medications, proper dosage is essential. Animals used for food or milk production must be off the medication for a while before its products can be used for food.

4–30. MA liquid medication is being administered with a dose syringe. (Courtesy, Education Images)

Other

Medications may occasionally be given in other ways. Some are in capsule or liquid forms and inserted into the rectum. Medications for some reproductive problems may be placed in the reproductive tract. Ear disease may involve placing medications in the ear canal. Eye problems may involve placing medications directly into the eye. Always read and follow directions on medicine. Get the assistance of a veterinarian or other qualified person.

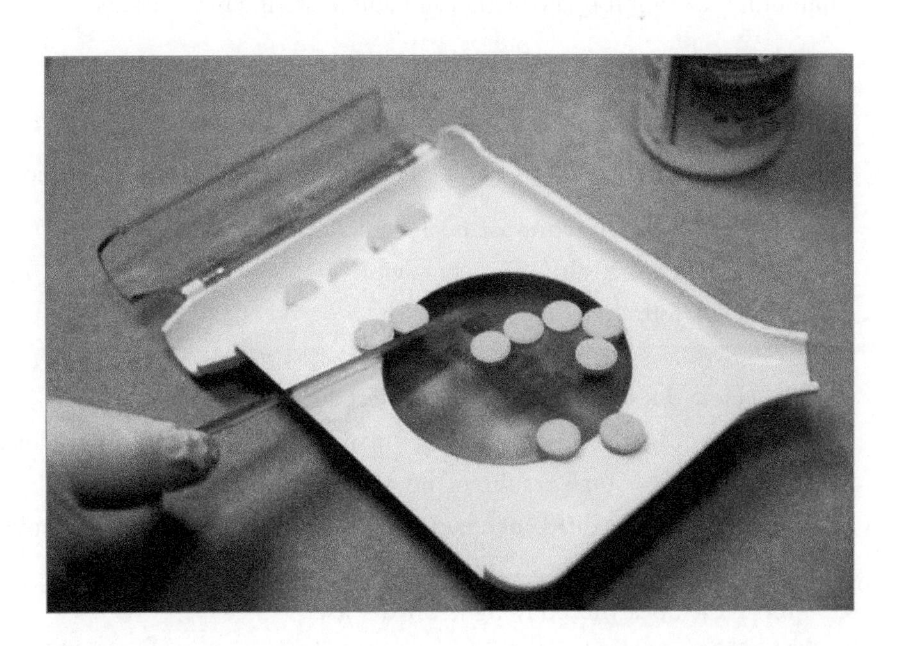

4–31. Veterinary pharmacies in clinics dispense drugs much like pharmacies people use when they are sick. (Courtesy, Education Images)

REVIEWING

MAIN IDEAS

Health is the condition of the body and a measure of how the functions of life are being performed. Health affects the behavior of the animal. Poor health causes personal and economic losses to the owner or producer. Death, low production, and human contraction of disease may result.

Signs of good health include good appetite, alert and content behavior, bright eyes, shiny coat, normal feces and urine, normal vital signs, and normal reproduction.

Signs of ill health include lack of appetite, sunken eyes or discharge from the eyes, discharge from the mouth or nostrils, inactivity, rapid breathing, rapid pulse rate, high temperature, full hair coat, lumps or protrusions on the body, open sores, seclusion, bloody urine or feces, and loss in production levels or weight.

Diseases may be contagious or noncontagious. Contagious diseases may come from viruses, bacteria, fungi, protozoa, and parasites. They are spread by direct or indirect contact with other animals. Noncontagious diseases are due to poor management practices.

Maintaining good animal health is easy if the well-being of the animals is considered. Animals must be placed in a good, clean environment that is free of debris. They should not be overcrowded and must have a clean, fresh, and nutritious supply of feed and water. Dead animals must be disposed of properly. Diseased and sick animals should be isolated. A disinfectant should be used to clean stalls, feeders, and waterers. Animals may be preconditioned for stress and vaccinated to develop artificial immunity. Use appropriate methods of administering medications.

QUESTIONS

Answer the following questions using complete sentences and correct spelling.

1. Compare and contrast health and disease.
2. How does health affect behavior?
3. Give four signs of good health.
4. Give four signs of ill health.
5. How does the environment affect health?
6. How can the environment be manipulated for the benefit of animals?
7. What are some results of poor health? Explain each.
8. Explain the difference between contagious and noncontagious diseases.
9. List three diseases of animals and describe each.
10. How does the body fight off disease?
11. What methods are used to prevent the spread of disease? Name and briefly describe at least six.
12. What kinds of medications are used?

EVALUATING

Match the term with the correct definition. Write the letter of the term on the line provided.

a. disease
b. health
c. immunity
d. disinfectant
e. contagious disease
f. isolation
g. shipping fever
h. preconditioning
i. balling gun
j. injection

_____1. Disease spread by direct or indirect contact among animals.
_____2. Substance that destroys the causes of disease.
_____3. Separating diseased animals from healthy animals.
_____4. The ability to resist disease.
_____5. Preparing animals for stress.
_____6. Caused by conditions animals encounter when hauled.
_____7. Disturbance in the functions of the body.
_____8. Condition of the body and how well life functions are being carried out.
_____9. Using hypodermic needles and syringes to get a substance into a body system..
_____10. Equipment for administering pills to large animals.

EXPLORING

1. Visit a local veterinary medical clinic and tour the facilities. Ask about the most common health problems that they deal with, including symptoms, causes, preventions, and treatments. Observe the treatment of an animal. Write a report on what you saw.

2. Conduct a job-shadowing project and follow a veterinarian, rancher, small animal producer, or animal health technician for at least a day. Write a report on your experiences.

3. Check the vital signs of an animal. Record its pulse, respiration rate, and temperature. Compare the data to normal animals. Check the vital signs over a period and observe the fluctuations. Be sure to follow the proper procedures so the animal is not injured.

4. Observe an animal that is in good health. List the signs that this animal exhibits. Observe an animal in ill health and list the signs observed that indicate ill health. Compare the two lists.

CHAPTER 5

Animal Reproduction

OBJECTIVES

This chapter covers the fundamentals of animal reproduction. It has the following objectives:

1. Explain the role of animal reproduction.
2. Name and describe the functions of the major reproductive organs.
3. Describe the phases of the estrous cycle.
4. Explain the phases of reproductive development in the life of an animal.
5. Describe the role of animal reproduction technology.
6. Evaluate breeding animals.

TERMS

accelerated lambing
anestrus
artificial insemination
breeding
cloning
conception
copulation
crossbreeding
egg
embryo
embryo transfer
estrous cycle
estrous synchronization
estrus
fertilization
fetus
gamete
genetic engineering
gestation
incubation
insemination
lactation
natural insemination
ovary
ovulation
parturition
performance testing
pregnant
production records
progeny testing
puberty
purebred animal
reproductive efficiency
scrotum
sexed semen
sexual reproduction
spawning
sperm
testicle
visual appraisal
zygote

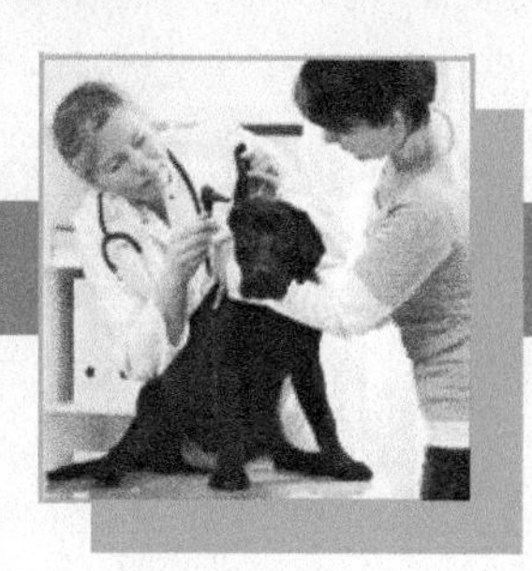

5–1. A mare with a minutes-old foal nursing for the first time depicts successful horse reproduction. (Courtesy, Education Images)

ANIMAL reproduction is essential in the production of animals. Without reproduction, there would no new members of a species. Further, producers depend on reproduction to produce products and gain profit from their work. Reproduction is a process that can and should be managed by successful animal producers.

The goal of animal reproduction is to produce numerous animals. These animals must have desired qualities including the ability to grow and be productive. If there are not enough animals, people will not have the products they want. Shortages might develop and prices would go up.

Producers pay close attention to the reproductive efficiency of their animals. Those that fail to efficiently reproduce are culled. Producers also manage reproduction to assure that animals produce the desired offspring. For example, a dairy producer wants cows that produce a lot of milk. To gain such cows, only animals with genetic potential for high milk production are allowed to reproduce. Several technologies are available to promote quality offspring. Understanding reproduction processes helps producers make better choices in animal management. There is a lot of difference in how cattle, chickens, and fish reproduce...knowing these processes for the species you have is essential!

ROLE OF ANIMAL REPRODUCTION

Reproduction is the process by which offspring are produced. The offspring are of the same species and have traits of their parents. For example, rabbits reproduce other rabbits though the offspring are not completely identical to their parents. We know them as rabbits when we see them!

Reproduction is essential if a species is to stay in existence. Without reproduction, there would be no new animals. Reproduction is not needed for an animal to live. Animal producers often neuter animals so that they cannot reproduce. Neutered animals have certain advantages from the production standpoint. Neutering advantages include that animals do not reproduce resulting in unwanted offspring; they do not engage in disruptive mating behavior; and, particularly with steers (castrated bull), the animals gain weight more rapidly when in a feeding facility. In addition, a steer that has been fed grain brings a higher per pound price than a bull. With hogs, a feeding facility would have almost every female pregnant if the males were not castrated. Prices are discounted for pregnant females plus feed nutrients are used for developing fetuses rather than growth. (If a female hog or sow, gave birth to 10 pigs weighing 3.5 pounds each that would be 35 pounds of babies, which requires a lot of nutrients.)

SEXUAL REPRODUCTION

Sexual reproduction is the union of a sperm and an egg. A new animal begins at the time of this union. Two parents are required: male and female. The ***sperm*** (spermatozoa) is the sex cell produced by male animals. The ***egg*** or ovum is the sex cell produced by female animals.

The union of sperm and egg is fertilization. Fertilization takes place inside the reproductive tract of female mammals, such as cattle, horses, and hogs. With birds, fertilization occurs within the reproductive tract of the female and the egg is laid (expelled) for incubation of the developing young. With some animals, fertilization is outside the reproductive tract, such as fish. However the process may occur, time is needed for the union to grow into a new individual.

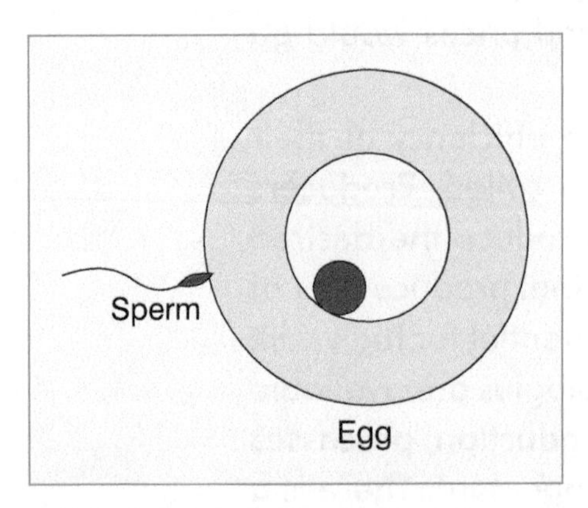

5–2. Sexual reproduction involves the union of a sperm and an egg.

With mammals, the fertilization of an egg to create new life is ***conception***. The new developing organism is formed in the reproductive tract of the female. A female that has a developing unborn baby in her reproductive tract is ***pregnant***. The period of pregnancy varies by species.

Insemination

Insemination is placing sperm in the reproductive tract of the female. Animal producers sometimes regulate insemination. Male animals may be kept separately from the females. New technologies are being used in animal reproduction to assure desired offspring.

Natural insemination is the process of the male depositing semen (fluid containing sperm) in the reproductive tract of the female. It occurs during copulation (mating). ***Copulation*** is the sexual union of a male and female animal. In copulation, the semen is released (ejaculated) by the male in the reproductive tract of the female near the eggs.

Artificial insemination is collecting semen from a male and placing it in the reproductive tract of a female of the same species. Special equipment and methods are used in the process. Artificial insemination gives the producer more control over the reproductive process.

Breeding

Breeding is promoting animal reproduction so the desired offspring result. Producers get the kinds of animals they want to meet market demand. Without some control over breeding, animal quality would not meet the demands.

Producers want animals that meet their needs. To do so, animal breeds have been developed. A breed is a group of animals of the same species that share common traits. For example, the Angus is a breed of beef cattle that is black and polled (some Angus are red and known as Red Angus). Breed traits are inherited. They are passed from one generation to the next. Traits offspring inherit are color, milk capacity, size, type, and presence of horns (in some species).

5–3. A purebred Hampshire boar meets all of the standards of the Hampshire Swine Registry. (Courtesy, Hampshire Swine Registry)

A ***purebred animal*** is one that is eligible for registry in a breed association. Its parents must meet the standards for the breed. The offspring of two purebred animals of the same breed will qualify for registration as a purebred. Some animal producers raise purebred animals.

To produce animals of a certain type, purebred animals of the same species, but of different breeds, are mated. This is ***crossbreeding***. For example, an Angus bull could be mated with a Hereford cow. Crossbreeding is used to improve the quality of the products produced by the offspring. In some cases, crossbred animals are mated to create even more desirable offspring. Other systems of breeding are also used.

5–4. This ewe with the newborn quadruplet lambs represents reproductive efficiency. Most ewes produce two young at each lambing. (Courtesy, Northwest Experiment Station, University of Minnesota, Crookston)

REPRODUCTIVE EFFICIENCY

Reproductive efficiency is the timely and prolific replacement of a species. It may result in some species flourishing and others declining. Reproductive efficiency is the difference between success and failure in animal production.

The union of a sperm and an egg may not always produce a new, healthy individual. The developing animal may die before birth or hatching. With mammals, the death of a fetus is natural abortion. Some are born or hatched at low birth weights or are defective in some way. Losses to abortion and other problems lower reproductive efficiency. For example, a cow that does not have a calf is not productive and should be culled from the herd.

Research findings can influence a species or breed so the cost of products is reduced. Artificial insemination is an important example. More new technology is being developed.

A good example of new technology is accelerated lambing. ***Accelerated lambing*** is a method that uses out-of-season lambing techniques to get three lamb crops in two years

CONNECTION

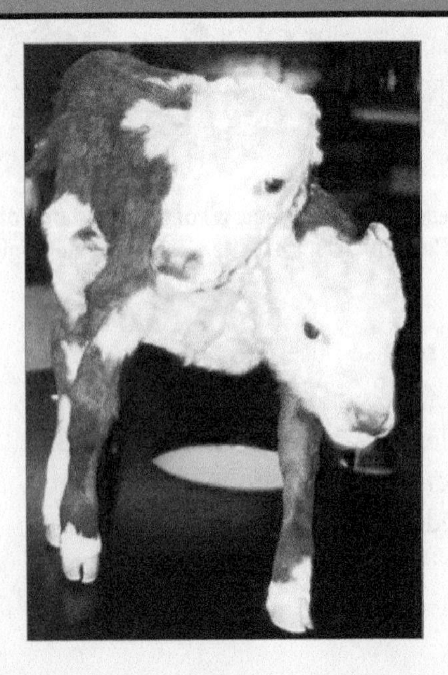

TWO HEADS ARE NOT BETTER THAN ONE

This day-old calf was born with two heads. Each head was joined by a neck to the body. Both heads appear to be well developed. The body was near normal except where the necks joined. The calf did not live.

Why was the calf born this way? No one knows for sure. Several reasons have been suggested. The two heads could have been the result of a mutation. Another reason is that the birth defect was due to pollution in the pasture where the calf's mother lived, such as a range with radioactive wastes present. Still another reason could be that the animal was from a diseased parent.

Two-headed calves do not usually live. They rarely occur. When they do, they represent reproductive failure. Producers prefer normal calves. An abnormal calf has cost the owner feed for its mother and delayed income from a marketable calf for about two years. (Courtesy, Education Images)

from an ewe. Most ewes have one lamb crop a year. Accelerated lambing allows sheep producers to increase their productivity by 30 to 50 percent.

REPRODUCTIVE ORGANS AND SYSTEMS

Reproductive processes must work properly. When they do not, new animals are not produced. A new calf being born or a chick hatching depends on the proper function of reproductive organs. Anyone planning to breed animals needs to know about the reproductive organs of animals and their functions.

FEMALE REPRODUCTIVE ORGANS AND SYSTEMS

The primary reproductive organ of the female is the ***ovary***. The ovary produces female gametes. A ***gamete*** is a sex cell that can unite with other sex cells. It is a mature sex cell capable of a union that produces an individual organism.

Female mammals have two ovaries. Some variation exists in other animals. Only the left ovary fully develops in chickens. Ovaries contain follicles, which are tiny structures that produce ova. A heifer has approximately 75,000 primary follicles in her ovaries at birth. A chick has about 4,000 miniature ova when hatched.

CAREER PROFILE

ARTIFICIAL INSEMINATOR

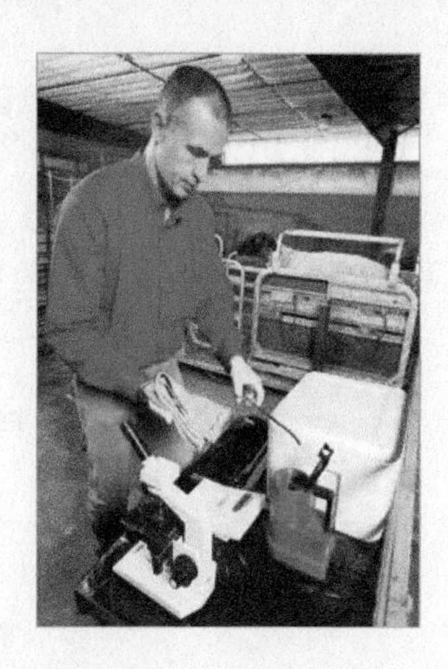

An artificial inseminator works on a farm or ranch to artificially inseminate animals. Some are entrepreneurs and own an artificial insemination (AI) business. The work involves storing and preparing fresh or frozen semen for insemination and performing the insemination process. The semen must often be observed with a microscope to determine sperm quality and presence of abnormal sperm. In some cases, the artificial inseminator will also collect semen from the male of the species.

Artificial inseminators must have a good knowledge of reproductive biology. Education and training include two to four years or more of college study plus workshops and short courses in artificial insemination. They need thorough knowledge of the female reproductive systems of the species they inseminate. Placing sperm in the reproductive tract to assure pregnancy requires knowledge of reproductive cycles, particularly determining when a female is in heat. Jobs for artificial inseminators are on farms and ranches where animals are artificially inseminated.

This image shows frozen sheep semen being thawed in a water bath for use in artificially inseminating an ewe that is in heat. (Courtesy, Agricultural Research Service, USDA)

The maximum number of ova a female will have are present at birth or hatching. Ova are slowly used throughout a lifetime. An older cow may have only 2,500 follicles containing ova when culled from the herd.

As an ovum matures, it moves to the edge of the ovary for release. Releasing an egg is **ovulation**. The egg is normally caught by the infundibulum (first segment of oviduct) and sent through the oviduct (fallopian tube).

In mammals, the oviduct moves the ova and spermatozoa (if present) for fertilization and early cell division. The embryo (developing young) remains in the oviduct an additional three to five days.

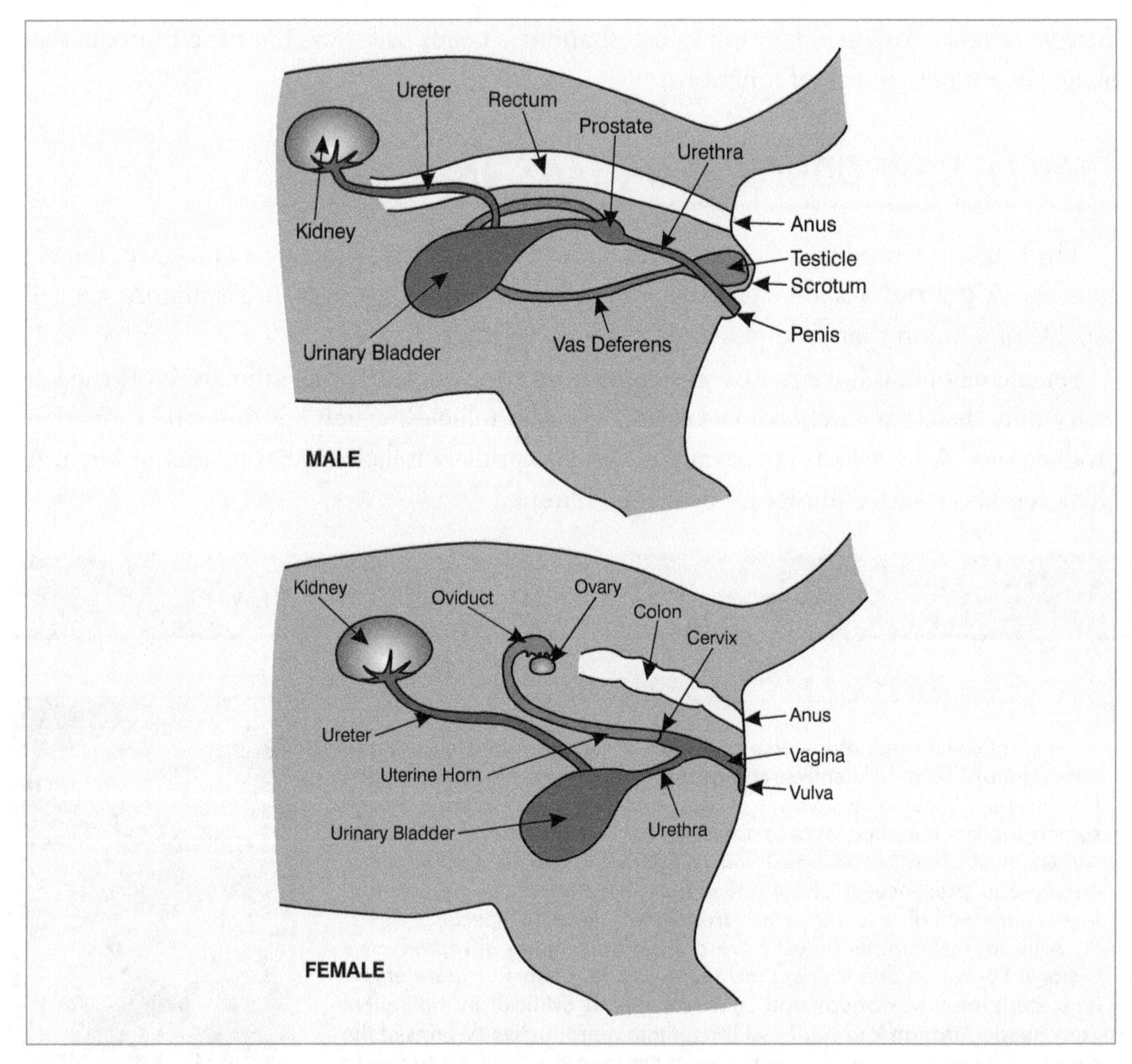

5–5. The reproductive systems of the cat.

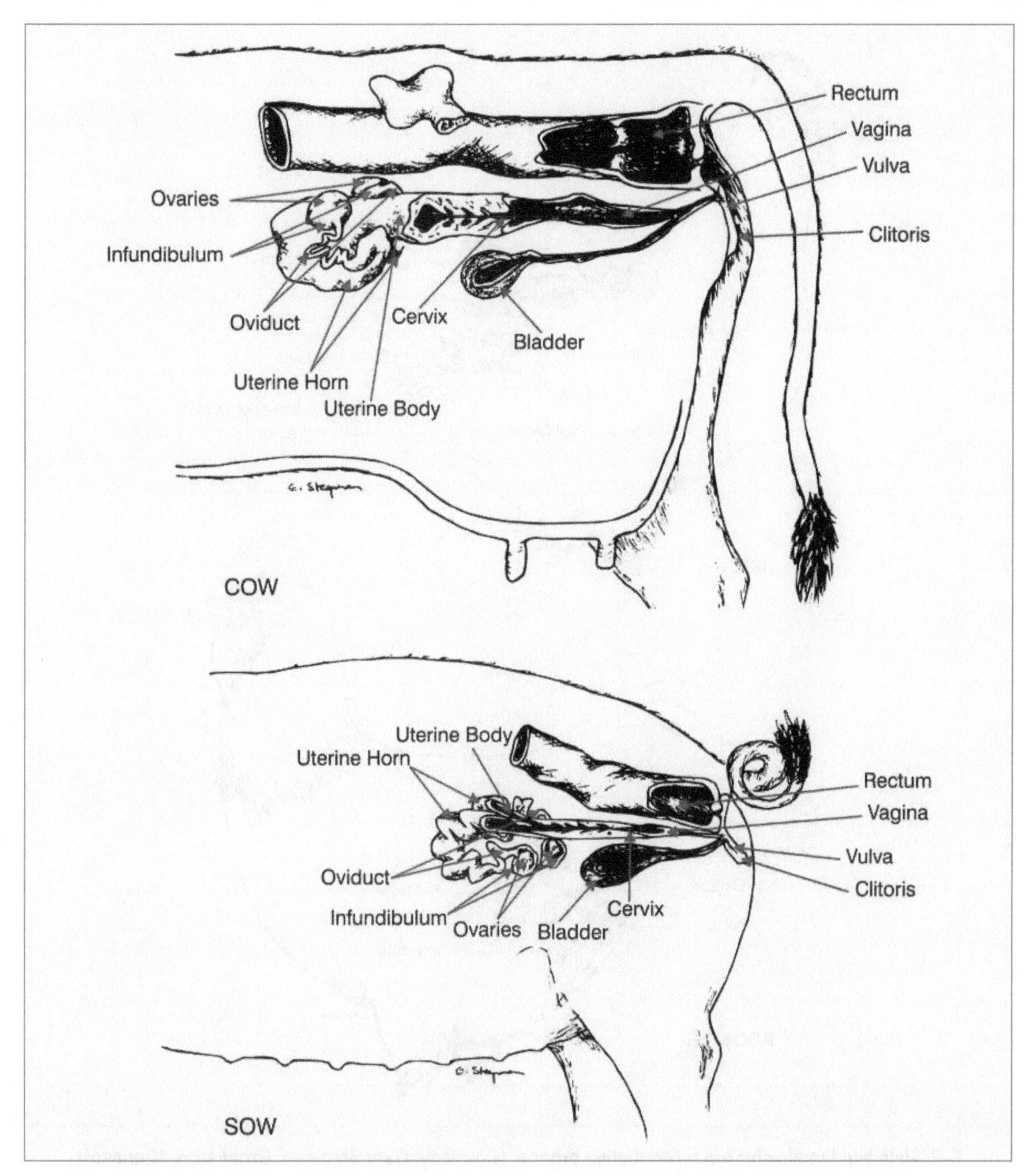

5–6. Cow and sow reproductive organs. (Courtesy, Gary Stegman, Crookston, Minnesota)

Fertilization in a hen occurs in the infundibulum (sometimes called a funnel). The oviduct of the hen has four segments. A developed ovum (egg yolk) goes through these segments. The magnum secretes the albumen (thick white) of the egg. As the egg travels through the isthmus, it receives shell membranes. The uterus adds the thin white, the outer shell, and the shell pigment. The completed egg then passes through the vagina and is laid

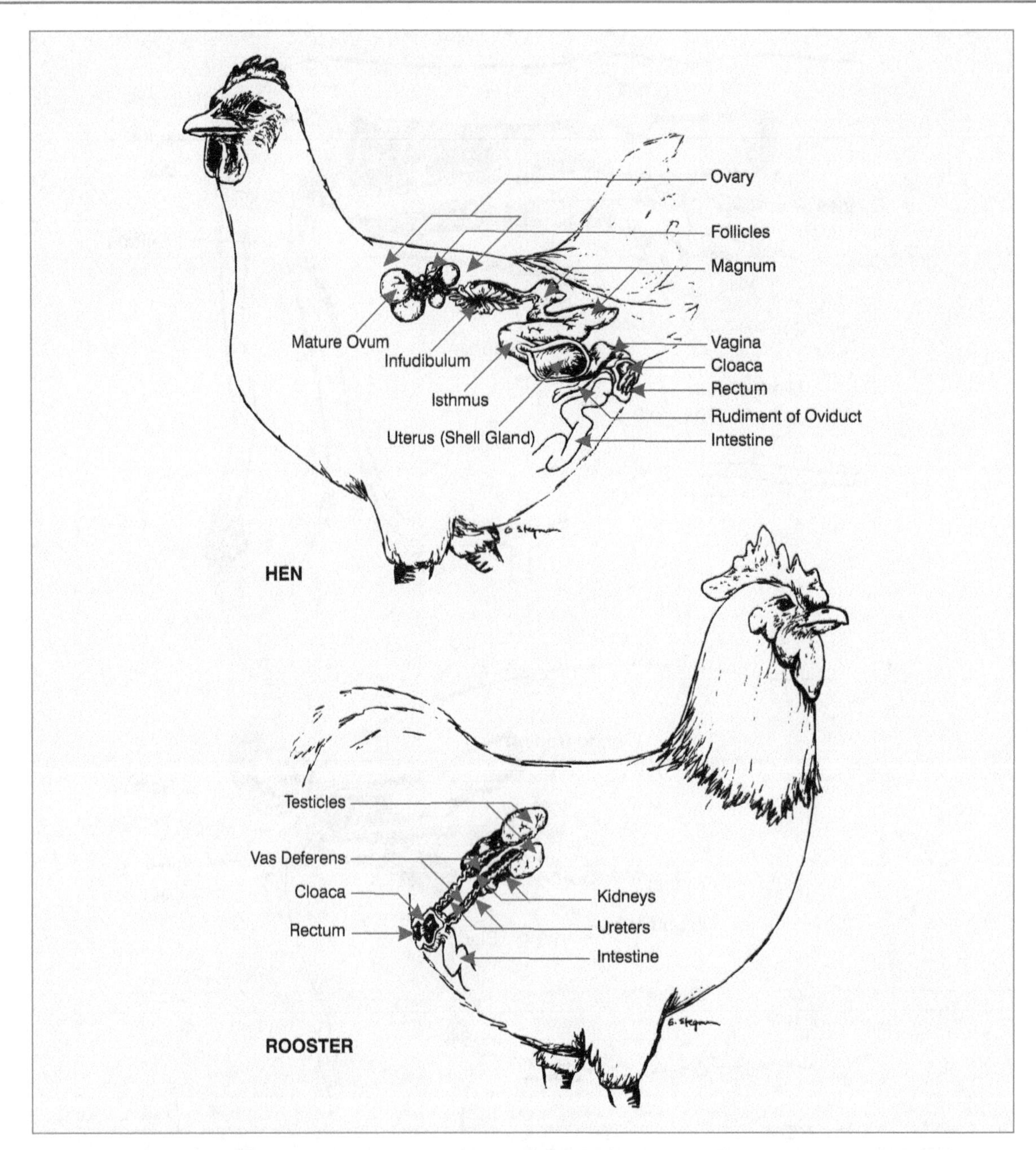

5–7. Male and female chicken reproductive organs. (Courtesy, Gary Stegman, Crookston, Minnesota)

through the cloaca. The cloaca is the opening for both the digestive and reproductive tracts. The entire time from ovulation to laying is slightly more than 24 hours.

In mammals, the embryo goes to the uterus. The uterus has two parts: uterine horn and uterine body. A placenta forms and attaches itself to the uterine wall. Cattle and swine differ from horses. A mare carries a foal in the uterine body, while a pig or calf is carried in the uterine horn.

The cervix is a thick-walled, inelastic organ (technically part of the uterus) that connects the uterus and the vagina. The cervix contains annular rings and cervical mucus that seal the uterus to keep out contaminants. During pregnancy, the mucus thickens to form a gel-like plug. Just before giving birth, hormones (chemical substances produced by the body) relax the cervix to allow the young to pass through.

The vagina is the passageway for reproduction and urine excretion. The vulva is the external opening of the reproductive and urinary systems. The clitoris is located just inside the vulva. It is a highly sensitive organ.

The reproductive organs of fish differ from mammals and poultry. The ovary of most fish species is a single large sac of eggs that appears as two ovaries that are joined. The eggs travel via the oviduct to the urogenital opening where they are excreted. Most fish eggs are fertilized outside the female's body.

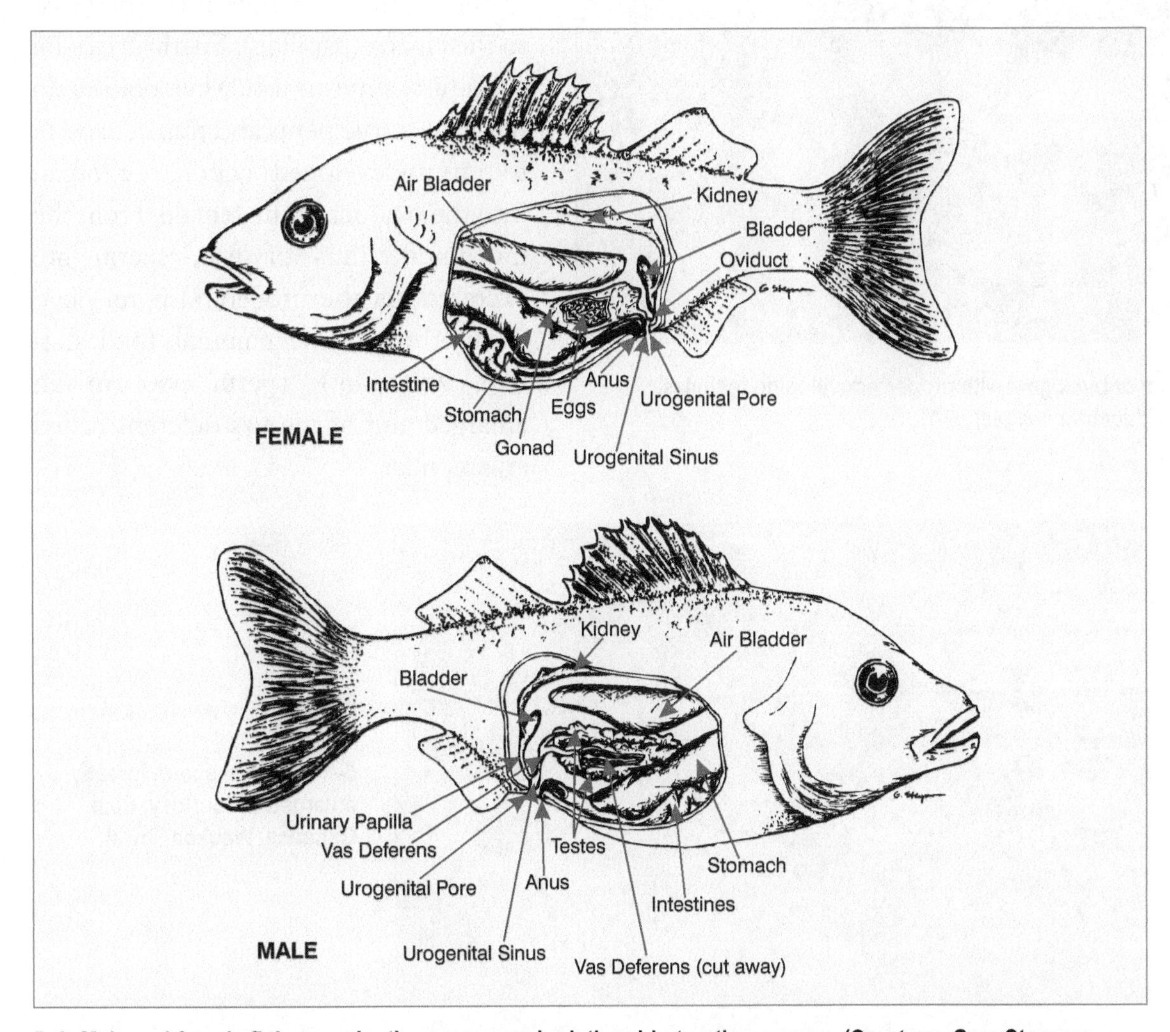

5–8. Male and female fish reproductive organs and relationship to other organs. (Courtesy, Gary Stegman, Crookston, Minnesota)

MALE REPRODUCTIVE SYSTEM

The primary reproductive organ of the male is the **testicle**. Testicles produce male gametes or sperm. Two testicles are held externally in the scrotum of mammals and internally in the body cavity of male poultry and fish.

Males begin producing sperm at puberty and do so throughout life. The **scrotum** is a two-lobed sac in mammals that protects the testicles and helps control the temperature. Sperm production is best if testicles are cooler than the body.

Sperm are formed in the seminiferous tubules of the testicles. Sperm pass through the epididymis into the vas deferens. The vas deferens carries the sperm and fluid (semen) for release. In mammals, it goes to the urethra. With poultry, sperm go to the cloaca (papillae). Sperm go to the urogenital sinus in fish. Male poultry do not have a true penis and transfer sperm to an undeveloped copulatory organ into the oviduct of the female. From the urogenital sinus of fish, sperm are excreted via the urogenital pore. Near the urethra of some mammals (bull, stallion, and ram), sperm pool in an enlarged end of the vas deferens called an ampulla.

5–9. Scrotum of buck goat with properly developing testicles. (Courtesy, Education Images)

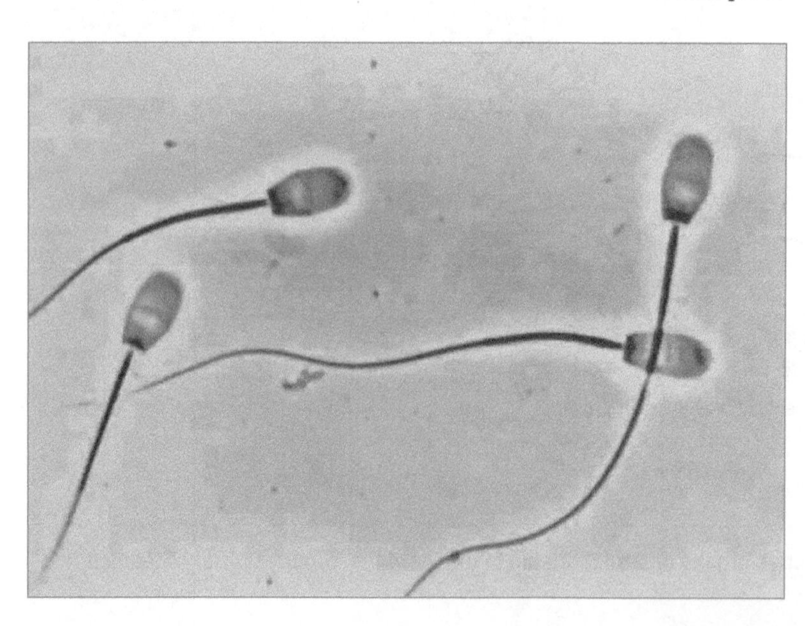

5–10. Normal sperm (greatly enlarged). (Courtesy, Elite Genetics, Waukon, Iowa)

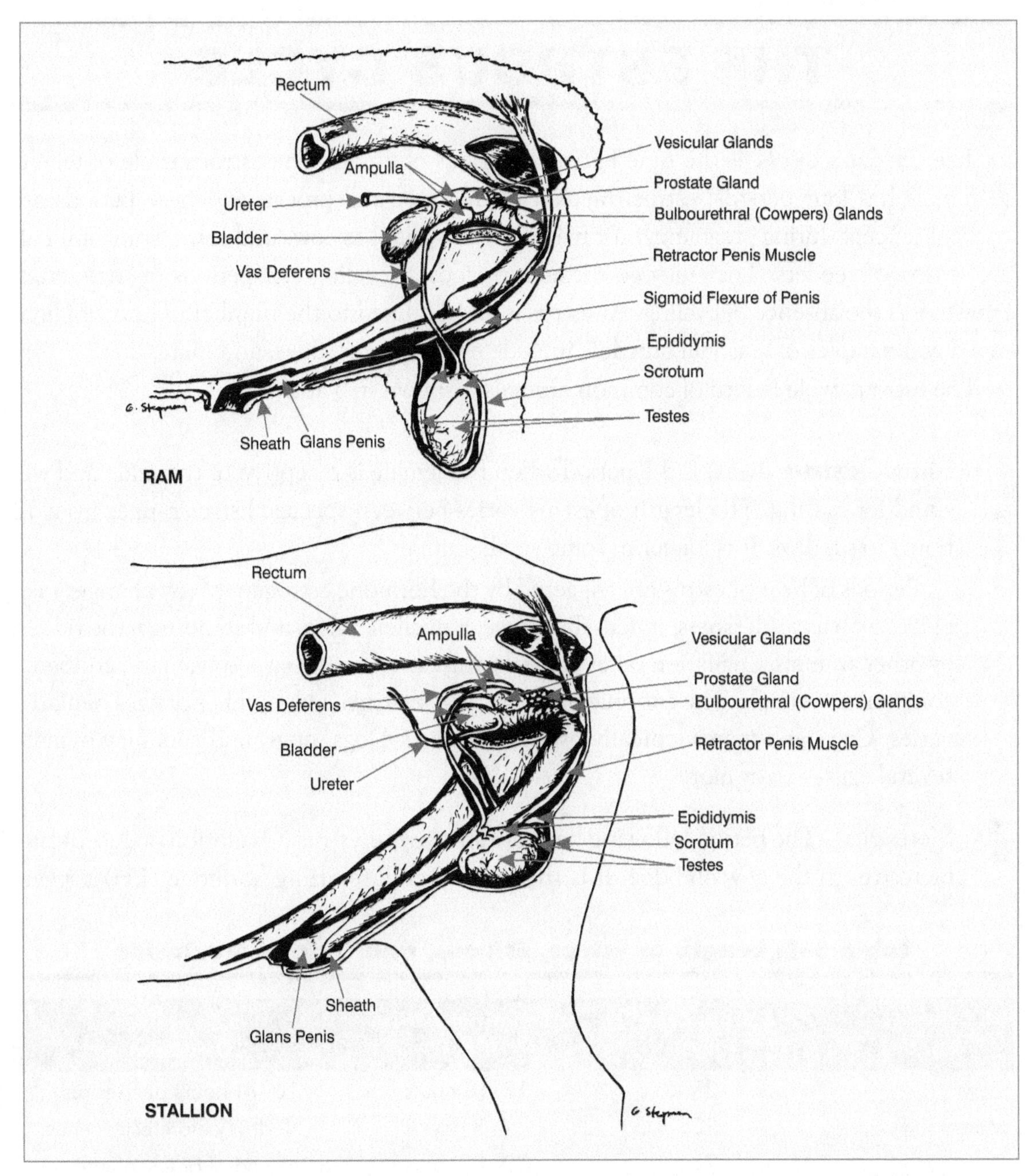

5–11. Ram and stallion reproductive organs. (Courtesy, Gary Stegman, Crookston, Minnesota)

At ejaculation in mammals, semen passes through the penis and is released in the reproductive tract of the female. The penis is normally relaxed but is erect during sexual arousal. The free end of the penis (known as glans penis) serves as the sensory organ similar to the clitoris in females. It also retracts beneath the protective sheath (fold of skin) in most animals when not aroused.

THE ESTROUS CYCLE

The ***estrous cycle*** is the time between periods of estrus. The estrous cycle of female mammals has four periods: estrus, metestrus, diestrus, and proestrus. These periods are cyclical (except during pregnancy) for many animals, such as cows and sows. Some animals are seasonal breeders. They may go through periods of cycling and periods of ***anestrus***. Anestrus is the absence of cycling. Anestrus is often related to the number of hours of light in a day. Examples of seasonal breeders include ewes, queens, does, and mares.

The estrous cycle length of common animals is shown in Table 5–1.

- Estrus—***Estrus*** (heat) is the period when the female is receptive to the male and will stand for mating. The length of estrus varies between species. Estrus ranges from 12 hours to six days. It is longer in some small animals.

  Periods of heat or estrus are triggered by the hormone estrogen. Many changes take place, such as restlessness, mucus discharge, a swollen vulva, and standing to be ridden by other animals. Ovulation takes place during estrus for the ewe, sow, mare, and some companion animals. The cat only ovulates after mating. The number of eggs ovulated varies. Cows and mares typically ovulate one egg. Dogs, hogs, and cats may ovulate several eggs—15 or more.

- Metestrus—The period following heat or estrus is metestrus. Ovulation occurs during metestrus in the cow and doe. It is the time when luteinizing hormone (LH) triggers

Table 5–1. Length of Estrus, Estrous, and Time of Ovulation

Species	Estrous Cycle (days)	Length of Estrus (heat)	Ovulation
Cow	21	12–18 hours	10–14 hours after estrus
Mare	22	6–8 days	1–2 days before estrus ends
Doe (goat)	21	30–40 hours	at the end of or just after estrus
Doe (rabbit)	Constant	Constant	8–10 hours after mating
Sow	20–21	40–72 hours	mid estrus
Ewe	17	24–36 hours	late estrus
Bitch (dog)	—	9 days	1–2 days after estrus begins
Queen (cat)	14–21*	5 days	24 hours after mating

*Estrous cycle influenced by length of daylight, with mating season typically when there are more than 12 hours of daylight in one day.

the corpus luteum (CL) to develop from follicular tissue that remains after release of the ova. The corpus luteum (yellow body) is important in maintaining pregnancy.

After ovulation, some capillaries break and release small amounts of blood. The blood is occasionally seen on the tail of a cow a day or so after estrus. This blood is not a sign that a cow is or is not pregnant.

- Diestrus—Diestrus is the period in each estrous cycle in which the system assumes pregnancy. It is characterized by a fully functional corpus luteum that releases high levels of progesterone (the hormone that maintains pregnancy). Diestrus is often 9 to 12 days in length. It is during diestrus that the uterus is prepared for pregnancy.
- Proestrus—Proestrus begins with the regression of the corpus luteum and a drop in the hormone progesterone. Follicle stimulating hormone (FSH) causes rapid follicle growth in preparation for estrus and ovulation. Late in proestrus, changes in behavior may occur as estrus approaches.

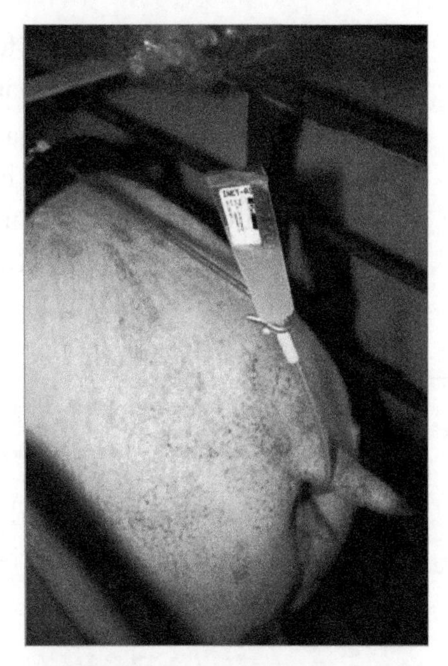

5–12. An in-heat (estrus) sow is being artificially inseminated. (The hollow spirette tube has been inserted through the vagina into the cervix. A tube is draining diluted semen through the spirette into the cervix so that fertilization of released ova can occur. Ovulation is 24-48 hours after the onset of estrus.) (Courtesy, Education Images)

PHASES OF REPRODUCTIVE DEVELOPMENT

Reproduction involves a series of events that must be properly timed. Its success is measured in various ways, such as the number of pigs per litter or calves per 100 cows. Hatchability and liveability are important to poultry and fish operations.

PUBERTY

Puberty is the time at which animals reach a level of sexual development that makes them capable of reproduction. Puberty in female animals is indicated by the first estrus with ovulation. Puberty in males is indicated by the first ejaculate with fertile sperm.

Neither males nor females are sexually mature at puberty. The female is often too small to bear young. The male is not highly fertile nor capable of breeding regularly.

Both environmental and genetic factors affect the age at which puberty occurs. Weight at puberty is affected by genetic factors, such as size of the parents. The primary environmental factor is anything that affects growth. Differences between breeds are the primary genetic factors within species. It is important to not breed animals until they reach the recommended weight.

FERTILIZATION

Fertilization is the union of a sperm and an egg. The sperm penetrates the egg and pairs of genetic material are formed. The fertilized ovum is a **zygote**.

GESTATION

Gestation is the period of pregnancy. It begins with conception and lasts until parturition or birth. The average length of gestation varies from 114 days for a sow to 337 days for a

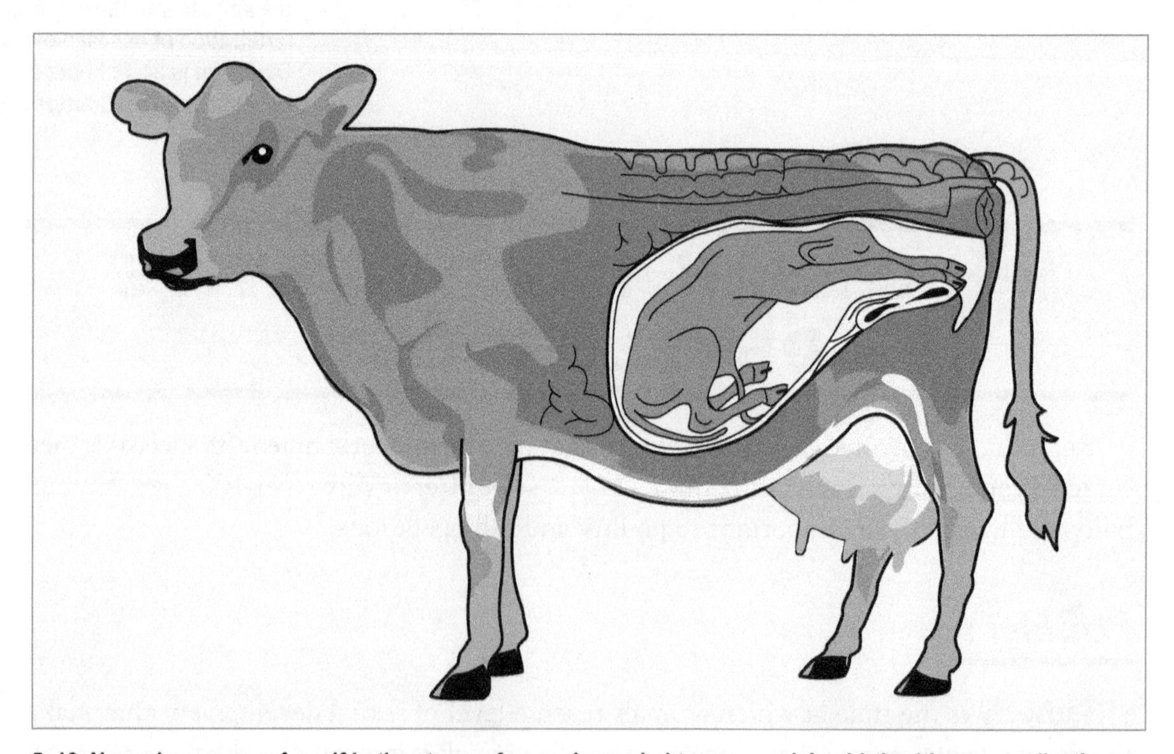

5–13. Normal presence of a calf in the uterus of a cow is needed to assure giving birth without complications.

mare. Fertilization starts gestation. A series of cell divisions without growth (cleavage) occurs while the embryo slowly migrates toward the uterus over a three- to four-day period.

The new animal is known as an **embryo** while the organs in the developing animal are forming, which is differentiation. Growth is rapid during this period. Differentiation is complete by day 28 in a sow and day 45 in a cow. After differentiation and until birth, the young animal is a **fetus**. The fetus gains weight and matures in preparation for birth.

PARTURITION

Parturition is the process of giving birth. Several hormone levels change and begin this process. These initiate uterine contractions and milk production.

Signs of approaching parturition can be observed. Females begin to exhibit a "nesting" behavior. They become restless and attempt to separate from the herd or flock.

The first stage of parturition includes dilation of the cervix and entry of the fetus into the cervix. This usually is the longest stage and ranges from one to 12 hours, depending on the species. The second stage completes the birth of an animal through strong uterine contractions.

The birthing process and the first hour after birth are critical. Animal producers may watch to ensure the fetal position is correct. They may observe the delivery for problems. They check to see that respiration has started and help the newborn nurse to get colostrum—the first milk that is high in antibodies.

The last stage of parturition is the expulsion of the afterbirth (placenta) from the uterus. This normally occurs shortly after giving birth. Sometimes the placenta is retained. This can interfere with future reproduction. The next gestation depends on a return to normal estrous cycling. The uterus must return to normal size and condition.

LACTATION

Lactation is the production of milk. Young mammals must have milk for food. The second purpose of lactation is to provide early disease resistance for the newborn. Colostrum contains important antibodies.

Lactation is a part of the reproductive cycle. Hormones that trigger the onset of lactation also play a role in parturition. Lactation requires nutrients that extend periods of anestrus following parturition. Weaning pigs or temporarily removing calves from cows can trigger estrus.

MATING BEHAVIOR

Conditions for domestic animals must be such that mating occurs. In addition to a properly cycling female, the male must be in proper condition and have the proper status (social rank). Social rank refers to status of the male within a herd or group. Males need both the desire to mate (libido) and the ability to mate. Both can be enhanced with proper nutrition that keeps the males from becoming overweight and with an exercise area/breeding pen that allows animals to move about. Males should be kept disease and injury free so mating is possible.

Management of rams in a sheep flock is an example of stimulating proper estrous cycling and mating behavior. Rams should be kept away from ewes (out of sight and smelling range) until near breeding time. Reintroducing rams near breeding time will stimulate cycling in ewes. Males of any species that are very dominant can keep other males from mating. Pasture and breeding pen design helps males that are lower in the social order breed their portion of females. This ensures a high pregnancy rate in the total herd/flock.

INCUBATION

Incubation is the development of a new animal in the fertile egg of poultry, fish, and other egg-laying species. Incubation occurs outside the body of the female. Proper incubation is needed to hatch eggs. Eggs are tended by species in different ways, such as birds "sit" on the nest. Eggs can be, and are, artificially incubated in hatcheries.

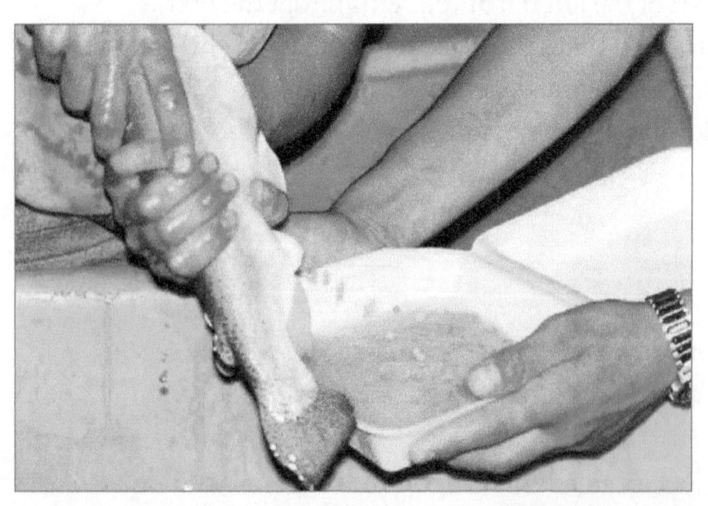

5–14. Eggs being artificially removed from a trout for fertilizing and hatching. (Courtesy, Education Images)

Four important factors in incubating bird eggs are temperature, humidity, oxygen, and egg rotation. Temperature should be maintained at 99 to 103°F (37 to 39°C). The humidity should be about 60 percent during the first 18 days and 70 percent during the last 3 days (assuming a 21-day incubation period). Eggs should be slightly higher in the large end and be rotated two to five times daily for the first 18 days. Sufficient air exchange to prevent carbon dioxide buildup, while maintaining a 21 percent oxygen level, will enhance hatching results. Incubation times range from 21 days for chickens to 42 days for ostriches.

5–15. Poultry eggs are incubated in large incubators that provide the ideal temperature for fertile eggs to develop. (These chicken eggs will hatch in about 21 days from the beginning of incubation.) (Courtesy, branislavpudar/ Shutterstock)

Early embryo development takes place in the body of the hen (107°F or 42°C) before the egg is laid. Several cell divisions take place. After laying, the egg cools and growth stops. Fertile eggs should be incubated shortly after laying. Some must be held for awhile before incubation. Eggs stored below 80°F (27°C) are more likely to hatch when incubated.

Incubation continues embryo development. Organ formation (cell differentiation) is completed in the first six to eight days. Two key changes take place near the end of incubation. The beak turns toward the air cell and the yolk sac enters the body cavity. This yolk sac provides nutrition and water for the first several hours of life after hatching. The yolk is gradually used during the first 10 days of life.

Table 5–2. Incubation Times for Selected Poultry Species

Common Species Name	Incubation Period (days)
Chicken	21
Pheasant	24
Duck	28
Turkey	28
Goose	28–32
Ostrich	42

SPAWNING

Spawning is the release of eggs by a female fish and the subsequent fertilization by the male. A female may lay thousands of eggs at one spawning, depending on the species. The male fish then fertilizes the eggs by releasing sperm on the mass of eggs. The eggs of some species are protected only by a gravel covering. The males or females of some species provide the eggs with protection or incubation. Spawning is covered in more detail in Chapter 13.

ANIMAL REPRODUCTION TECHNOLOGY

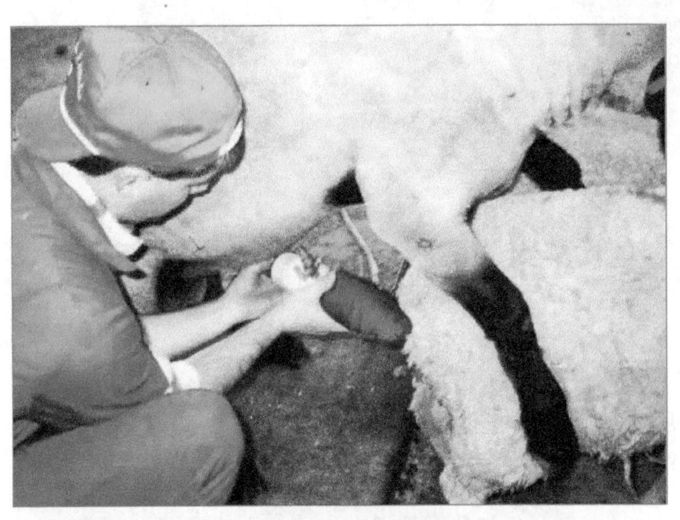

5–16. Ram semen is being collected with an artificial vagina and teaser ewe. (Courtesy, Elite Genetics, Waukon, Iowa)

5–17. Scientists are using a video microscope to view turkey sperm. (Their goal is to identify ways of improving the lifespan of stored turkey sperm.) (Courtesy, Agricultural Research Service, USDA)

The use of artificial insemination (AI) has increased. Today, AI is more common than natural service in some species, such as dairy cattle and turkeys. Synchronization and embryo transfer are also used but to a lesser extent.

ARTIFICIAL INSEMINATION

Artificial insemination is placing semen in the female reproductive tract by artificial techniques. AI has the advantage of using a quality male to breed many females. Superior traits can be extended to a large population. This has improved many species. AI can develop desired qualities in companion animal offspring. Meat animals can be produced that have high yields. Cows can be produced that give more milk. Hens can be produced that lay more eggs.

SEMEN COLLECTING

Artificial insemination requires that semen be collected from the male. A tom turkey or rooster requires manual stimulation and semen removal (sometimes called "milking of semen"). Sperm from a male fish can be obtained by applying gentle pressure on their abdomen. Semen from bulls, rams, and stallions is usually collected with an artificial vagina (AV).

Semen collection is improved if time is given for the male to become stimulated. The collected semen is evaluated, cooled slowly, and processed in preparation for freezing at temperatures of –320°F (–196°C). Semen stored in liquid nitrogen at –320°F (–196°C) can be thawed and used up to 40 years later.

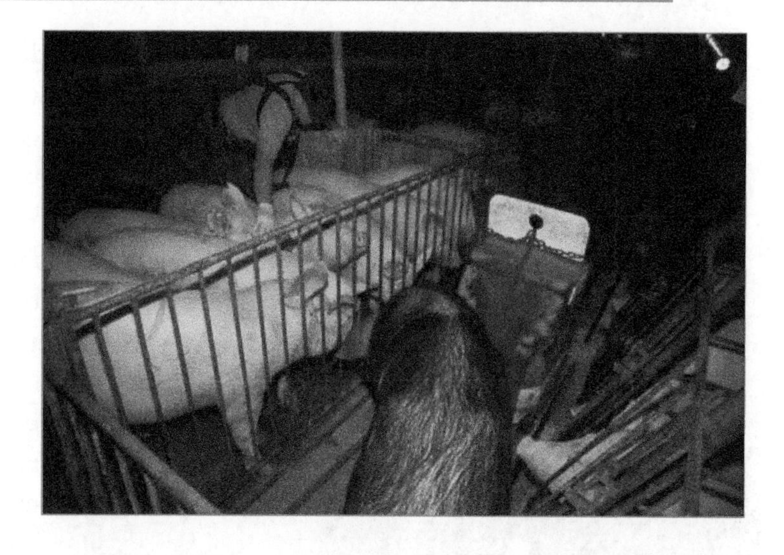

5–18. A non-breeding boar moves along the aisle in a sow unit to help the AI technician detect which sows are in estrus (heat). (A sow that is in estrus exhibits a desire to mate when she smells the presence of a boar. The technician sees this behavior and begins the AI process. The boar is attached to a device that slowly rolls along the aisle forming what is sometimes referred to as a "boar-bot.") (Courtesy, Education Images)

INSEMINATION

High conception rates require much more than quality semen. The female must be cycling and in estrus (heat). The semen should be thawed and inseminated after heat detection.

Detecting estrus usually requires keen observation. Failure to detect heat is the most common cause of AI failure.

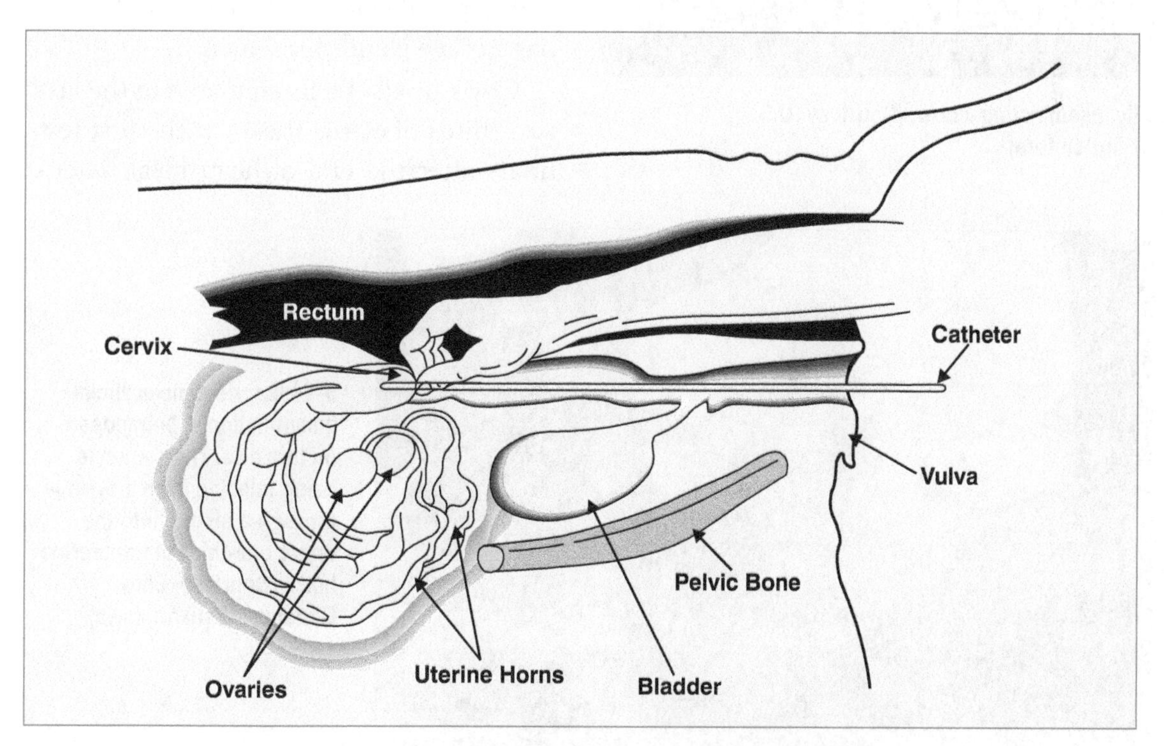

5–19. Using the recto-vaginal method to artificially inseminate a cow.

Estrus signs vary between species. The best indication of estrus for most species is standing heat. Standing heat is the stage of estrus when a female stands when mounted by another animal. Other behavioral patterns change when an animal is in estrus. Most animals become more active. Cows try to mount other animals and are generally restless. Mares urinate frequently and expose their clitoris in a process called "winking." Many animals display extra mucus and redness in the vulva. Sows may have redness in the vulva. The ewe requires a male being present before she will show signs of estrus. A ram that has had a vasectomy (made infertile) is often used to detect ewes in heat.

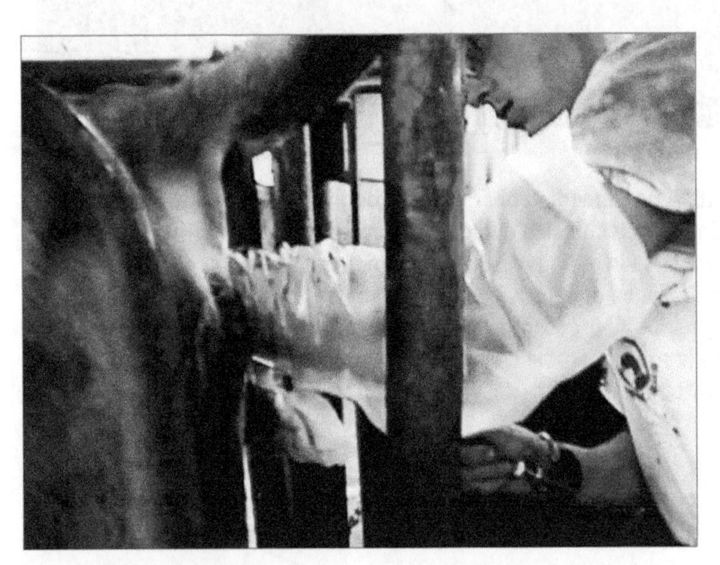

5–20. Artificially inseminating a cow. (Courtesy, U.S. Department of Agriculture)

The timing and placement of semen varies depending on the species. All require that frozen semen be thawed properly (95 to 98°F or 33 to 34°C). Fresh semen should be used within an acceptable time, depending on the species.

Most turkeys must be artificially inseminated due to their large size. Insemination does not need to be on a daily basis. A week is usually suggested between the first two inseminations and a slightly longer interval on subsequent inseminations. This is possible in poultry because the hen has storage glands for semen.

Cows should be inseminated in the last two-thirds of estrus (heat) or the first few hours after the end of heat. Heat detec-

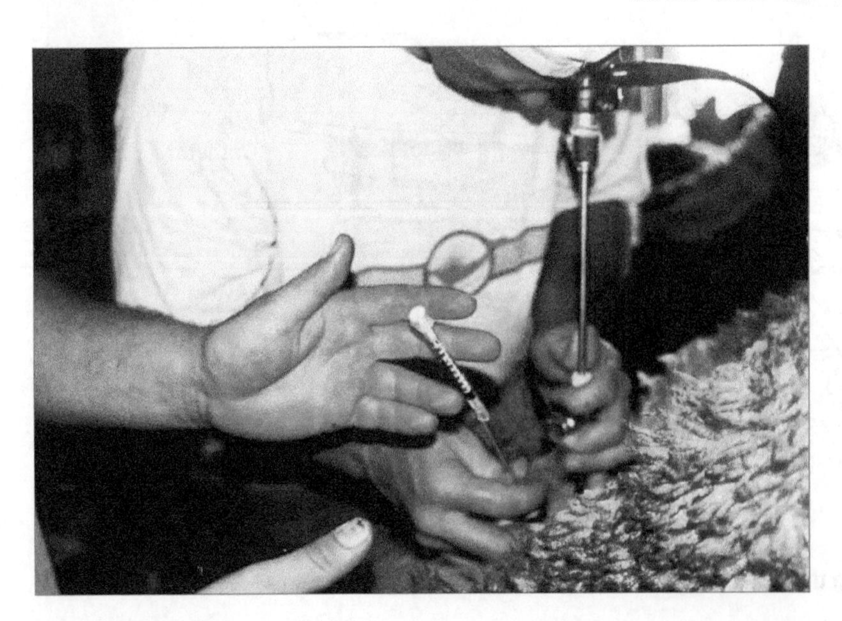

5–21. Lapriscopic artificial insemination is being used on this ewe. The semen is being injected from a syringe through a pipette into the upper one-third of the uterine horn. (Courtesy, Elite Genetics, Waukon, Iowa)

tion is most often done early in the morning or late at night, since most mounting activity occurs during or near the night hours. An approved practice is to breed cows detected in heat in the morning during the afternoon of the same day. Cows in heat in the evening should be bred early the next day. Insemination of cattle usually involves the recto-vaginal method where a gloved arm is placed in the rectum to palpate (feel) the reproductive organs.

Mares may need to be inseminated several times. This is because it is difficult to detect ovulation. Inseminations are made on the third, fifth, and seventh days of estrus. The AI procedure for mares differs from that for cows. Mares have a delicate rectum and other anatomical differences. The arm and syringe are placed in the vagina with a finger through the cervix. This requires special emphasis on cleanliness. The semen is deposited in the body of the uterus.

Sows should be bred 24 hours after the onset of estrus. Some people prefer to breed a second time, 40 to 48 hours after the onset of estrus. An inseminating tube is inserted into the cervix. The semen is injected with a plastic squeeze bottle syringe.

Artificial insemination of ewes is less common. Two inseminations will increase chances of conception and multiple lambs. The size of a ewe requires smaller equipment or the use of lapriscopic artificial insemination. In lapriscopic artificial insemination, semen is placed in the uterus through the ewe's abdominal wall.

SEXED SEMEN

The sex of offspring can be controlled with sexed semen. ***Sexed semen*** is semen that has been prepared to produce all male or all female offspring. It is collected in the manner as other semen used in artificial insemination. Sexed semen will predict sex with about 90 percent accuracy. The cost of sexed semen will likely be about four times the cost of unsexed semen.

ESTROUS SYNCHRONIZATION

Estrous synchronization is bringing a group of animals into heat simultaneously. This helps schedule animal breeding and birthing.

Synchronization usually involves the use of prostaglandin, progestin, or a combination of the two. Prostaglandin causes the corpus luteum to stop production of progesterone, allowing the animal to come into estrus. Progestin has the effect of keeping progesterone levels high, holding animals in an "extended" diestrus. When the progestin source is removed, the animal quickly comes into estrus.

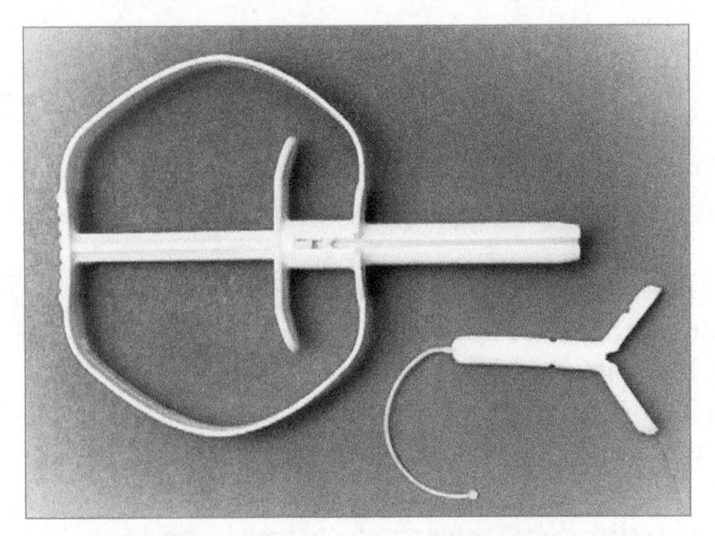

5–22. A controlled internal drug release (CIDR) dispenser for use in synchronizing ewes. (Courtesy, Northwest Experiment Station, University of Minnesota, Crookston)

CIDR (controlled internal drug release) dispensers are used with sheep to promote synchronization within a flock. CIDRs are placed in the vagina of ewes. When they are removed, ewes cycle shortly afterward. This method is used to breed ewes during times of the year when they do not cycle. This helps sheep producers get three lamb crops every two years.

EMBRYO TRANSFER

Embryo transfer is moving embryos from one female (known as a donor) to the reproductive tract of another female (known as a recipient). Donor females usually carry extraordinary genetics. Recipient animals have far less worth and are used as surrogate mothers.

Most embryo transfer work is done with beef and dairy cattle. Embryo transfer is used following superovulation. This assures a larger than normal number of eggs. Multiple breedings and careful administration of extra follicle stimulating hormone (FSH) will cause a donor to produce an average of five or six transferable embryos. Transfer of embryos can be done surgically or nonsurgically. The success rate is higher when transferring fresh embryos. Embryos can be frozen in liquid nitrogen and transferred later.

CLONING

Cloning is the production of one or more exact genetic copies of an animal. Cloning has been expanded and may be widely available soon. There are several methods of cloning animals.

One method of cloning in animals involves letting embryos grow to the 32-cell stage before splitting into 32 identical embryos. Scientists feel that this could be repeated producing 1,024 identical copies, 13,088 identical copies, etc.

A second method of cloning in animals involves taking a cell from an adult animal. This method resulted in the creation of Dolly from mammary cells of a six-year-old sheep (ewe).

Still another variation of the cloning technique involves taking cells from primordial germ cells during fetal development. These cells are more stable and can be cultured and frozen for indefinite periods of time. Bovine (cattle) calves have been born using this technique.

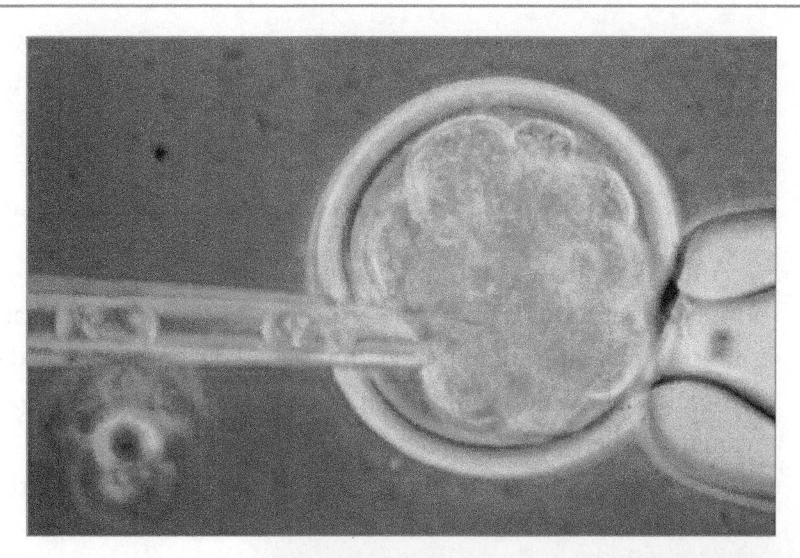

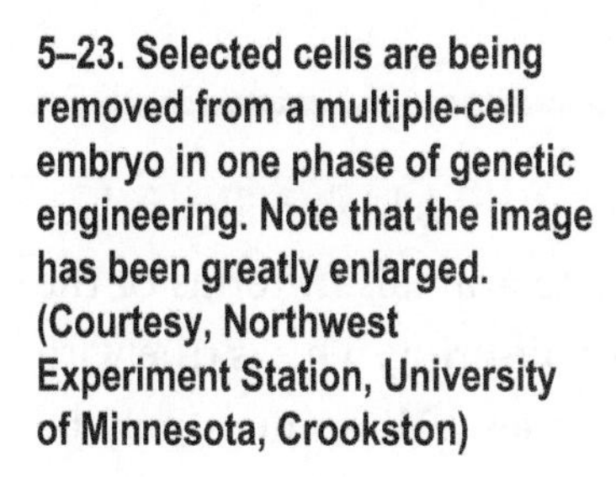

5–23. Selected cells are being removed from a multiple-cell embryo in one phase of genetic engineering. Note that the image has been greatly enlarged. (Courtesy, Northwest Experiment Station, University of Minnesota, Crookston)

GENETIC ENGINEERING

Genetic engineering is removing, modifying, or adding genes to DNA. Genetic engineering using recombinant DNA (gene-splicing) along with other reproductive technology has the potential to greatly change animal science.

Genetic marker technology is being used to detect the presence of certain genes. This technology, along with cloning, has the potential to increase further genetic progress. (Chapter 6 presents background information on genetic engineering.)

5–24. Turkeys have been bred to give large amounts of meat. Many are so fleshy that they cannot mate naturally. Some people view breeding turkeys so that they are unable to mate as an ethical issue. What do you think? (Courtesy, Jeff Banke/ Shutterstock)

REPRODUCTIVE TECHNOLOGY ETHICS

Reproductive technology is increasing rapidly. This raises ethical questions. Technology can be good or bad depending on how it is used. People must be honest. Fear of genetic engineering abuse is not a good reason to stop the potential benefits.

People need to be educated about the use of reproductive technology to prevent poor decisions from being made.

EVALUATING BREEDING ANIMALS

Animal producers reproduce animals to achieve desired goals, such as high milk production, good muscling, and gentle temperament. Producers evaluate animals in terms of the traits that breeding animals have that will likely be transmitted to offspring. This is true with livestock, companion animals, and animals used for other purposes. All animals may be evaluated when considered for reproduction purposes.

Approaches used in evaluating breeding animals include the following:

- Performance Testing—***Performance testing*** is the selection of animals on the basis of their individual merit. Data are collected on the traits of individual animals, such as rate of growth, and used in evaluating their potential. The notion is that the desired traits will be passed to offspring. An example of performance testing is efficiency of growth of beef bulls, with those that demonstrate most rapid growth likely to transmit that trait to their offspring.

- Production Records—***Production records*** include information about the growth and productivity of animals. This information can be very useful in selecting breeding animals. Dairy producers use milk production records of cows in choosing heifer calves to keep as herd replacements. Horse producers use race results in selecting mares or studs to reproduce if offspring are to be used for racing.

- Progeny Testing—***Progeny testing*** is the evaluation of an animal on the basis of the performance of its offspring. It is most often used in cattle with bulls. Progeny testing allows traits not evident in a bull to be evaluated. This method of evaluating is most often used with males kept for semen production used with artificial insemination. Progeny testing is expensive and somewhat long-term.

- Visual Appraisal—***Visual appraisal*** is the practice of a knowledgeable person evaluating an animal on the basis of its physical appearance. Some people refer to this as "judging." The judge must be well trained and aware of the desired traits in animals.

Combinations of evaluation procedures are often used. For example, a dairy producer may use visual appraisal and production records in selecting animals for reproduction.

More information about these evaluation methods is presented in the chapters on different animal species.

5–25. A judge is using visual appraisal of hogs in a junior market hog show. (Courtesy, Education Images)

REVIEWING

MAIN IDEAS

Reproduction is the process by which offspring are produced. Animals naturally reproduce sexually. Sperm and egg unite to begin a new individual. Producers also use artificial insemination (AI) to help assure productive animals.

Reproductive efficiency is important to animal producers. Animals must breed and regularly produce offspring. If they do not, the animals should be removed from the herd or flock.

Male and female animals have distinctive reproductive organs and systems. The testicle produces sperm and is the primary reproductive organ in the male. The ovary produces eggs and is the primary reproductive organ in the female.

Reproductive development goes through several phases. Puberty is the phase of development when animals can reproduce. Fertilization is complete when there is a union between sperm and egg. Gestation is the time between conception and parturition in the female. Parturition is the birthing process.

New technology is used in reproduction. Artificial insemination is widely used. Research is developing cloning, genetic engineering, and other areas. People must consider the ethics of the newer technology.

QUESTIONS

Answer the following questions using complete sentences and correct spelling.

1. What is reproductive efficiency and why is it important?
2. Describe the structure of the female reproductive system.
3. Describe the structure of the male reproductive system.
4. Where does mammal fertilization take place? Hen fertilization? Fish fertilization?
5. What are the four phases of the estrous cycle? What happens in each phase?
6. How is puberty defined in males? In females?
7. What four factors are essential to successful incubation?
8. What reproductive technology has had the largest impact on improving animal efficiency? Why is it useful?
9. Discuss two ways that reproductive technology of the future may influence animal and poultry production.

EVALUATING

Match the term with the correct definition. Write the letter of the term on the line provided.

a. sperm
b. egg
c. testicle
d. ovary
e. reproductive efficiency
f. gestation
g. parturition
h. conception
i. estrus
j. estrous cycle
k. progeny testing
l. puberty

______1. The timely and prolific replacement of a species.
______2. A male sex cell.
______3. The time between periods of estrus.
______4. Primary female reproductive organ.
______5. The union of sperm and egg.
______6. Birthing process.
______7. Heat or the period when females of some species are receptive to mating.
______8. Time between conception and parturition.
______9. Female sex cell.
_____10. Primary male reproductive organ.
_____11. Time in sexual development when an animal is capable of reproduction.
_____12. Using information on offspring in evaluation.

EXPLORING

1. Dissect the reproductive tract of a cow. Find all the important reproductive organs. Animal slaughter houses may provide the reproductive tracts to schools for educational purposes. Be sure to follow proper safety procedures and properly dispose of the reproductive tract after the dissection.

2. Tour an artificial insemination company and observe the correct procedure used to collect semen from a bull. Observe how the semen is handled in preparation for storage and shipment.

3. Hold a debate in class on reproductive technology. Consider the ethics of the technology and other areas that are important.

4. Observe heat signs in your herd or a friend's herd or flock. Determine the kind of animal and previous breeding history. What change in behavior did you see?

5. Producers feed non-productive animals even when they do not reproduce. How much does a rancher lose when cows do not calve? Assume that ZYX Ranch has 1,000 cows. How much income is lost if 5 percent of the cows do not have calves in a year? In your calculations, use a weaning weight of 500 pounds for each calf and a per pound selling price of 70 cents. (Clues: first determine the total number of calves not born and multiply by 500 pounds. This is the total weight of all lost calves. Multiply the total weight by 70 cents to get the number of dollars lost.) How would you overcome this loss if you owned ZYX Ranch?

6. Interview an animal producer. Determine the approaches used to evaluate animals kept for reproduction. Prepare a report on your findings.

5–26. Marsupials, such as the kangaroo, reproduce similar to other mammals with one major exception: the baby is born quite small, moves inside a pouch, and attaches itself to a nipple in the pouch for milk. The opossum and kangaroo are two familiar examples. This shows a female kangaroo with a joey in her pouch. As a young joey develops, it will leave and return to the pouch to nurse and for protection from perceived harm. (The name, kangaroo, is applied to four species of animals in the *Macropus* genus.) (Courtesy, idiz/Shutterstock)

CHAPTER 6

Animal Biotechnology

OBJECTIVES

This chapter introduces biotechnology in animal production. It has the following objectives:

1. **Explain biotechnology and how it is used.**
2. **Explain the role of genetics in animal production and biotechnology.**
3. **Distinguish between organismic biotechnology and molecular biotechnology in animals.**
4. **Describe molecular biotechnology, including genetic engineering and recombinant DNA processes in animals.**
5. **Explain examples of organismic biotechnology.**
6. **Discuss the meaning and potential uses of stem cells with animals.**
7. **Identify issues associated with animal biotechnology.**

TERMS

adult stem cell
allele
animal biotechnology
atom
autologous
biotechnology
chromosome
deoxyribonucleic acid (DNA)
dominant trait
DNA sequencing
embryonic stem cell
gene
gene transfer
genetic code
genetics
genome
genotype
heredity
heterozygous
homozygous
microinjection
molecular biotechnology
molecule
mutation
oocyte
oocyte transfer
organismic biotechnology
particle injection
phenotype
probability
Punnett square
recessive trait
recombinant DNA
stem cell
superovulation
synthetic biology
transgenic animal

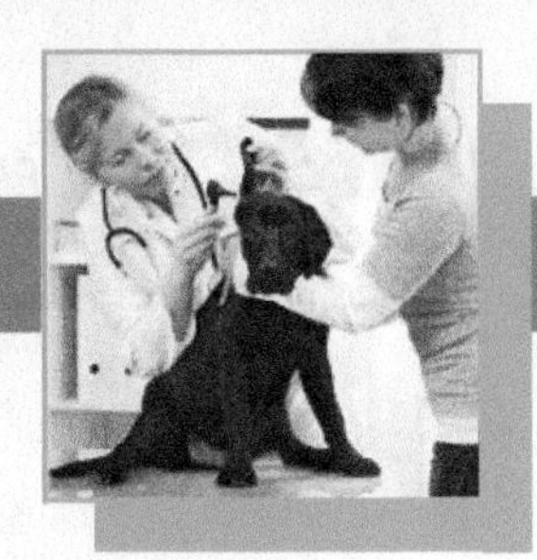

6–1. Molecular biologists load DNA to be sequenced into an automated DNA sequencer. (Courtesy, Agricultural Research Service, USDA)

ANIMAL producers want better ways of raising animals. They also want animals that grow fast and meet the needs of people. They also want ways to keep animals healthy, mobile, and free of pain. Animal producers are beginning to have an array of biotechnology approaches available. A highly promising area is that of stem cell technology.

Discoveries in biotechnology will change animal agriculture. Animal production may be quite different in the years ahead.

People have different opinions about biotechnology. Some people feel that it should be used. They feel that the benefits outweigh the possible problems. Other people feel that biotechnology should not be used. What do you think?

BIOTECHNOLOGY AND AGRICULTURE

6–2. Animal producers are using biotechnology to improve offspring.

Biotechnology is the management of biological systems for the benefit of people and their environment. This broad definition includes many areas of animal production. Biotechnology is often viewed as the application of science in food and fiber production. All plants and animals are included.

Biotechnology has been used for hundreds of years. Using yeast to make bread and bacteria to make cheese is biotechnology. However, today's biotechnology refers to new technologies developed since 1977. It is based on a thorough understanding of life, heredity (genetics), and the mechanics of living things. Today's biotechnology involves three broad areas: molecular biotechnology, organismic biotechnology, and stem cell technology.

Animal biotechnology is the application of biotechnology methods to improve animals. Livestock, companion animals, exotic animals, and other animals are included. The products are important to animal producers, companion animal keepers, and consumers.

MULTIPLE TECHNOLOGIES

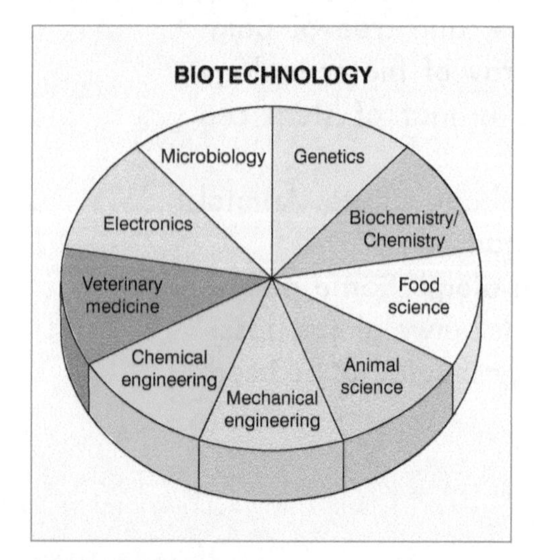

6–3. Many fields of study are involved in animal biotechnology.

Biotechnology involves many fields. Examples of important areas are biology, microbiology, genetics, biochemistry, and computer technology. These fields have technology that enables biotechnology to be carried out.

In animal biotechnology, specific fields related to animals are involved. Animal science, veterinary medicine, agricultural engineering, and other areas are important.

People working in biotechnology must be well educated in a specific area and must have broad preparation across related fields. They must be able to work with other people. Often, a team approach is used. This allows several people with in-depth education to bring their areas of specialization together to solve a problem.

SYNTHETIC BIOLOGY

Synthetic biology goes beyond biotechnology. It is the use of chemicals to create systems with some characteristics of living organisms. Now confined to laboratories, synthetic biology uses vesicles to create lifelike conditions. Vesicles are tiny cell-like structures that have external membranes.

All synthetic biology work is carried out using powerful microscopes with attached computer networks. Without computer capability, the research would be almost impossible.

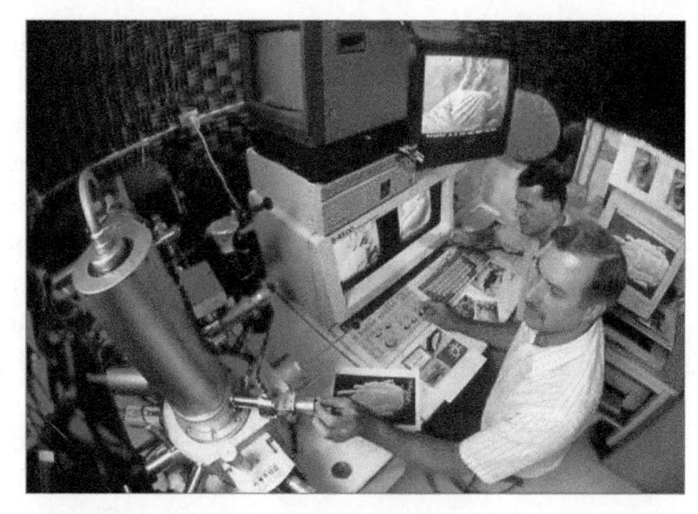

6–4. A low-temperature scanning electron microscope is being used to study tiny parasitic mites in great detail. (Courtesy, Agricultural Research Service, USDA)

Physical organic chemists and other highly trained scientists study the use of synthetic biology in animal health and production. The interaction of vesicles and living organisms is a top priority. Could the creation of living organisms from nonliving substances be next?

GENETICS

Genetics is the study of the laws and processes of biological inheritance. It is important to animal producers as well as in biotechnology. Parents have traits that are passed to their offspring. The passing of traits from parents to offspring is ***heredity***. The heredity of animals is important to producers.

6–5. The study of genetics can explain why the color of the calves is different from the color of the cows. (Courtesy, Agricultural Research Service, USDA)

CHROMOSOMES

All organisms are made of cells. A living cell consists of a cell membrane and the cytoplasm, which contains the nucleus. The nucleus has chromosomes. A ***chromosome*** is a tiny

Table 6–1. Number of Chromosomes for Selected Animal Species

Species	Number of Chromosomes
Cat	38
Cattle	60
Chicken	78
Dog	78
Donkey	62
Horse	64
Human	46
Mule	63
Sheep	54
Swine	38

threadlike part in a cell that contains the genetic material. This genetic material is known as the ***genome*** of the organism.

Chromosomes are the link between parents and offspring. When organisms reproduce sexually, the genome is the combination of the traits from the mother and father. The offspring receive chromosomes from both parents. This gives them pairs of "like" chromosomes. All of the cells within the organism are genetically identical. Each cell contains identical numbers of chromosomes. Table 6–1 lists the number of chromosomes for a few species.

Chromosomes are made of genes that consist of deoxyribonucleic acid (DNA). DNA forms genes that make up chromosomes. Chromosomes make up genetic information for cells. Cells make tissues that form organs. Organs form organ systems. These organ systems make up the organism.

GENES

A ***gene*** is a segment of a chromosome that contains the hereditary traits of an organism. Since chromosomes come in pairs, genes also come in pairs.

An ***allele*** is a different form of a gene. Forms of alleles may be similar or they may be different. An organism having similar alleles is said to be ***homozygous*** for a trait. A ***heterozygous*** organism is one having different alleles for a particular trait.

Genotype and Phenotype

6–6. Japanese shin puppies have very similar phenotypes. (Use close observation to see if you can note differences.) (Courtesy, Shchipkova Elena/Shutterstock)

The genetic makeup of an organism is its ***genotype***. The ***phenotype*** is the organism's physical or outward appearance. In basic genetics, a homozygous allele is called "AA" or "aa". A heterozygous allele is commonly noted as "Aa."

The transmission of traits by parents to offspring is important in breeding animals. Breeding practices use the concept of alleles to produce various phenotypes or outcomes. Outcrossing is used to develop heterozygous offspring and inbreeding produces homozygosity. The genotype and the environment determine phenotype.

Dominant and Recessive

Some traits are dominant while others are recessive. A ***dominant trait*** is one that covers up or masks the alleles for recessive traits. A ***recessive trait*** is a trait or gene that is masked by dominant traits. If an animal has a dominant allele, it will phenotypically show the dominant trait in complete dominance. If the animal receives two recessive alleles, then it will be homozygous recessive.

An example of dominant and recessive traits is the polled trait in cattle. The polled (PP) trait is dominant over the recessive (pp) trait of having horns. If a calf is homozygous dominant (PP) or heterozygous (Pp), they will be polled. If they receive an allele for the horned trait from both parents, they are homozygous recessive (pp) and will grow horns.

6–7. This Hereford bull received the genetic trait for horns from both parents. (Courtesy, Jim Guy/ Shutterstock)

Red Angus cattle are another example showing dominant and recessive alleles. Angus cattle have been black for hundreds of years. Occasionally, a

CONNECTION

TWO PARENTS?

How many parents does this pig have? In this case, when you say two you are correct. Biologists have long known that an organism that is sexually reproduced has two parents. The male sex cell (sperm) and the female sex cell (ovum or egg) unite to form one, unique individual offspring.

(Courtesy, janecat/Shutterstock)

New biotech methods could lead to individuals having more than two parents. Research announced in 2008 at Newcastle University may lead in this direction. Once an ovum is fertilized with a sperm, scientists have been able to remove selected mitochondria and insert other mitochondria. This changes the genetic makeup of an organism. It has the huge potential of removing genetic diseases from embryos in the very early stages of development.

With these new methods, an individual might have genetic inheritance of, for example, one father and two mothers...maybe more. Wow! (Don't forget the ethical issues of this.)

red calf is born. Black is the dominant allele and red is the recessive allele. When two red calves are crossed, they will always produce a red calf. These red animals are homozygous recessive.

Other examples of dominant and recessive traits are:

- Color is dominant to the absence of color in most animals. (The absence of color is known as albinism.)
- A white face is dominant to a colored face in cattle.
- Black is dominant to brown in horses.
- Rose and pea comb are dominant to single combs in chickens.

Incomplete Dominance

Most traits are not products of complete dominance. They are usually influenced by several alleles and by the environment. An example is incomplete dominance. Shorthorn cattle are either red (RR), roan (Rr), or white (rr). The dominant allele does not completely mask the white color. The Shorthorn animal appears to have red and white hairs.

Predicting Genotype

Prediction is based on probability. ***Probability*** is the likelihood or chance that a trait will occur. It is used by geneticists to predict the outcome of mating. Mating animals of particular traits does not guarantee that the trait will be expressed in offspring.

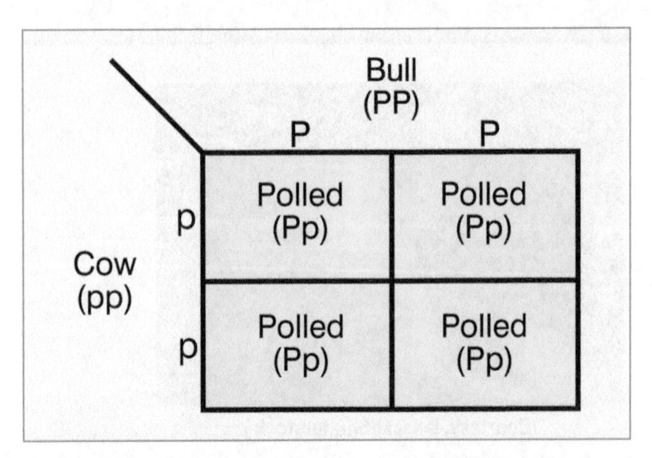

6–8. An example of the Punnett square prediction of the polled trait. A homozygous (dominant) bull is mated with a heterozygous (recessive) cow. This is noted as a PP x pp cross. All the F_1 have the genotype Pp (heterozygous) and will be polled.

Punnett square. The ***Punnett square*** is a technique for predicting genotype. It considers the dominant and recessive genes of the male and female parents for one trait. In a Punnett square, a parent is represented by the letter P and the offspring are sometimes represented by the letter F. Subscript numerals are used to designate the generation involved, such as F_1 is the first generation and F_2 is the second generation. Homozygous and heterozygous genes are considered. Figure 6–8 shows an example using the Punnett square.

Chi Square. Chi square is a statistical test used to determine if the data observed are a good fit with the data that were expected. In genetics, chi square is used to assess differences in offspring genotype. It answers the question, "Does the observed ratio of traits agree with the theoretically expected ratio of traits due to chance?"

The chi square formula is:

$$\text{Chi Square} = \Sigma \frac{(o - e)^2}{e}$$

o = observed traits
e = expected traits
Σ = Greek letter sigma meaning that all of the differences are to be summed

6–9. Probability can be used to select parents for breeding to achieve desired qualities in offspring, such as this Hampshire gilt. (Courtesy, Education Images)

The chi square formula is used to calculate a chi square value. This value is compared to those in a table of chi square values. These tables are published in statistics and genetics books. Using the table requires skill in understanding degrees of freedom (df) and how df are derived. Df is one less than the number of data classes being considered. It is often 1 in simple genetics calculations.

ROLE OF DNA

Deoxyribonucleic acid (DNA) is a protein-like nucleic acid on genes that controls inheritance. Each DNA molecule consists of two strands shaped as a double helix or spiral structure. These strands are nucleotides bonded together by pairs of nitrogen bases.

This double helix is similar to a tiny twisted ladder. The supports or backbone are made up of sugar molecules held together by phosphates. The rungs consist of four nitrogen materials: cytosine (C), guanine (G), adenine (A), and thymine (T). The rung structure allows DNA segments to be cut out and new ones inserted.

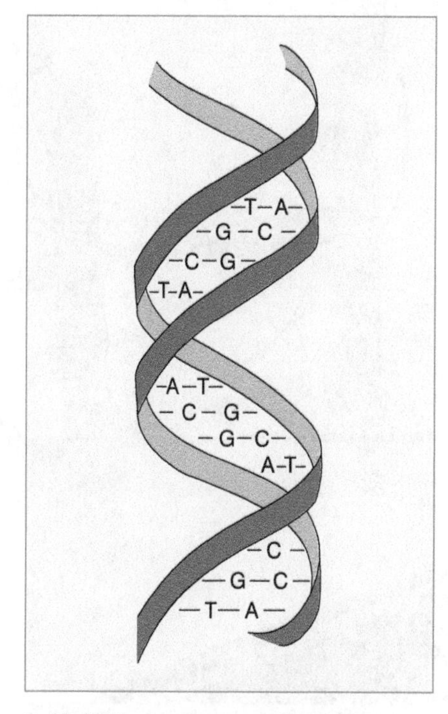

6–10. DNA structure consists of two nucleotide strands, which are bonded by the nitrogen bases cytosine, guanine, adenine, and thymine.

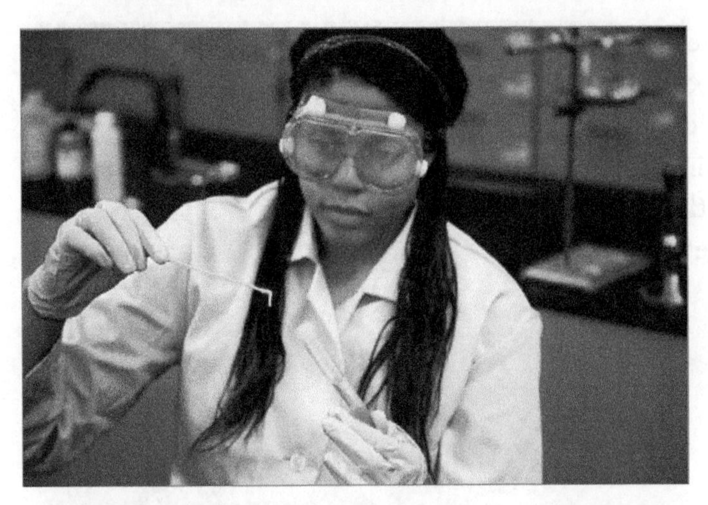

6–11. DNA can be obtained from several animal and plant tissues using relatively simple processes. (Courtesy, Education Images)

Genetic Code

The ***genetic code*** is the sequence of nitrogen bases in the DNA molecule. This sequence codes for amino acids and proteins. The unique ability of DNA to replicate itself allows for the molecule to pass genetic information from one cell generation to the next.

When a chromosome divides during mitosis, the two strands unravel and separate. Each strand contains the complementary base and will immediately become a new double strand identical to the original. A gene is structurally a triplet sequence of nitrogen bases (C, G, A, and T).

Genetic code determines the nature of an organism. Hogs have a different genetic code than do cattle or humans. Research is identifying the genetic code of animals.

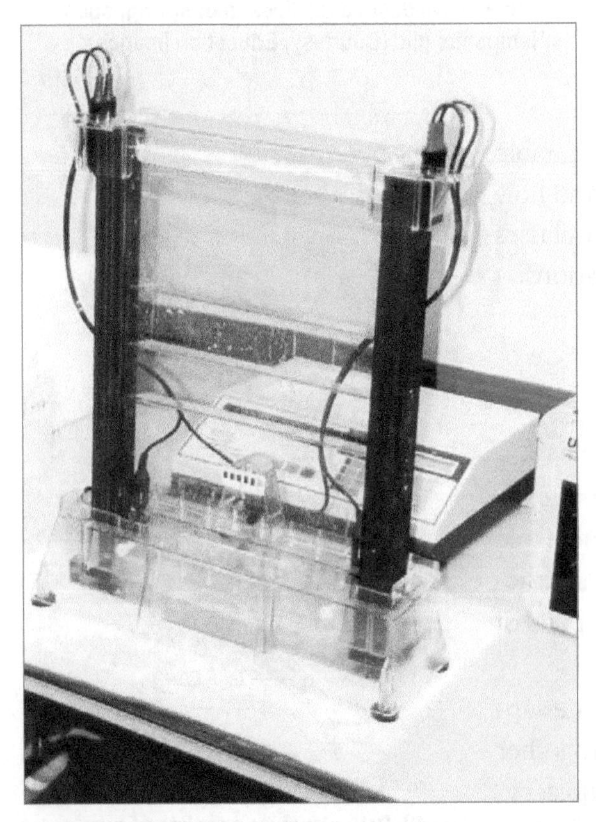

6–12. Sequencing gel apparatus used in electrophoresis. (Courtesy, Education Images)

DNA Sequencing

A tool being used in biotechnology is DNA sequencing. ***DNA sequencing*** is determining the order of nucleotides on a DNA fragment. It is being used to unravel the mystery of heredity and allow genetic engineering to be used. A good understanding of the DNA double helix is essential.

A process known as chain termination sequencing is used. Three steps are followed:

Step 1. DNA fragments are made using DNA synthesis.

Step 2. DNA fragments are separated according to length using gel electrophoresis. (Gel electrophoresis is a technique for separating substances, typically nucleic acids. An electrical field is used to

A C G T

Lane

6–13. Sample DNA sequencing ladder obtained with gel electrophoresis.

move the substances through the gel. Separation is by the size of the molecule.)

Step 3. DNA fragments are made visible by using a stain. (Once visible, these fragments can be studied and used in determining if an organism is likely to have a genetic disease, such as dwarfism.)

DNA sequencing is being used to understand the genetic makeup of animals and other organisms. The genetic information in one organism is huge. Large projects are being used to research the information. These are known as genome projects. Among the best known is the Human Genome Project.

With DNA sequencing information, heredity can be used in animal selection. The information will provide a genetic map of the characteristics of an animal. Methods of genetic engineering can be used to modify the genetic makeup of an organism. The use of DNA sequencing information is now largely limited to scientists. This is because of its complexity and the ethical issues surrounding some uses of biotechnology.

6–14. Molecular biologists are examining the DNA sequences of a virus isolated from a broiler breeder flock. (Courtesy, Agricultural Research Service, USDA)

MOLECULAR AND ORGANISMIC BIOTECHNOLOGY

The two major areas of biotechnology in animal science are molecular and organismic.

MOLECULAR BIOTECHNOLOGY

Molecular biotechnology is changing the structure and parts of cells to change the organism. It begins with the ***atom***, which is the smallest unit of an element. Elements are such things as hydrogen, oxygen, nitrogen, phosphorus, and sulfur. Two or more atoms make up a ***molecule***, which is the smallest unit of a substance. Molecules form cells in plants and animals.

Each cell of an organism has several molecules of water within it. Water is a molecule represented by the chemical symbol of H_2O. The elements in water are hydrogen (H) and oxygen (O). Water is made up of three atoms, two hydrogen and one oxygen.

Molecular biotechnology often changes the physical appearance of an organism. In some cases, undesirable traits may develop along with those that are desired. Thorough research and testing are always needed. Today's livestock operations are much more productive than earlier domestic animals. This is because of genetics and heredity information.

Genetics and molecular biology are important in both molecular and organismic biotechnology. Both areas are covered in more detail in this chapter.

6–15. A genetic analyzer is being loaded with endophyte samples to examine DNA sequence data as an aid in identifying the endophyte. (An endophyte is a fungus or bacterium that lives inside of plants, including forage plants that are consumed by cattle and potential causes of health issues.) (Courtesy, Agricultural Research Service, USDA)

ORGANISMIC BIOTECHNOLOGY

Organismic biotechnology deals with intact or complete organisms. The genetic makeup of the organism is not artificially changed. Organismic biotechnology is used to improve animals for the benefit of humans. This is the most widely used type of biotechnology.

Much genetic variation exists within a species. Hogs will always have piglets and cows will always have calves, but each piglet or calf is unique. Agricultural researchers have used this variation to improve animals using organismic biotechnology.

MUTATIONS

A ***mutation*** is a change that naturally occurs in the genetic material of an organism. The changes are not predictable. Mutations include changes in the chromosome number or chromosome structure. As animals grow and develop, thousands of cells are constantly dividing and millions of genes are being reproduced. This increases the risk of change resulting in a mutation.

Most mutations result in death of the cells, but occasionally mutations will result in a viable offspring that is genetically different. Mutations may be thought of as birth defects. It is likely that the first Polled Hereford beef animal resulted from a mutation.

6–16. The Polled Hereford breed probably originated as a mutation in the genetic material of horned Herefords. (Courtesy, Education Images)

MOLECULAR METHODS AND APPLICATIONS

Molecular methods have increased in the last few years. These have provided new applications to improve animal production.

METHODS

Much of the initial work in molecular biotechnology has involved previously unknown processes. The scientists were pioneers in their fields. Three areas are included here.

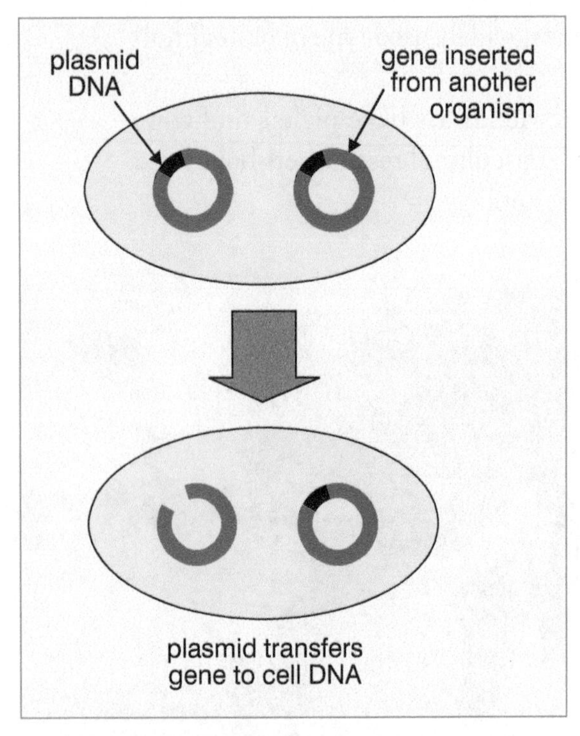

6–17. Recombinant DNA is moving a tiny part of one chromosome to another using plasmid DNA.

Genetic Engineering

Genetic engineering is a molecular form of biotechnology. The genetic information is changed to make a new product. Through this process, sections of the DNA strand are cut out and new sections are inserted.

Chromosomes contain thousands of genes. This makes it difficult to find the right gene to remove or add. A genetic map is used to locate the desired genes. This is needed for successful genetic engineering. ***Gene transfer*** is the moving of a gene from one organism to another. Recombinant DNA methods are used.

Genetic engineering is a means of complementing traditional breeding programs. It is not a replacement. Genetic engineering of plants is less controversial than of animals. The first release of a genetically engineered product was a tomato. This tomato has an altered gene to increase the time it is ripe without spoiling. This makes a higher-quality final product at a lower price. Improvements in animals will be made in the future using genetic engineering.

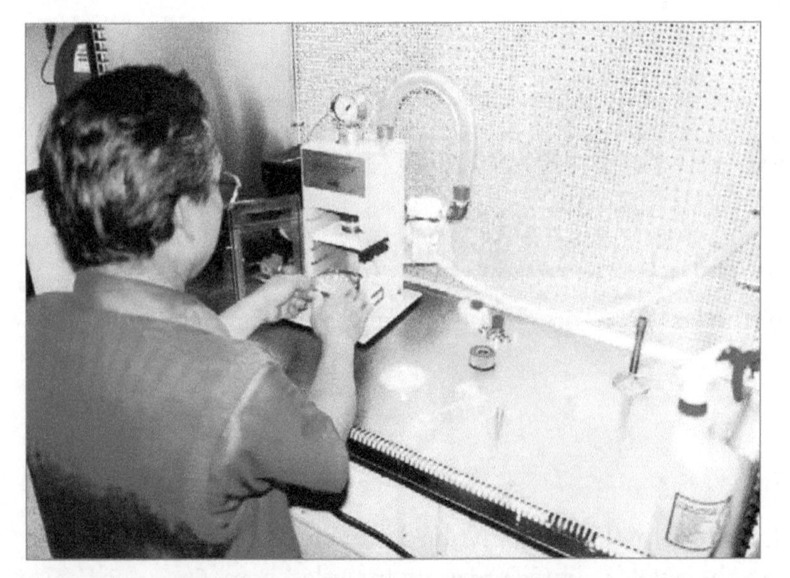

6–18. A microprojectile unit is used for particle injection. It transfers genetic material from one cell to another by shooting DNA on a small particle of gold. (Courtesy, R.J. Biondo, Naperville, Illinois)

Recombinant DNA

Recombinant DNA is gene splicing. Genes are cut out of a DNA strand with a restriction enzyme that works like a scalpel. They are then inserted into circular DNA molecules in bacteria plasmids. Plasmids are circular pieces of DNA found outside the nucleus in bacteria. The plasmid is inserted into the cell that is to be altered. This allows the DNA of two different organisms to be combined. Gene splicing is a challenging process because DNA is so small!

All cells have membranes to protect them. Adding DNA to the nucleus of a cell is not easy. Several techniques may be used in an attempt to insert DNA into the nucleus. DNA may be mixed in a solution and injected into the cell with microinjection. Most cells die in the process. Many surviving cells do not have the "new" DNA present in the nucleus.

Transgenic Animals

A ***transgenic animal*** is an animal that has incorporated a "foreign" gene into its cells. This organism can pass to its offspring this transgene (altered gene). All of the cells within the transgenic animal contain this transgene.

6–19. This sow is a transgenic animal named Genie. She was developed through genetic engineering. Her milk contains a human blood protein needed in human medicine. She has produced a litter of normal piglets with the same trait. (Courtesy, Virginia Tech University)

Transgenic methods involve microinjection and particle injection. ***Microinjection*** is a common method of producing transgenic animals. It is injecting DNA into a cell using a fine diameter glass needle and a microscope. Microinjecting a bovine growth hormone has increased growth rates in fish, chickens, sheep, cows, hogs, and rabbits. ***Particle injection*** is using a microprojectile unit to shoot tiny particles coated with DNA into cells. Microprojectile units are sometimes known as particle guns. These methods have relatively low success rates. Many attempts may be needed for one successful genetic transfer to occur.

APPLICATIONS

Molecular methods have been used to genetically alter animals. Results with animals can often be applied to humans. Medicines, nutrients, and animal quality are improved with molecular biotechnology.

Animals are used to study diseases in humans, such as Lou Gehrig's disease, sickle cell anemia, and cancer. The growth and development of animals, diseases, mutations, and the influence of male and female chromosomes can be studied in similar ways.

Insulin was once extracted from the pancreas of slaughtered cattle and hogs. This process is expensive and would sometimes cause allergic reactions. Today, high-quality insulin is artificially made in laboratories.

Milk composition is suited to using molecular biotechnology. Increases of beta casein decrease the time required for rennet coagulation and whey expulsion in making cheese. This reduces the amount of time needed to make cheese. Fat content in milk can be reduced to make fat-free cheeses and ice cream. If a substance known as kappa casein is increased 5 percent, milk is more stable and easier to ship.

Livestock quality is improving. Growth rate, efficiency, and disease resistance are increasing. Reproductive performance is being enhanced. The high birth rate among Chinese hogs is being studied to increase the proliferation of American breeds.

Transgenic and other engineered animals are expensive to produce. Research is often with mice or other small animals because of relatively lower cost.

ORGANISMIC BIOTECHNOLOGY

Organismic biotechnology deals with improving organisms without artificially altering their genetic makeup. Most animal management practices are in this area of biotechnology. Several examples of organismic biotechnology are presented here.

GREATER FERTILITY

Increasing the reproductive capacity of top animals is important to producers. Sexually reproducing more offspring has been a major limitation. Superovulation and embryo transfer are methods of increasing the number of young from a genetically superior individual.

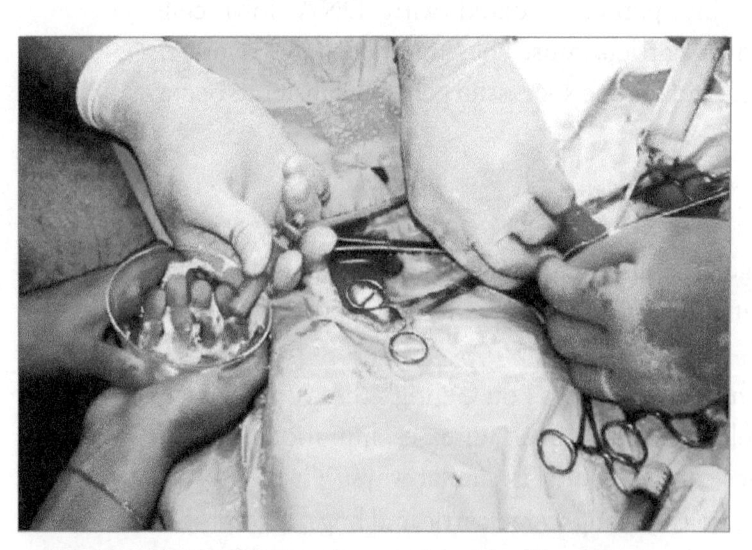

6–20. Harvesting embryos from a donor ewe. The embryos are collected in a small round dish at the end of the tube. (Courtesy, Elite Genetics, Waukon, Iowa)

- Superovulation—***Superovulation*** is getting a female to release more than the usual number of eggs during a single estrous cycle. Hormones are injected to assure more eggs. Superovulation is used with sheep, cattle, and a few other animals.

  Here is how superovulation is used with cows: Cows normally produce only one egg every 21 days. By injecting the cow with a hormone, such as gonadotropin, the cow can release up to 20 eggs. When in heat, the cow is inseminated and the eggs are fertilized. The embryos are flushed out of the cow seven days after breeding. A saline solution (saltwater) is used to wash the embryos out of the uterus. The embryos are no bigger than a match tip. They are either frozen or immediately placed in a recipient cow. Embryos can be frozen in straws similar to semen. After transfer, the embryos are carried to full term by the recipient. They do not receive any genetic material from their recipient mothers, only nutrients.

- Embryo Transfer—Embryo transfer is taking an embryo from its mother (donor) and implanting it in another female (recipient). The embryo completes development in the

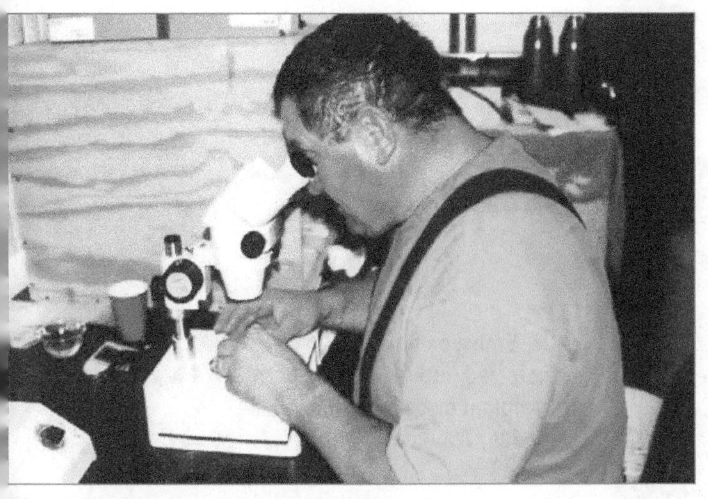

6–21. A microscope is used to search for harvested embryos in a dish. (Courtesy, Elite Genetics, Waukon, Iowa)

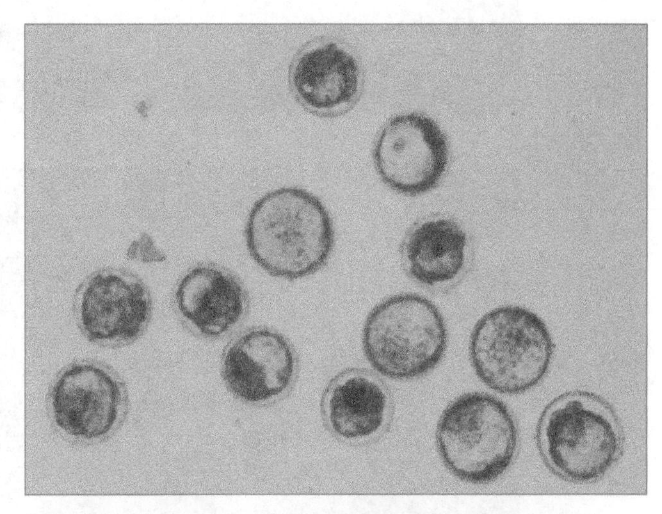

6–22. This shows the magnified appearance of harvested embryos. Unfertilized ova are identified and removed. Unfertilized ova show no cell division, with the cytoplasm filling the entire egg. Unfertilized ova have a lighter color. Fertilized ova show cell division, are darker, and have cell mass. Embryos are evaluated by stage of development and spherical shape. (Courtesy, Elite Genetics, Waukon, Iowa)

6–23. Potential recipient ewes are waiting to receive embryos. (Courtesy, Elite Genetics, Waukon, Iowa)

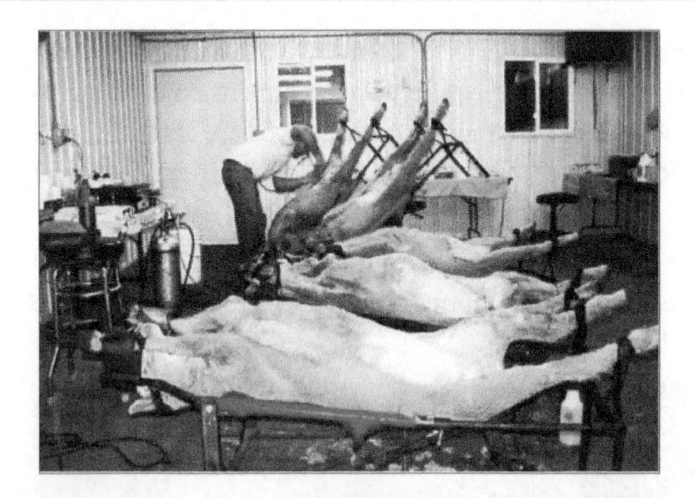

6–24. Recipient ewes are sedated, scrubbed, and ready for embryos. Their well-being is considered throughout the process. (Courtesy, Elite Genetics, Waukon, Iowa)

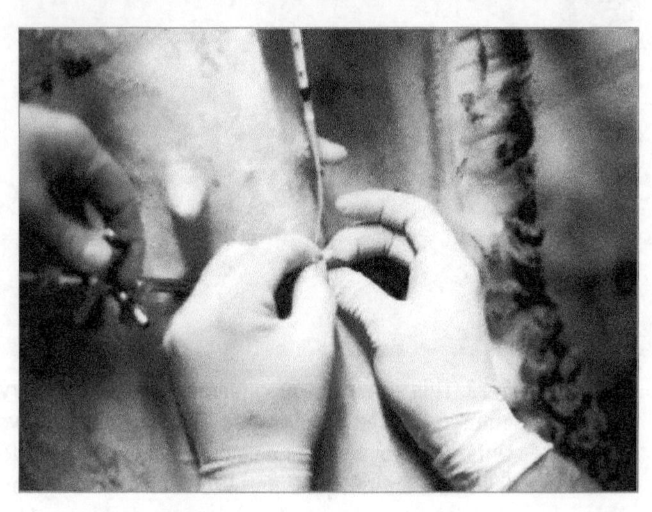

6–25. Embryos are being transferred into the uterine horns of recipient ewes. (Courtesy, Elite Genetics, Waukon, Iowa)

recipient. Embryo transfer usually follows superovulation, but superovulation is not necessary. With cattle, the valuable donor cow can continue estrus cycling and produce more eggs. Beef producers may implant embryos in dairy cows because they are larger and give more milk. These traits help the newborn calf grow fast.

- Oocyte Transfer—Oocyte transfer is emerging as another alternative in achieving efficiency in animal reproduction. An **oocyte** is an immature egg cell in the ovary of a female animal. With **oocyte transfer**, an unfertilized egg is collected from a donor female, placed in a recipient female, and the recipient female is artificially or naturally bred. Oocyte transfer varies from embryo transfer by where the egg is fertilized. Oocyte transfer is useful with females that cannot conceive, such as valuable mares. The donor female is the biological mother of the offspring. In the case of horses, the donor mare is the dam of the foal. Registry associations are beginning to accept offspring from oocyte transfer for registration. An example is the Arabian Horse Registry of America, which voted in 2002 to accept oocyte for registration.

Increased Production

Four examples of ways to increase production are included here. Some are widely used.

- Gene Alteration—By altering certain genes desired qualities can be obtained in animals. An example is the gene that produces myostatin (a protein). Researchers have found that if the myostatin gene is altered so that it produces an inactive protein, a beef animal will have a higher ratio of protein-to-fat in its meat. Though not widely used in animal production, this is just what many beef consumers want: less fat and more lean!
- Milk Hormones–Bovine somatotropin (bST) is an approved hormone product to promote milk production in cows. The hormone naturally occurs in cows. Injections can be used to provide additional hormone to gain greater milk production. It is now widely used in dairying.
- Meat Hormones–Porcine somatotropin (pST) can be used to gain larger pork chops and hams in hogs. Naturally produced in pituitary glands, pST causes hogs to produce more valuable muscle tissue.
- Growth Implants—Porcine somatotropin (pST) can be used to gain larger pork chops and hams in hogs. Naturally produced in pituitary glands, pST causes hogs to produce more valuable muscle tissue.

Enhancing Animal Nutrition

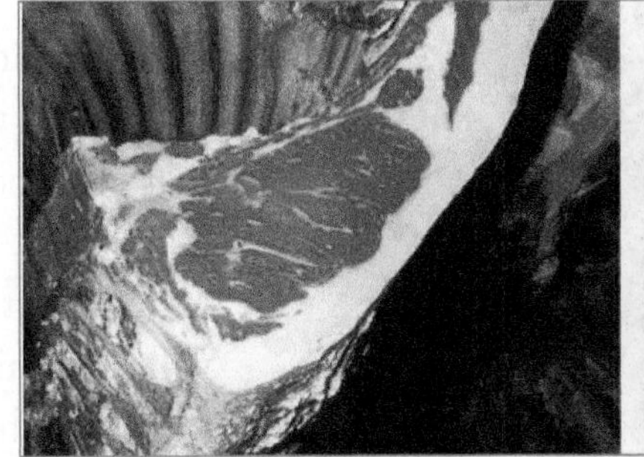
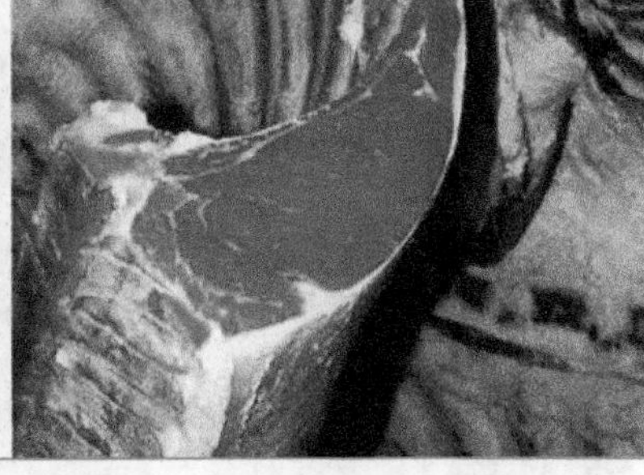

6–26. Differences in lean-to-fat ratios are apparent in the marbling (amount of fat) in these meat cuts. (The results were obtained by altering the gene that produces myostatin. On the left is a cut from an animal where the gene was not modified. The cut on the right is from an animal that received gene alteration.) (Courtesy, Agricultural Research Service, USDA)

Greater results have been in animal nutrition than any other area. Enhanced nutrition has increased feed efficiency, decreased time to market weight, increased milk production, and decreased overall production problems.

- Digestibility Testing—Digestibility of feeds varies. Tests are made by getting samples from the digestive system. Cattle, which are ruminants, have been used for most of this research. A small opening is made in the side of the cow and into the rumen (large compartment of the stomach). A rubber fistula is put in the opening. This allows taking partially digested feed samples from the stomach. Cannulas or small tubes can be placed into the duodenum (part of the small intestine) of cattle. The animal has no pain during these processes. Fistulas make it

6–27. A Holstein cow has been fitted with a fistula so scientists can sample partially digested food from the rumen. (The individual is observing the grazing behavior of the cow.) (Courtesy, Agricultural Research Service, USDA)

6–28. Sensors on chains around the necks of these cows are used to identify them to determine feed consumption and milk production. (Courtesy, polat/Shutterstock)

easy to monitor feed intake and digestion. This information is used to improve feed use and animal growth.

- Controlled Feeding—In a feedlot or pasture, many animals will over or under eat to meet their nutritional demands. One method of regulating this is with computer chips. The chips can be put in a monitor around the animal's neck, in an ear tag, or just under the skin in their ear. A computerized feeding station will read this chip and regulate the amount of feed that the animal gets. A daily feed intake sheet is used to detect health problems.

STEM CELLS

Stem cell technology is emerging in the animal industry as a major approach in promoting animal health and well-being. It is an application of biotechnology with considerable promise. Some of its potential is now being realized. The genetic makeup of an animal is not altered.

A ***stem cell*** is a cell with the ability to divide indefinitely and give rise to specialized cells such as those of the heart or cartilage. The goal is to have stem cells that become specialized cells. They will regenerate new cells to fulfill the need of an organism, such as new heart cells or blood cells. As unspecialized cells, stem cells can give rise to specialized cells. The genetic material in stem cells is not altered. Though not without some controversy, it appears that the benefits of "cell-based therapies" will far outweigh issues associated with stem cell use.

TYPES OF STEM CELLS

Stem cells are of two broad types: embryonic and adult.

An **embryonic stem cell** is harvested from a developing embryo. The cells form just following conception and are undifferentiated when harvested. Such stem cells are obtained from the inner mass of about 150 cells that are first formed as an embryo develops (at this stage, the new embryo is known as a blastocyst). Embryonic cells naturally divide, forming specialized cells and tissues that comprise organs and organ systems of an organism. In a laboratory, embryonic stem cells can multiply indefinitely in vitro (in glass). Scientists first used mice in research to study stem cells in 1981. It was at first felt that embryonic stem cells held the greatest potential.

An **adult stem cell** is harvested from an adult organism. The cells have differentiated to form various tissues of an adult organism's body. Adult cells can be used to form specific kinds of cells, such as stems cells from bone marrow can be used to generate new bone, cartilage, fat, and fibrous connective tissues. New procedures now use methods that "reprogram" adult stem cells to form specialized cells that are needed.

AUTOLOGOUS STEM CELLS

Autologous stem cells appear to have the greatest potential with animals. **Autologous** means that the stem cells were derived from the individual that will receive them back in some form. The animal is both the donor and recipient. Autologous blood has been used for many years. Now, the focus is on autologous stem cells. No diseases from other animals are transferred in the process. The match of cells should be the best "fit" for the animal, since the cells are its own.

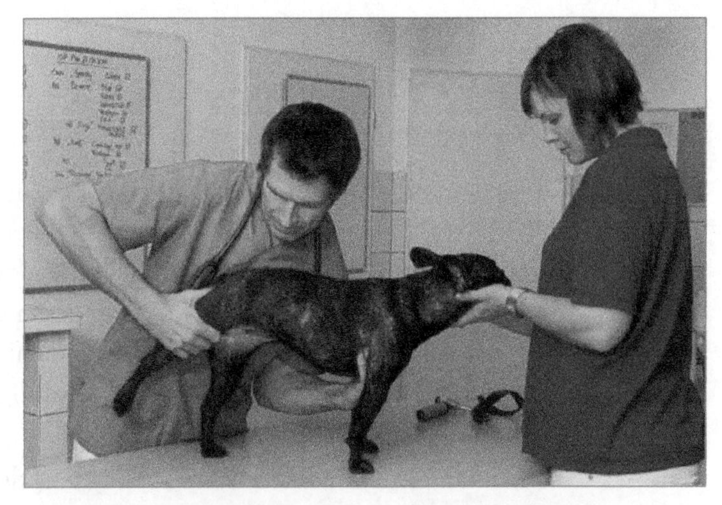

6–29. The function of hip joints in an arthritic dog is being investigated to determine potential benefit to be derived from stem cell therapy. (Courtesy, hightowernrw/Shutterstock)

Four sources of autologous adult stem cells are:

- Bone marrow–Marrow is the soft connective tissue that occupies the hollow spaces of most bones. The marrow cells can be harvested by drilling into a bone and extracted. The femur (large bone from hip to knee) is the most common bone for harvesting bone marrow for stem cell work.
- Adipose tissue–Adipose tissue refers to lipid cells (fat cells). These are harvested by making an incision into a location with fat, such as behind the shoulder of an animal. Only a small amount is needed. A procedure may involve only harvesting 30 grams (a little over 1 ounce) of adipose tissue. The tissue is sent to a laboratory for centrifuging to yield stem cells. These cells are mixed with platelet-rich blood plasma resulting in an injectable solution.
- Blood–Blood is drawn from the donor (and later recipient) in a manner similar to a blood donation or sample. It is passed through a machine that extracts the stem cells from it.
- Umbilical cord–Adult stem cells can be harvested from properly-preserved (usually frozen) umbilical cords. This requires keeping the cord with the identity of the animal from which it came and matching it back to the animal. If not correctly matched, the stem cells will not be autologous.

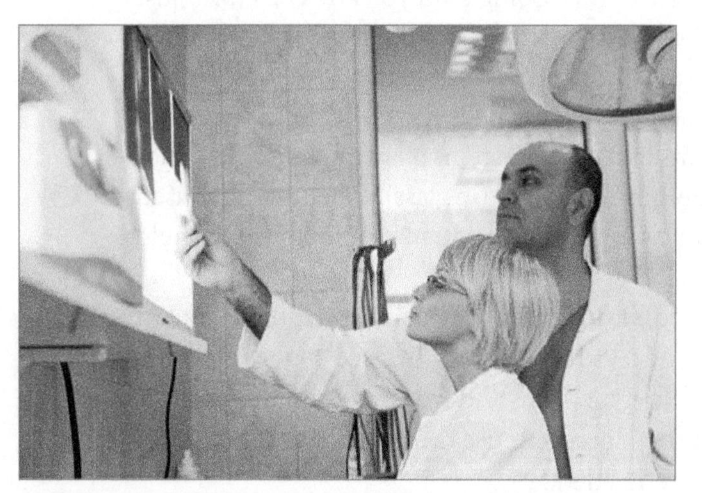

6–30. Veterinarians are reviewing radiographs of the joints of an arthritic dog in preparation for administering stem cell therapy. (Courtesy, .shock/Shutterstock)

USING STEMS CELLS

Once autologous stem cells have been isolated, they are mixed with the plasma containing platelets. The material is ready for use in treatment by injection into the site that needs therapy. Veterinarians use radiographs of the area and carefully study the anatomy of the animal. The animal is anesthetized. The stem cells are injected with a hypodermic needle and syringe precisely into the affected area. This results in the production of new specialized cells, such as cartilage in a deteriorated joint, ligament and tendon tissues, and bone in case of a fracture.

An example is the treatment of arthritis of dogs with stem cell treatments. Once injected and within a few weeks sufficient regeneration typically occurs for a severely affected dog to become active and move without pain. In effect, the dog has use its own resources (with a little help from science) to heal itself!

BIOTECHNOLOGY ISSUES

Some people are concerned about the use of biotechnology. They feel that it should not be used. Other people feel that biotechnology is tested and used in safe ways. This gives rise to controversy. Most concerns arise when molecular approaches are used.

The major areas of biotechnology where issues have developed are listed here.

- Threats to Animals—Molecular biotechnology alters the natural genetics of an animal. Some people fear that this unnatural intrusion is wrong. Other people feel that animals should be improved as long as it does not threaten the well-being of the animals.
- Threats to People—The risks with new, unknown products need careful study. Biotechnology has many benefits. It is not always dangerous. New products always have some risk to workers and users. Proper safety standards are set to protect people. The safety of food products and the effects on humans are questions that consumers may have. Some people are reluctant to accept biotechnology because it represents change. Change brings uncertainty.
- Environmental Concerns—Some people fear the release of altered life forms into the environment. That is not likely to happen. Researchers are careful with new life forms. Other people fear that approved organisms will have an unknown bad trait. The possibility exists for the organism to upset the ecosystem.
- Economic Gains—New feedstuffs and altered organisms present questions about patented products. Gaining a government patent on an item allows a company to make money. On the down side, patents can slow the process of scientific knowledge.
- Benefits to Developing Countries—Developing countries may not have the resources to benefit from new technologies. They may struggle to sustain a way of life. On the other hand, new technologies may make it possible for developing countries to increase production.
- Possible Threats to Personal Beliefs—Some people are concerned about biotechnology. They often need more information to help understand issues. This could change views about creation and evolution theories. New technology could change our view of the "perfect" child. Should new plants or animals be patented? Is it right to genetically engineer an organism?

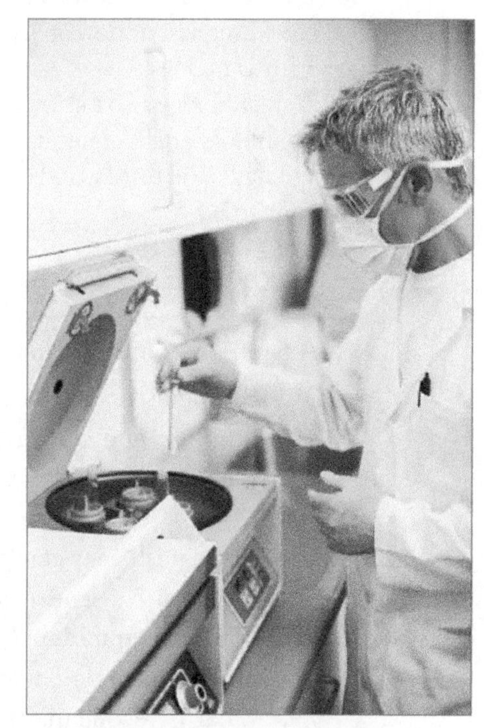

6-31. Developing new applications in biotechnology requires indepth knowledge and skills in laboratory work. (Courtesy, Yuri Arcurs/Shutterstock)

REVIEWING

MAIN IDEAS

Biotechnology is using biological systems for the benefit of humans. This uses molecular and organismic techniques.

Genetics is important in many areas of biotechnology. It is useful because it deals with heredity and how genes can be altered to change heredity.

Molecular biotechnology involves changing the structure and parts of a cell to change the organism. Genetics is a key to successful molecular biotechnology. A cell's nucleus contains chromosomes. Chromosomes are made of genes that are in turn made up of DNA. Alleles are the different forms of genes that code for a particular trait. DNA is the ultimate controller of inheritance. Recombinant DNA involves "cutting and splicing" new genetic material into a cell.

Methods of molecular biotechnology include genetic engineering, recombinant DNA, and transgenic animals. These are expensive techniques, but have great potential for helping society.

Organismic biotechnology deals with improving organisms as they are. Examples of organismic biotechnology include superovulation, embryo transfer, milk hormones, meat hormones, growth implants, digestibility testing, and controlled feeding.

Stem cell technology is making rapid advances. Both embryonic and adult stem cells are being researched. Four sources of adult stem cells are used: bone marrow, adipose tissue, blood, and umbilical cord. Emphasis is now on using autologous adult stem cells as therapy for certain conditions, such as arthritis in dogs.

Biotechnology also raises many questions. What damages could occur to individuals? How could biotechnology affect our environment? What economic gains will occur? Who will benefit from biotechnology? Will third world countries suffer because of biotechnology? Are people's personal beliefs threatened?

QUESTIONS

Answer the following questions using correct spelling and complete sentences.

1. Distinguish between the two main types of biotechnology.
2. Compare and contrast elements, atoms, and molecules.
3. What is genetics? Why is genetics important in understanding biotechnology?
4. Describe the relationship between DNA, genes, and chromosomes.
5. Distinguish between genotype and phenotype.
6. Distinguish between dominant and recessive.
7. What are mutations? How are mutations helpful?
8. What is genetic engineering?
9. What is recombinant DNA? How can recombinant DNA technology be used?
10. How is the genetic code related to recombinant DNA?
11. What is a stem cell? What two kinds are used?
12. What is an autologous adult stem cell? What are the sources of such cells?

EVALUATING

Match the term with the correct definition. Write the letter by the term in the blank provided.

a. biotechnology	e. chromosome	i. phenotype
b. DNA	f. transgenic animal	j. Punnett square
c. synthetic biology	g. mutation	
d. gene	h. molecular biotechnology	

______1. Changing the structure and parts of cells to change an organism.
______2. The part of a chromosome that contains heredity material.
______3. A protein that controls inheritance.
______4. Changes that occur naturally in the genetic material of an organism.
______5. Using biological systems for the benefit of humans.
______6. An animal that has a foreign gene in its cells.
______7. Threadlike parts of a cell that contain genetic material.
______8. Using chemicals to create systems with the characteristics of living things.
______9. A technique for predicting genotype.
_____10. The physical features or appearance of an animal.

EXPLORING

1. Invite an animal scientist or veterinarian to serve as a resource person on biotechnology in animal production. Ask the person to review superovulation and embryo transfer. Write a report that summarizes what you have learned.

2. Conduct a debate in class addressing the question "Is animal biotechnology good for society?"

3. Take a field trip to a research or biotechnology facility to study the work that is underway. Investigate and apply the principles of livestock nutrition in predicting the impact of current advances in genetics. Write a report on your observations.

4. Investigate the use of stem cell therapy. Use published, Internet, or other resources. An Internet site to visit is: **www.vet-stem.com/science.php**. Interview a local veterinarian or animal producer who has used stem cell therapy with animals. Prepare a report on your findings and give an oral report in class.

5. Design and conduct an experiment to support known principles of genetics. Since you are designing with animals and due to animal well-being, you may not be able to carry out your experiment. You can choose any species, small animal such as a dog, or large animal species such as cattle. You can experiment about characteristics such as color, size, or temperament. For example if you wanted black puppies, how would genetics influence selection of the breeding animals? In some cases, guinea pigs, mice, or other small animals may be used in breeding experiments related to color or other characteristics of offspring.

CHAPTER 7

Veterinary Technology

OBJECTIVES

This chapter introduces selected concepts of veterinary technology and practices that can be followed by animal producers. It has the following objectives:

1. **Describe the meaning of veterinary technology, and list important events in its history.**
2. **Identify major duties in a veterinary medical clinic.**
3. **Explain examples of veterinary medical terms.**
4. **Identify positions and direction of animal body parts.**
5. **Identify practices used in veterinary technology.**
6. **Explain the meaning and importance of veterinary pharmacology.**
7. **Identify and use appropriate physical restraint of animals.**
8. **Discuss the meaning and importance of asepsis.**
9. **Identify and perform common surgical skills in livestock production.**
10. **List major hazards and ways to avoid them in the veterinary workplace.**

TERMS

asepsis
aseptic technique
AVMA
care
diagnostic ultrasound
drug
electromagnetic radiation
heart murmur
hyperthermia
hypothermia
laboratory testing
penicillin
pharmacology
practice
prefix
pulse
restraint
root word
stethoscope
suffix
target tissue
triage
vector
venipuncture
veterinarian
veterinary assistant
veterinary technician
veterinary technologist
veterinary technology
zoonosis

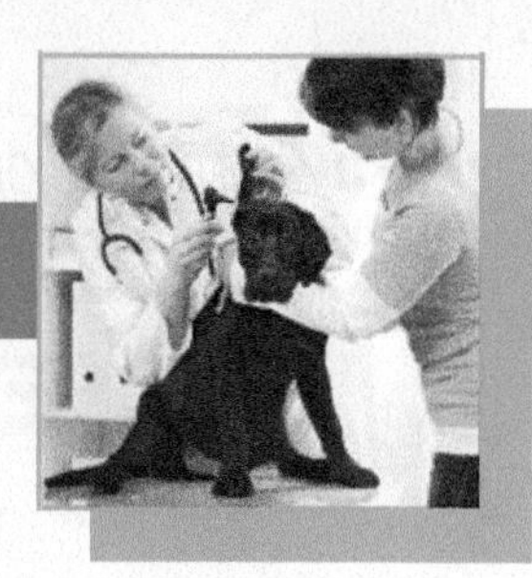

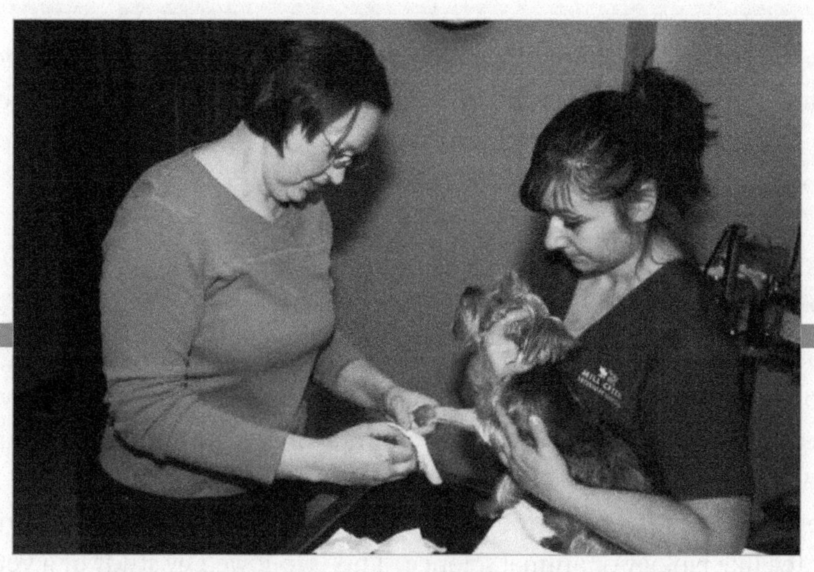

7–1. A veterinarian is preparing a small dog for anesthesia. (Courtesy, Education Images)

VETERINARY technology involves animal health care practices that require a team of individuals to carry out. Most importantly, it includes veterinary medicine. The producer or keeper of the animals has a major role. Veterinary medicine is the use of medical, diagnostic, and therapeutic practices to promote animal health. A veterinarian is the individual with medical training to provide these practices. Various individuals assist veterinarians in their work. These individuals work side-by-side with veterinarians. They assist in carrying out veterinary practices under the direction of a veterinarian.

The next time you hear, "It's time to take the dog to the vet," you know that veterinary medical care is at hand. Health check-ups, vaccinations, dentistry, lab tests, injury treatment, and other animal health services are provided by veterinarians and their assistants. These services cover a wide range of animal species: livestock, birds, fish, reptiles, companion animals, exotic animals, wildlife, and many others.

VETERINARY TECHNOLOGY AND ITS HISTORY

Veterinary medicine is the branch of medicine that deals with animals. The goal is to prevent disease and injury, diagnose problems, and treat animals to assure good health. Other services related to animal well-being may also be provided. Understanding the role of veterinary technology requires some knowledge of veterinary medical work.

THE VETERINARIAN

A ***veterinarian*** is an individual with a doctor of veterinary medicine degree from an accredited college. To practice veterinary medicine, a veterinarian must have a license. Gaining a license requires graduating from a veterinary medical school and passing an examination in the state in which the veterinarian wishes to practice.

Most veterinarians today have four years of pre-veterinary studies leading to a B.S. degree in an area like biology or animal science. This is followed by study at a veterinary college. Upon completion, they receive the Doctor of Veterinary Medicine (DVM or VMD) degree. The majority focus on small animals such as dogs and cats though some focus on large animals. Two areas of high need are: food animal veterinarian (works with animals on farms and ranches such as cattle, hogs, goats, and sheep) and food supply veterinarian (inspects meat processing and fulfills other roles to assure quality foods). Some veterinarians specialize with additional education.

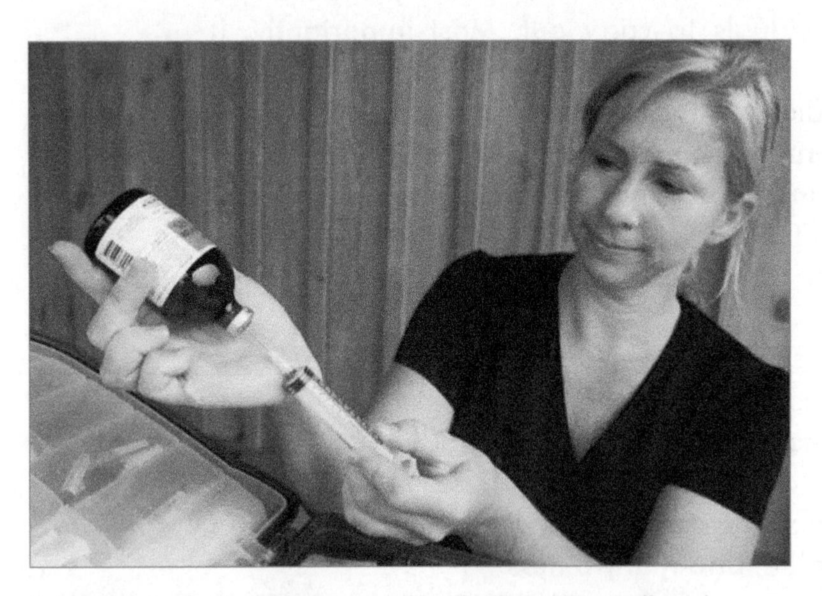

7–2. A veterinarian is preparing to pull medication with a needle and syringe. (Courtesy, Education Images)

Veterinarians may own small clinics or form larger veterinary hospitals, where several may practice together. Veterinarians may specialize in small animals, large animals, aquatic animals, or other areas. Some work with government agencies or companies may involve animal care, food quality, and research.

In veterinary technology, it is important to understand that the veterinarian is most highly trained. He or she is the individual in charge of providing veterinary medical care. Persons in veterinary technology support the work of veterinarians. They and the veterinarian(s) form a team of individuals with the overall goal of providing successful animal care.

VETERINARY TECHNOLOGY

Veterinary technology is the science and art of providing professional support service to veterinarians in their practice. Persons in the work must have basic skills in areas of animal health care. Jobs may involve veterinary assisting or veterinary nursing. Some veterinary clinics may also have grooming services.

New ideas, concepts, and techniques have been used to expand and improve veterinary medicine. Since the mid-1900s, veterinary medicine has become more sophisticated. Many practices used in veterinary medicine originated in human medicine. In some cases, the opposite is true. Increasing sophistication means that those who work in a clinic must be well qualified.

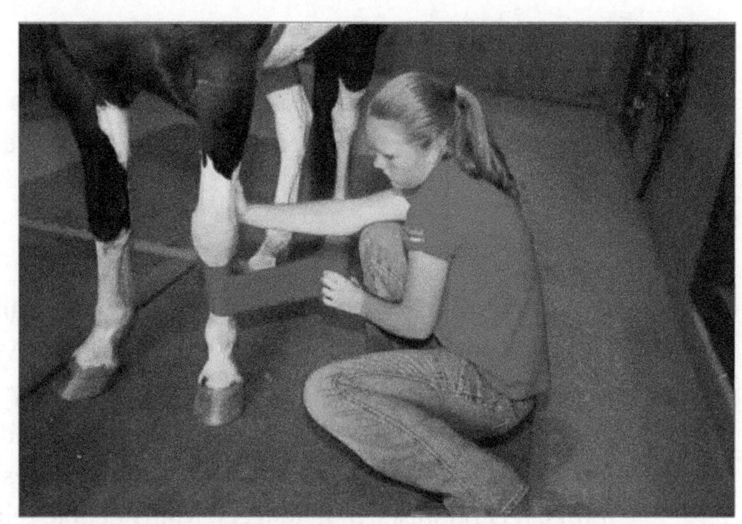

7–3. Correctly wrapping a horse's leg can help prevent injury during exercise. (Courtesy, Education Images)

The development of veterinary-centered television programs has given the public an inside look at the roles of veterinary hospitals. Thus, there is an increased awareness of veterinary medicine by the pet owner and the animal producer. The expectation is that the animals will receive excellent veterinary care.

Importance

Why is veterinary technology important? There are many reasons. Probably the most important reason would be the care of your animal(s). ***Care*** is the quality of appropriate maintenance of an animal, including food, water, shelter, and health upkeep. Since veterinary assistants often interact with the public, their ability to relate animal conditions and health to owners is very important.

If your pet were injured or sick, wouldn't you want it to have the best care? Of course! We love our pets as companion animals and as friends. We want them to live long and healthy

7–4. A veterinarian is preparing to inject a vaccine for bovine leptospirosis. (Courtesy, Agricultural Research Service, USDA)

lives. Farmers and ranchers also want their animals to live and be productive. Healthy animals are more productive and are more likely to help a farmer or rancher realize a profit from animal production.

Technology is improving animal health care. New techniques are coming out every day that are quicker and more efficient. Good veterinary medical service is important to animal owners. For example, horse racing is a very popular and expensive hobby. If you invested in a high-dollar thoroughbred colt that was a winning horse, you would want him to be sound and healthy. If anything happened to him, such as a leg or hoof injury, top-quality veterinary care would be needed to ensure that he would be able to race again and win you money!

History

Where would we be today without the veterinary medical field? Obviously, services to promote animal health would not be available. Most likely, it would be impossible to have the needed care to prevent disease and treat sick or injured animals.

Historically, many veterinarians practiced independently and performed many laboratory and nursing duties themselves. Often spouses and other family members served as assistants, receptionists, or managers. Many people relied on these family practices. A ***practice*** is the routine operations and procedures carried out by veterinarians. Many practices were similar to small businesses that used family help.

Veterinary technology began to take form in the early 1960s. The first organization of veterinary assistants and the first veterinary technician training program were initiated. Veterinarians began to rely on the services of assistants and technicians.

Today, many veterinary medical clinics or hospitals employ multiple veterinarians and require staffs of veterinary technicians, assistants, receptionists, and kennel workers to carry out the duties of running the clinics.

Approximately 90 American Veterinary Medical Association (***AVMA***)–accredited programs of veterinary technology are found in the United States. Thousands of students have graduated from these programs and gained jobs in the veterinary medical field. The number will likely continue to increase as the demand for educated, skilled assistants increases.

AREAS IN VETERINARY TECHNOLOGY

Jobs are available in the veterinary field for people who like to work with animals. Three overall classifications are used here based on education requirements. These are veterinary technologist, veterinary technician, and veterinary assistant.

Veterinary Technologist

A ***veterinary technologist*** is a graduate of a four-year AVMA-accredited veterinary technology program or a graduate veterinary technician with a bachelor of science (B.S.) degree in another program, with studies in supervision, leadership, management, or a scientific area.

Veterinary technologists know the scientific bases of many specialized applications. They plan and execute organized projects. They may be employed as

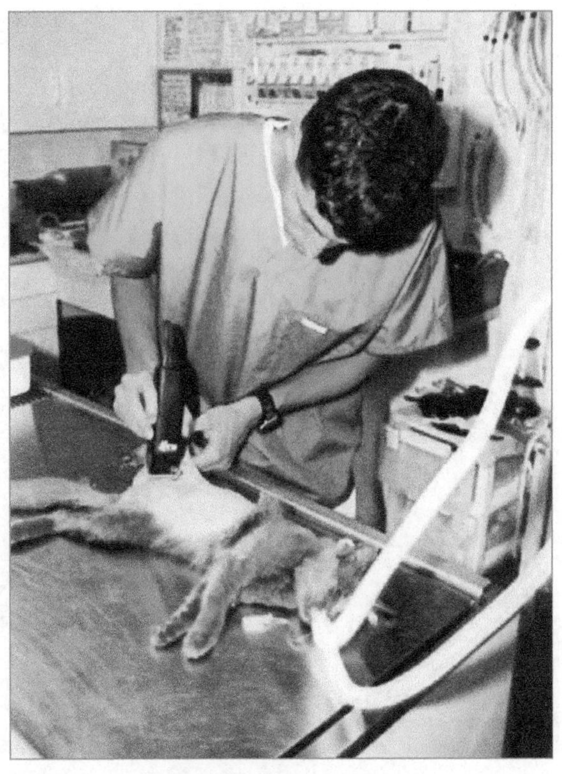

7–5. A veterinary technician is clipping hair from an animal in preparation for surgery by a veterinary surgeon. (Courtesy, Education Images)

CONNECTION

TRIAGE

Triage is a system for deciding how injuries in emergency situations will be handled. It is a French term that was first applied to sorting injured soldiers on a battlefield according to severity of injury and the medical treatment needed. Today, triage also applies to animal injuries.

Veterinary medicine uses triage to describe actions for an emergency. Triage involves assessing injuries and setting priorities on treatment. A veterinarian will make an assessment, recommend actions, and use professional skills for the well-being of the animal(s). In dealing with an animal owner, empathy is important in approaching and assessing the situation. Trust must be established with the owner. Prompt and swift attention is often needed.

How would you deal with the owner of a dog injured by an automobile? The owner loves the dog. The dog has been seriously injured. Broken bones may be evident. Internal injuries are unknown until a veterinarian makes an examination. Think about it. The accompanying photo shows a dog with a leg injury.

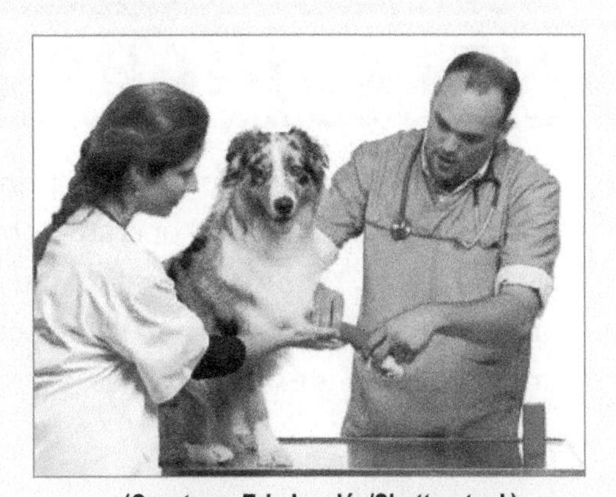

(Courtesy, Eric Isselée/Shutterstock)

teachers, research associates, group leaders, sales managers, or clinical technologists in specialty practices.

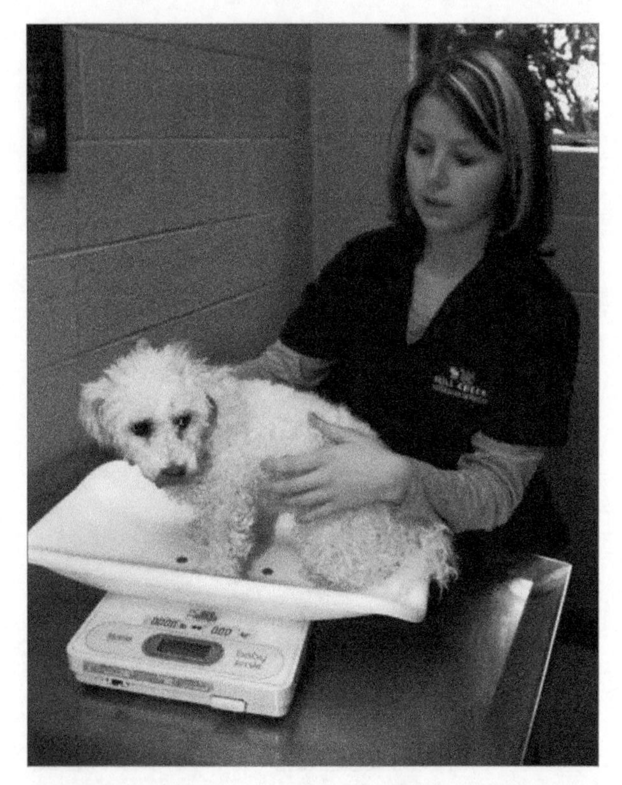

7–6. Weight of an animal is important in providing medications. (This puppy weighs only a few pounds and would receive far less dosage than a dog weighing 75 pounds.) (Courtesy, Education Images)

Veterinary Technician

A ***veterinary technician*** is a graduate of a two- or three-year AVMA-accredited program in veterinary technology. Technicians should be adaptable and know how to perform a wide range of technical tasks. They should also know why conducting the tasks is necessary. A veterinary technician can have a career as a certified veterinary technician on a veterinary team. He or she may also work in research, educational, sales, or governmental positions.

Veterinary Assistant

A ***veterinary assistant*** is a person with training, knowledge, and skills at the level of a clinical aide but below the level of a veterinary technician. Veterinary assistants can do general tasks, such as restraining, moving, feeding, and exercising patients, cleaning premises, and performing other clinical support tasks. Generally, veterinary assistants are trained on the job, but some graduate from six-month training programs. Little or no college education is required. Many high school students take veterinary science classes and work part-time in a veterinary facility to gain practical experience with animals and their care.

VETERINARY MEDICAL PRACTICE

Veterinary medical practice is primarily carried out in veterinary clinics and veterinary hospitals. These facilities operate as businesses and must conform to applicable regulations for businesses. A clinic or hospital must also have sufficient income to cover all costs of operation and provide a return to its owners, normally one or more DVMs. A good client base of animal owners is essential.

BUSINESS OPERATION

Veterinary clinics and hospitals may be established as sole proprietorships, partnerships, or corporations. A sole proprietorship is a clinic owned by one individual. A partnership is a clinic owned by two or more people. A corporation is a clinic owned by stockholders. It has advantages over a sole proprietorship or a partnership.

A corporation is viewed as a legal entity or "artificial person." A clinic that is a corporation must obtain a charter from the state in which it is located. Annual fees and taxes must be paid by the corporation. This means that financial records must be kept so that accurate reports can be provided.

7–7. An attractive sign identifies a veterinary hospital and promotes patronage by animal owners. (Courtesy, Education Images)

Facilities and Equipment

A clinic or hospital must have facilities and equipment. Office areas, examining areas, surgery rooms, and boarding facilities are needed. Most veterinary medical hospitals have separate pharmacies for storing and dispensing medicines. They may have facilities for grooming and otherwise caring for animals. Establishing the facilities and obtaining the equipment necessary to operate them requires an outlay of money. The facilities must be maintained, kept clean, and made presentable to the public.

7–8. A veterinarian is using a computer to record information on a patient. (Courtesy, Education Images)

Management

Management is needed for a veterinary clinic or hospital to be successful. In a small clinic, a veterinarian may be in charge. In a larger clinic or hospital, a hired manager may oversee the day-to-day operations of the business. Management is sometimes known as clinic administration.

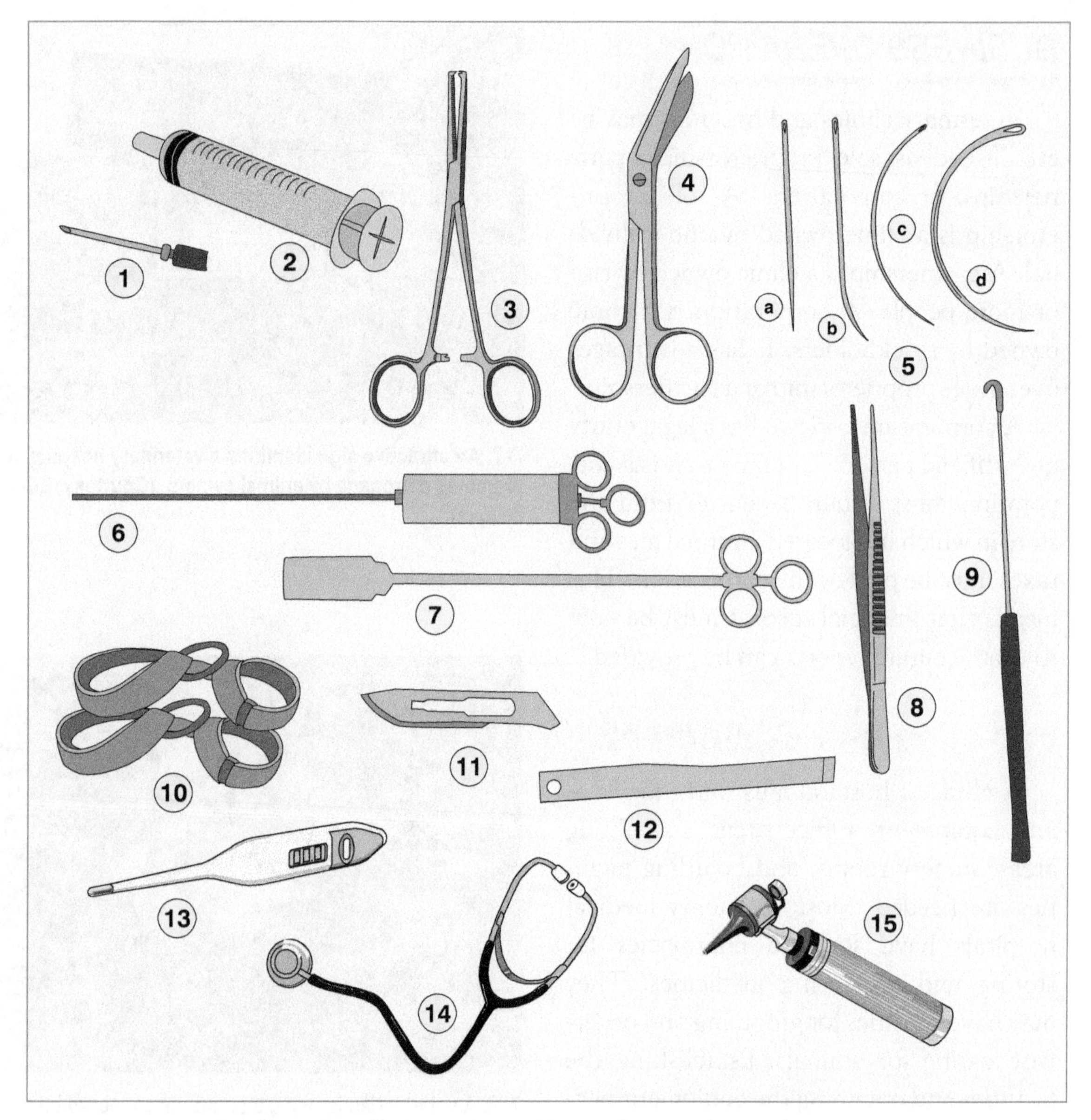

7–9. Examples of instruments used in veterinary work are shown here.

1. Hypodermic needle—used on a syringe to penetrate skin and release liquids.
2. Syringe—used with a hypodermic needle to give an injection.
3. Hemostatic forceps—used to control blood flow during surgery or after injury.
4. Bandage scissors—used to cut off bandages.
5. Suture needles—(a) straight, (b) half curved, (c) ᑯ circle, and (d) ½ circle—used to stitch wounds or incisions for healing.
6. Dose syringe—used to administer liquids through the mouth.
7. Balling gun—used to administer solid boluses (tablets) through the mouth.
8. Thumb forceps—used to grasp very small tissues or objects.
9. Spay hook—used through an incision to grasp ovaries and other reproductive parts within the body cavity.
10. Hobbles—placed around the ankles to restrain animals for treatment.
11. Scalpel blade—when attached to a handle, used to make incisions (cut skin and muscle tissue).
12. Scalpel handle—used to hold a scalpel blade.
13. Digital thermometer—used to measure the temperature of an animal.
14. Stethoscope—used to detect internal sounds, such as those of the heart and lungs.
15. Otoscope—used to examine (see into) ears and eyes.

Duties in clinic management include the following:

- Greeting customers (animal owners) to welcome them and determine their needs
- Answering the telephone
- Making appointments
- Answering questions
- Receiving mail and packages
- Collecting information from customers about themselves and their animals
- Keeping records (financial, patient/customer, and inventory)
- Scheduling workers around the tasks to be done, such as feeding boarded animals on weekends and holidays
- Preparing invoices and sending bills
- Receiving payments for services
- Preparing receipts for payments
- Making bank deposits
- Ordering supplies
- Paying bills in a timely manner
- Looking after the facilities, equipment, and supplies
- Doing public relations and advertising

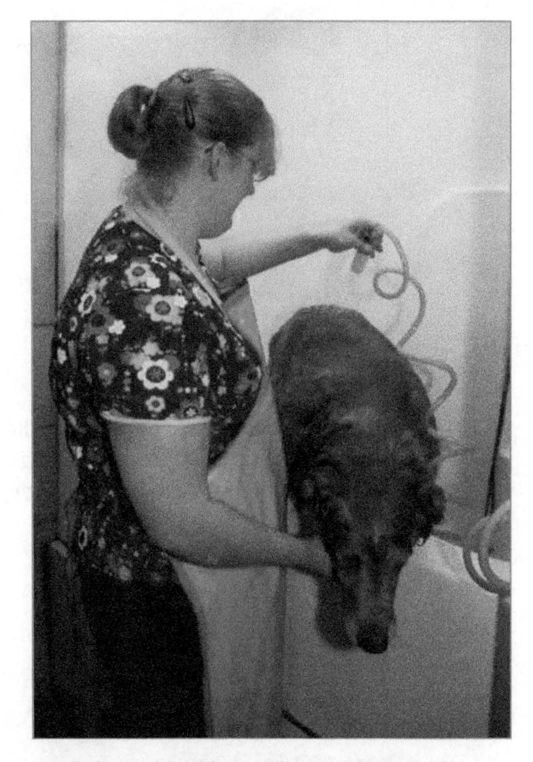

7–10. Animal bathing and grooming may be done at a veterinary clinic. (Courtesy, Education Images)

Office staff members are usually employed to handle many of these duties. A receptionist, a record keeper, and other individuals may be employed to work in the office.

Human Relations

Human relations deals with how employees in a veterinary clinic relate to each other and to customers. Common courtesies are important in such relationships. Being cooperative and working as a team member promotes good relations. Doing more than the minimum needed also promotes human relationships.

Relating well with the public may be the most important factor in the success of a veterinary practice. Everyone in a clinic must promote good relationships.

Here are a few skills that promote good human relations:

- Respect other people.
- Be cordial, and remember to say “Good morning,” “Thank you,” and “Please . . .”
- Smile and be happy.
- Communicate effectively.
- Listen carefully to animal owners.
- Be empathetic with animal owners.
- Commend other people for doing a good job.
- Live up to your word (do what you say you will do).
- Help other people feel good about themselves.
- Allow other people to help make decisions.
- Make eye contact when greeting others and discussing information.
- Be honest and fair.

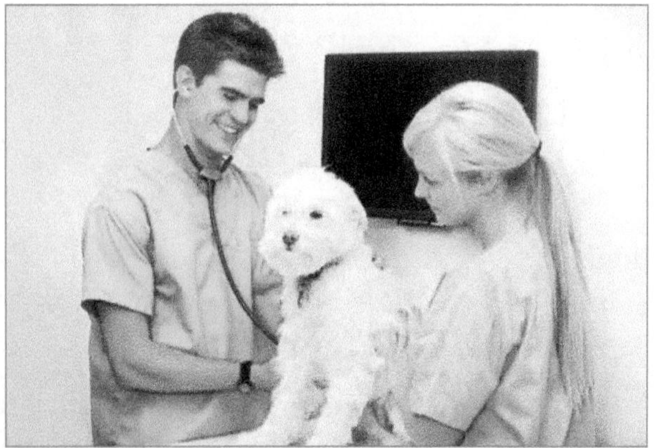
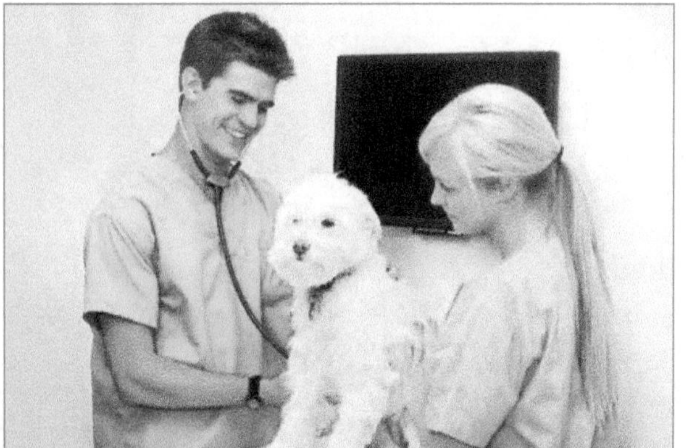

7–11. A pleasing personality promotes good human relations. (This veterinary assistant is holding a dog while the veterinarian does an examination.) (Courtesy, Tyler Olson/Shutterstock)

ANIMAL EXAMINATION AND CARE

The major function in a veterinary medical clinic or hospital is to examine animals and provide needed care and treatment to promote health and productivity. Some clinics focus more on preventive health care; others focus on treatment and surgery.

Major activities in a veterinary medical clinic or hospital including the following:

7–12. How to correctly lift and carry a large dog without injury to the dog or yourself.

- Admitting animals—This includes receiving animals for treatment and assuring their well-being. Small animals may be put in cages; larger animals may be placed in pens.

- Restraining animals—Most animals must be restrained to examine them and collect specimens. They must also be restrained to administer medicine and perform surgery.
- Performing general examinations—A general examination involves checking vital signs, assessing overall condition, checking the ears, listening to internal organs, and other protocols.
- Collecting specimens and conducting laboratory tests—This includes collecting samples of blood, urine, and other body fluids and tissues for analysis. Microscopes and other equipment and materials may be used in analyses.
- Performing euthanasia and necropsy (postmortem examination)—This includes humanely putting animals to death and examining deceased animals.
- Prescribing and administering medicines—Examination and laboratory tests may be needed to determine the kinds of medications needed. These medications must be administered properly to promote health and avoid injury to an animal or its products.

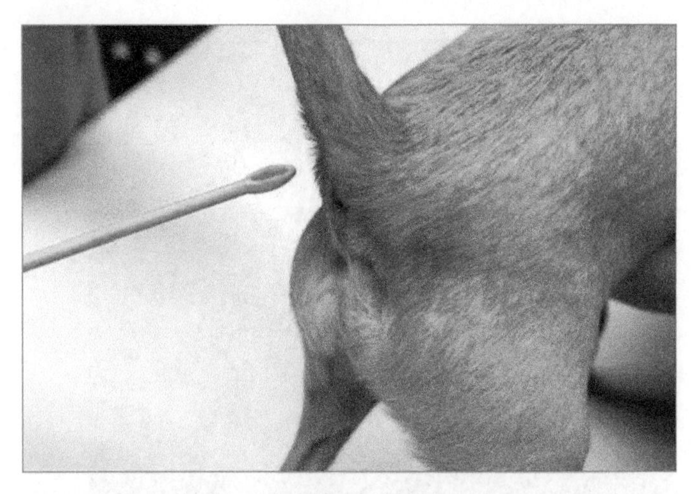

7–13. A fecal loop is being used to collect a fecal sample from a small dog to check for internal parasites. (Courtesy, Education Images)

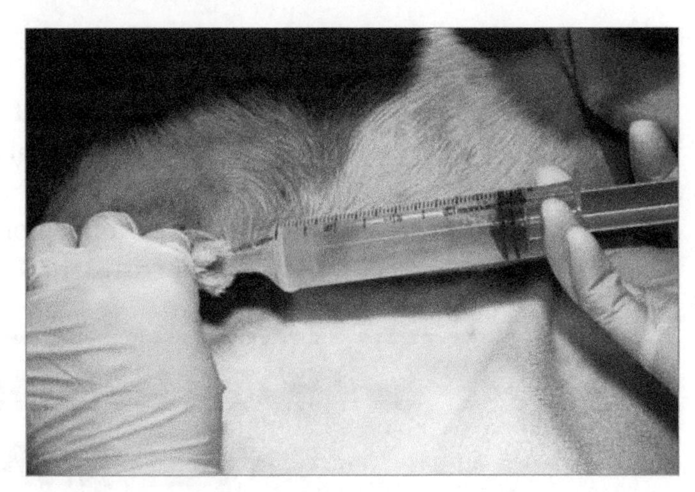

7–14. The prepuce of a male dog is being irrigated to treat for an infection. (Courtesy, Education Images)

- Using diagnostic imaging, such as ultrasound and X-rays—Such imaging is used to view internal organs and processes. It must be carefully done for accuracy and safety.
- Using anesthesia—Anesthesia is used in preparation for and during surgery. Anesthesia creates insensibility to pain.
- Performing surgery—Surgery uses manual methods to access, repair, or remove diseased tissues or organs. Most surgery involves making incisions through the skin with scalpels or other instruments.
- Treating wounds, fractures, and other injuries—Procedures may involve applying antiseptics, suturing ("sewing") wounds, and "setting" broken bones.

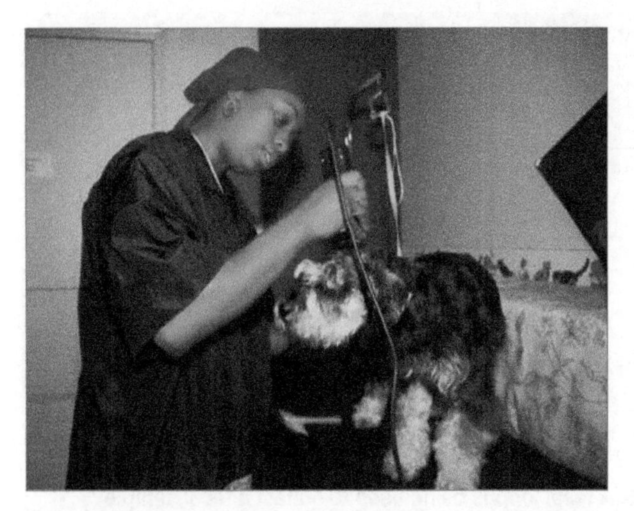

7–15. A groomer at a veterinary hospital trims the hair of a dog. (Courtesy, Education Images)

- Administering fluid therapy, including blood transfusions—Fluid therapy is the administration of a solution, such as dextrose solution, intravenously. Such a treatment is called an IV or a "drip." Fluid therapy also includes the administration of blood transfusions.
- Providing dental services—This includes examining and taking corrective measures with teeth.
- Caring for boarded animals—Many veterinary medical clinics have animal boarding facilities. One type of boarding facility is used to house recovering animals. Another type is used to keep animals while their owners are away.
- Discharging animals—After treatment and satisfactory recovery, animals are released or "checked out" of a clinic. All medical records must be completed before discharge.

Some of these functions are performed only by a veterinarian with the help of a veterinary technologist, technician, or assistant.

VETERINARY MEDICAL TERMINOLOGY

Veterinary medical terminology is the "language" of the veterinary profession. This language is used in everyday speech and recorded in medical records. You may see it in journal articles and veterinary books. You will want to learn the proper spelling and correct pronunciation of common medical terms. This will help you to "talk the talk"!

Spoken and written words should be used properly. They are important in communicating about animals and veterinary medicine. Understanding how words are formed promotes your ability to use them.

Many words are formed by adding prefixes or suffixes to root words. These are explained below.

- Root word—A ***root word*** is a word or part of a word that consists of a syllable, group of syllables, or word that is the basis for the meaning of the word.

Examples:

root word = horn

root word = cardi (the heart)

Table 7–1. Examples of Common Veterinary Medical Root Words and Meanings*

Root Word / Combining Form	Meaning	Root Word / Combining Form	Meaning
abdomin	abdomen	muscul, my, or myos	muscle
adren or adrenal	adrenal gland	myel	bone marrow or spinal cord
angi	vessel or blood vessel	neur	nerve
arteri	artery	ocul	eye
cardi	heart	orchi or orchid	testes
cholecyst	gallbladder	oste	bone
chondr	cartilage	ot	ear
col	colon	phleb	vein
crani	cranium (skull)	pneum	breath (lungs and air)
cyst	bladder	proct	rectum
cyt	cell	pulmo or pulmon	lung
dent	teeth or tooth	ren	kidney
derm or dermat	skin	rhin	nose
encephal	brain	spondyl	vertebra, spinal column
enter	intestines	thorac	thorax
esophag	esophagus	thyr or thyroid	thyroid gland
gastr	stomach	trache	trachea
gloss	tongue	ur	urine
hem or hemat	blood	urethr	urethra
hepa or hepat	liver	uter	uterus
hyster	uterus	vagin	vagina
lip	fat	ven	vein
mast or mamm	mammary glands	vertebr	vertebra
metr	inner lining of uterus	vulv	vulva

*This is a partial list. Reading and listening will help you learn these and other terms.

- Prefix—A ***prefix*** is a syllable, group of syllables, or word placed at the beginning of a root word. The prefix alters the meaning of the root word or creates a new word.

  Examples:

  prefix = de; root word = horn; new word = dehorn (to remove or take away the horns of an animal, such as a cow) prefix = hypo; root word = thermia; new word = hypothermia (lower-than-normal body temperature)

Table 7-2. Examples of Common Veterinary Medical Prefixes

Prefix	Meaning	Examples
a or an	not having or without	anemia (not enough red blood cells)
anti	opposed to or against	antibiotic (drug that acts against bacteria)
contra	opposed or does not exist	contraindicated (not found)
de	remove or take away	dehydrated (having lost body fluid); declaw (remove claws)
hyper	high, much	hyperthermia (above-normal body temperature)
hypo	low, lacking	hypothermia (below-normal body temperature)
peri	around	pericardial (around the heart)
un	some form of problem	unsound (such as lameness)

- Suffix—A ***suffix*** is a syllable, group of syllables, or word added at the end of a root word. The suffix changes the root word's meaning, gives it grammatical function, or forms a new word.

  Examples:

  suffix =ectomy; root word = hyster; new word = hysterectomy (removal of the uterus)

  suffix = iasis; root word = acar; new word = acariasis (infestation with mites)

Table 7-3. Examples of Common Veterinary Medical Suffixes

Suffix	Meaning	Examples
graph	device that writes or records	electrocardiograph (device that records impulses of beating heart)
itis	inflamation	mastitis (bacterial disease of udder)
meter	measuring instrument	thermometer (instrument for measuring body temperature)
oma	tumor	melanoma (tumor on skin, on eyes, or in mouth)
osis	degeneration or abnormal	necrosis (cell death that influences tissues, organs, and functions)
scope	an instrument for observing	microscope (instrument for magnifying very small things)
scopy	using a scope	laparoscopy (using an instrument to view the abdominal cavity)

- Compound words—Two or more words may be combined to form words such as black-leg and lockjaw. These are known as compound words. In a few cases, prefixes or suffixes may be used with compound words.

Table 7–4. Examples of Veterinary Medicine Abbreviations*

Abbreviation	Meaning	Abbreviation	Meaning
AAHA	American Animal Hospital Association	HBC	hit by car
AD	right ear	ICU	intensive care unit
Ag	antigen	IV	intravenous
AL	left ear	MLV	modified live virus
AMA	against medical advice; American Medical Association	NS	normal saline
ASAP	as soon as possible	OB	obstetrics
BAR	bright, alert, and responsive	OD	right eye
BLD	blood	OL	left eye
CBC	complete blood count	ppm	parts per million
CHOL	cholesterol	q2h	every two hours
CMT	California mastitis test	qh	every hour
CNE	canine distemper	RBC	red blood cells
C-S	coughing and sneezing	UA	urinalysis
CVS	cardiovascular system	VS	vital signs
DOA	dead on arrival	WBC	white blood cells
EKG or ECG	electrocardiogram		

*Only a few of the common abbreviations are listed here. Local clinics may develop abbreviations of their own, such as "CVH" for Cornelia Veterinary Hospital.

ANIMAL POSITIONS AND DIRECTION

Veterinary medical terms are used to describe the direction, position, and movement of animals' bodies. These help identify locations of injuries, diseases, and treatments.

Terms associated with position and direction include:

- Ventral—Ventral is the underside or belly area. It may be somewhat more toward the abdomen. Ventral is opposite of dorsal. Example: The ventral fins of a fish are attached to its belly side.

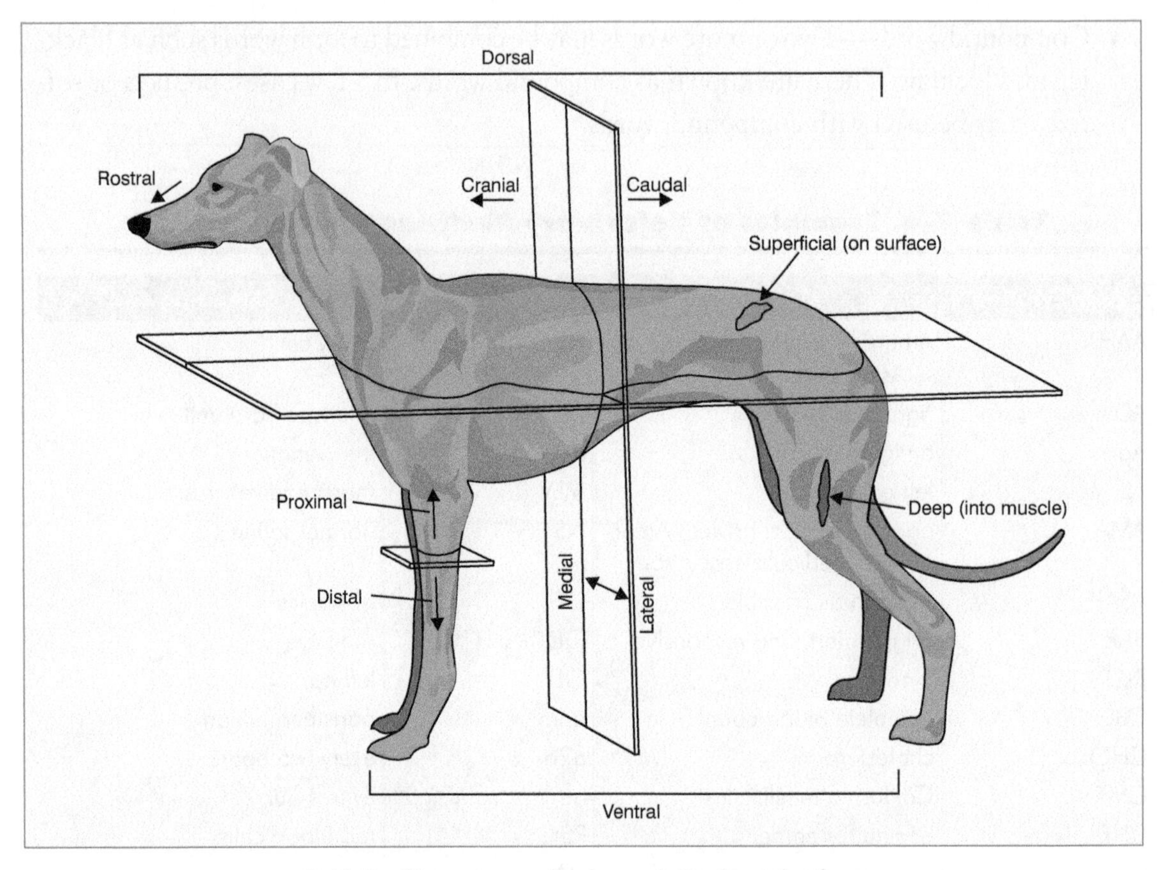

7–16. Positions on an animal are relative to each other.

- Dorsal—Dorsal is the upperside or back area and somewhat along the spine. It is opposite of ventral. Example: The dorsal fins of a fish are attached to its back side.
- Caudal—Caudal is the tail end of the body or toward the tail end. It is opposite of cranial. Example: The caudal fin of a fish is the large tail fin.
- Cranial—Cranial is the head area or skull. Cranial includes positions that are more toward the head. It is opposite of caudal.
- Superficial—Superficial is a location near the surface of the body. It is opposite of deep. Example: A superficial wound is a skin wound caused by a briar scratch.
- Deep—Deep is a location within the body away from the surface. Example: Muscles are deeper than the skin.
- Proximal—Proximal is a location nearer the center of the body as related to other locations. It may be deep within the body and is often thought of as opposite of distal. Example: A proximal injury may be to an internal body organ.

- Distal—Distal is a location farther from the center of the body as related to other locations. It is opposite of proximal. Example: A foot is distal to a thigh.
- Rostral—Rostral is a position that is toward, near, or at the nose of an animal. Example: The nostrils are rostral to the ears.
- Medial—Medial is a position that is more toward the middle of the body or limb. Example: The medial surface of a wing is the inside surface.
- Lateral—Lateral is a position that is farther from the median. It is at the outside of the body. Example: The lateral surface of a wing is the outside surface.
- Peripheral—Peripheral is a position toward the outermost part of an animal, organ, or other structure. Example: Enamel is peripheral to the root canal of a tooth.

COMMON PRACTICES IN THE VETERINARY FIELD

When it comes to health, animals can be similar to people. Some will go through their lives without any health problems; others seem to be prone to just about every illness that comes along.

Animals cannot tell people when they are feeling bad. Owners should know their animals and look for signs that they are not feeling well. It is a good idea to schedule annual or regular vaccinations, deworming, and other care for some animals. This will help prevent problems and ensure that they will live long, beneficial lives.

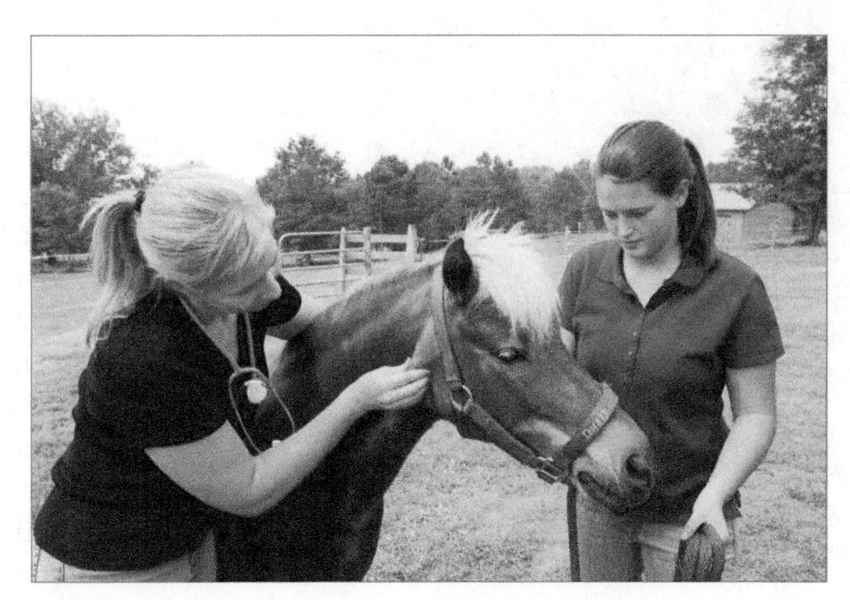

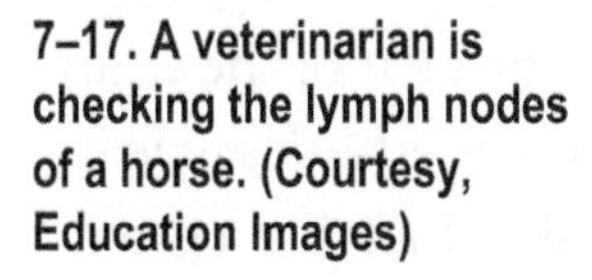

7–17. A veterinarian is checking the lymph nodes of a horse. (Courtesy, Education Images)

Some of the practices used in veterinary medicine are briefly described here. The practices included are not complete but should help you become familiar with the work in veterinary technology.

CHECKING VITAL SIGNS

The vital signs are temperature, external respiration rate and effort, heart rate and rhythm, blood pressure and flow (sometimes known as perfusion), level of consciousness, and urine production. (The vital signs of selected species were presented in Chapter 4.)

Temperature is important in indicating the presence of an infection. Temperature is measured using a thermometer of some type. With most animals, temperature is taken with a rectal thermometer. Ear probes are also used. Temperature may be taken at locations other than the rectum or the ears, but accuracy may be a problem. An above-normal body temperature is known as ***hyperthermia*** and is a sign of infection. An animal exposed to cold may have a below-normal body temperature and a condition known as ***hypothermia***.

7–18. An otoscope is being used to check the ears of a dog. (Courtesy, Education Images)

Respiration rate is the frequency of inhaling and exhaling air. The process exchanges gases in the lungs and body with those in the air. Oxygen is acquired, and carbon dioxide is given off. (Respiration also has other meanings, such as the process by which cells breakdown food substances to produce energy.) Observing movement of the diaphragm is used to count respiration for one minute. The number of inhalations/exhalations in one minute is compared with that of normal animals. Deviations are signs of health problems. Note: A watch with a second hand is needed for veterinary work.

Heart rate is checked by determining pulse rate and intensity. ***Pulse*** is the beat of the heart as felt through the walls of the arteries. Heart rate is commonly determined by locating a prominent artery in the leg or neck area and using finger tips to count the number of heartbeats per a minute. A watch with a second hand is needed. Accuracy of count and time is essential. The number obtained is compared with the acceptable range for a normal heartbeat of that species, age, and condition. Variation from normal is a sign of disease.

Blood pressure is the pressure of the blood in the blood vessels. The most common measure is arterial blood pressure. It involves placing a specially designed cuff around a limb or tail and measuring the air pressure required to obliterate the pulse wave. Measured blood pressure is compared with normal pressure for the species. Deviation from normal is a sign of disease.

Level of consciousness refers to how well the animal exhibits normal characteristics appropriate for its species, age, and gender. A good indication is to note how the animal relates to its surroundings. Is the behavior normal? Other things to note are clearness and alertness of eyes; discharge from eyes, nostrils, or mouth; how the animal stands; how it moves about; how it carries its body; whether it will consume food and water; appearance and quantity of urine and solid wastes discharged; and whether other vital functions are interfering with normal life processes.

Urine production and composition give information about the condition of an animal. Lack of urine production may indicate kidney failure, dehydration, blockages in the urinary tract, and other conditions. Urine may be collected for laboratory analysis.

LISTENING TO INTERNAL SOUNDS

Internal sounds are important in assessing animal health and identifying disease problems.

A **stethoscope** is an instrument used to hear and amplify the sounds produced by the heart, lungs, and other internal organs. The modern stethoscope has two earpieces and a flexible rubber cord leading to them from the two-branched opening of the bell. Through this, sound travels simultaneously through both of the branches to the earpieces.

7–19. A stethoscope is being used to listen to internal sounds of a donkey. (Courtesy, Education Images)

Stethoscopes help give accurate physical examinations. An example may be their use in detecting heart murmurs. A **heart murmur** is an irregular sound caused by the disruption of normal blood flow within the heart. A skilled veterinarian can identify which portion of the heart is affected and arrive at an accurate diagnosis.

LABORATORY TESTING

Laboratory testing is observing samples of fluids or tissues for abnormalities. Such testing often involves blood or urine. Lab tests may also be used to assess parasites, such as small mites in the ears or internal worms found in feces samples or collected from the anus. Tests may also be made on tissues (skin, muscle, bone, and others), cells, milk, and semen.

Blood is often used in laboratory testing. It is tested to determine the presence of disease in the organism. Several approaches are used, with most using whole blood. Fresh whole blood provides red blood cells (RBCs), white blood cells (WBCs), platelets, plasma proteins, and coagulation factors. The components can be used to determine a complete blood count (CBC). Blood types may vary between and within species. (Type is especially important if transfusions are used. A transfusion is the transfer of blood from one organism to another.)

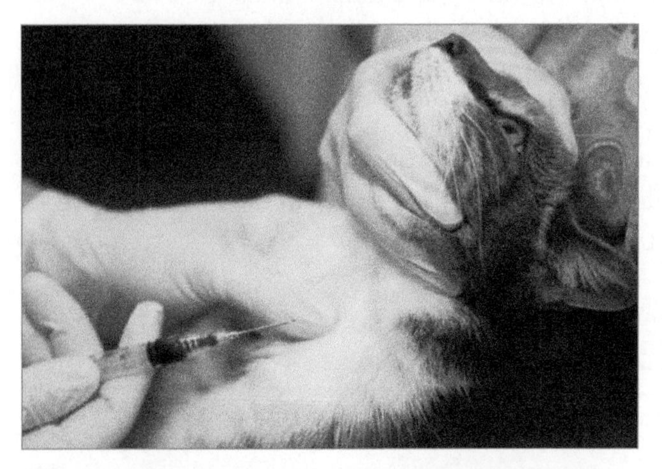

7–20. A blood sample is being drawn from a cat to test for feline AIDS and leukemia. (Courtesy, Education Images)

Blood samples are collected with venipuncture. ***Venipuncture*** is the surgical puncture of a vein. A needle and syringe are often used in venipuncture, but other methods of collecting in tubes are also used. The best site from which to collect blood in a small animal is the jugular vein. Large veins near the base of the tail are sometimes used. A 20- to 25-gauge needle on a syringe is used to withdraw blood from small animals. With large animals, the jugular vein can be used, but other veins are often prominent. A 16- to 20-gauge needle is used on larger animals, such as cattle, horses, and pigs. The needle should be removed before the syringe is emptied for testing.

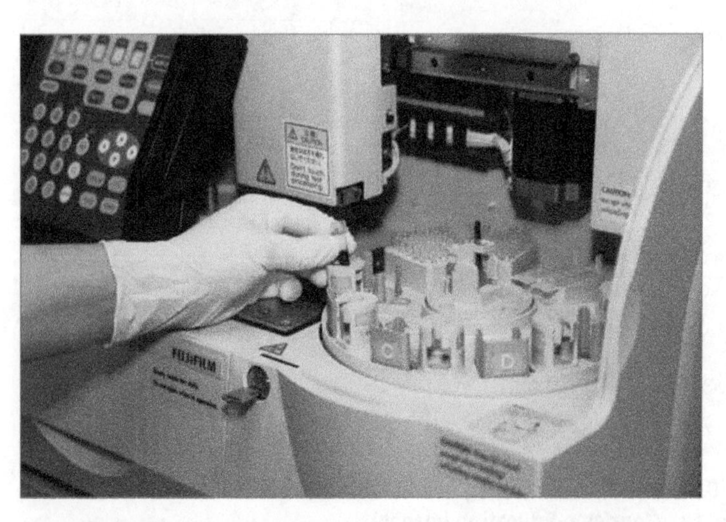

7–21. A blood sample is centrifuged as a step in the lab testing process. (Courtesy, Education Images)

Tests are best when fresh blood is used. Blood smears onto microscope slides are common. Special hemacytometers may be used to establish grids for making blood counts. Training and practice are needed to effectively draw blood samples and make the appropriate tests.

Samples must be properly collected and handled to avoid contamination. Collection must assure that a sample is representative. The sample must be kept free of alteration by dirty equipment or improper storage. It is best to submit a sample for testing as soon as possible. Many veterinary clinics routinely perform some analyses on the premises.

GIVING INJECTIONS

Some medicines work best if administered by injection. Sometimes, medicines cannot be given by mouth for various reasons, such as allergic reactions or inconvenience. Some medicines must be injected so they will act rapidly. An injection involves using a syringe and a hypodermic needle to force fluid into the skin, muscle, or bloodstream. According to the location on the body and the type of medication, different types of injections are used. Common types of injections include intramuscular, intradermal, and subcutaneous.

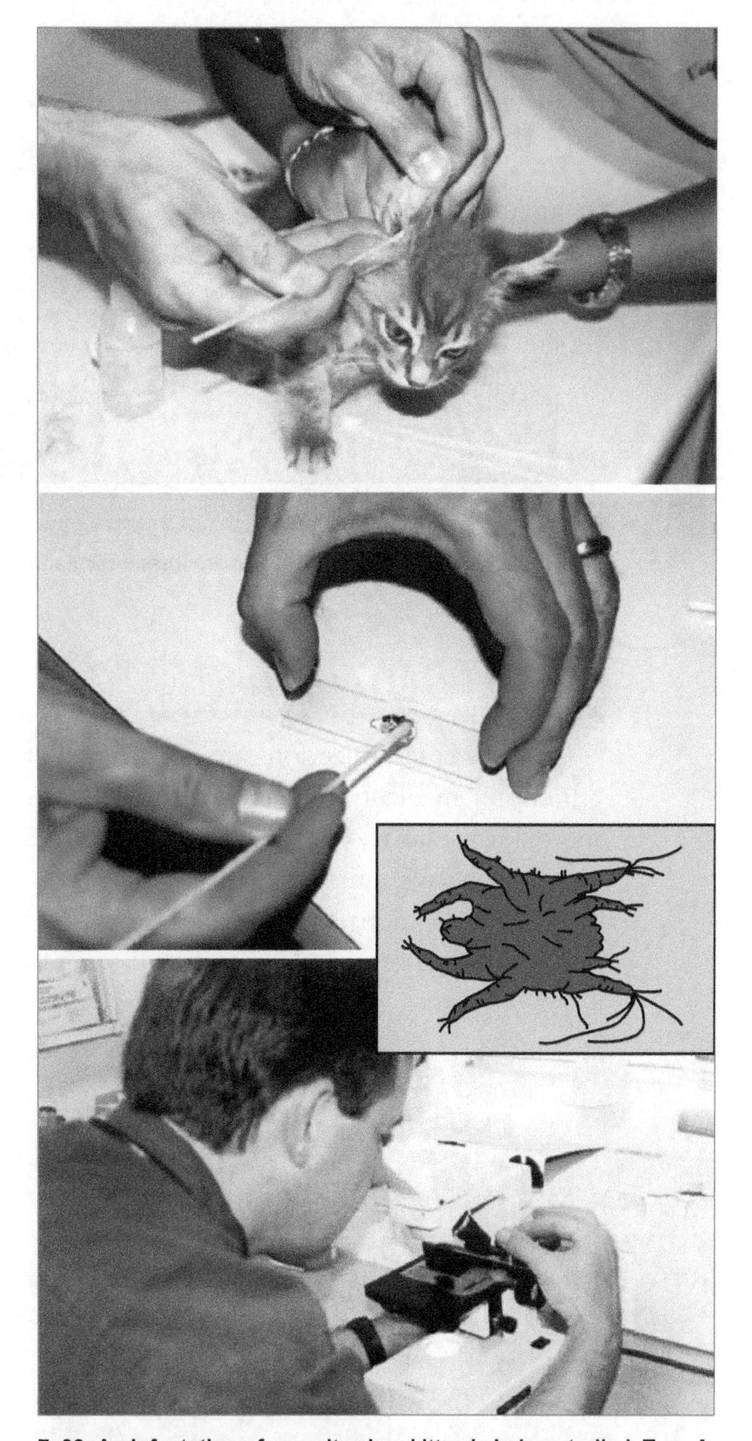

7–22. An infestation of ear mites in a kitten is being studied. Top: A swab is used to collect a sample of the brownish, waxy exudate (exuded substance) in the kitten's ear. Middle: A microscope slide is prepared by rubbing the swab in a drop of mineral oil on the slide. Bottom: The slide is observed under a microscope for the presence of mites. Inset: A greatly enlarged drawing of a cat ear mite (*Otodectes cynotis*). (Courtesy, Education Images)

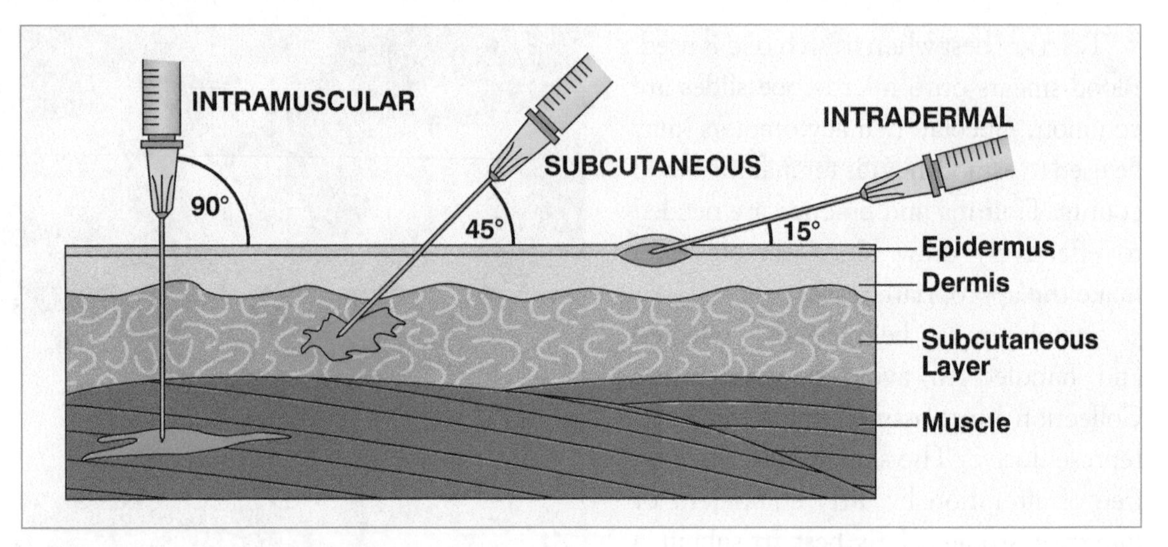

7–23. Three examples of injections are intramuscular (left), subcutaneous (middle), and intradermal (right).

USING ORAL ADMINISTRATION

In some cases, injections are not needed or may not be tolerated. It may be simpler to administer medication in through the mouth. Medications given orally are metabolized more slowly than those that are injected. The patient must be able to swallow and must have normal digestive functions.

Oral medications are available in capsule, tablet, and liquid form. If necessary, tablets can be crushed or capsule contents dissolved in water and given with a syringe or feeding tube. Pills for some animals, such as dogs, can be wrapped in meatballs or disguised in food, if the patients are willing and have good appetites. Sometimes a pilling device, such as a balling gun, is used. This is a plastic or metal rod with a rubber-tipped plunger that holds the medication. If the animal is correctly and securely held, the pill can be placed into the animal's throat and swallowed. Always check an animal's mouth to be sure that medications have been swallowed.

7–24. Using a dose syringe to administer a liquid medication to a donkey. (Courtesy, Education Images)

A liquid medication may be given with a dose syringe. To do so, tilt the animal's head back slightly, and pull the lips outward to form a pocket. Place the syringe between the lips and the back teeth so that the liquid flows between the molars and to the throat. It is important to administer slowly and in small amounts to allow the patient time to swallow and not spit up. Rubbing the animal's throat may promote swallowing.

USING DIAGNOSTIC ULTRASOUND

Diagnostic ultrasound, or ultrasonography, is a noninvasive way of imaging soft tissues in the body. It records the reflections (echos) of the ultrasonic waves into the tissues. The basic principle of ultrasonography is the same as that of depth-sounding in oceanographic studies. A transducer sends low-intensity, high-frequency sound waves into the soft tissues, where they interact with tissue interfaces. The sound waves that are reflected back to the transducer are analyzed by a computer to produce a gray-scale image.

The use of ultrasound equipment gives the veterinarian an excellent diagnostic tool. Ultrasound displays the findings of soft-tissue textures and the dynamics of some organs. It can be used to examine the heart, identify size and structural changes in organs, distinguish cancers from benign cysts, and diagnose pregnancy.

USING STOMACH TUBING

An injured or sick animal may refuse to swallow. A stomach tube may be used.

The placement of a stomach tube is a fast and effective way of delivering large volumes of liquid to the stomach. Both liquid medications and foods can be tubed. The tube can be entered nasally or orally. This depends on the animal and its condition.

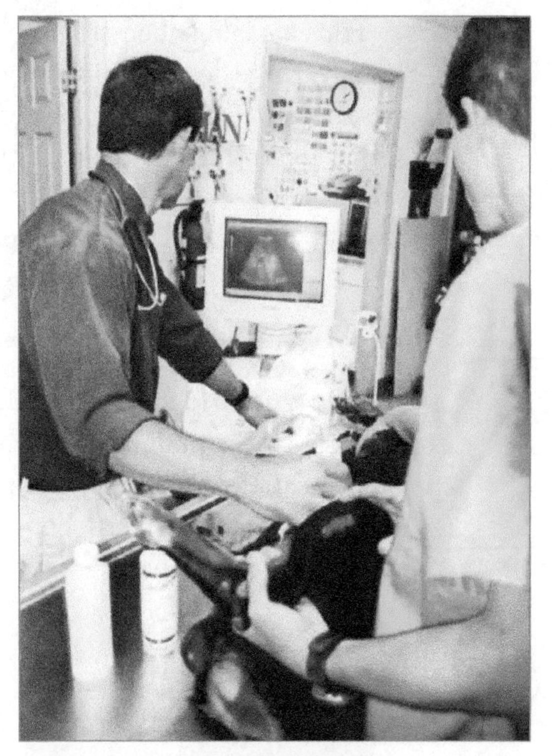

7–25. Using ultrasound to identify a tumor in a dog. (Courtesy, Education Images)

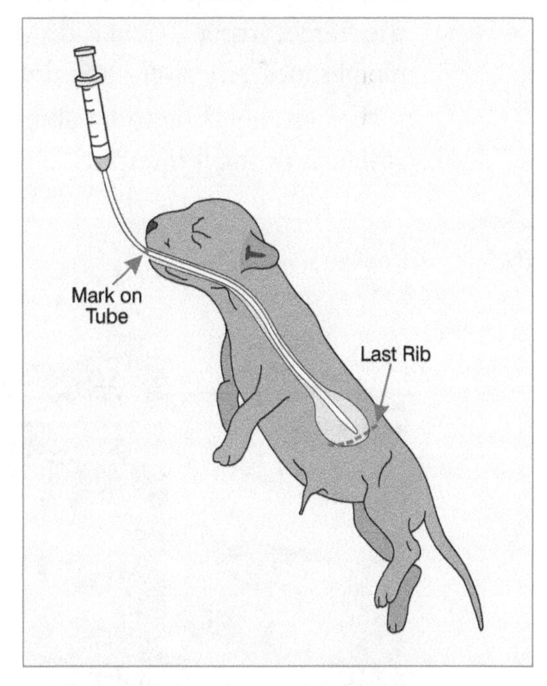

7–26. Correct placement of a feeding tube in a puppy. (Caution: Never insert the tube too far or feed too much.)

For example, when stomach tubing a horse for impaction colic, the tube is inserted nasally. Large doses of mineral oil are administered through the tube and into the gut. This has shown to be an effective treatment in horses that have eaten dry, hard-to-digest food. Other treatments may be necessary for other types of equine colic.

PHARMACOLOGY

Pharmacology is the study of drugs. A ***drug*** is any substance used to promote life processes in maintaining health and treating disease. Drugs are sometimes called medicines or medications. Most emphasis in pharmacology is on the action of drugs in living organisms to gain the desired effects. Functions of living organisms are not created by drugs, but they may be altered by them.

DRUG ACTION

A drug must reach the site where action is needed. This site is known as the ***target tissue***. How drugs reach the target tissue varies. Some can be applied to the skin directly on the target tissue. Other drugs must be absorbed and cross barriers created by cell membranes. Afterward, the drugs are deactivated and eliminated from the body.

For an oral drug to be absorbed into the body, it must pass through the lining of the stomach or small intestine. This means that the drug must dissolve in the digestive tract.

7–27. View of the pharmacy in a veterinary hospital. (Courtesy, Education Images)

Stomach contents can absorb or deactivate some drugs. Drugs of plant origin may be ineffective in ruminants because of deactivation by digestive microorganisms.

Drugs may be injected in several ways. Those injected subcutaneously or intramuscularly must be absorbed and transported to the target tissue. Subcutaneous injections are appropriate for small volumes of drugs; intramuscular injections are better for larger volumes. Drugs given intravenously do no go through an absorption phase.

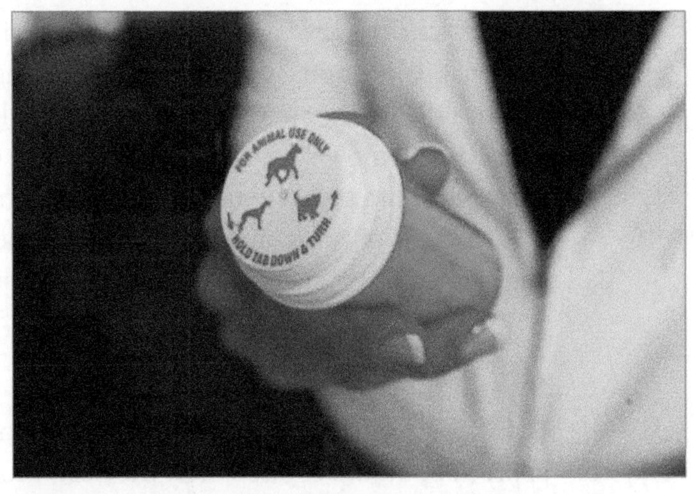

7–28. Be sure to keep animal medicines securely sealed and marked for "for animal use only." (Courtesy, Education Images)

Once in the bloodstream, much of a drug may become unavailable. This is because it has bound to plasma proteins. The unbound or free drug is available to reach the target tissue. Doses must be carefully administered to prevent an overdose as well as to assure that sufficient free drug is present. The formulation (concentration or strength) of the drug is also a factor.

Always read and follow instructions when administering any medication (drug). An overdose can cause a serious reaction—even death. An insufficient amount can result in no beneficial effects from the drug.

KINDS OF DRUGS

Several kinds of drugs are used. The kinds are often grouped on the basis of the disease that is targeted. This means that proper diagnosis of the disease is essential. As with all animal health care, a veterinarian should be consulted in the selection and use of drugs. Some drugs are controlled substances and, therefore, are not available to the general public. Five groups of drugs are discussed here.

Microbe Fighters

Drugs used to treat diseases caused by microbes (bacteria, etc.) are called antimicrobial agents. We know them as antibiotics. An antibiotic is a substance produced by a microbe that destroys or slows the growth of other microbes. Different antibiotics are used to treat different diseases. The species of bacteria affected by an antibiotic are known as its spectrum. Some antibiotics have broader spectrums than others.

Penicillin is the best known and most widely used antibiotic. Penicillin acts by blocking bacterial replication. It does so by destroying cell walls. There is no effect on animal cells because they do not have cell walls. Several kinds of penicillin are available, with Penicillin G being best known. The kind selected is the one appropriate for the animal and the disease. Most penicillin is safe to use, though severe reactions sometimes occur.

Examples of other antimicrobials include streptomycin and related drugs, cephalexin and related drugs, quinlones, tetracyclines, sulfonamides, and antifungal agents.

Parasite Fighters

Since a parasite is an organism that lives in or on another organism, treatment should not appreciably injure or discomfort the host. Proper identification of the parasite is essential. A treatment is then selected that is appropriate for the species of animal.

Substances used to control internal parasites, such as roundworms, are known as anthelmintics. These are commonly called wormers or dewormers. Several kinds of wormers are used. Most are given orally as liquids or boluses or in feed. Heartworms in dogs pose special considerations.

Substances to control external parasites are applied to the hair or skin of the affected animals. These kinds of drugs are known as topicals. Fleas, mites, and ticks are among the most common external parasites.

7–29. A bolus given to cattle to prevent digestive upsets when they are being moved or their feed is being changed. (Courtesy, Education Images)

Digestive System Drugs

Digestive system drugs are used for several purposes. Some are used to stop vomiting, others to induce vomiting. Some are used to treat diarrhea, others as laxatives to gain bowel movement. Ulcers in the gastric system may be treated with antacids and other appropriate medicines.

Body Process Regulators

Glands in the bodies of animals produce hormones. These hormones regulate certain life processes. It is sometimes necessary or useful to intervene in the regulation of these processes. Prostaglandins (PGs) are used to regulate the breeding cycle in order to obtain offspring at desired times and make more efficient use of artificial insemination. Insulin is another regulator used to alter the use of food, especially sugar, by the body. Thyroid drugs may be used with some animals when problems are found. Growth hormones are routinely used in gaining rapid and efficient growth. One such hormone is pST used with hogs.

Other Drugs

Many other kinds of drugs are available. Tranquilizers can be used to alter behavior, such as to calm an animal. Epinephrine is used to regulate body functions and treat shock. Anticonvulsants are used with animals that have seizures. In addition, anesthetics are used to gain a condition known as anesthesia, which is the loss of sensation. The animal has no pain during surgery or other procedures.

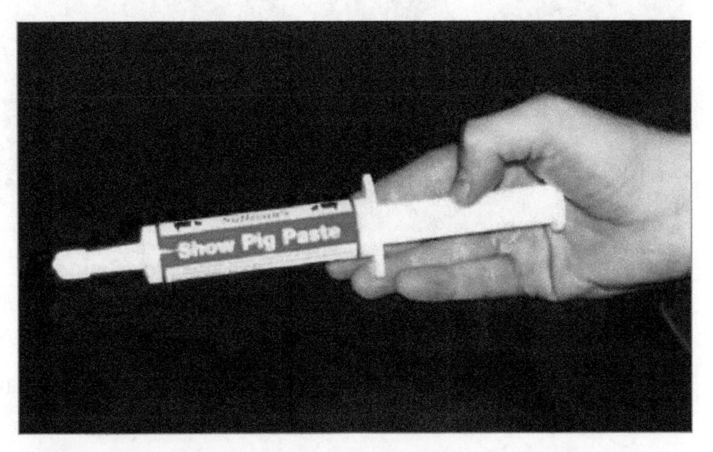

7–30. A product given orally to show hogs two to three hours before show time to promote calm demeanor in the show ring. (Courtesy, Education Images)

RESTRAINT

Restraint is the control of an animal so that it can be examined, treated, groomed, or otherwise managed. Without proper restraint, these activities are inefficient and may be impossible.

Restraint is needed with animals of all sizes and species. Improper restraint can injure an animal, injure the people who are trying to administer to it, damage facilities, or result in the animal escaping.

NEED FOR RESTRAINT

The need for restraint depends on the temperament, size, and species of animal, as well as the nature of the work to be done. Large animals are strong. They require restraint measures that are appropriate. Animals with teeth and claws may require restraint measures that prevent biting and scratching. Some animals have poisonous bites and stings; they must also be managed to prevent injury.

7–31. A gentle animal is easily restrained for grooming. (Courtesy, Education Images)

Animals often become frightened in different environments, such as when they are brought to a veterinary clinic, when they are corralled on a ranch, or when strangers and other animals are around. The animals have fear. Fear may create aggression. Fear-biting by animals is the most common injury to people in veterinary clinics.

Animals can become aggressive around others of the same species. Males may readily fight if not kept separate. Boars can be vicious fighters with each other. Bulls, stallions, and others may also be quite aggressive.

Mother animals with babies may become aggressive to protect their young. This means that a mother should be securely separated in a nearby pen from where work with her baby animals is occurring. An example is pigs. A sow should be well away from pigs when they are vaccinated or castrated. Their squeals arouse and incite aggression in the sow.

HOW TO RESTRAIN

Veterinary assistants are often responsible for helping restrain animals so that a veterinarian can closely observe a condition and/or provide a treatment. Restraint is also used by farmers and ranchers in performing certain procedures with animals.

Some common ways of restraining animals are

7–32. A feline bag can be used to restrain a cat. (Courtesy, Education Images)

- Feline bags—A feline restraint bag is used with a cat to reduce the likelihood of scratches. Restraint bags have openings and fasteners that allow access to certain parts of the body. In some cases, a cat can be wrapped in a heavy bath towel.
- Squeeze chutes—A squeeze chute is used with cattle, wild-mannered horses, sheep, and a few other spe-

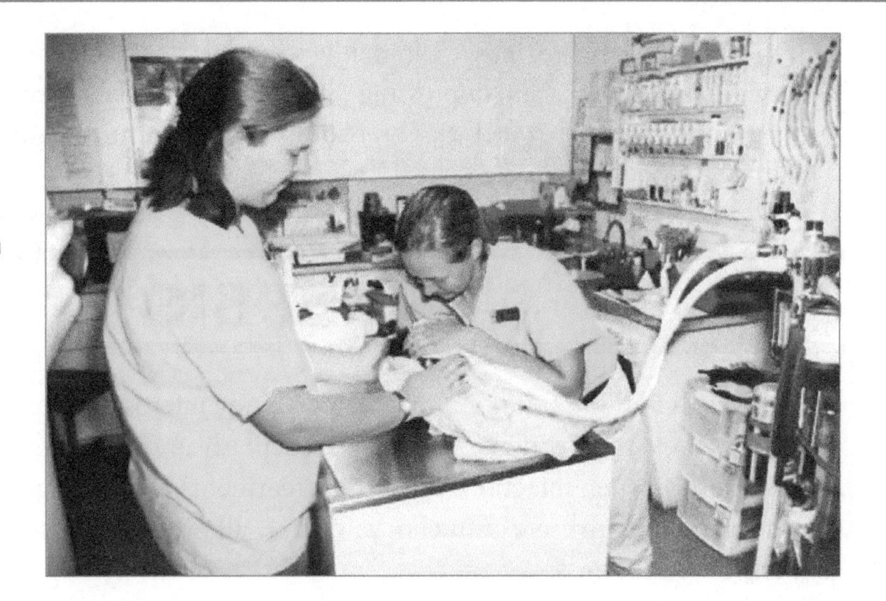

7–33. A small dog has been wrapped in a towel for restraint until anesthesia goes into effect. (Courtesy, Education Images)

cies. A chute allows animals to move through one at a time for administering medicine, checking condition, tagging or tattooing, and performing other activities. Cattle may need to be herded from a pasture into a corral equipped with a chute. Chutes are sometimes called restraint cages.

- Chemical restraint—Chemical restraint is the use of tranquilizers, sedatives, and other anesthetics. Such methods are most appropriate with wild animals or in the capture of horses or cattle that have escaped. A blow dart may be used to administer the chemical. A blow dart is a syringe-like instrument that contains a tranquilizer and is fired from a gun some distance away into the skin of an animal. The impact of the needle moving into the skin releases the chemical.
- Physical restraint—This involves physically holding an animal in place. Halters, hobbles, nose rings, and other instruments may be used. With a small animal, such as a baby pig, human strength is adequate to hold the animal for most procedures. This method of restraint is often used to vaccinate, castrate, dock, or dehorn baby animals.
- Diversionary restraint—This is used with large animals to divert their attention away from the pro-

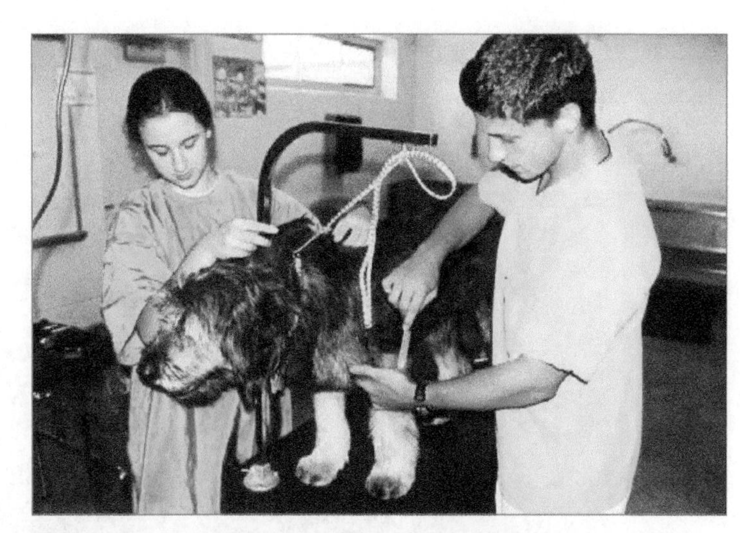

7–34. A dog has been restrained on a grooming table using a leash attached to its collar. (Courtesy, Education Images)

cedure. Horses are most commonly restrained using diversionary tactics, such as twitching the upper lip or pressing on a fold of skin. Sensation is such that the animal may not appreciably feel the insertion of a hypodermic needle for an injection or to draw a blood sample.

ASEPSIS

Asepsis is a sterile condition in which no living microscopic organisms are present. It is essential to prevent infections. Asepsis is particularly important where surgery is performed. It is less critical when injections and other treatments are being given. This is because surgery is much more invasive of the body and may allow infection-causing microorganisms to gain entrance deep inside.

ASEPTIC TECHNIQUE

Aseptic technique is taking action to assure asepsis. The environment is free of microorganisms. All who work in a veterinary clinic have a role in assuring asepsis.

Asepsis is very difficult to achieve. Bacteria and other organisms are naturally found on all surfaces, including tables, clothing, skin, hair, and instruments. Places where animals are congregated, examined, and treated may have a higher concentration of microorganisms unless steps are taken to reduce and/or eliminate them. Unfortunately, some are becoming resistant to the methods used.

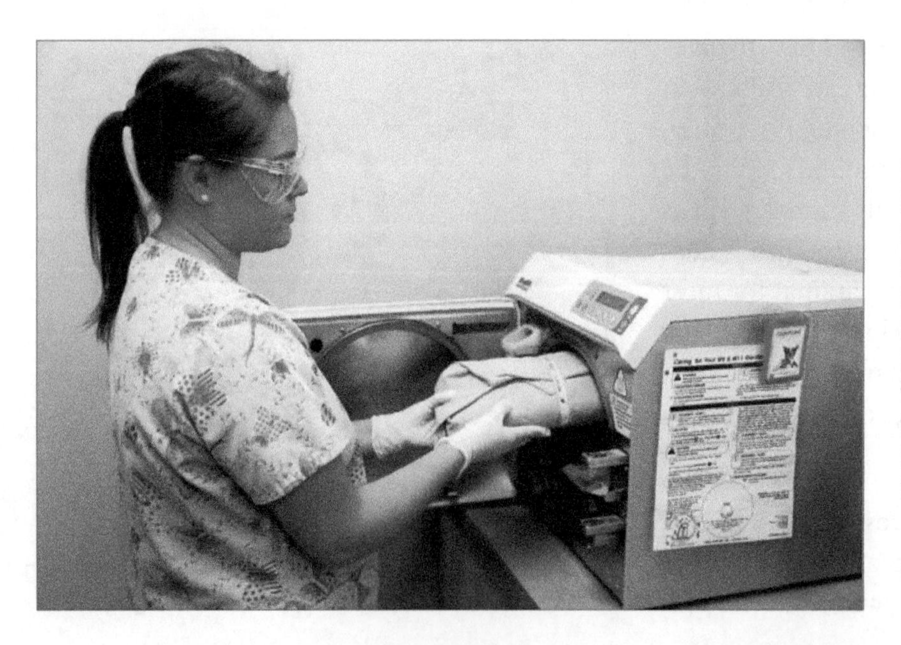

7–35. A surgical gown has been autoclaved to assure asepsis. The assistant is wearing sterile gloves to remove it from the autoclave. (Note how it has been folded to include a hand towel and sterilization indicator.) (Courtesy, Education Images)

The operating room and equipment should be washed and disinfected. The surgery table, legs on the table, casters, and all other parts should be disinfected. In most cases, thorough daily cleaning is recommended because cleaning activities cause dust to go into the air. Never use a dry mop, as it will stir up dust (including microbes). The goal is to have zero microbes present, but this is impossible to achieve totally.

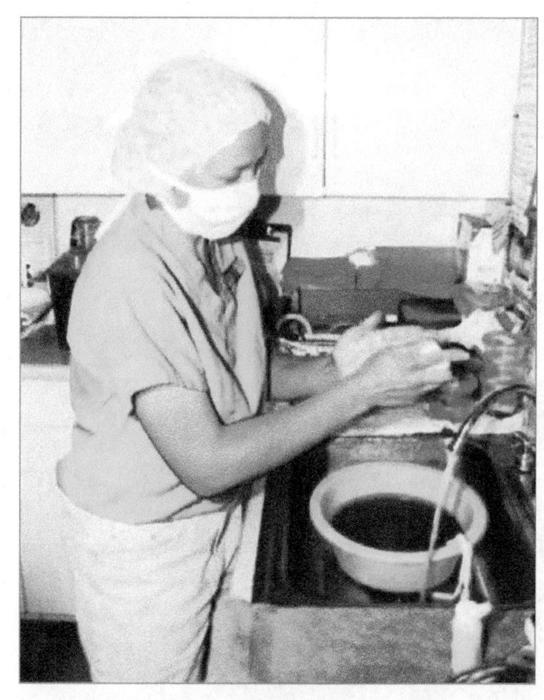

7–36. A veterinarian is scrubbing her hands in preparation for surgery. (Courtesy, Education Images)

Hand scrub is used. The hands and arms to slightly above the elbow are lathered for one minute and rinsed. They are lathered again, and a sterile scrub brush is used on hands, fingers (all sides), and wrist area. Once done, the process is repeated. In hand scrubbing, hold the hands higher than the elbows so excess water drips off the elbows and not the hands. Never touch anything that isn't sterile after a hand scrub.

People performing procedures on animals should practice appropriate asepsis. This includes all members of the surgical team. Proper surgical attire should be worn, including aseptic gowns, gloves, caps or hair nets, and masks. Street clothing and shoes should not be worn, as these may collect contaminants from the air.

METHODS USED

Chemicals, heat, filtration, and radiation are used to gain asepsis. A combination of these is often used in a veterinary clinic.

Chemical methods include a range of products. Some are disinfectants. These destroy bacteria but not certain spores and other forms of microorganisms. Disinfectants are used to wash and mop. Other chemicals are applied to the surgical area of the animal to make the area sterile. Iodine compounds and ethyl and isopropyl alcohols are common and effective against most infectious agents.

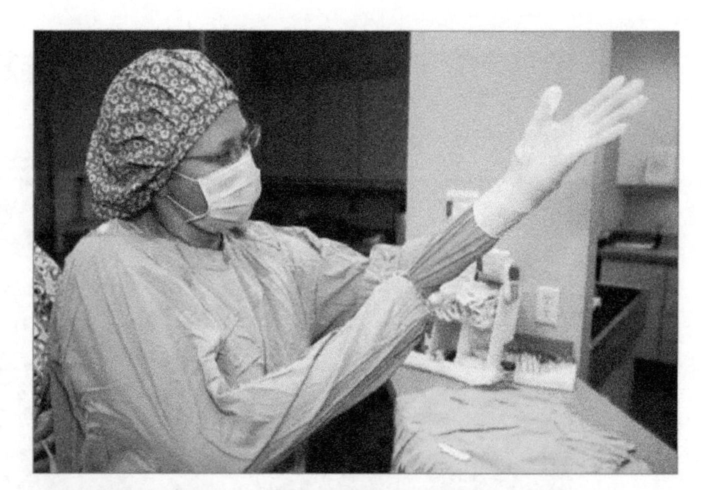

7–37. Sterile gloves complete the surgical attire. (Courtesy, Education Images)

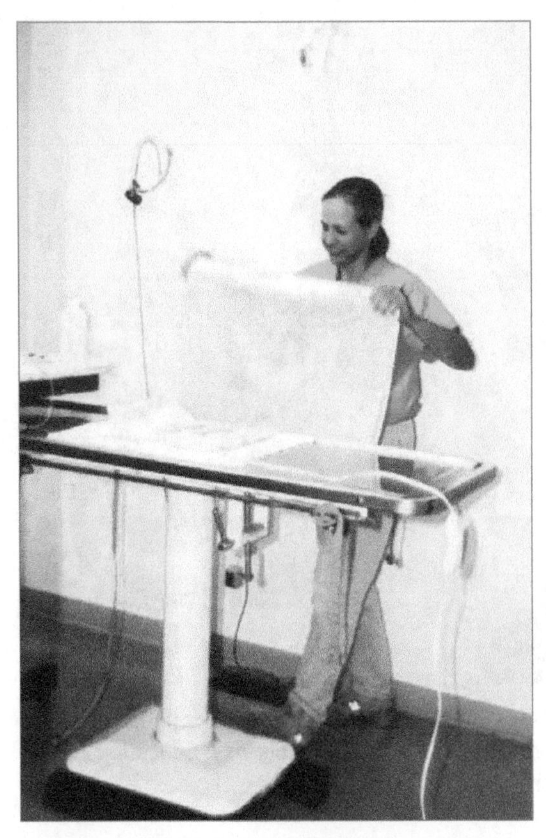

7–38. A surgical table for small animals is being prepared for surgery by placing a sterile covering over a heating pad. (A heating pad is used to prevent hypothermia during surgery.) (Courtesy, Education Images)

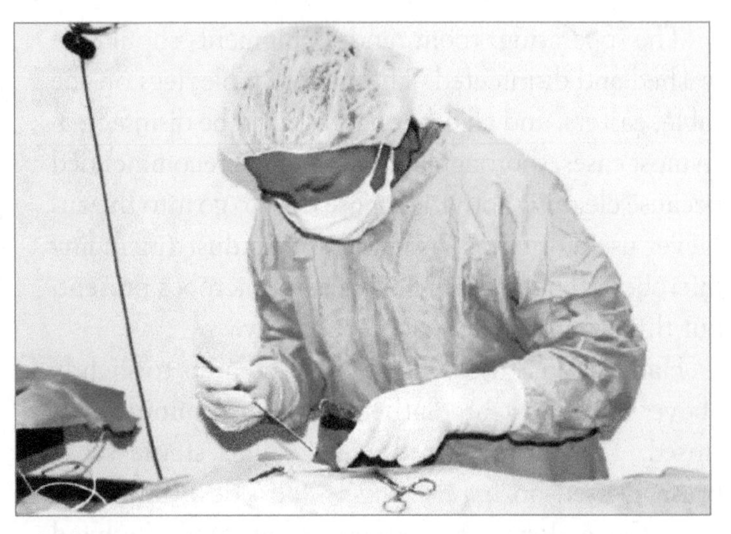

7–39. A spay hook is being used to remove the ovaries of a cat. (A small incision is used in an aseptic environment.) (Courtesy, Education Images)

Heat is used to destroy microorganisms. Instruments may be heated in an autoclave. An autoclave is a steam-pressure sterilizer with somewhat the appearance of a kitchen pressure cooker. Forceps, scalpels, scissors, and other instruments are placed in an autoclave for 15 minutes at 121°C and 15 psi (pounds per square inch). Always use caution, as steam can quickly cause burns. After heating, the autoclave is allowed to cool. When it is opened, protect the instruments from contamination.

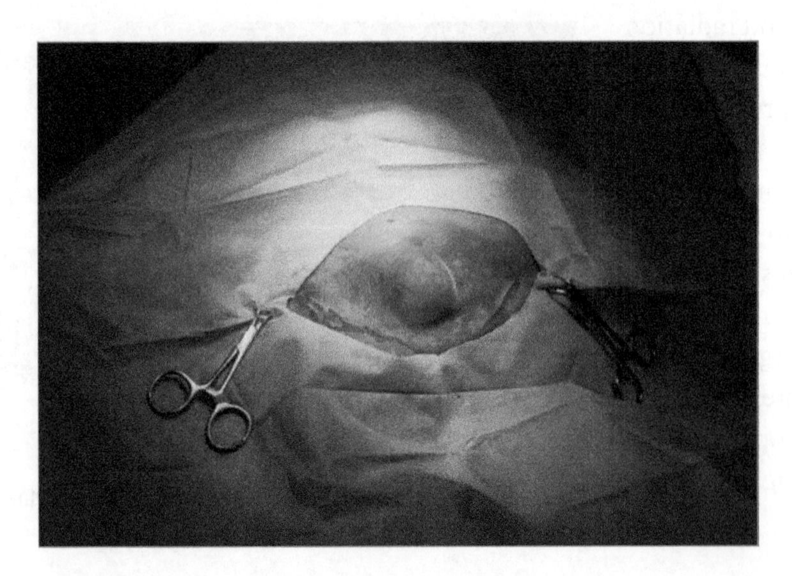

7–40. A sterile surgical drape covers a dog where surgery is beginning to remove a tumor. (Courtesy, Education Images)

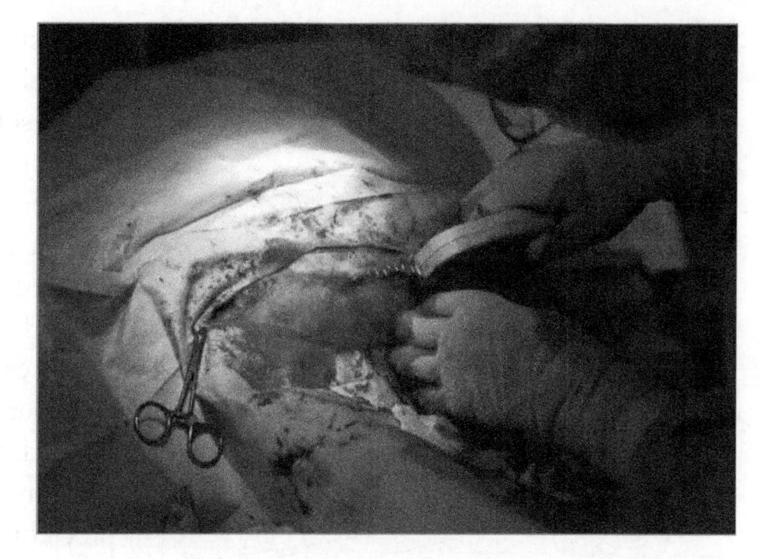

7–41. Following surgery to remove a large tumor from a dog, staples are being used to secure the closed and sutured wound. (Courtesy, Education Images)

Filtration is used to assure that the air in the area is as free of dust particles as possible. Filters should be changed regularly to assure they are working properly.

Radiation can be used with certain kinds of instruments. Gloves and suture material can be radiated.

SURGICAL SKILLS IN LIVESTOCK PRODUCTION

Veterinarians are highly trained to perform surgery on animals. Their assistance should be used as available and needed. Most animal producers develop skill in performing surgery often associated with efficient livestock production. This is performed on the farm or ranch rather than in the aseptic environment of a veterinary operating room.

Common livestock surgical skills are introduced here. Each is also included in the chapters where specific species are covered.

GENERAL PROCEDURES

Docking, castrating, and dehorning should be done when animals are young and with a minimum of stress to the animals. Doing the work properly and using disinfectant promotes fast healing and minimizes the risk of infection.

Some methods used are bloodless; others involve cutting and opening the body and skin. Any area where cuts are made should be disinfected. Common disinfectants used are iodine surgery spray, a 7 percent iodine solution, isopropyl alcohol, and a Lysol® solution. These

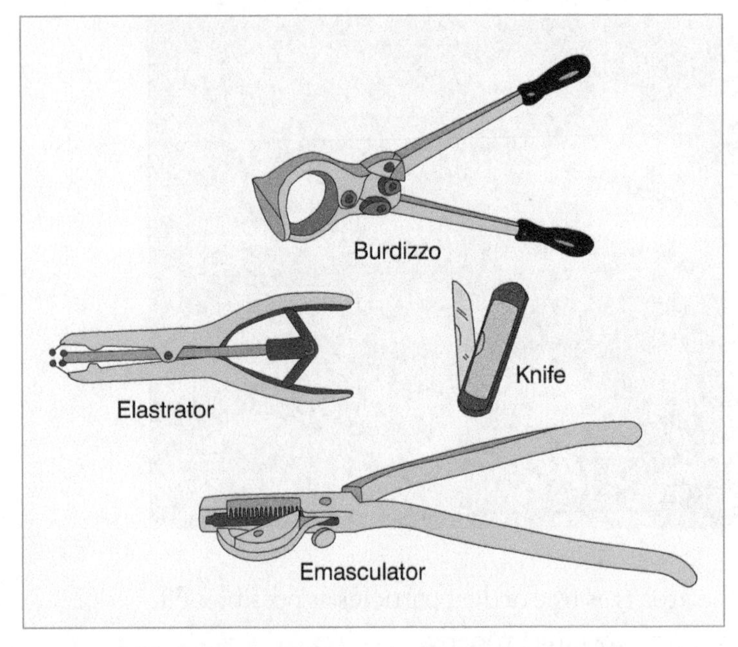

7–42. Common docking and castrating instruments.

7–43. An elastrator is being used in docking a lamb. (Courtesy, Education Images)

products are available from veterinary supply houses or farm and ranch stores. Weak solutions may be poured on the area of the surgery; slightly stronger solutions may be wiped or brushed on with a cloth. It is sometimes recommended that the scrotum area be washed with soapy water and dried before castration.

Several instruments are available for use in docking and castrating. The same instruments can often be used for both procedures. The instruments should be dipped in disinfectant before use and between uses on different animals.

DOCKING

Baby lambs and pigs are docked. The procedure removes their long tails to promote efficient production. Docking may appear to be a harsh thing to do. However, it is a management technique that is widely accepted by producers.

With a lamb, the tail is removed so that it will not collect filth (feces) that would contaminate the wool and provide a place to attract flies and for maggots to grow. Docking should be done before a lamb is 14 days of age. A few breeds of sheep are not docked, especially those that have hair rather than wool, such as the Katahdin.

Pigs raised in confinement are docked to prevent tail biting, a problem in which pigs bite each other's tails. The procedure with pigs is similar to that with lambs and should be done when the animals are a few days old.

The methods of docking are based on the instrument or process used to remove the tail. The elastrator uses a strong, specially made rubber ring known as an elastrator band. When the ring is placed tightly around the tail, blood circulation is stopped, and the tail falls off in two to three weeks. The elastrator is a bloodless method of docking. An emasculator can also be used on a lamb by placing the device 1 inch from the animal's buttocks and mashing the tail off. The clamping action seals blood vessels and reduces bleeding. Docking can also be done with a knife, axe, or chisel. With cuts, bleeding may be a problem.

The general procedure in docking a lamb is:

1. Select the method of docking to use, and organize the supplies and work area.
2. Restrain the lamb.
3. Apply antiseptic to the place where the cut will be made. (Not needed with the elastrator.)
4. Place the elastrator on the tail, or cut the tail off leaving about a 1-inch stub.
5. Apply antiseptic to the tail stub. (Not needed with the elastrator.)
6. Release the lamb.
7. Observe the lamb for several days for signs of complications, such as an infection. (Observe a lamb on which an elastrator has been used for two to three weeks.)

CASTRATING

Castrating is removing or destroying the testicles of male animals. It is used with certain species to prevent unwanted breeding and to promote growth. Since testicles produce sperm, a male without them cannot create a pregnancy. Testicles also produce testosterone, which is a male hormone that influences development and behavior.

Male animals are best castrated when quite young. They are easier to restrain, and the procedure is less traumatic to them. Castration may appear to be a harsh process to use with young animals. It should be done so that stress is kept to a minimum and to assure success. Producers of sheep, hogs, cattle, goats, horses, and other species use castration. (A neutering process is used on some females to make them incapable of reproduction. It is an invasive process that involves making an incision into the abdomen and removing the ovaries or other reproductive organs. Neutering is more common with companion animals. With female dogs, neutering is known as spaying.)

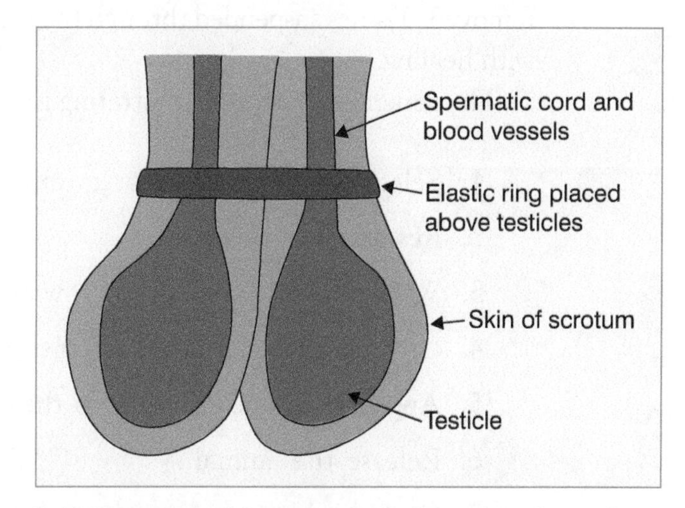

7–44. Placement of elastrator band in castration of a small bull calf.

Most of the instruments used in docking can be used in castrating. Two bloodless methods are used: the elastrator and clamping (pinching). These can be used with species that have testicles that are suspended in the scrotum. With the elastrator, the rubber ring is applied to the scrotum well above the testicles and cuts off the blood supply. After a couple of weeks, the lower part of the scrotum containing the testicles falls off.

Clamping involves using the Burdizzo or other instrument to pinch off the blood vessels to the testicles. Without blood, the testicles shrink, fail to mature, and do not produce sperm. The scrotum does not fall off. With pinching, do not mash the divider in the scrotum. The blood vessels to each testicle must be pinched separately. Use the left hand to find the blood vessels through the scrotum wall, and use the right hand to apply the pincher. Properly pinch the blood vessels to assure castration success. A blood vessel that is not properly pinched may allow blood flow and continued development of a testicle. An animal needs only one testicle to produce sperm. Do not mash the testicles with the clamp. Bruised testicles may cause swelling and shock in the animal.

Surgical methods involve using a sharp knife, scalpel, or emasculator. With cattle and sheep, these methods cut off the bottom of the scrotum, and the testicles are removed. This assures successful castration. All cords connected to the testicles should be pulled from the body cavity. If some cords remain, it is possible that testicles can be regrown, resulting in a fertile mature animal. Always apply a mild antiseptic before and after surgical castration. Some producers also use a fly spray to keep flies away from the wound. (Do not apply the fly spray directly to the wound.)

With pigs, testicles are held more tightly to the body cavity. Clamps and elastrators will not work; surgical procedures are needed. A sharp knife or scalpel is used. The scrotum is washed with mild, soapy water and dried. Antiseptic may be applied to the scrotum. An incision is made vertically in the scrotum over each testicle and the testicle is pulled out with cords attached as far as it will go. Sometimes the cords break off in the body cavity. If not, cut them as close to the body as possible. All tissues attached to the testicles should be removed. Tissues suspended through the incision opening may cause infection and interfere with healing.

The general procedure in castrating is:

1. Select the method of castration to use, and get all supplies ready.
2. Restrain the animal.
3. Wash the scrotum with soapy water, and apply a mild antiseptic to the scrotum.
4. Use the chosen method of castration.
5. Apply antiseptic solution to the scrotum area.
6. Release the animal..
7. Observe the animal for several days for infections or complications that need attention.

DEHORNING AND DESCENTING

Several animal species grow horns as they mature. The horns can cause injuries and increase safety issues. Livestock with horns include cattle, goats, and sheep. Other species may develop scent glands that produce undesirable odors as the animals mature. Among livestock species, goats are more likely to have scent glands removed. Ferrets and skunks are other species with scent glands.

Dehorning is removing the horns or horn "buds" so that horns do not develop. It is used primarily with cattle, particularly dairy cattle. Goats may also be dehorned. Dehorning is best done while calves are young, before horns show much development. Healing time is less than with mature animals.

Dehorning is done with elastrators (rubber-band method), caustic paste, electric dehorning irons, or gougers. With older cattle, saws may be used to remove horns, but this is no doubt painful to the animal. You will need to get the assistance of a qualified individual to learn the procedures to use.

Producers can eliminate the need to dehorn with some species by selecting breeds that do not develop horns, such as Polled Hereford or Black Angus cattle. "Polled" means that the animals are hornless.

Only male goats are descented. Mature bucks have an offensive, musty odor. Descenting is done at the same time as dehorning (five days of age) and involves destroying the scent glands. The scent glands are located behind the horns and toward the middle of the head. Groups of yellow scent gland cells are easy to see when the hair is clipped off this area. If a hot iron is used to dehorn, it can also be used to destroy the musk cells. It is often difficult to remove all the scent glands. The assistance of an experienced goat keeper may be useful.

AVOIDING HAZARDS IN THE VETERINARY WORKPLACE

Working in the veterinary field can be exciting and rewarding. By the very nature of the work, veterinary personnel are exposed to potential safety hazards. There are many hazards with both animals and equipment. Steps must be followed carefully and correctly to avoid injuries in the workplace. Some potential hazards are included here related to animals, the workplace, and radiation. (Additional safety considerations are presented elsewhere in the book.)

ANIMAL-RELATED INJURIES

When being treated, an animal can be aggressive and unpredictable. It is important to know the animal or ask the owner about its behavior around other people. Even after that, it is critical always to keep an eye on the animal. At any time, an animal may bite, kick, scratch, paw, step on, knock down, or even trample someone. Appropriate knowledge about the type of animal helps in any situation. For example, when a feline is being given medication orally, an assistant should securely hold the animal so that no one gets scratched. A judgeable distance is always recommended for those not involved with a procedure.

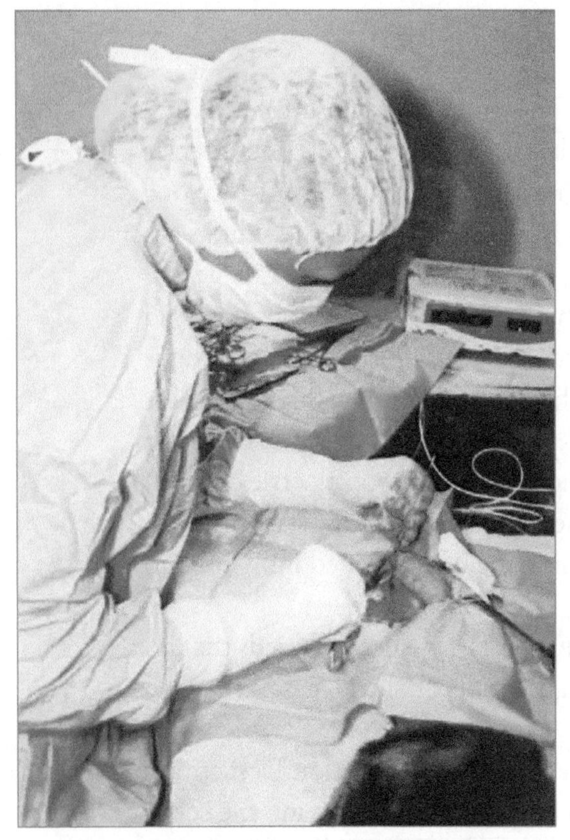

7–45. Proper attire protects the animal and the veterinarian during surgery. (Courtesy, Education Images)

LABORATORY ACCIDENTS

There are many responsibilities when working in a veterinary laboratory. It is important to learn the safety hazards and the approved procedures for avoiding injury in the workplace. Some hazards in a laboratory may be:

- Slippery floors resulting in falls. (Wear proper shoes or boots.)
- Steam burns from autoclaves. (Know how to use autoclaves.)
- Spills caused by loose lids. (Keep all containers tightly closed.)
- Poisoning. (Always know the qualities of the substances you use.)
- Spread of disease through body fluids. (Wear rubber gloves.)
- Infection from unclean hands. (Wash hands thoroughly.)
- Fire from chemicals or flammable materials.
- Expired medications and chemicals.

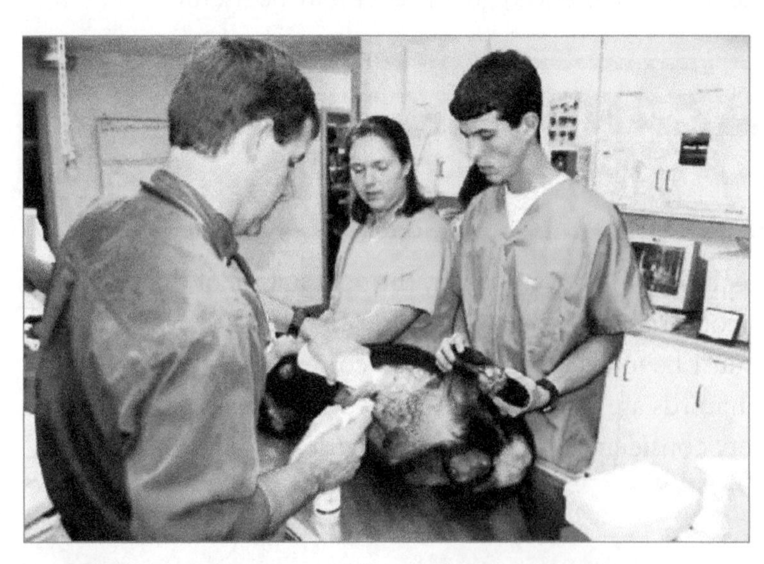

7–46. Working as a team to restrain animals promotes safety. (Courtesy, Education Images)

- Cuts from improperly disposed of glass or needles.
- High noise levels.
- Cuts from sharp instruments.

7–47. A container for proper disposal of sharps, such as needles, scalpel blades, and broken glass. (Courtesy, Education Images)

RADIATION

Persons working in veterinary hospitals may be exposed to electromagnetic radiation produced by X-ray machines, lasers, fluoroscopes, and ultraviolet lamps.

Electromagnetic radiation consists of photons of energy traveling at the speed of light and used to produce an image on film or digitally. Different forms of radiation have different wavelengths. Some forms of electromagnetic radiation have the potential to damage cells, even in small doses. Note that the image made with X-ray technology is known as a radiograph.

Here are some precautions that should be taken to prevent unnecessary exposure:

1. Do not take more X-rays than are necessary.
2. Use the least amount of radiation possible for a useable radiograph.

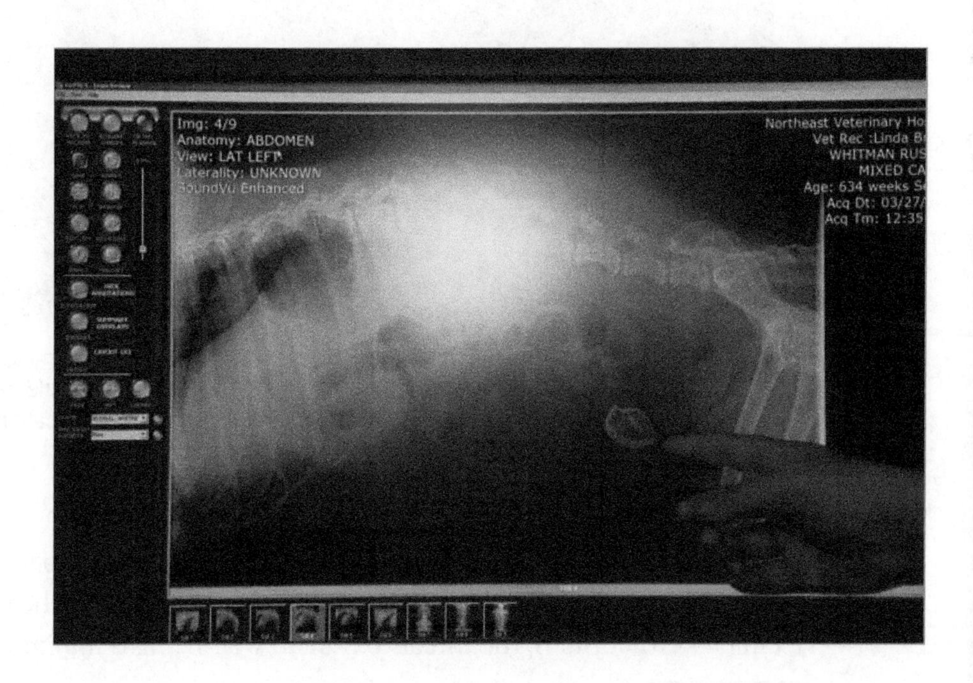

7–48. A digital radiograph reveals a beverage bottle cap causing a blockage in the digestive system of a dog. (Courtesy, Education Images)

3. When the radiograph is being taken, stay as far from the X-ray beam as possible.
4. When restraining an animal for radiography, keep your hands out of the primary beam.
5. Everyone in the room where an X-ray is being taken must wear protective clothing, including lead-lined gloves and a lead-lined apron, unless standing behind a lead barrier or control booth.
6. If a portable X-ray unit is used, take special precautions to prevent exposure.
7. Monitor X-ray exposure.

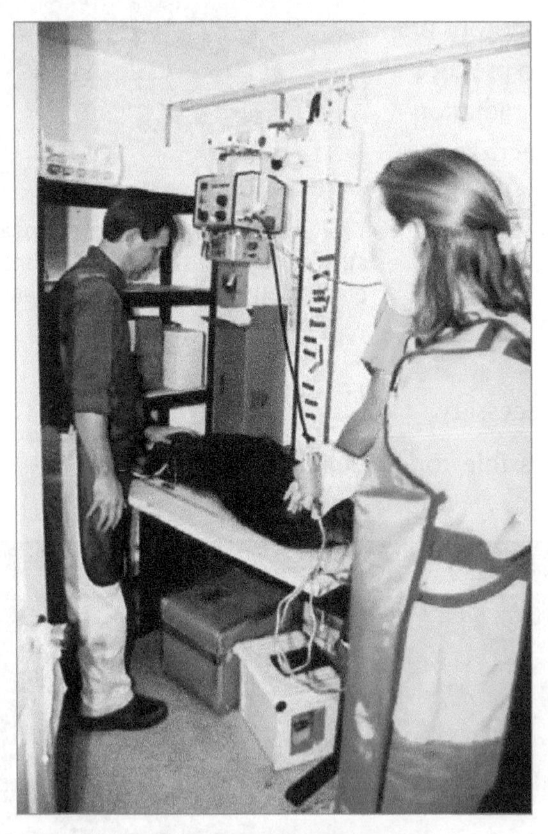

7–49. A dog is being prepared for an X-ray. (Courtesy, Education Images)

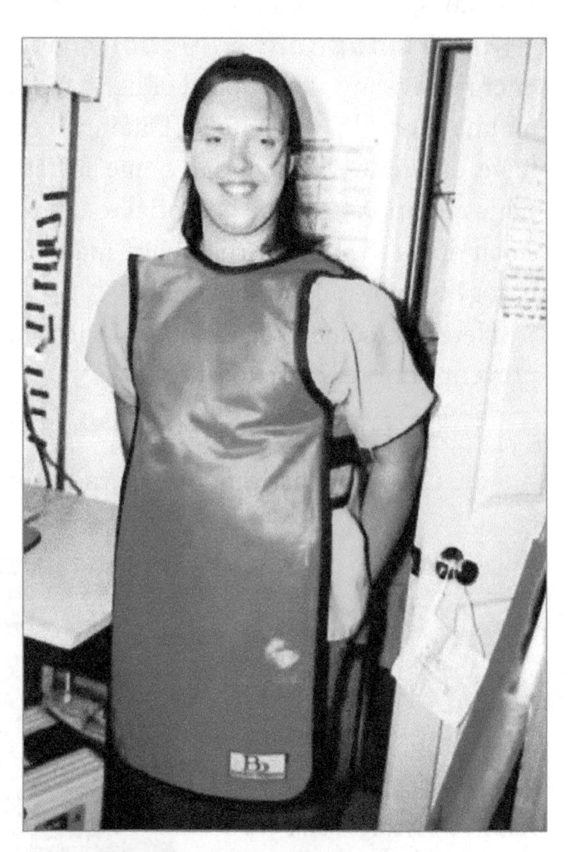

7–50. A lead-lined apron and other protective equipment should be worn when taking X-rays. (Courtesy, Education Images)

ZOONOSIS

Diseased animals may pose threats to the humans who care for them. Some diseases that are contracted from animals are major threats to human health, with rabies being an example. Other times, the threat to human well-being is minimum though a problem, such as ringworm.

An infectious disease that can be transmitted from an animal to a human is known as a **zoonosis**. In some cases, a vector may be involved. A **vector** is an intermediary such as a bite from a dog flea that might transmit a disease from the dog. Other vectors are bats, birds, cats, flies, lice, mice, and snails. About 60 percent of known pathogens are zoonotic. These pathogens include bacteria, fungi, viruses, pria (incomplete organisms), and parasites, including protozoans.

In a few cases, reverse zoonosis occurs. Reverse zoonosis is the case of a disease being transmitted from a human to an animal, with internal parasites such as the tapeworm being an example.

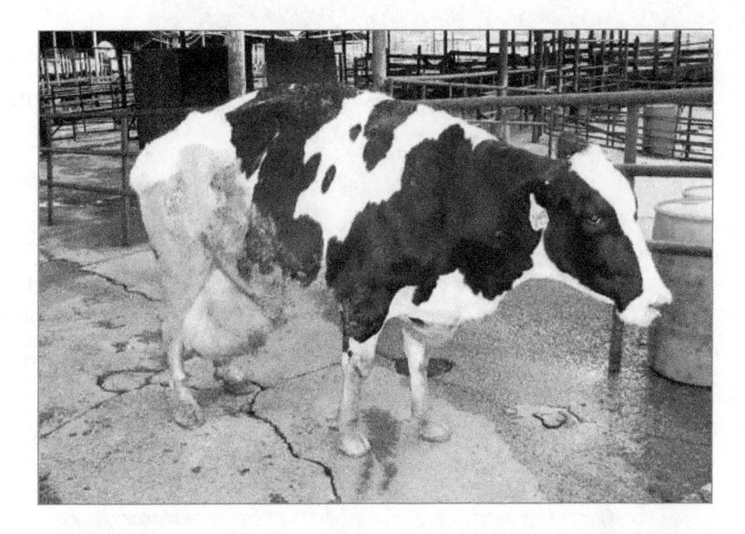

7–51. This cow is obviously not in good health but what is the problem? Seek the services of a veterinarian for an accurate diagnosis and the treatment procedures to follow. (Courtesy, Nancy S. Jackson, DVM, Mississippi)

Here is a partial list of zoonoses: anthrax, brucellosis, lyme disease, bovine tuberculosis, *E. coli* (foodborne), equine encephalitis, *Giardia lamblia*, helminth, plague, rabies, salmonellosis (foodborne), and trichinosis. There are many others.

Protecting yourself against zoonosis involves several important sanitation and prevention practices. Here are a few examples:

- Innoculate animals–Prevention of disease is highly important. Have a plan of vaccination to prevent disease problems. A veterinarian can help with such a plan.
- Seek professional services–Use a veterinarian to aid in diagnosing and treating diseased animals. The veterinarian can also provide information on the procedures to follow and how to protect yourself from zoonoses.
- Avoid direct contact with diseased animals–Do not touch diseased animals with unprotected hands. Avoid contact with feces, urine, body fluids, and flesh of a diseased animal. Some food products can transmit disease, such as undulant fever from a cow infected with brucellosis.

- Use personal protective devices–Gloves, masks, and other protective devices help prevent the spread of zoonoses.
- Practice sanitation–Clean areas, including examination instruments and equipment, where diseased animals have been kept and examined. Use a disinfecting agent to minimize populations of zoonotic agents.
- Sterilize instruments–Sterilizing instruments helps prevent spread to humans and to other animals. Autoclaves are widely used for this purpose.
- Isolation and quarantine–Keep infected animals away from other animals and in locations where human contact is minimized.
- Keep informed–Good information will help you to avoid being exposed to zoonotic agents. It will provide information on how to prevent exposure as well as provide an awareness of zoonotic agents.

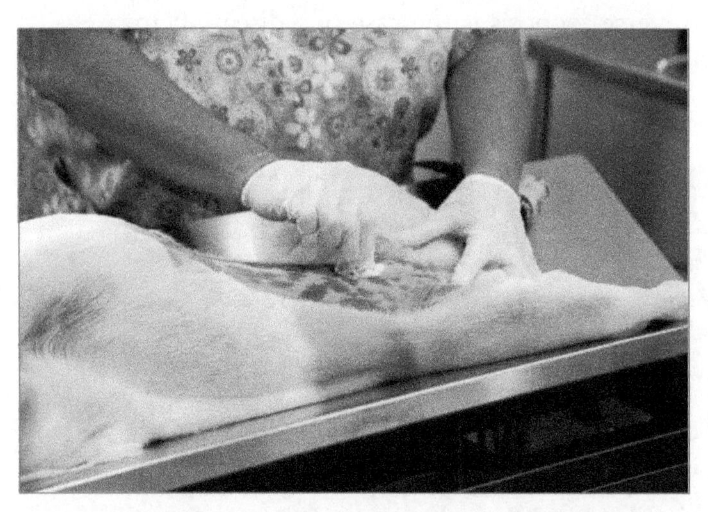

7–52. Wear proper gloves when preparing a patient for surgery. (Courtesy, Education Images)

REVIEWING

MAIN IDEAS

Veterinary technology is an important part of today's society. Many people rely on technology to improve methods of helping sick or injured animals. Every day, new and exciting procedures are developed. Veterinary technology has improved notably from what it used to be. Historically, many veterinarians practiced independently and performed duties by themselves. Now, practices are carried out with multiple veterinarians. Veterinary assistants, technicians, and technologists are needed to help in practices.

Medical terminology is important in the veterinary profession. Proper pronunciation and correct spelling help assure accurate communication. Words may contain important parts, such as prefixes, suffixes, and root words. These parts have universal meanings that are understood by veterinarians.

There are many practices carried out in veterinary medicine. Sick or injured animals may require veterinary attention. Some procedures include checking vital signs, giving injections, giving drugs orally, using diagnostic ultrasound, and inserting stomach tubing.

Pharmacology is the study of drugs as related to health management. A knowledge of drug action and how drugs are effective helps assure that the proper drugs have been used.

Animals must be restrained for examination and care. The way they are restrained is based on their size, species, and temperament. Always restrain animals to prevent injury to them, yourself, people around the area, and facilities.

Asepsis is an important part of successful animal care. It involves having the veterinary work area free of microorganisms that cause infections. If not kept clean, a veterinary facility can harbor disease agents and spread them to uninfected animals.

Animal producers on farms and ranches often perform routine surgery on animals. Docking, castrating, and dehorning are most common. Proper procedures should be followed to assure success of the procedures and prevent infections.

Safety precautions must always be taken whenever you are around animals or equipment. Hazards can be prevented when proper safety measures are taken.

Individuals around animals need to be aware of zoonotic diseases. Sanitation, using personal protective equipment, and isolating new animals are useful practices. Always obtain the services of a veterinarian if in doubt about zoonosis.

QUESTIONS

Answer the following questions using correct spelling and complete sentences.

1. Why is veterinary technology important?
2. When did veterinary technology begin to take form?
3. What is involved in the business operation of a veterinary medical clinic?
4. What are the qualifications of a veterinarian?
5. Why is it important to learn proper veterinary medical terminology?
6. What is the use of a stethoscope?
7. What are three common types of injections?
8. What use is made of diagnostic ultrasound?
9. What are some possible hazards in a veterinary clinic or hospital?
10. What is electromagnetic radiation?
11. What is involved in animal examination and care?
12. How are the vital signs of an animal checked?
13. What is pharmacology?
14. What is restraint? Why is it needed?
15. What methods of restraint are used?
16. What is asepsis? Why is it important?
17. What instruments may be used in docking? Castrating?
18. What steps are followed in docking a lamb?
19. What is zoonosis? Name two diseases that may be transferred to humans.
20. What is a vector? Name two examples with animal diseases.

EVALUATING

Match the term with the correct definition. Write the letter by the term in the blank provided.

a. care
b. penicillin
c. veterinary technology
d. target tissue
e. suffix
f. veterinary technologist
g. prefix
h. veterinary technician
i. root word
j. veterinary assistant

______1. The site in an animal where drug action is needed.

______2. The quality of appropriate maintenance of an animal, including food, water, shelter, and health upkeep.

______3. A graduate of a four-year AVMA-accredited program in veterinary technology.

______4. A graduate of a two- or three-year AVMA-accredited program in veterinary technology.

______5. A widely used antibiotic.

______6. A syllable, group of syllables, or word added at the end of a root word to change its meaning or create a new word.

______7. A syllable, group of syllables, or word added at the beginning of a root word to change its meaning or create a new word.

______8. A person with training, knowledge, and skills at the level of a clinical aide but below the level of a veterinary technician.

______9. The part of the word consisting of a syllable, group of syllables, or word that is the basis for the meaning of the word.

_____10. The science and art of providing professional support service to veterinarians in their practices.

EXPLORING

1. Search the Internet for recent articles on a topic in veterinary technology. Choose an article that interests you. Prepare a report on your reading.

2. Tour a veterinary clinic. List five possible hazards. Beside each, write a prevention method that might help someone avoid such a mishap.

3. Study common medical prefixes and suffixes. Look in a veterinary handbook and see how many words you can identify. Try making up your own veterinary terms.

4. Call your local veterinarian, and schedule an interview. Ask the veterinarian about the nature of the work and technology in his or her field. Report your findings to the class.

5. Sketch an outline of the body of an animal, such as a horse or a cat. Label the following positions on the body: dorsal, ventral, cranial, rostral, caudal, distal, and superficial.

6. Investigate becoming a food animal veterinarian. Refer to: **http://www.foodanimalvetservice.com/** or **http://www.avmf.org/whatwedo/food-animal-veterinarian-recruitment-and-retention/**. Prepare a report on your findings.

UNIT THREE

(Courtesy, Phillip W. Kirkland/Shutterstock)

Food Animal Technology

CHAPTER 8

Beef Cattle Production

OBJECTIVES

This chapter introduces beef cattle production. It has the following objectives:

1. Explain the importance of beef production.
2. Explain how to select beef animals.
3. Name and identify the common breeds of beef animals.
4. List the advantages and disadvantages of beef cattle production.
5. Describe the types of beef production systems.
6. Describe the promotion of herd health.
7. Assess methods of animal identification.
8. Describe two common herd management practices.
9. Describe facility and equipment needs with beef cattle.
10. Discuss common feed resources for beef cattle.
11. Identify selection, feeding, grooming, and training practices with show animals.

TERMS

anthelmintic
backgrounding system
Bos indicus
Bos taurus
branding
conformation
cow-calf system
creep feeding
cutability
dehorning
dual-purpose breed
elastrator
expected progeny difference (EPD)
finishing system
forager
grade cattle
grass-fed beef
halter broken
heterosis
marbling
polled
purebred
steer

8–1. Research has a big role in producing better beef. Here, scientists are comparing marbling and yield grade of carcass images in assessing the role of feed materials. (Courtesy, Agricultural Research Service, USDA)

PEOPLE enjoy raising and watching beef animals. Their production serves as a major source of farm and ranch income in the United States. Products made from their meat, known as beef, are popular foods. Steaks, roasts, ribs, and ground beef meat are a few of the most widely consumed beef products.

The production of beef animals is big business that takes time and money. A female will have a calf at two to three years of age. Most have only one calf a year. With good care, the calf will reach a weight of 500 to 700 pounds in its first year. The female calves may go back into herds to produce more calves or be fattened for meat harvest. With male calves, most are harvested for meat though the best animals are used for breeding.

Producers are guided by the Beef Quality Assurance (BQA) program. This is a national program that provides guidelines for beef cattle production. The goal is to use best management practices that produce quality beef for consumers and a profit for the producer. More details on BQA are presented in this chapter.

IMPORTANCE OF BEEF CATTLE PRODUCTION

8–2. Beef cattle should be regularly observed on the range. (Courtesy, Beefmasters Breeders Universal and B. E. Fichte, Rosebud Communications, Inc.)

Beef cattle are raised for their meat, known as beef. The animals are selected to yield a high amount of good meat. This is contrasted with dairy cattle, which are used for milk production.

Beef production is a large part of agriculture in North America. Providing quality beef requires good animals and efficient marketing. Products must reach the consumer in wholesome condition.

ORIGIN OF BEEF CATTLE

The origin of cattle is not fully known. It is assumed that they were first domesticated in Europe during the Stone Age. Scientists initially classified cattle into three separate species: *Bos indicus* (cattle of Indian origin—have humps on shoulders and droopy ears), *Bos taurus* (cattle of European origin), and *Bos primigenius* (now extinct). Today, after greater knowledge related to classification, all cattle are classified as *Bos primigenius*. This includes those kept for meat, milk, and draft power of oxen. The *B. primigenius* is further classified into subspecies of *B. primigenius taurus* and *B. primigenius indicus*.

8–3. Cattle of the *B. p. indicus* subspecies have droopy ears, loose skin, and humps over their shoulders. (Courtesy, Jackiso/Shutterstock)

Bos primigenius taurus are cattle common to the temperate regions, such as the midwestern United States. They are often known as the European breeds. Examples include Angus and Hereford and the dairy breeds such as Jersey.

Bos primigenius indicus cattle are those cattle that are descendants of the Zebu cattle with humps on their necks. They also have large, droopy ears and loose skin. *B. p. indicus* cattle are common to warmer tropical areas, such as the southern part of the United States. They are sometimes said to be more resistant to insect pests and certain diseases.

Christopher Columbus introduced beef cattle in the United States. Beef animals were on his second voyage to America. Early settlers from the British Isles and other European countries brought cattle with them as they settled new land in America. As the settlers moved west, so did the cattle. Abundant pasture was found in the great plains areas.

THE BEEF CATTLE INDUSTRY

Today, the cattle business is an important part of American agriculture. Sales of cattle amount to a large portion of all farm markets. Though the number has declined somewhat in recent years, about 90 million head of cattle are on U.S. farms and ranches. Of these, 9 million are dairy cows. Ultimately, most dairy animals wind up being processed into meat products. (Source: ERS/USDA. Accessed on April 14, 2012, at: **www.ers.usda.gov/news/BSECopverage.htm**) There are 33 million head of beef cows on farms and ranches in the United States.

Cows will be bred to produce calves for human use and serve as herd replacements. (Detailed statistical information about cattle production is compiled by the Economic Research Service, USDA: **www.ers.usda.gov/**.)

There is also a global market for beef. The United States exports beef to countries such as Japan, Canada, Republic of Korea, Mexico, and Hong Kong. Japan is by far the largest foreign customer of the United States' beef industry, including variety meats and hides.

Consumption

The consumption of beef is part of a balanced diet designed for good nutrition and health. Beef continues to be one of America's favorite foods. The per capita consumption of beef is 57.4 pounds per year in the United States.

8–4. Beef is an important part of many restaurant-prepared meals. (Courtesy, michaeljung/Shutterstock)

Beef helps consumers meet daily dietary requirements in protein, vitamin B-12, iron, zinc, and other essential nutrients. In recent years, beef has received criticism because of the amount of fat and cholesterol it can add to the diet. New research, improved genetics, and better management techniques are reducing

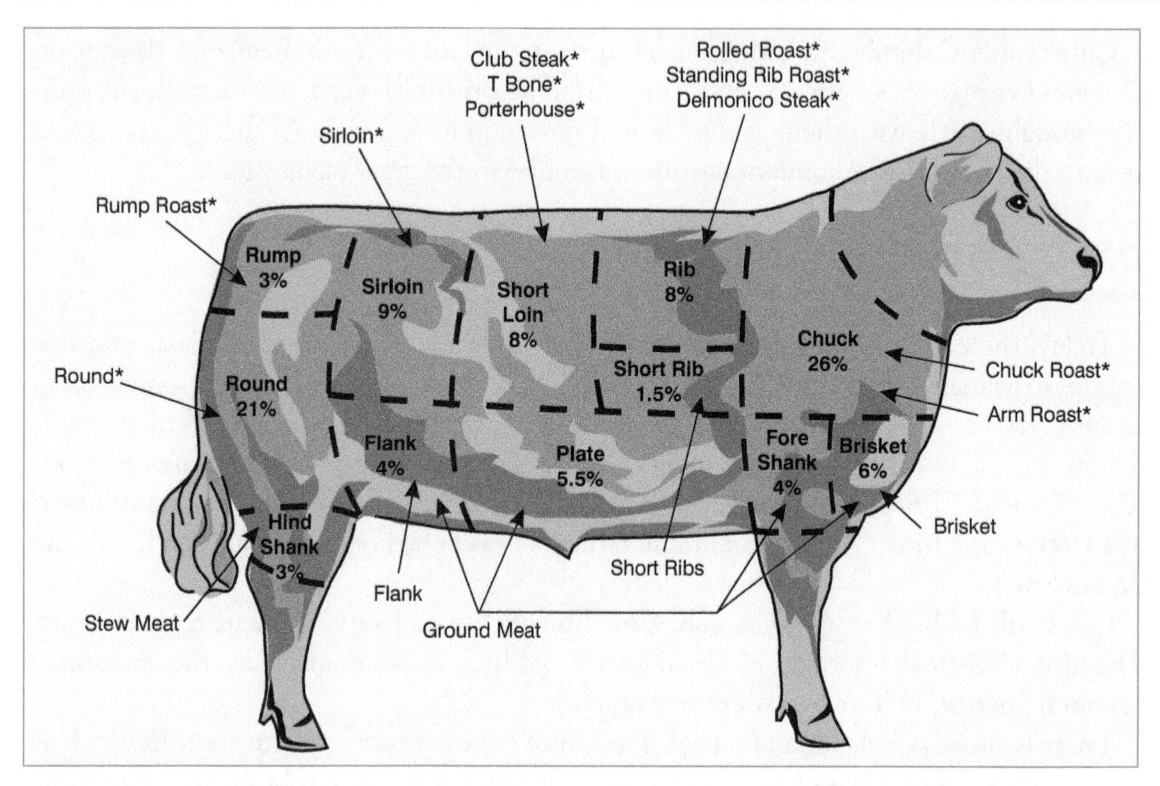

8–5. Locations of the beef cuts on an animal. (An asterisk * indicates the most desirable cuts that sell for a higher price.)

these concerns. Moreover, the key to a healthy diet is in the balance, variety, and moderation of consumption.

CHECKOFF PROGRAM

The U.S. Beef Promotion and Research Act of 1985 established a checkoff program to fund needed research in beef production and gain funds for promoting beef consumption. The Act established a structure for a national $1 per head of beef animal sold to provide funds for research and promotion. The amount is deducted from payment that the producer receives. National and state boards oversee the program. Half of the money is retained in the states for use by the State Beef Councils and the other half goes to the Cattlemen's Beef Board.

Funds generated have been used in a range of activities. Some funds have been used to develop materials on the care and handling of cattle. Other funds have been used to promote beef consumption, including international activities that promote beef exports. Research activities are also carried out by universities and others using checkoff funds.

For more information, go to: **www.beefboard.org/**.

Consumer Demands

Consumers want good, wholesome beef. They will pay top prices for steak and loin; less for organs and bony pieces. Consumers want beef to be produced and handled in a safe manner and free of bacteria, pathogens, and foreign matter.

Producers need to be aware of where the more valuable cuts are located on an animal. They also need to produce the type of animal that has larger amounts of meat in these cuts.

QUALITY ASSURANCE

In order to meet consumer demand and provide a product that is safe to eat, the beef industry has developed and implemented the Beef Quality Assurance (BQA) program. BQA is a national program that provides guidelines for producers to follow in the production of beef cattle. The program is coordinated at the state level. Overall, it is said to be a common sense approach to beef husbandry so that cattle are produced with optimal management and environmental practices.

BQA includes the use of the best practices to protect herd health. These are documented by records kept by the producer. Beef processors are assured of getting a quality animal when they buy from a BQA producer. Consumers are assured of getting quality foods that are safe to eat when the meat they buy is from a processor who has used BQA animals.

The major guidelines followed with BQA are:

- Feedstuffs—Animals are to receive only quality feeds that are free of mold and chemical toxins, meet nutritional needs, and do not infect with diseases.
- Feed additives and medications—Only feed additives approved by the Food and Drug Administration (FDA) should be used. If antibiotics or other medicines are used with animals, proper withdrawal times must be followed.
- Processing/treatment and records—The processors of beef products have a major role in this regard. The identification of animals is important in tracing them and their products from the producer to the consumer.
- Injectable animal health products—Regulations in the use of health products require proper administration according to instructions including subcutaneous (SQ), intramuscular (IM), and orally. Limitations on dosage and frequency of treatment are also to be followed, such as SQ injections are to be only in the neck and and IM injections should involve no more than 10 cc of product at one site.

- Care and husbandry practices—Good veterinary and husbandry practices are to be followed, such as moving animals to minimize stress, keeping feed and water equipment clean, and maintaining bio-security (restrict access to animals).
- Guidelines for the care and handling of cattle—A manual on the care and handling of beef animals details a wide range of guidelines to be followed.

More details on BQA are available: **www.bqa.org/**.

BEEF CATTLE SELECTION

Selecting beef cattle is an important decision. Much money and other resources are involved. Some people want a specific breed. Other people are not concerned about specific breeds. The strengths and weaknesses of each breed should be considered. This will allow producers to meet their goals.

8–6. Grade cattle are being herded on a New Mexico ranch. (Courtesy, Agricultural Research Service, USDA)

PUREBRED AND GRADE CATTLE

A ***purebred*** is an animal of a particular breed. It is entitled to be registered as a purebred animal. Breed associations set rules that purebred cattle must meet. Producers must follow the rules and certify breeding.

Some cattle producers decide to raise cattle that are not purebreds (entitled to registration). They may have crossed breeds to get certain desired traits. For example, bulls of one breed may be bred to cows of another. The animals that result are crosses. Other producers raise grade animals. ***Grade cattle*** are not registered. Some have a purebred background, but most are mixed breeds.

Crossbreeding occurs when individuals of two breeds are mated. By crossing two breeds, heterosis may result. ***Heterosis*** is the tendency of offspring to perform better than the average of their parents in areas such as milk production and rate of growth.

SELECTION GUIDELINES

Eight guidelines to follow in selecting a breed are:

1. No one breed is the best for all traits. Producers need to have priorities on the traits that are most important to them.

2. All breeds of beef cattle will have variation within their genetics. A decision has to be made by the producer on the genetic traits that are most important.

3. Selection of superior animals and sound breeding practices are as important as the breed. The cattle will only be as good as their selection and management.

4. Select a breed that has the desired conformation as a beef animal. **Conformation** is the type, shape, and form of an animal. It is related to muscle (meat) in the valuable cuts and the ability of the animal to be productive. Over the years, the desired conformation has changed to moderate-framed animals that can provide much meat.

5. The breed needs to be selected on its production capabilities in the specific environment where it will be raised. The producer needs to consider climate, feed, and forage resources. Some breeding animals are performance tested. This

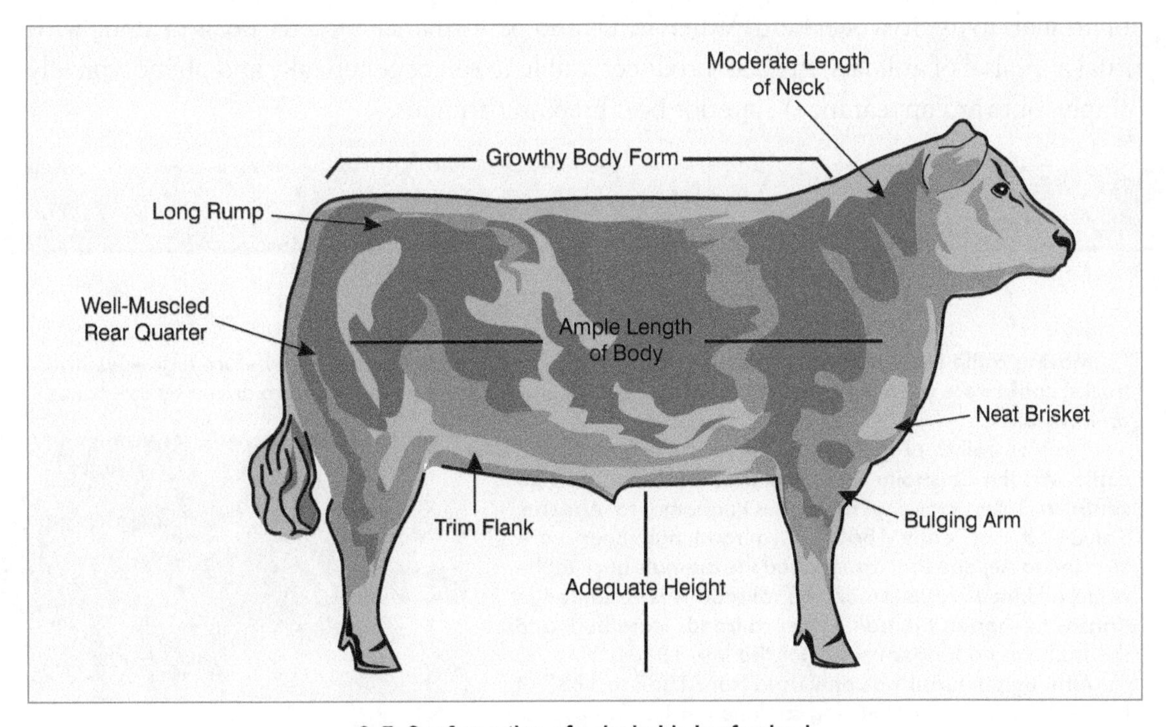

8–7. Conformation of a desirable beef animal.

involves careful growth and feeding evaluations of young animals, particularly bulls.

6. Markets for the breed should be evaluated. Can the offspring be readily sold in the area in which they are produced? Breeding stock needs to be available at a price that makes production economically feasible. Buy the best animals at the lowest price.
7. The producer should consider personal preference and select a breed with which they are comfortable and familiar. This is probably the most commonly considered criterion in the selection process.
8. Buy animals from established producers with a good reputation.

PERFORMANCE RECORDS: USING EPDS

Performance records are important in selecting beef cattle. Performance records serve to predict how an animal comparatively performs with other beef animals. Performance records that are commonly used in selection are birth weight, weaning weight, yearling weight, maternal milking ability, and carcass performance.

A common tool used to aid in selecting on the basis of performance rate is information on expected progeny differences. An **expected progeny difference (EPD)** is a prediction of traits a parent passes to its offspring (progeny). EPDs are used to select genetically superior animals to use in a beef herd. When EPDs and performance records are used along with visual appraisal of animals, the beef producer is able to select genetically and phenotypically (simply, outward appearance) superior beef breeding animals.

CONNECTION

GREAT CATTLE ROUTE

Moving cattle to market is a major step in meeting consumer demands. Well before highways and trucks, cattle were moved by herding them along overland trails. The animals were driven by cowpokes on horseback.

A major trail to provide an outlet for an abundance of cattle was the Chisholm Trail. This trail extended from far south in Texas crossing through Oklahoma to Abilene, Kansas. By the early 1860s, a railroad had been constructed to Abilene that transported the animals back to the eastern United States. Later, the railroad was extended in Kansas to shorten the trail. Other railroads were built and the trail was no longer used after the late 1880s.

Although the trail was only used from 1867 to 1887, it had a major impact on moving cattle to market. Today, the significance of the Chisholm Trail is commemorated by a longhorn monument in a park in Yukon, Oklahoma. (Courtesy, Education Images)

BEEF BREEDS

The breeds of beef cattle common in the United States are briefly described here. These were developed in Europe, Great Britain, or the United States. Some genetics of Indian cattle may be included. Animals in the breeds conform to standards set by the breed associations. Many are purebred and registered. Others may be purebred but have not been registered.

ANGUS

The first Angus cattle were imported from Scotland by George Grant around 1873. Angus are the most popular purebred cattle. There were 281,965 purebreds registered in the United States with the American Angus Association in 2002.

8–8. A modern Angus bull. (Courtesy, American Angus Association)

The breed is black in color and has a smooth hair coat and black skin. A few Angus cattle carry a recessive gene for the color red, which are described below as Red Angus. Nearly all purebred Angus carry the dominant gene for being polled. ***Polled*** means that the animal is naturally without horns.

Angus are cattle with moderate frames. They produce a high-quality carcass that has good marbling. ***Marbling*** is the presence of intramuscular fat. Marbling increases the quality of the carcass and makes the meat cuts tastier.

RED ANGUS

The Red Angus breed developed from Black Angus in the United States about 1945. Black Angus carrying the recessive red gene for color were mated to produce red offspring.

Red Angus are similar to Black Angus in many respects, except for color. However, red does not absorb as much heat as black, so Red Angus are somewhat more heat tolerant.

ANKOLE WATUSI

The Ankole Watusi breed originated in the Ankole district of Uganda, Africa. It is not widespread in the United States but has striking features. Colors include dark red, black,

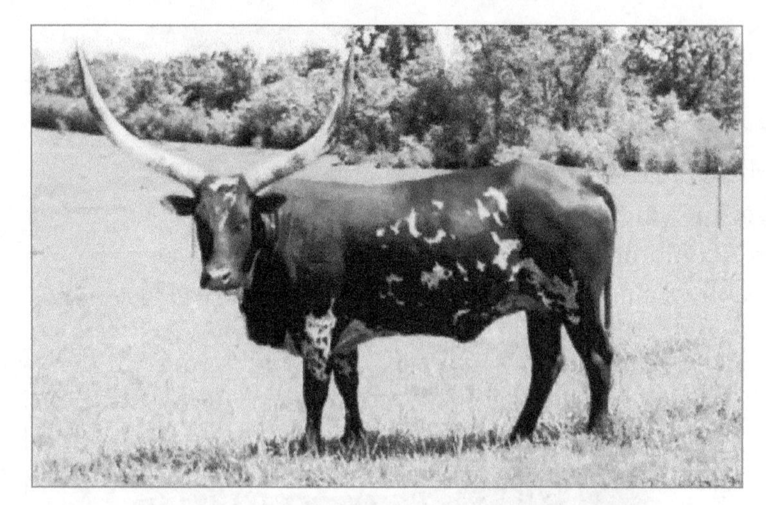

8–9. An Ankole Watusi cow. (Courtesy, Ankole Watusi International Registry)

8–10. A Charolais bull. (Courtesy, American International Charolais Association)

8–11. A Chianina bull. (Courtesy, American Chianina Association)

white, gray, and yellow. This breed has the largest horns of any beef cattle. The horns may reach 8 feet in total length, with a diameter of 9 inches at the base. The horns are said to serve the animal by casting-off excess heat in hot climates. The animals have a slight hump. Many recessive genes result in a wide variety of animals, including some with dwarfism and some that are polled.

CHAROLAIS

The Charolais is an old breed developed in central France. The first Charolais were introduced in the United States in 1936 by the King Ranch in Texas. The Charolais has continued to increase in popularity since its introduction.

Charolais cattle are white or off-white in color with pink skin. They are a large-framed, heavy-muscled breed. Charolais may be either polled or horned. This breed is commonly used in crossbreeding programs to increase frame size and muscle.

CHIANINA

The Chianina breed originated in the Chiana Valley of Italy. It is, perhaps, one of the oldest breeds in the world. Chianina were introduced to the United States in 1971, when semen was first imported.

Chianina are large framed and are commonly used in crossbreeding programs. They have good maternal traits and generally improve the growth rate of the offspring. The color of Chianina cattle varies because of the influx of crossbreeding. Original Chianina cattle were white with a black tail switch and black skin pigment.

8–12. A Gelbvieh cow and calf. (Courtesy, American Gelbvieh Association)

GELBVIEH

The Gelbvieh breed was developed by crossing four yellow breeds of cattle. These four breeds were Glan-Donnersburg, Yello Franconian, Limburg, and Lahn. Together they form the Gelbvieh breed, which originated in 1920.

Gelbvieh cattle vary in color from cream to reddish yellow. In recent years, black Gelbvieh lines have been developed. These cattle are medium framed, produce an acceptable carcass, and have a high milking ability, when compared to other beef breeds.

8–13. A Hereford bull deemed exemplary on its genetic merit, DNA, and profitability information. (Courtesy, Agricultural Research Service, USDA)

HEREFORD

Hereford cattle originated in Hereford County, England. The first Hereford cattle were imported by Henry Clay to Kentucky. These cattle were mated to other native cattle and the first purebred herd was believed to have been established in 1840 in New York.

Hereford cattle have white faces and red bodies. Additionally, they may have a white feather or stripe on the top of their necks, which is commonly called a "feather neck." They are predominately white on the belly, legs, and switch. Herefords are horned, usually very docile and easy to handle, and are moderate in frame size.

Hereford cattle are extremely good foragers. A ***forager*** is an animal that makes especially good use of grasses and plants for food. They will search out food plants on rangeland. Furthermore, this breed is very hardy. The foraging ability and hardiness make this breed

8–14. A Longhorn. (Courtesy, Texas Department of Agriculture)

8–15. A Polled Hereford cow. (Courtesy, Education Images)

8–16. A modern Limousin bull. (Courtesy, North American Limousin Foundation)

popular to the grazing regions of the western United States. It is common to hear Hereford cattle referred to as "whiteface."

LONGHORN

Longhorn cattle were brought to the United States from Mexico by Spanish explorers in the 1800s. The fertility of Longhorn cattle is usually better than many other breeds. They are commonly used in crossbreeding because of their light birth weights.

Longhorn cattle do not grow as fast as some other breeds, but are very durable and adapt well to rough range conditions.

POLLED HEREFORD

The Polled Hereford breed originated in Iowa in 1901. An Iowa breeder named Warren Gammon was instrumental in developing the breed. Gammon located four Hereford bulls and ten Hereford cows that were carrying the polled gene and mated them to develop the foundation for the Polled Hereford breed. Polled Herefords have the same basic traits as Herefords, except that they do not have horns.

LIMOUSIN

Limousin cattle originated in France. Limousin cattle were introduced in the United States in 1968, when semen was imported from Canada.

The Limousin will be either a light yellow or an orange color. The area around the muzzle will generally be a lighter version of the body color. Recently, black lines of Limousin cattle have been developed. Both horned and polled cattle are found in this breed.

Limousins have a moderate, heavy-muscled frame. The breed is noted for its carcass leanness and large loin areas. Therefore, this breed's carcass is usually very high in cutability. ***Cutability*** is the amount and quality of salable retail cuts that may be obtained from a carcass.

MAINE-ANJOU

The Maine-Anjou breed originated in France in the early 1800s. It is thought to have developed through crossbreeding English Shorthorns and French Mancelle breeds. Maine-Anjous were originally used as draft animals. Beef traits were developed in the breed through selective breeding. Maine-Anjou cattle were introduced to the United States in 1970, when semen was imported from Canada.

Maine-Anjou cattle are dark cherry red and white in color and have light pigmentation in the skin. They are commonly used in crossbreeding. From this crossbreeding, black Maine-Anjous have been developed. The breed is usually horned and moderate framed. They are generally docile and easy to handle, grow quickly, and have good marbling in their carcasses.

8-17. A young Black Maine Anjou bull. (Courtesy, Lindsey Broek, Maine Anjoy Voice)

SALERS

The Salers breed was developed in central France and recently brought to the United States. Initial herds were established in 1973, when semen was imported.

Salers are a deep cherry red and, sometimes, will have white on their bellies. Recently, black Salers cattle have been developed. Salers cattle are noted for their calving ease, primarily because of their smaller heads and long slender necks and bodies. Most Salers females milk quite well.

8–18. Salers bull. (Courtesy, American Salers Association)

8-19. A Shorthorn cow. (Courtesy, American Shorthorn Association)

SHORTHORN

Shorthorns originated in northern England around 1600. Shorthorns were originally developed as a **dual-purpose breed**, meaning they were used for both milk and meat production. The breed was introduced to the United States in 1783 in Virginia.

Shorthorn cattle are red, white, or roan. Shorthorns may be either polled or horned. Frame size of this breed is moderate. The females are good mothers and milk exceptionally well. Shorthorns are generally gentle and have good dispositions.

SIMMENTAL

The Simmental breed was developed in western Switzerland and is still quite popular in Europe today. Originally, they had many purposes, including milk and meat production, and use as draft animals. Simmentals were brought to the United States in 1969 from Canada.

8–20. A Simmental cow with her calf. (Courtesy, American Simmental Association)

Simmental cattle usually have a cream-colored face with a light yellow, red, or spotted body. Recently, black strains of the breed have been developed. Simmental cattle may have some color pigment around their eyes. Horned and polled cattle will be found in this breed as well. The Simmental is large-framed, heavy-muscled, and produces a lean carcass.

SOUTH DEVON

8–21. A South Devon cow and calf. (Courtesy, North American South Devon Association)

South Devon cattle were developed in the western region of England. This breed of cattle is believed to have originated from a large, red breed in France, which was brought to England during the Norman Invasion. The South Devon breed is related to the Devon breed. This breed was developed for dual purposes of meat and milk production.

South Devons were first imported into the United States in the late 1930s and early 1940s. However, the popularity of the breed increased in 1970 when a group from Stillwater, Minnesota, started using South Devons for crossbreeding purposes. The South Devon breed will be a medium-red color and will usually be horned.

BRAHMAN

8–22. A modern Brahman cow. (Courtesy, American Brahman Breeders Association)

Brahman cattle were developed in the southwestern part of the United States. The breed is thought to have developed between the mid 1800s and early 1900s and originated from the cattle from India. Bos indicus cattle have a hump on the shoulders and usually have long floppy ears and loose hide. These cattle may also be called Zebu.

Breeders select the Brahman breed because of their hardiness, resistance to insects and diseases, and their heat tolerance, which makes them very popular in the warmer climates of the south and southwestern United States. Furthermore, the females are good mothers, and Brahman cattle do quite well on poor-quality forages common to range conditions.

Brahman cattle are somewhat unpredictable in their temperament. The Brahman breed develops skeletal maturity early and often is used in crossing with other breeds of cattle. Brahman cattle will vary in color from gray to red to black.

8-23. Red and Black Brangus bulls. (Courtesy, International Brangus Breeders Association)

BRANGUS

The Brangus breed is a cross between the Brahman and Angus. This breed was developed at a U.S. Department of Agriculture Experiment Station in Louisiana in 1912.

Brangus cattle will have many of the same characteristics as the Angus and Brahman. They are solid black in color and are polled. (There is also a Red Brangus, which is a different breed from the Brangus.) They also have long floppy ears (not as long as the Brahman) and loose hide. Brangus cattle are very adaptable to various climates and have good mothering abilities.

8–24. A Santa Gertrudis bull, cow, and calf. (Courtesy, King Ranch)

SANTA GERTRUDIS

Santa Gertrudis cattle were developed around 1918 on the King Ranch in Texas. This breed was developed by crossing Brahman and Shorthorn cattle.

Santa Gertrudis are a deep cherry red and may be horned or polled. Because the Santa Gertrudis are part ancestors of the Brahman breed, they have many of the same characteristics, including loose skin and long ears. They are also very adaptable to the warmer climates and are more resistant to disease and insects.

8-25. A Beefmaster cow. (Courtesy, Tommy Perkins, Beefmaster United)

BEEFMASTER

Beefmaster cattle were developed in 1931 in Texas. The breed is the result of crossing Herefords, Shorthorns, and Brahmans. The breed varies in color, and reds and yellowish-whites are common. Beefmaster cattle are hardy and milk well.

TARENTAISE

8–26. A Tarentaise bull. (Courtesy, American Tarentaise Association)

The Tarentaise originated in the French Alps in 1859. The breed was brought to North America in 1973 with importations into Canada. They are a hardy breed. Tarentaise are relatively free of eye cancer and udder burn because of dark pigment around the eyes and on the udder.

Tarentaise are wheat-colored, ranging from cherry to dark blonde. Black pigment is found on the muzzle, eyes, and udder. Females calve easier than most breeds. Bulls are well suited for use with first-calf heifers. Calves are somewhat smaller and vigorous at birth. Bulls may weigh up to 1,800 pounds and cows up to 1,150 pounds. Adults are more active than other breeds. The breed is growing in popularity. Research by Montana State University animal scientists has found that Tarentaise show particularly good performance at higher mountain elevations.

ADVANTAGES AND DISADVANTAGES OF BEEF CATTLE PRODUCTION

Producing beef cattle has advantages and disadvantages. The producer must analyze these and decide if beef production is the right thing to do.

ADVANTAGES

Producing beef cattle may have the following advantages:

- Beef cattle can consume roughages, including most grasses and several legumes, which other livestock may not eat or be able to use. Therefore, little forage is wasted with beef cattle.
- Labor requirements are usually not as great with beef cattle as with other livestock production. Because beef cattle are good foragers, they do not necessarily have to be fed daily.
- Death losses are relatively low if good management practices are followed.

- Beef cattle adapt well to various sizes of operations. Productive beef herds can be from one or two head to several thousand head.
- There is a good demand for meat. Beef is a mainstay in the American diet. Additionally, meat is an important export to other countries.
- The beef industry creates many jobs. It relies on services, such as herdsmen, trucking firms, processors, researchers, and feed suppliers.
- Beef cattle may provide avocational interests, as well as youth projects through FFA and 4-H, which allow new experiences for both young and old.

DISADVANTAGES

Producing beef cattle has the following disadvantages:

- If cattle are fed to slaughter weight, the producer incurs risk. Grain prices tend to fluctuate and available forage is dependent on precipitation for a specific area. An example of this would be whenever a region suffers a drought. Grain costs more because yields are lower and forages do not grow readily because of the lack of precipitation.
- Beef cattle do not convert feed and forages into meat as efficiently as do other animals. (However, beef cattle do make efficient use of forages that have little use for other purposes.)
- Because of the 283-day gestation period of beef cattle, it takes longer to increase herd size. The producer will usually only get one offspring a year from each cow.
- Capital investment is high to start a modern beef operation. Not only does the producer have to invest in the cattle, but also equipment, facilities, and herd management.
- Droughts can result in little forage growth on land. The result may be that cattle have short or no grazing. The animals must be fed "bought" feed or sold to prevent starvation.

TYPES OF BEEF PRODUCTION SYSTEMS

Beef cattle production varies among producers. The kind of system depends on the producer's goals. It is also influenced by the resources available and the market in the local area.

COW-CALF SYSTEMS

A ***cow-calf system*** is keeping adult cattle to raise calves for sale to other growers. Most cow-calf systems involve high-quality grade cattle. The cows are bred to calve in the late winter or early spring. The calves are sold in the fall. This takes advantage of the warm summer growing season. A few producers use fall calving in September to November.

Natural breeding is typically used with grade cow-calf systems. The bull may be put in the pasture with the cows at breeding time. A producer schedules breeding to assure calving at the desired time. One three-year-old bull will breed 25 to 40 cows a year.

Some beef producers use artificial insemination (AI). AI is more important with expensive, purebred females. This allows the producer to have access to a wider variety of bulls.

Heifers need to be the proper size and age before breeding. Most heifers will have their first calf at two years of age. Since the gestation period of a cow is approximately 283 days, a cow should have one calf a year. Occasionally, cows will have twins or multiple births, but this is not prevalent in beef.

The cow-calf system uses forages for feed. Several kinds of plants are grown in pastures for forage. Both grasses and legumes are used. Pasture grasses include fescue, orchard grass, or Bermuda grass. Legumes include several kinds of clovers, lespedeza, vetch, and alfalfa. Pastures should be kept free of weeds, especially those that are poisonous. If pasture is not available, hay or silage can be fed as a forage, but this is not typical with beef cattle production.

An additional mineral supplement should be fed free-choice to meet the need for minerals, which are vital to beef cattle. Forage alone may not provide sufficient nutrients. Calves that are a few months old may be supplementally fed with grain. A method known as creep feeding is used. ***Creep feeding*** is using equipment that only allows small animals to enter. Creep feeding will allow the calf to gain more weight while it is still nursing the cow. Creep feeding is usually done when the calves are not receiving enough milk from the cow, the supply of forages is limited, or when grain prices are low and it is economically feasible.

Producers of purebred cattle follow a somewhat different cycle. They raise herd replacements and animals for other cattle breeders. Purebred cattle are not raised to go to a feedlot.

8-27. This cow-calf producer has grade cattle of Hereford breeding. Note the portable watering system, back rubber, and other management practices. (Courtesy, Natural Resources Conservation Service, USDA)

8-28. Finishing systems involve providing feed concentrates to achieve rapid growth and fattening. (Courtesy, Education Images)

Backgrounding Systems

A ***backgrounding system*** is an approach that takes a calf from the time it is weaned from the cow to the feedlot phase. Backgrounding is essential to add more weight to the calf before it enters the final feeding phase.

During the backgrounding, forages, such as annual grasses, are commonly used to add more weight to the calf. Supplemental grain may also be fed to the calves to increase the rate of gain and prepare the calves to enter the finishing phase earlier. Calves that are backgrounded will also need a free-choice mineral supplement.

Finishing Systems

A ***finishing system*** is an approach that takes a calf through the complete grow-out phase of beef production. The calf is fed to harvest weight, which is usually 1,100 to 1,300 pounds, depending on the breed.

Some beef producers feed calves to harvest weight, but most cattle will be finished by commercial feedlots. The feedlots are located in the western and southwestern United States. During the finishing period, calves are fed feed concentrates. These are feeds high in

CAREER PROFILE

RANCH MANAGER

A ranch manager oversees the operation of a ranch. Ranches vary in production; many focus on beef cattle. A ranch manager often supervises other workers, observes cattle, schedules feeding, performs health practices, and carries out other duties in the successful production of beef animals.

Ranch managers need practical experience with cattle or other species being produced. A high school education is essential. Many have college degrees in animal science or a related area.

Jobs are found on ranches and other facilities that have cattle such as a research station. This photo shows a ranch manager and ranch hands branding a calf. (Courtesy, U.S. Department of Agriculture)

8-29. The goal of finishing is to produce beef carcasses with desired quality—fat deposits, including marbling (presence of intramuscular fat which are observed as tiny fat particles in lean muscle tissue). (Courtesy, Pavzyuk Svitlana/ Shutterstock)

energy from grains, such as corn, milo, and oats. Free-choice minerals may be provided if the minerals are not part of the finishing ration.

HERD HEALTH

A good herd health program is essential in all phases of beef production. Prevention, including good sanitation and vaccination programs, is key to maintaining good herd health. Diseases and parasites greatly reduce the profitability for the beef producer. A good health program will also aid in marketing beef animals. (Chapter 4 covered animal health in more detail.)

8-30. Animal specialists examine a cow for the presence of external parasites. (Courtesy, Agricultural Research Service, USDA)

PARASITES

Beef cattle are affected by both external and internal parasites.

Common external parasites are flies, mosquitoes, lice, mites, and ticks. Good sanitation practices with facilities and equipment will reduce problems with these external parasites. In addition, external parasites may be controlled by using insecticides.

Only approved products should be used in controlling external parasites. Some producers consult veterinarians for parasite control. In general, insecticides may be applied by spraying or dipping the cattle. Backrubbers may also be used when spraying or dipping are not options. A backrubber should be placed in a high-traffic area, such as near feeders and waterers. The insecticide is poured on the backrubber. As the cattle walk under the backrubber, the insecticide is applied.

Internal parasites, such as roundworms and flatworms, must be controlled. Common symptoms of internal parasites are weight loss, poor weight gain, rough hair, and diarrhea.

An important way to prevent worm infestation in cattle is to provide good nutrition and good grazing conditions. Pasture management is a key with cow-calf and other production systems. Cattle with worms will excrete worm-forms in their feces. Worm larvae are then on and in the top soil. Cattle that graze short pastures are more likely to contact the ground with their mouths. Overgrazed pastures have short forage so that the mouths of cattle contact the ground when grazing.

A management practice is to provide low-risk pastures. These are pastures with little or no infestation from infected cattle. Counts of larvae in a pasture can provide some indication but performing counts is not always easy. Scientists sometimes place a worm-free animal on a pasture to graze for a while and then determine if eggs or larvae are being excreted.

8-31. Estimating the number of brown stomach worms in a pasture may involve researchers placing worm-free calves old enough to graze for a measured time on the pasture. The feces is then checked for parasite eggs. (Courtesy, verityjohnson/Shutterstock)

Anthelmintics are chemical compounds used for deworming animals. Anthelmintics may be administered to cattle orally, by injection, by pour-on, or by adding to the feed or minerals. A regular worming schedule should be followed. Best results are achieved if cattle are dewormed once or twice a year. A veterinarian may be consulted to decide which anthelmintics to use.

PATHOGENIC DISEASES

Preventing disease is important in beef cattle production. Several diseases may be problematic with beef cattle. These include blackleg, Bangs (brucellosis), bovine virus diarrhea (BVD), foot and mouth disease, anaplasmosis, mastitis, and shipping fever. Careful management helps to assure good herd health. A veterinarian may be used to guide this phase of management. (Most of diseases were covered in Chapter 4.)

In the 1970s, concern developed in the beef industry about Mad Cow Disease, which is properly known as bovine spongiform encephalopathy (BSE). The word, "spongiform," is used because the brain of cattle with the disease has a spongy appearance when viewed under a microscope. It is also an issue with dairy cattle that are used as beef. Worldwide, thousands of head of cattle were euthanized and millions of pounds of harvested beef were destroyed because of possible contamination. Beef cattle producers in the United Kingdom were particularly hard hit. Since a peak in the 1990s, the number of BSE cases has declined markedly in recent years. Very few have been in the US.

Infected animals have degenerative central nervous systems that affect body movements and mobility. Cattle acquire BSE by eating feed contaminated with the infectious BSE agent; therefore, do not use feed containing meat scraps or tankage. The Food and Drug Administration has enacted rules prohibiting the use of mammalian protein in ruminant feed.

Preventing BSE in cattle is essential as humans may get forms of the disease from eating beef from infected animals. Evidence is that cooking beef will not destroy the BSE agent. There is no treatment or cure for BSE. Infected animals must be disposed of by approved veterinary personnel. Research into BSE is continuing.

Infectious bovine rhinotrachetis (IBR, red nose, IPV) has several forms. Symptoms may include fever, discharge from the nose, rough breathing, coughing, weight loss, inflammation of the genitals, watery eyes, convulsions, and incoordination. Since the symptoms are quite varied, a veterinarian should be consulted if IBR is suspected.

Leptospirosis is caused by several strains of bacteria. Symptoms may include a rise in temperature, rapid breathing, stiffness, loss of appetite, bloody urine, and abortion. All animals should be vaccinated twice a year to control the disease.

Campylobacteriosis or vibriosis may occur in intestinal or venereal form. The intestinal form usually does not have harmful effects on beef cattle. The venereal form is more serious. The symptoms may include infertility, abortion, erratic heat periods, and poor conception rates due to early embryonic death. Animals should be vaccinated before breeding and the vaccination should be repeated annually.

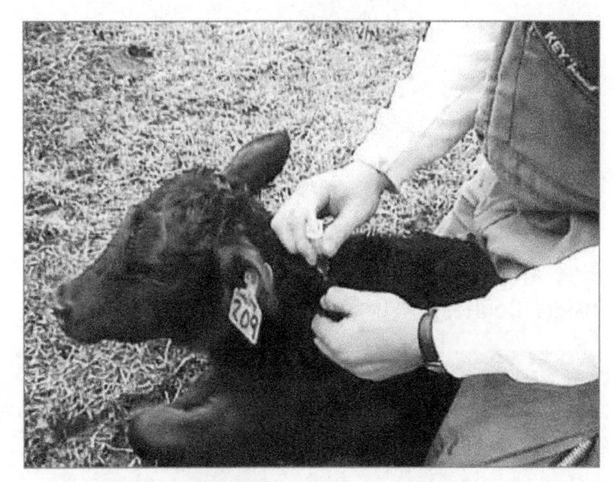

8–32. Baby calf receiving its first injection. (Courtesy, Thunderbolt Angus)

ANIMAL IDENTIFICATION

Most beef animals need to be identified in some way. Proper identification is important in managing the beef herd, as well as specifying ownership. It allows the producer to identify all animals and keep records on production. Identification helps follow an animal from the

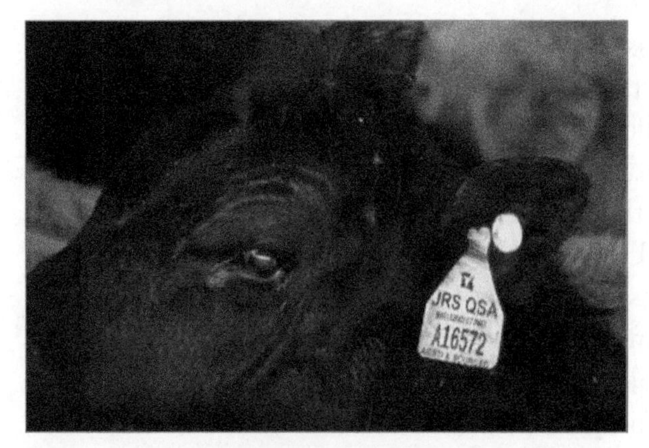

8–33. A source verification tag aids in identifying the ranch that produced this calf. (Courtesy, Joann Pipkin, Joplin Regional Stockyards)

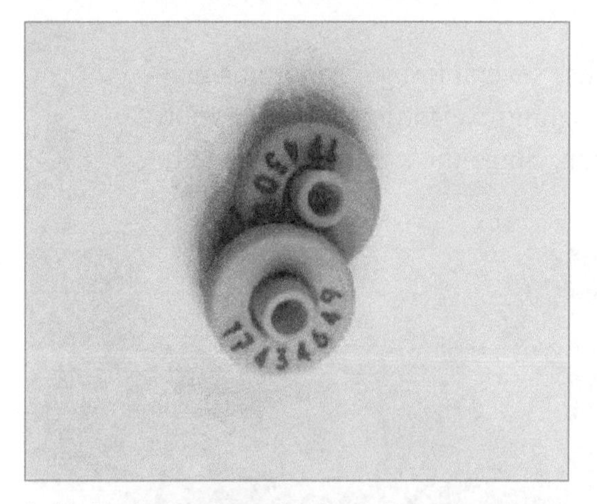

8–34. An electronic ear tag has information stored inside. (Courtesy, Allflex USA)

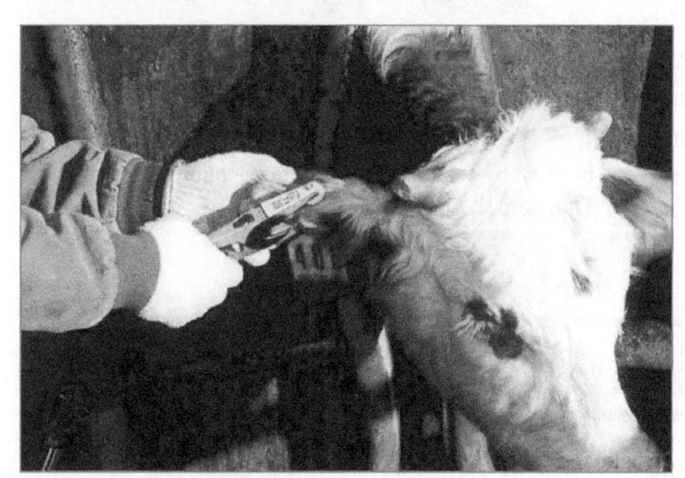

8–35. Using an applicator to insert an electronic ear tag. (Courtesy, Allflex USA)

producer through marketing. Newer methods are being developed for this purpose. A leader in this effort is Allflex USA. This company operates a research facility and collaborates with producers.

Some beef animals will have permanent tags or other permanent identification. When the animal has a permanent tag, it can be tracked from the producer all the way through to the eventual harvesting of the animal. Being able to track an animal completely through the marketing system allows for greater food safety, greater quality control, and disease control. Several methods may be used.

- Tags—Tags are usually placed in the ears of animals, though some are on chains around their necks. Typically made of plastic, tags have identifying numbers or letters. Tags are an aid in record keeping. Some tags allow electronic tracking of animals using electronic codes that are sensed by a scanning device hooked to computers. Source verification tags may also be used to identify origin of the animal. This record keeping helps in accountability of animal health and quality.
- Branding—***Branding*** is a permanent method of identification. Branding may be with a hot iron or by freeze branding. The producer needs to select a brand design that will represent ownership of the cattle. Brands are usually registered with a state agency. A hot iron brand will burn the surface of the skin, leaving a permanent mark. When freeze branding, liquid nitrogen is used with the iron, which freezes the skin pigment. This depigments the hair follicles and allows the hair to grow back white in color. In white-haired cattle, the liquid nitrogen is

left on a longer time. This produces a bald spot the shape of the branding iron. (See brand example in 8-37.)

- Earmarks—Earmarks may be as simple as using a knife or an earnotcher to cut a notch in the ear. Tatoos may also be placed in the ears of some animals. With earnotches, a system is developed based on location in the ear using the left and right ears for the notches.
- Neckbands—Neckbands or chains may also be placed on animals for identification. Again, animals may tear the bands or chains or someone other than the owner may remove them. In rare cases, an animal may catch the chain or band on an object, which could possibly choke the animal to death.
- Tatoos—A tattoo is a method of identification frequently used among purebred breeders. An instrument applies letters and numbers by piercing the skin. Ink is then rubbed into the area. A common place for tattoos on beef cattle is the inside of the ear. A drawback is the animal must be confined for the tattoo to be read. Also, some tattoos are hard to read on dark-skinned cattle.
- Other methods—Several other methods of identification are available, including nose prints, DNA testing, and microchips. Nose prints and DNA testing are used to establish the precise identity of an animal. These procedures are often used with show animals. Microchips may be placed beneath the skin of an animal. A scanner will decode the signal from the microchip. The information identifies the animal. Microchips are not appropriate in animals to be slaughtered. Removing the microchip from the carcass may pose a problem.

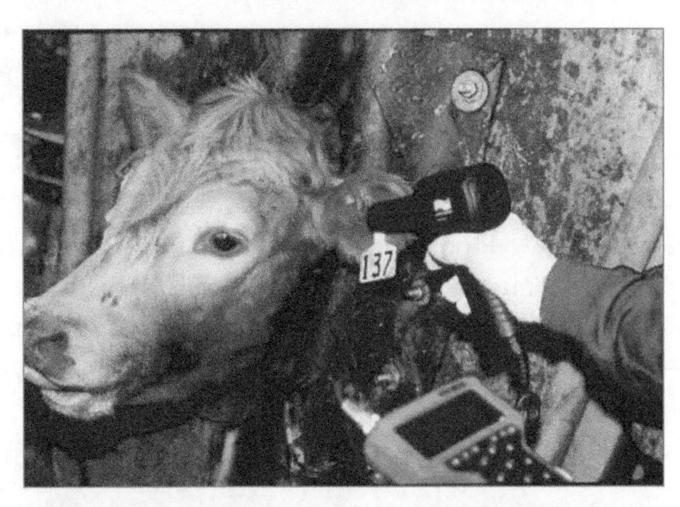

8–36. A handheld reader emits a radio signal that detects the identification number programmed into an electronic ear tag. (Courtesy, Allflex USA)

8–37. Hot brand. (Courtesy, Tommy Perkins, Southwest Missouri State University)

COMMON HERD MANAGEMENT PRACTICES

Producing quality beef animals is more than turning them out in a pasture. The animals must receive the proper care. Care is known as management practices. Several areas of care include providing feed and water, breeding, disease control, worming, and animal identification. Animals should be observed each day. This is to check for injury and disease as well as calving by pregnant females. Two additional practices are included here.

8–38. A calf dehorned with a caustic chemical that prevents the growth of horn tissues. (Courtesy, U.S. Department of Agriculture)

DEHORNING

Dehorning is used with horned breeds of cattle to remove horns. Large horns can be dangerous and injure other animals, as well as people.

Several dehorning methods are used. It is preferable to dehorn young calves. Dehorning larger animals is more difficult as well as traumatic to the animal.

With calves, caustic chemicals may be placed on the location where horns grow. This prevents horn growth by destroying tissue. It is widely used on younger animals.

Horns may also be sawed or clipped off. This is a surgical procedure that removes the horns. When using this method, all bleeding needs to be stopped afterward. A repellent should be used to keep flies off the wound.

CASTRATION

The beef producer will usually not want to keep all male offspring as bulls. These calves are castrated.

Castration is removing the testicles or destroying their development. Castration makes male calves easier to handle and manage. A calf castrated at a young age is a ***steer***. Except for breeding animals, steers often bring higher prices than bulls. Castration also prevents bulls from breeding any females that are present once puberty has been reached.

A common method of castrating cattle is with an ***elastrator***. This stretches a specially made rubber band over the scrotum, which will cut off circulation. The scrotum will essentially dry up and fall off. This method should be done on younger, smaller calves.

A knife or a scalpel may be used to cut the testicles from the scrotum. This is a popular method. Bleeding must be stopped and flies controlled. Additionally, castrated calves need to be watched closely for infection.

A Burdizzo clamp is also used. No bleeding occurs with this method. The clamp crushes and severs the cords and blood vessels going to the testicles. This causes them to dry up. When using the Burdizzo, it is important that the cord does not slip out of the clamp and miss being crushed.

8-39. Using a Burdizzo to castrate a calf by severing cords and blood vessels to the testicles without bleeding. (Courtesy, Claudia Otte/Shutterstock)

FACILITIES AND EQUIPMENT

The facilities and equipment needed for beef cattle depend on the system of production and climate. Factors that assist in planning include the number of cattle, amount of space, location, kind of facility, herd and property security, environmental conditions, amount of money to invest, and availability of other facilities and equipment.

FACILITIES

Facilities may range from simple to elaborate. Producers should allow for flexibility so they can expand or change beef production systems.

Because beef cattle are primarily range animals, it is not necessary to have a great amount of shelter. However, barns and loafing sheds should be available to provide shade in the summer if trees or other shade sources are not available. Additionally, barns and loafing sheds may be used to protect cattle from harsh weather.

All beef cattle producers need fences to keep animals confined. These fences must be kept in good condition. Some fences are of barbed wire while others use a meshed wire. Electrically charged fences are used on some farms and ranches. Board and vinyl material are also used, especially in public areas.

8-40. High tensile wire fences are gaining in popularity for cattle and horses. The wire has no injurious barbs. It can be used in conjunction with an electric fence system. The inset shows a wire strainer that is used to maintain tension on the wire. (Courtesy, Education Images)

EQUIPMENT

Cattle production requires various equipment. The kind needed depends on the type of animal produced. Water tanks and water fountains need to be available for all cattle.

Because most producers supply salt and mineral free-choice to beef cattle, feeders need to be available. These feeders should be large enough to allow several head to consume salt and mineral mix at the same time.

8–41. A properly designed corral. (Courtesy, University of Missouri Extension Service)

Feed storage is needed to assure high-quality feedstuff and reduce the amount of feed wasted by spoilage and pest damage. Feed may be stored in upright silos, trench silos, metal storage bins, or feed rooms inside barns.

Feeders for hay, forages, and grain need to be large enough to accommodate all cattle at once. This may require the producer to use more than one feeder. This alleviates the dominant, aggressive animals from keeping other cattle away.

Equipment to handle cattle is needed in all beef operations. A good corral makes it easier to work cattle. Corrals should also decrease labor and time, reduce stress on cattle, decrease chance of injury and weight loss, and increase safety for the producer.

Corral designs vary. In planning a new corral, allow for expansion in the future. Sharp turns and corners should be avoided. The corral should be built so cattle can be easily moved and trailers loaded.

Working chutes and headgates are used to hold and restrain cattle for tagging and vaccinating. There are many models of chutes and headgates. Select the design that best allows efficient cattle handling.

8–42. A squeeze chute for working cattle. (Courtesy, University of Missouri Extension Service)

FEED RESOURCES

Beef cattle must receive proper nutrition in order to grow, reproduce, and be healthy. Nutritional requirements vary with age and other factors in an animal's life. With most producers, animals are fed as a herd or group rather than individually. Chapter 3 covered animal nutrition and feeding. The focus here is on specific practices in beef cattle production, particularly the use of forages in pastures.

The types of feed vary for beef cattle depending on the beef operation or system. Grains and forages are the primary feeds for beef cattle. Common grains used to feed beef cattle are corn, oats, and barley. Soybean and cottonseed meals may be used to add protein to the grains fed to the cattle. Most rations for beef cattle will ranges from 12 to 16 percent protein. Cattle being finished for market in a feedlot will typically receive a diet high in grain and other concentrates to promote growth. Some roughage will also be available.

FORAGES, SUPPLEMENTS, AND WATER

Beef cattle need forages. And, they have digestive systems that are uniquely equipped to handle forages—the ruminant digestive system. The success of a beef operation, particularly cow-calf operation, is tied directly to the amount and quality of forage available. Common forages utilized by beef cattle are grasses, legumes, hay, and silage. Beef cattle can utilize these forages as their sole source of nutrients or a grain or protein supplement may be fed to

8–43. Cows grazing a quality grass pasture. (Courtesy, Natural Resources Conservation Service, USDA)

assure balanced nutrition. Beef cattle consume forages by grazing in pastures on grasses such as fescue, orchard grass, rye grass, wheat, and native grasses. Clovers and other legumes may be planted with the grasses. During times when grasses are limited (such as in the winter), beef cattle need to be fed hay or silage as a source of roughage.

In some cases, a mineral supplement may be provided. Such supplements are used if cattle do not gain adequate minerals from their forages. The minerals needed most are calcium, phosphorus, iron, iodine, magnesium, sodium, and potassium. Though not needed in large amounts, deficiencies of minerals may lead to poor performance and health deterioration of beef animals.

Most cattle producers use permanent pastures or ranges. These pastures may be seeded to improve varieties adapted to the local area. Sometimes a cool season forage variety is seeded into permanent pastures in the fall to help provide late winter and early spring grazing.

As with all animals, quality drinking water is needed by beef animals. The water should be free of harmful substances such as pesticides and of disease agents that could be ingested with the water and cause health problems. The average adult animal will drink 10 to 20 gallons of water a day. It should be available free access so that animals can drink as they wish.

GRASS-FINISHED BEEF

Grass-finished beef is meat from an animal beef animal that has been raised on a forage diet its entire life. The animal has not been fed grain to enhance finish. Concentrates are not fed. Grass-finished beef is also known as grass-fed beef or pasture finished beef.

Pasture quality is extremely important with grass-finished beef. Keeping young, tender forages may prove to be challenging to the producer. Therefore, increased attention to pasture management is needed. The vegetative growth stage of forage plants is a factor in the nutrition gained by grazing animals. Overseeding with temporary or seasonal forage plants may be beneficial. Seasonal growth of forage plants must be considered. In some areas, forage plants are tender and more nutritious in the spring and early summer. They lose nutrition in the summer and early fall as they develop greater amounts of dry matter.

Rotational grazing may be used. This is the practice of allowing animals to graze in a pasture for a few days or weeks before moving the animals to another pasture where the forages have been growing without grazing. This requires cross-fencing of pasture areas.

Most finished beef animals go to market at weights of 1,000 to 1,250 pounds. Grass-fed animals will require longer to reach this stage of development than with animals in feedlots receiving grain and other concentrates.

Before initiating a grass-finished beef production operation, be sure that a market is available for the animals. Of course, this is true with any beef production enterprise.

SHOWING BEEF CATTLE

Showing beef cattle can be fun. Showing beef cattle allows producers to promote their cattle to prospective buyers and other breeders. It also allows breeders to compare their beef cattle with other beef breeders' cattle. For 4-H, FFA, and other junior members, showing beef cattle (usually haltered heifers or steers) allows them to learn about the promotion of their individual breed of beef animal. Furthermore, showing beef cattle creates a sense of responsibility by properly caring for, grooming, and showing the animal.

The well-being of show animals is very important. Never abuse an animal. Always keep the animal in conditions that are favorable to its well-being.

Learn more about animal well-being and showing by attending local seminars or field days designed for people who show animals.

ANIMAL SELECTION

Show animals need to be carefully selected. Use a highly qualified show person to help in the selection process. Get the animal from a producer who consistently has quality cattle. Some cattle producers will hold special sales of calves that have the genetic potential to be good show cattle. A quality animal is needed to be competitive in the show ring.

In selecting an animal for show, you need to know the dates of the shows you wish to enter and the deadlines for ownership of the animal. Consider how long it will take to get the animal ready for showing. Most steers are shown at 16 to 20 months of age. Heifers may be shown up to 24 months of age.

Steers and heifers are typically on feed for 270 days. This means that if a show is in January, the animal would go on feed about May 1 of the previous year and be 6 to 10 months of age. Typically, the animal chosen would weigh 400 to 600 pounds. Further, the animal would have been born the previous summer to reach the age for feeding. Beef animals gain 2.0 to 3.5 pounds a day while on feed. The animal you show will likely weigh 1,100 to 1,300 pounds.

ANIMAL CARE

An animal being raised for showing must receive proper care. Feed must be appropriate to the desired gain. Facilities should be used that provide for the well-being of the animal. Get the assistance of a person who is an authority on caring for show animals. Local feed dealers often have the proper feed.

In general, feeds high in energy are needed to prepare a steer or heifer for showing. Corn is the best grain. Barley, oats, and sorghum grain may partially substitute for corn. Grain should be of high quality. It should not be infested with insects or spoiled. The amount of grain to feed per day is about 2.25 pounds, with above this amount leading to increased fat on the animal. Somewhat less will result in a leaner animal.

A protein supplement will also be needed. Cottonseed meal, soybean meal, and linseed meal are often used. Small amounts of fish meal, dried blood meal, and others may be used to have a better balance of amino acids. Younger, lighter animals need slightly more protein in their rations.

8–44. Shelled corn is a major ingredient in show calf feeds when cracked, crimped, or rolled. (Courtesy, Education Images)

Beef animals will need roughage. Cottonseed hulls are often suggested for animals being raised for showing. Some good quality grass hay can be fed. Hay tends to keep down digestive system problems. It may be placed in a rack and available for the animal day and night.

A little molasses is sometimes used to make the concentrate ration more palatable. Vitamin supplements, mineral supplements,

8–45. Cottonseed hulls may be used as a roughage with calves. (Courtesy, Education Images)

and feed additives may be used. Be sure to get the help of a local expert. Do not feed anything that is harmful to the animal or restricted.

Beef calves will typically consume 2 to 3 percent of their body weight in feed each day. The weight of the feed is calculated without water (dry matter basis). Feed intake on a percentage of weight basis tends to decrease as an animal grows. The feed can be weighed and fed individually once a day (or half the amount if twice a day) in a trough protected from rain and contamination. Feeding allows you to check on the animal and bond with it for training and showing purposes. It also helps see how well the feed is consumed and if the amount given is sufficient. You want your animal to have plenty.

Growing animals need good, clean water. Water should be available all of the time on a demand basis so that animals are able to drink all they want.

Most show calf growers find it easier to use a locally-available commercial feed.

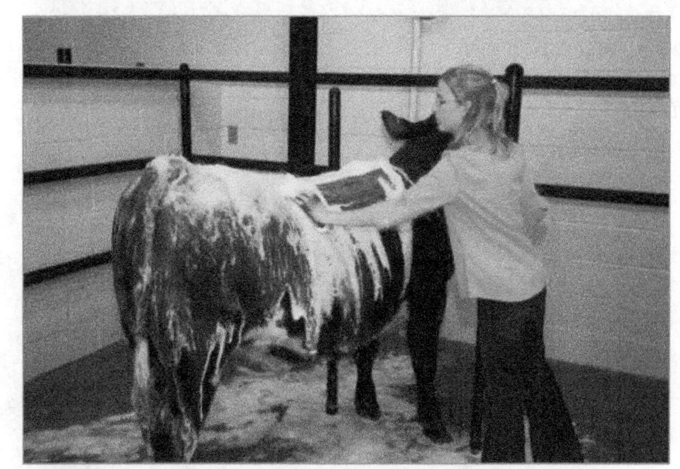

8–46. A first step in preparing an animal for showing is to wash the animal. (Use an approved cattle shampoo mixed with water.) (Courtesy, Education Images)

HALTER BREAKING

For an animal to be shown, it must be halter broken to stand and lead. ***Halter broken*** refers to an animal that is comfortable with a halter on its head and responds to the commands of the person who is leading it. Many hours may be needed to halter break an animal.

Begin halter breaking by tying the animal with a rope halter for short periods. Brushing the animal while it is tied up will help to calm the animal. It is important not to leave the animal unattended during this period to avoid injury to the animal. After the animal has been tied up several times, the next step is to lead the animal to water or to feed. This should be done in a secure area of a corral so, if the animal gets away, it can be caught again and tied. After the repetition of being led back and forth to feed and water, the animal will soon be accustomed to being led. In most cases, younger animals will be easier to break to lead.

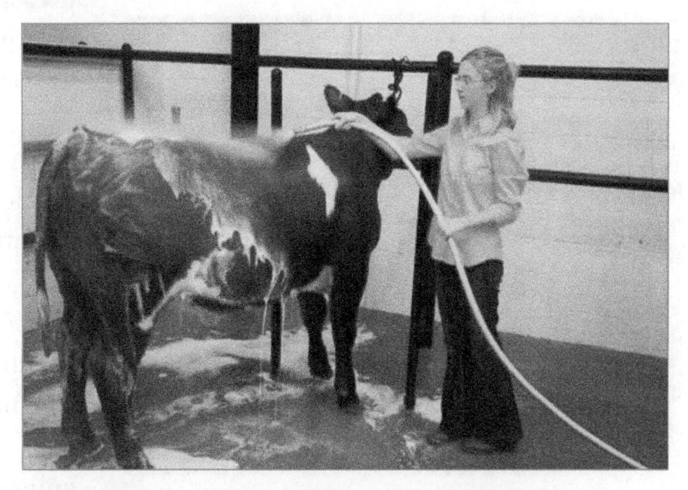

8–47. Use clean water to thoroughly rinse the shampoo out of your animal's hair. (Courtesy, Education Images)

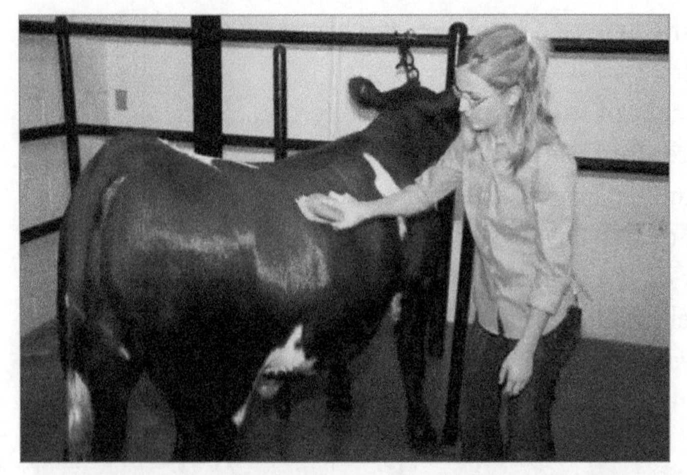

8–48. After washing and while wet, comb and brush the hair of your show beef animal. (Follow this by blow-drying the hair.) (Courtesy, Education Images)

GROOMING

Grooming, or fitting, the animal is a very important part of showing beef cattle. The initial step is to train the hair of the animal. This may be done by washing the animal at least once a day, then brushing and drying the hair on the animal. The next step is to clip the animal. This involves clipping the long hair off the animal to highlight its strengths. The areas of the animal that may be clipped are the head, neck, underline, body, and tail. The hair may be left longer on the hind quarters and the body of the animal to make it look thicker and bolder. Products to hold hair in place, such as adhesives, may not be used at most shows.

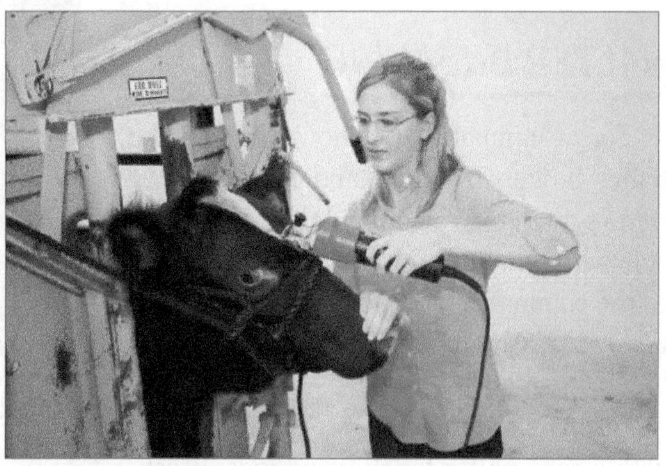

8–49. Use clippers to remove unwanted hair and gain desired characteristics. (Courtesy, Education Images)

SHOWING

8–50. A Charolais heifer is being fitted for showing. (Courtesy, J&S Cattle Company)

How an animal is managed in the show ring is important. When the animal is fitted and ready to go to the show ring, the rope halter should be replaced with a leather show halter. Upon entering the show ring, the exhibitor should pay attention to the judge of the animal. When asked to stop, position the animal's feet with the show stick. The animal should stand squarely and evenly distribute its weight on all four legs. Exhibitors should always be aware of where the judge is positioned and how their animals are positioned. Considerable practice is needed to develop skills in showing an animal.

8-51. This young man's heifer was selected best in its class at a county livestock show. (Courtesy, Education Images)

8–52. Heifers competing in a show ring. (Courtesy, American Gelbvieh Association)

REVIEWING

MAIN IDEAS

Most beef cattle breeds originated in Europe or Great Britain. Beef cattle are produced by a series of systems: the cow-calf, backgrounding, and finishing. All of these systems are beneficial to the beef producer.

The producer has to follow sound management practices to be successful in producing beef cattle. These practices will increase profitability, decrease loss from disease, and reduce stress to the animals. Proper management practices will allow the beef producer to have higher-quality beef cattle.

The promotion of these quality animals may be through showing the animals at local, district, and national level shows.

Proper facilities and equipment are necessary for a beef cattle operation. Properly planned facilities will decrease labor and make beef operations more efficient. Designing facilities that will allow for expansion is important to facility design.

Equipment varies from one operation to another but is needed in all beef production. Corrals make it safer and easier to handle cattle, as well as reduce stress on the cattle. Watering equipment is important for most operations. Feeders and feed storage equipment will insure higher-quality feed.

Additional detailed information on beef cattle production is available in Beef Cattle Science available from Prentice Hall Interstate.

Beef cattle are ruminants and make good use of forage. Quality pastures and hay crops are quite useful in their production. An increasing area of demand is for grass-fed beef.

Some individuals enjoy showing beef animals. It allows animal owners to compare their animals with the animals of other producers. Showing requires attention to animal selection and providing proper care to assure growth and development. Halter breaking (training) and grooming skills are needed. Managing an animal in the show ring requires an understanding of animal behavior and expectations of the judge.

QUESTIONS

Answer the following questions using complete sentences and correct spelling.

1. Distinguish between the Bos taurus and Bos indicus cattle.
2. List advantages and disadvantages of raising beef cattle.
3. What should be considered in selecting a breed of cattle?
4. What beef production systems are used?
5. Name and describe two important herd management practices.
6. What facilities and equipment may be needed in beef cattle production?
7. How are beef cattle identified?
8. Select any three breeds of cattle and provide a brief description.
9. What steps should be followed when preparing a beef animal for a show?
10. What is the purpose of using source verification tags?
11. What is the importance and use of EPDs?
12. What is grass-fed beef?

EVALUATING

Match the term with the correct definition. Write the letter of the term on the line provided.

a. dual purpose
b. cutability
c. polled
d. anthelmintics
e. forager
f. halter broken
g. steer
h. Bos taurus
i. Bos indicus
j. backgrounding

______1. Chemical compounds used for deworming animals.

______2. Beef cattle that naturally do not have horns.

______3. Male calf castrated at a young age.

______4. Cattle used for both meat and milk production.

______5. Cattle that are from India and have long floppy ears and loose hide.

______6. The amount of salable retail cuts obtained from a beef carcass.

______7. A system that takes the calf from weaning to the feedlot phase.

______8. Cattle common to the United States of European origin.

______9. An animal that is comfortable with a halter on its head.

_____10. Makes good use of grasses and other plants for food.

EXPLORING

1. Prepare a computer-based presentation on a breed of beef cattle. Include pictures and descriptions of the breed. Each student in class should choose a different breed and provide a report to the class.

2. Construct a bulletin board that depicts the components of the beef industry. All components should be identified and pictures should be included.

3. Tour various beef operations in your area and prepare a report on your observations.

4. Make a field trip to a livestock show. Identify the breeds that you see. Describe the conformation of the animals. Note how they are fed, bedded, watered, and groomed.

5. Use the following Web site to explore different breeds of beef cattle: www.ansi.okstate.edu/breeds.

6. Investigate Beef Quality Assurance. Determine the "BQA Code of Cattle Care" and other details. Include *The Cattle Industry's Guidelines for the Care and Handling of Cattle* in your investigation. Identify the BQA Coordinator for your state. Prepare a report on your findings. For details, go to the BQA Web site: www.bqa.org.

7. Review the registration papers and EPDs of two different bulls of the same beef breed. Identify which bull would increase growth rates in a beef herd. Prepare a brief report on your findings.

CHAPTER 9

Swine Production

OBJECTIVES

This chapter has basic information on swine production. It has the following objectives:

1. Explain the importance of swine production.
2. Describe swine as organisms.
3. List and describe common breeds of swine.
4. Explain the possibilities of pork production.
5. Describe pork production systems.
6. Explain important management practices in swine production.
7. List nutritional requirements of swine.
8. Explain additives and withdrawal time.
9. Explain health management practices for swine.
10. Describe facility and equipment needs for swine.
11. List and discuss major factors in raising a show pig.

TERMS

additive
barrow
boar
contract production
farrowing
feeder pig
finishing
gilt
meat-type hog
needle teeth
pedigree
piglet
porcine somatotropin
porcine stress syndrome
probe
prolific
sow
specific pathogen free
tail docking
type
ultrasonics
withdrawal time

9–1. Video equipment is being located above a sow with pigs. (This will help the producer to remotely observe, record, and manage behavior.) (Courtesy, Agricultural Research Service, USDA)

HOW hogs are raised has changed and will continue to change. Why? A big reason is consumer demand. In order to meet consumer demand, new bloodlines, feeds, vaccines, equipment, and other developments are being used in swine production. These new developments are also intended to make production more efficient and profitable for the producer.

Swine are popular because of the meat, known as pork. Pork consumption has leveled off and slightly declined since the year 2000. Per person consumption is now approximately 47.2 pounds of harvested pork a year in the United States. Among nations, the U.S. is 12th in pork consumption. Denmark is the leading nation followed by Spain, Hong Kong, and Germany.

New approaches with swine production help to assure a high quality product. The Pork Quality Assurance Program (PQA) is a management education program for adults and youth. Pork harvest facilities are increasingly only buying hogs from producers who have completed PQA certification. Practices are followed to assure a quality pork product.

IMPORTANCE OF SWINE PRODUCTION

9–2. A roasted pork loin (with potatoes) represents the quality product available today in the United States. (Courtesy, ElenaGaak/Shutterstock)

Swine have been produced for a long time. They were first domesticated in Asia about 9,000 BC. Swine were brought to America by Christopher Columbus in 1493. North America already had wild hogs, often known as boars. They had been used only as hunted game animals.

Swine brought to America were of European and Asian breeding. Many changes in these animals have been made to provide desired products. New breeds were developed. Today, the breeds have given way to crossbred animals.

SWINE INDUSTRY

The swine industry closely parallels the production of corn. Three-fourths of the hogs produced in the United States are produced in the Corn Belt. When corn yields are high, corn prices are low and hog production increases.

There are 67,000 pork production operations in the U.S. with slightly over half producing 5,000 or more pigs a year. Nearly three-fourths of these hogs are produced in confinement. These are often under contract with a buyer or harvest company. Hogs are the second largest livestock population in the United States, with 64.6 million head. Iowa, North Carolina, Minnesota, Illinois, and Indiana are the leading states in swine production. These states also tend to produce considerable grain, which is used in swine feed.

9–3. Baby pigs (sometimes called piglets) will rapidly grow if properly managed to produce quality pork products. (Courtesy, Phant/Shutterstock)

Several sources provide influence on the swine industry. Pork product purchases by large restaurant chains tends to dictate some pork production practices. An example is the decision by Burger King to only purchase and sell pork products that have been produced following certain practices. Another influence in the industry is that of the associations that represent producers, such as the National Pork Producers Council (NPPC). Certainly, government regulations also play a factor in the pork industry, though these may be targeted at sustaining the environment or providing protection of workers in pork industry jobs. Demand for pork for international markets is also a factor. Currently, Japan followed by Mexico and Hong Kong are the major purchasers of U.S. pork.

Favorable Factors

Swine production has many factors that make it a favorable enterprise. Swine are efficient in converting feed to meat. Fewer pounds of feed are needed to produce a pound of pork than for a pound of beef. Less than 5 pounds of feed are needed for a pound of pork. Nine pounds of feed are needed for a pound of beef.

Swine are very prolific. **Prolific** means that they will produce a large number of young. Sows will usually farrow (give birth) 7 to 12 piglets twice a year. (Gestation is 114 days.) They also excel in dressing percentage. They will yield 65 to 80 percent of their live weight. Comparably, cattle only dress out around 50 to 60 percent, and sheep at 45 to 55 percent.

Labor requirements are lower because hogs are good at self-feeding. The capital investment (money) is low because they require little land or buildings, depending on the type of operation. The approaches used on modern farms require greater investments in buildings and waste disposal. However, the level of production is higher. The time to get a return on investment is relatively short. A return can be made within 10 months.

9–4. Some hogs are finished in parlor-type facilities though this type of facility is being replaced in some areas with intensive confinement. (This shows a lagoon at the top right for managing wastes.) (Courtesy, U.S. Department of Agriculture.)

Unfavorable Factors

Several factors make the swine enterprise unfavorable. Hogs are very susceptible to disease and parasites. Since hogs have a simple stomach, they must have a large amount of concentrates and a minimum of forages. Hogs also require special attention at farrowing time. Because of the relationship between corn and hog prices, economic conditions can become unfavorable at times. The disposal of wastes is a major issue in hog production. Communities may refuse permits for hog farms and protest plans to build them.

9–5. A modern swine facility in Pennsylvania. (Courtesy, Education Images)

CORPORATE SWINE PRODUCTION

Swine have typically been produced on small, family-operated farms. The farm might also grow grain crops, primarily corn, as feed for the swine. In recent years, changes have taken place that have become major issues with some traditional pork producers.

More swine are being grown in large, farm systems. Large numbers of hogs are produced in a relatively small amount of space using intensive methods of production. The producer may be under contract with a company that provides the piglets, monitors production, supplies feed, and markets the hogs. This approach is often known as vertical integration.

The swine are bred for the most efficient growth and yield. During growth, careful attention is given to nutrition and disease control. People are not allowed in facilities where the hogs are growing. Producers enter only after bathing and wearing carefully laundered clothing. These precautions allow faster growth to market size.

CONNECTION

QUALITY ASSURANCE PROGRAM

The Pork Quality Assurance (PQA) program of the National Pork Producers Council (NPPC) promotes good management to produce a quality pork product. Some pork packers have made participation in PQA mandatory for producers who sell to them.

The PQA has four goals:

- To improve management practices in hog production.
- To avoid violative drug residues in pork.
- To decrease the costs of hog production.
- To increase food safety awareness among hog producers.

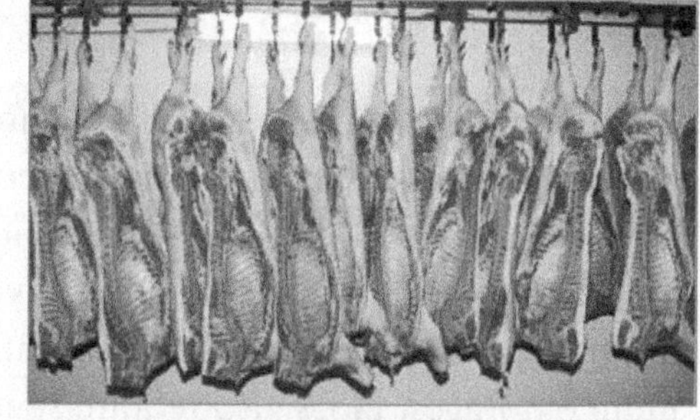

(Courtesy, LOYISH/Shutterstock)

Overall, the goal is to assure continued consumption of pork food products by providing quality products. Information on the PQA is available from the National Pork Producers Council, P.O. Box 10383, Des Moines, IA 50306. The Web site is **www.nppc.org**.

SWINE AS ORGANISMS

Swine have organ systems similar to most animals. They are mammals, with sows giving birth to litters of pigs. Swine differ from cattle in that they have a monogastric stomach. This means that the stomach has a single compartment. They must be fed feed that is more concentrated. Swine make poor use of forage.

Swine are classified as mammals with hoofs. The scientific name for the species produced in North America is *Sus scrofa domestica*. Other species include pygmy hogs (*Sus salvanis*), bearded pigs (*Sus barbatus*), and the warthog (*Phacochoerus aethiopicus*).

Swine used for food products are typically young hogs only a few months of age. They have grown rapidly and been protected from disease. Older hogs, especially boars, may develop a strong objectionable flavor in the meat.

CLASSIFICATION BY AGE AND SEX

Swine are classified by age and sex. A young swine is known as a **piglet** or baby pig. At a young age, male pigs are castrated, unless they are kept for breeding. The male castrated at a young age is a **barrow**. A young female that has not farrowed is a **gilt**. An older female is a **sow**. A male hog is a **boar**. Sometimes, male hogs are castrated after they reach sexual maturity. If so, they are known as stags.

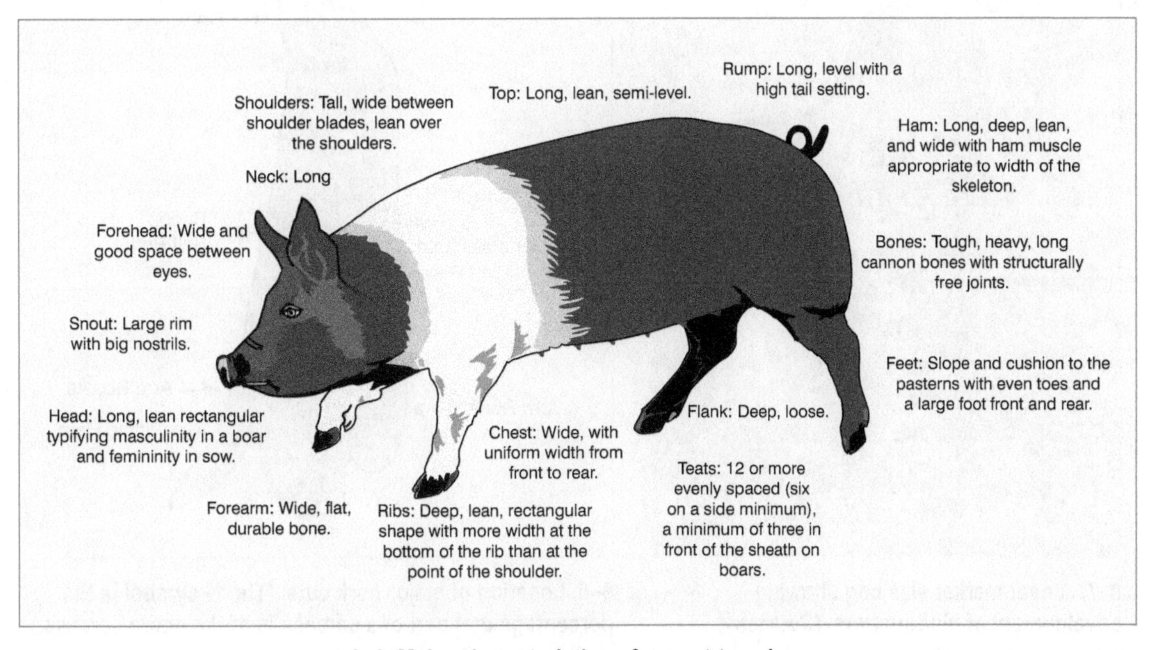

9–6. Major characteristics of a meat-type hog.

The hogs used for high-quality meat are barrows and gilts. Sows and boars may be made into lower-quality meat, such as sausage or cooked food products. Remember, lower-quality meats are wholesome but may not have the same flavor as the higher quality.

SWINE TYPE

Swine are produced for meat. People want lean meat without much fat. This reflects a change from years ago when fat was used to make lard. Consumer demand has resulted in a ***meat-type hog***. Lard-type and bacon-type hogs have declined in use. The meat-type gives the greatest amount of lean meat in high-value cuts, such as the ham.

Producers need to know the desired features of a hog. These are useful in describing the desired conformation. Conformation is the type and shape of an animal.

Meat-type hogs need to grow fast and efficiently. They need to have plenty of muscle tissue that will be used for lean food products. In general, a meat-type hog is a long animal with deep, well-developed muscles in the hams. Breeding animals should have good bone structure to be able to carry their weight. A meat-type hog will not have thick layers of fat.

The major meat products from hogs are ham, bacon, loin cuts, and roasts. Other parts produce cuts that are of lower value. Some low-value cuts are ground into sausage.

9–7. A near market-size hog showing development of hindquarters. (Courtesy, Education Images)

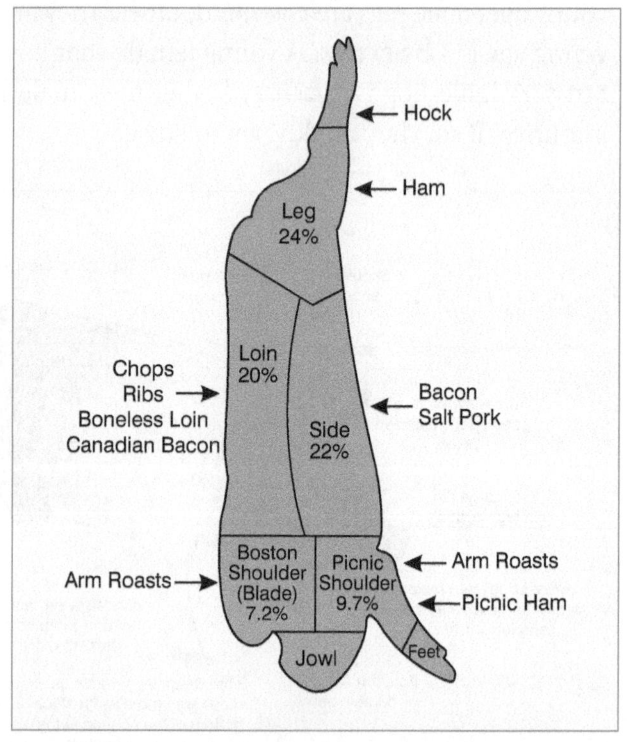

9–8. Location of major pork cuts. (The % symbol is the percentage that part of a carcass is of the entire carcass.)

BREEDS OF SWINE

Several breeds of swine are found in the United States. Swine breeds are very important to the purebred producer. Hogs grown for meat are typically crosses of the purebreds. The breeds of hogs include Duroc, Hampshire, Yorkshire, Hereford, Berkshire, Poland China, Landrace, Tamworth, and Chester White.

Some producers have detailed breeding systems to get the type wanted. For example, a Duroc boar may be bred to a Yorkshire female. A Hampshire boar may then be bred to female offspring from the Duroc-Yorkshire cross. A Yorkshire boar may then be bred to the females from the second cross.

Intensive production systems often use "hybrid" hogs. These result from crossing two or more breeds and selecting animals based on specific traits. Specific breeds are not important to many producers. The conformation of the animal is important.

9–9. A barrow of Duroc breeding. (Courtesy, National Duroc Swine Registry)

DUROC

The Duroc is light to dark red and has ears that droop over the eyes. It is a popular breed because of prolificacy and hardiness. Durocs are known today as good meat-type hogs. Durocs should not have any white on their bodies. This is a popular breed of hog.

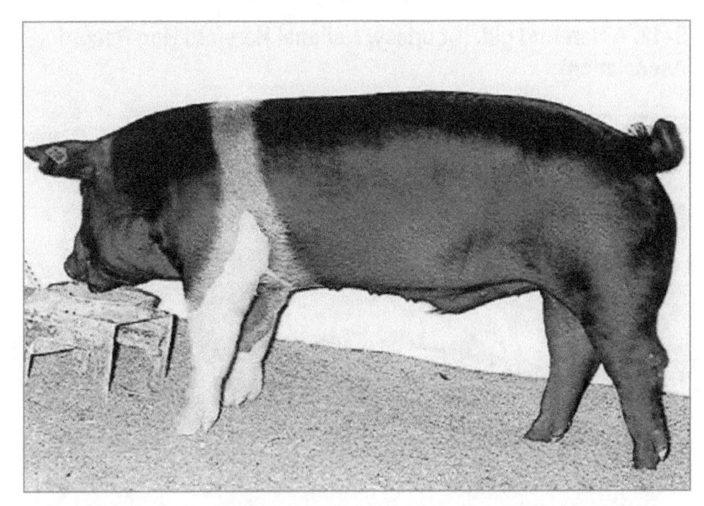

9–10. A barrow of Hampshire breeding. (Courtesy, Seedstock Edge, West Lafayette, Indiana)

HAMPSHIRE

The Hampshire has the most striking appearance as a black hog with a white band or belt encircling the shoulders and front legs. The ears stand out and do not cover the eyes. Hampshires cannot be reg-

9–11. A Yorkshire sow with litter. (Courtesy, American Yorkshire Club, Inc.)

istered if the band does not encircle the body and if there is too much white on the head and hind legs. Hampshires have been popular as show animals.

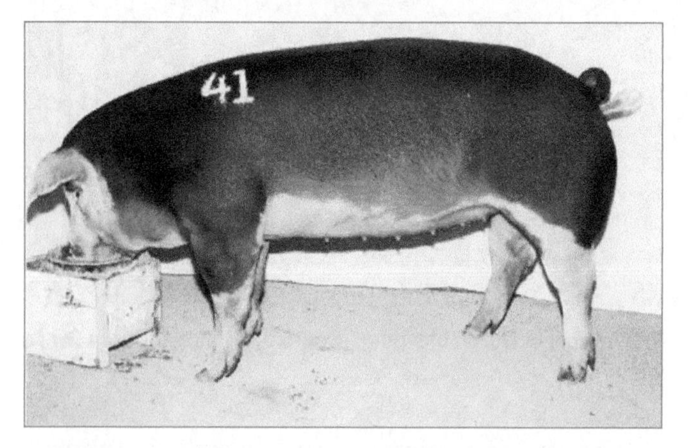

9–12. A Hereford gilt. (Courtesy, National Hereford Hog Record Association)

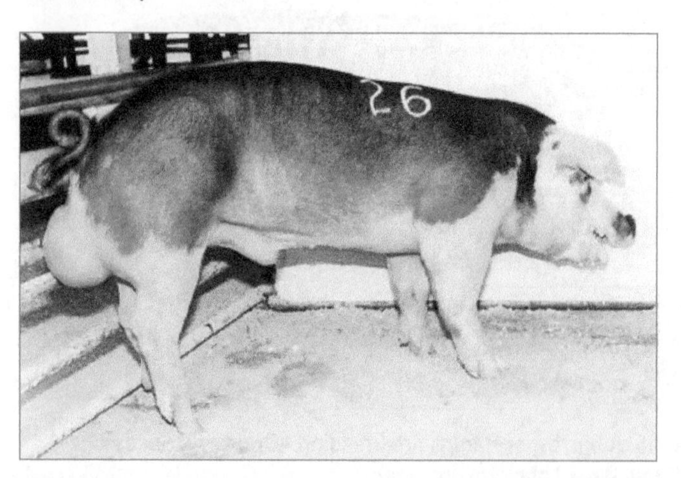

9–13. A young Hereford boar. (Courtesy, National Hereford Hog Record Association)

YORKSHIRE

The Yorkshire is a large, white hog. The females are known for their mothering ability. Yorkshires have long bodies that may tend toward a bacon-type of hog. They are sometimes faulted for small hams.

HEREFORD

The Hereford is a newer breed. The breed was founded by R. L. Webber of Missouri by inbreeding Chester Whites, Durocs, and hogs of unknown origin. The color markings are similar to those of Hereford cattle. They should be red with a white face and white underneath the body. At least 2/3 of the body must be red. The ears can be red or white. At least two of the four feet must be white. Herefords are primarily found in Indiana, Illinois, Wisconsin, and Ohio, though some are found throughout the United States.

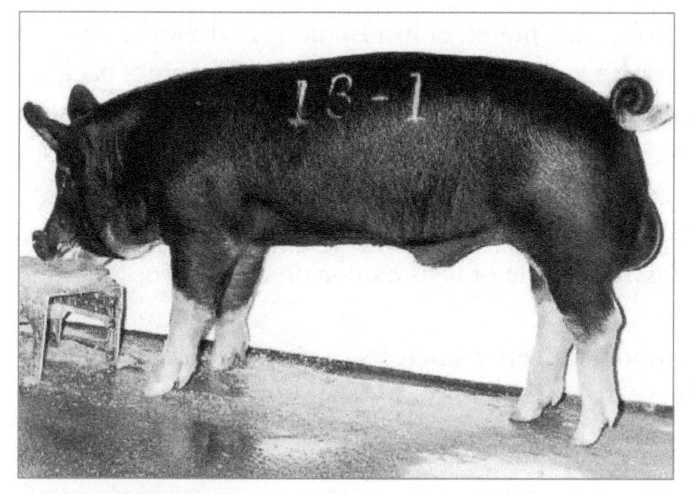

9–14. A Berkshire boar. (Courtesy, Seedstock Edge, West Lafayette, Indiana)

BERKSHIRE

The Berkshire is easily identified by its short, upturned nose. The hog is black with six white points: the feet, some white on the face, and white tail switch. Splashes of white may be in other places on the body. Berkshires are lean and typically have less fat than other hogs.

POLAND CHINA

The Poland China is black with white spots on the feet, the tip of the tail, and the nose. In the past, Poland Chinas were known as large, fat hogs. Producers have selected breeding stock based on more lean and less fat. Selection efforts have resulted in the breed being more the meat-type.

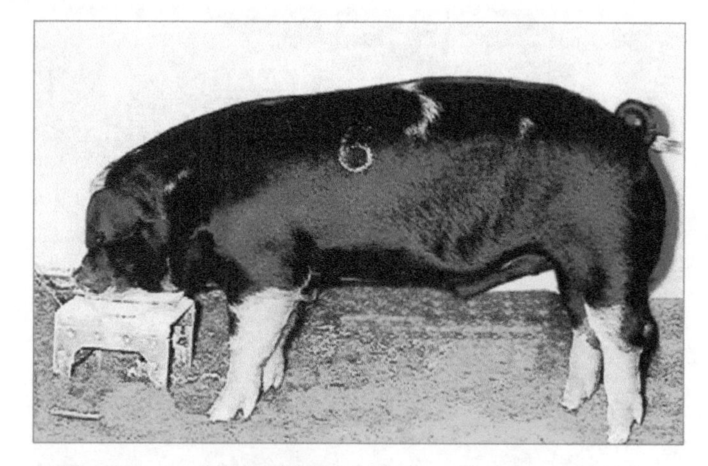

9–15. A barrow of Poland China breeding. (Courtesy, Poland China Swine, Peoria, Illinois)

PORK PRODUCTION POSSIBILITIES

Carefully select hogs to establish a herd. Selecting the wrong hogs can result in failure as a hog producer. Three important considerations are type and pedigree, health, and production testing.

TYPE AND PEDIGREE

Selection based on ***type*** means to select an animal that is as close to the ideal as possible. A ***pedigree*** is a record of an individual's heredity. In selecting for type, test for meatiness, porcine stress syndrome, and genetic defects.

Meatiness can be measured with a probe, lean meter, or ultrasonics. All show the amount of backfat present on an animal. The **probe** is a tool for measuring the thickness of backfat. The lean meter is based on the difference in conductivity of electricity in the fat and muscle. Muscle and blood are better conductors than fat. **Ultrasonics** use bursts of high frequency sound. It measures the reflection of the pulses and the time of reflection.

Porcine stress syndrome (PSS) is a nonpathological disorder in heavily muscled animals that results in sudden death loss. A couple of tests can be used to detect PSS, and they are 100 percent accurate.

Litter mates and parents should be noted for genetic defects to avoid. Commonly inherited genetic defects include scrotal hernia, cryptorchidism, hermaphrodites, and umbilical hernia.

9–16. Swine shows help producers to refine breeds for the desired type. (Courtesy, Hampshire Swine Registry)

HEALTH OF THE ANIMAL

Selected stock should be disease free. Specific swine diseases will be discussed later in this chapter.

Many producers purchase hogs that come from a **specific pathogen free** (SPF) herd. SPF pigs are free of disease at birth. They are raised in an aseptic environment. The diseases that SPF animals should be free of are brucellosis, leptospirosis, lice, mange, pneumonia lesions, pseudorabies, swine dysentery, and turbinate atrophy (atrophic rhinitis) or snout distortion.

PRODUCTION TESTING

Several purebred swine registries, as well as producers, have a production registry and a meat certification program. The most important way to evaluate the productivity of potential animals is through production testing.

- Boar selection—One boar should be selected for every 15 to 20 sows. Because they will be producing so many offspring, their evaluation is extremely important. Table 9–1 summarizes the major characteristics of a good boar.

  The boar should also be docile and have an easy temperament, yet be aggressive enough to breed the herd. The boar should be purchased at six to seven months of age.

Table 9–1. Characteristics of a Good Boar

Trait	Standard
Litter Size	from litter with 10 or more farrowed, 8 or more weaned
Underline	12 or more well-placed, undeveloped nipples
Feet and Legs	medium to large boned; wide stance both in front and in the rear; free in movement; good cushion to both front and rear feet; equal sized toes
Age at 230 lbs.	155 days or less
Feed/cwt gain	275 lb/cwt
Daily Gain	2.00 lb/day or higher
Backfat Probe	1.0 inch or less

Use of boars in breeding may begin at eight months of age. Boars should be of good health and have a masculine appearance. The testicles should be well developed. With purebred boars, the standards of the breed should be met.

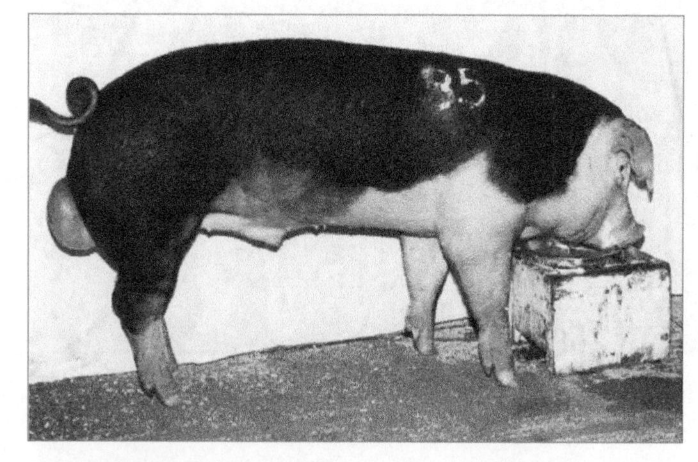

9–17. A champion Hereford boar that has the characteristics of a good boar. (Courtesy, National Hereford Hog Record Association)

- Female selection—Gilts and sows should be carefully selected. Only gilts from large litters should be selected. Their backfat should be less than 1.2 inches. The gilt or sow should be sound and free of any flaws or defects that would interfere with normal production. They should be the fastest growing and the leanest of the bunch. Gilts in purebred herds should meet the standards of the breed.

Gilts are often selected from the animals that are being fed for meat production. When gilts reach 180 to 200 pounds (81.6 to 90.7 kg), they should be evaluated for their reproductive potential. If selected as a replacement, they should be removed from the meat animals and fed separately.

9–18. A Hampshire gilt that has the characteristics of a good gilt. (Courtesy, Hampshire Swine Registry)

PRODUCTION SYSTEMS AND CONTRACTS

A production system is the method by which swine are raised for a specific purpose. It is based on the age and weight of the pigs when they are obtained and marketed.

9–19. Feeder pigs being produced on pasture at an Iowa farm. (Courtesy, National Hereford Hog Record Association)

9–20. Hogs nearing market size in an automated and temperature controlled facility. (Courtesy, Natural Resources Conservation Service, USDA)

SYSTEMS

Four traditional production systems have been used. New, integrated approaches are replacing traditional systems. The traditional systems are

- Feeder pig production—A ***feeder pig*** is a pig weighing about 40 pounds (18.2 kg) and has been weaned. This system involves maintaining a herd of brood sows and having the necessary facilities for successfully breeding and farrowing litters of pigs. Boars may be maintained for natural breeding or for semen collection. Artificial insemination of the sows may be used.

  Feeder pigs are often produced in confinement though some are with sows on improved pastures. Pastures work better in warmer climate. Pastures may be seasonal in cold climates. Feeder pigs are sold to farms that grow them to market weight.
- Finishing—***Finishing*** is feeding feeder pigs until market size is reached. Most hogs are marketed for slaughter at 240 to 260 pounds (109–118 kg), which is an increase over the 220 to 240 pounds (100-109 kg) of a few years ago. Some are lighter; others are

heavier. The trend is toward a heavier market hog. Many packers may have specific weights, and a producer should target these weights. The price paid for pigs that vary from the weight more than a few pounds may be less. For example, the per pound price paid for a pig weighing more than 270 pounds (122.5 kg) may be less. The goal of producers is to get the best possible weight gain on the minimum investment in feed.

9–21. Pigs being finished to market weight in a less intensive environment without temperature control. (Courtesy, Education Images)

- Farrow-to-finish—Pigs are farrowed and fed to market weight on the same farm. The producer must have farrowing facilities as well as facilities for feeding out the pigs. This type of system is less specialized than the feeder pig or finishing systems.
- Purebred—Purebred hog production is raising hogs that are subject to registration by the breed associations. All standards of the breed association must be met. If not, the pig cannot be registered. The goal is to produce breeding stock. The breeding stock is sold to other producers who are producing pigs.

CONTRACTS

An agreement between two or more people is a contract. It is usually in writing and signed by all parties. ***Contract production*** involves a contract between a producer and a buyer before hogs are raised. The contract specifies the kind of hog to be raised and other details. Contracts may be used for feeder pigs or market hogs. Feeder pig contracts are with growers who will provide feeder pigs to finishing operations. Market hog contracts are with finishing operations that may buy feeder pigs or farrow their own feeder pigs.

Contract production is increasingly popular. This is because of the high capital investment, the difficulty in financing an operation, and the willingness to forego large profits in return for security. The producer is assured of a market.

PRODUCER RESPONSIBILITIES AND PRACTICES

The pork industry is dedicated to assuring quality food products. Producers have an important role in providing quality animals. The husbandry (care) used with hogs shapes the quality of the finished product—all the way to the pork products in a supermarket.

Hog producers often participate in a checkoff program. This involves making a payment of a few cents per head of hog sold to support research, promotion, and other activities. Benefits to producers usually far outweigh the cost. One use of the checkoff funds is to promote quality products for consumers.

The National Pork Board has used to checkoff funds to develop information for producers. Emphasis is on swine care practices for safe, humane, and efficient pork production. The

PORK PRODUCER CODE OF PRACTICE

Producers take pride in providing proper care to the swine on their farms. They consider management and husbandry practices for good swine care to include the following:

- Providing facilities to protect and shelter pigs from weather extremes while protecting air and water quality in the natural environment.
- Providing well-kept facilities to allow safe, humane, and efficient movement of pigs.
- Providing personnel with training to properly care for and handle each stage of production for which they are responsible with zero tolerance for mistreatment of swine in their care.
- Providing access to good quality water and nutritionally balanced diets appropriate for each class of swine.
- Observing pigs to make sure basic needs for food and water are being met and to detect illness or injury.
- Developing herd health programs with veterinary advice.
- Providing prompt veterinary medical care when required.
- Using humane methods to euthanize sick or injured swine not responding or not likely to respond to care and treatment in a timely manner.
- Maintaining appropriate biosecurity to protect health of the herd.
- Providing transportation that avoids undue stress caused by overcrowding, excess time in transit, or improper handling during loading and unloading.

9–22. Pork Producer Code of Practice (used with permission of the National Pork Board).

Board stresses that care is related to the interaction of the producer with the animals. This includes feeding, facilities, disease prevention, and all other aspects of hog production.

A goal is for producers to take pride in providing proper care of the swine on their farms. To this end, a Pork Producer Code of Practice has been developed. This code is to guide producers in managing and caring for their swine. It also promotes the need for producers to keep updated on advancements and make decisions based on sound production practices and welfare of the pigs.

SWINE MANAGEMENT

Hogs are unique livestock. Except for a few breeds of sheep, they are the only livestock that naturally produce offspring twice a year. Because hogs reproduce at a younger age and more quickly than other livestock, genetic gains can be made quite rapidly. But, this requires good management. Good management is important because there is little difference between producing a litter of five pigs and a litter of ten pigs.

Selection of the animals to be used as breeding stock is very important. Some producers only keep females and buy semen for breeding from reliable sources. In all cases, the genetic potential of the animals is a high concern. Pedigree information is useful. Obtain stock from a reliable source. Choose breeding animals with structural correctness, good muscle shape, a wide body, and good rib shape. If under contract with a processor, use the breeds and crosses that are recommended. With females to be used for breeding, consider the number of pigs born in litters of previous generations and number of pigs that were weaned.

NORMAL BREEDING

Swine reach puberty from four to eight months of age. This wide variance is due to sex, environment, breed, and breeding lines. Gilts can generally be bred to farrow at 11 to 12 months of age. They should weigh at least 225 pounds when bred.

The heat period or estrus lasts from one to five days and averages two to three days. Signs of heat include restless activity, mounting of others, swelling and discharge from the vulva, increased urination, and loud grunting. The gestation period is 114 days.

CROSSBREEDING

Most meat hog producers use crossbreeding, which is the mating of different breeds. Crossbreeding allows for heterosis or hybrid vigor. Heterosis is a biological phenomenon that causes crossbreds to outperform either of their parents.

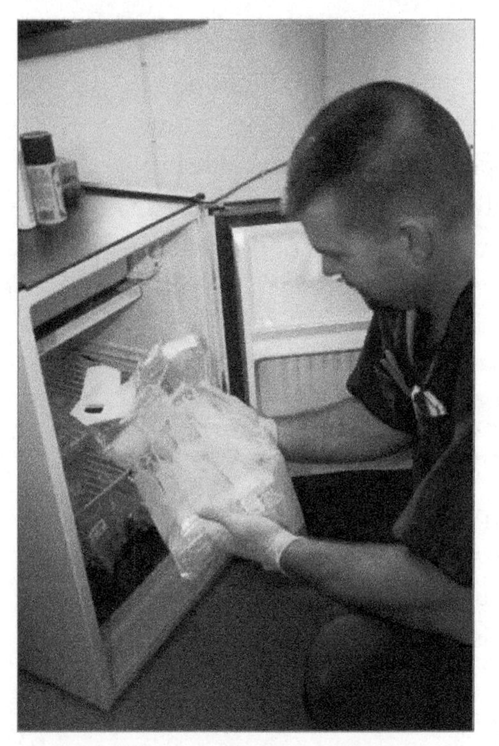

9–23. Semen vials are kept in a plastic bag in a refrigerator to retain sperm viability. (Courtesy, Education Images)

Crossbreeding is widely used because it allows for increased production and profits. Roughly 90 percent of commercial hogs are crossbred.

THE BREEDING PROCEDURE

Sows can be bred naturally by a boar or artificially. Natural breeding results when the boar and female are placed together for breeding. This is sometimes known as hand mating, which is the most common method. A breeding crate is used when a heavy boar is to be mated with a young gilt. This reduces the risk of injury to both.

Artificial insemination is increasing in use with hogs. It has advantages and disadvantages. AI decreases the risk of disease and increases genetic improvement. However, the main problem with AI is that swine sperm cannot be frozen, thus decreasing the opportunities for use.

Thirty to forty-five days after breeding, the sow or gilt should be pregnancy checked. The use of ultrasonics has increased detection up to 95 percent accuracy.

CONNECTION

(Courtesy, dyoma/Shutterstock)

LITTER SIZE

The number of living pigs farrowed by a sow is important to a producer. There is profit in number! Obviously, a sow that farrows 12 or more pigs is more valuable than one that farrows fewer. What makes the difference?

Scientists have determined that uterine size is a major factor in the number of pigs farrowed. Small litter size is not highly related to the ovulation of too few eggs or the use of poor quality semen resulting in eggs not being fertilized. Up to 40 percent or more of all fertilized eggs may suffer fetal death due to limitations in uterine capacity. A sow that farrows 12 pigs weighing 3 pounds or more each must have a large uterus!

Genetic markers related to uterine capacity are now being investigated at the U.S. Meat Animal Research Center, Nebraska. Until more is known about the associated genetic markers, researchers are suggesting that average pig weight at birth may be the best measure. Pigs below 2.2 pounds (1.0 kg) at birth have poor survival rate, while those over 3 pounds (1.4 kg) have a much better chance of survival.

CARE OF THE PREGNANT SOW

The trend in caring for pregnant sows in confinement is to house them in groups, often referred to as group housing. Each sow should have at least 15 square feet (1.4 square meters) of floor space if housed in groups. Each group must have feeding and watering equipment.

Many producers have used individual confinement to care for pregnant sows. The sows are put in small pens or farrowing crates. Some producers are moving individual confinement to larger pens with several sows in each pen. Advantages of group confinement are a decrease in labor requirements, the use of automatic feeding, freeing pasture land for other purposes, a controlled environment, improved control over disease and parasites, and better management.

Some disadvantages to confining pregnant sows are the increased need for facilities, higher initial investment, possible delay in sexual maturity, lower conception rate, and the requirements of better management.

Sows should receive feed with more bulk shortly before parturition. A drug to induce labor may be given 111 to 113 days after breeding. The sow will usually farrow within 18 to 36 hours.

9–24. Facility for managing pregnant sows showing overhead feeding systems. (Courtesy, U.S. Department of Agriculture)

CARE OF THE BOAR

Boar care is important. A boar will be expected to service dozens of gilts and sows and produce healthy offspring. The boar should have plenty of room to exercise, which is usually in a pasture. The pasture should be well fenced. If kept in confinement, boars should have clean pens with slotted floors or concrete. Boars should be kept in individual pens.

Boars that are not being used should be kept separate and away from the rest of the herd. Also, never allow them to grow tusks as this drastically increases their dangerousness.

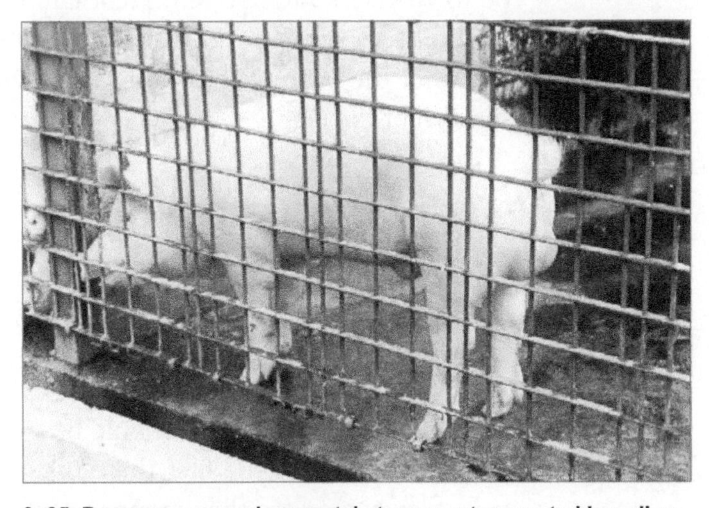

9–25. Boars are penned separately to prevent unwanted breeding. (Courtesy, Education Images)

9–26. Sow with litter in heated farrowing facility. (Courtesy, Education Images)

CARE OF SOW AT FARROWING

Giving birth to pigs is ***farrowing***. Care of the sow at farrowing is important to have a strong, healthy crop of pigs. Only 70 percent of the piglets born reach weaning, thus increasing the importance of good, sound management practices at farrowing time. This means that 30 percent die due to disease, mashing by the sow, bad weather, or other conditions. Increasing the survival rate makes a farm more profitable.

- Signs of parturition—Signs of parturition include nervousness, uneasiness, enlarged vulva, and mucous discharge. Milk will be present in the teats and the sow will begin to make a nest for the young. She should be placed in a crate or pen by at least the 110th day of gestation. Farrowing crates are commonly used because they reduce the number of young that are crushed.
- Sanitation—Sanitation is essential. Before the sow is moved to the crate or pen, she should be scrubbed to remove dirt, manure, and any parasites. Bedding, if used, needs to be clean and fresh. Good sources of bedding are wheat, barley, rye, or oat straw; chopped hay; corncobs; and shavings. The farrowing unit should be scrubbed and cleaned with a disinfectant between uses and left empty for five to seven days.
- Environmental factors—Sows and piglets are sensitive to their environment. Good ventilation and protection from extreme weather are needed. Their quarters should be 60 to 70°F (15.5 to 21.1°C). Heat lamps or mats are placed in the farrowing crate or pen to keep piglets warm. An attendant should check on the sows frequently during and after the birth of the young. This will decrease loss and increase the health and survival of the pigs.

THE SOW AND LITTER

Getting the pig litter off to a good start is important for economic gains. Sanitation is necessary because swine suffer heavy losses due to disease and parasites. A diseased herd can be wiped out leaving the producer in debt and the facilities highly contaminated.

- Feeding the sow—The sow should be fed liberally before parturition to stimulate milk production. During the first three days after giving birth, she should be fed minimal feed. Slowly increase the amount to full feed at two weeks. Use a manufactured feed designed for lactating sows.

- Starting piglets on feed—Newly born pigs need to be observed to be sure they nurse within a few hours. If not, a pig may need to be placed near a nipple. Pigs will often begin learning to eat solid feed within 7 to 10 days. Use a manufactured baby pig feed. Feeds for pigs may have protein content of 20 percent. The feed should be in meal form. Using feed containing dried whey has shown positive signs in research trials.
- Clipping needle teeth—When pigs are born, they have eight **needle teeth**. Two are located on each side of the upper and lower jaws. These teeth are very sharp and may injure the sow's udder when pigs nurse. Clip these teeth off with pliers or forceps on the day of birth. Take care not to injure the pig. This is a good time to treat the umbilical cord with an approved iodine disinfectant solution. An iron shot may also be given at this time.

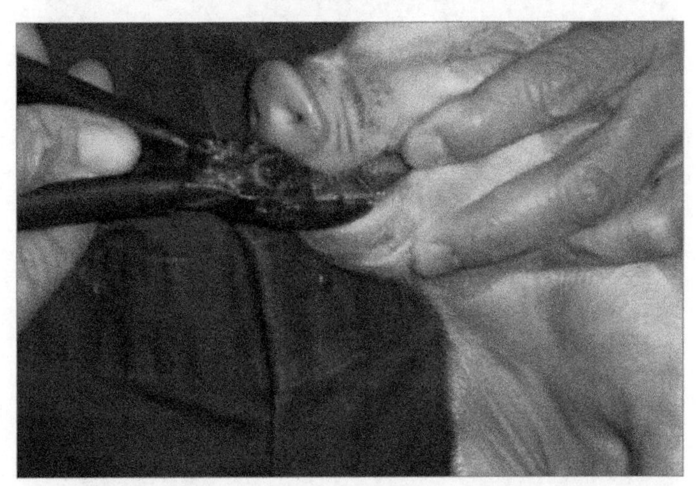

9–27. Clipping needle teeth prevents injury to the teats of the sow by the nursing pigs. (Courtesy, Nancy S. Jackson, DVM, Mississippi)

- Tail docking—***Tail docking*** is clipping the tail from baby pigs. It is done at the time that needle teeth are snipped. The tail is cut about 1 inch or slightly less from the bone of the tail. Tail docking prevents tail biting that may happen in hogs as they grow in confinement.
- Ear notching—The most common method of identifying hogs is to use a special V-notcher to notch their ears. This is especially important in herds with thousands of hogs. This enables producers to exactly identify the animals, which is necessary when selecting breeding stock and replacements. Plastic ear tags are sometimes used. A disadvantage is that they will tear out easily. Branding and tattooing are also used to some extent but are hard to read.

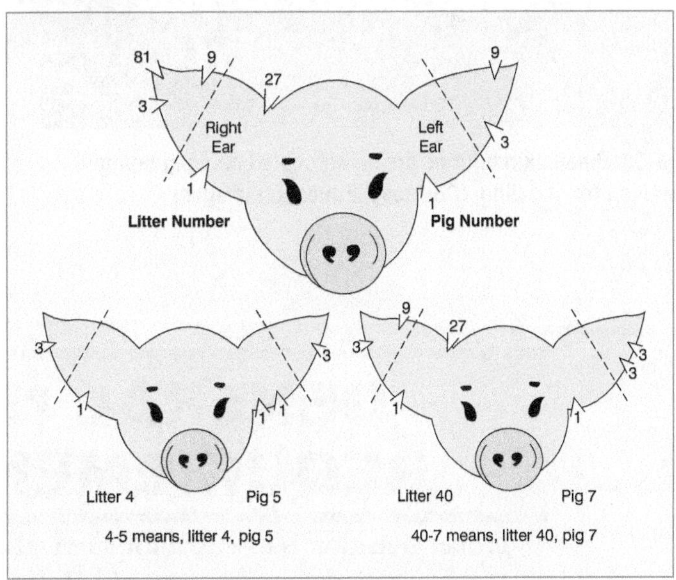

9–28. Ear notching is done on baby piglets as a permanent identification system. The right ear is the litter number. The left ear is the pig number within the litter. The position of each notch has numeric value.

- Castration—All male pigs being raised for meat should be castrated a couple of days after birth. Use a sani-

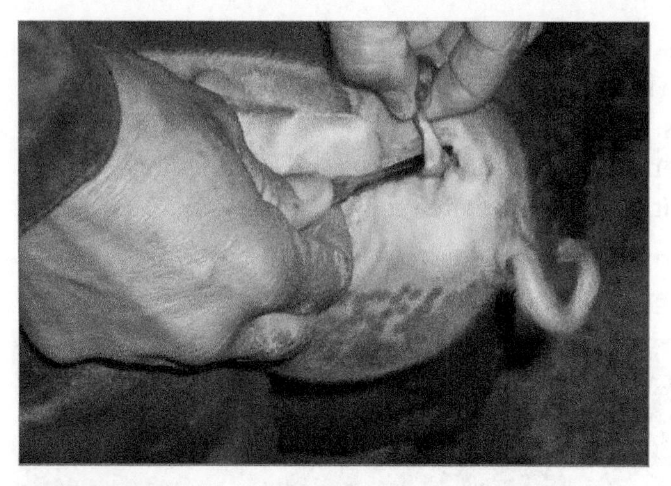

9–29. Castrating a young pig. (Courtesy, Nancy S. Jackson, DVM, Mississippi)

9–30. Small skin injuries are being treated on a pig being raised for showing. (Courtesy, Education Images)

tary and safe surgical process. Sanitation in the area of confinement promotes healing without infection. Of course, males kept for breeding should not be castrated.

- Vaccination—Newborn pigs should receive an iron shot (1 ml of 100-200 mg iron) within a week of birth to prevent anemia. Vaccinate for atrophic rhinitis before weaning. Vaccinate for erysipelas at 40 pounds.

GROWING FEEDER PIGS TO MARKET SIZE

Feeder pigs are pigs that can be finished to market size in a few months with proper feeding and management. Feeder pigs are typically produced in confinement though some are raised on pasture with supplemental feed. Finishing to market size is typically done in confinement. The pigs are given complete rations for their stage of growth. From 40 to 125 pounds, the feed is 16 to 18 percent protein. For pigs over 125 pounds, the feed is 14 to 16 percent protein. Main ingredients are typically yellow corn and soybean meal. Calcium, phosphorus, salt, and vitamins are added. A manufactured feed is often used.

GENERAL NUTRITIONAL REQUIREMENTS OF SWINE

Proper nutrition is essential for survival and growth. Hogs raised in confinement must receive a complete feed. Costs need to be kept down to assure a profit. About 75 percent of the total cost of hog production is feed.

Each individual varies in the kinds and amounts of nutrients needed. These requirements are influenced by age, function, disease level, environment, and other factors.

- Carbohydrates and fats—Carbohydrates and fats provide energy for the various functions of the body. Cereal grains are the biggest source of energy, mostly carbohydrates. Roughages are a good source of energy but are not suitable for nonruminants, such as swine. Roughages are too bulky for the restricted size of the digestive system. Corn is the most widely fed concentrate supplying energy. Other sources of energy fed to hogs include barley, sorghum, wheat, and fats and oils.
- Protein—Protein is needed to build and repair tissue. Protein is broken down in the stomach and small intestine into amino acids. These amino acids are then recombined into proteins that the animal needs for muscle development and repair of other tissue. Hogs, as nonruminants, are unable to synthesize their own protein, thus increasing the importance of having a well-balanced diet. Soybean meal is the most widely fed protein. Because of its high palatability to hogs, it should be mixed with grain to prevent overeating. Young, growing hogs are fed feed higher in protein than older hogs. Protein supplements of 35 to 40 percent protein may be used with corn and other feedstuffs. Commercial feeds with 20 percent protein are used to start pigs at three weeks of age. Growing and finishing feed may have 12 to 16 percent protein. Pregnant sows are fed 12 percent protein feed. Lactating sows and boars are fed 13 to 16 percent protein feed.
- Minerals—Hogs are the most common farm animals to suffer from mineral deficiency. Those minerals needed in the largest quantities include salt, calcium, and phosphorus. Cobalt, copper, iodine, iron, manganese, selenium, and zinc are needed in smaller amounts but are still essential.

CAREER PROFILE

HOG PRODUCER

A hog producer raises hogs using one or more production systems. Some produce feeder pigs; others have finishing operations, farrow-to-finish, or purebred farms. The work involves tending animals, building and repairing facilities, obtaining feed, preventing disease, supervising other workers, marketing hogs, and assuring the wastes are properly disposed of to prevent environmental damage.

Hog producers need hands-on experience and training in the skills of hog production. A high school education is needed though many may have technical or college degrees. Experience working with hogs is essential.

Jobs are found where hogs are raised. Many are entrepreneurs and own their farms. The trend is toward contract production for large feed and processing companies. This photo shows a hog producer sampling water in the lagoon on the farm used for the disposal of wastes from the hog operation. (Courtesy, Natural Resources Conservation Service, USDA)

Mineral supplements are usually mixed with the feed ration. Newborn pigs are given iron shots shortly after birth.

- Vitamins—Vitamins are needed in small amounts, yet are essential for performance of normal body functions. Hogs also suffer from more instances of vitamin deficiency than any other livestock. This is due to confinement and the restricted variety of feeds that they can consume. Fat-soluble vitamins A, D, and E and water-soluble vitamins biotin, niacin, pantothenic acid, riboflavin, and vitamin B-12 are the most likely to be deficient.
- Water—Hogs will generally need $^1/_4$ to $^1/_3$ of a gallon of water for every pound of dry feed consumed. Ideally, they should have access to automatic waterers or be hand-watered at least twice a day. Water needs are based on size. Pigs weighing 60 pounds need $^3/_4$ gallon of water a day. Larger pigs need more water. Pigs weighing 60 to 100 pounds need about $2^1/_2$ gallons a day while those weighing 100 to 250 pounds need about 4 gallons a day. A mature boar needs about 5 gallons a day. Water needs of sows vary with condition. A pregnant sow needs about 5 gallons a day. A lactating sow may need 7 gallons.

9–31. Pigs will quickly learn to drink from automatic waterers. (Courtesy, Education Images)

ADDITIVES AND WITHDRAWAL

Additives are substances added to feed to meet a particular need. Some additives are standard in hog rations. They do not provide nutrients but provide other advantages. Anthelmintics may be added as wormers. Antibiotics are used to stimulate growth, improve feed efficiency, and control infestations. Piglets do not begin producing antibodies until they are five to six weeks old. Under ideal conditions, antibiotics can increase gain by 10 percent, while decreasing feed consumption by 5 percent.

Porcine Somatotropin (pST) is a growth hormone for swine. It increases protein synthesis and growth in most tissues. Research shows that pST can increase feed efficiency by 20 to 30 percent, average daily gain by 15 to 20 percent, and improve muscle mass by 10 to 15 percent.

Feeds with most additives should not be fed for a period before slaughter. ***Withdrawal time*** is the period before slaughter when an animal should not receive a particular medication or other product. The days of withdrawal allow the hog's system to purge itself of the medications. This assures pork that is free of medication residues above the amount that may be allowed.

Withdrawal time applies to feed additives and medications given orally, by injection, or in other ways. The label on medicines and growth products gives withdrawal information. For example, feed containing sulfathiazole should not be used within seven days of slaughter. Other additives have similar withdrawal times, ranging from 2 to 70 days. It is essential for producers to follow recommended withdrawal times before marketing hogs.

HEALTH MANAGEMENT PRACTICES

Healthy animals vary widely. Some animals may appear healthy but may actually be suffering from pain or disease. Knowing the normal behavior of animals helps detect and correct any ill health. A producer or animal owner should learn to diagnose good and ill health.

9–32. Keeping visitors away from hog production facilities helps prevent the spread of disease. (Courtesy, U.S. Department of Agriculture)

9–33. This pig has conjuctivitis, which is more likely to occur in intensively housed swine. (Note the pus at the bottom of the eye. This case was most likely caused by *Chlamydia sp.* bacteria. Precise diagnosis would require a swab of the infected area and microscopic examination. Treatment involves the administration of a topical medication to the eye and an injection. Sanitation and fly control help prevent spreading.) (Courtesy, Nancy S. Jackson, DVM, Mississippi)

PARASITES

Parasites live in and on hogs. The parasites feed on the blood, flesh, and other tissues of a hog. This reduces the rate of growth and overall health.

- Internal parasites—Internal parasites live inside a hog. They may be taken in through the mouth when eating contaminated feed or when eating feed on a dirty floor. Most internal parasites are obtained through feces in the food. Pastures with very short grass also increase the likelihood of ingesting parasites. Hogs on pasture are more likely to have internal parasites. Minimize the amount of feed eaten on the ground. Roundworms, tapeworms, and other worms are parasites that can infect all animals. They are found in the intestines and the stomach. Tapeworms can grow several feet long. When they become this large, they use much of the feed that the animal has eaten. Animals with low infestations of small-sized worms may not show symptoms. As the number and size of worms in the digestive tract increases, the animal loses weight, becomes anemic, and may have diarrhea. Sanitation is a good control measure.

- External parasites—External parasites live outside on a hog's body. They get their food from the blood and tissue of the animal. External parasites can transfer contagious diseases or other parasites to other animals. Lice are small insects that suck their host's blood. Animals may become anemic. Lice cause animals to itch. This is why hogs with lice are often seen scratching or rubbing against trees or posts. Lice usually appear in the winter. Back rubbers and other means can be used to administer pesticides to control lice. Mange is a dermatitis (skin disease) caused by one of several types of mites. Mange results in intense itching around the head and neck and wrinkling of the skin. It spreads rapidly. If untreated, it can be debilitating and even fatal. Dipping or high-pressure sprays may be used for treatment.

NUTRITIONAL DISEASES

Most nutritional diseases are caused by an unbalanced diet. Animals require specific amounts of nutrients. Too high or too low amounts of any of these may result in a nutritional disease that decreases profitability. Mineral and vitamin deficiencies are quite common in swine.

COMMON INFECTIOUS DISEASES

Swine infectious diseases can often be prevented by vaccination, sanitation, and good management. Vaccinate boars and prebreeding gilts and sows for leptospirosis, parvovirus, and erysipelas. Prefarrow females should usually be vaccinated for *E. coli* and atrophic rhinitis. Baby pigs may be vaccinated for atrophic rhinitis. Vaccinate feeder or grower pigs for erysipelas.

Some swine diseases were discussed in chapter 4; others are included here.

- Erysipelas—Erysipelas is primarily a disease of growing pigs. It is caused by the bacterium, *Erysipelothrix rhusiopathiae*. Symptoms include arthritis, heart disease, and skin discoloration. Death may result. Survivors may have stunted growth. Vaccinate

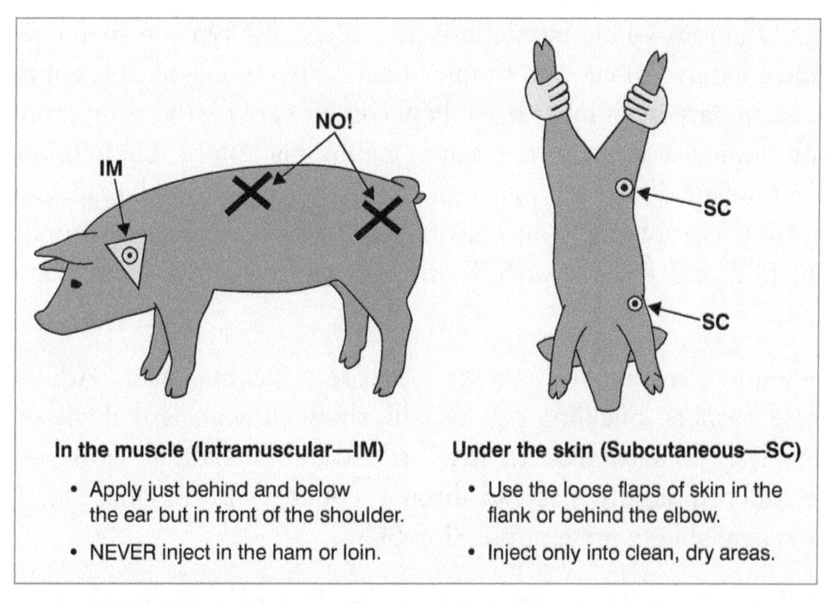

9–34. Injection sites for pigs and hogs. (Injections into the ham or loin damages the product for meat use.)

sows before breeding so that immunity is passed to pigs. Pigs from non-vaccinated sows may be vaccinated.

- Atrophic rhinitis (AR) —AR is the deterioration of internal nose tissues, including the septum, causing the snout to twist. An early sign is sneezing by pigs. AR is caused by one or more kinds of bacteria. Adult hogs may carry the disease. Vaccinate sows before farrowing so their colostrum will carry immunity to pigs. Practice good sanitation and bring only disease-free pigs into a herd.

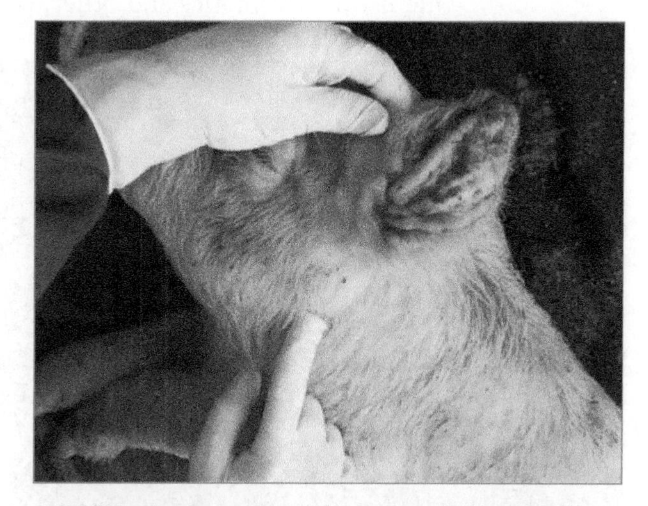

9–35. This pig shows signs of hematoma, which results when blood builds up in the ear canal due to excessive head shaking. (Note the abscess below the ear and the dried blood on the ear.) (Courtesy, Nancy S. Jackson, DVM, Mississippi)

- Porcine Parvovirus Infection (PPV) — A disease of swine, PPV only affects females. It causes reproductive failure, including small litters, failure to breed and farrow, and lack of reproductive cycling. It is best to vaccinate breeding females to prevent PPV. There is no effective treatment for a PPV infection.
- Leptospirosis—Leptospirosis is a bacterial disease. Symptoms include high fever, poor appetite, bloody urine, and females will abort their fetuses. Antibiotics are sometimes used to treat leptospirosis. Animals are often vaccinated against this disease.

- Pneumonia lesions—Pneumonia lesions is a specific type of pneumonia that affects swine and can have a major economic impact. Pneumonia itself is not that significant, but the secondary infections that follow can be very costly. Symptoms are a chronic cough and lung lesions, thus the name pneumonia lesions. Lung lesions can result in increasing the incidence of lungworms and migrating ascarid larvae. Control is usually unsuccessful. Use of sulfas and antibiotics helps to decrease secondary infections. Depopulation and the use of SPF animals will lower the incidence of pneumonia lesions.

- Pseudorabies—Pseudorabies is a viral disease attacking swine. Adults may serve as unaffected carriers. Suckling pigs usually show clinical signs. Piglets will develop a fever, paralysis, coma, and death, in as little as 24 hours. It may cause adults to abort or to have stillborn pigs. It is spread through contact with nasal and oral secretions. A vaccine is available to prevent pseudorabies.

- Diarrheal diseases—Pigs are more likely to be affected by diarrheal (intestinal) diseases than adult hogs. Some of these diseases are caused by bacteria and others by viruses. Swine dysentery is a type of diarrheal disease of the large intestine of pigs. First signs are soft feces and fever. Later stages include a condition known as bloody scours where pigs are weak and emaciated. Another rather common diarrheal disease is *E. coli* (*Enteric colibacillosis*). It is most common with nursing pigs and shortly after weaning. A disease of the small intestine, *E. coli* is passed through feces from a sow or other infected animal. Vaccinate pregnant females before farrowing. These and other diarrheal diseases cause stunting, loss of feeding efficiency, and, possibly, death. Treatment with antibiotics and other medications may be beneficial.

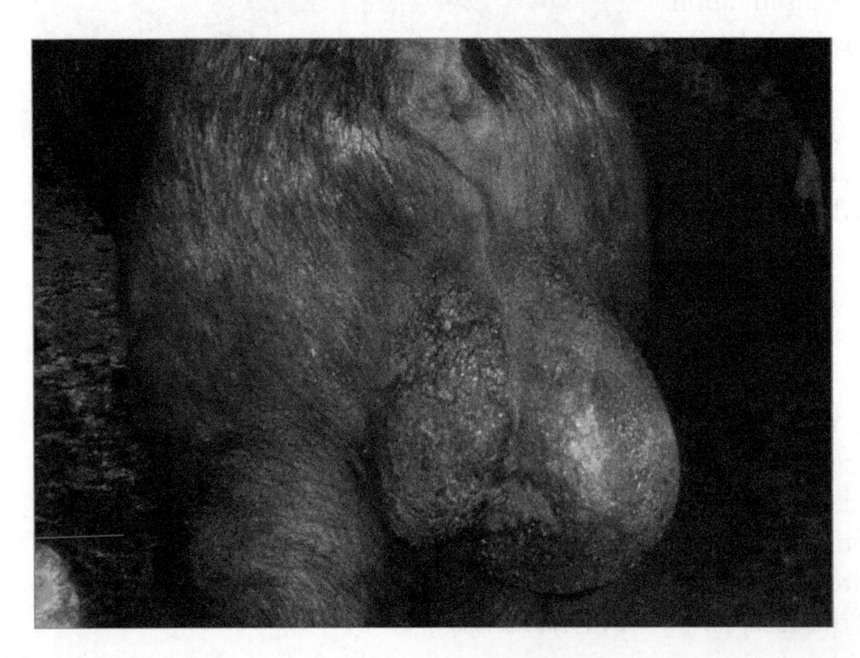

9-36. A boar with orchitis, as evidenced by the swollen right testicle. (Note the condition may result from a fight injury or an infection.) (Courtesy, Nancy S. Jackson, DVM, Mississippi)

FACILITY AND EQUIPMENT NEEDS

Hog production requires good facilities and equipment. The facilities must be properly located, constructed, and maintained to provide an appropriate environment. Climate and size of the pig should be considered. In cold weather, baby pigs need an enclosed environment with a temperature around 80°F. A temperature of 60 to 75°F is satisfactory for larger pigs, though they can tolerate 50°F.

GENERAL FACILITY TYPES

Production facilities determine how hogs are raised. In general, three types of approaches are used:

- Pastures and drylots—Hogs are placed in fenced pastures or drylots. Some protection from the weather should be provided. Overall, the trend is away from pastures and drylots except for brood stock on some farms. Sows are pastured at about 10 per acre or only 7 per acre if they have pigs. Each boar needs about $^{1}/_{4}$ acre of pasture.
- Buildings with concrete slab floors—The buildings may have roofs that cover half to all of the concrete area. The concrete should have a finish that is not slippery and a gradual slope to allow easy cleaning. Bedding is typically not used. Automatic waterers and feeders are used.

9–37. A free-range hog production enterprise. (Courtesy, EMJAY SMITH/Shutterstock)

9–38. A sow unit that produces 35,000 pigs a year. (Careful management is used to ensure a quality environment and well-being of sows and pigs. Air movement is created by the fans on the end of the building. Air goes through panels on the side of the building.) (Courtesy, Education Images)

- Enclosed buildings with slotted floors—These facilities provide a completely controlled environment. Manure collects below the slotted floor and is washed out each day into a tank or lagoon. Automatic feeders and waterers are used. This type of facility works well in cold climates. Producers exercise control over the total production environment.

9–39. An enclosed facility. (Courtesy, Education Images)

LOCATION

A good site must be selected for hog production facilities. Location is more important as fewer and fewer people live on farms and understand what is involved. Neighbors may dislike the odors and noise coming from the facilities. Locating buildings downwind of housing areas would be beneficial. Protection from wind and snow and access to good roads are also things to consider in site selection.

BUILDINGS

Swine facilities usually include buildings for farrowing, growing, and finishing. The farrowing barn may contain a nursery. Pasture may be used with gilts and sows depending on the climate, availability, and time of year. Attention should be given to three areas:

- Ventilation—Well-ventilated facilities help to remove extra moisture in the wintertime. During the summer, it helps to control temperature. Hogs are generally healthier when fresh air is present. Diseases are reduced.
- Manure disposal—Hogs create much waste. Disposing of manure is a major problem for producers with large numbers of hogs. Waste management is needed to maintain good health, avoid air and water pollution, and comply with government regulations. Manure pits are usually located under confinement buildings for temporary storage.
- Cleaning—Facilities should be able to be cleaned easily. Construction materials should resist damage from water and be durable.

9–40. A modern swine building with good ventilation. (Courtesy, Education Images)

9–41. This producer is checking the pump on a waste lagoon. (Courtesy, U.S. Department of Agriculture)

- Space requirements—Pigs kept in enclosed housing need adequate space. Needs increase as pigs grow. Pigs weighing less than 30 pounds should have about $2^1/_2$ square feet of floor space each. Other space requirements are: 30 to 60 pounds, 3 to 4 sq. ft.; 60 to 100 pounds, 5 sq. ft.; 100 to 150 pounds, 6 sq. ft.; and 150 pounds to market weight, 8 sq. ft.

FENCING

Some hogs are raised in fenced pastures and other areas. Sows and boars are sometimes put in pastures. Strong, woven wire fencing should be used to make the fence at least 3 feet high. Temporary fencing may be made of electric wire. Gates need to be made of wood or metal, be of sturdy construction, and have good hinges and fasteners.

HANDLING EQUIPMENT

Swinging gates make holding and handling swine easier. A series of gates throughout the building allow for easier sorting. Loading chutes are best located at the end of the building.

9–42. Portable scales for weighing individual hogs. (Courtesy, Education Images)

FEEDING AND WATERING EQUIPMENT

Hogs are hard on feeding and watering equipment. The equipment must be sturdy and not easily damaged. The equipment should also be easy to use and clean. Good automated feeders and waterers make a more efficient hog operation.

RAISING A SHOW PIG

Youth and adults enjoy showing pigs. County fairs, state livestock shows, regional and national expositions, and others afford opportunities for showing. Most youth enter the junior division of a show. Some advance to the open division. You might even have the grand champion barrow, gilt, market hog, or other animal at a show!

Success in raising a show pig involves several important activities. These require careful planning, some financial investment and facilities, and giving the animal good care.

- Kind of Animal—You will need to decide if you will raise market animals or breeding animals. Most youth focus on market hogs. This means that your pig should weight 230-270 pounds at the time of the show. If you raise breeding hogs, you will want animals that conform to breed standards and that have registration papers with the breed association. Market animals are typically sold for harvest after showing; breeding animals may be sold to an animal producer or returned to your facility for breeding. And, if you are a beginner, you will likely want to start with one or two market pigs.
- Facilities—You must have a place to keep your pig. A pen with board or wire fencing is needed to keep a pig confined. The floor may be concrete or earthen. Select a site that is well-drained and easy for you to get to so you can feed and care for the pig. A barn or

small shelter building is needed to protect the pigs from inclement weather, including preventing the sunburn of pigs that are white. Sometimes an entire small pen is covered with a roof to keep out rain and sun. The facility should not be muddy and should be cleaned regularly to remove manure. A pig needs at least 60 square feet of floor space on concrete and double or more that amount if on earth or sand. Two pigs will require nearly twice that amount of space. A pig kept in a small space will need a larger exercise pen. The facilities should also include a waterer and feeder, which with a small set-up could be a stationary pan or tub and trough. A chute and ramp for loading and unloading pigs should be convenient. About two weeks before placing a pig in a facility, clean it and spray with a disinfectant, such as 1 part bleach to 4 parts water–20 percent mixture. Be sure there are no nails or other objects that would injure your pig or you as you care for it.

9–43. Sample facility for a small show pig facility. (Size varies with number of pigs. For example, a place for one to three pigs could have a shed that is 8 x 14 feet with an outside exercise area of 20 x 14 feet or larger. Locate the feeder and waterer inside the shed. In cold climates, additional protection from the weather may be needed.)

- Choosing a Pig—Selecting a pig to raise involves at least three important considerations. First, will you raise a market hog or a breeding hog? Most youth begin with market hogs. Today's market hogs are typically crossbreeds that manifest the desired conformation. Crosses of Yorkshire and Hampshire are popular. Raising breeding hogs is more of a long-term enterprise. Secondly, what is the target date for having your pig show-ready? With market hogs, finish is gained at 6-7 months of age. This means that if a show is in July, you would choose a pig that was farrowed in January or February. If you buy a feeder pig, it should weight 40 pounds or so and be weaned from its mother. You would likely buy your pig in March or April. For example, if you got a 50–pound pig in March it would need to gain at least 170 pounds before the show in August. Thirdly, what is the desired conformation? You will want a pig that is structurally correct, have good muscle shape and definition, and wide body and rib cage. Market demands sometimes change and you would want to adapt accordingly. Always seek the help of an informed hog person in selecting your pig. (Refer to Figure 9-6 for pointers on desired conformation.)

- Providing Nutrients—Your pig will need proper nutrients in order to grow and meet desired weight and finish for the show. A self feeder is preferable, as a pig can eat at any time. Experienced hog growers know how to regulate feed intake to assure rate of gain needed for a show. With a feeder pig, you will feed a growing ration and a finishing ration. Growing rations have more protein, calcium, and phosphorous, while finishing rations have more carbohydrates. Transitioning from the growing to the finishing ration begins when a pig weighs about 110 pounds and may take a few days. When transitioning, feed half growing and half finishing rations for 7 to 10 days in changing to all finishing ration. A pig needs 4 to 5 pounds of feed for each pound of gain in body weight. Fresh water should always be available upon demand.

Table 9-1. Feeding a Show Pig*

Stage of Growth	Weight (lbs.)	Daily Feed Intake (lbs.)	Average Daily Gain (lbs.)
Growing	44-110	4-5	1.5
Finishing: Lighter	110-240	5-6	1.8
Finishing: Heavier	240-280	7-8	1.7

*Source: Dani Peters. *The Market Hog Guide to Success.* Oregon State University Extension Service, January 2011.

- Managing Health—Good health is essential for a pig to grow. Establish a relationship with a local veterinarian to develop a health management plan. This also helps you to have one available if a problem arises. Provide routine vaccinations, such as for erysipelas, rhinitis, dysentery, and parvovirus (see earlier section in this chapter). Control internal and external parasites. Deworm a pig every 30 to 45 days and use a different wormer each time. Regularly observe for signs of a health problem. Here are some signs that may indicate that a pig has a health issue:
    - Goes off feed (stops eating or greatly reduces feed consumption).
    - Has scours (this is a sign of a digestive system problem).

- Does not drink water as it previously did.
- Goes off to one side and lies down.
- Other abnormal behavior (obviously, you should observe your pig during normal health so you will be able to identify abnormal behavior).

If signs of disease develop, use the services of a veterinarian. Be sure to follow rules and regulations of your state or the show that you will be entering your pig.

9–44. Brushing a mixed-breed pig that is being finished helps to prepare for showing. (Note: The barrel contains an automatic waterer that makes water available any time the pig wishes to drink.) (Courtesy, Education Images)

- Training and Showing—Regular contact with your pig helps tame it and establish a bond. This can involve providing exercise for the pig (be careful not to do so in hot weather as pigs do not have sweat glands). Teach commands by tapping gently on the side of the head with a show stick or piece of PVC pipe. Do not tap on the ribs, back, or hindquarters. Use exercising of the pig as a time to further teach commands. Use exercise as time to pretend you are in a show ring and direct the pig around. Regardless of a pig's behavior, do not hit or slap it. Hard hits cause bruises and, possibly, other injuries. Also, a pig remembers abuse and will not respond well the next time you wish to drive it. Note: Pigs that have not been exercised will tire when in a show ring. Begin washing the pig once a week a few weeks before the show.

Follow all rules of a show. Arrive on time with your pig. Put your pig in the pen assigned to you. Be sure the pen is secure and the gate closed properly–you never want a pig to escape! Provide feed and water, with water available all of the time. Provide a comfortable temperature with a fan in warm weather or protection from the cold in the winter. Use the same shampoo, feed, and other supplies at the show as you use at home. Know the time you will be showing and have your pig ready. Follow directions when called to the show ring. Keep your eyes on the judge and position your pig for the judge to get the best view.

9–45. In the show ring, keep your eyes on the judge and keep your pig between you and the judge. (Courtesy, Education Images)

REVIEWING

MAIN IDEAS

The swine industry is the second largest livestock enterprise in the United States. More than 60 percent of hogs are raised in confinement facilities.

Hogs are good to grow because they are prolific, labor requirements are low, the capital investment is low, and the return can be fast. Hogs are susceptible to numerous diseases and parasites. A farrowing enterprise requires more labor because of the attention to sows and piglets.

Four production systems are used—feeder pig, finishing, farrow-to-finish, and purebred. The goal of the first three is to produce high-quality pork. The goal of the latter is to produce quality breeding animals.

Good breeding management of swine is needed to attain high numbers of pigs twice a year. Keeping facilities clean is essential.

Because hogs have a simple stomach, they require concentrates and eat very little forages. Cereal grains, soybean meal, and mineral and vitamin supplements are used in a total mixed ration. Additives, such as antibiotics, sulfas, and porcine somatotropin, may be used to increase efficiency. Always follow withdrawal time guidelines when using additives and administering medicines in other ways.

Hogs require good facilities. Selecting the location for a hog facility is an important decision. Locate the facilities downwind of large populations. Building requirements vary based on the type of operation and the management system adopted.

QUESTIONS

Answer the following questions using complete sentences and correct spelling.

1. Why is the swine industry a good or favorable enterprise?
2. What is a meat-type hog?
3. Name five breeds of swine and distinguish between them.
4. List and explain the three different systems of the swine industry.
5. What are the characteristics of a good boar? Good sow?
6. How should the sow be managed at the time of parturition?
7. Describe management practices with newborn pigs.
8. Describe the nutritional requirements of swine.
9. Name three diseases of swine and describe how each affects hogs.
10. Describe the facility and equipment needs of swine.
11. What six activities are involved in raising a show pig? Briefly tell what each is about.
12. What are the different stages of growth in raising a show pig? How does feeding vary with the stages?

EVALUATING

Match the term with the correct definition. Write the letter of the term on the line provided.

a. prolific
b. tail docking
c. pedigree
d. probe
e. specific pathogen free
f. gilt
g. barrow
h. feeder pig
i. needle teeth
j. withdrawal time

_____1. A record of heredity.

_____2. Free of disease and raised in an aseptic environment.

_____3. Tool for measuring thickness of backfat.

_____4. Pig weighing 40 pounds.

_____5. Clipping or cutting the tails on newborn pigs.

_____6. Young female swine that have not had pigs.

_____7. Produce a large number of offspring.

_____8. Male swine castrated at a young age.

_____9. The period before marketing in which medications and other additives should not be given to hogs.

_____10. Sharp teeth in baby pigs that injure the teat of the sow.

EXPLORING

1. Visit a swine farm. Note the management of reproduction, nutrition, and health. Interview the manager or a worker about the farm. Determine the breed or cross being raised. Take a camera and make photographs to prepare a poster about the farm.

2. Write a report on the use of porcine somatotropin. Use reference material to help in preparing the report. Contact the animal science department at the land-grant university in your state for information. Your agriculture teacher can provide the name and address of the individual to contact.

3. Investigate the Pork Quality Assurance (PQA) program. Prepare a report on your findings. Identify food safety and HACCP issues. List and describe the 10 Good Production Practices. (A good place to begin your investigation is: **www.porkboard.org/PQA/**.)

CHAPTER 10

Sheep and Goat Production

OBJECTIVES

This chapter provides basic information on sheep and goat production. It has the following objectives:

1. **Explain the importance of the sheep and goat industry.**
2. **Describe sheep and goats as organisms.**
3. **List and describe the major breeds of sheep.**
4. **List and explain the groupings of goats.**
5. **Explain sheep selection and production systems.**
6. **Describe breeding management of sheep.**
7. **Explain nutritional needs and feeding practices for sheep.**
8. **Explain important health management practices used in sheep and goat production.**
9. **Describe facility and equipment needs with sheep and goats.**
10. **Describe fitting and showing sheep and goats.**

TERMS

billy
browse
buck
chammy
confinement method
docking
drenching
ewe
farm flock method
kid
kidding
lamb
lamb feeding
lambing
mutton
nanny
orphaned lamb
purebred flock
ram
range band method
wether
wool

10–1. A young woman enjoys her Boer goat. (Courtesy, Education Images)

SHEEP and goats look a lot alike. They are closely related and have been domesticated for a long time. Both are ruminants but eat different plants. People who know a little about sheep and goats can easily tell them apart.

Here are a few distinctions: goats have a beard, sheep have foot glands, and male goats have a strong smell. Slight differences exist in the horns and skeletons. Goats are more intelligent, independent, and have a better ability to fight and protect themselves.

Goats were one of the first animals to be domesticated. They were once used to plant seed by trampling them into the ground. They were used as a source for food and fiber, and their skin was used for bottles. Sheep and lambs have also long been used for food and fiber.

10–2. Observe the difference in the head of a goat (top) and sheep (bottom). (Courtesy, Education Images)

THE SHEEP AND GOAT INDUSTRY

Sheep and goats are raised for food and clothing. They provide many important products. Goats were domesticated about 9,000 years ago, and sheep more than 8,000 years ago.

Sheep originated in Asia and Europe. Goats came from the eastern Mediterranean area and Asia. The first were brought to America by settlers some 400 years ago. The numbers of both grew until the mid 1900s in North America.

WHERE SHEEP AND GOATS ARE RAISED

Australia and New Zealand are the leading producers of sheep. The per capita consumption of sheep and lamb products in New Zealand is 57 pounds (27.2 kg) a year. Australia is second at 30 pounds (13.6 kg) of sheep and lamb products a year. The U.S. is way down the list of countries, with per capita consumption of lamb and mutton being 0.8 pound (0.36 kg). The world per capita consumption is about 3 pounds (1.4 kg) a year. Several European countries produce large numbers of sheep.

10–3. Sheep are often pastured on rangeland unsuited for other uses. (These are being herded by a herder on horseback and a Border Collie.) (Courtesy, Agricultural Research Service, USDA)

China and India are the leading producers of goats. These two countries have more than 460 million head. The U.S. has had considerable increase in goat keeping in recent years. Today, there are nearly 4 million head in the U.S., with fewer than 1 million being milk goats. California has the largest dairy goat population. Obviously, meat goats are far greater in number than milk or Angora goats. Some authorities feel that the number of goats will soon surpass the number of sheep and lambs.

The number of sheep in North America decreased between 1940 and 2012. This decline was due to lower returns, higher risks than cattle or crops, increased loss due to predators, scarcity and high wages of good sheep herders, and uncertainty in price. Today, there are 5.6 million sheep and lambs in the United States.

On a global basis, the United States ranks twenty-seventh in sheep numbers. Over half of the sheep are found in ten western range states and Texas. Farm flocks range from a few head to thousands.

PRODUCTS FROM SHEEP AND GOATS

Sheep and goats provide many products that are important to people. The major products are food and clothing.

Food

The food products from sheep and goats are meat and milk. The meat from sheep is lamb and mutton. ***Lamb*** is the meat from a young sheep that is less than one year old. Lamb includes both male and female sheep. Lamb is a delicate meat prepared in many delicious ways. ***Mutton*** is the meat from a sheep that is more than one year old. Mutton has a stronger flavor than lamb and is less desirable as a meat in the United

10–4. Lamb carcasses. (Courtesy, Education Images)

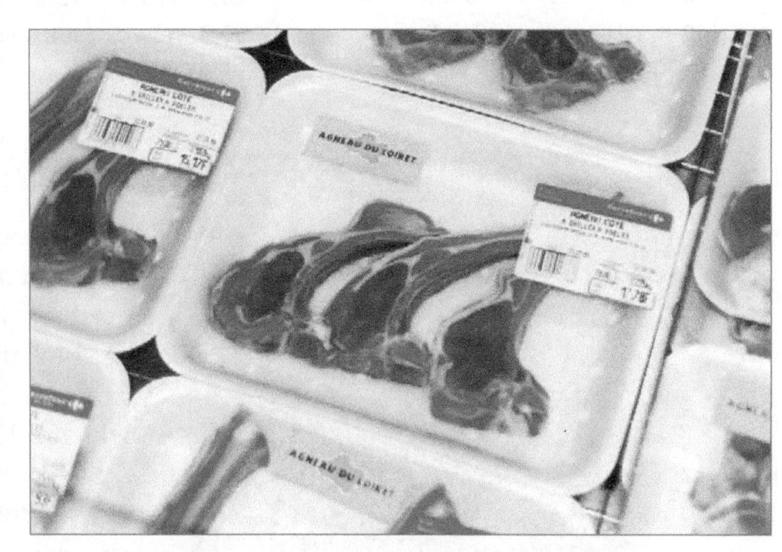

10–5. Lamb chops in a supermarket display case. (Courtesy, Education Images)

States. Goat is referred to as goat except for younger goat meat, which is known as kid. Many people enjoy barbecued goat meat.

Of goats and sheep, milk is primarily from goats. Goat's milk is easier for some people to digest than cow's milk. Goat's milk contains more vitamin A than cow's milk and has smaller fat particles (known as globules) than cow's milk. Goat and sheep's milk are used to make cheese, such as Roquefort cheese from sheep's milk.

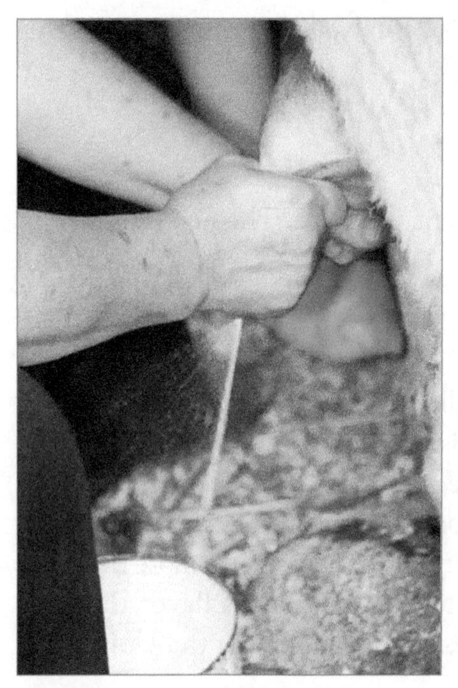

10–6. Milking a goat. (Courtesy, frantab/Shutterstock)

10–7. Wool is shorn from a sheep. (Courtesy, Gail Johnson/ Shutterstock)

10–8. Bags of wool loaded for hauling to market. (Courtesy, Education Images)

Clothing

Clothing is made from the hair or wool and hides of goats and sheep. ***Wool*** is the soft coat of sheep and is a major fiber used in making clothing. Mohair and cashmere are made from the fiber covering of specific breeds of goats. Sheep yield 5 to 15 pounds of wool each shearing.

Sheep and goat skins are used for leather products. ***Chammy*** is a soft, pliable leather made from sheep and goat skin. It is used in cleaning and polishing—often on automobiles.

ADVANTAGES AND DISADVANTAGES OF SHEEP AND GOATS

Many factors make sheep and goat production favorable. Compared to other types of livestock, they are better suited to more arable land for grazing. They are excellent scavengers. Sheep are more efficient at converting feed to meat than cattle. They are dual-purpose animals producing wool and meat. Lambs are usually marketed in eight months yielding a fast return on investment.

10–9. Sheep have been corralled from a pasture. (Courtesy, U.S. Department of Agriculture)

Goats and sheep can be pastured together because they eat different plants. Goats eat **browse** (woody plants) and broadleaf plants. Sheep graze short grass and some broadleaf plants.

Several factors make sheep and goat production unfavorable. The price of wool is low and unstable. Competition from synthetic fibers has hurt the industry. Consumption of lamb is very low. Sheep are susceptible to disease and parasites with little resistance to keep themselves healthy. Sheep and goats are also susceptible to attack from predators. Management of sheep and goats is important because they are dual-purpose animals. About 80 percent of sheep production is lamb for meat purposes, while only 20 percent is for wool production.

Sheep and goats are often raised by youth. They are relatively small animals and require less elaborate facilities than cattle and do not create the waste disposal issue as with hogs. Lamb and goat shows have increased considerably in recent years.

10–10. Spanish goats thrive on brushy rangeland. (Courtesy, Texas Department of Agriculture)

SHEEP AND GOATS AS ORGANISMS

Sheep and goats are alike in many ways. The differences provide advantages in different production situations.

Sheep and goats are mammals with ruminant digestive systems. They have cloven (divided) hoofs. Both are scientifically classified in the Bovidae family. Domesticated sheep are Ovis aries. Goats vary in scientific name, with the Spanish goat being Capra pyrenaica and other domestic goats being Capra hircus.

Sheep are far more important economically than goats in the United States. Greater emphasis in this chapter will be on sheep.

DIFFERENCES BETWEEN SHEEP AND GOATS

Management of sheep and goats is basically the same. However, there are a few differences. For example, sheep herders work behind the herd. Goat herders work ahead of the animals to lead them. (Chapter 2 presents the major external parts of a sheep.)

Size

Goats and sheep are of similar size. Variations exist within the species as well as between.

Mature goats range from 20-pound (9 kg) dwarfs to 150 pounds (68 kg) or more for common domestic species. Heights range from 1.5 to over 4 feet (45 to 120 cm). Goats live 8 to 10 years.

Mature sheep range from 100 to 225 pounds (45 to 102 kg) or more. The wool on a large sheep may weigh 15 pounds (7 kg) or more. Sheep live 7 to 13 years.

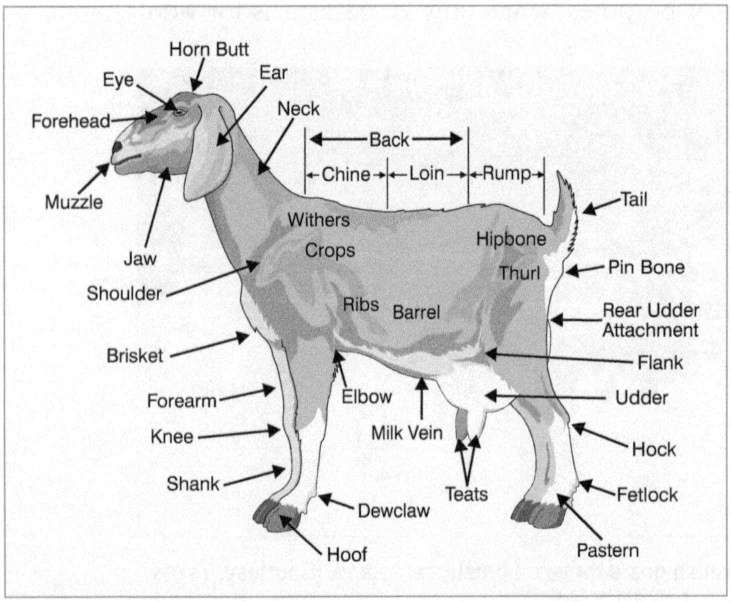

10–11. Major external parts of a dairy goat.

Age and Sex Classification

The names for sheep and goats vary based on sexual classification and age. A lamb is a young lamb of either sex that is less than one year old. A **ewe** is a female sheep of any age. A

ram is a male sheep kept for breeding purposes. A male sheep castrated before sexual maturity is a ***wether***. Its secondary sexual characteristics have not developed at the time it is castrated.

Goats have similar life cycles to sheep, but the terms applied to age and sex classifications are quite different. A female goat is a ***nanny*** or doe. A male goat is a ***buck*** or ***billy***. A young goat under a year of age is known as a ***kid***. Most all kids reach puberty by a year of age.

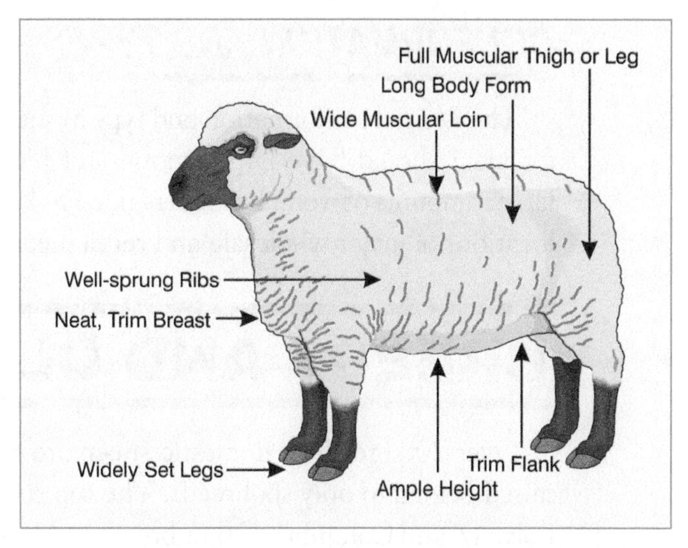

10–12. Desired conformation of a lamb for meat.

Reproductive Traits

The breeding season of goats is regulated by climate and weather. In hot climates, goats may mate year round. In climates away from the equator (including most of the United States), goats typically mate in the late summer and fall. Kids are born in the early spring the following year.

Does are bred to have their first kids at two years of age. The process of giving birth is known as ***kidding***. Most kids weigh about 5 pounds at birth, but weight depends on the breed of goat. Goats typically have two or three kids.

The gestation period of goats averages 151 days, whereas 148 days is common in sheep. The estrus period for goats lasts 18 to 19 days, which is one to two days longer than sheep. Newborn kids require greater care than lambs. Kids are either staked near a small A-shaped box or put in pens or corrals. Goats require little, if any, shelter. If they are well fed, they will not suffer in cold weather.

More information on sheep reproduction is presented later.

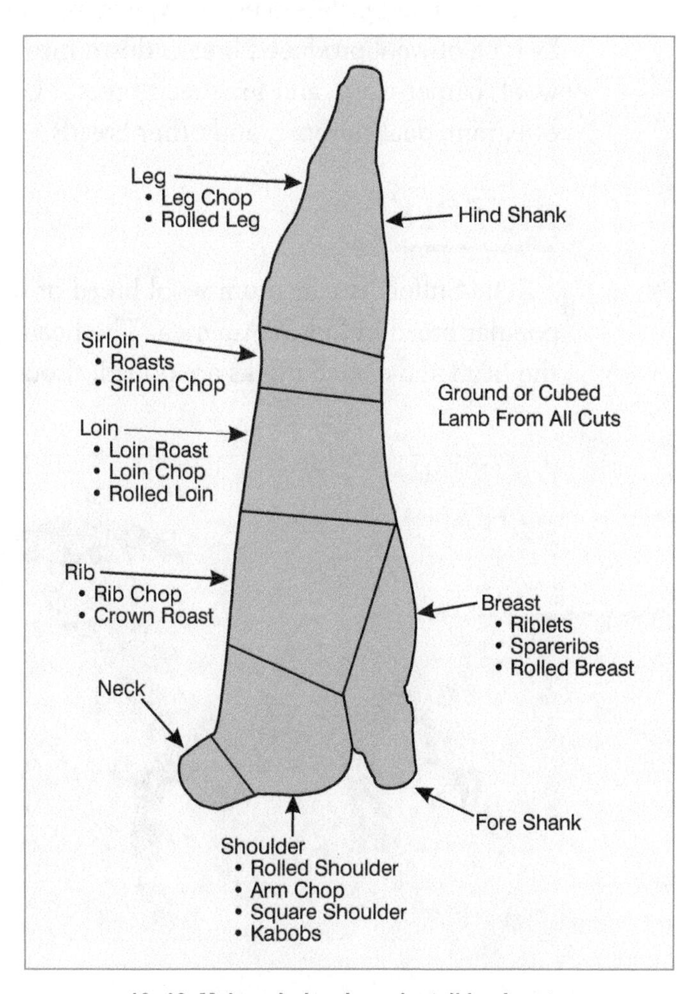

10–13. Major wholesale and retail lamb cuts.

CONFORMATION AND TYPE

The desired conformation and type in sheep vary with the product to be produced. Sheep for meat should have large amounts of higher-valued cuts. Sheep for wool should have larger amounts of wool. Producers need to know the major external parts of a sheep and the locations of major wholesale and retail meat cuts.

BREEDS AND CLASSES OF SHEEP

Over 200 breeds of domestic sheep are in existence today. Three-fourths of the sheep raised belong to only six breeds. The top six are Suffolk, Dorset, Hampshire, Rambouillet, Polypay, and Columbia. Other breeds in North America include Oxford, Southdown, Corriedale, Montadale, Shropshire, Cheviot, St. Croix, and Katahdin.

Sheep may be classed by the type of wool produced or by their breeding use. Classification by type of wool produced breaks down into fine-wool, medium-wool, long-wool, crossbred wool, carpet-wool, and fur sheep breeds. Classification by breeding use breaks down into ewe, ram, dual-purpose, and other breeds.

SUFFOLK

The Suffolk is a medium-wool breed of sheep. It originated in England and is the most popular breed in North America. The head, ears, and legs are black, without any wool on the head and ears. Suffolks are polled, though some males may have scurs.

10–14. Suffolk ewe. (Courtesy, Education Images)

DORSET

The Dorset is a medium-wool breed. The sheep may be polled or horned, with most being polled. The Dorset is entirely white and has no wool on its face. Ewes breed out of season and are known as good milkers. The Dorset breed originated in England.

10–15. Dorset ewe. (Courtesy, Education Images)

HAMPSHIRE

The Hampshire originated in England and is a medium-wool breed. The face, ears, and legs are dark brown and often nearly black. Both males and females are hornless, though males may have scurs. The Hampshire is known as an early-maturing, large sheep.

10–16. Hampshire ewe. (Courtesy, Education Images)

10–17. St. Croix sheep have many desirable traits, such as heat tolerance and parasite resistance. (Courtesy, Agricultural Research Service, USDA)

OXFORD

The Oxford is a medium-wool breed that is popular with youth in lamb shows. Originating in England, the Oxford is a cross of Hampshire and Cotswold. The face, ears, and legs may have brownish or grayish coloring. It has a topknot of wool (wool cap) on its head. At maturity, the Oxford is among the largest breeds.

KATAHDIN

The Katahdin is a breed of sheep with hair rather than wool. Originating in the Virgin Islands, the Katahdin was first brought to Maine along with the English breeds of sheep. Colors vary from white to cream or red. The coloring may be spotted or solid. Most Katahdins are polled, though some males may have small horns or scurs. The Katahdins are not tail-docked as are the wool breeds. Docking is not necessary because the hair does not collect feces, as does the wool on other sheep.

10–18. Katahdin sheep. (Courtesy, Education Images)

RAMBOUILLET

The Rambouillet is a descendant of Spanish Merino sheep. Half of all sheep in the United States have some Rambouillet blood. Rambouillets are good dual-purpose sheep. They are white, have blocky bodies, and may be horned or polled. All ewes are polled.

10–19. Rambouillet lamb following a successful show. (Courtesy, The American Rambouillet Sheep Breeders Association)

GROUPINGS OF GOATS

Some 300 breeds of domestic goats are known. Wild goats are still found in many areas, such as the Rocky Mountains of North America.

Goats should be selected based on your goals. Goats should be healthy and disease free. Milk goats should have a well-developed udder. Color and the presence of horns may also be factors. A discussion of the five groups of domestic goats follows.

ANGORA GOATS

There are more than a million Angora goats in the United States. However, few people really know what they look like. Texas is the leading producer of Angora goats. They are well adapted to living in areas where the grazing is not suitable for other livestock.

Angora goats are almost totally white. Occasionally a black one will be found. Kids may be red but usually shed their hair, and the new growth is white. The Angora goat's coat is of long locks of hair, known as mohair, and covers the entire body except the face. The average Angora produces up to 7 pounds of fine-quality mohair each year when sheared twice.

DAIRY GOATS

Dairy goats produce 1.8 percent of the world's supply of milk. An average doe may produce five pounds of milk per day during a ten-month lactation.

Goat's milk has a sweeter taste and contains more minerals than cow's milk. During digestion, goat's milk forms into a soft curd. This makes it easier for children and the elderly to digest than cow's milk. The presence of bucks around does causes the milk to take on an unpleasant odor. Major breeds of milk goats are described below.

10–20. Nubian doe. (Courtesy, American Dairy Goat Association)

- Nubian—The Nubian is considered a dual-purpose breed for milk and meat. It has larger, drooping ears than other breeds. The record milk production in a year is 4,420 pounds. Though the amount of milk is less than some other breeds, the milk has a higher fat content.
- Lamancha—The Lamancha breed originated in California. A major characteristic is the absence of external ears. The Lamancha has higher milk production than the Nubian.

10–21. Lamancha doe. (Courtesy, American Dairy Goat Association)

10–22. Alpine doe. (Courtesy, American Dairy Goat Association)

- Alpine—The Alpine is a good milker, with annual production up to 5,700 pounds a year. It has distinctive coloring, with white and darker colors forming unique patterns.

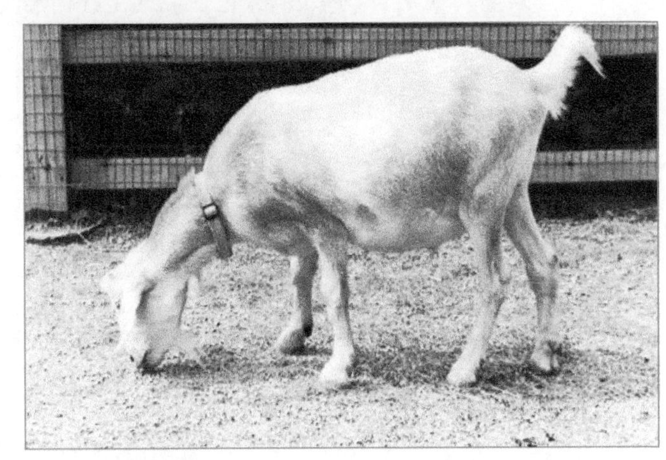

10–23. Spanish goat. (Courtesy, Education Images)

10–24. Boer goats have distinctive coloring with brown heads and white bodies though the color pattern may vary. (This shows a young female Boer in a playful mood.) (Courtesy, blewisphotography/Shutterstock)

MEAT GOATS

Some meat goats were derived from the Mexican Criollo; thus, they are called Spanish goats. The coloring varies from white to black, with some being orange to sandy.

The Boer goat is emerging as an important meat goat. It is an improved breed that is sometimes pastured with cattle. Boer goats prefer browse over pasture. They do not compete with cattle for grass.

Boer goats arrived in the U.S. in the early 1990s from New Zealand but originated in South Africa. They have rapidly become popular, with some people being Boer goat enthusiasts. They have a high resistance to disease and are tolerant of heat. Mature bucks weigh 240-300 pounds (110-135 kg); does 200-220 pounds (90-100 kg). Boer goats have a fast growth rate and excellent carcass qualities.

10–25. Pygmy goat. (Courtesy, Education Images)

CASHMERE GOATS

Cashmere goats have a fine down undercoat. This is the finest animal fiber used. Cashmere produces clothing that is light and soft, yet warm. It also has three times the insulating value of wool.

PYGMY GOATS

Pygmy goats are miniature goats used for research and as pets. People find them enjoyable to watch.

SHEEP SELECTION AND PRODUCTION

The management of sheep is more complex than the management of most animals. The complexity is the result of trying to get two products (wool and meat) from one animal, instead of one product, which is the case with most animals. Sheep meat makes up about 80 percent of the market and wool only 20 percent.

SELECTION

When selecting sheep, consider the production goals. Is the goal to raise purebred, crossbred, or grades? Breeds are selected for their characteristics. Also consider the size of the flock or herd; the time of year to start; and the uniformity, health, age, soundness of the udder, and price of the animal.

Usually only experienced breeders need to buy purebred animals. Beginners should start with a crossbred ewe. Selection of breed or class should be based on the purpose for raising

10–26. Herding sheep on the Ramah Navajo reservation in New Mexico. (Courtesy, U.S. Department of Agriculture)

them. Size of the flock should be based on the space and facilities available. The animal itself should be healthy, sound, and uniform for its age and breed.

PRODUCTION SYSTEMS

Five types of sheep production systems are used. Success in any of these types depends on managing a healthy and productive flock—that is, economically marketing lambs and wool.

The Farm Flock Method

The ***farm flock method*** is the most popular method used in the central, southern, and eastern United States. Production of market lambs is the focus of this method with the production of wool being secondary. Usually, the producer raises other livestock that compete with the sheep for pasture.

10–27. Farm flock in good pasture with shade structures and good fences. (Courtesy, Education Images)

The Purebred Flock Method

Few farms raise a ***purebred flock***. Their main purpose is the sale of rams and ewes. This requires a high level of management ability. The producer must raise sheep of ideal type. A full-time caretaker is usually employed by large producers.

The Range Band Method

With the ***range band method***, each band of sheep has its own herder who moves them over a large area of land. The emphasis on lamb or wool depends on the region's rainfall and vegetation. In arid regions, such as in the southwest, wool production is emphasized because there is not enough grass to finish a lamb.

10-28. Lambs are fed concentrates to promote growth and good health. (Courtesy, National Lamb and Wool Grower Magazine)

The Confinement Method

The ***confinement method*** is becoming more popular. Raising animals in confinement means that they are raised totally indoors. Reasons for the popularity of this method include the success of raising other types of livestock in confinement, high land prices, fewer parasite problems, and the increase in growth rate of the young.

The confinement method has several unfavorable factors for most producers. Equipment and building costs are higher. This type of system is more restrictive than the other methods. More skill and better management practices are required.

CAREER PROFILE

SHEEP HERDER

A sheep herder cares for sheep. They are sometimes called shepherds. The work involves moving, feeding, protecting, helping with births, and treating diseases. Duties depend on the nature of the farm or ranch and the kind of production system.

Sheep herders need practical experience with sheep. They should like sheep and enjoy working outside. A high school education is needed. Working under the direction of an experienced sheep herder helps develop the needed skills.

Jobs are found where farms and ranches produce sheep. Some jobs are with research facilities and demonstration farms. This photo shows a sheep herder moving a flock of sheep from one pasture to another. (Courtesy, U.S. Department of Agriculture)

Lamb Feeding

Lamb feeding is a specialized method of sheep production. After weaning, lambs are sold to feedlots. These feedlots are usually located near good, cheap food supplies. Feedlots feed out the lambs to a desired slaughter weight.

BREEDING MANAGEMENT PRACTICES WITH SHEEP

Commercial and purebred breeders alike strive to have a high lamb crop. The average lamb crop is 106 to 108 percent with the mortality as high as 25 percent from birth to weaning. Breeding and selecting ideal animals is important in meeting these objectives.

NORMAL BREEDING

Ewes reach puberty at 8 to 10 months of age, while rams reach puberty at 5 to 7 months. Ewes are usually bred to give birth (lamb) the first time at 24 months of age. Giving birth is known as ***lambing***. The time of the year when ewes have lambs is known as the lambing season.

The heat period lasts an average of 30 hours. Ewes show no visible signs of heat other than the acceptance of the ram. Estrus occurs every 16 to 17 days. The gestation period of sheep lasts around 148 days but varies among breeds.

RAM CARE AND MANAGEMENT

Rams are normally kept separate from ewes. A dry barn or lot with room to get plenty of exercise is adequate. Good hay and pasture are needed. Concentrate supplement may be needed in the winter and if the pasture is short. Rams do not require much, if any, grain. Many rams are pastured. Excess fat is harmful to a breeding ram.

EWE CARE AND MANAGEMENT

A pregnant ewe needs feed, water, shelter, and exercise. Quality hay and pasture have important roles in ewe rations. Supplemental concentrates and minerals may be used. A ewe that does not lamb is not profitable because lambs are more profitable than wool. Because of this, pregnancy testing is very important. Detection of one or multiple fetuses affects nutritional and health management practices.

10–29. Ewe with nursing lamb. (Courtesy, Education Images)

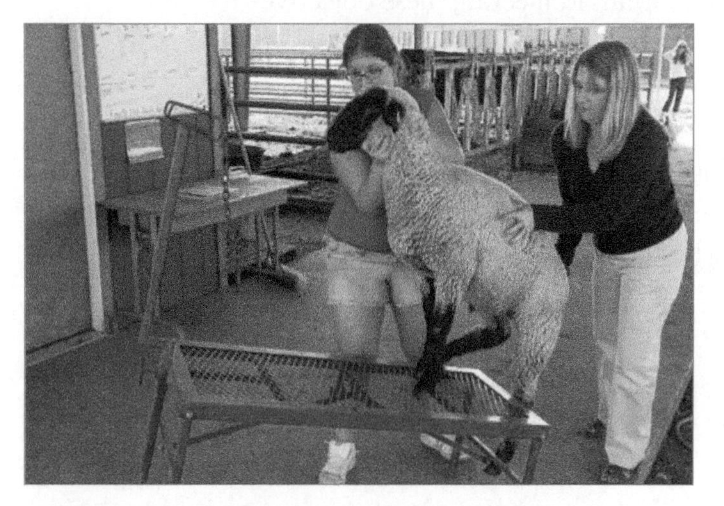

10–30. A lamb is being placed on a blocking stand as part of fitting for show. (Courtesy, Education Images)

Care at Lambing

Most death loss of lambs occurs in the first few days after lambing. As parturition approaches, the ewe should be sheared around the udder, flank, and dock; placed in a dry, roomy pen; and the amount of grain is reduced. Water is available free-choice. Proper care of the ewe and her lamb(s) include use of lambing pens, helping lambs if they are being born in the wrong position, nursing chilled and weak lambs, and examining and treating any health problems.

Orphaned Lambs

If a ewe dies or is unable to nurse, her lambs are orphaned. An ***orphaned lamb*** needs a foster mother. A ewe who is very healthy may accept an additional lamb. A ewe who has lost her own young may accept an orphaned lamb, if the orphan is first rubbed with the dead lamb. A more effective method is to tie the skin of the dead lamb onto the orphaned lamb. The skin may then be removed a piece at a time after a few days. If the lamb is not accepted, cow's milk or milk replacer may be used to feed the lamb.

MANAGEMENT FROM LAMBING TO WEANING

Between the time that lambs are born until they are weaned, they should be docked and castrated.

Docking

Docking is cutting off all or part of the tail. This is one of the first things done after lambing. It should be done when they are 3 to 10 days old. Docking may be done with a knife, hot docking iron, electric docker, or elastic band. The Katahdin lambs are not docked because they have hair rather than wool. Goats are not docked.

Castration

Ram lambs should be castrated when they are young. Many producers dock and castrate simultaneously. Castration may be done by similar methods as docking. Remember to be as sanitary as possible to prevent infection. Goats and Katahdin sheep are not docked.

NUTRITION AND FEEDING OF SHEEP

Success in the sheep enterprise can be measured by the percentage of lambs raised and the pounds of lamb marketed. Nearly 100 percent of a sheep's diet is made up of roughages. Sheep have a ruminant digestive system that allows for use of pasture and hay. Sheep are not fed grain except just before and immediately following lambing.

NUTRITION NEEDS

Sheep, like other animals, require the six basic nutrients of carbohydrates, fats, proteins, minerals, vitamins, and water.

CONNECTION

FEEDING AN ORPHAN

Occasionally a doe or ewe will die or be unable to care for her kids or lambs. With care, you can bottle feed the orphan. A bottle with a nipple and patience on your part is essential. Hold the bottle upward so that the milk is swallowed rather than getting into the lungs, as is being done in feeding this orphaned kid.

An orphan needs 3 ounces of milk per pound of body weight within the first 24 hours. Use milk from another doe or ewe that has just given birth so it contains colostrum. An artificial colostrum product is sometimes used. After the first day, colostrum is not so important. Increased amounts of milk will be consumed as the baby grows. Cow's milk can sometimes be used but it isn't as high in some nutrients as goat or sheep's milk. (Courtesy, Education Images)

Carbohydrates and Fats (Energy)

Energy, which comes from carbohydrates and fats, is provided chiefly by pasture, hay, and silage. Corn, oats, wheat, barley, or grain sorghums may be added to the feed ration during times of drought, overgrazing, or snow-covered pastures. During the fall, they may have access to stalk or stubble fields.

Protein

Sheep need high levels of protein because they produce wool, which is made up of protein. Alfalfa, clover, soybeans, and green pasture are all good sources of protein. These crops are best when harvested before the protein quantity begins to decline.

When these are not available, a protein supplement may need to be added. Soybean, cottonseed, linseed, canola, peanut, and sunflower meals are good sources.

Range sheep may develop a protein deficiency. Protein blocks may be placed in the pasture to alleviate this problem.

Minerals

Sheep require the macrominerals salt (sodium chloride), calcium, phosphorus, magnesium, potassium, and sulfur. The microminerals that they require include cobalt, copper, fluorine, iodine, iron, manganese, molybdenum, selenium, and zinc.

Phosphorus deficiency may be experienced by range sheep. Self-feeding salt and mineral mixtures may be placed in pastures to alleviate any mineral deficiencies.

10–31. Commercial feed concentrate is often used with lambs being raised for show. (Courtesy, Education Images)

Vitamins

All of the fat-soluble vitamins (A, D, E, and K) are required by sheep. Bacteria in the rumen produce adequate quantities of the water soluble (B) vitamins. Vitamin A is the most likely to be deficient. Vitamin mixtures may be fed free choice.

Water

The average sheep needs one gallon of water each day. Sheep obtain water not only from watering facilities, but also from feed, dew, and snow. When feeding on lush, green pasture, sheep may go for weeks without drinking water.

FEEDING

Sheep and goats benefit from pasture and other roughage. They are suited to rangeland where cattle would not thrive well.

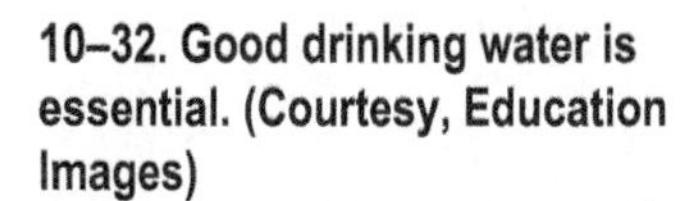

10–32. Good drinking water is essential. (Courtesy, Education Images)

Sheep do best on quality pasture and hay. They may also be given supplements with protein, minerals, and vitamins. Lambs may be creep fed while still nursing.

Goats tend to thrive in more rugged terrain. Milk goats are often kept in improved pastures and provided quality hay and concentrate supplements.

HEALTH MANAGEMENT

Keeping sheep and goats healthy means developing a preventive health program. When an animal gets sick, the expense of treating the animal usually outweighs the value of the animal. Therefore, prevention is clearly best.

GENERAL PRACTICES

Some recommendations for prevention of ill health of sheep and goats include monitoring the animals, regularly checking for any signs of illness; providing a well-balanced ration; reducing stress whenever possible; controlling outside traffic from other animals and vehicles; rotating pastures to reduce parasite problems; and keeping facilities and equipment sanitary.

A good vaccination and checkup program should be established with the local veterinarian. If an animal becomes ill or has just been purchased, isolate the animal from the rest of the flock or herd. New animals should be isolated for at least 30 days.

DISEASES

Several diseases afflict sheep and goats. Prevention is especially important. Chapter 4 has general information on diseases. Examples of sheep and goats' diseases follow.

- Actinobacillosis (lumpy jaw)—Actinobacillosis causes economic loss to the producer. It affects the jaw and parts of the animal's head. Tumors or lumps of yellow pus appear on the jaw. The muscles and organs may also become infected, causing parts of the animal to be condemned at slaughter. The animal will usually not die.
- Actinomycosis (wooden tongue)—Actinomycosis causes lesions on the animal's head and swelling of the lymph glands. The tongue may have lesions on it as it becomes hard and immobile. The animal finds it more difficult to eat and will lose weight. They will

10–33. A healthy lamb is being groomed for showing at a county fair. (Courtesy, U.S. Department of Agriculture)

eventually die. Animals that have actinomycosis should be isolated from the rest of the herd or flock. Contaminated feed usually spreads this disease.

- Blue tongue—Blue tongue is a viral disease that is spread by gnats. Blue tongue weakens the immune system of sheep and goats. Thus, most deaths occur due to secondary diseases. Signs of blue tongue include a high fever, loss of appetite, and sluggish behavior. The lips, muzzle, head, and ears will become swollen. The mouth will turn red or blue and will develop ulcers. The animal will have trouble eating. Blue tongue does not have any available treatments. Vaccination at shearing time is the best method of prevention.
- Enterotoxemia—Enterotoxemia is a bacterial disease that usually affects lambs and kids. It is commonly known as overeating disease. Enterotoxemia is more likely to affect lambs being fed for show because of the large amount of grain they are given. Bacteria normally present in the intestines can grow rapidly and produce a powerful toxin. Affected animals are commonly found dead with their heads being arched up. This occurs because of the convulsions that they experience shortly before death. There is no treatment for enterotoxemia but practices can be followed to prevent it. Vaccination, good management, and proper feeding are keys to preventing this disease. Ewes should be vaccinated two and six weeks before giving birth. Lambs should also be vaccinated and a booster given about three weeks later. Increasing the amount of roughage and chlortetracycline may be necessary.
- Foot abscess—Foot abscess affects the foot's soft tissue. It is usually found when conditions are wet and muddy. Bacteria enter the foot causing abscesses or pockets of pus. The joints and tendons may also be affected. The animals should be placed on clean, soft bedding, abscesses drained, and antibiotics used.
- Foot rot—Foot rot is caused by bacteria, which are different from the one that infects cattle. It is extremely contagious and occurs when the animals are in wet conditions. Foot rot causes animals to lose weight and requires added labor to treat these animals. Animals will become lame and will have an unpleasant odor about the feet. Keeping their quarters dry and clean and isolating infected animals will help reduce losses from foot rot. A vaccination is available.
- Johne's disease—Johne's disease causes the intestinal walls to become thicker. Animals will appear to have diarrhea, lose weight, and eventually die. There is no effective treatment of Johne's disease. Purchased animals should be checked for a history of this disease.
- Lamb dysentery—Lamb dysentery is caused by bacteria and affects lambs when they are only one to five days of age. They will lose their appetites, become depressed, have diarrhea, and will suddenly die. Death loss can be very high. Sanitation and good management practices are the best methods of prevention.

- Polyarthritis—Polyarthritis may be a problem with 3 to 5 week old lambs. The lambs are inactive, lose weight or fail to gain weight, and are reluctant to move about. If made to move, they appear to be in pain. Tetracycline antibiotics prescribed by a veterinarian are used in treatment.
- Soremouth—Soremouth is a virus disease that appears as scabs on the mouth and lips of lambs. It is a contagious disease that can be transferred to humans. Wear rubber gloves when handling lambs with sore mouth. Rubbing iodine into the lesions after removing the scabs tends to promote healing. A vaccine is available that helps prevent soremouth.

PARASITES

External parasites cause losses to sheep and goat producers because of the decrease in the quality of wool, mohair, meat, and milk. Controlling of external parasites is done through good sanitation and proper use of pesticides. Common external parasites of sheep and goats include blowflies, bot flies, lice, mange, and sheep ked or tick.

Ringworm can be an external parasite problem with lambs. It is a contagious disease caused by several species of fungi and can be transferred to humans. As with all diseases, prevention is best. Several treatments are available such as Fulvicin® powder, Novasan®, and a 10 percent bleach solution sprayed on lambs, equipment, and pens.

The biggest problem of raising sheep and goats is the loss caused by internal parasites. Infested animals lose weight and lose milk and meat production. They have poor-quality wool and decreased reproductive efficiency. Signs of internal parasites include weight loss, rough hair coat, loss of appetite, diarrhea, and anemia. Animals should be wormed regularly and rotated between pastures frequently. Common internal parasites of sheep and goats include coccidia, liver fluke, lungworms, and stomach and intestinal worms.

Drenching is giving a liquid medication by oral means to control internal parasites. A dose syringe is used in drenching. Proper dosage must be used. Only use approved products to control parasites.

OTHER PROBLEMS

Sheep and goats may experience other disease problems. With lambs being fed for show, rectal prolapse is sometimes a problem. Rectal prolapse is a protrusion of the rectum. It is associated with feeding high amounts of concentrate and dusty conditions that cause a lamb to cough. Treatment by a veterinarian is needed.

FACILITY AND EQUIPMENT REQUIREMENTS

Sheep do not require expensive shelter. They are relatively hardy animals. Some producers raise sheep in confinement facilities using automatic feeders and waterers. This increases costs but decreases labor requirements. The average farm or ranch does not need confinement facilities.

HOUSING

Barns or sheds for sheep usually open to the south. This provides protection from winter weather. The barn should be well bedded, dry, and well ventilated. Where lambing is to take place, there should be electricity to hook up heat lamps. Troughs, feeders, and waterers should be placed in an easily accessed location.

Milking does are commonly kept in free stalls. Loose housing may be used for kids and yearlings. Both require extra bedding and care during the cold winter months.

10–34. Ewe and lamb in warm facility during the winter. Note the heat lamp is located to allow the lamb to enter but keep the ewe away. (Courtesy, Northwest Experiment Station, University of Minnesota, Crookston)

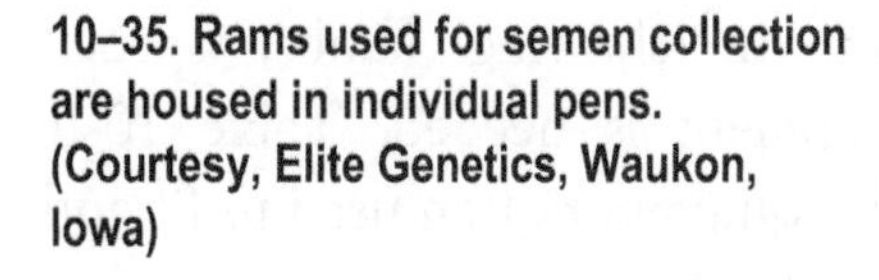

10–35. Rams used for semen collection are housed in individual pens. (Courtesy, Elite Genetics, Waukon, Iowa)

10–36. A specially designed corral for working sheep. (Courtesy, Education Images)

10–37. Judging is underway in a local lamb show for youth. (Courtesy, U.S. Department of Agriculture)

Table 10–1. Equipment Used to Fit and Show Lambs*

- Trimming table (45" long, 20" wide, and 18" tall)
- Electric clippers (and extension cords)
- 20- and 23-tooth combs with cutters
- Wool card or poodle comb
- Syringes and needles
- Lamb blankets and/or socks
- Rope halters
- Hoof trimmers
- Hand shears
- Bolus gun (for giving medicine)
- Drench-gun (for giving medicine)
- Small portable feed trough
- Shovel to clean pen
- Stiff brush for cleaning water trough
- Bucket or hose for water
- Hot air blow dryer
- Show box to hold equipment
- Portable livestock scales
- Electric fans (to cool lambs in barn)
- Muzzle (to protect lamb from eating bedding)

*You may not need to own all of this equipment. In some cases, you can borrow it from your FFA chapter, 4-H club, or managers of the show. However, you will need a halter and a few items for grooming the lamb.

Goats kept in loose housing or tie stalls must be dehorned.

FENCING

Fencing is needed to keep predators out, more than keeping sheep and goats in. Sheep need a 60-inch, or higher, fence. It could be made of woven wire or barbed wire with only 4 to 5 inches between each strand.

EQUIPMENT

Loading chutes, corrals, portable shelters, weigh crates, and pregnancy testing cradles may also be needed, depending on the operation. In showing lambs, special equipment is needed. Table 10–1 lists equipment often used to fit and show lambs.

10–38. An exercise track in which lambs are run with a four-wheeler. (Courtesy, Education Images)

EXERCISE TRACK

Lambs kept confined in barns or small areas should be exercised. An exercise track is often used. Tracks should be 5 to 7 feet wide and long like a corridor and have rounded corners. The outside fence should be at least 4 feet tall. The size of the oval area is typically 40 to 100 feet. The pathway should be free of rocks and trash that can injure lambs. The lambs are run in the oval using a sheep dog or a four-wheeler.

SHOWING

Lambs, sheep, and goats are often shown. With youth, lambs are most widely shown. Goat showing is increasing in some areas. As with all animals, good practices help promote success in a show ring. As with all animals, using ethical practices and promoting well-being of the animals are essential.

10–39. A muzzle is sometimes used on goats and lambs to protect them from eating unwanted materials such as sawdust, wood shavings, and unhealthy bedding material. (Courtesy, Education Images)

SHOWING LAMBS

Success in showing lambs involves several important practices. You can learn skills in many areas by observing competent people as they prepare a lamb for showing.

Selection

Selecting the lamb is likely the most important decision in showing. A person skilled in lamb selection should assist a beginner in choosing the lamb. You may buy a lamb or select one from your flock. Of course, you want to have a good quality lamb.

Consider the following when selecting a lamb:

- Classification—Classification includes the color markings, skeletal shape, physical structure, and softness of the pelt. Show managers set classification guidelines. If showing in a purebred class, be sure the lamb is ideal for the breed and is entitled to registration. Market lambs also have classes.
- Muscling—Muscling is the amount of muscle a lamb has. A lamb with good muscling feels firm to the touch. A lamb should have a long, level, square rump that is wide at the dock. A lamb that stands with legs wide apart is likely to have more muscle.
- Structural correctness—This refers to the skeletal system or bone development. A lamb should hold its head erect and stand with a strong and straight top. Other factors are also considered by the experienced lamb judge.
- Style and balance—A lamb should appeal to the eye when viewed. The hide should be tight and free of wrinkles. A level rump, trim middle, and straight legs are also desired. Again, experienced judges look for other factors.
- Growth potential—Select a lamb that has the ability to grow rapidly. Lambs with larger frames and longer necks may have this ability.

10–40. A lamb is being weighed to determine its rate of growth. (Courtesy, Education Images)

Feeding and Nutrition

Good practices are needed to assure growth. Clean, fresh water is essential for a growing lamb. Protein is needed to assure growth. Young lambs grow best on rations of 16 to 18 percent protein. Older lambs may have rations of 11 to 12 percent protein. Lower protein rations tend to promote fattening and slower growth. Carbohydrates and fats, minerals, and vitamins are also needed. In addition to grains and other concentrates, lambs may also consume small amounts of high quality hay. Using a commercially available feed designed to promote lamb growth is a common practice and is advised for the inexperienced lamb grower.

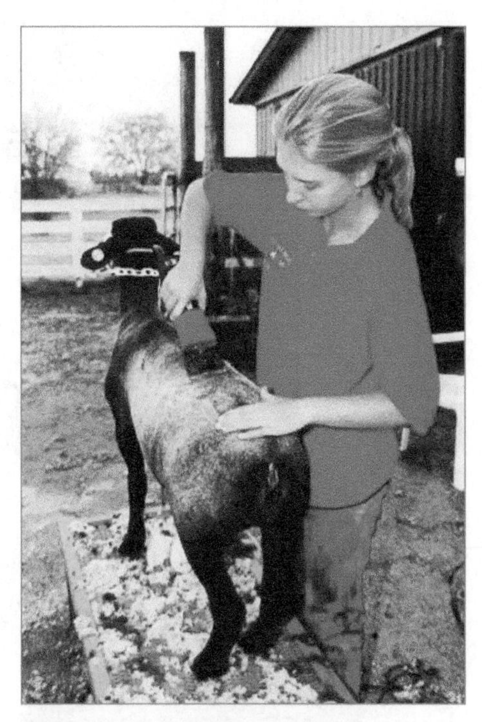

10–41. A lamb is being sheared in preparation for showing. (Courtesy, Education Images)

Health

A lamb needs good health to grow well and perform properly.

If you buy the lamb, it is probably a good idea to assume that it has had no immunizations or treatments for potential problems. (The earlier section of this chapter on health will provide useful information.)

Feeding and Management

Feeding a show lamb requires extra attention to the quality of the ration. Commercially prepared feeds or those you mix yourself may be best. In some cases, a local feed mill will specially manufacture a feed for the local area.

Feed is provided in two ways: self-feeding and hand-feeding. Self-feeding is using a trough with feed in it all of the time. Hand-feeding is providing feed once or twice each day at a fairly routine time.

Exercise is a part of management. Careful attention to exercise should begin two to three months before the show. Lambs should be exercised hard and fast for short distances or 350-450 yards each day. The benefit of exercise is that it causes increased adrenalin in their bodies that helps develop muscle. Lambs that have not been exercised may need to gradually build up to a rigorous exercise program.

10–42. A lamb sock has been used on this lamb after shearing and washing to protect it from dirt. (Courtesy, Education Images)

Fitting

Fitting is to get the lamb in top physical appearance for showing. An experienced show person can be very helpful to beginners in lamb showing.

Lambs are washed first. Washing is usually before arrival at the show, though some facilities allow wash-

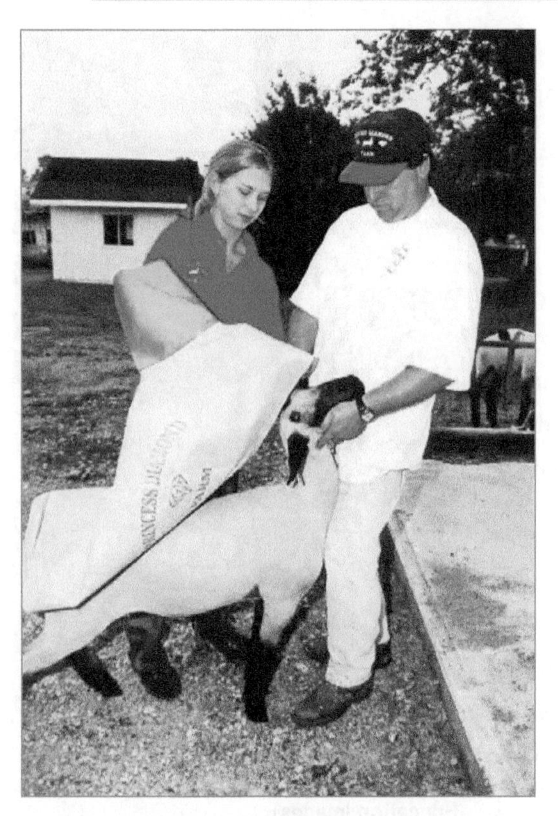

10–43. A blanket and hood are being placed on this lamb. (Courtesy, Education Images)

ing on the show grounds. Water with a mild soap, such as livestock soap, is used. The soap must be thoroughly rinsed from the fleece. Washing cleans the fleece before shearing. A blow dryer may be used to speed drying. Shearing is with clippers fitted with a 20-tooth goat comb and point cutter or a 23-tooth comb and 9-point cutter. Lambs are sheared smooth and parallel with the length of the body. Wool below the knees may be left on the lamb. Wool around the eyes, ears, and delicate areas may be removed with a small clipper. After shearing, a lamb is fitted with a blanket and hood or a lamb sock. The lamb is then kept in a clean, dry pen with good bedding.

Showing

Training is needed to assure that a lamb performs properly in a show. It should begin no later than two to three weeks before a show. Halter breaking is an important part of training. A lamb can be haltered and tied to a fence. Be sure someone is present while the lamb is tied. This is also a good time to place a lamb's feet so that it can get accustomed to setting up.

Once the lamb is gentle, teach it to lead. Put one hand under its chin and another hand on the back of its

10–44. Setting up a lamb. (Courtesy, Education Images)

head. A reluctant lamb may need to be pushed by another person from behind. With experience, you can lead the lamb to show its best features and to be led without a halter. Set up the lamb hind legs first followed by the front feet. Keep the body and neck straight and the head in a high, proud position with the ears up and forward. Always stand while setting up a lamb. Once set up has been mastered, teach the lamb to brace. A lamb braces when it pushes as pressure is applied to its neck or chest. Front feet should remain on the ground when bracing. Again, learning these skills under the direction of an experienced sheep handler is beneficial.

10–45. Bracing a lamb. (Courtesy, Education Images)

All of your hard work is presented in the show ring. Dress neatly and have a good appearance. Do not wear a hat or cap or use a halter on the lamb in the show ring. Position your lamb to do its best and be seen by the judge. Set up the lamb and be sure the body, neck, and head are in a straight line. Always know where the judge is in a show ring. Do not block the judge's view of your lamb. Brace the lamb as the judge handles it. Follow the judge's instructions. Always move a lamb forward and never backward. Be courteous to other exhibitors and have a pleasant expression on your face.

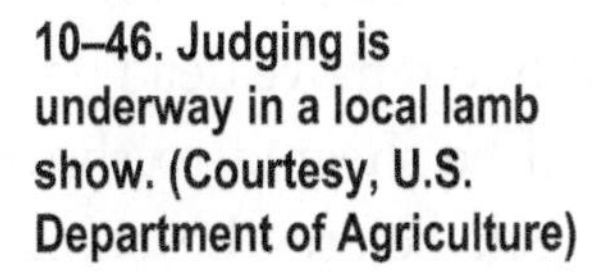

10–46. Judging is underway in a local lamb show. (Courtesy, U.S. Department of Agriculture)

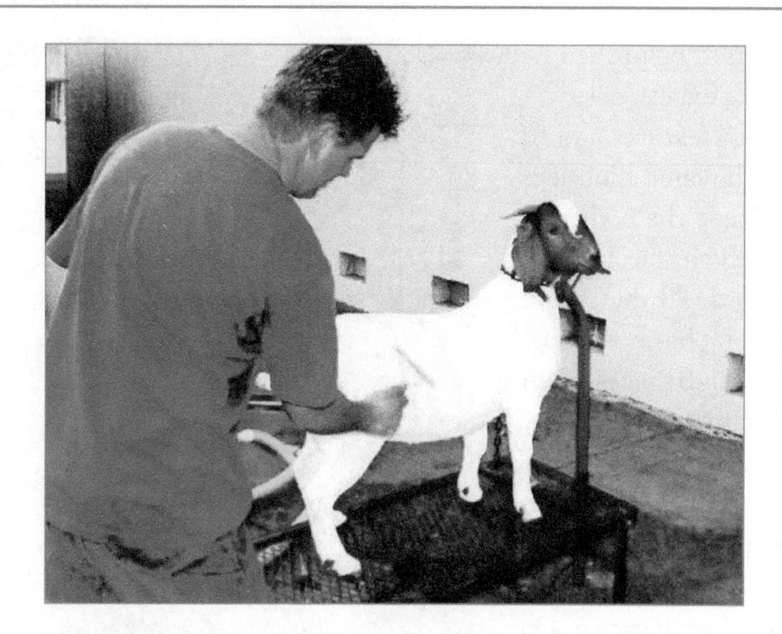

10–47. A Boer goat is being fitted for showing. (Courtesy, Education Images)

GOATS

Goat shows publish rules and regulations. Always read and follow these in selecting, fitting, and showing a goat. Vaccinate a goat for overeating disease, tetanus, and soremouth. Control scours with antibiotic boluses or electrolyte drench. Control stomach worms by deworming kids four to eight weeks after they get old enough to graze. Be sure the goat gets plenty of feed and water. Use a commercially available feed that meets the age requirements and desired growth of the goat. A 16 percent protein feed in pellet form is often used.

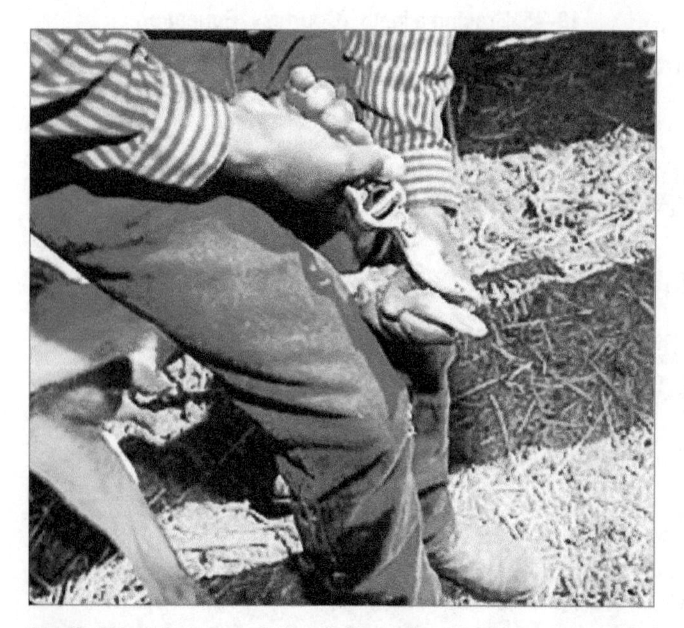

10–48. Trimming a goat's hoof. (Courtesy, U.S. Department of Agriculture)

Fitting involves several practices. Goats are usually washed using a mild soap and thoroughly rinsed. The coat is brushed on a regular basis with a stiff brush to remove dead hair and dirt. A goat should be sheared to show regulations about a week before the show. Hooves may need to be trimmed. Be careful not to cut too much away and cause bleeding. In cold weather, cover a goat with a lamb sock or blanket after shearing. Provide clean bedding to keep the goat clean and dry.

Some training is needed to get a goat ready for a show. A halter, collar, or chain is used. Training a goat to lead may involve haltering and tying to a fence. Always tie a goat where it cannot injure itself. After the goat is gentle, teach it to lead by having someone push from behind when the goat stops. Hold the head in a high and proud position. Practice setting up the goat, hind legs first.

In the show ring, set up your goat properly to make it look its best. Be sure the judge can see your goat. Make sure the legs are set and that the body, neck, and head are in a straight line with its head held up. Use both hands, with one holding the halter and the other helping keep the goat's head and body straight. Follow instructions of the judge and ring stewards.

As you develop skills in showing goats, get the assistance of an experienced goat shower. Observe a goat show in progress noting how individuals show their goats to best advantage.

10–49. Training a goat for showing. (Courtesy, Education Images)

10–50. A Boer goat is being prepared for showing. (Courtesy, Education Images)

REVIEWING

MAIN IDEAS

Sheep and goats are closely related. They have relatively minor differences in physical characteristics and management practices. The number of sheep in the United States has been declining since the 1940s.

Sheep are used for their meat and for their wool. Meat production is more profitable. Goats are raised for their meat, milk, mohair, and cashmere, or as pets.

Selection of sheep and goats should be based on many things. They should be healthy and without defects. Sheep and goats should fit into your facilities and management abilities.

Sheep and goats may be raised by various means. They may be part of a farm flock, a purebred flock, a range band, a confinement system, or they may be fed out in a feedlot.

With the mortality of lambs as high as 25 percent, proper breeding and health management are essential. Unbred ewes that do not have lambs are unprofitable. Lambs should be docked and castrated when young. They consume mostly roughages and very little concentrates. Drenching is a common practice used to prevent internal parasites.

Minimum facilities and equipment are required for raising sheep and goats. However, a good fence is needed to keep out predators.

Showing lambs and goats is a fun and interesting experience. Properly selecting, feeding, managing, and training an animal will provide rewards in the show ring. Get the assistance of an experienced show person in learning the first time.

QUESTIONS

Answer the following questions using complete sentences and correct spelling.

1. How are goats and sheep similar in appearance and in management practices?
2. Describe how sheep are classified.
3. Describe how goats are grouped.
4. What should be considered when selecting a sheep or goat?
5. What are the five types of sheep production systems? Describe each.
6. Which type of production system would best suit the area in which you live? Explain.
7. How should a ewe be managed in terms of reproduction, nutrition, and health?
8. Describe a good health management system.
9. Describe facilities needed on an average sheep or goat farm/ranch.
10. What are the major steps in showing a lamb?

EVALUATING

Match the term with the correct definition. Write the letter of the term on the line provided.

a. ram
b. nanny or doe
c. kidding
d. wether
e. mutton
f. chammy
g. lamb
h. ewe
i. browse
j. docking

_____1. Soft, pliable leather made from a sheep skin.
_____2. Sheep less than one year old.
_____3. Male sheep castrated before sexual maturity.
_____4. Male sheep.
_____5. Female sheep.
_____6. Female goat.
_____7. When a nanny gives birth.
_____8. Meat from a sheep that is more than a year old.
_____9. Removing all or part of the tail of a lamb.
_____10. Woody plants that goats often eat.

EXPLORING

1. Visit a local sheep farm. Note the management of reproduction, nutrition, and health. Take a camera and record major observations. Prepare a poster or bulletin board that depicts what you have learned.

2. Visit a local goat farm. Note the management of reproduction, nutrition, and health.

3. Fit a lamb for showing. Use your own lamb or assist someone who owns a lamb. Determine the different procedures in fitting and why these are followed. Write a report on your experiences. Give an oral report to the class.

4. Investigate caprine arthritis encephalitis (CAE) as a disease of goats. What species are affected? What causes it? What are its signs? Can it be prevented? Treated? Prepare a report on your findings. Here are Web sites for beginning your study:

- Veterinary Medicine at Washington State University: **www.vetmed.wsu.edu/depts_wadd/caefaq.aspx**
- Iowa State University: **www.cfsph.iastate.edu/factsheets/pdfs/caprine_arthritis_encephalitis.pdf**
- Onion Creek Ranch: **www.tennesseemeatgoats.com/articles2/CAE.html**

CHAPTER 11

Dairy Production

OBJECTIVES

This chapter provides general information on dairy cattle and dairying. It has the following objectives:

1. Describe the dairy industry.
2. Describe dairy cattle as organisms.
3. Explain dairy conformation and type.
4. Identify common breeds of dairy cattle.
5. Explain important management practices in dairy production.
6. Explain reproduction in dairy cattle.
7. Describe dairy feeding and nutrition.
8. Describe how the environment is modified for dairy cattle.
9. Explain health management practices with dairy cattle.
10. Describe facility and equipment needs with dairy cattle.
11. Identify general considerations in showing dairy cattle.

TERMS

alveoli
animal model
cold housing
culling
Dairy Herd Improvement Program (DHI)
dry cow
functional type
homogenization
immunoglobulins
ketosis
linear evaluation
management intensive grazing
mastitis
metabolic disorder
milk fever
milking system
nutrient dense
pasteurization
predicted transmitting ability
progeny
selection
total mixed ration (TMR)
type production index (TPI)
udder
warm housing

11–1. Milking technicians use automated equipment to milk and record information on dairy cows. (Courtesy, Agricultural Research Service, USDA)

THE goal of dairy production is to efficiently produce large amounts of cow's milk. Milk and the foods made from milk, such as cheese and ice cream, are "nutrient dense." This means that they contain large amounts of essential nutrients compared with calories. This has made milk and dairy foods popular with consumers.

Milk is secreted by the mammary glands of female mammals. Its purpose is to feed her babies. As the primary food for baby animals, milk is also a good source of human food. Dairy cows are selected based on their capacity to produce large amounts of milk. Years of research and development have resulted in vast improvements in the productivity of dairy cows.

Humans use milk from several animal species. Cows are most important in the United States. Goats, camels, llamas, reindeer, sheep, and water buffalo are also sources of milk for human consumption.

THE DAIRY INDUSTRY

11–2. A modern milking facility. (Courtesy, U.S. Department of Agriculture)

The dairy industry provides milk and other dairy foods to consumers in North America and many foreign markets. Its success can be attributed in part to the ability of the dairy cow to efficiently convert feed to milk. As a ruminant, the cow converts plant nutrients from plants into forms humans will eat. The cow then converts the plant nutrients into the nutrient-dense product called milk. ***Nutrient dense*** means that the food has large amounts of nutrients relative to calories.

BEGINNINGS

Dairy cows were first brought to the Jamestown Colony in 1611. Early farms had only one or two dairy cows for their own use. Lack of refrigeration made it more difficult for people living in large cities to obtain milk.

Pasteurization, refrigeration, and bottled milk were developed in the second half of the nineteenth century. This allowed milk to be stored and transported to population centers.

11–3. A dairy facility in California. (Courtesy, Education Images)

Pasteurization is a process that destroys harmful bacteria and other tiny organisms. It involves heating the milk to 161EF for 15 seconds. This keeps people from getting disease, such as undulant fever, when drinking milk from diseased cows.

Small family dairy herds began to develop in response to improved storage and handling technology. The dairy industry continued to evolve throughout the twentieth century. Compulsory pasteurization, homogenization, and addition of Vitamin D to milk became accepted marketing practices in the United States before World War II.

Table 11-1. Major Events in the Dairy Industry

Year	Event
1611	first cows arrived at Jamestown
1841	unrefrigerated milk shipped to New York by rail
1851	first cheese factory in Oneida, New York
1890	Babcok Butterfat Test introduced
1895	pasteurization available on a commercial basis
1919	homogenized milk marketed
1932	vitamin D fortification of milk introduced
1936	artificial insemination successfully used with cattle
1948	ultra-high temperature pasteurization first used
1975	embryo transfer began to gain acceptance
1994	bovine somatotropin (bST) first used on commercial basis

Homogenization is a process that is used to keep the fat and milk liquid from separating. Cream rises to the top of milk that is not homogenized. In homogenization, the fat droplets are broken into very small particles so they stay in suspension.

More changes occurred in the mid 1900s. Dairy farmers started using artificial insemination on their cows. Commercial dairy farms became market oriented. This pushed production, per cow, upward.

TODAY

There are 9.2 million milk cows on farms in the United States. These cows annually produce a total of 185.6 billion pounds of milk. California, Wisconsin, and New York are the leading states in dairy production. Each dairy cow, on the average, produces 20,209 pounds of milk a year. This is a big increase over the last half century. Today, one cow provides dairy products for almost 30 people!

The number of farms with dairy cows has declined. There are 95 percent fewer dairy farms today than in 1950. Each dairy farm is larger today. Most herds range from 35 to 500 cows, and some herds have as many as 10,000 cows.

Dairy farms use high-tech milking equipment. The hand labor of milking has been replaced with milking machines.

Biotechnology has contributed to the changing dairy industry. The use of rbST (bovine somatotropin) began in 1994 and has steadily increased. Bovine somatotropin is a growth hormone naturally found in cows, but at a higher level in high-producing cows. Giving addi-

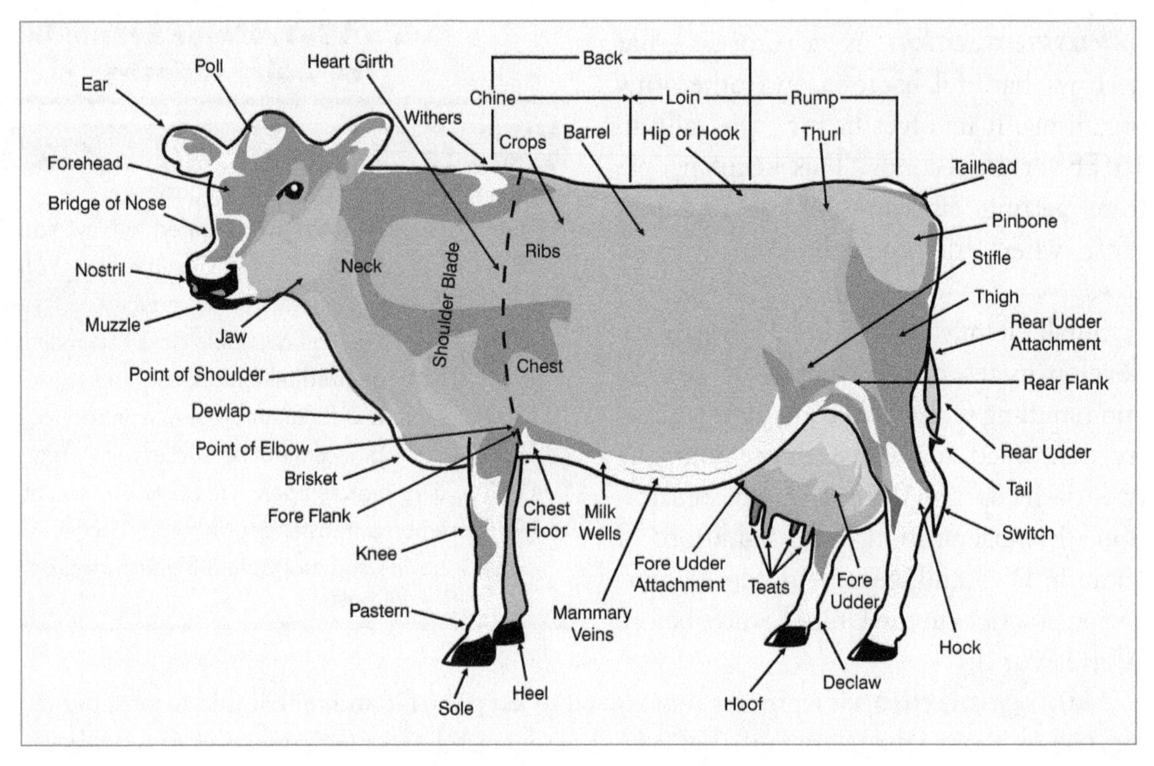

11–4. Major external parts of a dairy cow.

tional bovine somatotropin to cows causes them to milk at higher levels for a longer period of time in well-managed operations.

Dairy production is concentrated near large cities. California leads all states in total milk production. Wisconsin is second, followed by New York, Pennsylvania, and Minnesota. All states have some dairy cattle.

Dairy cattle also contribute to the beef supply. Calves are used for veal. Older cows are used for processed meat products, such as bologna and hamburger.

DAIRY CATTLE AS ORGANISMS

Dairy cattle belong to the family bovidae, which includes ruminants with hollow horns. Members of this family also chew their cuds. The following shows the scientific classification of the dairy cow:

Kingdom Animalia: The animal kingdom

Phylum Chordata: Animals with either a backbone (in the vertebrates) or the rudiment of a backbone (in the chorda).

Class Mammalia: Mammals are warm-blooded, hairy animals whose offspring are fed with milk produced by the mammary glands.

Order Artiodactyla: Even-toed, cloven-hoofed mammals.

Family bovidae: Ruminants have numerous placental attachments and hollow, nondeciduous, upbranched horns.

Genus bos: Four-footed ruminants, including wild and domestic cattle, distinguished by a stout body and hollow, curved horns standing out laterally from the skull.

Species: *Bos primigenius*

DAIRY CATTLE CONFORMATION AND TYPE

Dairy cattle are milk producers. The body structure should contribute to high milk production.

CONFORMATION

Dairy cows appear angular and often lack muscle development in high-value meat cuts, compared to beef cattle. In cows, the udder should be well attached and have the capacity to

CONNECTION

MASTITIS

Mastitis has long been a major health problem in dairy production. As an inflammation of the udder, continual care is needed to prevent it from occurring. Mastitis costs the U.S. dairy industry $1.7 billion annually.

New developments using bioengineering have resulted in the potential of a transgenic dairy cow that resists mastitis. The heifer shown here is a cloned purebred Jersey whose cells have been genetically modified to produce lysostaphin, a protein that kills *Staphylococcus aureus* bacteria that cause mastitis. Using such animals to produce will ensure milk that is free of these bacteria. (Courtesy, Agricultural Research Service, USDA)

hold 50 to 70 pounds of milk. Four teats should be shaped and spaced uniformly for machine milking.

Since much feed is needed for the cow to produce milk, the body must be capable of eating and digesting feed. A cow needs a good appetite and the strength to compete with other cows for feed. They also need good body capacity to hold feed for digestion.

Dairy cows should regularly reproduce. Milk production is associated with the reproductive cycle. Lactation is the secretion of milk by the mammary glands. It begins at parturition and ends when the cow is dried up.

A **dry cow** is one that has stopped producing milk. Most dairy cows produce milk for 305 days. They are bred to calve each year. With a gestation period of 283 days, some rest from milk production is needed. Milking is stopped and they are dried-up 50 to 60 days before the next calving.

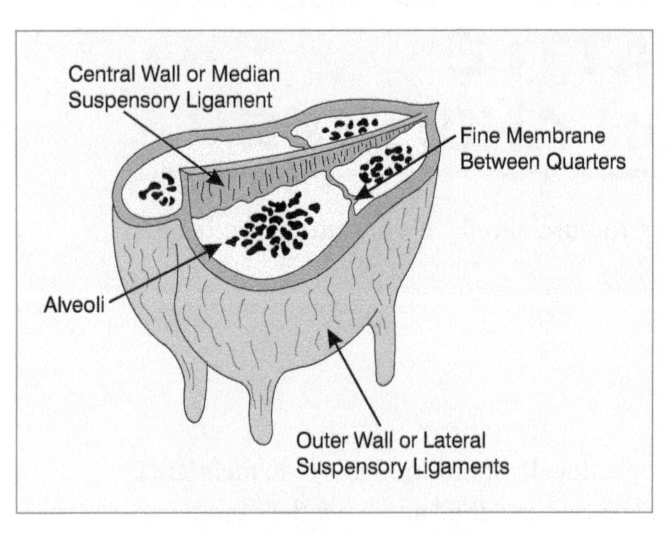

11–5. Structure of a cow's udder.

UDDER STRUCTURE

The **udder** is the bag-like structure that contains the mammary glands and to which teats are attached. Cows have four mammary glands or quarters. Each quarter has a teat. A canal in each teat allows removal of milk.

The udder should be large and strongly attached to the body. Udders sometimes break away from the cow. This shows problems in producing a large amount of milk. The milk is produced in the ***alveoli***. These are tiny structures that remove nutrients from the blood and convert the nutrients into milk.

Milk is 87 percent water. Fat and protein make up a little less than 4 percent each. It has a little more than 4 percent lactose (milk sugar). Cows must drink much water to carry on life functions and secrete milk.

DAIRY CATTLE BREEDS

There are seven major breeds of dairy cattle. Differences in breeds include size, color, milk yield, milk fat, and milk protein.

Variation exists within each breed. Holsteins produce the most milk by volume. They are followed by Red and White, Brown Swiss, Ayrshire, Milking Shorthorn, Guernsey, and Jersey. The breeds ranked in order based on percent of milk fat and protein are Jersey, Guernsey, Brown Swiss, Ayrshire, Red and White, Holstein, and Milking Shorthorn.

11–6. A Jersey cow. (Courtesy, Pete's Photo, Wykoff, Minnesota)

JERSEY

The Jersey is from the Island of Jersey and was brought to North America in 1850. Jerseys are grayish to fawn and near white in color. They are smaller than other breeds. Jersey milk has the highest fat and protein content. The amount of milk produced is less than other breeds. Jersey numbers have been increasing in recent years.

HOLSTEIN-FRIESIAN

The Holstein-Friesian (commonly shortened to Holstein) breed came to North America from the Netherlands in 1621. It is the most popular breed of dairy cattle. Most Holsteins

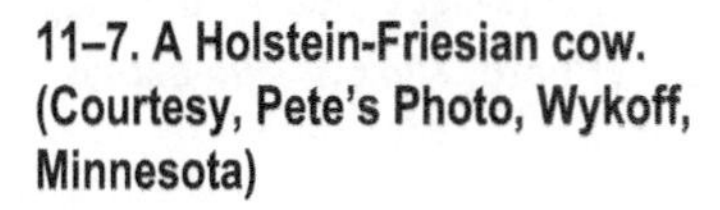

11–7. A Holstein-Friesian cow. (Courtesy, Pete's Photo, Wykoff, Minnesota)

11–8. A Brown Swiss cow. (Courtesy, Pete's Photo, Wykoff, Minnesota)

have distinctive black and white color. They produce the largest amount of milk though low in butterfat and protein. Holsteins are large, with cows weighing up to 1,500 pounds, and bulls, 2,200 pounds.

BROWN SWISS

Brown Swiss were brought to North America in 1869 from the Alps of Switzerland. The color is solid, light to dark brown. The nose and tongue are black. The Brown Swiss has a calm disposition. The milk has a favorable protein to fat ratio for the present milk market.

GUERNSEY

The Guernsey was introduced into North America in 1831 from the Island of Guernsey. The color is fawn with white markings. Guernseys are quiet cattle and easy to work with. Their numbers have been declining. They are sometimes referred to as "golden Guernseys" because of the slight golden color in their milk.

11–9. A Guernsey cow. (Courtesy, Pete's Photo, Wykoff, Minnesota)

AYRSHIRE

The Ayrshire was brought to North America from Scotland in 1822. Color varies from light to dark red, brown, and white. The Ayrshire is known for well-attached udders and good grazing ability.

11–10. An Ayrshire cow. (Courtesy, Pete's Photo, Wykoff, Minnesota)

MILKING SHORTHORN

The Milking Shorthorn breed was designated as a dairy breed in 1968. The coloring is similar to the Shorthorn, with red, white, and combinations of these. The breed is adaptable to a variety of situations.

RED AND WHITE

The Red and White evolved from the Holstein. It is similar to other Holsteins in milk production, size, and disposition. As the name suggests, the color is red and white.

11–11. A Red and White Holstein cow. (Courtesy, Pete's Photo, Wykoff, Minnesota)

11–12. A Milking Shorthorn cow. (Courtesy, Pete's Photo, Wykoff, Minnesota)

MANAGEMENT OF DAIRY HERDS

Good management is essential to success in dairy production. Dairying is a daily venture. Cows must be milked at least twice each day. Some cows in a herd produce milk every day of the year.

STARTING A DAIRY ENTERPRISE

Some people like dairy farming. The decision to start a dairy enterprise is a long-term one. Once started, it is difficult to change without serious consequences. As a result, the final decision should be carefully thought out.

11–13. This Wisconsin dairy farm required a large financial investment. (Courtesy, Nancy Gill/Shutterstock)

Dairying has been one of the most stable agricultural enterprises. It has returned a reasonable profit and allowed long-term financial growth. Personal pride and satisfaction of ownership are additional rewards for most dairy farmers.

The rewards of dairying also come with risks and challenges. The capital investment is sizeable and carries with it risks just as any investment does.

The final decision is to choose the type of operation best suited to resources and personal goals. This can include choosing between a commercial (grade) and purebred operation, a family-size (100 or less) or large herd, a total herd (cows and replacement heifers) or a milking herd, and whether the feed will be raised or purchased.

SELECTION AND CULLING

A key to success in dairying is the ability to select and cull cows and heifers. ***Selection*** is choosing or picking the animals for the herd. ***Culling*** is choosing the cows or heifers to remove from the herd. Criteria in selection and culling are production factors, genetic makeup, and type.

Dairy Herd Improvement

Milk production can be assessed by the ***Dairy Herd Improvement (DHI) Program***. DHI is a national dairy production testing and record-keeping program. USDA personnel work with dairy producers to get information and compare a herd with others in the area, state, and nation.

The state and local Dairy Herd Improvement Associations (DHIA) are the most common official testing plans. All cows must be properly identified and all DHI rules enforced. Cows are tested every 15 to 45 days. The results are published and used by the USDA to evaluate sires (bulls).

The Dairy Herd Improvement Registry (DHIR) includes the standard DHIA requirements plus the added requirements of breed associations. Additional testing is needed when an individual cow's milk production exceeds specified standards. The records for registered cows in herds enrolled in DHIR are sent to the respective breed registries.

These rules apply to both the DHIA and DHIR programs:

1. All cows in the herd must be entered into the official testing program.
2. All animals in the herd must be permanently identified.
3. Copies of pedigrees of all registered cows must be made available for DHIR.
4. Testing is done each month with not less than 15 days nor more than 45 days allowed between test periods.
5. Testing is conducted over a 24-hour period.

CAREER PROFILE

DAIRY FARM MANAGER

A dairy farm manager oversees the operation of a dairy farm. The duties may vary with the size and nature of the farm. Dairy farm managers often supervise others in caring for animals, milking, and performing other duties. They often keep records, make purchases, build and repair facilities, select cattle, artificially inseminate cows, and care for calves. Some dairy farm managers are entrepreneurs and own their farm.

Dairy farm managers need practical experience on a dairy farm. A high school education is essential. Two-year college or a baccalaureate degree in dairy science or a related area are often needed.

Jobs are found where dairy farms are located. Increasingly, the jobs are with large dairy producers. (Courtesy, Education Images)

6. An independent supervisor must be present for supervising the weighing of the milk and the sampling of the milk for milk fat and other determinations.

7. Milk or milk fat records that are above values established by the breed association require retesting of the cow to assure that an error was not made. An owner may request retesting if he or she feels the test does not properly reflect the production of the cows.

8. Surprise tests may be made if the supervisor suspects that they are needed to verify previous tests.

9. Any practice that is intended to, or does, create an inaccurate record of production is considered a fraudulent act and is not allowed.

Several unofficial DHI testing programs also exist. These programs often involve the owner doing the sampling rather than a neutral supervisor. They may only involve the recording of milk weights. Unofficial records cannot be published or used to promote the sale of cattle.

Progeny Information

Progeny information is useful in selecting dairy animals. ***Progeny*** refers to the offspring of animals. Information about the offspring is important. Calf identification is the cornerstone to successful progeny data. Progeny information identifies the genetic worth of sires and dams.

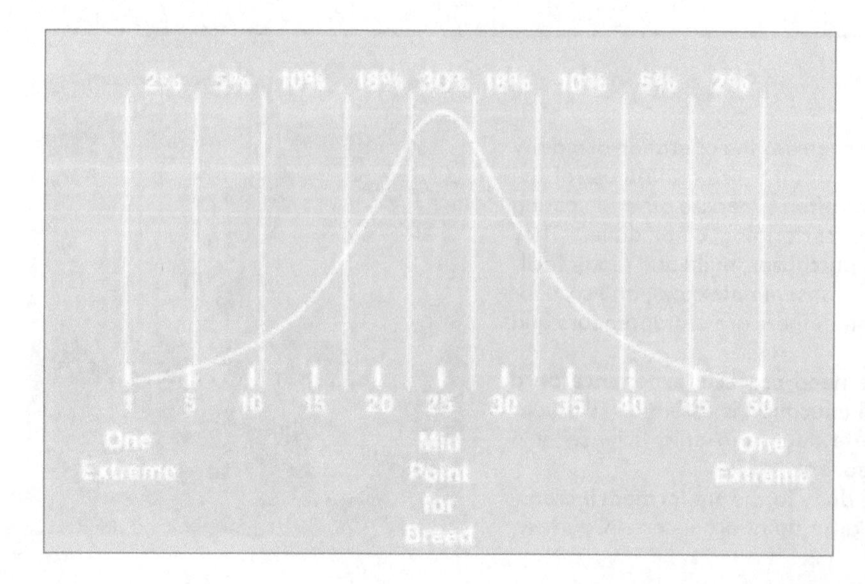

11–14. Linear evaluation uses 16 traits of dairy cows between biological extremes. This shows typical frequencies on a 50-point scale. (Courtesy, Lake-Plains Education Materials, Crookston, Minnesota)

The dairy industry has benefitted from progeny information. The USDA gathers DHI information in a large, computerized data bank. Sire and dam predicted transmitting abilities are calculated using the animal model.

Predicted transmitting ability is an estimate of the traits an animal will transmit to its offspring. It includes both genetic superiority and inferiority. In other words, both good and bad traits are transmitted.

The ***animal model*** is the genetic evaluation for dairy cattle production used by the USDA. It uses pedigree data and production-related factors.

Linear Evaluation

The type or physical appearance of a dairy cow may be used in culling or selection. Purebred producers want outstanding type. This improves the market value and pedigree of an animal. Commercial herds are more interested in functional type.

Functional type refers to the traits of a cow that are good enough to allow her to complete a useful life in a herd. ***Linear evaluation*** is the coding of 15 primary traits of dairy cows. It was started in 1978 and has been well received. Nearly one-half of the herds that use artificial insemination use linear evaluation.

With linear evaluation, 15 primary traits of cows between biological extremes are coded. Several numerical scales have been used, but the most common is a 50-point scale with 25 being the midpoint. The most limiting biological extreme is usually given a score of "1."

Most experts agree on the use of linear evaluation. A dairy farmer should select a group of sires that meets specified production standards (goals) for their operation. The selected group of bulls should then be mated to each cow in the herd to maximize type improvement.

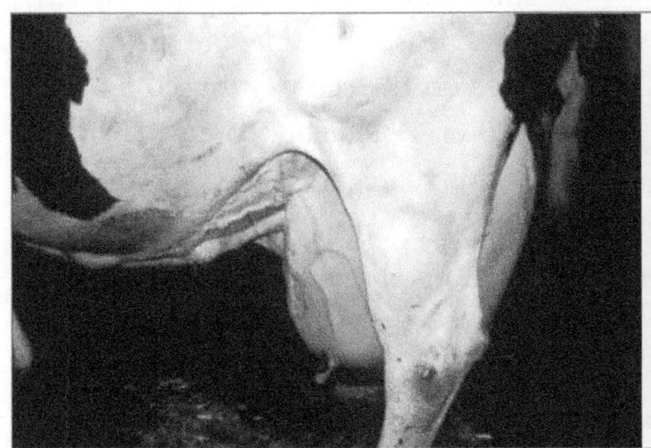

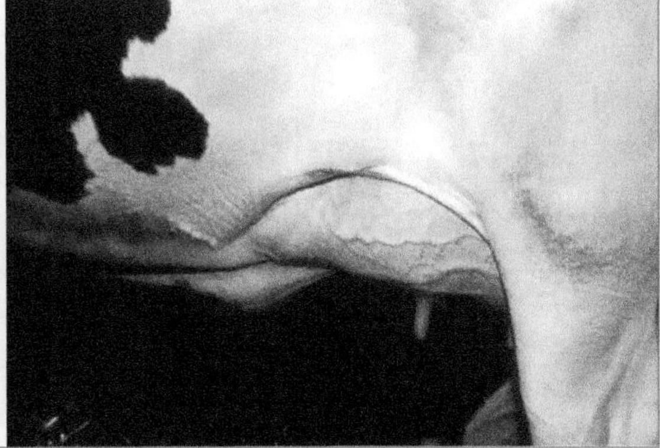

11–15. Biological extremes of fore udder attachment are shown here. (Linear scores of 1–5 are given for the cow on the left and 45–50 for the cow on the right.) (Courtesy, Lake-Plains Educational Materials, Crookston, Minnesota)

Breed associations also calculate predicted transmitting abilities for type (PTAT). A higher PTAT indicates to a dairy farmer that it has a potential to improve the type classification of a particular mating. Dairy farmers may use PTAT together with linear scores. PTAT is also included in the ***type production index*** (TPI), which combines several type and production factors into a single figure. The type production index ranks cows based on overall performance. Higher type production indexes are a popular sire selection tool.

Judging Dairy Cattle

The ability to judge dairy cattle is often practical with both commercial and purebred herds. Judging is comparing animals as a whole based on their body type. Judging competition usually includes four animals in a class. The animals are compared and ranked according to their desirable physical characteristics, commonly called type. Functional type refers to the physical characteristics that are important for a cow to efficiently produce well with acceptable longevity.

11–16. Dairy judging at the World Dairy Exposition in Madison, Wisconsin. (Courtesy, Dairy Club, University of Minnesota, Crookston)

1234	1243	1324	1342	1423	1432	2134	2143	2314	2341	2413	2431	3124	3142	3214	3241	3412	3421	4123	4132	4213	4231	4312	4321

JUDGING CARDS

No. ______

Name ______________________

Class ______________________

Date ______________________

SCORE

P ______

R ______

11–17. Sample dairy judging placing card.

Judging competition is popular throughout the United States. Youth who are members of FFA and 4-H compete at local, county, state, and national competitions. The Dairy Unified Scorecard is considered the guide for students to select one animal over another. Frame, dairy character, body capacity, feet and legs, and udder are the five categories on the scorecard used for judging cows. A similar scorecard without the udder category is used for heifers.

Many contests include giving oral reasons. Oral reasons provide a way for judging contestants to defend or justify their decisions (placing) to an official judge. Students are scored both on their place and their oral reasons. Official judges at a county or state fair may use a modified reasons format to explain their placing at a show.

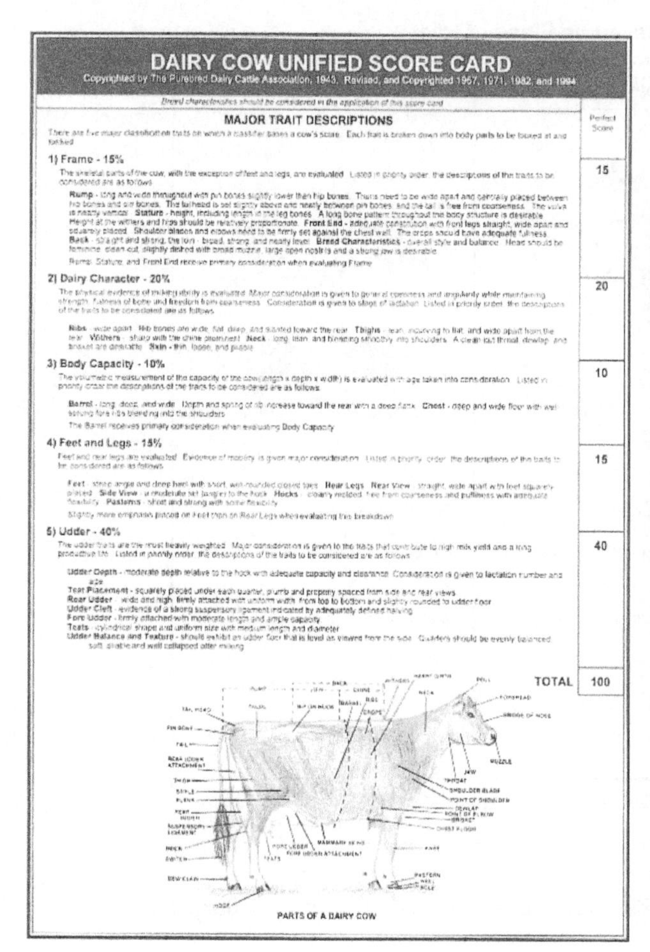

DAIRY COW UNIFIED SCORE CARD

Copyrighted by The Purebred Dairy Cattle Association, 1943. Revised, and Copyrighted 1957, 1971, 1982 and 1994

MAJOR TRAIT DESCRIPTIONS

Trait	Perfect Score
1) Frame - 15%	15
2) Dairy Character - 20%	20
3) Body Capacity - 10%	10
4) Feet and Legs - 15%	15
5) Udder - 40%	40
TOTAL	100

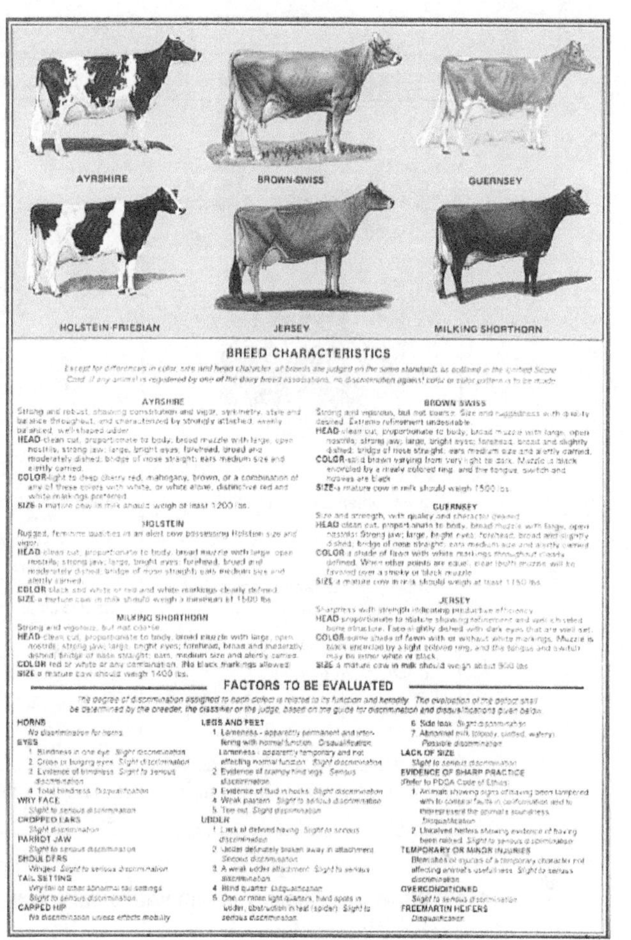

BREED CHARACTERISTICS

AYRSHIRE

BROWN SWISS

GUERNSEY

HOLSTEIN

JERSEY

MILKING SHORTHORN

FACTORS TO BE EVALUATED

11–18. Dairy Cow Unified Judging Scorecard.

DAIRY CATTLE REPRODUCTION

A dairy cow must reproduce annually while under the stress of high milk production. A combination of proper nutrition and good health are necessary. (The basics of reproduction were covered in Chapter 5.)

REPRODUCTION

Reproductive goals include (1) a 12 to 13.5 month calving interval, (2) less than 2 units of semen per pregnancy, (3) heifers first calve at 23 to 25 months, (4) a live and healthy calf

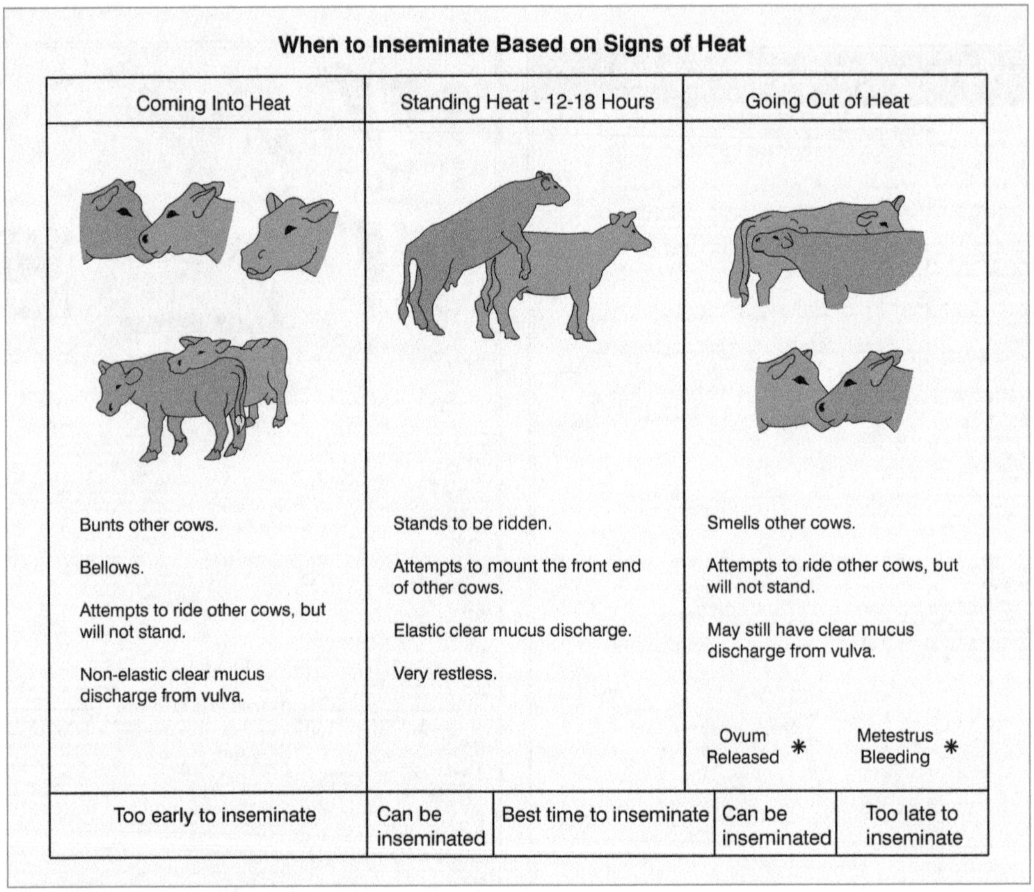

When to Inseminate Based on Signs of Heat

Coming Into Heat	Standing Heat - 12-18 Hours	Going Out of Heat
Bunts other cows. Bellows. Attempts to ride other cows, but will not stand. Non-elastic clear mucus discharge from vulva.	Stands to be ridden. Attempts to mount the front end of other cows. Elastic clear mucus discharge. Very restless.	Smells other cows. Attempts to ride other cows, but will not stand. May still have clear mucus discharge from vulva. Ovum Released * Metestrus Bleeding *

Too early to inseminate	Can be inseminated	Best time to inseminate	Can be inseminated	Too late to inseminate

11–19. When to artificially inseminate a dairy cow based on signs of heat. (Courtesy, Gary Stegman, Crookston, Minnesota)

crop of 95 to 98 percent, and (5) a minimum of reproductive tract disorders. Embryo transfer is used. Some consideration is being given to cloning and the use of sexed semen.

Achieving reproductive goals requires a good knowledge of animal health. Many diseases that may cause reproductive problems, such as bovine virus diarrhea (BVD), brucellosis (Bang's disease), and leptospirosis, must be included in a herd vaccination program. Other problems, such as metritis and retained placentas, should be analyzed and treated by a veterinarian.

The danger of many genital diseases, such as trichomoniasis, vaginitis, and vibriosis, can be virtually eliminated with artificial insemination. (Other diseases are presented in more detail elsewhere in the book.)

Heat detection in cows and keeping good reproductive records are important. Breeding programs using synchronization methods may be used for one of two major reasons: to bring cows in heat or to group the heat cycles of many animals. Most dairy cows are artificially inseminated.

SELECTING SIRES

Selection of quality sires is important. Dairy farmers have only five or six cow generations to achieve production goals that have been set. Breeding artificially with bulls that are in the top 20 percent of the breed could be the difference in making or not making a profit.

Semen from five to eight bulls should be adequate for small herds of 60 cows or less. Intermediate size herds (61 to 200 cows) may use 7 to 10 bulls in their breeding program. Large herds (over 200 cows) may use 10 to 15 bulls. Semen is available from many sources. Bulls that meet the standards established by a given herd can easily be selected using predicted transmitting ability data.

FEEDING A DAIRY HERD

Dairy animals have inherited genetic potential for performance. How well that potential is realized depends on the environment of the animal. Nutrition is the most important environmental factor.

A great need for energy and protein is created by lactation. A cow that weighs 1,400 pounds and gives 80 pounds of milk per day needs 2.5 times more energy for milk production than for maintenance.

Modern dairy technology is used by dairy farmers to feed many times per day. Computerized feeders, mechanized track feeders, or auger-type delivery systems all allow for multiple feedings. Total mixed rations are increasing in use. A ***total mixed ration (TMR)*** is one in which all feedstuffs are mixed together. They provide all feed ingredients in each mouthful a cow eats.

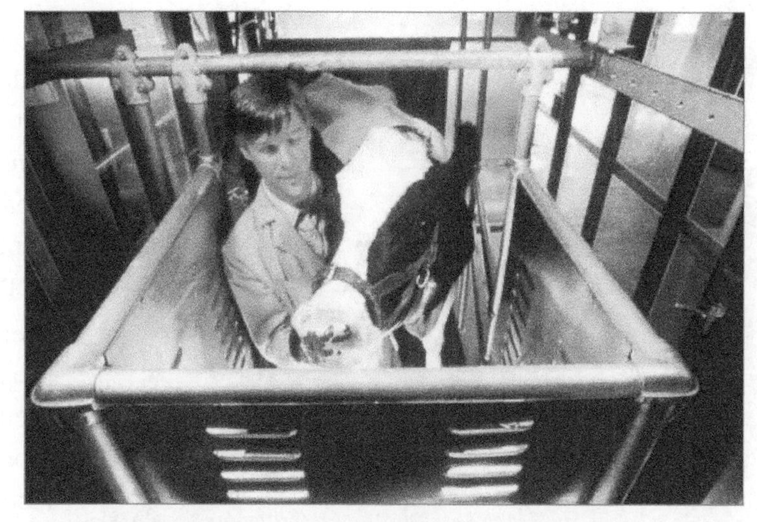

11–20. Evaluating a ration in feeding dairy cows. (Courtesy, University of Minnesota and Northwest Experiment Station, Crookston, Minnesota)

RATIONS

Rations for dairy cattle begin with high-quality roughages. Roughages are usually the lowest-cost sources of nutrients. High-quality roughages can be bought or grown. Proper soil fertility and harvesting will be needed. The common roughages for dairy cattle are

11–21. Dairy cattle, such as these Jersey cows, may graze on improved pasture. (Courtesy, U.S. Department of Agriculture)

hay, silage, green chop, and pasture. Roughage can be restricted, fed free choice, or given as part of a total mixed ration.

High-producing dairy cows cannot get an adequate supply of nutrients from roughages alone. While 50 to 60 percent of a ration may be roughages, the other 40 to 50 percent must be obtained from concentrates. Concentrates refer to high-energy feeds, such as grains or protein supplements.

A higher percent of concentrates is usually fed to a dairy cow during the first four months of lactation. The nutrient demand is greatest at this time. Lower-producing cows or cows in late lactation are fed a reduced percent of concentrates to reduce feed cost.

NUTRIENT REQUIREMENTS

The nutrient requirements of dairy animals can be summarized easily. Maintenance needs are related to body size but imply no weight gain or loss. Most dairy animals under five years of age also require nutrients for growth. Lactation requires large amounts of nutrients.

Growth rates and nutrients needed for growth vary according to the stage of maturation. Bred heifers and cows require nutrients for reproduction, including both gamete production and fetal growth. The greatest nutrient requirement is for a pregnant lactating cow.

Body condition scores (1 = thin, 3 = average, 5 = fat) are often used to monitor proper nutrition. Ideally, a cow should spend her entire life with a body condition score that fluctuates from 2.0 to 4.0. Peak lactation (60 to 90 days into lactation) requires a cow to use her body reserve to supplement her ration. Her body condition may drop to 2.0 or 2.5 during that phase of

11–22. Dairy cattle on a ration that includes silage. (Courtesy, University of Minnesota, Crookston)

lactation. A cow should replenish her body reserve during the last half of lactation, entering the dry period with a body condition score ranging from 3.5 to 4.0.

FEEDING CALVES

Heifer calves must be healthy and well-nourished. These are vital for the calf to become a high-producing dairy cow. Colostrum fed to a calf ensures that it can use the immunoglobulins. ***Immunoglobulins*** are antibodies in the milk that help the baby calf develop passive immunity. Colostrum is the first milk given by a female after giving birth. One-half gallon of colostrum should be fed within the first hour of birth and a full 6 percent of body weight within the first six hours of life. An additional 6 percent of body weight should be fed between the sixth hour and one day of age for a total of 12 percent of body weight during the first 24 hours.

Colostrum is usually fed to a calf from its mother, but there are times when colostrum from other older cows in the herd should be fed. This is necessary when cows have been purchased shortly before calving or bred heifers have been raised separately from the older cows. Isolated heifers do not develop immunoglobulins to the diseases in the herd.

Preferably, a calf should be fed whole milk from cows not carrying Johne's disease for about five days. From five days until weaning, it can be fed whole milk or a high-quality milk replacer. A high-quality milk replacer should contain 20 percent fat, 20 percent protein, and less than .5 percent fiber. In addition to milk or a milk replacer, a calf should be fed a calf starter as soon as it will begin to eat. At four to six weeks of age, a calf should be eating 1.5 pounds of calf starter per day.

Forages should not be fed until after weaning. If hay is fed to a preweaned calf, it should be a leafy, high-quality alfalfa or alfalfa-grass hay fed only after four weeks of age.

11–23. Dairy calf receiving proper nutrition. (Courtesy, University of Minnesota and Northwest Experiment Station, Crookston, Minnesota)

FEEDING YOUNG STOCK

Feeding postweaned calves becomes easier due to a functional rumen. Some basic feeding guidelines are still necessary.

Forage can be fed free choice, with concentrates being supplemented at a rate up to 6 pounds per head. This depends on forage quality. Pasture can be the only forage used. Pasture quality should be carefully monitored so an appropriate supplemental grain mix can be fed.

Just as with lactating cows, body condition can be a useful tool in monitoring nutritional needs. If the nutritional level for young stock appears correct, but the body condition level and growth rate are less than desired, look for related problems. A common problem is parasites, which should be controlled.

FEEDING DRY/TRANSITION COWS

Concentrates are not fed when a cow is dried off. For two to three weeks, average to good roughage may be the only feed. Concentrates, as a special dry cow formulation, are often reintroduced to a dry cow about midway through a standard 50 to 60 day dry period at a rate of 4 to 7 pounds per day. Toward the end of the dry period, some producers challenge or lead feed. Lead feeding refers to feeding concentrates that more closely resemble the lactating ration (7 to 14 pounds per day).

11–24. Jersey heifers are being raised to replace older dairy cows. (Courtesy, U.S. Department of Agriculture)

A few basic rules should be followed for a dry cow ration: (1) the calcium:phosphorus ratio should fall in the 1:1 to 1.5:1 range to reduce the incidence of milk fever; (2) a maximum of 35 pounds per day of corn silage should be fed; (3) alfalfa hay or haylage is not a recommended dry cow forage, but if it must be fed, it should never make up more than 50 percent of the forage dry matter; and (4) salt should be reduced to .25 percent of the ration (dry matter basis).

FEEDING THE LACTATING COW

The most challenging ration to develop is that for a high-producing cow. Each cow is unique and should be fed that way when possible. At a minimum, a cow should be fed a ration that is being fed to a group of cows that closely resembles her needs.

Several cow factors must be considered when developing lactating rations. These are age, size, stage of lactation, milk composition, and labor capability. Feed quality, availability, cost, and palatability to the cow must be considered.

High-producing cows typically require 1 pound of concentrate (grain) for each 2.4 to 3 pounds of milk produced. Concentrates rarely make up more than 60 percent of the total dry matter (a 50:50 ratio of concentrates to roughages is more desirable). Forages make up the balance of the ration and usually are fed at a rate of 1.5 to 2.0 percent of body weight. Feed protein in a ration should be 19 percent in early lactation and decreased slightly in later lactation. A higher level of bypass protein (undegradable) is necessary in early lactation.

11–25. Feedstuffs stored in bulk on the farm are becoming more common to reduce feed costs. (Courtesy, University of Minnesota, Crookston)

The fiber level in a ration should be 19 to 21 percent acid detergent fiber (ADF) and 25 to 28 percent neutral detergent fiber (NDF). Acid detergent fiber is the residue of the fiber content of feed. Neutral detergent fiber is cell walls and other feed content that are not readily digested. A maximum of 5 to 6 percent fat level in the ration is suggested. Salt should be fed at .5 percent of the total ration, a calcium/phosphorus mineral source at 1 to 2 percent of the grain mix, and vitamins as needed.

ENVIRONMENTAL MODIFICATION

Modification of a dairy cow's environment began with domestication. Most environmental change is to provide for the well-being of the cow. It includes people who handle the cow, the climate in which a cow lives, facilities, and economic considerations. Modifications are to improve a cow's environment. Conditions should provide for the well-being of the cow.

ANIMAL/HUMAN INTERACTION

The major environmental modification for most domesticated animals is increased interaction with humans. This is definitely true with dairy cattle.

A good dairy herdsperson is patient and understands animal behavior. Dairy producers should understand normal behavior to detect and treat animals that are acting abnormally. Dairy animals bond to kindness.

11–26. A dairy facility with an open ridge and curtains allow for environmental control at Windgate Dairy in Minnesota. (Courtesy, University of Minnesota, Crookston)

STRESS MODIFICATION

Many people feel that stress should be eliminated from both human and animal lives. A better viewpoint is that stress should be modified to a desirable level. It should be at a level that prevents boredom but does not create unnecessary fear. Finding the right level of stress requires knowledge of the past experiences of an animal. Feeding time is an example. Dairy animals accustomed to eating at a specific time are stressed by a one-hour delay.

Some potential stressors are under human control. These include regular care, space allocation, transportation, and presence of strangers. These stressors can be easily managed with proper planning.

Other potential stressors cannot be controlled by humans. However, some can be modified. Examples of this type of stressors include weather changes, social-order fights among animals, and physiological cycles, such as heat or calving.

Keeping dairy animal stress at an appropriate level requires knowledge of animal behavior, animal history, and coming changes in the life of an animal. A bred heifer raised on a large pasture with little human interaction will require a longer acclimation period to the dairy cow housing facility. A bred heifer raised with close human interaction in a facility will usually adapt quickly.

CLIMATE MODIFICATION

Four climate factors affect dairy animals: (1) temperature, (2) humidity or precipitation, (3) wind, and (4) radiation. While dairy cattle have a fairly wide tolerance range to individual factors, combinations of extremes can quickly have an adverse effect on milk production, conception rates, and growth.

A combination of heat and humidity with little wind is especially difficult for dairy animals to cope with. Shade and misting are used to reduce the negative effects. While dairy animals are more tolerant of cold, protection against severe cold, wind, and snowstorm conditions is beneficial.

Understanding an animal as it reacts to the weather is useful. For example, cows eat more before storms and less during hot, humid weather.

Dairy animals prefer to travel in groups or herds to water while grazing if the water is further than 500 to 700 feet away, while they travel individually at closer distances. Hot weather requires greater water intake and disrupts grazing if the water is at a distance that encourages herd travel.

DAIRY CATTLE HEALTH

Good health is essential for milk productivity. A good manager can use practices that promote health.

The primary objective of a health program is to prevent and control disease. Knowing normal and abnormal dairy animal behavior is the first step in management. Noticing a cow in heat, a calf with dull eyes, or a cow whose body condition is declining makes it easy to respond.

Several diseases are important in dairying. Prevention and fast action when problems arise help keep a herd healthy. Many of the diseases of beef cattle may also afflict dairy cattle. Chapter 4 has basic information on disease.

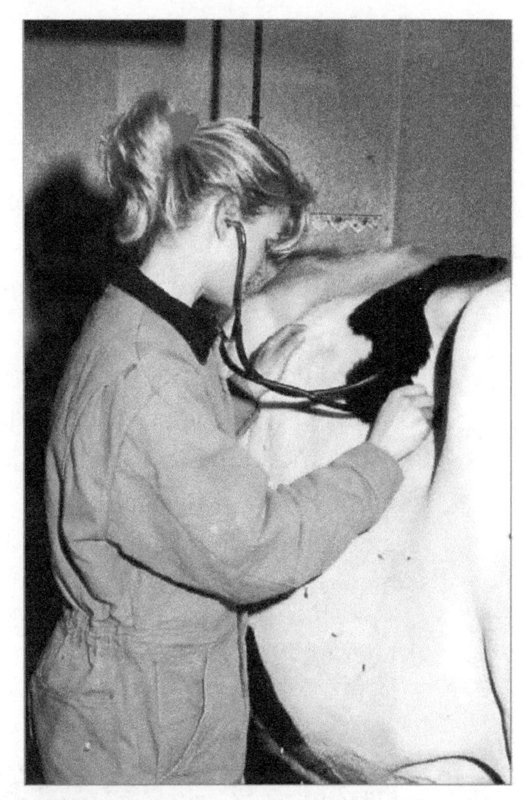

11–27. Checking for rumen movement in a dairy cow. A lack of rumen movements may indicate a displaced abomasum, which is often referred to as a "twisted stomach." (Courtesy, University of Minnesota, Crookston)

METABOLIC HEALTH DISORDERS

Nutrition overlaps most areas of dairy cow health and management. Similarly, several metabolic disorders are directly or indirectly related to nutrition.

A ***metabolic disorder*** is a chemical transformation of energy by the body. Most occur at or shortly after calving. This is due to the stress associated with high milk production. The three most common metabolic disorders are milk fever, ketosis, and fat cow syndrome. Other metabolic disorders, such as bloat, retained placentas, and grass tetany, are not as specific to dairy cows.

Milk fever is a metabolic disorder characterized by low blood calcium and paralysis. It is caused by overfeeding calcium during late lactation or the dry period. Milk fever most often occurs near calving time, when the need for calcium increases rapidly. A low calcium to phosphorus ratio and an adequate amount of vitamin D in the ration reduce milk fever problems.

Ketosis is characterized by a poor appetite and dullness. Blood sugar is low and should be treated with propylene glycol. While ketosis can be a primary problem, it is more often a secondary problem. It results when a cow is struggling with a health problem early in lactation.

Fat cow syndrome develops when cows become overfat in late lactation or the dry period. This leads to other health problems, such as ketosis. Avoid fat cow syndrome by proper feeding.

MASTITIS

Mastitis is inflammation of the mammary gland. Mastitis costs $200 or more per cow annually. Mastitis is sometimes visible to the human eye; it is often not visible. Laboratory tests are needed to measure somatic cells. A cowside California Mastitis Test (CMT) may be used.

Controlling mastitis requires clean yards, barns, and milking equipment. Spread of mastitis is reduced with good sanitary milking practices. Common practices include using individual towels for each cow, pre- and post-teat dipping, drying of teats, and elimination of overmilking. Cows with mastitis that do not react to treatment during lactation should be dry treated. If a cow is a problem after reasonable treatment, it may be wise to sell her.

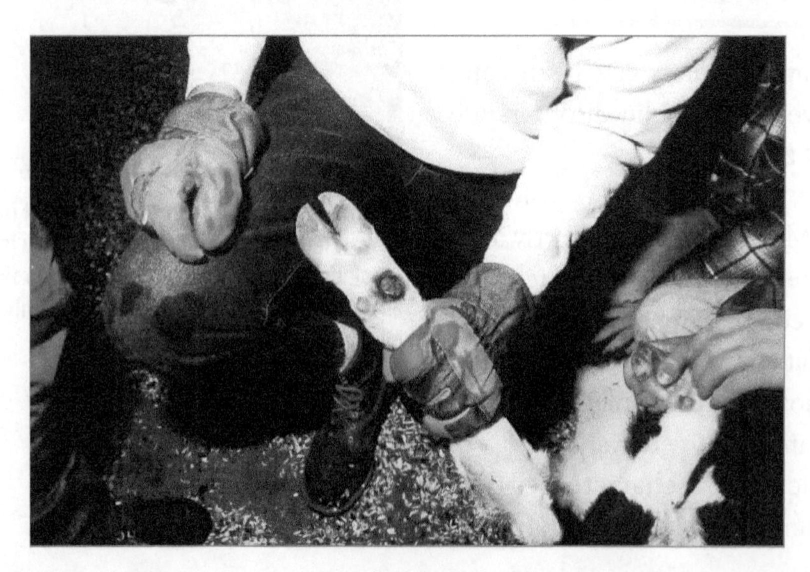

11–28. Mastitis is reduced by removing the inside dew claws of heifers. (Courtesy, University of Minnesota, Crookston)

FACILITIES AND EQUIPMENT

Facilities and equipment help make efficient dairying possible. These also consider the well-being of the cows.

HOUSING

Housing protects dairy cattle and the humans who work with them. The goal is a system that does not allow conditions to get outside the "acceptable" range. Each dairy operation has a level at which additional expenditures will not increase profits.

11–29. Calf hutches provide a healthy environment for young calves. (Courtesy, Education Images)

Cold housing is an unheated building kept cold during the winter. Natural air movement keeps these buildings cold and removes moisture. ***Warm housing*** refers to buildings kept warm in the winter. Body heat from the animals provides the heat source while insulation helps retain it in the building.

Cold housing for cows usually takes the form of loose housing. Loose housing allows individual cows freedom, while the herd is handled on a group basis. Loose housing may be an open area with a manure pack or a free-stall system in which cows can enter individual stalls to rest and ruminate. Free stalls often reduce bedding requirements by 75 percent.

Warm housing for cows can also include a free-stall barn with insulation. Northeastern and midwestern United States dairy farms commonly use stall barns. Stall barns usually consist of two rows of cows confined in tie stalls, comfort stalls, or stanchions. Each cow has an area (usually 4 feet by 6 feet for large breeds) where it spends most of its life. Most often, cows are milked in their stall.

Calves, young stock, and dry cows are commonly housed in a natural, cold environment. Calf hutches for 0- to 6-week-old calves and super hutches or open-sided sheds provide shelter for older calves, young stock, and dry cows. Other sheds are occasionally used, but frequently they lack proper ventilation and temperature control to provide the calf with a healthy environment.

Cattle on open lots in the southern United States or on pasture should have shade and a clean, readily available water supply. Northern pastures may require wind protection or shelter during the early and late grazing season.

MILKING SYSTEMS

A ***milking system*** is the way the cows are milked. This includes how the cows are brought to the milkers, how the milking equipment is attached, and how the harvested milk is stored to assure good quality.

Four kinds of systems predominate: robotic, rotary, herringbone, and parallel. Systems are increasingly integrated with computers to maintain records and provide feed.

Robotic milking systems are rapidly gaining popularity in 40 to 250 cow herds. A robotic milking system eliminates the need for human labor to milk the cows. The cows come to the robotic system on their free will, the udders are washed, and the milking unit automatically goes onto and removes itself from the udder by a system of sensors. Feed is electronically delivered to cows during milking.

11–30. A robotic milking system in operation. (The inset shows milker attachment.) (Courtesy, Craig Roerick, University of Minnesota Extension)

Rotary milking parlours involve the cows loading onto an area that turns much like a merry-go-round. The cows are milked as the system turns. Most finish milking in a revolution and exit the facility. A slow milking cow can go around another time.

Some cows are still milked within their stall in a flat milking barn system. This requires a pipeline to carry the milk from the stalls to the bulk tank.

Milking parlors are used on some dairy farms. A milking parlor is a concrete platform raised above the parlor floor (pit). Cows enter the parlor in groups or individually and stand on the platform for milking. The way the cows stand may resemble other systems. A wide range of milking parlor designs are used.

11–31. A diamond milking parlor. (Courtesy, Joshua Johnson, University of Minnesota-Crookston)

11–32. A rotary milking facility. (Courtesy, polat/ Shutterstock)

FEEDING SYSTEMS

Nearly twenty tons of feed are fed to a cow in one year. Labor efficiency, ability to individualize to each cow or groups of cows, and cost must be considered in selecting a feeding system. Storing and feeding forages can require different equipment from feeding concentrates or can overlap in the case of total mixed rations (TMR).

Storage is needed for silage, haylage, dry hay, and straw. Silages can be stored in traditional upright silos, bunker silos, oxygen-limiting silos, or silage bags. Dry hay and straw can be baled in small square bales, large square bales, or round bales and can be pelleted or put up in loose stacks.

Fresh or ensiled forages are usually fed with mechanized delivery systems or a wide range of silage/green chop wagons. Dry forages are moved mechanically with a tractor or manually in flat stall barns. Mechanical chopping of hay is required in certain feeding systems.

Management intensive grazing is a system that allows the cows to harvest their own forage. Cows are placed in a pasture that can be eaten down to the desirable height in a 24- to 48-hour period. The cows are then moved to another pasture. This concept encourages proper nutrition by forcing the cow to eat both leaves and stems. It also maximizes the number of cows that a pasture can support. Management intensive grazing has the disadvantage of requiring more labor since cattle and waterers must be moved each day.

11–33. Cattle are fed outside during the summer on this Minnesota dairy farm. (Courtesy, Education Images)

Concentrates are stored in grain storage bins. A wide range of mechanical feeding equipment is available. Simple push carts, mechanized feeding carts, augers, cable or track feeders, computerized feeders, and lever-operated storage boxes are all available to meet a wide range of housing/feeding arrangements.

11–34. Cattle are being fed harvested forage from a drive-by vehicle. (Courtesy, Natural Resources Conservation Service, USDA)

Total mixed rations are gaining acceptance as they provide rumen microorganisms with a stable food supply. Chopped hay and silage are weighed and mixed with concentrates in a mixer wagon. Cows can be fed in groups according to production level.

MANURE HANDLING SYSTEMS

Dairy animals produce urine and feces at a rate of about 8 percent of body weight daily. This is about 15 to 20 tons a year for a mature dairy cow. Manure from one cow has nitrogen, phosphorus, and potassium worth over $100.

A well-designed manure disposal/handling system serves three functions: (1) keeps dairy animals clean through frequent removal of manure/waste; (2) provides efficient and economical collection of manure/waste; and (3) disposes of manure/waste in an environmentally sound manner.

11–35. Dairy heifers are kept in a facility with slatted floors to allow liquid wastes to escape. (Courtesy, Education Images)

11–36. A dairy farm lagoon being agitated prior to removal of manure. (Courtesy, University of Minnesota, Crookston)

Solid manure systems usually require the lowest investment in equipment and facilities. Manure can be hauled on a daily basis or stored and hauled periodically. Daily disposal reduces odor and fly problems. All solid manure systems are labor intensive. Some require daily labor and others require intensive labor for a few days.

Liquid manure handling requires a larger investment in equipment. These systems meet most environmental standards and preserve nutrients. Conventional liquid systems may have above or below ground storage tanks/basins. Water is frequently added to the waste material to allow it to be handled by pumps. A minimum of four to five months storage capacity is needed. Adequate land for manure application must be available to avoid leaching of the nutrients to the underground water supply.

An alternate liquid handling system uses lagoons (a type of pond). Lagoon systems can be aerobic (require oxygen), anaerobic (do not need oxygen), or a combination. Aerobic systems are relatively odor-free but require larger surface area. Anaerobic systems can handle more waste and require less land but can cause a bad odor problem.

Waste disposal is a concern. Manure should be viewed as an asset and considered as such when designing a dairy operation.

SHOWING DAIRY CATTLE

Showing dairy cattle has been a popular tradition throughout the United States. Dairy cattle are divided into lots based primarily on age, sex, and breed. Animals in each lot are then shown against each other with judges using type qualities for making their placing. Animals that win each lot or place high enough within a lot can qualify for overall categories of competition.

The Fitting and Showing Scorecard developed by The Purebred Dairy Cattle Association (PDCA) is often considered the official guide for dairy showmanship competition. Appearance of the animal, appearance of the exhibitor, and how the animal is shown in the ring are all factors in the final decision made by the judge. A unique difference in showing dairy cattle is that the exhibitor walks backward while an animal is being evaluated by the judge. New rules allow an exhibitor to walk forward when entering the ring or when the judge is clearly not evaluating the animal.

Skills learned and experience gained from showing dairy animals can be useful for employment in the dairy industry. Showing can also provide enjoyment to the exhibitor. Participants who follow PDCA guidelines will grow in dairy industry knowledge while building a network of people who share a common bond.

11–37. A student has a new heifer to raise for showing. (Courtesy, Education Images)

SHOW DETAILS

Most dairy cattle shows have specific rules and regulations. You should prepare to follow these as you plan to participate in the show. Many shows have a junior division and an open division. Only youth under a certain age participate in the junior division. Youth and adults may show in the open division. Regulations deal with the breeds, ages of animals, how animal well-being is promoted, and general show behavior.

Investigate the date of a show well ahead of time so that you can select and prepare your animal. You will need several weeks or months to get your animal ready. As the show date approaches, you also need to know the time that your animal's class will be judged. Several hours are needed to wash and otherwise prepare an animal for the show ring.

Be sure your application is turned in on time to show officials. Have all immunizations and other health information ready. Get the equipment and materials you need to take with you to a show.

SHOWING YOUR ANIMAL

Showing a dairy animal involves planning well ahead of time. Some of the activities you will need to address are included here. (You can learn the skills associated with these by going to a show and observing how animals are prepared. You will also want to get the assistance of a person trained in showing dairy animals.)

Training

Training is teaching an animal to lead, stand, and be calm. This will begin weeks before a show. Animals are gentle if treated kindly, praised, brushed, and accustomed to a halter while young. Training an animal to lead requires patience. A halter is placed on the head of the animal. The halter should be comfortable for the animal. A short lead rope on the halter helps you direct the animal. Some people tie a calf and take it to get water at least twice a day. Getting the water serves as a reward for leading and reinforces leading behavior.

11–38. A Holstein heifer is being prepared for showing. (Courtesy, U.S. Department of Agriculture)

Washing

Washing is used to clean the hair coat of an animal and make the hair have a good appearance. The first washing is often six weeks before a show. Regular washings occur until the show. A mild soap and plenty of water should be used. It is best to use products prepared for use on dairy cattle or livestock.

After washing, brush and dry the hair. This helps the animal be gentle when these are done away from home in a show barn. After an animal is dry, some people put blankets on them to protect from dust and other dirt.

Grooming

11–39. Brushing with a grooming brush cleans the hair and brings out natural oils. (Courtesy, Education Images)

Grooming is brushing down an animal each day for several weeks before a show. A rubber curry comb is used first to loosen the hair. A brush is then used to remove dust and dirt. Grooming helps bring out the natural oils in the hair and adds gloss to the animal's coat. Learn the techniques involved by observing an experienced groomer.

Clipping

Clipping is shortening hair in some places to enhance the appearance of the animal. Electric clippers are typically used. Hair is typically clipped from the tail, head and neck, legs, and udder-belly a day or so before showing. Only the rough hair is clipped on the body. The switch of the tail is not clipped but is teased to make it fluffy.

Trimming

Trimming is used on the hooves to give each hoof the proper shape. This helps the animal stand properly. Trimming removes excess growth and prevents lameness. It also removes chipped areas and gives hooves a neat appearance. Use the proper tools for trimming. The animal will need to be appropriately restrained to do the work. Never trim too deeply so that the hoof bleeds or develops a sore.

In the Ring

11–40. Students at a New Hampshire school practice showing dairy cows. (Courtesy, Education Images)

Practice with your animal ahead of time helps promote calmness in the show ring. Be sure the animal is groomed and ready before the time of the show. Use a halter that is in good condition and properly fits the animal. Lead from the left side of the animal and hold the halter lead securely in your hand. Arrive early for your class or at least be prompt when the class is called. Wear appropriate clothes for showing. Walk confidently and handle the animal with relaxed skill. Observe the judge and respond to any instructions. When the show is over, promptly leave the show ring and return your animal to its stall. Always demonstrate good sporting behavior and courtesy to others.

REVIEWING

MAIN IDEAS

Dairying is concerned with providing wholesome milk. Producers follow many practices to keep costs down and earn a profit.

Milk production is associated with the reproductive cycle. Cows are milked about 305 days following parturition. They are allowed to be dry 50 to 60 days before the next calving. Cows normally have a calf each year and complete the reproductive cycle.

Dairy cows are selected based on their milking potential. Some differences exist among the breeds. The Holstein is the most popular breed and gives the largest amount of milk. The Jersey gives the milk with the highest butterfat and protein content.

Selecting and culling dairy cows may be based on Dairy Herd Improvement records, linear evaluation, and progeny information. Cows that do not produce or are diseased should be sold for slaughter.

Providing a good environment improves efficiency. The environment should meet the needs of the cows and make high production possible. A ration with high-quality feed is needed. An abundance of good water is essential. Preventing and controlling disease requires continual observation of the herd. Reproductive management and a herd health program are needed. Feed is the greatest cost in dairy production.

Showing is an important activity for youth and adults interested in dairy cattle. Preparing to enter show competition begins months before the show in selecting the animal. Fitting includes properly feeding the animal as well as training, washing, grooming, clipping, and trimming. Poise in a show ring comes with observing the desired skills and practice.

QUESTIONS

Answer the following questions using complete sentences and correct spelling.

1. How has dairying changed since dairy cows were first brought to the United States?
2. What are the important breeds of dairy cattle? Distinguish between the breeds.
3. What should be considered before starting a dairy farm?
4. How are cows selected and culled?
5. How does linear evaluation differ from judging?
6. Describe the ration for a lactating dairy cow.

7. How would you react to the statement that "stress is bad for animals"?
8. What are four climatological factors that affect the performance of dairy animals?
9. Identify three types of dairy cattle housing. How does each modify the weather?
10. What is a milking parlor?
11. What is a total mixed ration (TMR)? What are its advantages?
12. What are the functions of a manure disposal/handling system?
13. What is the overall objective of a health program?
14. Why is dairy cattle reproduction important?

EVALUATING

Match the term with the correct definition. Write the letter of the term on the line provided.

a. immunoglobulin
b. alveoli
c. dry cow
d. homogenization
e. pasteurization
f. progeny
g. animal model
h. Dairy Herd Improvement Program

______1. A cow that is not lactating.
______2. The genetic evaluation for dairy cattle production.
______3. Antibodies in colostrum that develop passive immunity in a calf.
______4. Breaking the fat globules in milk into smaller pieces so they stay suspended in milk.
______5. Offspring.
______6. Heating milk to destroy microorganisms.
______7. Industry-wide dairy production testing and record-keeping program.
______8. Structures in the mammary gland that convert blood nutrients to milk.

EXPLORING

1. Tour a dairy farm. Discuss with the dairy manager how the environment is modified for various age dairy animals.

2. Ask a breed association classifier or an AI organization cow evaluator to discuss the linear traits and how they relate to functional type.

3. Collect several forage samples and send them to an infrared laboratory for analysis. Discuss the analysis with your instructor.

4. Contact a local veterinarian about preventive health visits. How do these visits differ from emergency calls?

CHAPTER 12

Poultry Production

OBJECTIVES

This chapter provides an overview of poultry production. It has the following objectives:

1. **List and describe the major kinds of poultry.**
2. **Explain the poultry industry.**
3. **Describe poultry as organisms.**
4. **Explain production systems for raising poultry.**
5. **List facility and equipment needs for poultry production.**
6. **Describe sanitation and disease control programs.**

TERMS

albumen
broiler
candling
capon
cock
cockerel
debeaking
down
drake
duckling
egg injection
gaggle
gander
gizzard
gosling
incinerator
layer
litter
molting
peacock
peahen
poult
poultry science
pullet
ratite
roaster
spent hen
tom
vertical integration
yolk

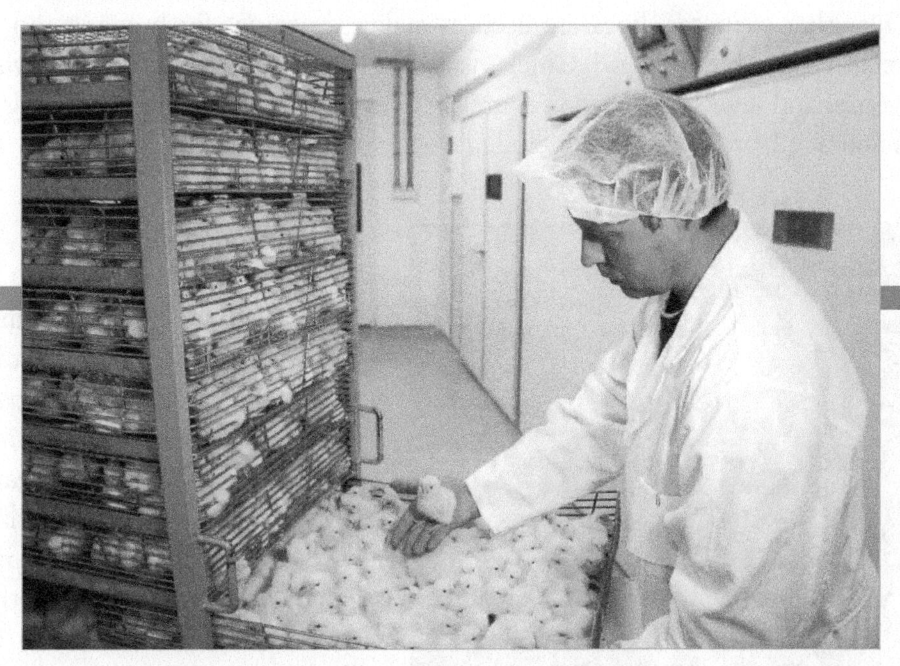

12–1. A hatchery supervisor is checking newly-hatched chicks in the incubator. (Courtesy, branislavpudar/Shutterstock)

POULTRY products are increasingly important in the diets of Americans. Chicken and turkey are most important. Genetic improvements and innovations in production have resulted in better products.

Poultry consumption has been increasing in recent years. Chicken is most popular with an average annual per person consumption of 80.0 pounds. Turkey consumption is 16.8 pounds per person annually. In addition, the average person consumes 242.5 chicken eggs a year either as egg entrees or as ingredients in other dish recipes.

Poultry provide many products. Eggs are important in many dishes. Feathers are used in making pillows, clothing, and fishing lures. Eggs are used in laboratories to make human medicines, including vaccines. Some poultry are kept for ornamental uses and as a hobby. Much of the increased use of poultry is due to changes in how they are raised and marketed.

KINDS OF POULTRY

Poultry are domesticated birds raised primarily for meat, eggs, and feathers. Poultry products are used primarily by humans. Some products, often by-products, are used as pet food and for other purposes. Eggs are used in making vaccines for humans and animals. Poultry are also called birds or fowl. Chickens, turkeys, ducks, geese, quail, peafowls, swans, ratites, pigeons, and several other species are classified as poultry. A few game birds, such as pheasants and quail, are sometimes included.

12–2. Day-old chicks are being placed in a broiler house for growing. (Courtesy, U.S. Department of Agriculture)

12–3. Broilers nearing harvest size in a modern poultry house. (Courtesy, U.S. Department of Agriculture)

CHICKENS

Chickens are the most important poultry species. They are raised for meat and eggs. The type raised depends on the product wanted. A few other uses include game and fancy show chickens. The latter are used mainly for hobbies. Groups of gamecocks may be seen tethered to small individual houses on a range. They account for only a tiny part of the poultry industry.

The meat from a chicken is based on its age and sex. Most chicken is from broilers. A **broiler** is a young chicken six to seven weeks of age that weighs about 4 pounds (2.8 kg). Some growers using chicks with improved genetics will reach this weight in 36 days in a poultry house. Broilers are tender and easy to cook. Chickens of either sex are used as broilers, which are sometimes known as fryers. About 9 billion broilers are raised each year in the United States.

Other chickens used for meat include roasters, capons, and spent hens. A **roaster** is a young chicken that is older and slightly larger than a broiler. A ***capon*** is a male chicken that has been neutered (castrated). Most capons are five to seven months of age and weigh about 6 pounds. A ***spent hen*** is a hen that is no longer laying. Spent hens go into processed foods, such as soup, or are baked. Older male chickens have tough flesh and are not used much in cooking.

12–4. White Leghorn cock and hens. The White Leghorn is primarily for egg production. (Courtesy, U.S. Department of Agriculture)

Eggs are produced by mature female chickens, known as hens. A ***layer*** is a mature female chicken kept to produce eggs. Most hens in laying flocks produce almost 300 eggs in the year after they mature. A ***pullet*** is a young female chicken that is being raised for laying. Often, layers are force-molted so they can be recycled to lay for another year.

A mature male chicken is a rooster or ***cock***. Young male chickens are known as cockerels. A ***cockerel*** is less than one year of age.

12–5. New Hampshire hens. (Courtesy, Mississippi State University)

Breeds of chickens are not important in commercial poultry production. Varieties of chickens have been developed by poultry integrators to meet particular needs.

The common breeds of chickens are White Leghorn (used for egg production, white egg shell); Barred Plymouth Rocks (meat and eggs, brown egg shell); New Hampshire (meat and eggs, brown egg shell), and the White Rock (meat and eggs, brown egg shell). The White Leghorn is the smallest, with mature cocks weighing 6 pounds (2.7 kg) and hens weighing 4 pounds (2 kg). The White Rock is among the largest, with cocks weighing 9 pounds and hens weighing 7 pounds (3.4 kg).

TURKEYS

Turkeys are raised primarily for meat. Consumers want birds that have a high proportion of white breast meat. Nearly 300 million turkeys are raised each year in the United States.

12–6. White turkeys being raised on a range. (Courtesy, U.S. Department of Agriculture)

12–7. Bronze turkeys of breeding size. (Courtesy, Ed Phillips/ Shutterstock)

Turkeys are being raised on fewer farms, but the size of the farms has increased. Many commercial producers raise more than 100,000 birds each year on a single farm.

A young turkey is a ***poult***. It has not grown to the point where its sex is easily determined. A mature male turkey is a ***tom*** or gobbler. A female is a hen. Younger and smaller turkeys are roasted whole. Larger turkeys are often made into boneless breasts and other further processed products.

Most turkeys are raised in confinement under contract with a processing plant. Many turkeys are marketed after 20 weeks of age. Toms weigh considerably more than hens, with the market for further processing requiring heavier and heavier turkeys. Good producers can have feed conversion ratios of 2.2 pounds (1 kg) of feed to 1.1 pounds (0.5 kg) of gain. Turkeys have been bred to be larger and have broad breasts. Colors vary from white to bronze, with the broad-breasted white being essentially 100 percent of those in commercial turkey production.

DUCKS

Ducks are raised for meat, eggs, and down and feathers. (***Down*** is the soft feathery covering that grows under the feathers.) Some are kept as hobby or ornamental ducks.

Young ducks are covered with down. Feathers develop as they grow. A young duck that still has down rather than feathers is a ***duckling***. A mature male duck is a ***drake***. Mature female ducks are known as hens.

Duck and duck products are not nearly as widely used as chickens. Some 15 million ducks are raised in the United States each year. White Pekin ducks are most widely found.

Ducks grow faster and heavier than chickens. They can also swim, which is not the case with chickens and turkeys. Most commercial ducks are now raised indoors.

12–8. Pekin ducks. (Courtesy, Mississippi State University)

GEESE

About 1 million geese are raised in the United States each year. Geese are raised for meat, eggs, and feathers and down. Many are also kept for ornamental purposes. Geese like to eat tender grass in pastures and parks. Some are used to control weeds and grass and are known as "weeder geese." Geese are hardy birds and resist many poultry diseases. A baby goose of either sex is a **gosling**. A mature male goose is a **gander**. A mature female is a hen or goose. A flock or group of geese that are not flying is known as a **gaggle**. Wild Canada geese have been encouraged around lakes and resorts. However, in some places, they have become pests.

12–9. Embden is a breed of goose that has a striking white appearance. (Courtesy, Kimberley McClard/Shutterstock)

PEAFOWL

Peafowl are raised for their large, beautiful feathers. A male is known as a **peacock** and a female as a **peahen**. Sometimes, either sex is known as a peacock. Male peafowls may spread their feathers, known as a train. The feathers may be five times the length of the body. Most peafowls are raised for ornamental purposes. The hens lay a few eggs each year.

12–10. A peacock. (Courtesy, Education Images)

SWANS

Swans are similar to ducks and geese in their preference for water. Several kinds of swans are found, with colors ranging from white to black. Most swans are kept for ornamental purposes. They are frequently used with pools of water at resorts and similar locations.

12–11. A white swan (left) and an ornamental black Australian swan. (Courtesy, Education Images)

RATITES

No other birds have created more excitement in recent years than a group known as ratites. A ***ratite*** is a group of flightless birds that have some commercial uses. Ratites include the ostrich, emu, rhea, cassowary, and kiwi.

12–12. The ostrich is the best-known ratite. (Courtesy, Education Images)

The largest and best-known ratite is the ostrich. An ostrich can weigh as much as 350 pounds and stand 10 feet tall. Ostriches have the longest life span of most birds, living up to 70 years. These birds are raised for feathers, meat, skin, and oil products. The feathers are known as plumes and are used in decorations.

Interest in ratites was high in the late 1980s but declined by 2000. Many growers paid high prices for breeding animals but were unable to sell offspring at a profit. The novelty of ratites has worn off.

GUINEA FOWL

Guineas are raised for food, as novelty birds, and to stock game preserves. They are hardy and often roam free. Their eggs are smaller than chicken eggs and have a thicker shell. Because the eggs are not as easily broken, they are often dyed and decorated in various ways.

12–13. A guinea fowl. (Courtesy, Bogdan VASILESCU/Shutterstock)

THE POULTRY INDUSTRY

The poultry industry has changed from a family farm with a few birds for home use to large commercial producers that raise hundreds of thousands of broilers each year.

EARLY POULTRY PRODUCTION

The first poultry was raised more than 5,000 years ago in Asia, and in Egypt about 3,500 years ago. Poultry were brought to North America by the early settlers. A few people in Jamestown had small flocks of chickens as early as 1607.

In colonial times, many families kept small flocks that ran about and often nested in brush and thickets. They ate whatever they could find—insects, seed, and tender leaves. Fox and other animals often preyed on them. Most poultry were kept for eggs, with young males and older hens used for meat. People often raised only what they needed, with extra eggs or chickens traded to neighbors for a ham or vegetables.

The turkey is native to the Americas. The Aztecs domesticated the wild turkey. Native Americans used the turkey for food and its feathers for decoration. Turkeys were exported to Europe before chickens were brought to Jamestown!

Dramatic changes began to occur in poultry production in the mid-1900s. These changes led to the modern poultry industry.

12–14. Modern poultry houses have controlled environments and automated feeding systems. (Courtesy, Education Images)

MODERN POULTRY INDUSTRY

Poultry production is a large commercial industry. Thousands of birds may be raised in confinement in one house. Science and technology are used to assure their well-being. On the other hand, some consumers want free-range chickens that have been allowed outside and not confined to a house.

The number of small flocks has declined considerably since 1950. Most people could not raise poultry even if they wanted to because of where they live. Urban life is not well suited to having a small flock of poultry. People prefer to go to the supermarket and buy poultry products that have been carefully inspected and are ready to cook.

Providing an abundance of poultry products at low costs required major changes in how poultry were raised. Good information was needed. A system to market products had to be developed for consumers to have a year-round and uniform supply. Much has happened in the poultry industry since Harland Sanders opened the first KFC restaurant in 1956! Today, there are other speciality restaurants that serve chicken.

The leading states in broiler production are Georgia, Arkansas, Alabama, North Carolina, and Mississippi. California is the largest producer of eggs. North Carolina, Minnesota, and California are leading states in turkey production.

Poultry Science

Poultry science is the study and use of areas of science in raising poultry. It includes breeding, incubation, rearing, housing, feeding, sanitation, marketing, and other areas. The goal of poultry science is to provide consumers wholesome poultry products at a reasonable price.

Poultry science has resulted in improved production. A hen now lays almost 300 eggs a year (compared to 100 per year in 1900). Automation allows one worker to care for many birds. A single broiler house may have 40,000 birds. One worker can care for more than 100,000 broilers or 40,000 laying hens in cages. Most of the information used in poultry science has been developed in the last 50 years. It is based on biology and related areas of science, including environmental science.

Vertical Integration

12–15. Inspection to assure quality food products is carried out in processing plants. (Courtesy, Agricultural Research Service, USDA)

Raising, processing, and distributing poultry has developed into one continuous chain. ***Vertical integration*** is one agribusiness being involved in more than one step in providing poultry products. The steps are linked together to assure a continual supply of poultry. The poultry industry has more vertical integration than other areas of agriculture. All areas of poultry are vertically integrated.

The best example of vertical integration is in broiler production. One company vertically integrates by carrying out all of the steps in the process of poultry production. A poultry company may sign agreements with growers, provide the chicks and feed, supervise the grower, process the broilers, and distribute the chicken to buyers. The grower (farmer) provides a poultry house and looks after the chicks. The company may pay the grower a fee for each bird raised. The amount is based on how well the birds grow or other conditions.

POULTRY AS ORGANISMS

Poultry are terrestrial animals with feathers, a backbone, and wings that have limited use in flying. Modern poultry do not need to fly, nor do they have the ability to do so. Some species can swim, such as ducks and swans. Chickens and turkeys cannot swim. Poultry are classified in the Aves class of vertebrates. Each species has a scientific name. The scientific names of common poultry species are shown in Table 12–1.

Table 12–1. Scientific Names of Common Poultry Species

Kind of Bird	Scientific Name
Chicken	*Gallus domesticus*
Duck	*Anas domestica*
Goose	*Anser domesticas*
Guinea fowl	*Numida meleagris*
Ostrich	*Struthio camelus*
Pigeon	*Columba domestica*
Swan (black)	*Cygnus atratus*
Turkey	*Meleagris gallopavo*

LIFE PROCESSES

Poultry carry out the same life processes as other animals. The structures for doing so vary. Two major areas of difference are food digestion and reproduction.

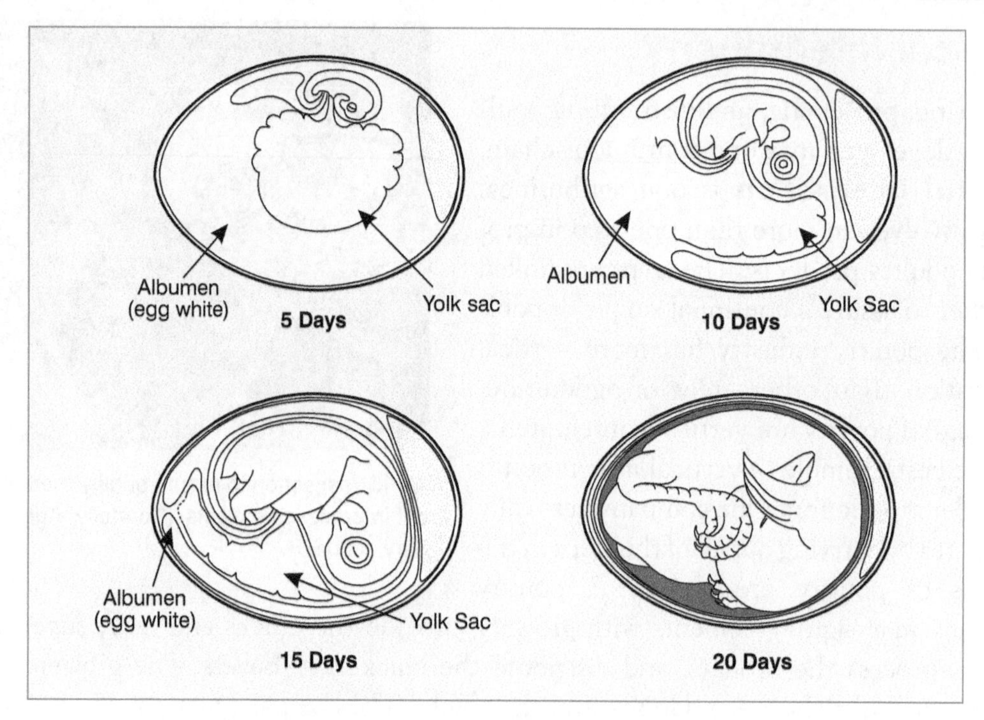

12–16. Development of a chick embryo in an incubating egg.

Digestion

Birds take in food with their beak. A beak is a strong mouth part made of material similar to a person's fingernails. It can be used for pecking and breaking foods apart. In large poultry houses, birds may be debeaked to reduce damage they can inflict on each other. ***Debeaking*** is removing the tip of the beak so they cannot peck other birds.

Poultry do not have teeth and rely on a ***gizzard*** to grind food. The gizzard is a strong, muscular organ that may contain grit that the bird has eaten. Grit (e.g., grains of sand) helps in the food-grinding process.

12–17. Chicks are beginning to hatch from these eggs following incubation. (Courtesy, U.S. Department of Agriculture)

Reproduction

To reproduce, a female bird must mate with a male for eggs to be fertile. After laying, eggs are incubated. The length of time for the embryo to develop varies by species: chickens need 21 days; geese, 29 to 31 days; turkeys, 27 to 28 days; and ducks, depending on their species, need 28 to 35 days. The incubation period is also influenced by temperature and humidity conditions. Eggs go through distinct changes in incubation.

12–18. A modern incubator gently rotates eggs much as a hen does sitting on a nest. (Courtesy, U.S. Department of Agriculture)

12–19. Hatching eggs in a small incubator involves: placing fertile eggs in tray (top), adding water to incubator tank to assure sufficient humidity at 82 to 84 percent (middle), and observing temperature and humidity to be sure the desired environment is maintained during incubation (21 days). (Courtesy, Education Images)

Most eggs sold in supermarkets are infertile. Hens do not need to mate to produce an egg. There is usually no difference to the consumer between fertile and infertile eggs if they are fresh. A fertile egg that was not refrigerated quickly and stored in a warm environment may show signs of embryo development.

Artificial insemination is used in the poultry industry. Some birds, particularly turkeys, are unable to mate because their mass of flesh presents a mechanical barrier. Natural fertilization is still predominant with chickens.

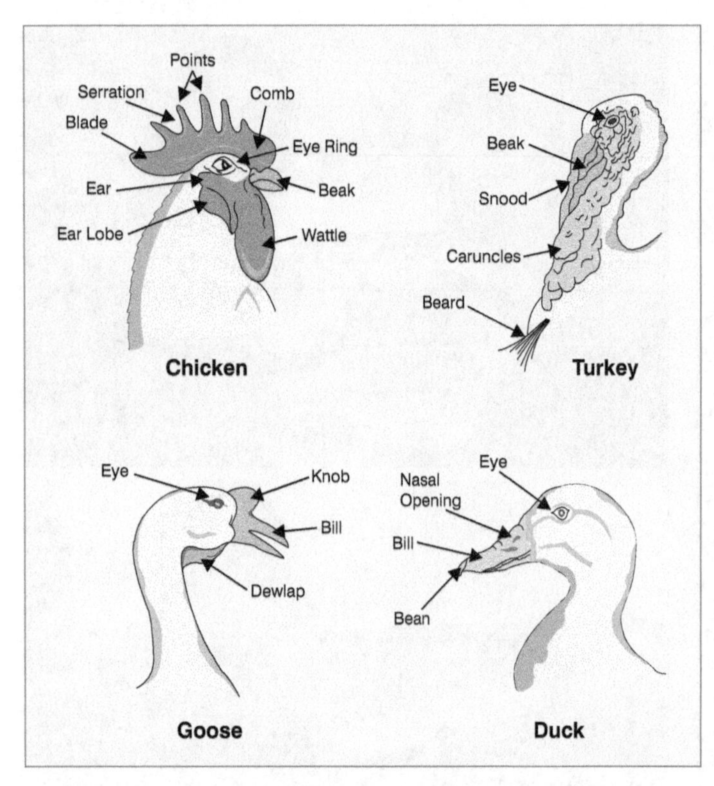

12–20. Head and neck features help identify common poultry.

APPEARANCE

The external parts of most birds are similar. Some, such as the game chicken, have long spurs near their feet. Chapter 2 provides additional information on biology.

The health and condition of a bird are evident in its exterior appearance. For example, some breeds have yellow shanks; others have white shanks. The disappearance of the yellow pigment in those with yellow shanks is a sign that the female has been laying eggs. A large, red comb is a sign that the bird is in good health. Ragged feathers may indicate poor condition.

It is fairly easy to distinguish between most types of poultry. Some, such as ducks and geese, live in and around water. Chickens and turkeys cannot swim. Turkeys have beards and other fleshing structures from their heads. Some chickens have serrated combs and wattles. Geese and ducks have rather plain heads. Looking at the heads of poultry is a good way to tell them apart. The heads also vary with the sex of the poultry. Males tend to have larger head features, such as the comb.

POULTRY PRODUCTION SYSTEMS

Chickens are the dominant form of poultry in the United States. Most chickens are produced on large poultry farms that specialize in birds for a particular use. With vertical integration, producers usually have little investment in the chickens, and large investments in the land for the farm, housing and equipment, electricity and water, and labor. The poultry company provides the chicks, feed, medications, and field services in an effort to assure that the birds receive proper care.

The chicken industry has four areas: broiler production, egg production, pullets for egg production, and pullets and cockerels for broiler egg production.

BROILER PRODUCTION

The goal of broiler production is to produce the most meat as quickly and efficiently as possible. Growers take extra care to keep chicks healthy and growing. Mortality (death) is kept to a minimum—no more than 5 percent while the broilers are growing.

Most broilers are raised in large poultry houses. About six weeks are needed to raise a broiler to market size, or about 4 to 4.5 pounds (2 kg). The time required is declining as genetic stocks and diets are continually improved. Housing is designed to meet the needs of young chickens.

12–21. Newly hatched turkeys closely resemble chicks. (They will soon become poults, the term for young turkeys.) (Courtesy, Willmar Poultry Company, Minnesota)

CONNECTION

QUICK CHICKEN

Broilers grow fast when done the "Fieldale way." Fieldale farms is a large poultry integrator in Georgia with a number of growers under contract. Here are the events in quickly raising top-quality broilers.

One grower with four modern, computer monitored and controlled houses would likely:

- Have a total of 80,000 genetically improved birds.
- Grow day-old chicks to market-size (4 lb.) broilers in 36 days.
- Have 97 percent or more of the chicks live to market size.
- Have nearly 80,000 birds weigh 320,000 lbs. at market.
- Have birds eat 800,000 lbs. of feed (2:1 feed conversion ratio).
- Use chicks that are ***in vivo*** vaccinated on day 18 of incubation.
- Re-vaccinate chicks again at 8 and 21 days of age in drinking water.
- Have lighting for 24 hours a day the first week; cut back to 6 hours a day the second week; and increase it 4 hours a day each week thereafter until it increases to 22 hours a day.
- Produce six (or maybe more) batches a year.

(Courtesy, branislavpudar/Shutterstock)

Amazing! Every production day counts. Similar schedules are repeated by poultry producers across the nation with other integrators.

Growers attempt to provide an environment that is ideal for broiler production. A day or so in the time required to raise a broiler makes a big difference in feed cost. The goal is to get broilers to grow as fast and efficiently as possible. Most growers get a pound of growth on less than two pounds of feed. Young chicks are fed a finely ground meal feed that has 18 to 20 percent protein. As they get older, the percentage of protein is reduced to 12 to 15 percent. Protein in feed is typically from soybean meal and fish and other meat-meal ingredients. The feed also contains ground yellow corn, other grains, minerals, vitamins, and, in some cases, medication.

A small niche market exists for broilers raised on a range. They are outside and allowed to run about inside fenced areas. The efficiency of feed conversion and control over product quality is not as good as that in a modern house.

EGG PRODUCTION

The goal of egg production is to produce high-quality eggs for human use. Hens are given an ideal environment to produce. Producers strive to get one egg each day from hens. The eggs used for food are not fertile. They will not hatch if incubated.

Eggs are graded and placed in cartons designed to protect them from damage. Eggs are graded by size, such as small, medium, and large. The eggs are put in a carton with the small end downward. The larger end should be up to help maintain the quality of the egg. The large end has the egg's air cell. Dirty, misshaped, or cracked eggs are removed.

Managing hens for egg production requires facilities where their well-being is met and the eggs are of high quality. The facilities must keep the eggs clean and prevent breakage. Regularly collecting the eggs or using automated egg collecting systems is needed. Most layers are kept in specially designed cages. Nearly 90 percent of the layers are in cages. Other layers are in houses with slatted or litter floors. A few layers are on a range, and the eggs they produce are for niche markets. A nest must be provided for the hen in laying.

12–22. An egg production facility. (Courtesy, Tomas Sereda/ Shutterstock)

Feed and Water

Laying hens need proper feed. The formulation of a complete commercial feed should vary according to the hen's age, stock, and time of laying. Diets are supplemented with calcium because egg shells contain substantial calcium. A good supply of water is critical because laying hens will drink 2 to 3 pounds (1 to 1.5 kg) of water for each pound of feed.

An average chicken egg weighs 2.2 ounces (0.062 kg). It is made of 66 percent water, 10 percent fat, 13 percent protein, and 11 percent other material, primarily calcium in the shell. The fat is mostly in the egg yolk. Hens that do not get feed with the needed nutrients cannot be productive.

Egg Quality and Color

Egg quality is determined by external appearance and interior condition. External appearance includes size, shape, color, cracks, and blemishes or dirt.

Egg color includes both the shell and internal parts of the egg. The shell is a protective layer of material, primarily calcium. Egg shell color varies. Nearly 95 percent of the market eggs in North America are white. In recent years, markets for brown eggs have increased in areas other than the New England states, which have long had a great demand for brown eggs.

The inside of the egg has two major parts: albumen and yolk. The **albumen** is the white part that surrounds the yolk. It should be firm and cling near the yolk when broken out. Albumen that spreads out indicates deterioration and aging. The **yolk** is the center part of an egg that varies in degree of yellow color. Hens fed diets high in carotene lay eggs with yellower yolks.

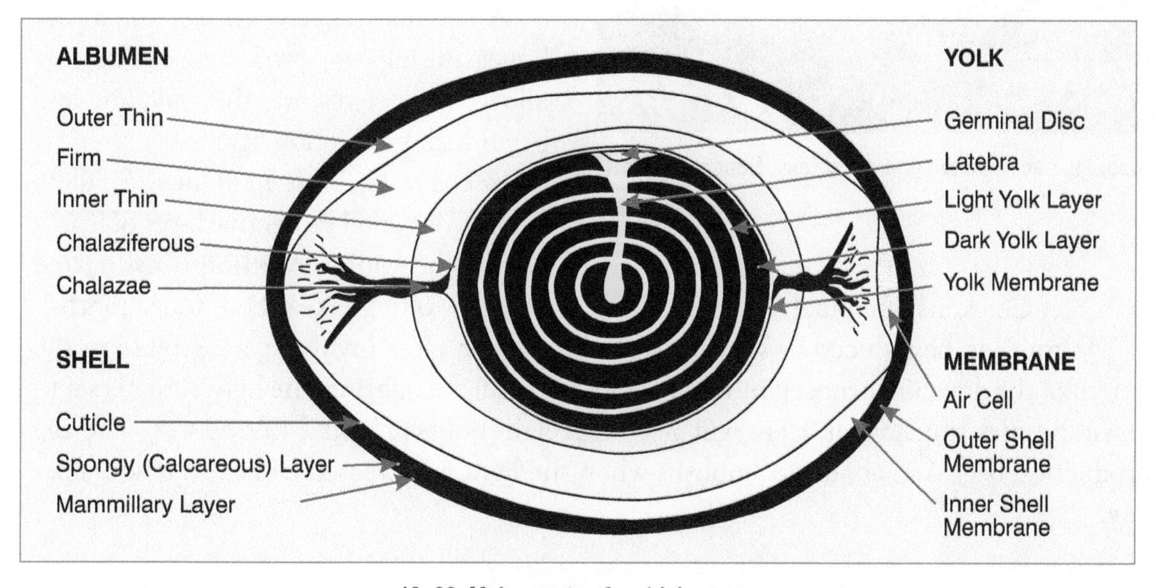

12–23. Major parts of a chicken egg.

12–24. Eggs being candled. (Light shining through eggs makes any spots or imperfections visible.) (Courtesy, chalabala/ Shutterstock)

Inside structure indicates egg quality. The presence of meat or blood spots is objectionable. An air cell is located inside at the large end of an egg. As an egg ages, the air cell gets bigger. Fresh eggs have small air cells. ***Candling*** is shining light through an egg to see the inside features. Most of an egg is fairly transparent. Dark spots are signs of problems.

Cracked shells lower the quality of eggs. In most cases, cracked eggs are unacceptable to consumers. Breaking plants use cracked eggs. The eggs are frozen and used in food products.

12–25. Consumers expect quality eggs. (Courtesy, Education Images)

Molting

Molting is the process of shedding and renewing feathers. Hens usually stop laying during molting. If not, the laying rate is greatly reduced. Some growers use forced molting or recycling. This gives the hen a rest before returning to production. Forced molting is becoming popular.

A hen may deplete the calcium in her bones during egg production. The rest allows her to restore the calcium and begin high production again.

In the wild, birds molt in the fall so they have a good protective layer of feathers for the winter. Molting is associated with age. Chickens often molt at about one year of age. Molting takes about four months.

Molting can be induced by various procedures, with most involving a decrease in the length of the day. Since most poultry houses have artificial lighting, the light timer is set to make the dark time longer. Hens will molt with eight hours of light. Molting stops and egg production begins in about two months when the light is increased to 14 to 16 hours per day.

PULLETS

Hens for egg production are raised by pullet producers. The grower takes specially bred, day-old chicks and raises them similar to broiler chicks. The chicks are sexed, so only females are raised. The sexing is done prior to delivery to the grower. The male chicks are raised as broilers for meat. The birds are bred for maximum egg production. The pullets are known as replacement hens.

Pullets are raised to 20 weeks of age by the pullet producer. They then go to laying farms as starter pullets. Pullets will begin egg production a few weeks after delivery. Pullets start laying regularly at 24 weeks of age and reach maximum production at 30 to 35 weeks.

BREEDER BIRDS

Some growers specialize in raising pullets and cockerels for fertile egg production. The fertile eggs are hatched to become broilers. Chickens are carefully selected for their potential in producing broilers with large amounts of meat. Many producers raise pullets and cockerels for breeding under a contract with a poultry company.

The pullets and cockerels are raised separately. A typical ratio is to raise one cockerel for each eight to ten pullets. At maturity, the cockerels and pullets are placed in a combined flock so mating can occur and the hens will lay fertile eggs.

CAREER PROFILE

FERTILE EGG PRODUCER

A fertile egg producer keeps hens and roosters for producing fertile eggs for a hatchery. Fertile egg producers are under contract with large processing commercial hatcheries and follow specific directions with the chickens. General duties include maintaining houses and equipment, caring for chickens, managing house environments, gathering eggs, and observing for health problems.

Fertile egg producers need practical experience with chickens. A high school diploma is beneficial. Land and adequate facilities are required to accommodate birds. Facilities for storing feed and providing water are required. Methods for disposing of dead birds are needed. A producer must be able to follow instructions provided by field representatives of the hatchery.

Most fertile egg producers are entrepreneurs who work under contracts with poultry hatcheries and integrators. A few small, independent producers may be able to serve niche markets. This photo shows a producer checking automated equipment in a chicken house. (Courtesy, U.S. Department of Agriculture)

The initial practices with breeder chicks are different from those with broilers in many ways. Growing conditions are used to prepare pullets for high egg production, and feed intake of both males and females is carefully controlled. In the breeder house, separate facilities for males and females are common, in order to control body weight of each sex.

HOUSING AND EQUIPMENT

Housing and equipment used in poultry production varies with the system used and the species being produced. Some poultry is produced inside buildings under environmentally controlled conditions. Other poultry is produced with partial freedom to have access to the outdoors. Some poultry is almost completely produced on range conditions.

This section of the chapter addresses housing and equipment used in broiler production.

12–26. The floor of this house is covered with clean wood shavings used as litter. (Courtesy, Education Images)

12–27. Economical lighting is needed to promote efficient growth. (Courtesy, Education Images)

HOUSING

Houses for broilers are designed to meet the needs of newly hatched chicks through market-size birds. A day-old chick needs between one-quarter to one-third of a square foot (0.02 to 0.03 m^2) of floor space. Space requirements expand as the chickens grow. A four-week-old chicken needs about three-quarters of a square foot (0.06 to 0.09 m^2) of floor space. Older, larger birds would need more space. Growers often use a portable fence to isolate baby chicks to a small part of a house. The fence is removed as they grow.

Poultry houses have a controlled environment. The controls provide for the well-being of the chickens. Environmental control characteristics:

- Litter—***Litter*** is the floor covering in a poultry house. Litter is often made of wood shavings. The litter absorbs moisture from the manure. Litter is normally changed between batches of chickens.

- Lighting—Modern poultry houses rely almost exclusively on electric lights. No natural light may be allowed in some houses. The lighting is timed to give chickens the light needed to eat and grow. Shorter days are made longer by artificial lighting. Most poultry houses use a specially designed bulb that provides adequate light on a low amount of electricity.
- Temperature—Baby chickens are especially sensitive to temperature changes. Many growers keep house temperatures in the 85 to 95°F (29 to 35°C) range for small chicks. Large hood-type brooders with electric or gas heat are used. By six weeks of age, a temperature of 70°F (21°C) is adequate. It is lowered as the chickens grow. In warm weather, cooler outside air may be pulled into the house for ventilation.
- Humidity—Humidity control is used to provide the ideal level of water vapor in the air of chicken houses. Heat tends to dry the vapor from the air. Water vapor may be added to the air with mist systems. The humidity level should be 50 to 75 percent in broiler houses.
- Ventilation—Ventilation involves using large fans to bring fresh air into a chicken house. The fans are often

12–28. A thermostat controls the temperature of this gas-powered brooder. The large, round, reflective hood assures chicks of adequate heat. (Courtesy, Education Images)

12–29. Humidity and ventilation are regulated by large fans that automatically turn on when needed. (Courtesy, Education Images)

12–30. Sensors above the level of the chickens in a poultry house help control the environment. (Courtesy, Education Images)

12–31. An emergency generator is on standby if the electricity fails. (Courtesy, Education Images)

on controls to help regulate temperature. Ventilation also removes excess moisture from a house.

- Alarm systems—Modern poultry houses should have systems to alert people to problems. Failures of feeders, waterers, ventilation systems, and other areas are monitored.
- Standby electrical generators—Poultry farms need standby electrical generators in case of power failure. Chickens cannot live long without feed, water, and environmental control.

FEEDING AND WATERING EQUIPMENT

Automated feeding and watering equipment are used. Young chicks may require feed in trays or pans, and waterers that are easy to use; however, they soon learn to eat from automated feeders. After a few days, most producers rely only on automated systems.

Feeder systems may consist of tubes or troughs located throughout the house. Feed bins and weighing equipment are located outside at the middle of a long house. A conveyor chain moves the feed throughout the house. A large flock will eat a lot of feed. Plan ahead to be sure there will be plenty. Adequate feed storage must be provided. Feed is usually stored in bins designed to keep the feed from damage and contamination. Bins should be easy to use with access for delivery by trucks.

12–32. Automated feeding systems carry feed throughout a house. (Courtesy, Education Images)

12–33. Feed is stored, weighed, and metered into the feeding system by these automated bins. (Courtesy, Education Images)

12–34. Two types of watering systems are shown here: watering nipple (left) and watering trough. (Courtesy, Education Images)

Watering systems may involve troughs and nipples attached to tubes. The water is carefully metered. The amount consumed is an indication of the health of the flock. Water consumption drops when chickens are sick or in less than optimum environment. Medications may be added to water. Waterers need to be cleaned each day.

Most producers use water from wells or municipal systems. Many have standby sources in case one source fails.

SANITATION AND DISEASE CONTROL

Sanitation and disease control are important in successful poultry production. A disease outbreak can quickly wipe out a flock. It is far better to prevent disease than it is to try to treat an outbreak after it occurs.

SANITATION

Practicing good sanitation is important in preventing the spread of disease. Many diseases are spread by diseased birds, insects, people, and vehicles that travel from one farm to another. Carcasses of dead birds and contaminated feed, water, and litter also spread disease. Keeping a poultry farm clean is a major part of sanitation.

12–35. Disposable boots and shoe dip are both used at the entrance to this poultry facility. (Courtesy, Education Images)

Removing all litter from a poultry house and spraying the building and equipment with a disinfectant helps prevent disease. This is done after a batch of chickens has been moved to processing or other farms, and prior to bringing a new batch of chickens to a farm.

Many poultry farms restrict the access of vehicles and people. Tires, shoes, clothing, and equipment can transport a disease from one farm to another. Tires of a vehicle must be washed with a disinfectant solution before going onto a poultry farm. People dip their shoes in disinfectant solution or wear disposable boots.

12–36. Day-old chicks being debeaked. (Courtesy, Education Images)

12–37. An automated *in vivo* system to vaccinate embryos on day 18 of incubation. (Courtesy, Embrex, Inc.)

VACCINATION

Vaccination is used to help poultry develop immunity to disease. The common diseases vaccinated against include Newcastle, Marek's disease, and infectious bronchitis.

Recent advances have allowed for vaccination in vivo. This is into the embryo of the incubating egg by **egg injection**. Egg injection is used on the eighteenth day of incubation to help the chick be immune to certain diseases at hatching. An egg injection system is used in the hatchery at a high rate of speed. A specially designed needle penetrates the egg shell with a minimum of damage to the shell or embryo. The dosage is delivered inside the egg. One machine and two operators can inject 20,000 to 30,000 eggs an hour. Vaccinating during incubation results in the chicks hatching with immunity. Handling and stress are reduced. The chicks are said to be ready for market about two days earlier if vaccinated before hatching.

If chicks are not vaccinated in vivo, they are vaccinated in the debeaking process. The vaccine is placed in the nose, eye, or wing. Medications are also placed in drinking water.

DEAD BIRD DISPOSAL

Any birds that die should be promptly disposed of to prevent the spread of disease to others in the flock. Dead birds are usually handled in one of three different ways: incineration, burying, and composting.

An ***incinerator*** is equipment that burns chicken carcasses so only ashes remain. Incinerators are made of steel and heated with propane, natural gas, or fuel oil.

Composting is a low-cost, self-contained method of layering the carcasses with straw or other litter, and leaving them to decompose naturally. A special compost bin is used. The nutrient-rich remains, or compost, can be used as fertilizer on crops.

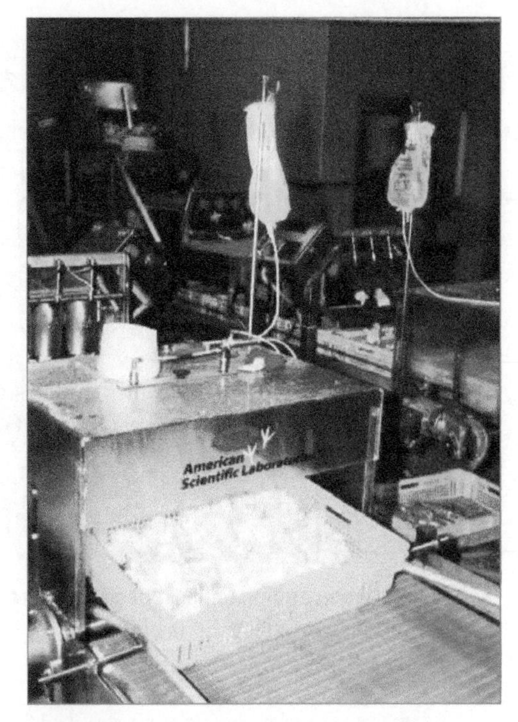

12–38. Day-old chicks are automatically vaccinated at the hatchery in preparation for delivery to a poultry farm. (Courtesy, Education Images)

DISEASES

Poultry are affected by some 33 pathogenic diseases and 10 parasites. In addition, nutritional diseases and deficiencies affect poultry. These can cause huge losses. In some cases, the birds die. In others, the birds are less efficient. Either way, the owner has a loss of money from the disease. Further, who wants to eat a diseased chicken?

Diseases

Pathogenic diseases are caused by bacteria, viruses, or other living agents. Some examples follow.

- Marek's disease—Marek's disease is also known as range paralysis. It is caused by a herpes virus. Symptoms include diarrhea, paralysis of legs or wings, loss of weight, and death. No treatment is available. Prevent Marek's disease by vaccination of day-old

12–39. An approved incinerator for a poultry farm. (Courtesy, Education Images)

12–40. Incubation equipment is kept clean to prevent disease in newly hatched chicks. (Courtesy, Education Images)

chicks or embryo vaccination. Some genetic resistance has been developed.

- Newcastle disease—Newcastle disease is caused by a virus. Affected birds gasp, wheeze, twist necks around, may be paralyzed, and lay soft-shell eggs, if any are laid. No treatment is available. Vaccination prevents Newcastle disease.
- Infectious bronchitis—Infectious bronchitis only affects chickens. Young birds wheeze and gasp and have nasal discharge. Older birds stop laying. Sanitation and isolation are effective in prevention. A vaccine is available. Infectious bronchitis is caused by a virus.
- Fowl cholera—Fowl cholera affects many different birds. Caused by bacteria, birds get a fever, purple-colored heads, yellowish droppings, and sudden death may result. Some growers have success in treating fowl cholera with antibiotics and sulfonamides. Vaccination can be used to prevent fowl cholera.

12–41. Records related to chick health and hatching rate are carefully kept in a hatchery. (Courtesy, Education Images)

Parasites

Sanitation and control of pests, such as flies and wild birds, help in preventing parasites in poultry. All poultry may be affected. Some examples of parasites:

- Coccidiosis—Coccidiosis is an internal protozoan parasite disease. It is transmitted by the droppings of affected birds, including wild birds. The droppings may be bloody, and affected birds may be sleepy, pale, and listless. Anticoccidials may be put in feed or water.

- Large roundworms—Birds with large roundworms are droopy, emaciated, and may have diarrhea. The worms may reach lengths of 3 inches (7.62 cm) in the intestines. Sanitation and rotation of range and yards help control large roundworms. Dewormers can be used to treat affected poultry.
- Mites—Mites are external parasites that cause birds to be droopy, pale, and listless. Approved insecticides can be used to treat mites. Mites are prevented by sanitation and controlling access of wild birds to the flock.
- Tapeworms—Tapeworms are internal parasites that use snails, earthworms, beetles, and flies as intermediate hosts. Pale head and legs and poor body flesh indicate tapeworm infestation. Dewormers are available.

12–42. Healthy broilers are nearing market size. (Courtesy, branislavpudar/Shutterstock)

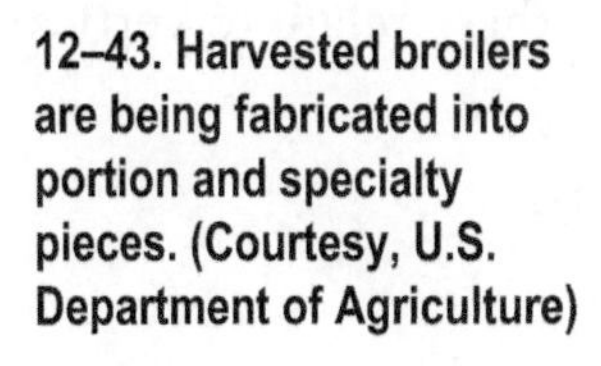

12–43. Harvested broilers are being fabricated into portion and specialty pieces. (Courtesy, U.S. Department of Agriculture)

REVIEWING

MAIN IDEAS

Poultry are domesticated birds raised for meat, eggs, and other products. Chicken is by far the most important poultry. Turkey is second in importance.

Broilers are the major source of chicken meat. A broiler is a young chicken 6 to 8 weeks old. Most commercial broiler producers can grow broilers to the weight of 4 pounds or more in six weeks or a little less. Egg production, pullet production, and breeder pullet and cockerel production are also specialized areas of the chicken industry.

Vertical integration is used to link the poultry industry. Most producers specialize in one kind of bird, such as broilers or layers. They get the baby chicks or pullets from specialized producers.

Housing and equipment are used to provide a good environment for chickens. The well-being of birds is a major part of the environment. Birds are managed to efficiently grow and produce.

Sanitation is an important part of reducing disease problems. A method for properly disposing of dead birds is essential on every poultry farm. Many farms use incinerators.

Chicks are vaccinated for disease. In vivo vaccination is now widely used. It involves egg injection of the embryo on the eighteenth day of incubation. This reduces the handling stress newly hatched chicks often experience.

QUESTIONS

Answer the following questions using complete sentences and correct spelling.

1. What are the major kinds of poultry? Briefly describe the two most important.
2. Explain how the poultry industry has changed chicken production.
3. How does the digestive system of poultry vary from other animals?
4. What is the reproductive process in poultry?
5. What are the distinguishing features of the heads of chickens, ducks, turkeys, and geese?
6. What are the four production systems with chickens? How do these relate to vertical integration?
7. What are the housing and equipment needs in raising broilers?
8. How does the feed vary from young chicks to older broilers?
9. How are layers managed for production?
10. What are the major parts of an egg?
11. What is molting? Forced molting?
12. Why is sanitation important in controlling disease?

EVALUATING

Match the term with the correct definition. Write the letter of the term on the line provided.

a. egg injection	e. debeaking	i. layer
b. incinerator	f. gosling	j. capon
c. candling	g. tom	
d. litter	h. cockerel	

_____1. A mature female chicken kept for egg production.

_____2. A castrated male chicken.

_____3. A mature male turkey.

_____4. A young male chicken.

_____5. The floor material in a chicken house.

_____6. Removing the tip of the beak.

_____7. Vaccinating chicks as embryos through the egg shell before hatching.

_____8. Equipment to burn the carcasses of dead poultry.

_____9. A young goose of either sex.

_____10. Shining a light through an egg to grade quality.

EXPLORING

1. Invite a poultry technician or other qualified person to serve as a resource person in your class. Have them explain the nature of the poultry industry, particularly their role in providing poultry products.

2. Incubate a small batch of eggs. Use a laboratory educational incubator. Get the fertile eggs from a local producer or person with a small farm flock. Carefully monitor temperature, humidity, and other conditions. Follow instructions with the incubator to be sure the eggs are properly incubated. Keep records on all processes and activities.

3. Use a whole broiler from a local supermarket. Carefully cut the broiler into different pieces based on the natural joints of the bird. Study muscling and bone structure. CAUTION: Be careful with sharp knives. Practice proper sanitation. Wash your hands and all equipment thoroughly. Bacteria, especially salmonella, may grow in water or blood from the chicken.

CHAPTER 13

Aquatic Animals

OBJECTIVES

This chapter provides background information on aquaculture. It has the following objectives:

1. Define aquaculture and list requirements of aquacrops.
2. Distinguish between types of water on the basis of salinity and temperature.
3. Describe fish, mollusks, and crustaceans as organisms.
4. Identify fish production systems.
5. Describe management practices in the fish production cycle.
6. List the nutritional and feeding requirements of fish.
7. Explain important health and predator management practices.
8. Describe harvesting procedures with fish.

TERMS

aquacrop
aquaculture
bivalve mollusk
brackish water
broodfish
cage
crustacean
dissolved oxygen
expressing
fingerling
fish
food fish production
freshwater
fry
harvesting
hatchery
mariculture
mollusk
operculum
parts per million
parts per thousand
pen
plankton
pond
predator
production cycle
production intensity
raceway
sac fry
saltwater
smolt
spawning nest
spring
surface runoff
swim bladder
water facility
water quality
well water

13–1. A fish farmer is guiding a basket with 2,000 pounds of fish into a transport truck. (Courtesy, Agricultural Research Service, USDA)

ANIMALS are sometimes grouped as terrestrial or aquatic. Terrestrial animals live on land, with examples being goats, chickens, and horses. Aquatic animals, on the other hand, live in water environments. Examples of aquatic animals include trout, clams, and shrimp.

Fish and other water-based animals are popular for their food products. Consumer demand is too great to be satisfied by the wild stocks in rivers, lakes, and oceans. Annual per person consumption of fish and seafood products is about 15 pounds in the United States. More and more aquatic species are being cultured (farmed) for food. Some are raised in ponds. Others are raised in tanks, cages, and similar facilities.

Farming aquatic animals helps meet consumer demand while, at the same time, protecting species that live wild in streams, lakes, and oceans from over-harvest. Fortunately, research and demonstrations are helping us learn more about how to raise these species. Raising an aquatic animal requires a somewhat different set of skills than those involved raising terrestrial animals.

AQUACULTURE

Aquaculture is the production of aquatic plants, animals, and other species, such as algae. Aquatic plants and animals are species that grow naturally in water. Most aquaculture in North America is the production of aquatic animals, primarily fish, a few shell fish, and crustaceans, such as shrimp and crawfish.

Important skills are used in aquaculture. Each species is different. Species are reproduced and raised to grow-out size. Appropriate water must be provided and managed to keep it in good condition. Feeding fish and controlling disease are important. The crop must be harvested and marketed for a profit.

13–2. Pompano fish are being harvested from grow-out tanks used in nutrition studies. (Courtesy, Agricultural Research Service, USDA)

AQUACROPS

An ***aquacrop*** is a commercially produced aquatic plant, animal, and other species. "Commercial" means that the aquacrop is to be marketed for income to the producer. This is different from a pond used for personal sport fishing where no money is gained.

13–3. Culture pens located in a saltwater ocean. (Courtesy, Vladislav Gajic/ Shutterstock)

Aquacrops are produced for several purposes. The most important purpose is for human food. Other uses of aquacrops include bait for sport fishing; recreation sport fishing; ornamental fish or pets; and feed, which is using a fish, such as goldfish, as feed for other fish crops.

The kinds of aquacrops vary with location. Climate, water, and consumer demand are important factors. A few examples:

13–4. In addition to food, some aquatic species provide other valuable products such as this pearl in an oyster. (Courtesy, Education Images)

- Oysters are grown in protected saltwater areas, such as Wilapa Bay off the coast of Washington.
- Catfish are raised in "pure" freshwater in warmer areas, such as the delta area along the Mississippi River.
- Trout are cultivated in cool freshwater in areas with mountain streams or in the north, such as Pennsylvania and Idaho.
- Tilapia, an imported species, is being raised in warm tank systems.

Important Species

Different aquatic animals are raised, with some more important than others. Species chosen are adapted to the available water and market. Some species have more demand than others. Species with high demand include trout, shrimp, oysters, salmon, and catfish. Examples are in Table 13–1.

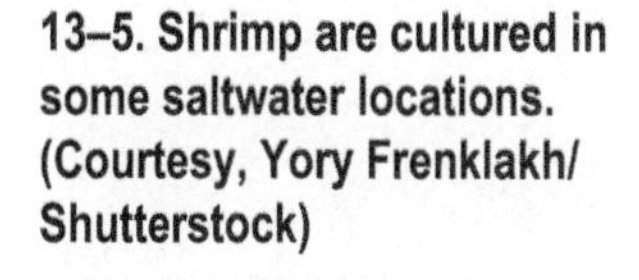

13–5. Shrimp are cultured in some saltwater locations. (Courtesy, Yory Frenklakh/ Shutterstock)

Table 13–1. Important Cultured Aquatic Species in North America

Species (scientific name)	Climate/Culture	Use
abalone (red) *Haliotis rufescens*	cool, saltwater	food, decorative shell
alligator *Alligator mississippiensis*	warm, freshwater	food, leather
bullfrog *Rana catesbiana*	warm, freshwater	food, sport
catfish (channel) *Ictalurus punctatus*	warm, freshwater	food
coho salmon *Oncorhynchus kisutch*	cool or cold, saltwater/freshwater	food
crawfish (red swamp) *Procambarus clarkii*	warm, freshwater	food
prawn (freshwater) *Macrabrachium rosengbergii*	warm, freshwater	food
minnow (fathead) *Pimephales promelas*	cool or mild, freshwater	food
oyster (American) *Crassostrea virginica*	cool to warm, saltwater or brackish	food, pearl
red drum (red fish) *Sciaenops ocellatus*	warm, saltwater	food, sport
salmon (chinook) *Oncorhynchus tshawytscha*	cool or cold, saltwater/freshwater	food
shrimp (western white) *Penaeus vannamei*	mild, saltwater	food
hybrid striped bass *Morone chrysops x M. saxatilis*	mild, freshwater	food, sport
sea scallop *Patinopecten yessoensis*	cool, saltwater	food
tilapia (blue) *Tilapia aurea*	warm, freshwater	food
trout (rainbow) *Oncorhynchus mykiss*	cool or cold, freshwater	food, sport

13–6. Four species of fin fish cultured in North America are (clockwise from top left) channel catfish, hybrid striped bass, coho salmon, rainbow trout. (Top left, top right, and bottom left: Courtesy, Education Images; Bottom right: Courtesy, American Fisheries Society)

General Requirements

Aquacrops differ from terrestrial crops. A major difference is the water environment. Farmers who are very good in producing terrestrial crops will find that water crops require new skills.

Six requirements must be met:

1. A market must be available. (Large crops require mechanical harvesting equipment and automated processing plants. A way must exist to get products to consumers. Profitability and success require a market.)

2. A suitable species must be selected. (The species should be adapted to the water and the available growing conditions.)

3. Water appropriate for the species must be available. (Without the "right" water, production will be difficult or impossible.)

4. The nutritional needs of the aquacrop must be met.

5. Practices must be followed to keep diseases under control.

6. The water must be properly managed for the aquacrop to grow.

13–7. Large aquaculture ponds may require aeration to assure adequate dissolved oxygen in the water. (Courtesy, Mississippi State University)

Production Intensity

A stream, lake, or ocean naturally supports a low density of production. Foods for aquatic species grow in most bodies of water. These food sources include tiny plants and animals, known as **plankton**, and larger plants and animals. Producers want a higher production level than exists naturally. Ponds will naturally support the growth of 100 to 200 pounds of fish a year. This is far below the amount needed in fish farming.

Production intensity is the number or weight of aquatic species in a volume of water. Scientists call intensity biomass. Ponds may be stocked at 6,000 to 8,000 or more fish per acre in intensive production. This level of stocking may produce 8,000 pounds of fish per acre in a year. Tanks may have several fish per gallon, depending on the species and system. Carefully manage the water, feed, and control disease.

WATER ENVIRONMENTS

Aquatic animals vary in the water environment they need. Water is the limiting factor with most aquatic animals. Two features of water are its salt content and temperature.

WATER SALT CONTENT

Aquaculture is practiced in three different water environments based on salt (salinity) content. Species must be carefully selected to match the available water.

- Freshwater—***Freshwater*** is water with little or no salt content. Salt is measured as ***parts per thousand*** (ppt) or the number of parts of salt in a thousand parts of water and salt solution. Freshwater typically has less than 3.0 ppt salt. The water from most wells, streams, and surface runoff is freshwater. Species that are adapted to freshwater will not thrive, and may die, when put into water that is not freshwater. Most aquaculture in North America has been with freshwater species, such as trout, catfish, and hybrid striped bass.

13–8. Preparing to test the suitability of water in a lake for aquaculture. (Courtesy, Education Images)

- Brackish Water—***Brackish water*** is a mixture of freshwater and saltwater. It is found where freshwater streams run into saltwater oceans and seas. The salt content is higher than freshwater, but lower than saltwater. Only a few species are cultured in brackish water. Two examples are mullet and crab.
- Saltwater—Much of Earth's surface is covered with saltwater in oceans, seas, and lakes. Water that has more than 16.5 ppt salt is ***saltwater***. The salt content of oceans and seas is typically 33 to 37 ppt.

  Many different species grow in saltwater. Salmon, oysters, shrimp, and red fish are a few examples. Some of these are produced in large cages that confine them. Producing aquacrops in saltwater is known as ***mariculture***.

WATER TEMPERATURE

Some species will live only in water that has a temperature within a suitable range. Species can be classified as warm water, cool water, and cold water. Most cultured species grow in either warm or cool water.

Water temperature is important in the metabolism of fish. For example, if the temperature is too cool or too warm, the fish will stop eating and fail to grow. The species selected must be appropriate for the water temperature.

- Cool Water—The predominant cool water species cultured in the United States are trout and salmon. These species grow best at a water temperature of 50 to 68°F (10 to 20°C). The rate of growth slows below and above this temperature range. The species

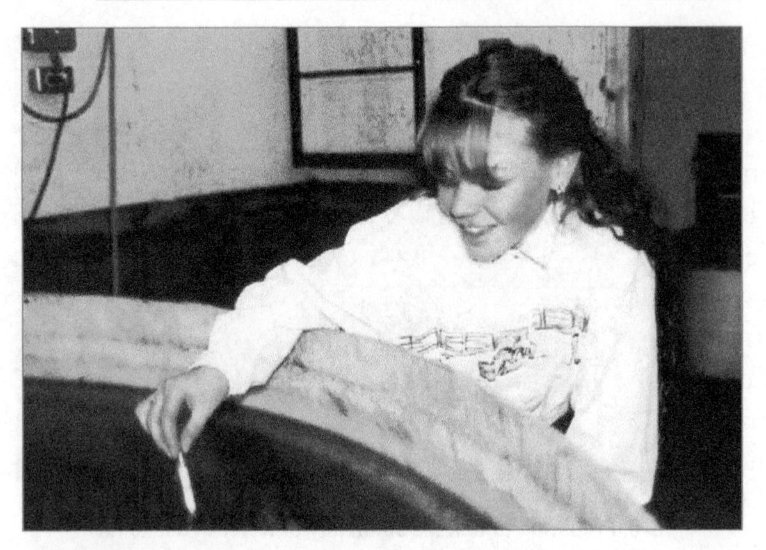

13–9. Water temperature can be measured with a thermometer. (Courtesy, Education Images)

will not survive long in water that is warmer than 80°F (26°C). Some research has been done with arctic char–a species that survives in cooler water.

- Warm Water—Catfish, tilapia, crawfish, and hybrid striped bass are the predominant species of cultured warm water fish in the United States. Catfish grow best in water that is 75 to 85°F (24 to 29°C), but will survive in water below or above the temperature range. Tilapia prefer a similar temperature and will not survive in water below 50°F (10°C). Crawfish prefer water that is 65 to 85°F (18 to 29°C). Hybrid striped bass prefer 77 to 88°F (25° to 31°C), but will survive in water as low as 40°F (4°C).

OTHER WATER FACTORS

Besides salinity and temperature, water must be suitable in other ways. Water quality is affected by pollution, pH, mineral content, gases, and the presence of other living plants and animals.

AQUATIC ORGANISMS

Aquatic organisms carry out the same life processes as other organisms. They do so in water rather than on land.

FISH

Fish are vertebrates, or animals with backbones. Fish are classified in the phylum of Chordata, meaning that they have a spinal cord. As organisms, they typically have bony skeletons, a skull, teeth (though not always well developed), nasal openings, swim bladder, and gills. A major distinction from other animals is that fish do not have lungs. Fish use gills to get oxygen from the water.

External Anatomy

Most fish have a streamlined body shape. The color of the fish matches their natural surroundings. Fins help fish move and keep their balance in the water. Some fins are sharp and can inflict painful wounds in human skin.

The integumentary system consists of skin and scales. Some species have scales, such as trout and tilapia. Other species do not have scales, such as catfish. All species have skin.

13–10. Tilapia are increasing in popularity as a cultured species. (Courtesy, Education Images)

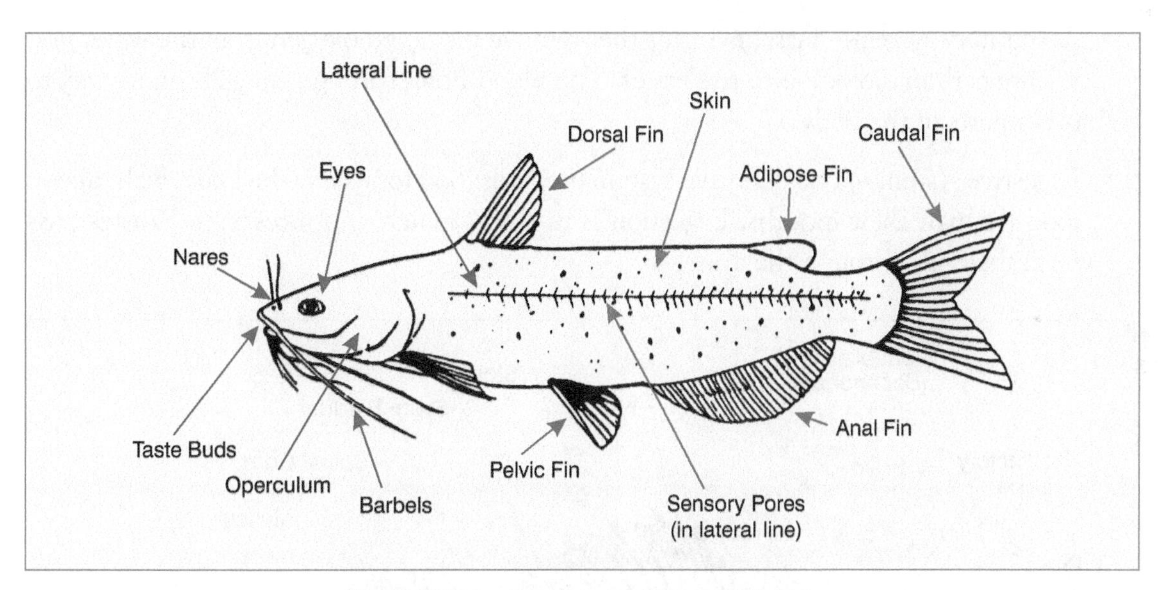

13–11. Major external parts of a typical fin fish.

The body of a fish is divided into three parts: head, trunk, and tail. The head extends to the back of the ***operculum*** (gill cover). The trunk begins at the back of the operculum and extends to the anus. The tail extends from the anus to the tip of the caudal fin.

Internal Anatomy and Systems

Fish have internal anatomy and systems similar to most other vertebrates. The major systems:

- Nervous system—Consisting of the brain and spinal cord and branching spinal nerves, the nervous system sends impulses or commands throughout the body.
- Circulatory system—The circulatory system consists of the heart and blood vessels. Blood contains red and white cells, similar to other animals.
- Sensory system—The sensory system collects information from the fish's environment. Skin and eyes are important to fish. Some fish also have barbels, or feelers, about the mouth to help in sensing danger, food, and movement.
- Skeletal system—The skeletal system consists of bone and cartilage. The skeleton gives the body shape and protects the internal organs.
- Muscular system—The muscular system is the largest system in the body of fish. Muscles make it possible for fish to swim and move in other ways. The large muscles along the trunk and tail are highly desired as food.
- Respiratory system—Fish have gills that remove dissolved oxygen from the water and release carbon dioxide into the water. The blood flows through the gills on its way to other parts of the body.
- Digestive system—The digestive system prepares food for use by the body. Fish take in food through their mouths. Digestion is in the stomach and intestines. Wastes pass from the body through the anus.

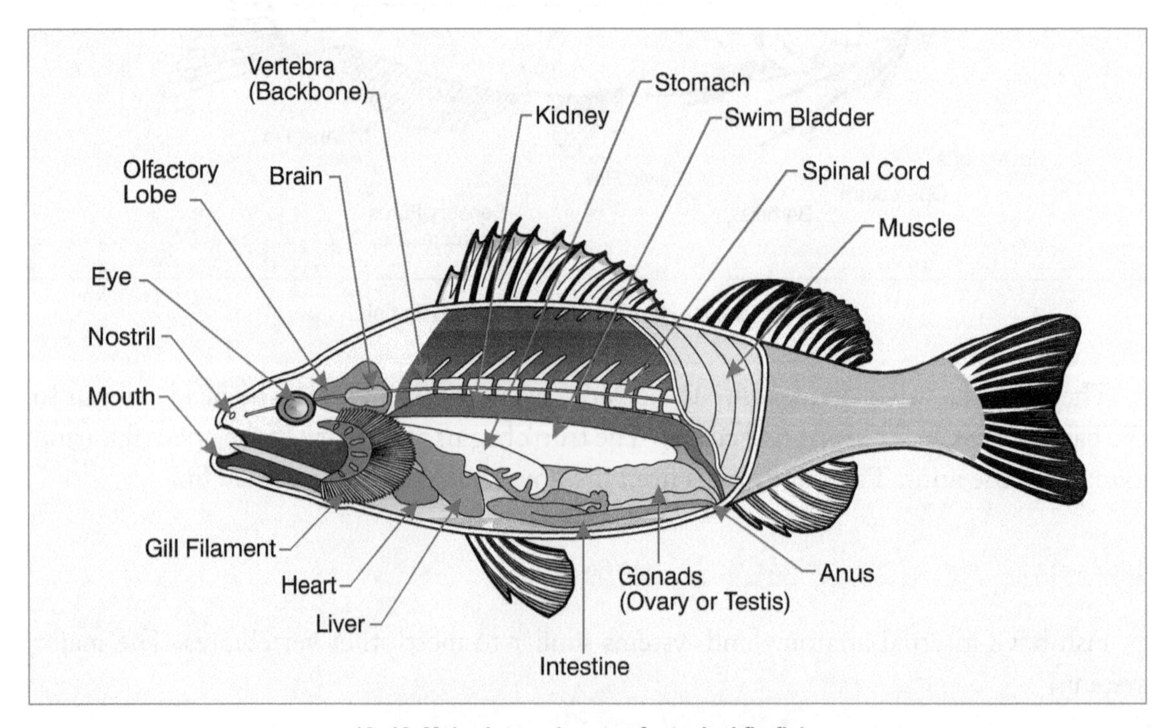

13–12. Major internal parts of a typical fin fish.

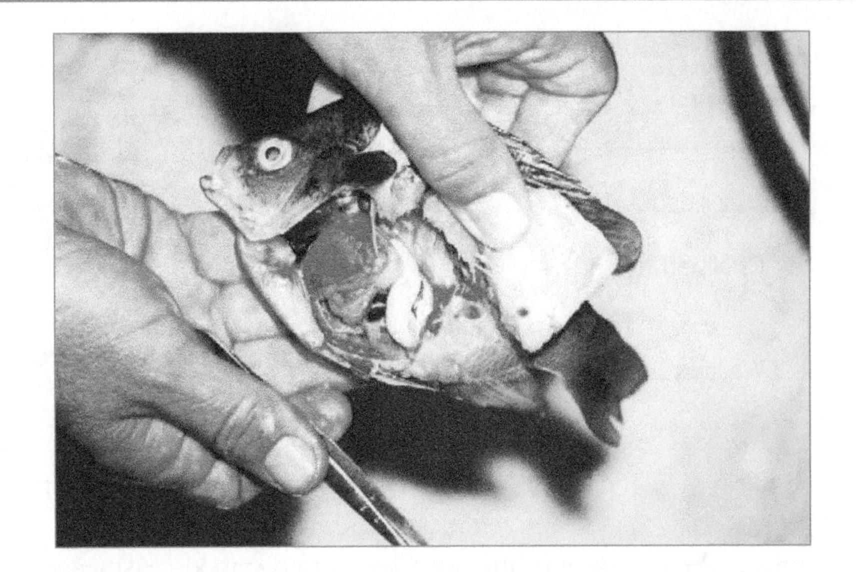

13–13. Dissection is used to study the internal organs of a fish. (Courtesy, Education Images)

- Reproductive system—Most fish spawn. Spawning (fish reproduction) is the process under which new fish are created. The female fish lays eggs that are fertilized by sperm from the male. After a few days of incubation, the eggs hatch as tiny young fry. A few species give birth to live young.
- Swim bladder—The ***swim bladder*** is a small balloon-like structure that helps a fish with buoyancy. Buoyancy allows a fish to stay at a particular depth in the water. The deeper a fish swims, the smaller its air bladder becomes.

CRUSTACEANS

A ***crustacean*** is any species in the Arthropoda phylum. Arthropods have exoskeletons, segmented bodies, and jointed legs. The species used for human food include shrimp, prawns, crawfish, crabs, and lobsters. Crawfish may also be known as crayfish or crawdads. Shrimp and crawfish are more common as aquacrops. New methods have been developed for freshwater prawn culture. Lobsters are primarily harvested from the wild though some young are cultured in Maine and released into the ocean to

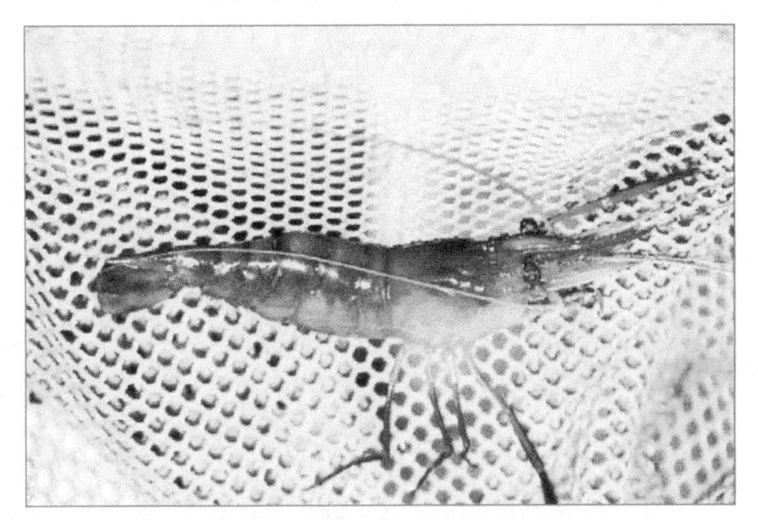

13–14. A freshwater prawn. (Courtesy, Education Images)

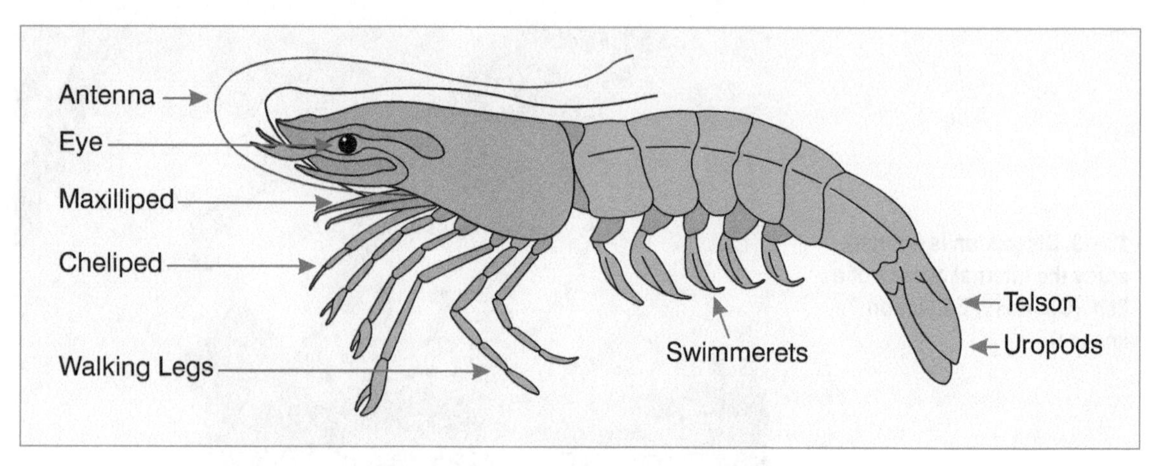

13–15. Major external parts of a shrimp.

grow. The crustaceans included here are decapod crustaceans, or decapods, because they have five pairs of limbs.

Crustaceans have exoskeletons that cover their bodies. The material is much like cartilage or the human fingernail. Crustaceans are in the same scientific order as insects. They have many other body features similar to insects, including three major body parts: head, carapace (similar to the thorax in insects), and abdomen. Decapods have five pairs of limbs, some of which are used for walking. A few decapods form one pair of strong pincers, such as the prawn and crawfish. Crustaceans have body systems similar to fish, but the systems are not as complex.

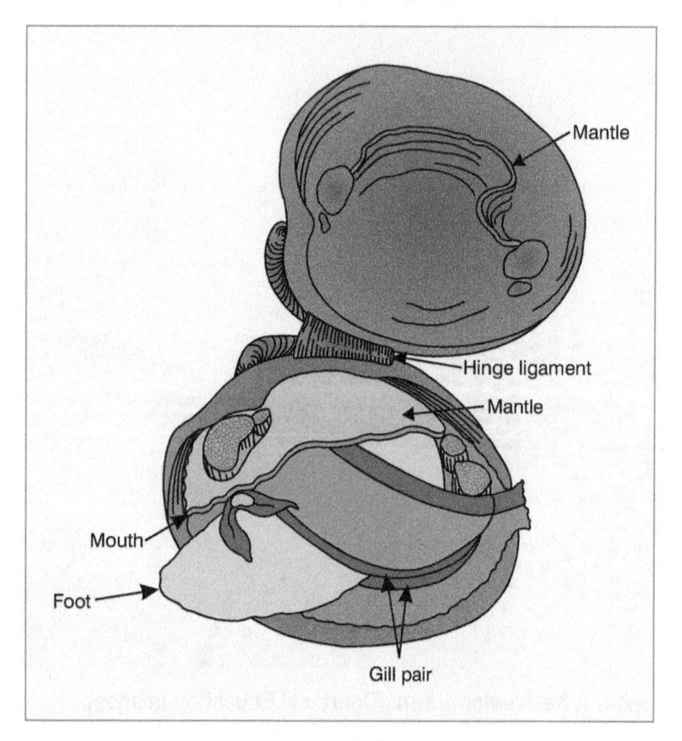

13–16. Anatomy of an oyster.

MOLLUSKS

A ***mollusk*** is any species that has a thick, hard shell. They are in the Mollusca phylum. Some shells are hinged and easily opened for the organism to capture food.

A mollusk with a hinged shell is a ***bivalve mollusk***. Oysters, mussels, and clams are examples. The major muscle in a bivalve is the abductor muscle, which opens and closes the shell. A mollusk with a one-part shell is a gastropod or univalve mollusk. The abalone and snail are examples of gastropods.

Mollusks have simple organs and systems. They have gills for respiration. A heart forces blood over the gills and throughout the body. Most mollusks are filter feeders, meaning that they remove food particles from the water and use them for food. Quality water is essential; mollusks will eat pollution if it is in the water!

Pearls are a valued crop from mollusks. Formed naturally by oysters, pearls are cultured by implanting a tiny grain of sand or other irritant in the oyster.

13–17. The oyster is a bivalve mollusk that is being cultured in brackish water (combination of freshwater and saltwater). (Courtesy, Agricultural Research Service, USDA)

PRODUCTION SYSTEMS

Production systems vary with the water facility. A ***water facility*** is the structure in which aquacrops are grown. Several kinds of water facilities are used; they include ponds, raceways, tanks, vats, aquaria, and pens and cages. Pens and cages are primarily used in natural bodies of water.

PONDS

A ***pond*** is a water impoundment made by building earthen dams or levees. Ponds may range from less than

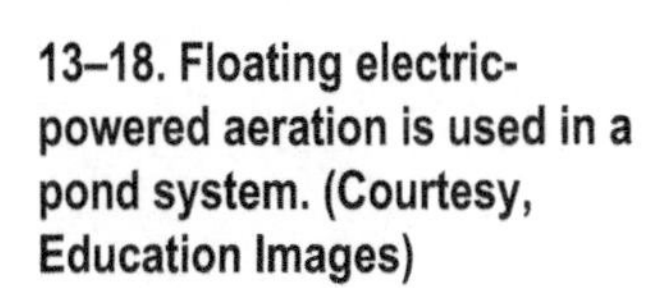

13–18. Floating electric-powered aeration is used in a pond system. (Courtesy, Education Images)

an acre to more than 50 acres in size. Many producers prefer ponds that are about 20 acres in size. The water in ponds does not flow and must be managed to keep fish growing.

Ponds should be designed for easy management and harvesting. Trees, stumps, and other obstacles are removed from ponds. The bottoms of ponds must be smooth, which makes harvesting easy. Levees should be wide enough for vehicles to travel on them and free of trees and utility poles. Ponds are typically rectangular and have water that is 3 to 5 feet (1 to 1.7 m) deep.

Careful management is needed to assure that a pond has a good environment for the fish.

RACEWAYS

A ***raceway*** is a water structure that uses flowing water. Raceways are typically long and narrow. The water may be 3 to 4 feet (1 meter) deep. More fish can be stocked in flowing water than in still water. Flowing water has more oxygen and carries waste away. Raceways vary considerably, with some built of earth, and others, of metal or concrete. Using pumps to keep the water flowing is expensive because of the cost of electricity to run the pump motors. Raceways also require more water and a method for disposing of used water.

13–19. A raceway used for salmon rearing. (Courtesy, Education Images)

TANKS, VATS, AND AQUARIA

Tanks are made of many different materials, such as fiberglass, concrete, and metal. Large round tanks are often used for intensive fish culture. Vats, often made of concrete, are similar to tanks and are often known as tanks. Aquaria are smaller tanks made primarily of glass.

The water in tanks may flow, much like a raceway, or it may be managed, as a pond. Intense production requires flowing water that has plenty of oxygen.

13–20. A tank system for tilapia. (Courtesy, Pan Xunbin/ Shutterstock)

PENS AND CAGES

Pens and cages confine aquacrops in small areas in large bodies of water. A ***pen*** is a confinement structure made so it is joined with the earth in the pond, lake, or stream. Posts that support the pen may be imbedded in the earth. Pens are often used with broodfish where the nesting container rests on the mud at the bottom of the water. Water where pens are located is often only 2 to 3 feet (0.6 to 0.9 m) deep.

A ***cage*** is a confinement structure that floats on the surface of the water. Cages do not touch the earth at the bottom. They have frames covered with net-type material, such as wire or plastic. Cages can be in water that is deep and open, such as a large lake. Many other species may be growing in the water. Cages confine the aquacrop so it can be managed and harvested.

13–21. A round net pen for raising salmon. (Courtesy, Education Images)

13–22. A square cage for raising salmon. (Courtesy, Education Images)

MANAGING FISH PRODUCTION

The water environment of fish must be carefully managed. This is more challenging than first believed. Water environments can be complex.

PRODUCTION CYCLE

A ***production cycle*** is the complete production of a crop. It begins with seedstock and ends with the crop reaching market size. In nature, these cycles often follow the seasons of the year. Reproduction may be in the spring, with the young growing all summer. A couple of years, or more, may be needed for the young to reach adequate size.

The production cycle described here is primarily for freshwater fish. It has four phases: broodfish and spawning, hatching and raising fry, producing fingerlings, and growing fish to food size.

Broodfish and Spawning

Young fish are needed to stock food fish growing facilities. ***Broodfish*** are the sexually mature fish kept for reproduction purposes. Female fish produce eggs and the males produce sperm that fertilize the eggs.

Broodfish grown in warm water, such as catfish, are often kept in separate ponds from young, growing fish. Sometimes, broodfish will eat young fish if they are together. Broodfish are usually larger than food fish and three or more years old. At spawning time, pairs of broodfish may be isolated in pens or left in small ponds that contain spawning nests. A ***spawning nest*** is a container large enough for the pair of fish and provides some protection for the spawn (egg mass).

13–23. A broodfish-size trout. (Courtesy, Education Images)

The size of eggs varies with the species of fish. Some eggs may be the size of a mustard seed or small pea. The red drum eggs are about the size of the head of a straight pen. Other species, particularly the smaller bait fish, may have tiny eggs.

Female fish may produce thousands of eggs at one spawning. For example, a female channel catfish may produce 2,000 or more eggs per pound (0.45 kg) of body weight. This means that a fish weighing 5 pounds (2.27 kg) might produce 10,000 or more eggs at one spawning.

Spawning varies among the species. Catfish spawn when the weather begins to warm in the spring. Trout spawn a little later in the spring or in the early summer. Some species spawn only once and die, such as the coho salmon, while others can be used in repeated years, such as the tilapia.

CONNECTION

LOBSTER BABIES

Lobsters have an under-shell covering that provides a good place for eggs to incubate. One female produces many thousands of eggs. The fertile eggs are attached under the curve of her tail for nearly a year before they hatch.

A baby lobster appears little more than a piece of trash in the water. They are about a-inch long. Other aquatic animals prey on the babies. For the first four weeks, the babies drift and swim about. Afterward, they sink to the bottom of the ocean where they live about 15 years unless captured or they die from disease.

Artificially hatching lobsters is limited. The Oceanarium at Bar Harbor, Maine, has had success with its lobster hatchery. This shows a female in the hatchery with many thousands of attached incubating eggs. (Courtesy, Education Images)

How fish respond to spawning also varies. In natural spawning, some females leave the eggs in a nest where the male tends them for 5 to 7 days. Other fish, such as some tilapia, incubate the eggs in their mouths. This is why they are called mouth brooders. Producers of mouth-brooding fish check the mouths of females for eggs and remove them for artificial hatching.

Spawning can be encouraged by regulating the environment of fish, such as water temperature, and by injecting the fish with a hormone. Of course, these are only practical in laboratories.

Eggs may be artificially removed from some female fish. With fish that die after spawning, the abdomen may be cut open and the eggs removed. With some species, particularly trout, the abdomen of the female is carefully squeezed to force the eggs out into a container. This procedure is known as **expressing**. Females that have eggs forced from them by gentle pressure on their abdomens have been expressed. Sperm, also known as milt, may be expressed similarly from the male. The milt is placed on and mixed with the eggs in a container so fertilization can occur.

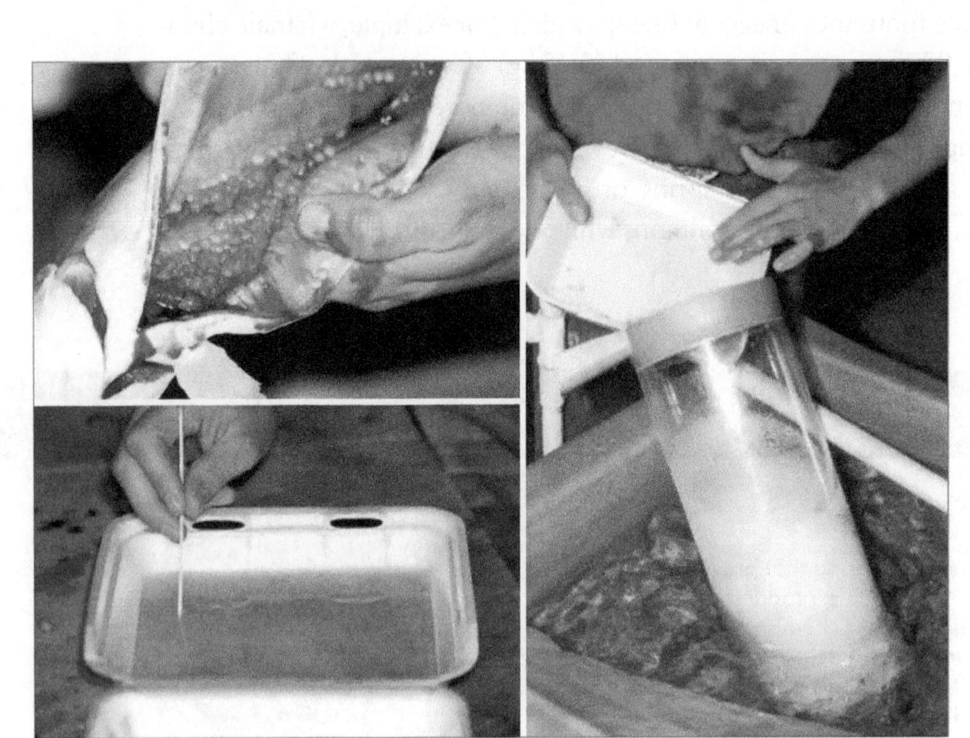

13–24. Eggs are being removed from a sacrificed female fish (top left), mixed with sperm expressed from a male (bottom left), and placed in a hatching jar for artificial hatching (right). (Courtesy, Education Images)

Hatcheries and Fry Rearing

A ***hatchery*** is a place where eggs are artificially incubated and hatched. Incubation is the time between spawning and when the young fish are hatched from the eggs. A favorable

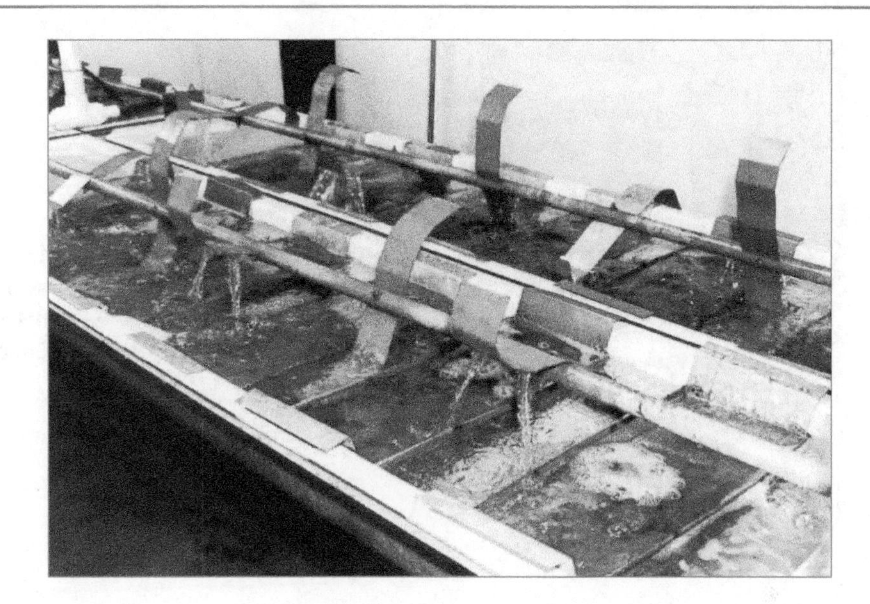

13–25. Paddle agitation is provided egg masses in a catfish hatchery. (Courtesy, Education Images)

environment must be maintained for the eggs during incubation. The water must be within an appropriate temperature range and the water around the eggs must typically be moving.

Hatcheries must have the proper equipment. Eggs are incubated in jars, trays, or troughs, depending on the species. The conditions provided must be ideal for maximum hatch and survival of the young fish. Red drum eggs hatch in about 30 hours at a water temperature of 72°F (22°C). Channel catfish eggs hatch in 5 to 9 days at water temperatures of 70 to 85°F (21 to 29°C).

Young fish are ***fry*** or larvae when they hatch. Fry are very small and delicate. Conditions must be provided so fry survive and grow free of disease. At first, a fry may have the yolk sac from the egg still attached for a few hours or a day. These are ***sac fry***. (Newly hatched salmon are known as alevin.)

13–26. Thousands of tiny catfish fry are being removed from a hatching trough. (Courtesy, Education Images)

13–27. An incubator tray system is used in trout hatcheries. (Courtesy, Education Images)

Feeding fry is essential. Since they are very small, they cannot eat feeds made for larger fish. Some feed manufacturers produce a finely ground meal for fry. Fry may be fed very finely ground organ meat (heart or kidney), yeast, or natural foods. Plankton is sometimes grown for fry. The yolks of chicken eggs may be made into a slurry and used as fry feed with some species. Once the fry have grown sufficiently, they are moved to other growing facilities. With food fish, fry are about 1 inch (2.54 cm) long when moved to a fingerling area.

13–28. A basket of coho salmon eggs in a hatching trough. Incubation is about 60 days. White eggs are infertile (not developing a fish) and should be removed. (Courtesy, Education Images)

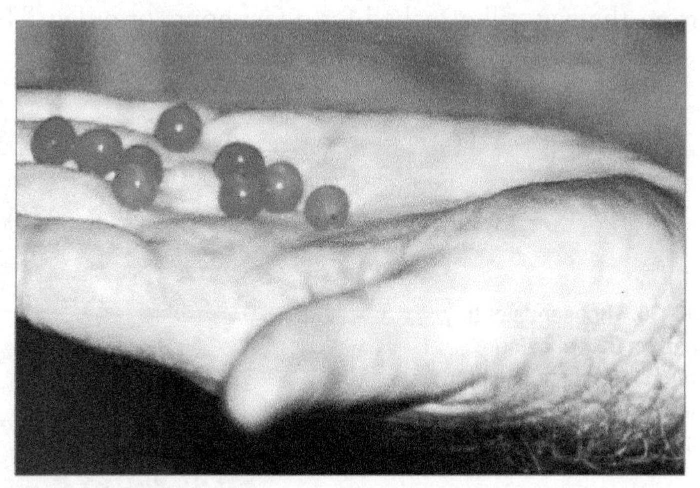

13–29. Eyed salmon eggs that will soon hatch. (Courtesy, Education Images)

Fingerling Production

A ***fingerling*** is an immature fish. They are between the size of a fry and the size of a fish for stocking in a food fish growing facility. A salmon this size may be known as a ***smolt***. A smolt has other characteristics, such as moving from the freshwater where it hatched to the saltwater where it will grow to maturity. A smolt's color changes from dark to light as it grows.

Fingerlings may be grown in tanks, small ponds, or other facilities. The rate of stocking is high because they will not grow to maturity in the facility. Fingerlings are fed a nutritionally complete feed. Since fingerlings are young and rapidly growing, their feed must have a higher percentage of protein (about 36 percent). A specially formulated commercial fingerling feed is convenient. Feed particle size is small so the young fish can eat it. Disease control is needed for fingerlings to survive.

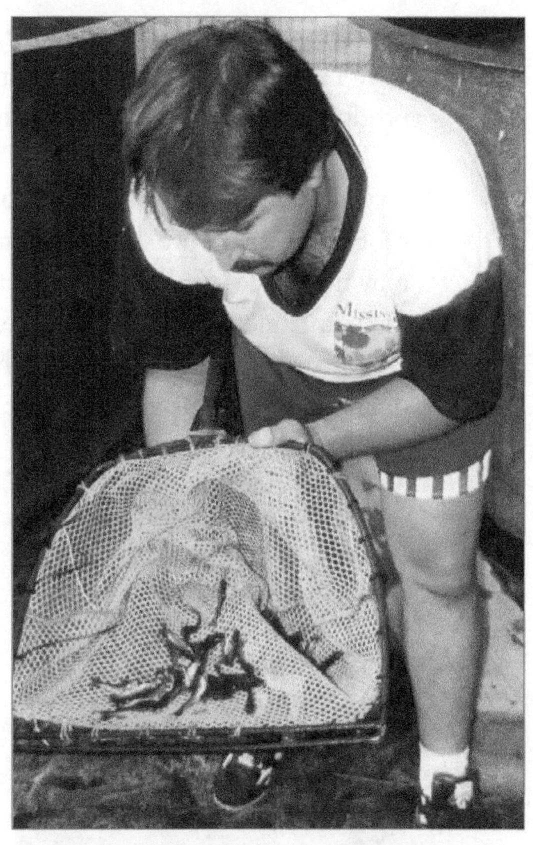

13–30. Channel catfish fingerlings. (Courtesy, Education Images)

13–31. Fish that have been corralled by a seine are being removed from a pond. (Courtesy, U.S. Department of Agriculture)

A channel catfish fingerling is from 1 to 8 or 10 inches (2.54–20 or 25 cm) in length. One growing season is needed for fingerlings to reach adequate size. Fingerlings should be at least 5 inches (12.7 cm) when stocked in growing ponds.

Food Fish Production

Food fish production is raising fingerlings to food fish size. Fingerlings are stocked in growing ponds, raceways, or tanks. Appropriate feed is needed at the rate of about 3 percent of the weight of the fish each day. The feed may have a protein content of 28 to 32 percent. Of course, the intensity of production has a big impact on the rate of feeding and other management details.

Fish may reach harvest size in six months or more depending on the climate and other conditions. Fish are considered food size when they are three-fourths pound or larger. Most processing plants are designed for fish within a certain size range. The fish that are too large may not process well with the equipment in the processing plant. With catfish, the processing plants most efficiently handle fish that are 1 to 2 pounds in size.

All areas that affect the growth of a fish must be considered. This includes feeding, water quality, and disease and predator control.

WATER MANAGEMENT

Water must be well managed for an aquacrop to live and grow. This involves having a good source of the water, maintaining oxygen in the water, and disposing of used water.

13–32. Tank-cultured rainbow trout are being used in developing a genetic map of the species. (Courtesy, Agricultural Research Service, USDA)

Sources of Water

Water from several different sources may be used. This depends on the climate and requirements of the species.

Surface runoff is excess water from precipitation. It may collect behind dams or in streams, lakes, or oceans. Runoff may be okay if it is free of pollution and dependable in supply. Water must be available throughout the growing season for an aquacrop. Runoff tends to be seasonal and available only during certain times of the year. Legal regulations may restrict pumping water from a stream.

Well water is water pumped from aquifers deep in the earth. The water is often of high quality, but is expensive to get. Deep wells, pumps, and electricity to operate the pumps are costly. Occasionally, the water may be too cool for pumping directly into fish facilities. It may need to be pumped and stored in open ponds for a few days so the sun can warm it. Permits are needed to drill water wells for aquaculture.

A ***spring*** is a natural opening in the earth that provides water similar to well water. The water may be of good quality for aquacrops. Some springs dry up at certain times of the year and are not dependable sources of water for aquaculture.

Other sources of water are industrial waste water and municipal systems. Waste water must be free of dangerous substances and have the right temperature. Municipal water systems may treat the water with chemicals that are harmful to fish. This source of water is often too expensive for large-scale aquaculture but can be used for small aquaria.

Water Quality

Water quality is the suitability of water for a particular use. It must have sufficient dissolved oxygen and be free of pollutants. In addition, acidity, alkalinity, hardness, nitrogen compounds, and carbon dioxide are important. The major emphasis in this chapter is on dissolved oxygen.

Dissolved oxygen (DO) is oxygen in the water that is available for living organisms to use. It is in a free, gaseous form and not the oxygen in a molecule of water. DO is measured as ***parts per million*** (ppm) in water. For example, 5.0 ppm DO in water means that there are 5.0 parts DO in 999,995.0 parts water, or a total of 1 million combined parts.

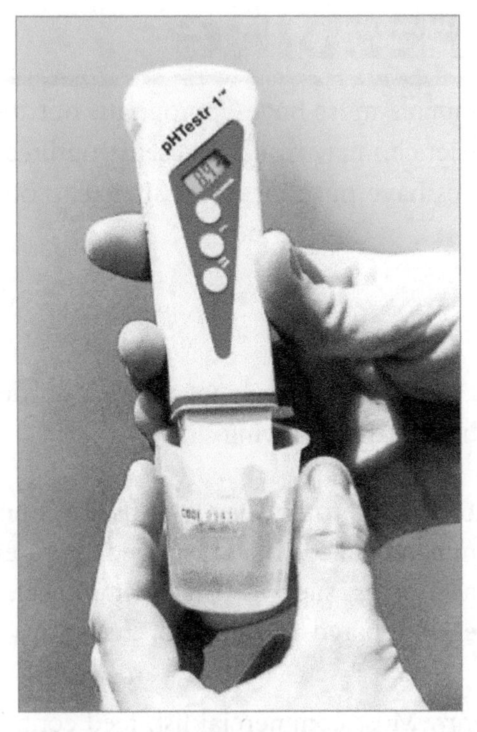

13–33. A pH meter is used to measure water pH. (Courtesy, Education Images)

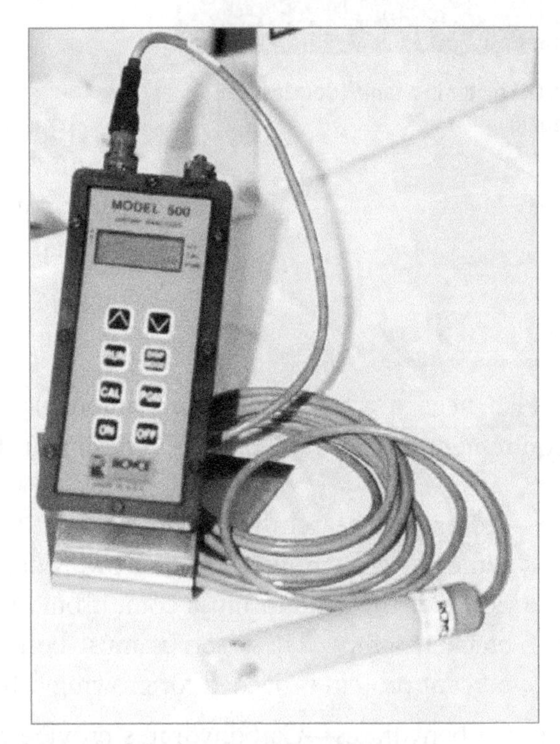

13–34. A DO meter for measuring oxygen levels. (Courtesy, Education Images)

13–35. Using a DO meter in a tank. (Courtesy, Education Images)

Cooler water holds more DO than warmer water. Fish are more active in warm water and produce more wastes. Water must be carefully monitored in warm weather. In ponds, DO is highest about noon on a bright, sunny day and lowest just before sunrise. This is because tiny plants in the water, known as phytoplankton carry on photosynthesis and produce oxygen. At night, when there is no sun, the phytoplankton does not produce much oxygen. Water facilities inside buildings respond differently.

Most aquacrops need 5.0 ppm DO or above to grow, though some will survive below that level. DO is frequently measured with an oxygen meter. If DO is low, the water is oxygenated by splashing it into the air, injecting pure oxygen into it, or in other ways. Fish can die quickly if the DO level drops too low.

NUTRITION AND FEEDING

Aquatic animals must have appropriate nutrition to grow. Deficiencies may result in disease outbreaks. Most aquacrops have nutrient needs similar to other animals.

NUTRITION

Fish need a ration (daily feed amount) that meets nutritional needs. The nutrient requirements of fish vary, but are very similar. Fish need the following:

- Protein—Research has given information about the protein needs of fish. Without sufficient protein, fish do not grow. Most commercial feeds have the needed protein. Some of the protein must come from animal sources, such as fish, shrimp, beef, or chicken scraps. These scraps must be properly prepared in the feed. Feed with 28 percent protein is needed for growing fish.
- Carbohydrates—Carbohydrates provide energy. Most commercial fish feed contains grains, such as corn and soybeans. These are good sources of carbohydrates. The amount of carbohydrate in the diet of a fish is about 10 percent.

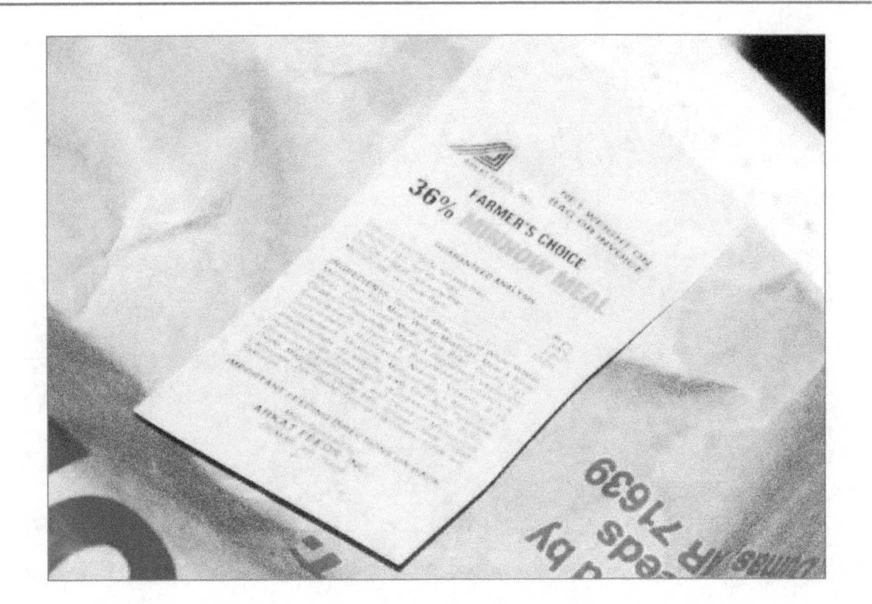

13–36. Bag with a label showing minnow feed analysis. (Courtesy, Education Images)

- Fats—In fish rations, fats usually come from animal wastes, such as by-products from a processing plant. The fat content of feed is 4 to 15 percent.
- Minerals—Fish need about the same minerals as other animals. These include calcium, iron, silicon, manganese, sodium, and seven others.
- Vitamins—Fish need vitamins or they will develop severe deficiency symptoms. Eleven vitamins are essential for fish.

FEEDS AND FEEDING

The intensity of production is a big factor in the amount of feed. If only a few fish are in a pond, adequate natural food may be available. This is not true with aquaculture, however. High stocking rates require appropriate feed.

Natural Foods

Plankton naturally grows in ponds, streams, and other water impoundments. It serves as food for fish and other aquacrops. Depending on the stocking rate, plankton can provide some of the needed food. The use of natural foods also depends on the species of aquacrop.

Crawfish eat vegetation in the water. Most crawfish growers plant rice in the crawfish ponds. The rice (stem, leaves, and grain) serves as food for the crawfish.

Manufactured Feed

Commercial mills produce feed for aquacrops. These products are often specially designed for a species, such as trout or catfish.

13–37. Floating pellets remain on the surface of the water for a few minutes. (Feeding activity of fish helps assess their condition.) (Courtesy, Education Images)

13–38. Floating pellets (left) and meal fish feed (right). (Courtesy, Education Images)

Feeds are manufactured in several forms:

- Meal—A meal is a finely ground feed used with small fish.
- Pellets—A pellet is a feed product made by binding meal into larger particles. Pellet sizes vary for different fish. Pellets are made to float, sink, or remain neutral in the water. Many food-fish farmers prefer floating pellets because the fish have to come to the surface of the water to get food. This helps see fish that otherwise may be unseen.
- Other forms—Crumbles, blocks, and other forms may be used with fish feed. These forms are for specific species of aquacrop.

Frequency and Rate of Feeding

Frequency of feeding ranges from feeding every hour to once or twice daily or once a week. How often fish are fed depends on the size of the fish, climate, and rate of growth.

13–39. A mechanical feeder that blows feed out into a pond. (Courtesy, Education Images)

Fry are fed several times during the day, often each hour. Fingerlings may be fed less often. Food fish are fed once or twice a day. Broodfish may be fed once a day or two or three times a week. In the winter, broodfish are fed less often.

Automatic feeders are sometimes used. Some provide feed on demand, such as when the fish touch a "trigger" that releases feed. Most automated feeders are on a timer and feed is provided at regular intervals.

The rate of feeding depends on fish size and water temperature. Small fish are fed at a higher rate than larger fish. A general rule is to feed fish 3 percent of their body weight each day. This depends on the water temperature and other factors. Most fish eat more in water as it warms. Too much feed is wasteful. The fish do not eat it. Over feeding fouls the water. Uneaten food particles settle to the bottom. Feed only the amount the fish will eat in 15 minutes.

HEALTH AND PREDATOR CONTROL

Diseases and predators can rapidly wipe out a fish population. Good management promotes health and well-being. Steps can be taken to control predators.

HEALTH

Health is the absence of disease. Fish that get adequate feed and have quality water are less likely to have health problems. Prevention of disease is the best control. Fish are difficult to treat once an outbreak of disease occurs.

Kinds of Disease

Diseases are infectious and noninfectious. Infectious diseases are caused by germs or pathogens. Bacteria, fungi, viral, and parasitic diseases may occur. Parasites include worms, flukes, and protozoa. These diseases attack the tissues of aquacrops and cause loss of growth and, sometimes, death.

Noninfectious diseases are from a variety of causes. Some are caused by organ systems that do not work properly. Others are caused by the environment in which an aquacrop lives and the food it gets. Sometimes the organs may not operate properly. This may be related to water quality, such as a sudden change in pH may cause blood problems. Nutritional disease results when the aquacrop has an inadequate diet. Environmental diseases result when the water contains gases, chemical residues, or other materials that are damaging to fish.

Disease Control

The best disease control is to keep fish healthy. Only disease-free fish should be brought on a farm or put in a tank. One fish with a disease can infect an entire population!

A few chemical treatments are available for fish. Some involve putting medications in the feed of the fish. Other treatments involve dipping fish into a chemical solution or bathing the fish in a solution. Valuable broodfish can be injected with medications. The U.S. Food and Drug Administration (FDA) regulates the medications that can be used on fish that will be used for human food in the United States. Always follow the regulations.

13–40. Double-crested cormorants are major pests on some fish farms. (Courtesy, Education Images)

PREDATORS

A ***predator*** is an animal that hunts and kills other animals. Large birds can prey on aquacrops in ponds or other growing facilities. For example, the cormorant (a water bird) can cause large losses in a fish production facility by eating a pound or so of fish a day. Other predators include alligators, turtles, frogs, other fish, snakes, and insects.

With ponds, keeping the grass mowed around the edge of the water will discourage some predators. Air cannons are sometimes used to frighten law-protected birds away. Small ponds and tanks can be covered with netting to keep predators away. Constructing ponds with a minimum area of shallow water where birds can stand will help keep some of them away.

HARVESTING

Harvesting is capturing the aquacrop so it can be hauled and sold to a processing plant or other facility. The methods used must result in a product free of injury.

How an aquacrop is harvested depends on the crop and the water facility. Large seines may be put around ponds and the fish confined to a small area in the pond. This is accomplished by drawing the seine through the water. After seining, the fish are dipped and loaded

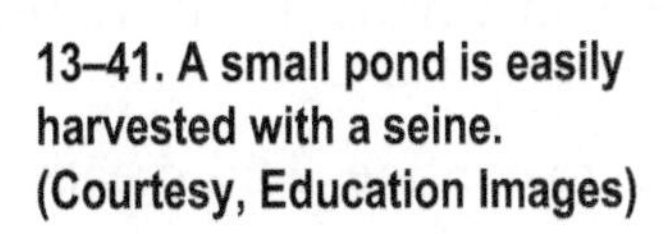

13–41. A small pond is easily harvested with a seine. (Courtesy, Education Images)

13–42. Powered seines are used in harvesting large ponds. (Courtesy, Education Images)

13–43. Modern and clean processing plants prepare fish. (Courtesy, U.S. Department of Agriculture)

into a haul truck. Frequently, the basket used to dip the fish has a scale for weighing the fish as they are being loaded.

Fish in tanks or raceways may be harvested similarly to those in ponds. A net or grading device may be put across the facility and moved toward one end where the fish are confined. The confined fish are dipped from the water.

Sometimes, the water level is lowered so the aquacrop is easier to confine. This creates the need for additional water to again fill the facility. Water is expensive!

Harvested aquacrops must be kept alive until they reach the processing plant. This requires hauling in a tank of water with oxygenation equipment.

13–44. After harvest, crops are processed. This shows cultured oysters being shucked, which is opening the shell and removing the oyster in one piece. (Courtesy, Education Images)

REVIEWING

MAIN IDEAS

Aquaculture is the production of aquatic organisms. Plants, animals, and other species, such as algae (kelp), are produced. The major emphasis in the United States is on aquatic animal production.

The most widely produced aquacrops are fish, crustaceans (including shrimp, prawns, and crawfish), and mollusks (including clams, oysters, and mussels). Freshwater fish farming has grown most rapidly in recent years. The species grown are catfish, trout, tilapia, salmon, hybrid striped bass, and a few others.

Water is important in aquaculture. Freshwater, saltwater, and brackish water are used, as appropriate to the species. Water is obtained from runoff, wells, and waste from industry. Water quality is the suitability of water for a particular aquacrop. A critical factor in any aquaculture is dissolved oxygen (DO). Most aquacrops prefer a minimum DO level of 5.0 ppm. The common water facilities (impoundments) include ponds, raceways, tanks, vats, and aquaria.

Production cycles of aquacrops include broodfish and spawning, hatcheries and fry rearing, fingerling production, and food fish production. Nutrition and feeding are important at all stages. Controlling diseases and predators is essential. Harvesting is a part of the final production cycle with food fish.

QUESTIONS

Answer the following questions using correct spelling and complete sentences.

1. What species of aquacrops are commonly produced in the United States? List any five and give one item of descriptive information about its requirements.
2. What are the general requirements for aquaculture?
3. What is production intensity? Why is it important in the production of aquacrops?
4. Distinguish between the kinds of water based on salinity and temperature.
5. Draw a fish and label its external anatomy.
6. What kinds of water facilities are used? Describe each.
7. Name and briefly describe the four phases in the production cycle of aquacrops.
8. What are the major sources of water?
9. What is water quality?
10. What is dissolved oxygen? Why is it important?
11. What are the major areas of nutrition and feeding of aquacrops?
12. What are the major areas of health management with fish?
13. How do predators cause losses?
14. What is harvesting? How is it done?

EVALUATING

Match the term with the correct definition. Write the letter by the term in the blank provided.

a. predator
b. aquaculture
c. aquacrop
d. pond
e. production intensity
f. freshwater
g. brackish water
h. saltwater
i. fish
j. crustacean

______1. Arthropoda phylum.
______2. Contains little or no salt.
______3. A mixture of freshwater and saltwater.
______4. Water with more than 16.5 ppt salt.
______5. Production of aquatic plants, animals, and other species.
______6. The number or concentration of aquatic species in a volume of water.
______7. Commercially produced aquatic plants, animals, and other species.
______8. Chordata phylum.
______9. A water facility made with an earthen dam or levee.
_____10. Animals that hunt and kill other animals.

EXPLORING

1. Set up an aquarium or fish tank in the classroom. Select a species that will survive in the environment. Goldfish are very hardy. Develop a plan for maintaining the water facility and caring for the fish. (A good Web site is the Aquaculture Network Information Center – http://aquanic.org.)

2. Tour an aquafarm or hatchery in the local area. Determine the species produced and the management needed for successful production. Also, learn how the aquacrop is marketed.

3. Invite an aquaculturist or fisheries specialist to serve as a resource person in class and discuss the locally adapted species and how to culture them.

UNIT FOUR

(Courtesy, Daniel Rajszczak/Shutterstock)

Pleasure and Draft Animal Technology

CHAPTER 14

Horses

OBJECTIVES

This chapter has basic information on the care and management of light horses. It has the following objectives:

1. Describe horses as organisms.
2. Distinguish between the types of light horses.
3. Explain breeding practices with horses.
4. Discuss the nutritional and feed requirements of horses.
5. Explain important health management practices with horses.
6. Describe facility and equipment requirements of horses.
7. Identify and describe equitation skills.

TERMS

draft horse
driving horse
equitation
farrier
filly
floating
foal
foaling
frog
gait
gallop
gelding
hand
horsemanship
hunting and jumping horse
jog
light horse
lope
mare
paddock
plug
polo mount
pony
racehorse
riding horse
stallion
stock horse
stud horse
walk

14–1. Horses are among the most popular large animals. (Courtesy, Education Images)

EVERYONE admires a sleek horse that has received good care. Some people are horse enthusiasts. They like to ride, care for, and train horses. These people do so for fun. Other people train, care for, and provide additional services as a source of income.

In the 1800s and early 1900s, horses were important sources of power and transportation. They pulled plows and wagons. In 1900, there were 20 million horses in the United States. With the emergence of gasoline engines, the number dropped to fewer than 6 million in the late 1980s. The number had increased to near 9.5 million head in 2012.

Horses are now primarily owned for their recreational and personal value. Riding, showing, racing, and grooming are all fun. As this personal interest in horses began to grow, more horses were produced. Horse owners spend millions of dollars each year on their care.

Horses also have many other uses. Horses are used in law enforcement and to work cattle on ranches. Some are used to produce medicines that are important in promoting human health. And, what about the horses in a parade?

HORSES AS ORGANISMS

14–2. Horses are used for recreation, such as riding on a calm afternoon. (Courtesy, auremar/Shutterstock)

Horses are similar to other animals in many ways, yet they have differences that make them unique. The scientific name of the horse is Equus caballus. Horses are sometimes called equine, which is based on the scientific name. The riding and managing of horses is ***equitation***.

AGE AND SEX CLASSES OF HORSES

Several terms are used to describe horses. A mature female horse four years of age or over is a ***mare***, while a mature male horse four years of age or over is a ***stallion***. In Thoroughbreds, both classes begin at five years of age. A mare that has never been bred is a maiden horse. A male parent is a sire. A female parent is a dam.

The gestation period in horses is longer than other livestock at 336 days. The act of giving birth (parturition) is ***foaling***. A young horse of either sex that has not been weaned is a ***foal***. A ***gelding*** is a male horse castrated before reaching sexual maturity. A ***filly*** is a female horse under three years of age (four years of age for Thoroughbreds). A ***stud horse*** is a male kept specifically for breeding purposes. A stag is a male castrated after reaching sexual maturity.

14–3. A Thoroughbred racehorse parades before a race. (Courtesy, U.S. Department of Agriculture)

Horses are known as intelligent animals. They can be taught and directed in many ways. A horse sometimes fails to meet the notion of a beautiful, intelligent animal. A horse that has an ugly head or behaves oddly is known as a jughead. A horse with a large, coarse

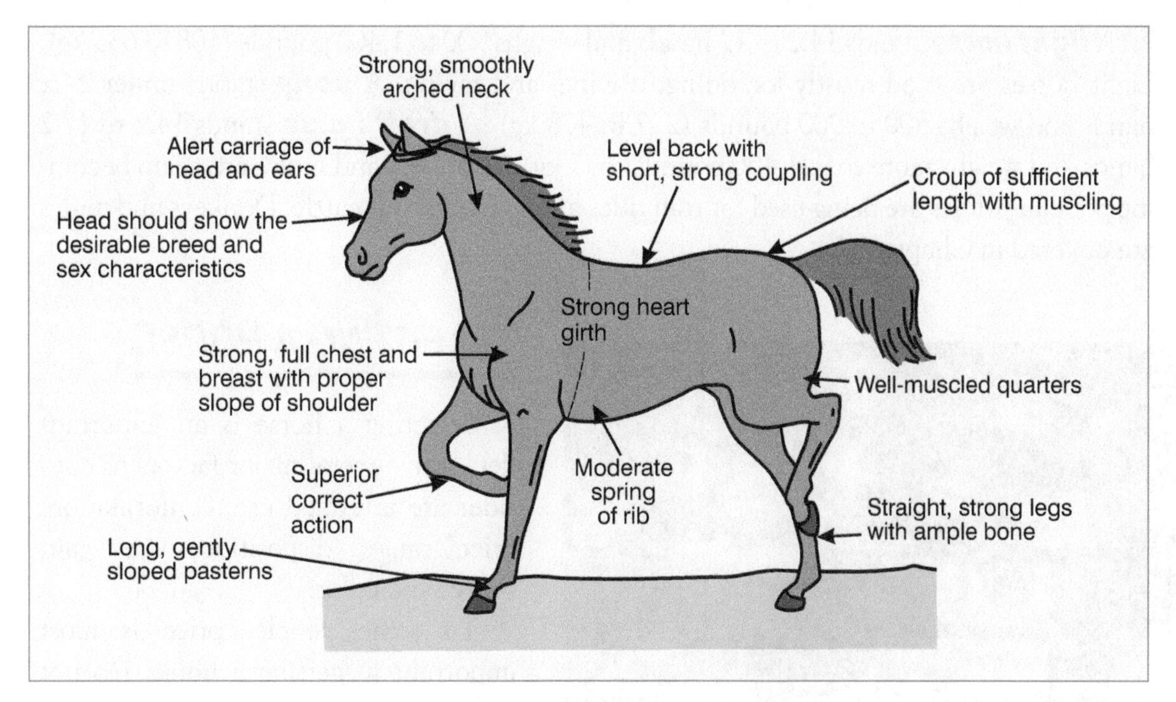

14–4. Desirable conformation of a horse.

head is called a hammerhead. A **plug** is a horse with poor conformation and common breeding. Older horses that show signs of aging may also be known as plugs.

Conformation tells much about the desirability of a horse. Healthy, disease-free horses should be selected. Avoid horses with defects that detract from your intended use. Figure 14–4 shows the desired conformation of a horse.

SIZE CLASSES OF HORSES

Horses are classified by size as light horses, ponies, or draft horses. Draft horses are suited to pulling loads and performing heavy work. This chapter includes light horses. Draft horses are covered in Chapter 15.

Horse classification is based on height. Height is the distance from the highest point of the withers to the ground. Horses are measured in hands. One **hand** is equal to 4 inches (10.2 cm). If a horse is 15.2 hands (meaning 15 hands, 2 inches, not $15^2/_{10}$ hands), then it is 62 inches (157.5 cm) tall.

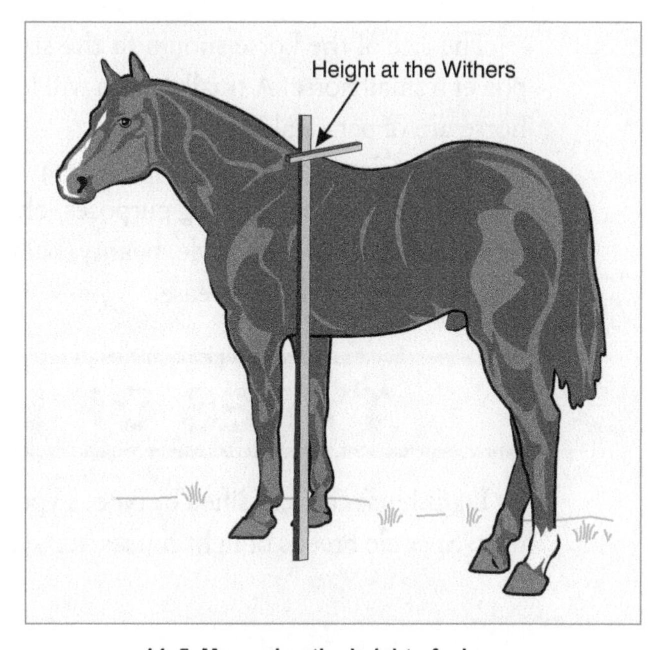

14–5. Measuring the height of a horse.

A ***light horse*** stands 14.2 to 17 hands and weighs 900 to 1,400 pounds (408 to 635 kg). Light horses are used mostly for riding, driving, and racing. A ***pony*** stands under 14.2 hands and weighs 500 to 900 pounds (227 to 408 kg). A ***draft horse*** stands 14.2 to 17.2 hands and weighs more than 1,400 pounds (635 kg). (Donkeys and mules are again becoming popular. Mules are being used for trail rides and working with cattle. Donkeys and mules are covered in Chapter 15.)

SELECTING A HORSE

14–6. Get the assistance of an experienced horseperson in selecting a horse. (Courtesy, Education Images)

Selecting a horse is an important decision. Several major factors to consider are intended use, conformation, price range, disposition, size, gait, breed, and color.

To some people, price is most important in getting a horse. Do not spend more than you can afford. Money is also needed for feed, housing, and other care. An amateur or child needs a quiet, gentle, and well-trained pony or horse. Younger horses tend to be more spirited and less trained than older horses.

The size of the horse should fit the size and weight of the rider. A tall person will overpower a small horse. A small person will look tiny on a tall horse. The breed and color of the horse are of personal preference.

When selecting a horse, bring an experienced horseperson along for his or her opinion. When selecting for breeding purposes, check the horse's progeny and pedigree. Other factors to consider include style, beauty, balance, symmetry, energy level, good wind, suitable age, and freedom from disease.

TYPES OF LIGHT HORSES

Light horses are classified by type. Type refers to the best use of a horse based on its attributes. Specific breeds of light horses are available for each type.

14–7. Four breeds of horses are (clockwise from top left) Quarter Horse, Paint, American Miniature, and Arabian. (Quarter Horse Courtesy, EquiPhotos by Cynthia, Georgia; Paint Courtesy, BarLink Paint Horses, Oregon; American Miniature Courtesy, Education Images)

RIDING HORSES

A **riding horse**, as the name implies, is a horse ridden for pleasure or work. Riding horses have a definite utility value. Types of riding horses include gaited horses, stock horses, polo mounts, hunters and jumpers, and ponies. The most popular riding horses are the Quarter Horse, Arabian, Appaloosa, Morgan, Thoroughbred, American Saddle Horse, and Tennessee Walking Horse.

Gaited Horses

Gait is the way a horse walks or runs. It is associated with the rhythmic movement of the feet and legs. The run or gallop is a fast gait. The two hind feet leave the ground at different times, followed by the two front feet. The walk is a gait in which each foot leaves and touches the ground at different intervals.

Three-gaited horses are known for their gaits of walking, trotting, and cantering. They are used chiefly for pleasure riding, showing, and performing. Gaited horses lack some of the action of stock horses, yet they have their own unique style.

Five-gaited saddle horses must also perform a slow gait called the rack or single-foot. The slow gait is the running-walk, fox trot, or slow pace. These horses were developed on the southern plantations by people who spent long hours on horseback. They chose animals that had an easy, springy step. Gaited horses are ideal for amateur riders because of the animals' easy disposition and gait.

14–8. A stock horse is used to herd cattle. (Courtesy, Stephen Mcsweeny/Shutterstock)

Stock Horses

A ***stock horse*** is a type of horse used in managing cattle. It is usually of mixed breeding and a descendant of the Mustang. The stock horse is the most popular type of horse in the United States. It is short coupled, well muscled, and deep bodied. The stock horse needs to be hardy, agile, sure footed, fast, short coupled, deep, powerfully muscled, and durable, and it must have good feet and legs. Stock horses must also possess "cow sense."

Polo Mounts

The ***polo mount*** is a type of horse used in playing the game of polo. Polo is played by four people on horseback, who try to drive a wooden ball between goal posts. A polo mount is of smaller size than the hunter or Thoroughbred type. It must be quick and clever in turning. It must be able to dodge, swerve, and turn while running. The sport of polo is very expensive because up to six years are required to train a polo mount and a player may use four to six mounts during a single game.

Hunters and Jumpers

A ***hunting and jumping horse*** is a type used in fox hunting. Hunters and jumpers are large, clean-cut horses that perform well in cross-country riding and jumping. Again, Thoroughbred blood is predominantly found in these breeds. Hunters and jumpers must be

14–9. Horses may be trained to jump and perform in competitive events. (Courtesy, U.S. Department of Agriculture)

of good size and height to jump tall fences and wide ditches. They also require stamina and good conformation to keep up with the pack as the hounds track the fox.

Ponies

A pony is a type of horse typically raised for children. Not only are ponies unique in their small size, but they must also possess many other characteristics. They must be gentle, be sound in feet and legs, be symmetrical, have good eyes, and possess endurance, intelligence, patience, faithfulness, and hardiness. The most popular breeds of ponies are the Welsh, the Shetland, and the Pony of the Americas.

14–10. A Shetland Pony is used for a fun activity.

RACEHORSES

A ***racehorse*** is a type of horse used for racing. It may be either a running racehorse or a harness racehorse. Running racehorses are almost exclusively Thoroughbreds or Quarter Horses. They are also used for other purposes, such as for polo mounts, hunters, and calvary horses. From the earliest recorded history, racing has taken place between horses.

Running racehorses must be extremely refined, with oblique shoulders, well-made withers, heavily muscled rear quarters, and straight hind legs. Harness racehorses need to be fast and light. The two main breeds of harness racehorses are the Morgan and the Standardbred.

DRIVING HORSES

A ***driving horse*** is of little utility value but is important in the show ring. Showing is competitive and entertaining to some people. They take great pride in their driving horses.

BREEDING PRACTICES

Conception rates among mares are the lowest of all farm animals. The average conception rate is about 50 percent. Some general recommendations for breeding are discussed here.

NORMAL BREEDING OF MARES

Mares will begin coming into heat at 12 to 15 months of age. The general recommendation is to breed a mare at three years of age so that she will foal at four years of age. This way,

CONNECTION

KENTUCKY DERBY MUSEUM

On the first Saturday in May each year, more than 100,000 people gather at Churchill Downs in Louisville, Kentucky, for a horse race known as the Kentucky Derby. It is the most famous horse race in the United States.

The Kentucky Derby Museum is at Churchill Downs. This museum features a wide range of exhibits about horse racing. Famous horses and their riders are featured in the exhibits. The museum also has educational exhibits about horse care and job opportunities in the field.

Lifelike exhibits have strong appeal. Horse enthusiasts, as well as novices, will find the exhibits interesting. This photo shows one of the exhibits in the museum. (Courtesy, Education Images)

she will not be training while heavy in foal, and she will be more mature, taller, and more fully grown. Mares will continue to foal up to 14 or 15 years of age.

Heat periods are around 21-day intervals and last 4 to 6 days. Signs of estrus include relaxation of the external genitals, increased frequency in urination, teasing of other mares, desire for company, and a slight mucous discharge from the vagina.

The average gestation period of a mare is 336 days, or slightly more than 11 months. Pregnant mares should be kept separate from others because they are usually more sedate. The pregnant mare needs exercise and is best kept in an open pasture.

14–11. A mare seven months into gestation is allowed pasture exercise and grazing. (Courtesy, Education Images)

Parturition

The most obvious sign of nearing parturition is a distended udder. The distension may occur from two to six weeks before foaling. Foaling is the act of a horse giving birth. About 7 to 10 days beforehand, there will be a falling away of the muscular parts of the top of the buttocks, near the tailhead, and in the abdominal area. The vulva will become loose and full. As foaling comes even closer, milk will fill the ends of the teats, and the mare will become very restless, often breaking into a sweat and urinating frequently. These signs are not 100 percent foolproof. Always watch the mare, and note changes in her behavior.

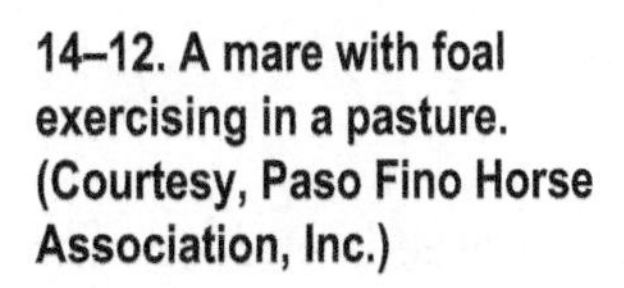

14–12. A mare with foal exercising in a pasture. (Courtesy, Paso Fino Horse Association, Inc.)

When the weather is warm, foaling may take place in a clean, open pasture away from other animals. There is less chance of injury and infection. During poor weather, the mare should be kept in a roomy, well-lighted, ventilated, quiet stall.

Just before parturition, the mare will exhibit extreme nervousness and uneasiness, biting of her sides and flanks, switching of the tail, sweating, and frequent urination. When the outer fetal membrane breaks, a large amount of fluid will flow out. (The water bag has broken.) Foaling should take no more than 15 to 30 minutes.

A foal should come front feet first, with heels pointing down. The nose should follow as it rests on the forelegs. Then the shoulders and the rest of the body should come. If the foal is presented in any other manner, call a veterinarian.

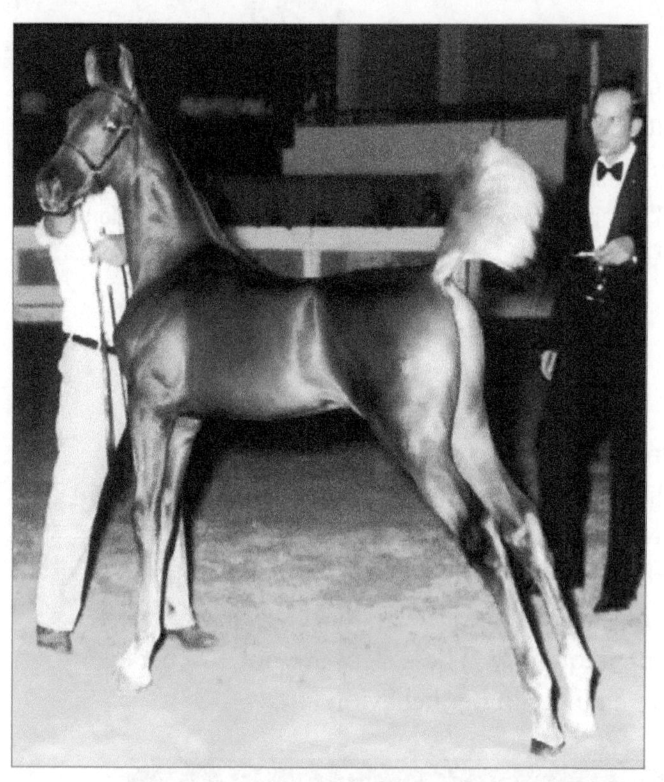

14–13. An American Saddlebred weanling with the traits of becoming a champion adult. (Courtesy, American Saddlebred Horse Association, Inc.)

Care After Parturition

Once the foal is out, check to see that it is breathing and that no membranes are covering the mouth or nostrils. Dip the navel cord in iodine to prevent infection. Feed the mother lightly, and give small amounts of lukewarm water at frequent intervals. Clean the stall often, and observe both mother and offspring for signs of illness.

Feeding colostrum to the foal immediately after birth is very important. Colostrum contains antibodies that protect the foal against certain infections. The healthy foal should be ready to nurse within 30 minutes to 2 hours but may need assistance and coaxing.

CARE OF THE STALLION

Any stallion bought or used for breeding purposes must be a guaranteed breeder. The number of healthy foals a stallion has sired is more important than the number of times the stallion has serviced mares. Some farms periodically check the semen of stallions to measure the fertility of the animals. Semen must be of good volume and have a high sperm count. The sperm must move about, and they must be morphologically correct.

14–14. A stallion in a teasing unit where semen is collected for artificial insemination. (Courtesy, Education Images)

Because stallions are more nervous, they need to be kept in separate quarters. The best arrangement for a stallion is a roomy box stall and a two- to three-acre pasture. The stallion must receive daily exercise to maintain thriftiness.

BREEDING METHODS

The mare should be in good condition at breeding time. She should not be too thin or too fat. She can be bred by hand breeding, corral breeding, or pasture breeding.

The most accepted breeding method is by hand. This will decrease the chance of injury to the mare or the stallion. Artificial insemination may also be used.

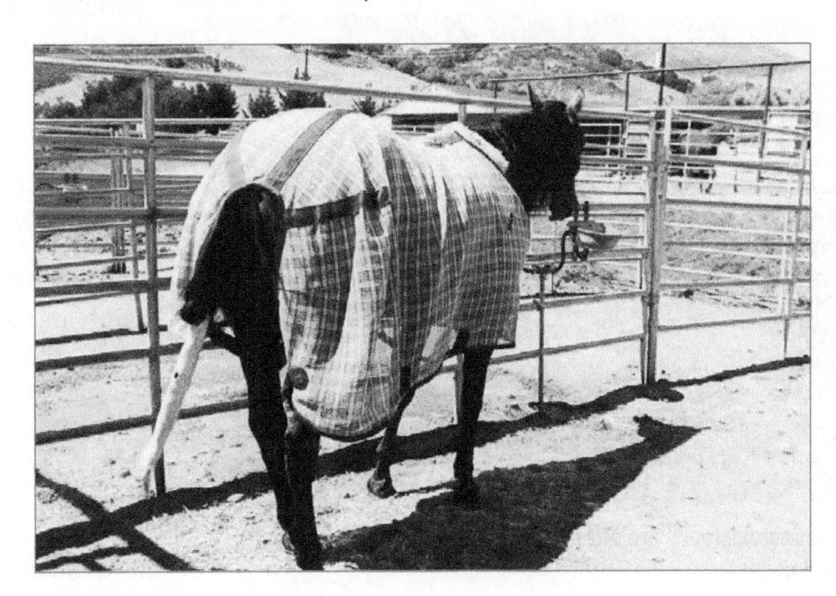

14–15. Tail wrap is used on this ready-to-breed mare to keep hair out of the reproductive tract during mating. (Courtesy, Education Images)

14–16. A mare in good condition. (Courtesy, Paso Fino Horse Association, Inc.)

With corral breeding, the mare and the stallion are turned loose in a corral. It is best to stand in a spot where you can see but not be seen. Do not stand in the corral. Once the mare is bred, she and the stallion are returned to their separate quarters.

Pasture breeding is hard to control and monitor. Valuable animals should not be pasture bred, as there is an increased incidence of injury. With pasture breeding, the stallion is turned out with a band of mares. As the breeding season progresses, the stallion may become sterile.

After breeding, both the stallion and the mare should be checked for injury. The mare should be pregnancy checked shortly thereafter.

FEEDING

Horses require nutrients for maintenance, growth, reproduction, lactation, and work. Like other animals, they require all six of the basic nutrients: carbohydrates, fats, water, protein, vitamins, and minerals. They require energy, or carbohydrates and fats, in the largest amounts. Horses receive these nutrients from concentrates, forages, free-choice vitamins and minerals, and water.

14–17. Commercial feeds are often used with horses. (Courtesy, Education Images)

CONCENTRATES

Concentrates consist of various grains and supplements. They mainly supply energy and protein.

Grains

Grains, such as oats, corn, or barley, are concentrates that provide energy.

Grains also provide protein. Protein is necessary for the formation of bones, ligaments, hair, hooves, skin, organs, and muscles. Horses are monogastric, so they cannot significantly synthesize their own protein. A young horse, especially, needs high-quality protein because its cecum is not fully developed. Maintenance requirements are about 1 pound of protein for every 1,000 pounds (453 kg) of weight.

50 Pounds NET
Johnson's Hi-Pro Horse Feed
HEAVY GRAIN
14% Manufactured By JOHNSON MILLING Clinton, Mississippi 14%

Guaranteed Analysis

Crude Protein, not less than 14.00%
Crude Fat, not less than 2.50%
Crude Fiber, not more than 9.50%

INGREDIENTS: Grain products, plant protein products, processed grain by-products, animal protein products, forage products, molasses products, ground limestone, di-calcium phosphate, salt, animal fat, fenugreek seed, anise oil, vitamin A supplement, D-activated animal sterol (source of vitamin D-3), vitamin B-12 supplement, vitamin E supplement, riboflavin, niacin supplement, calcium pantothenate, sodium selenite, iron oxide, manganous oxide, zinc oxide, iron carbonate, iron sulfate, copper oxide, potassium iodide, cobalt carbonate, magnesium oxide, sulfur, potassium chloride, calcium propionate and ethoxyquin (preservatives).

14–18. Label from a commercial horse feed. (Courtesy, Education Images)

Oats are the most commonly used grain for horses. Oats are low in energy but are easier on a horse's sensitive stomach. Dusty oats, however, can cause colic. Colic is the number one cause for loss of horses.

Corn is usually mixed with the oats, especially in winter, because it is higher in energy. Corn should be cracked, rolled, coarsely ground, or shelled. Care should also be taken to avoid dusty or moldy corn. Moldy corn is very dangerous for any animal. If barley is fed, it must be ground or rolled, because it is hard.

CONNECTION

HOUSING FOR HORSES

Horses need protection from the weather. They need a place for resting, eating and drinking, being groomed, and receiving other care. Having the horse barn you want requires planning. Size depends on the space needed for horses, feed storage, and a tack room. Space is also needed for grooming, and that may be with crossties in the aisle.

This illustration shows a barn under construction. You will note that its location is on a level site. The structure has been carefully framed and placed for efficient use. Electricity, water, and other needs are being included. Safety and comfort are being built-in. (Courtesy, Education Images)

14–19. Horses should receive quality hay if not pastured. (Courtesy, Education Images)

Supplements

The most common supplements are soybean meal and linseed meal or oil. Soybean meal is very high in protein content. Linseed meal or oil puts a shine on a horse's coat.

FORAGES

Forages for horses include pasture and hay. Forages should be fed at 1 to 2 pounds (0.91 to 1.82 kg) for each 100 pounds (45.36 kg) of body weight each day. Hay should be of good quality and free of dust or mold.

Two questions to answer in using pastures are: Will the pasture be used primarily for exercise or nutrition? What is the carrying capacity of the pasture? Carrying capacity is the number of animals an acre of pasture can support. It is an important factor in determining the amount of pasture a horse needs. With pastures planted to improved varieties of forage plants and that receive sufficient moisture, 2 to 4 acres of pasture will be adequate. With unimproved pastures and pastures of the dryland western U.S., each horse may need to 10 to 20 acres. Determine local recommendations on forage plants to use in horse pasture, maintenance practices, and carrying capacity.

14–20. Horses need ready access to clean drinking water. (Courtesy, Education Images)

MINERALS AND VITAMINS

Minerals are necessary for the growth and development of bones, teeth, and other tissues. The most essential and most likely to be deficient minerals are calcium and phosphorus. Steamed bone meal, dicalcium phosphate, salt, and trace minerals are the best sources of minerals.

Vitamins are needed for proper growth and development, health, and reproduction. Vitamin deficiencies are not a problem if the horse is fed a well-balanced diet. Supplemental vitamin A should be fed if a horse is not in a pasture to consume green grass. Vitamin D may be needed if the horse is kept inside.

WATER

Water is the most important nutrient that horses and all animals need. An average horse will consume 10 to 12 gallons (37.9 to 45.5 l) of water each day. Horses may require more while working or in the summer. They should not drink excessive amounts of water during the summer because they will founder.

HEALTH MANAGEMENT

Daily health management is imperative. Prevention is the key. Work closely with a veterinarian to devise a vaccination schedule. This will result in fewer health problems, lower veterinarian fees, and decreased animal losses. Cleanliness is necessary to prevent disease and for the well-being of the animal.

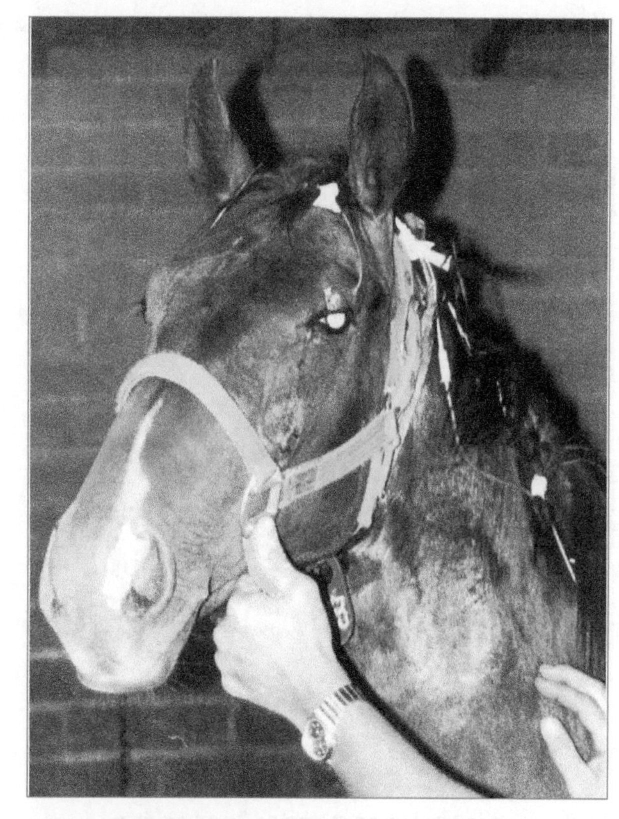

14–21. Diseased eye being treated. (Courtesy, Education Images)

COMMON AILMENTS

Several common ailments that may afflict horses are listed here. Refer to Chapter 4 for more information on animal health.

- Colic—Colic is the leading cause of death in horses. It may be caused by internal parasites, improper feeding, or excessive water intake. Signs of colic include a distended abdomen, kicking or rolling, heavy perspiration, constipation, and refusal to eat or drink. Walk the horse, and call the veterinarian immediately.
- Encephalomyelitis (sleeping sickness)—Encephalomyelitis is caused by a virus carried by mosquitos. Vaccinating and applying insecticides are the best preventives for this

14–22. A face mask is used to protect the eyes and face from flies. (The horse can see through it!) (Courtesy, Education Images)

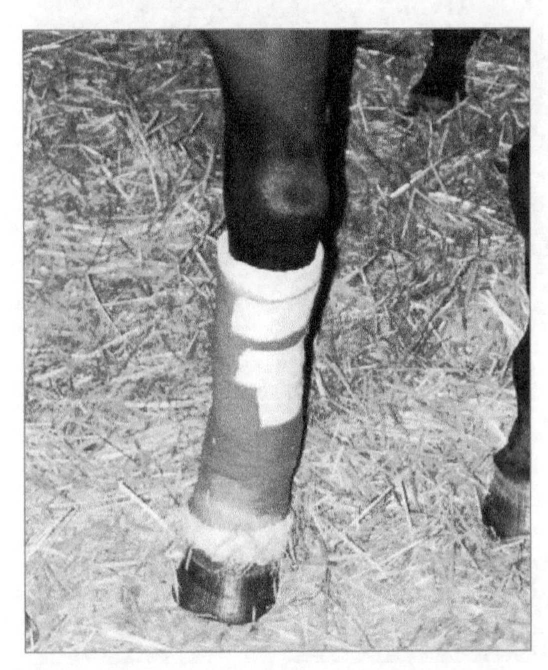

14–23. A leg injury has been treated by an equine veterinarian. (Courtesy, Education Images)

disease. Signs of encephalomyelitis are a sleepy attitude and localized paralysis of the lips and bladder.

- Equine infectious anemia (swamp fever)—Equine infectious anemia (EIA) is another viral disease transmitted by biting insects. It can also be spread by infected needles. There is currently no vaccine for EIA. Signs of EIA include a high, intermittent fever, stiffness, weakness, loss of weight, anemia, and swelling.
- Equine influenza—Equine influenza is highly contagious. Symptoms include fever, loss of appetite, depression, rapid breathing, cough, weakness, and eye and nasal discharge. If influenza is suspected, consult the local veterinarian; there is a vaccine available.
- Founder—Founder affects the tissue connecting the hoof wall to the foot. Severe founder cannot be cured. Founder is caused by the overeating of grain or forage, consumption of too much cold water, overwork, rapid change in diet, or inflammation of the uterus following foaling. Signs of founder include pain in the feet, fever, and a reluctance to move.
- Tetanus (lockjaw)—When an open wound becomes infected with Colostridium tetani, a bacterium, tetanus will result. These bacteria are found in manure, making it necessary to keep stalls and barns clean. Symptoms include stiffness about the head and slow and awkward chewing and swallowing. An annual vaccination is recommended.
- External parasites—External parasites common to horses include flies, mosquitos, mites, lice, ticks, and ringworm. They are not only annoying to the animal and owner, but they also carry diseases. Application of insecticides and regular cleaning of stalls will greatly decrease the external parasite problem.

- Internal parasites—Internal parasites include roundworms, bots (flies), pinworms, and strongyles (bloodsucking worms). To prevent internal parasites, do not allow a horse to eat off the ground or drink from puddles. Rotate pastures frequently. Regular worming may be needed. If internal parasites are suspected, get a veterinarian to examine the feces of the animal. The extent and type of infestation determine the plan of action.

CARE OF THE TEETH

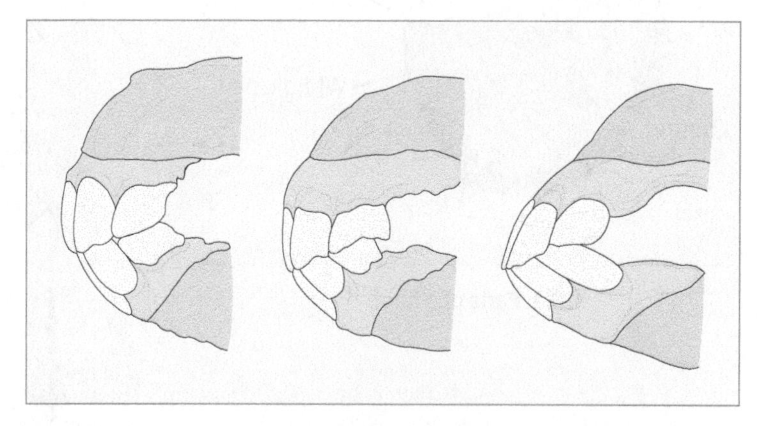

14–24. Teeth slant forward as horses age. This side view shows the slant in teeth of horses that are 5, 7, and 20 years old.

Determining Age by Teeth

Horses live to be 20 to 25 years old. Their best years are from age 3 to age 12. Age can be determined by examination of the teeth. The time of appearance of temporary and permanent teeth, the shape, and the degree of wear are important in aging horses.

Temporary teeth are smaller and whiter than permanent teeth. A mature male has 40 teeth, while a mature female has 36. As a horse ages, the cups of the teeth wear on the inside and center. After 12 years, the teeth change from an oval to a triangular appearance, and they slant forward more.

Floating the Teeth

Because a horse's upper jaw is wider than its lower jaw, teeth will wear unevenly. This causes sharp edges to form. Teeth should be floated each year to reduce the discomfort of the sharp edges. ***Floating*** is using a file to smooth the sharp edges of the teeth. Only a trained person should attempt floating.

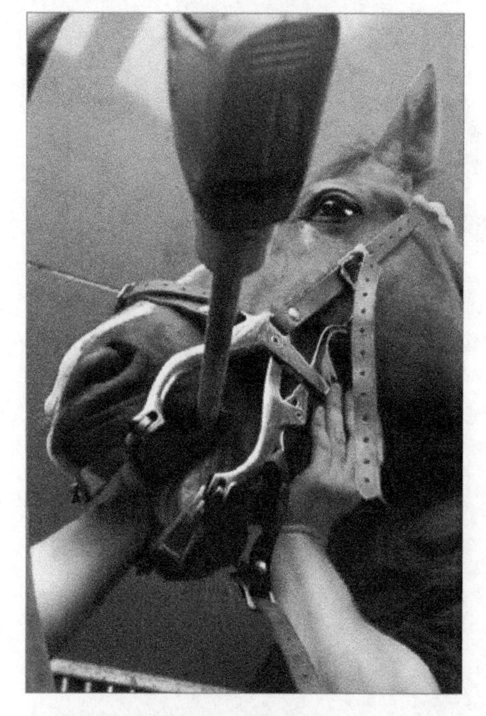

14–25. Floating the teeth of a horse to remove sharp edges and hooks. (Courtesy, kiep/Shutterstock)

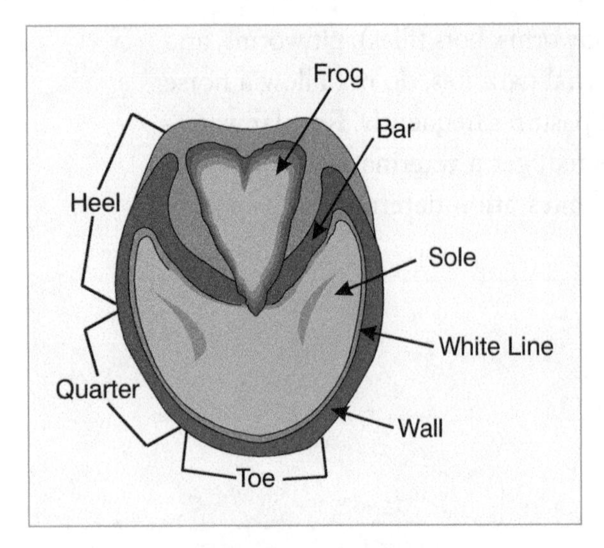

14–26. Parts of a hoof.

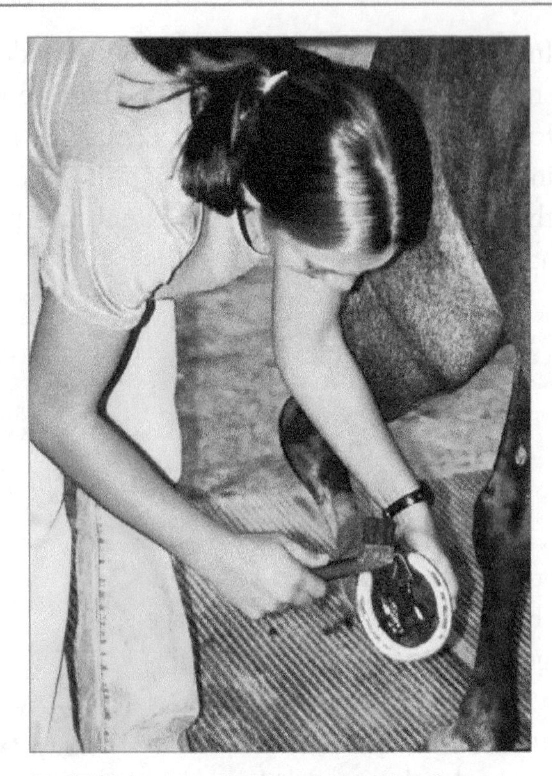

14–27. Using a hoof pick and brush to clean a foot. (Courtesy, Education Images)

14–28. A farrier is dressing a horse's foot. (Courtesy, Paso Fino Horse Association, Inc.)

HOOF CARE

The ability of a horse to move will determine its worth. Proper care and maintenance of the hooves are very important. Through the domestication of the horse and the change in its environment, the care the hooves require has significantly increased. The hooves should be kept clean, prevented from drying out, maintained at the proper length and in the proper shape, and protected from the hard surfaces with which they come in contact.

A hoof pick should be used daily to clean out a hoof. Carefully clean around the frog. The **frog** is the V-shaped elastic-like pad in the middle of the sole of the hoof. Keeping the hooves wet may be done by simply keeping the area around the water tank wet, applying dressing to the hooves, or attaching burlap to the hooves.

A ***farrier*** (a person who puts shoes on horses) should trim the hooves every four to six weeks. The sole of a hoof and jagged edges of the frog should be trimmed with a hoof knife. The sole of the hoof should not be trimmed unless necessary. Protection from hard surfaces is accomplished with shoes. Shoes should be either replaced or reset after the hooves are trimmed. Horses that are seldom ridden or that spend little time on hard surfaces may not need shoes.

FACILITY AND EQUIPMENT REQUIREMENTS

Horse facilities do not need to be expensive or fancy. Again, no more money should be spent on facilities than can be afforded.

BARN

The facilities need to protect the horse from the wind, sun, and poor weather. A pole-type structure over concrete may be used, though some barns have earth or clay floors. Obtain a plan by an architect to assure structural soundness and proper sizing for horses. The barn should have enough space for equipment and feed. Within the barn, the horse is given a stall. Standard recommendations for stall size are at least 10 × 12 feet (3 × 3.8 m), with a height of 8 feet (2.62 m). Tie stalls should be 5 × 12 feet (1.6 × 3.8 m). The bottom 5 feet (1.6 m) should be solid boards that can withstand kicking. Some horse farms use round pens or exercisers to maintain health of horses. A round pen is exactly what the name indicates: round (or close there to). Most are 50 to 60 feet in diameter. They are constructed of pipe, wood, or other materials that are strong and offer safety for the horse and trainer.

14–29. A horse barn at a boarding and training facility. (Courtesy, Education Images).

Keep the stalls clean and dry. If clay is used as flooring, remove and replace the top foot yearly. This decreases the incidence of diseases and parasites.

14–30. A cross-tie should be included in a barn to help restrain a horse during clipping, examination of the teeth, or other work. (Courtesy, Education Images)

14–31. An assortment of tools and equipment may be needed to clean and maintain a horse facility. (Courtesy, Education Images)

14–32. A round pen may be used to exercise and train horses. (Courtesy, Education Images)

FENCES

Safely confining and protecting a horse is essential. Providing an area for exercise and, possibly, grazing is also needed. Many horse keepers have paddocks for this purpose. A ***paddock*** is a small enclosure or area of pasture or dry lot where a horse can wonder, get exercise, and, possibly, graze. Most paddocks are located near the barn or stable.

14–33. Leading a horse into a paddock should be simple and safe. (Courtesy, Education Images)

Fencing material should be durable, strong, and safe. Wood, metal, and vinyl materials are used. These may include electric applications. Vinyl may be flexible or as rails. In very cold areas, rails tend to snap when a horse pressures against them. Barbed wire should not be used with horses. Some manufacturers have horse fencing systems that include gates and other features.

Gates should be strong and properly installed. Closures should be secure and have locking capability.

4-34. Flexible fencing materials and electric wire ave been used together in this fence. (Courtesy, ducation Images)

14-35. Exercising and training a horse in a paddock with a rigid vinyl fence. (Courtesy, Education Images)

FEEDING EQUIPMENT

Hay or other forage should be fed in a hayrack or manger positioned above the ground. Grain should be fed in a box, pail, or tub. Watering is best with buckets or automatic waterers. Do not allow horses to consume large amounts of water at one time.

14–36. Good facilities are needed with horses. These photographs show the exterior and interior of a horse facility with an alley between rows of stalls. (Courtesy, Education Images)

GROOMING

Horses should be groomed regularly. The grooming equipment needed includes a hoof pick, a body brush, and a mane and tail comb. A curry comb and a sweat scraper work well when bathing horses.

Each horse should be provided with its own halter. Make sure that the halter fits the animal and that the throatlatch is not too tight. When a horse is let out to pasture, remove the halter. Otherwise, the horse may get caught and strangle itself.

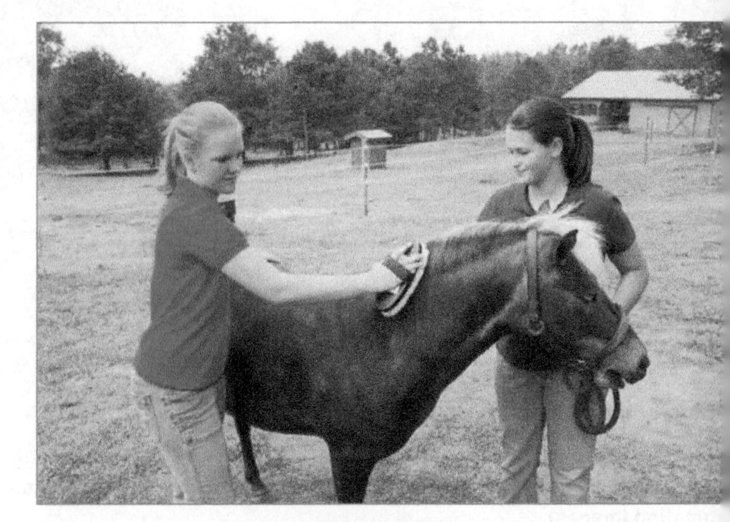

14-37. Brushing a horse removes dirt and gives the coat a healthy appearance. (Courtesy, Education Images)

SADDLE AND BRIDLE

There are two types of saddles: Western and English. The type chosen depends on the type of riding that will be done. The saddle should be comfort-

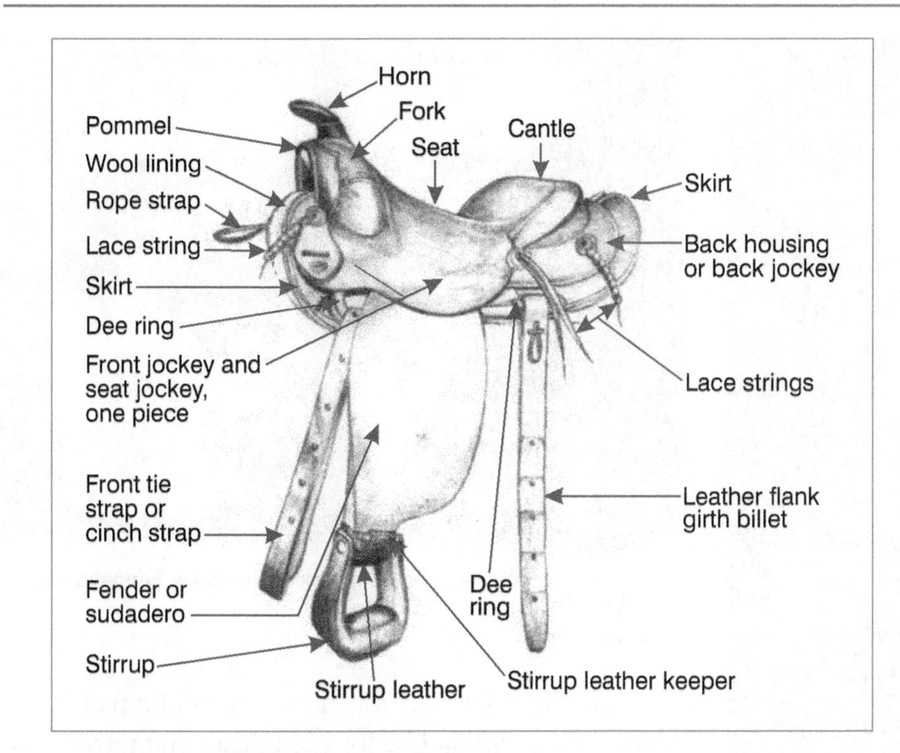

14–38. Parts of a Western saddle.

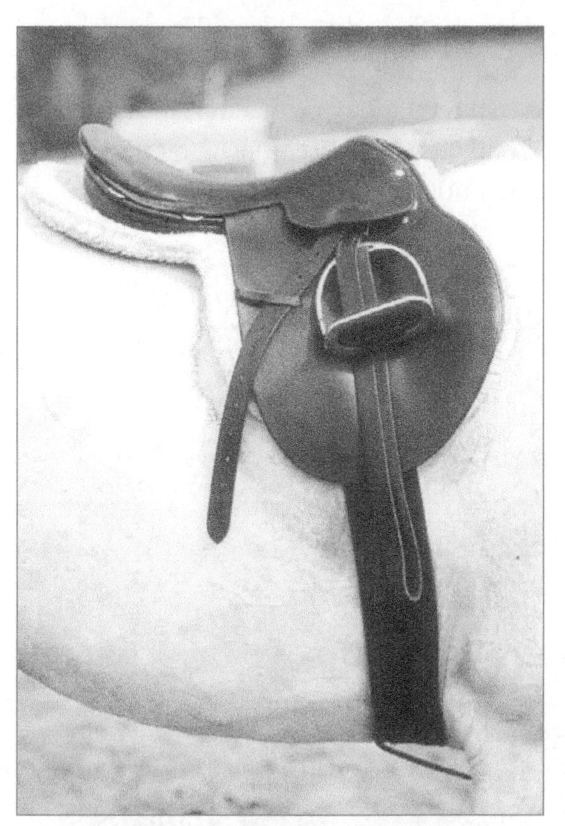

14–39. An English saddle. (Courtesy, Education Images)

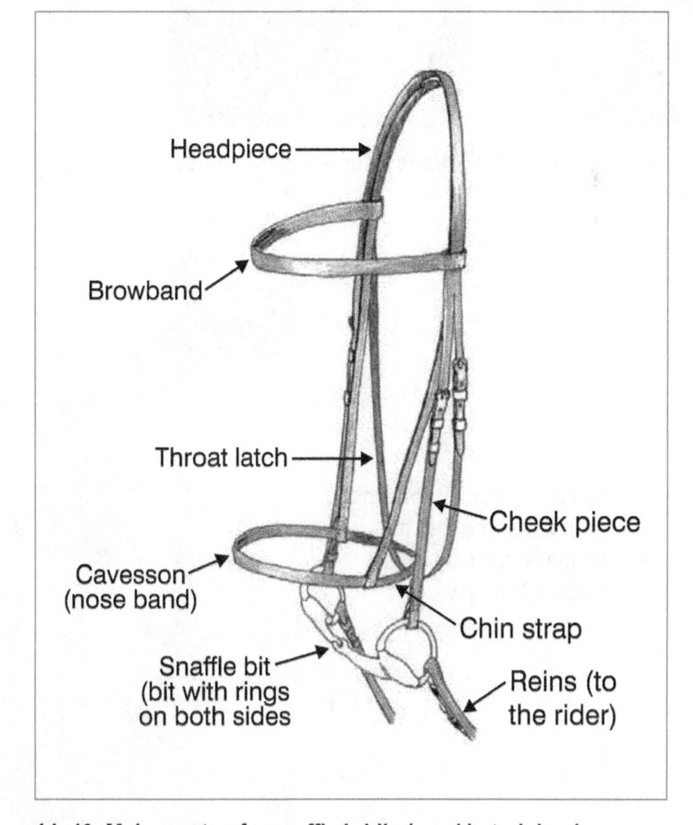

14–40. Major parts of a snaffle bridle (used in training horses and on sport horses).

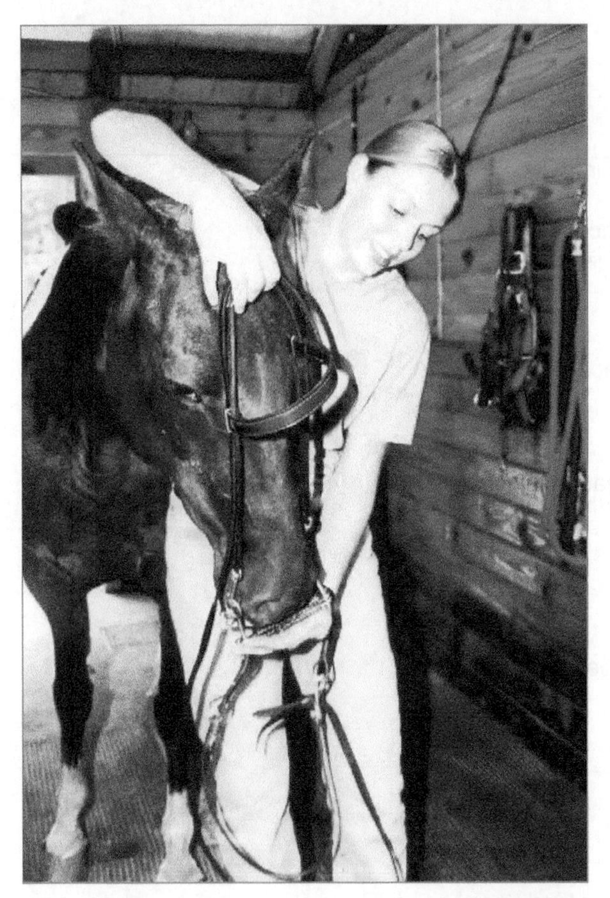

14–41. Putting a bridle on a horse. (Courtesy, Education Image)

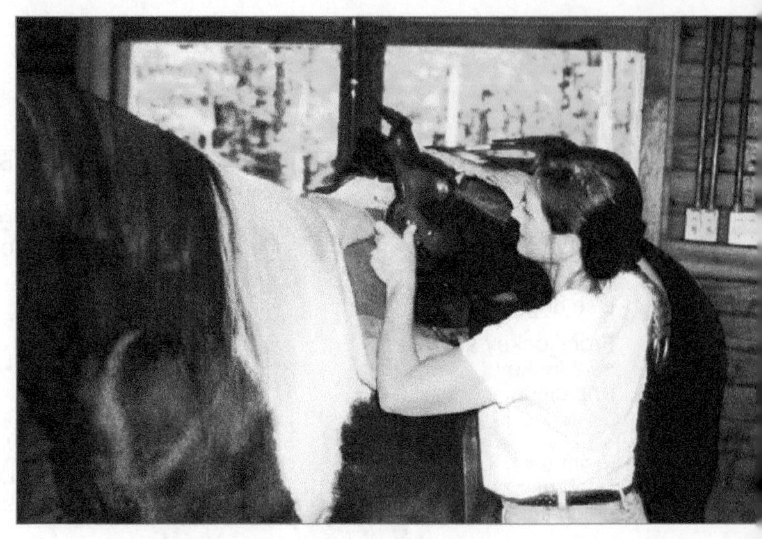

14–42. Putting a saddle on a horse. (Courtesy, Education Images)

able for the rider and the horse. A saddle pad is placed under the saddle for the horse's comfort.

A bridle is also needed when riding. The bridle is the part of the harness placed on the horse's head. It is used to direct and control a horse. There are many styles and bits to choose from. A bit needs to be wide enough not to pinch the cheekbones.

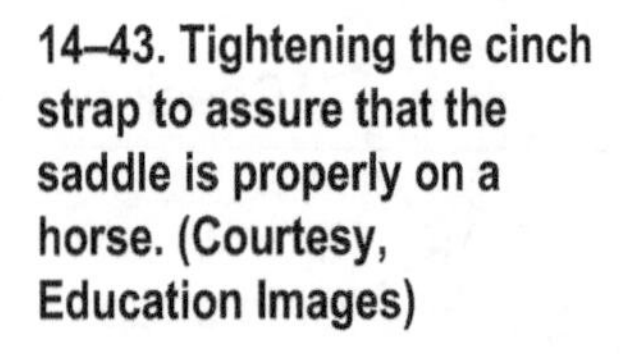

14–43. Tightening the cinch strap to assure that the saddle is properly on a horse. (Courtesy, Education Images)

EQUITATION

Equitation is the riding and managing of horses. It is often called ***horsemanship***, which is a word that has grown in disfavor in recent years with the increasing use of equitation. Good equitation and patience are essential for working with horses. Each animal should be treated as an individual. Horse and rider should work together as a team.

MOUNTING AND DISMOUNTING

A horse should be approached, mounted, and dismounted from the left side. When mounting, stand facing the saddle squarely. Then, place the left foot in the stirrup. Hop and swing up and into the saddle, using momentum of the right leg.

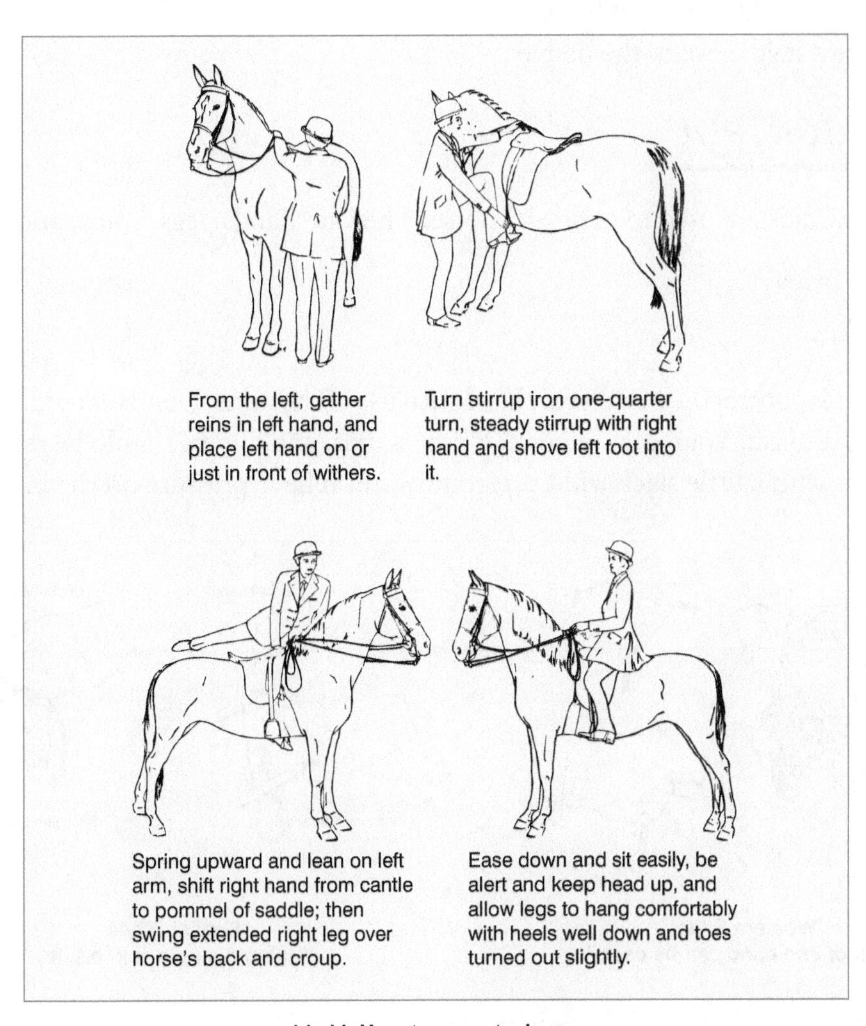

14–44. How to mount a horse.

When dismounting, hold the saddle horn with the right hand, and balance with the left hand on the horse's neck. Making sure the horse is steady, loosen the left foot in the stirrup, and shift your body weight to the left foot. Free the right leg, and gently let down.

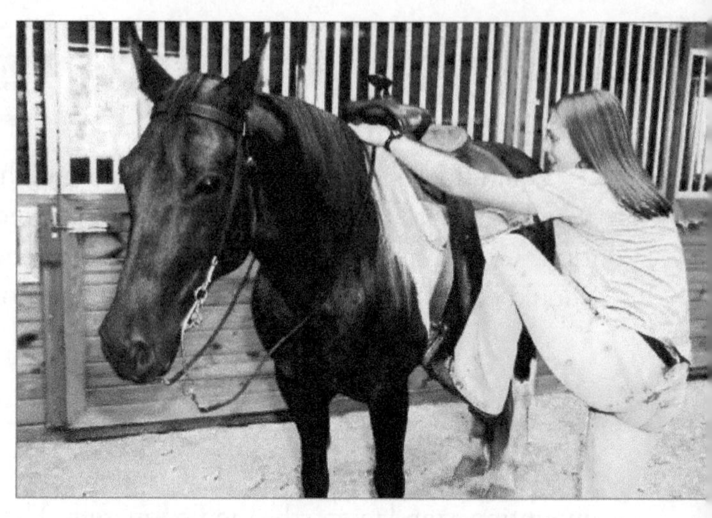

14–45. Mounting a horse. (Courtesy, Education Images)

SEAT POSITION

Sit tall in the saddle. Your back should be erect, not slouched. Your position in the saddle should allow for comfortable control of the horse and use of aids. Your shoulders should be back, and your arms held close to the body.

HORSE CONTROL

Four basic aids are used to control a horse. They are hands, legs, voice, and weight.

Hands

Your hands control the horse's forehead with use of the reins. Hands should be soft, light, and steady, yet firm. Your arms, elbows, wrists, hands, and fingers should be relaxed. Holding the reins with a little slack will keep control, yet relieve pressure on the bit.

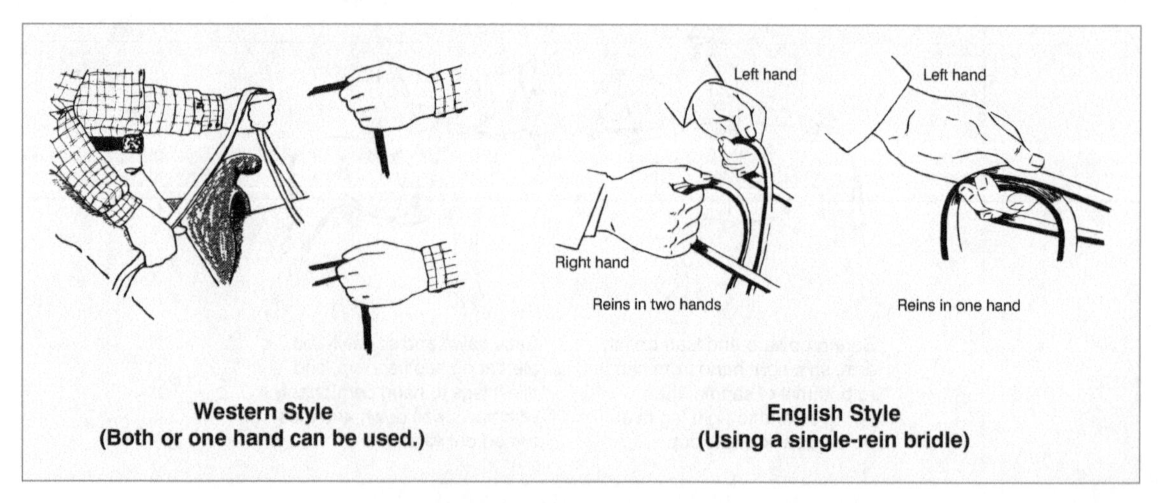

14–46. How to hold the reins.

Legs

Your legs are used to control the hindquarters and the forward movement of the horse. The horse should learn to move when your legs are squeezed.

Voice

Successful use of the voice means being consistent. Frequent commands are "Back," "Easy," and "Whoa." The words used, as well as the tone of the voice, are important. Remain calm, and do not yell at the horse.

Weight

Shifting your body weight signals the horse to shift its weight. This causes the horse either to be free to move or to be held in place. Signaling a horse by use of weight can be done by shifting your seat, moving weight from one stirrup to the other, or being in rhythm with the horse by leaning your upper body forward.

14–47. A rider with skills in horse control. (Courtesy, Education Images)

MOVEMENT

Movement is initiated by using the four aids. Overuse of these is a common fault of some beginning riders. There are various types of movement. The horse will become ready to move when you squeeze your legs. Movement is governed by the actions of the reins.

Walk

A **walk** is a four-beat gait. During a walk, both you and the horse are relaxed. Slightly squeeze your legs and move the reins to begin walking.

Jog

A **jog** is a two-beat gait in Western riding. You should be sitting deep in the saddle, with just enough weight on your ankles to absorb some of the motion. Your upper body should be leaning forward to maintain balance. The trot is the similar gait in English riding.

Lope

A **lope** is a three-beat gait. You should sit very deep in the saddle and lean forward to maintain balance. The three-beat gait in English riding is the canter.

Gallop

The **gallop** is a four-beat gait in both Western and English riding. It is sometimes known as the run. The four feet strike the ground separately. For a brief interval, all four feet are off the ground. The gallop is the fastest gait.

14–48. A horse can be temporarily restrained by tying to a sturdy post or rail. (Courtesy, Education Images)

Backing

To back a horse, sit erectly in the saddle, with your weight forward. Collect the horse, and squeeze with your thighs. Tell the horse, "Back." Alternately put pressure on each rein.

Stopping

For proper stopping, a horse should be cued ahead of time by saying "Whoa," squeezing with your thighs, tugging on the reins, or gently touching the horse's neck. To bring the horse to a stop, squeeze with your thighs and firmly pull on the reins. Sit slightly erect in the saddle, grip with your thighs, and position your body weight evenly in the saddle.

Some inexperienced riders are frightened by not being able to stop a horse quickly. Until you gain good experience in handling a horse, always ride one that readily stops.

REVIEWING

MAIN IDEAS

Horses are chiefly used today for their recreational and personal value. They are classified as light horses, ponies, or draft horses. Three types of light horses are riding horses, racehorses, and driving horses. Riding horses include gaited horses, stock horses, polo mounts, hunters and jumpers, and ponies.

Selection of a horse should be based on an affordable price range; the animal's disposition, size, gait, breed, and color; and the work to be performed. Purchase a horse that suits its rider.

Breeding requires careful management. Conception rates among mares are the lowest of all farm animals. Careful and precise management of horses is a must.

Horses require concentrates in their ration to provide energy and protein, minerals, and vitamins. Good water is needed. Oats, corn, and barley are the most common sources of carbohydrates and fats. Supplements widely used include soybean meal and linseed meal and oil. Mineral and vitamin supplements are mixed into the total feed ration.

Horse health should be checked each day. A horse's teeth may need to be floated. Its hooves should be cleaned daily, and if shoed, they should be trimmed every four to six weeks.

Facilities for horses do not have to be expensive and extravagant. A barn with a good-sized stall is needed. Basic equipment includes a hayrack or manger raised off the ground, a hoof pick, a body brush, and a mane and tail comb.

Equitation (horsemanship) is skill in riding and managing horses. A beginner must learn how to properly mount, dismount, sit in the saddle, use the aids (hands, legs, voice, and weight), coordinate movement, and stop a horse.

QUESTIONS

Answer the following questions using correct spelling and complete sentences.

1. What is the purpose of horses?
2. How are horses classified?
3. How are light horses broken down by types?
4. What factors should be considered when selecting a horse?
5. Why is good management of utmost importance in horse reproduction?
6. Name six things that should be considered in the breeding management of horses.
7. What does an average horse's diet consist of?
8. When managing the health of a horse, what should the owner look for?
9. How should hooves be cared for?
10. What facilities and equipment are needed for horses?
11. What is equitation? Why is proper equitation important?
12. How are the four aids used in equitation?

EVALUATING

Match the term with the correct definition. Write the letter by the term in the blank provided.

a. mare
b. gelding
c. plug
d. lope
e. farrier
f. gait
g. foal
h. stud horse
i. gallop
j. hand

_____1. A young horse that has not been weaned.
_____2. A male horse kept for breeding.
_____3. A person who puts shoes on horses' feet.
_____4. A slow, three-beat gait in which the head is carried low.
_____5. A horse with poor conformation and common breeding.
_____6. A sexually mature female horse.
_____7. The way a horse moves, walks, or runs.
_____8. A male horse castrated before reaching sexual maturity.
_____9. A horse measurement equal to 4 inches.
_____10. A four-beat gait.

EXPLORING

1. Visit a horse farm. Explore the breeding, feeding, health-care, and facility management of the farm. Prepare a written or oral report for the class on what you learn.

2. Study common management practices used with horses, and explore how they are changing. Interview authorities on horses. Use several references, Web sites, and other types of information in your study. Prepare a report on your findings.

3. Attend a horse show. Observe the breeds of horses, how the animals are groomed and ridden, and other areas related to horse management. Use a digital camera and prepare an electronic report on your observations.

CHAPTER 15

Draft Animals

OBJECTIVES

This chapter provides basic information on draft animals. It has the following objectives:

1. Describe draft animals and list major kinds and breeds.
2. List the possibilities in draft animal production.
3. Explain important management practices with draft animals.
4. List the nutritional requirements of draft animals.
5. Describe the facility and equipment needs with draft animals.

TERMS

automated feeding
breeding period
curry comb
draft
draft animal
feather
gregarious behavior
hand-feeding
harness
mule
ox
power
regurgitate
roan
ruminate
self-feeding
social ranking
tri-purpose animal
yoke

15–1. A draft horse is used to skid logs. (Courtesy, U.S. Department of Agriculture)

IF you have a heavy load that needs moving, a draft animal might do it for you! No gasoline or tires are needed. Hay, grain, and care will handle routine maintenance. Not many people use draft animals to do work anymore. They are used as hobbies, in animal-pull events, and as companions.

Draft animals had a major role in the settlement of America and farming of the land. They are still important in some countries. In early America, draft animals supplied the power to farm and clear land. Draft animals were replaced with engine-powered machinery and equipment.

Draft animals are still around today. They have only limited use in farming in the United States. They are no longer major power sources in the military. When military tactics changed and gun powder was developed, the horse was no longer vital in battle. Today, draft animals are largely kept for fun.

KINDS OF DRAFT ANIMALS

15–2. A camel is being used to plow land in India.

Several different animals are used for draft. A ***draft animal*** is an animal that has been trained and is used for pulling heavy loads. Most often, draft animals are horses or oxen. However, donkeys, mules, camels, and buffalo have been used as draft animals and still are used today in some countries.

When a draft animal is selected, consider the climate, the availability of the animal, the cost of the animal, the type of work to be done, and the social and religious traditions, which might limit the ownership or use of some animals.

15–3. A few farms in the United States continue to use draft animals, such as this team of horses on an Amish farm. (Courtesy, U.S. Department of Agriculture)

OXEN

The most popular draft animal, the **ox** is any animal of the bovine (cattle) family. Oxen are primarily castrated males or bulls, but sometimes females are used. Oxen are preferred over other draft animals because they are ruminants and use their feed better, but they are much slower than the other animals. Chapter 8 gives you more information on beef cattle

15–4. A team of steer oxen is being used to pull a plow on a farm in Cuba. (Courtesy, U.S. Department of Agriculture)

and Chapter 11 gives you more information on dairy cattle. Oxen are classified as a ***tri-purpose animal*** because they served three purposes: work, milk, and meat.

DRAFT HORSES

Draft horses are common draft animals in the United States. Some breeds of horses are defined as draft horses. They are larger and heavier than the horses used for riding and pleasure. All horses and horse-like animals are from the equine family. Draft horses were developed to be 1,500 to 2,500 pounds (680–1,134 kg); have a low center of gravity; and be wide, deep, compact, strong, and large boned. Other members of the equine family, such as mules, are used because they are cheaper to buy than horses.

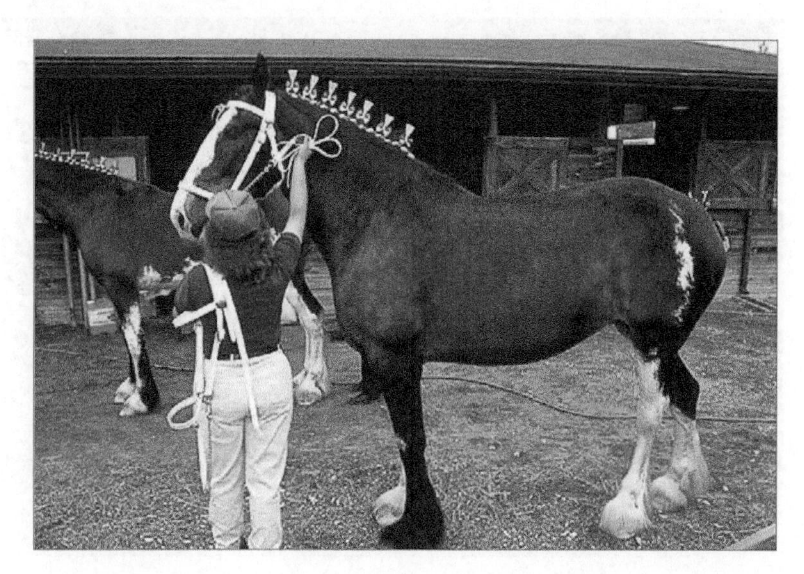

15–5. Cyldesdales are among the best-known draft horses. (Courtesy, U.S. Department of Agriculture)

15–6. Belgian horses plowing on an Ohio farm. (Courtesy, U.S. Department of Agriculture)

Five breeds of draft horses are found in North America: Belgian, Percheron, Shire, Clydesdale, and Suffolk.

Belgians, which are a Flemish breed, are brown to light blonde and sometimes even **roan**, which is a mixture of white and colored hairs, and became popular because of their size, endurance, tremendous power, excellent muscling, and style. Belgians are the number one breed of draft horses in the United States today, replacing the Percheron that was previously number one.

Percherons, which are from France, are more refined and balanced than some of the other draft horse breeds. Fifty percent of Percherons are black and the other 50 percent are black and gray.

15–7. A Percheron is pulling a cart at the Maryland State Fair. (Courtesy, U.S. Department of Agriculture)

Clydesdales are the third most popular breed in the United States. They are brown or black with white faces; have long white hair on their legs, known as ***feathers***, which sometimes extend all the way up to their bellies; and black manes. Clydesdales originated in Scotland and are well known for pulling heavy wagons.

Shires are black in color and, like Clydesdales, have feathers on their feet, but they have greater bulk than the Clydesdale. Shires were developed in England.

Suffolks were also developed in England. Unlike the other four breeds, their development was solely for agricultural reasons. They are chestnut color (sometimes referred to as sorrel) and have very little white on them. They have very little or no feathers. Suffolks are easy keepers, which means they stay in good condition with less feed than other horses require for the same condition. Suffolks have a very fast walk, great stamina, longevity, and a willingness to work.

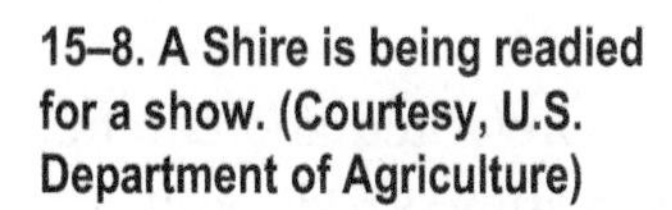

15–8. A Shire is being readied for a show. (Courtesy, U.S. Department of Agriculture)

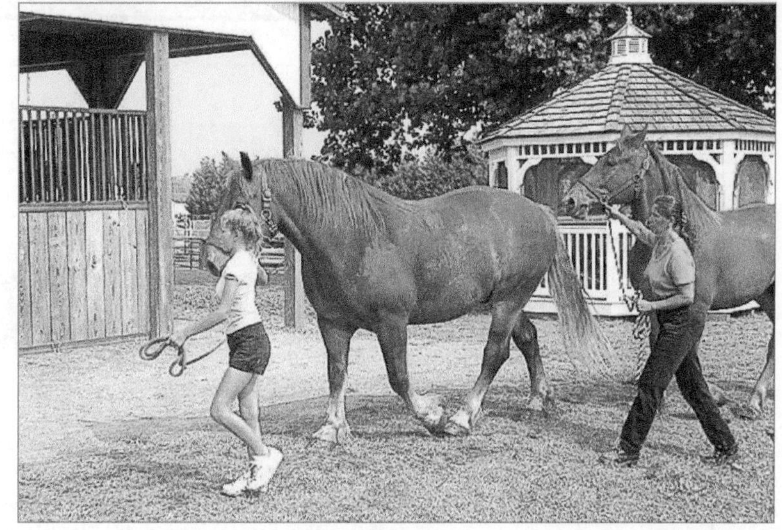

15–9. Suffolks are known as "easy keepers." (Courtesy, U.S. Department of Agriculture)

15–10. A donkey. (Donkeys are mated to horses to gain mule offspring.) (Courtesy, Education Images)

MULES

A ***mule*** is the hybrid offspring of a male ass (donkey, jack, or jackass) and a mare (female horse). Mules physically resemble both donkeys and horses. They have the large ears of donkeys, are strong like a horse, and tend to pace themselves for long-term power like a donkey. Horses tend to do work at a fast pace and do not hold up for the same long hours as mules. Mules are usually larger in size than donkeys, but smaller than draft horses. As hybrids, mules do not reproduce, though they are born as males and females. Male mules are gelded (castrated) as young animals.

Mules have advantages over other draft animals. Mules endure heat better than horses. They are less sensitive to their feed than horses and have fewer feed-related diseases. Mules eat lower-grade feed, which saves money.

Mules are physically sound, have fewer defects than horses, are surefooted and careful, and live a long time. When treated with force, mules tend to have stubborn dispositions somewhat like a donkey. Some people like mules and get along well with them. Other people prefer horses to mules. In some cases, mules will turn aggressive toward other animals, such as cattle, and chase them through a pasture.

CONNECTION

HORSEPOWER SCIENCE

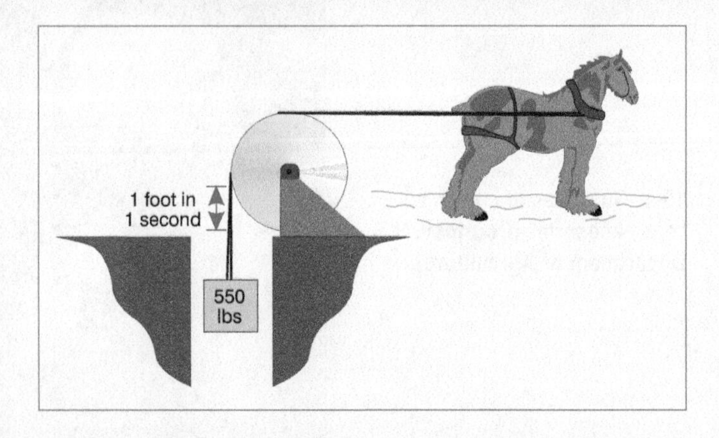

Horsepower is used in many ways today. Engines have horsepower ratings, as do electric motors. But what does the rating of engines and motors have to do with horses?

Horsepower was developed in Britain as the unit of power. It refers to work done at the rate of lifting 550 pounds a distance of 1 foot in a time of 1 second. A strong, fast horse would be needed to meet this definition of horsepower. Animal pull competitions use a different measure of power, as stated in the chapter.

Horsepower has been converted to other measurements. Electricity, for example, may use horsepower. One horsepower is equal to 746 watts of electricity.

15–11. A team of mules is pulling this covered wagon. (Courtesy, American Donkey and Mule Society, Texas)

15–12. Mules may be trained for showing and competitive events. (Courtesy, Lucky Three Ranch, Colorado)

Many farmers in the United States relied on mules to provide power for long hours of the day doing tedious crop cultivation. In the late 1800s and early 1900s, the large cotton plantations of the South often used hundreds of mules for power.

Today, some people view mules much as horses. They train them for show and dress them in fancy ways formerly reserved only for horses. The mules are pets!

DRAFT ANIMAL PRODUCTION

After a period of decline, draft animals are increasing in numbers. Special events feature draft animals performing power events as well as skill events.

Many enthusiasts show their draft horses in halter classes. The animal is judged on conformation. Hitch classes are used where one to six horses may be used in a class. There are also clubs of draft horse owners who put on old-fashioned farming demonstrations for the public. Loggers in the north are starting to use the draft horse again. Horses cause much less damage to the environment than tractors, especially crawler-type tractors. Teams of horses also compete in horse pulls. The team is hooked either on a boat or on a machine and they pull the weight. The team that pulls the most weight is the winner.

DRAFT ANIMAL REPRODUCTION

Reproduction is a key component in the draft animal industry. Strong, well-bred animals are needed.

Estrus is the time when the female animal is most receptive to be bred by the male animal or artificially. The time of the year when the female has her estrus periods and estrous cycle is the ***breeding period***.

Oxen

Oxen are raised similarly to beef cattle. The females reach puberty at 8 to 15 months of age. The duration of their cycle is 12 to 20 hours every 21 days year round. Breeding must occur during this time.

Even though cows cycle year round, most producers prefer to have the females bred about the same time so they calve about the same time of the year and all of the young are close to the same age. When it comes time to make management decisions, such as vaccinating or castrating of the male calves, it is easier when the whole group can be done at once. Oxen in estrus stand to be mounted by herdmates, lack an appetite, act nervous, restless, and are very vocal. After the cow has been bred, it takes about 9 months or 270 to 295 days for her to calve.

15–13. Horse mare with mule colt. (Courtesy, Lucky Three Ranch, Colorado)

Draft Horses

Draft horse females reach puberty at 12 to 15 months of age. Unlike oxen, females have a cycle that lasts 4 to 6 days and comes every 21 days during the estrous cycle, which begins in the spring and stops in midsummer. Females must be bred during this time of the year.

When females are experiencing estrus, they are very vocal, get nervous, restless and irritable, squat and urinate frequently, and seek out the stallion. Mares should be bred on the sec-

15–14. Foals are often pastured with mares, such as these Belgians. (Courtesy, U.S. Department of Agriculture)

ond day of estrus and every other day for two to four days. After the mare has been bred, either artificially or by the stallion it takes 11 to $11\frac{1}{2}$ months, or 330 to 340 days, for her to foal.

15–15. A Clydesdale yearling is being trained for showing. (A yearling is a horse of either sex between 1 and 2 years of age.) (Courtesy, U.S. Department of Agriculture)

YOUNG ANIMAL CARE

Both female horses and oxen clean their young with their tongues after giving birth. This gets blood circulating in the newborn and encourages the newborn to stand and nurse. Instinct results in mothers being aggressive in protecting their young. Even a calm pet may become ornery with a baby.

DRAFT ANIMAL MANAGEMENT PRACTICES

Understanding animal behavior is essential for good management of draft animals. First of all, ***draft*** is the pulling of an animal to move any object. For an animal to move an object, it must exert a force equal to or greater than the weight of the object.

DRAFT CAPACITY

The draft capacity of an animal increases with the weight of the animal. The rule is that an animal can exert a constant pull at its normal pace of one-tenth its body weight. If an animal is required to exert a concentrated effort, the time the animal can work will greatly diminish. Horses, mules, and oxen are preferred because they can pull loads at a normal speed over long time periods and exert extra energy when needed.

Power is the combination of pulling capacity and speed. A draft horse will pull a 150-pound (68 kg) load at a steady rate of $2^1/_2$ miles per hour, which is defined as one horse power. One horse power may also be defined as pulling 150 pounds (68 kg) out of a 220-feet (67 m) deep hole in one minute. An ox pulling the same weight will travel at $1^1/_2$ miles per hour, so at the end of the day the horse has traveled further. Also, shorter animals have a lower center of gravity, so they must be hitched lower so less work is required for them to get the task accomplished.

15–16. These Belgians are being used to rake hay. They are well conditioned for the work.

All animals have certain behavior patterns, which must be recognized for proper health and overall well-being of the animal. Some animals pull better and reach maximum capacity when worked alone, but not when worked with other animals. Animals work well only when they have been properly trained and properly outfitted with the correct harness (for more information on the harness see the equipment section of this chapter).

An ox can deliver 25 to 50 percent more horsepower when harnessed with a collar instead of a yoke. This is because of the ox's lower center of gravity and the wider pushing area of the collar.

DRAFT ANIMAL SELECTION

The type of draft animal selected depends on the job and the time in which to do the job. For example, horses work much faster than oxen, but oxen use their feed more efficiently because they are ruminant animals. If horses are considered, there are other animals in the equine family that may be cheaper to purchase than a horse, such as a mule. All of these factors need to be taken into account by the owner when a draft animal is selected.

When selecting animals, it is also important to note that horses and oxen in herds develop social rankings. ***Social ranking*** is the order the animal falls within the herd. Social rank is determined by the age, size, strength, genetics, if they have horns or are polled (which means they are bred to be born without horns), and previous experience of the animal.

There are many sources available to assist producers of draft animal owners and producers. Veterinarians, the Cooperative Extension Service, state universities, state experiment stations, feed suppliers, other animal owners, animal associations, and libraries are all important sources for people to gather information pertinent to their animal and situation. Since animals and situations are different, it is impossible to answer every question that may occur.

NUTRITIONAL REQUIREMENTS

Animals vary in their feed needs, just as they vary from breed to breed and horse to oxen. A good feeding program is vital in maintaining an animal's health and strength. In determining the feed ration for a draft animal, first the species, development stage of the animal,

15–17. Pack animals, such as this donkey with a load of bananas, must receive adequate nutrition to do work. (Courtesy, Agricultural Research Service, USDA)

weight and size of the animal, if they are or are not pregnant, and how often and hard they are worked must be decided. Many people use a commercial feed prepared especially for draft horses.

Water is important for all animals and should be available to them at all times, except during and right after working. Draft horses and oxen both require approximately 8 gallons (30.3 l) of water per head per day. Some may drink more, depending on activity.

All draft animals need protein. It may be fed in the form of alfalfa hay to both horses and oxen and/or cottonseed or cottonseed meal to oxen. The feed ration needs to be complete. The cost of the ration must be considered.

High-fiber and low-fiber feeds are also important to draft animals. In the high-fiber category, horses and oxen both benefit from oat hay. Oxen also benefit from corn silage and haylage. In the low-fiber category, both oxen and horses eat corn, grain, barley, and oats. By-product feeds are also used, such as pulp from citrus or beets, barley malt, and others. The availability and types of by-products depend upon the area in which the producer lives.

FEEDING OXEN

Oxen are ruminant animals, which means they have four compartments in their stomachs. They swallow their food as soon as they bite and salivate on it. Later, often while lying under a shade tree, they **regurgitate** the food. This is returning eaten food to the mouth for chewing. The food in the mouth is a cud. Oxen graze from four to nine hours per day and **ruminate** (chew their cud) for four to nine hours per day.

Extreme heat or cold lessens the amount of food consumed by the animal. Oxen are similar to beef cattle in many nutritional needs. An exception is that they need feed with higher energy when they are being worked.

FEEDING DRAFT HORSES

A horse is a single-stomached (monogastric) animal. It eats and chews its food immediately and then the food is used to give the animal energy. Ruminants are sometimes preferred because it has been said they use their food more efficiently.

Vitamins and minerals are important to assure the draft animal remains healthy and strong. The three minerals often added as supplements to the feed ration are salt (sodium chloride), calcium, and phosphorus.

Draft horses need the following seven major elements: calcium, phosphorus, magnesium, sodium, chloride, potassium, and sulfur. There are also some trace minerals that are required: chromium, nickel, silicon, vanadium, tin, copper, cobalt, and zinc. The vitamins needed by horses are A, D, E, and C. In most cases, good-quality hay will eliminate the need for vitamin supplements.

Horses should be allowed to cool down after working before they are fed. They may be given some water after they have cooled, but horses will get lame or founder if they eat or are watered when they are too hot. Give a horse an hour or so to cool down before feeding or watering it. Also, give a horse a ½ hour to 45 minutes to digest food before harnessing the horse and making it do hard work. When feeding horses, make sure the hay is dust and mold free to prevent illness. The harder the horse works, the more feed the horse should be fed.

FEEDING SCHEDULES

In addition to feeding the correct ration, it is important to develop a feed schedule so animals are fed at the same times each day. The animals may be fed by ***hand-feeding***, which is giving animals the same amount of feed either one or two times daily. This would entail their grain and hay. For example, hay may be given two or three times daily and grain only once or twice.

Self-feeding is having feed in front of animals at all times. In this system, the producer needs to ensure some way of having feed available at all times. Many times, hay is available in the self-feeding system and grain is given through the hand-feeding system. This tends to work well since draft horses will overeat and make themselves sick on sweet feeds, such as grain.

The third option of feeding is ***automated feeding***. Animals are fed mechanically so no one needs to haul feed at feeding time. Usually, an auger is used to bring the feed to the animals. This is a common way of feeding silage to cattle.

Draft animals ready for weaning from their mother should have their mother's milk supplemented with hay and grain. When the weaning takes place, the young animal will do better if it is used to eating food along with its mother's milk. It will cause less stress to the young animals if two or more of them are weaned at a time. The companionship is important to these animals and they will do better if they are not alone.

HEALTH PRACTICES

Animals respond differently to their environment. Some things may cause stress in one animal while not in another. It is vital that owners understand their animals' attitudes and behaviors. If an animal is behaving strangely, it may be ill, under stress, ready to be bred, or have other problems that may require attention.

There is no one who knows an animal better than the person who cares and works with it everyday. This person must always pay attention to the actions and attitudes of the animal.

CONDITIONING

An important health practice is to condition a draft animal before it is used for heavy pulling. If the animal does not work hard over the winter, start the spring with a smaller work load and shorter days until the animal has gotten back into condition.

A draft animal must physically train much as a human athlete. Because they do not have the power to reason and decide to stay in shape on their own, it is the responsibility of the owner to ensure that an animal is in shape before hard work. Animals may pull a muscle or physically hurt themselves if they are not conditioned. The least that will happen will be a break in their spirit because they will experience defeat.

15–18. Draft horses need to be in good physical condition to pull heavy loads. (Courtesy, U.S. Department of Agriculture)

CONSIDER HEALTH

Health begins with animal selection. To have a good, strong, sound draft animal, buy that type of animal. The stock purchased will make a big difference in the amount of work actually done. It is an accepted practice to take a veterinarian to inspect an animal before it is bought. The buyer is usually the one who has to pay for the veterinarian's visit, but the buyer then knows the soundness of the animal.

General Health

A draft horse has a normal heart rate of 28 to 50 beats per minute and an average temperature of 100.5°F (38°C) with ranges of 99.0 to 100.8°F (37.2–38.2°C). An ox has a normal heart rate of 40 to 70 beats per minute and an average temperature of 101.5°F with ranges of 100.4 to 102.8°F (38–39.3°C) temperature. When the owner has a feeling the animal is not feeling well, a higher temperature is a good indication that something is wrong. There are many possibilities of what can go wrong with an animal; and a veterinarian is a good person to call for assistance.

Nutritional diseases may affect draft animals. If animals are not fed the proper ration or supplements, they may have deficiencies. Too much feed can also cause problems. Metabolic disease may result from erratic feeding times and feeds themselves.

Parasites

Internal parasites may cause major problems in animals. They live at the expense of the animal they are living inside. External parasites get their nourishment from the outer surface of the animal's body.

Deworming and vaccinations of draft animals are important to stop or prevent parasites. Vaccinations are the injection of an agent into the animal for the purpose of developing resistance to what the animal is treated for. Animals pick up worms through their feed and pasture. As long as the animals are treated for parasites periodically, they will be okay.

Small precautions, such as brushing and grooming draft animals, will cut down on any manure on their coats. Manure attracts flies and other insects that may carry disease. A curry comb and a brush are both needed to groom the draft animal. The ***curry comb*** is an oval-shaped plastic or metal device used to loosen sweat, manure, and other foreign materials from the animal's coat. The brush is used to remove the materials from the animal's coat. If an animal is not bothered by flies and insects, it will be happier and healthier.

15–19. The foot of a mule is being examined. (Courtesy, Lucky Three Ranch, Colorado)

Foot Care

Another important area of health care is the feet of the draft animal. Since the animals are used to do heavy work, their feet must be maintained in good condition. If an animal has sore feet it will work much less and put forth minimal effort. Feet must be trimmed about every six months, depending upon the growth of the animal's hoof.

Horses have metal shoes formed to fit their feet. The shoes, which are nailed on, give the animal better traction for pulling loads. Proper shoes assist the horse in digging its feet into the ground to pull. Proper shoes and trimming lessen the possibility of the animal falling or pulling a muscle or developing a foot disease.

FACILITY AND EQUIPMENT NEEDS

To develop proper housing for a draft animal, the personality and characteristics of the animal must be understood.

Gregarious behavior is the instinct of animals to flock or herd. When an animal that normally is in a flock or herd cannot be with other animals, this causes stress in the animal and makes it difficult to deal with. Although draft animals prefer to be in herds, they also need to have enough space to live comfortably, because overcrowding of animals is also a source of stress.

When constructing windbreaks or barns, make sure the builder takes into account that the area needs to be easily cleaned and well drained. By doing so, animals can stay dry and out of the mud.

Draft horses need shelter or a windbreak, especially in the winter. Mares who are going to foal need to be kept separate from barren mares and need to be in an area where they can get plenty of exercise to stay in peak physical condition. If they are bred to foal in the winter, they need a warm, dry place to give birth. Some horses do foal outside, but the survival rate of the newborn is better and there are fewer complications if it is born inside.

Draft horses prefer to be in the company of other horses. If a person plans to work a horse—one that is usually worked as part of a team—alone, he or she may have trouble. Many times, when horses are worked as a team, it is difficult to separate the team without the horses getting unruly or upset. Horses tend to be naturally curious. Owners may put toys in the horse's stall to keep it entertained or leave it out in the pasture longer so it does not get bored in a stall. Remember, oxen are ruminants; they spend many hours per day laying around and chewing their cuds. Horses have to search for pasture longer since they only have one stomach.

Horses may be kept out on pasture where they have access to a windbreak, or kept in box stalls or tie stalls. Tie stalls are okay for short periods. After long periods in a tie stall, a horse will get bored and start causing trouble, such as chewing on the boards it can reach, pawing or whinnying, and being vocal. These are all signs of boredom in the horse. If horses are pas-

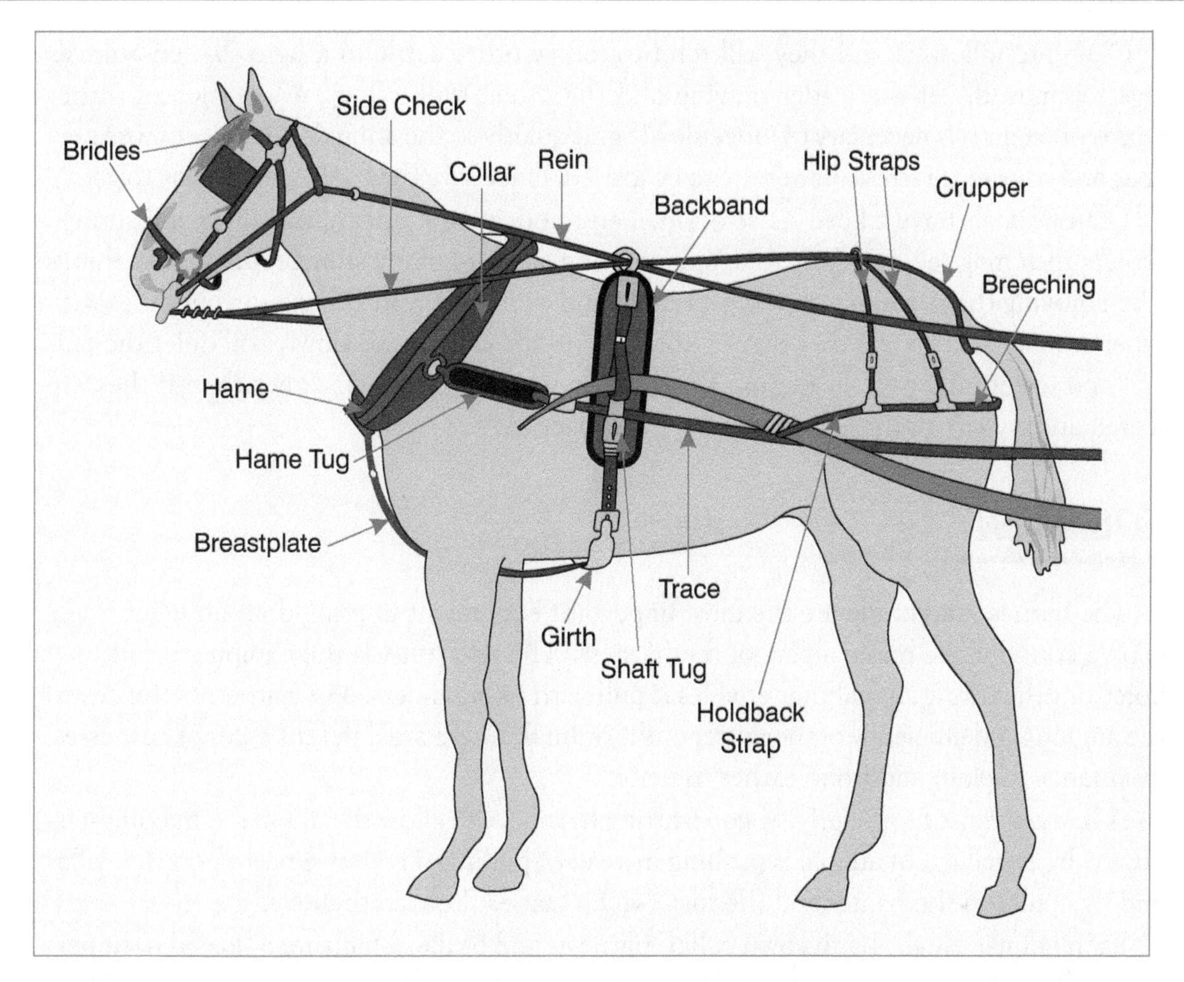

15–20. Parts of the harness of a draft horse.

tured in lots next to each other, they may run the fence together and be very vocal with one another.

STRONG FENCES AND STALLS

Oxen, on the other hand, because they are so big, tend to break through fences to get with other cattle if they are separated and penned alone. Oxen prefer shady areas in the warm weather and gather under shelter or windbreaks in the cold weather.

A cow ready to calve in extremely cold weather should be kept inside. The newborn has a better survival rate if it is born inside, out of the elements. A cow and calf require about 300 square feet (27.9 m^2) of space and a steer needs 150 to 200 square feet (13.9 to 18.6 m^2) of space. The fences should be 50 to 60 inches (1.3 to 1.52 m) high; however, this depends on the size of the cattle.

Oxen are followers and they will tend to follow other cattle in a herd. When animals must be moved, get the leader moving and the others will follow. When moving either horses or oxen, it is necessary to move slowly and quietly so the animals do not get more nervous and have extra stress. If they are to be loaded, make sure they only see one way to go.

Draft animals have a keen sense of smell and notice loud, unfamiliar noises and strange objects they may see. It is best to avoid exposing them to many strange, unfamiliar things when moving them. Many animals get scared and even unruly in strange surroundings with other animals and people they do not know. Working calmly and slowly will quiet the animal and eventually pay off because they will do what is expected sooner than if they are scared and try to run through things and over people.

EQUIPMENT

The harness and bridle are the most important equipment in using draft animals. Some people consider the bridle a part of the harness. The **harness** is the equipment put on a horse or other draft animal that enables it pull with its shoulders. The harness is the key to the amount and efficiency of the power of the animal. There are different types of harnesses, from fancy to plain and from leather to nylon.

A harness must fit properly. A good-fitting harness will allow the animal to maximize its power. In actuality, an animal is pushing instead of pulling. The horse pushes on the collar and this pulls on the hames and the tugs, which causes the load to move.

Each animal usually has its own collar, harness, and bridle, which are adjusted to fit perfectly. The collar goes over the horse's head and rests around the neck and on the shoulders.

15–21. A Percheron team is harnessed and ready for work.

15–22. A two-mule team is pulling a light wagon. (Courtesy, patti jean_images & designs by patti jean guerrero/ Shutterstock)

The collar should fit snugly, but not bind on the horse. If the horse is worked, and a sore results, the collar is not properly fitted.

The harness is put over the top of the horse and the hames are attached around the collar. It is vital the hames are a good fit with the collar. If the fit is improper, a sore will once again result. All of the power comes from the collar, hames, and to the tugs, which run the length of the horse and are attached to the load at the back end of the horse. The bridle goes over the head of the horse and is hooked to the lines, which are used to steer and stop the horse.

After the horse is harnessed, it is hooked onto a pair of whipple trees. The tugs hook onto the whipple tree, which is then hooked onto the load. It is imperative to hook the animals the proper distance from the load; otherwise, they may get their heals scrubbed or be so far out and away from the load that they lose the advantage of being hooked low and close.

If one has never harnessed an animal before, it is best to find someone who knows how and ask for assistance. Many times, the people the animal is bought from can give good advice as to the collar size and the way it was hitched before.

A harness can be a very expensive piece of equipment and must be cared for. A properly cared for harness will last virtually forever. All that is required is to keep the harness clean and well oiled with a good harness oil. Some communities have harness shops that can dip harnesses in oil, eliminating the owner's mess and problems of what to do with the old oil.

Just as horses push on their collars, oxen push through their yoke to move their load. The **yoke** is a wooden bar that hooks two animals together with a bar between them. The oxen are hooked around their necks or horns with a bar between their heads. They hold their heads low and since that is where their strength is, they can move more hooked this way.

Oxen do not have bridles like horses do. Some oxen are steered with lines hooked to a halter and some have nothing on their faces.

REVIEWING

MAIN IDEAS

Draft animals have been trained to pull heavy loads. Once used for power in agriculture, they have been replaced in North America by machinery. Today, people enjoy having draft animals as pets and for competitive events.

Draft horses, mules, and oxen are the most common draft animals. The horses and mules are often trained for showing and groomed in fancy ways.

To effectively work with and raise draft animals, many considerations should be taken into account. The appropriate breed needed for the work to be accomplished, good management practices, a nutritionally complete diet, proper health program, and the correct facilities and equipment are all important to healthy draft animals.

A healthy draft animal will work hard to accomplish the tasks that need to be done. Diseases, parasites, and nutritional problems affect most draft animals in one form or another.

Appropriate facilities and equipment are needed. Since draft animals are very strong, all pastures, pens, and other structures around them should be built to withstand their power.

QUESTIONS

Answer the following questions using correct spelling and complete sentences.

1. What is a draft animal? Name three and compare their performance.
2. What are the major differences between reproducing oxen and draft horses?
3. What are the power and drafting potential of animals?
4. What are the nutritional needs of draft animals?
5. What are the major health practices with draft animals?
6. What facilities and equipment are needed with draft animals?
7. In constructing facilities, why is the power of the animal important?
8. How is an animal or team harnessed to its load?

EVALUATING

Match the term with the correct definition. Write the letter by the term in the blank provided.

a. mule
b. power
c. draft
d. harness
e. social ranking
f. regurgitate
g. yoke
h. gregarious behavior

______1. The pulling of an animal.
______2. Hybrid of ass and horse.
______3. The combination of pulling capacity and speed.
______4. The instinct of animals to flock or group together.
______5. A wooden bar that holds two harnessed animals together.
______6. Returning undigested food to the mouth for chewing.
______7. The order an animal has in a herd.
______8. Attachments put on an animal so it can pull.

EXPLORING

1. Attend a draft horse, mule, or oxen pulling event. Observe how the animals are handled and the kind of work they do. Give a report in class on your observations.

2. Assist in harnessing a team of animals. Study the different parts of the harness and how they are connected. Be very careful; draft animals can be dangerous.

CHAPTER 16

Dogs

OBJECTIVES

This chapter presents basic information on companion animals. It has the following objectives:

1. Describe the biology of a dog.
2. List and describe the classes of dogs.
3. Identify factors to consider in selecting a dog.
4. Describe reproductive practices with dogs.
5. Describe nutrition and feeding of dogs.
6. Identify important health practices with dogs.
7. Describe facility and equipment needs of dogs.
8. Identify grooming practices with dogs.
9. Assess sanitation and well-being practices with dogs.

TERMS

bitch
grooming
herding dog
hound
mixed-breed dog
nail bed
non-sporting breed
orphaned puppy
sporting dog
stud dog
terrier class
toy breed
training
weaning
whelp
whelping box
working dog

16–1. Dogs often bond with human families and become an integral part. (Courtesy, Andresr/Shutterstock)

Do you have a dog? Do you know of other people who have a dog? Some people have more than one dog. Why do some people keep dogs as companion animals? Dogs provide fun as well as make the lives of their owners better. An enjoyable dog helps people relax. A strong human-dog bond may develop. But, dogs need care and training on a daily basis in order to be healthy and happy.

Just as there are people who love dogs, there are those who are afraid of them. Dogs tend to sense that a person is fearful of them. Anyone who has been attacked or bitten by a dog is always cautious around one that is an unfamiliar. If you own a dog, always respect the feelings and wishes of those who may be fearful.

There are 77.8 million dogs in the United States. The number is increasing each year. Most of these dogs are kept for companionship although some are "working" dogs. These dogs provide service beyond companionship, such as guarding property, hunting game, and sniffing for drugs or explosives.

16–2. An owner with a well-trained dog who sits on command. (Note that the dog is on a leash to help keep it under control.) (Courtesy, Education Images)

THE BIOLOGY OF DOGS

Dogs are among the animals that have been pets the longest. Their use as companion animals began more than 10,000 years ago. In the centuries since, many breeds and varieties have been developed.

Dogs are carnivores, meaning that they are flesh eaters. Dogs are also mammals and monogastric animals. The scientific name for the domestic dog is Canis familiaria. Dogs are often called canines (or K-9s) because of their family (Canidae) and genus (Canis). Understanding their nutritional needs and digestive systems helps us provide better care.

Dogs are endothermic animals. They have well-defined, sturdy skeletons. Many dogs are adapted to specific uses, such as hunting and retrieving.

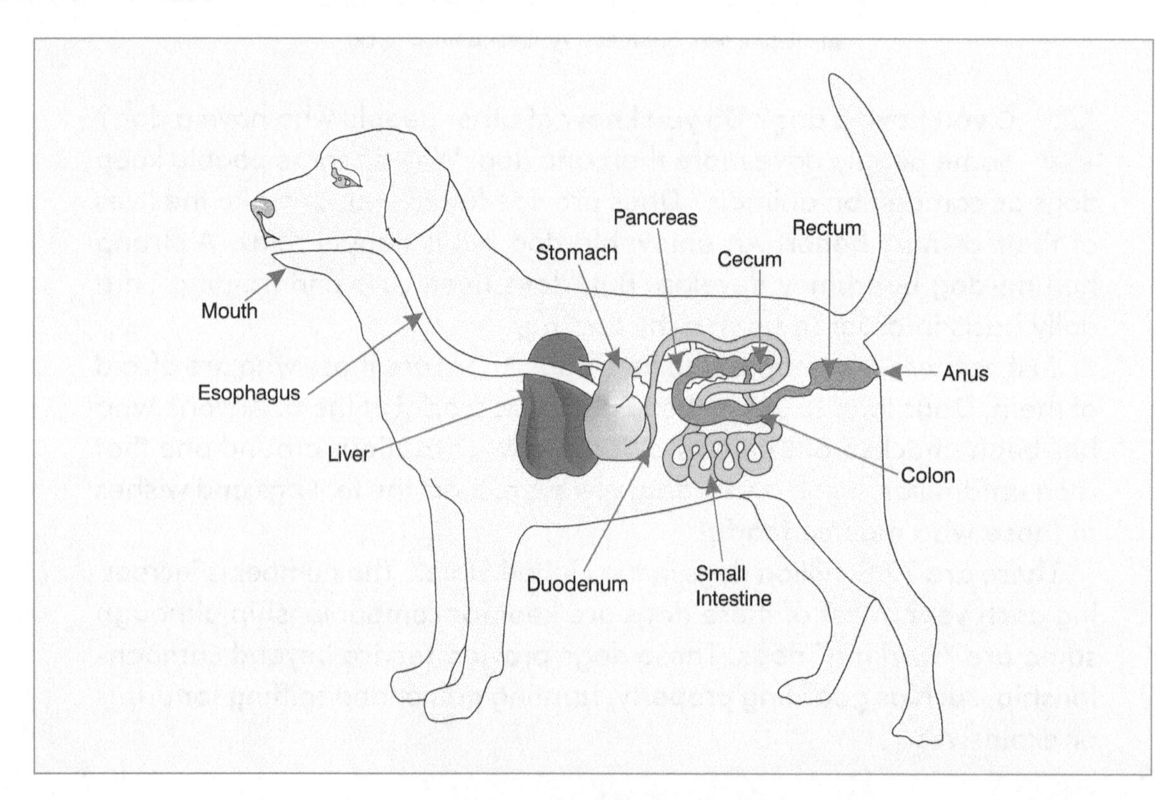

16–3. Gastrointestinal system of a dog.

The body structure of a dog varies with the size and breed of the animal. The weight of dogs varies. A mature Chihuahua may weigh less than 6 pounds (2.7 kg). A mature Saint Bernard may weigh as much as 200 pounds (90 kg). Most dogs have similar skeletons of about 320 bones. The major difference is in the size of the bones.

The smaller breeds of dogs may live 15 to 18 years. The larger breeds, such as the Saint Bernard, have shorter life spans, often only 8 to 10 years. Baby dogs are born in groups of 1 to 10 puppies, known as litters. The female, known as a ***bitch*** or dam, has a gestation period of about 63 days, depending on the breed. The male is the sire or ***stud dog***.

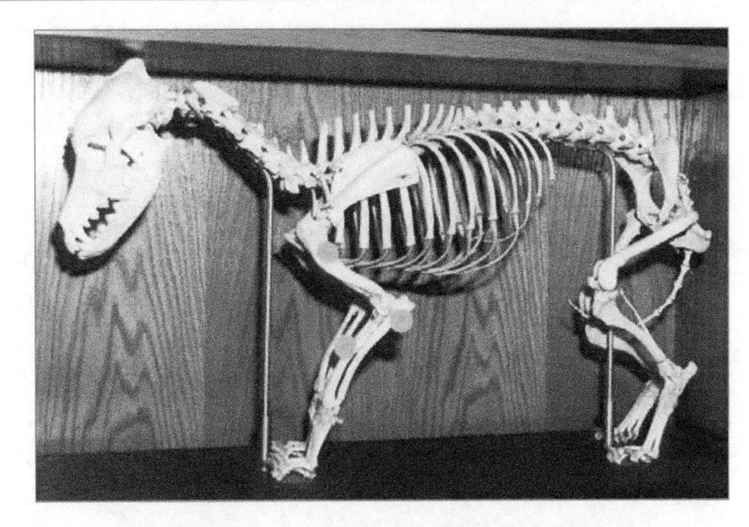

16–4. Skeleton of a dog. (Courtesy, Education Images)

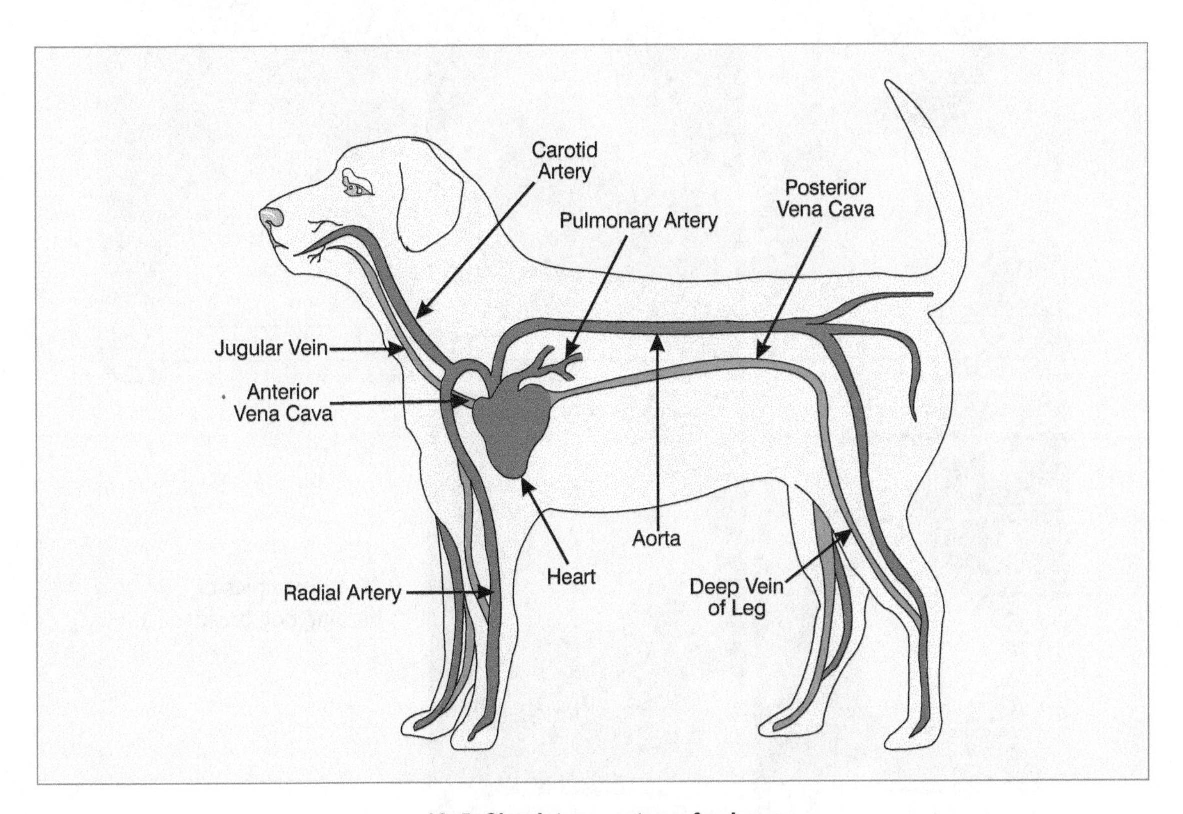

16–5. Circulatory system of a dog.

CLASSES AND BREEDS

16–6. This Australian cattle dog is working in a barn. (Courtesy, Education Images)

About 500 breeds of dogs are found throughout the world. In the United States, the American Kennel Club (AKC) classifies and maintains records on 150 dog breeds. Eight classes or types have been identified. These are based on the use and characteristics of the breeds. This information is helpful in selecting a breed. There are about 100 designer breeds such as the pubble (pug + beagle) and pomapoo (pomeranian + poodle).

16–7. Examples of herding dog breeds.

- Herding dogs—A ***herding dog*** is in a class that is popular with many people who have sheep and cattle. Herding dogs are easily trained to assist in herding animals into pastures and barns. The common herding species include the Australian Cattle Dog, Shetland Sheepdog, collie, and German Shepherd dog. Select these dogs as pets only if space is available for them to run and have fun. A herding dog does not usually like to live in confinement.
- Sporting dogs—A ***sporting dog*** is in a class that is used for sporting purposes. Breeds include pointers, setters, spaniels, and retrievers. Pointers and setters are hunters that run and then stop when they find their prey. Spaniels run and scare their game out of the cover. Retrievers swim and retrieve game both in the water and on the land.

16–8. The cocker spaniel (left) and the golden retriever (right) are examples of sporting dogs.

- Tracking dogs (hounds)—A ***hound*** is in a class of dogs used for tracking. Hounds have good abilities to follow the scent left by animals or people. Hunters use hounds to locate game. Law officials use them to find escaped prisoners or lost people. The most common breeds are the beagle, basset, dachshund, and greyhound. Some breeds run quickly to track, and others work slowly and thoroughly to find their catch.

16–9. The beagle (left) and the basset hound (right) are examples of tracking dogs.

16–10. This female Alaskan Malamute (also known as a husky) has been trained as a sled leader dog in a Tok, Alaska, kennel. (Courtesy, Education Images)

- Working dogs—A **working dog** is in a class used by people to get work done. Examples of the work done by these breeds include pulling sleds, protecting property and other animals, and sniffing bombs and drugs. Some working dogs are used to guard property and provide human protection. The breeds of working dogs include the Alaskan Malamute, Doberman Pinscher, Rottweiler, Saint Bernard, and Samoyed. These dogs are sometimes classified as service animals.
- Terrier dogs—The **terrier class** is a group of 25 different terrier breeds recognized by the AKC. The name terrier is from the Latin word terra, meaning "earth." The breeds are noted for following the animals they are chasing down into the earth. They dig in the earth to capture their prey. Miniature Schnauzers,

16–11. Examples of working dog breeds.

6–12. A Bedlington terrier (left) and a Scottish terrier (right) are 2 of the 25 terrier reeds. Scotties can be white or black.

16–13. A proud owner enjoys her Yorkshire terrier, a very small dog. (Courtesy, Education Images)

Airedales, bull terriers, and Scottish terriers are representatives of this group.

- Toy breeds—A ***toy breed*** is a class of small dogs that weigh between 4 and 16 pounds. Dogs in this class are known for their companionship and their long lives. Examples of toy dogs are Yorkshire terriers, toy poodles, English toy spaniels, and Chihuahuas.

CONNECTION

DOG BONDING

Dogs and their owners often develop a close bond. This bonding typically begins when the dog is a young puppy. The owner and dog share a long-term mutual attachment to each other. Dogs miss their owners when apart; owners miss their dogs when apart. Many life routines are built around each other.

Bonding involves close proximity of dog and owner. The dog provides and seeks affection. People may talk to their dogs much as they do to other people. In some cases, a dog will defend its owner or other family member if danger is perceived. Of course, there are situations in which some dogs turn on their owners and attack them. Be sure to select a breed that does not have this quality. You don't want your dog to cause injury to others!

This illustration shows a young woman continuing the bonding process with her young dog. No doubt, a long-term relationship is developing.

(Courtesy, Alena Ozerova/Shutterstock)

16–14. Examples of toy dog breeds.

- Non-sporting breeds—The ***non-sporting breed*** class of dogs was developed for specific purposes. Dogs in this class are now primarily pets. Examples of these are Dalmatians, bulldogs, and poodles (miniature and standard).

16–15. The Dalmatian (left) and the English bulldog (right) are examples of non-sporting breeds.

- Miscellaneous—Breeds not included in other groups or types are placed in the miscellaneous category. Inclusion in the miscellaneous group allows a dog to compete in scheduled AKC events.

SELECTING A DOG

Selecting a dog is an important decision. Good information is needed. Careful consideration should be given to the traits of various breeds and an individual's personal wishes.

If a puppy is obtained, it should be between 8 and 12 weeks of age. During this time, the puppy is old enough to be removed from its mother and is still forming its personality. If the owner wants to have a dog that is an average size for the breed, then an average-size puppy should be selected.

Some considerations in selecting a dog are discussed here.

- Mixed breed or purebred—Many people have mixed-breed dogs. A ***mixed-breed dog*** is one of unknown ancestry. In some cases, owners know the breeds of the parents, while in other cases, they do not. Some people want purebred dogs whose ancestry is known in terms of breed. In some cases, the dog may be bred to another of the same breed and a litter of puppies produced that is entitled to registration. A problem with a mixed-breed dog is knowing how large the mature dog will be. A young puppy can change significantly as it grows and matures. The mixed-breed dog cannot be shown in breed shows, nor can pups be sold as purebreds. Of course, with purebred dogs, keeping records and making application for breed registration requires effort.

16–16. Mixed breeds often have desirable traits. This puppy was selected for farm life because of the breeds of its parents. (Courtesy, Jody Pollock, Michigan)

- Space requirements—Dogs need good environments. Some are kept inside; others roam outside. The size and nature of the dog may determine whether it is kept inside. In most cases, a large dog in a home is much more difficult to manage. A hunting dog is best kept outside. In cities and restricted residential areas, most dogs must be confined. Non-dog owners do not like the dogs of other people on their property or threatening them. Decide whether the dog will live inside or outside before selecting it and bringing it home. Most breeds can live healthy lives outside if they have shelter. A dog should not be kept inside for a while and then be put outside. Temperature changes make it difficult for the dog to adjust. Also, a dog that is allowed outside part of the time will bring dirt into a home. Inside dogs may require more training than outside dogs. This is because dogs have to learn what they can and cannot do in the house. Owners must be involved in the training. Outside dogs need proper housing and may need to be in a pen appropriate to their needs.

- Hair length—Hair length is important in selection. Disease, parasites, and dirt are associated with longer hair. Dogs with short hair are not as likely to develop diseases of the skin and bring dirt inside. Long-haired breeds of dogs are more likely to collect plant seeds, feces, mud, and other dirt. The owner must note the extra time required to properly care for and groom a long-haired dog. A short-haired dog does not need to be brushed as often, and less time is required when the dog is brushed.
- Gender—The sex of a dog should be considered in the selection process. Most dog breeds reach sexual maturity between 9 and 10 months of age. There are positives and negatives with both males and females. If the dog is a female, steps are needed to keep it from getting pregnant. A male dog may pick up the scent of a female that is outside and breed her. To avoid breeding, the female can be spayed. A male dog will try to break out of a pen or off a leash if a female in heat is nearby. In some cases, a male may run away for several days looking for females. Male dogs also fight, so the dog may come home injured. Male dogs may be castrated to avoid breeding. Neutered or castrated dogs cannot be shown in breed shows. However, they can be shown in obedience classes.

CONNECTION

DOGS SAY "NO" TO CHOCOLATE

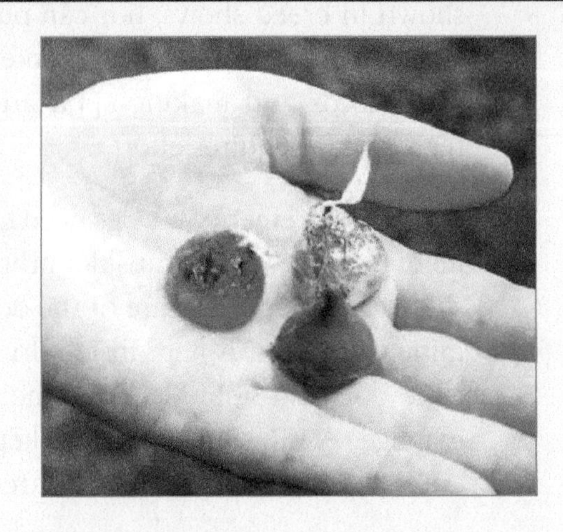

Dogs like chocolate. They welcome small pieces shared by admiring human chocolate-lovers. But chocolate is dangerous to dogs!

Never allow a dog to eat chocolate. The animal may suffer chocolate poisoning. A small dog can die within a few hours from eating chocolate. Larger dogs may have vomiting, diarrhea, and increased heart rate. These symptoms may be followed by tremors, seizure, and a coma.

Prevention: Don't feed chocolate. Treatment: Induce vomiting or administer activated charcoal to prevent absorption of the toxin by the body. (Courtesy, Education Images)

REPRODUCTION

Dogs should breed and raise puppies only if there is a demand for the puppies. Many dogs are put to sleep (humanely killed or euthanized) each year at shelters because they are not wanted.

BREEDING

Female dogs come into their first heat at 6 to 12 months of age in small breeds and 8 to 18 months of age in larger breeds—they mature slower. The owner can tell when a bitch is in heat because blood is discharged from the vulva. After the discharge, the bitch should be exposed to a male in 10 or 11 days. If puppies are not wanted, the female should be confined, because male dogs in the area will breed her if she is outside.

When selecting a male to breed the female to, choose one whose weak points complement those of the female. Although the bloodlines of registered dogs are on their papers, about 50 percent of the genetic make-up of a puppy comes from one parent.

WHELPING

The bitch will give birth, or **whelp**, 63 to 67 days after she is bred by the male. Most bitches will take care of their puppies with little or no assistance. A **whelping box** should be used. This is a box specially designed for the birthing process. It can be constructed about two or three weeks before whelping so the female can get used to it.

The box should be large enough for the female to lie down and stretch out comfortably. Bedding from the female's bed should be put into the whelping box, along with newspapers, shredded paper, or carpet for her to use at whelping time. The box should be placed in a warm, quiet, secluded place where there is very little if any traffic.

16–17. A five-day-old bulldog puppy with eyelids sealed. (Courtesy, Education Images)

CARING FOR PUPPIES

Puppies should be handled as little as possible for the first 14 to 21 days. For the first few days, they need a temperature of about 85°F (29°C).

Newborn puppies can neither see nor hear very well for 10 to 15 days following birth. Their ears and eyelids are sealed. Puppies begin walking at about two weeks of age when their ears and eyes open.

Newborn puppies cannot urinate or defecate except when their mother licks under their tails. Puppies begin to wag their tails and bark at about three weeks of age. When they are so young, they need warmth, their mother's milk, and plenty of sleep.

NUTRITION AND FEEDING

16–18. Use a container suitable for the size of the dog and designed so that it is hard to turn over. (Courtesy, Education Images)

Dogs and puppies need food that meets their nutritional needs. Puppies need foods higher in protein than adult dogs. Food intake is also regulated by a dog's activity. Working dogs need more energy than those that get little exercise.

FEEDING AN ORPHANED PUPPY

An ***orphaned puppy*** is a puppy whose mother has died or is not able to nurse. Orphaned puppies will quickly starve unless fed. Puppies can be fed a milk formula. Bitch milk is richer than cow or goat milk. Therefore, cow milk does not provide sufficient nutrients. A suggested formula is 1 cup of homogenized milk, 3 egg yolks, 1 tablespoon of corn oil, and 1 dropper of liquid vitamins (pediatric). The ingredients should be thoroughly mixed. Store the formula in the refrigerator. Warm the formula until a drop on your wrist cannot be felt. Overheated formula should never be given to a puppy.

Orphaned puppies can be fed with an eye dropper, a bottle and nipple, or a feeding tube. An eye dropper is used with the smallest and youngest puppies. To feed from a bottle, open the puppy's mouth, insert the nipple, and hold the bottle at a 45-degree angle. This will keep air out of the stomach. Using a feeding tube involves attaching a small plastic tube about the size of a small straw to a hypodermic syringe. The formula is released from the syringe.

A puppy's stomach is quite small and may hold only a small amount. It is usual practice to feed a puppy four times a day. The rate of formula is approximately 60 ml per pound of weight. As an example, an 8-ounce puppy would need 30 ml per day. Divided into four feedings, the amount per feeding is 7 to 8 ml. At a couple of weeks of age, the number of feedings can be reduced to three per day. Overfeeding can cause formula to get into the lungs, resulting in pneumonia.

WEANING

Weaning is withdrawing the need or opportunity for a puppy to nurse. Puppies can be weaned at the age of three to six weeks, depending on the species. A puppy should be eating solid food. Its digestive system should be developed sufficiently to handle the food. Special puppy foods should be provided.

A bitch will nurse puppies for several weeks, and sometimes longer than necessary. Most puppies will begin to lap food from a bowl at about three weeks of age. The puppies should be introduced to foods suitable to them. A commercial food for young puppies is likely appropriate. Puppies should be eating well at the time of weaning.

In some cases, the puppies may need to be kept in a separate place from their mother. This will encourage weaning. Puppies are often sold at weaning age.

16–19. Example of dry dog food. (Courtesy, Education Images)

FEEDING A NEW DOG

Dogs and puppies are fed differently. For a new owner, the key to feeding a puppy is to get a small amount of food from the person you are getting the puppy from. The same feeding schedule and the same food should be used the first day your puppy is in its new home, because it has enough other new things to get used to. Starting the second day, gradually switch your puppy over to the food you will be feeding it. After about four days, the puppy should be completely switched over.

Feed the puppy three times a day. The amount of food should not exceed what a puppy can consume in 15 minutes. A high-quality, balanced puppy ration is the best food for puppies, and there are many commercial types on the market. If the puppy is eating dry food, moisten

16–20. Before buying, read the label on dog food and choose the best product for your animal. (Courtesy, Education Images)

it with water so it is easier for the animal to eat. It is important to get the puppy on dry food now, so it will be content eating dry food the rest of its life.

FEEDING THE GROWING AND MATURE DOG

Dry food for dogs and puppies is much more economical than other foods. Dogs get more nutrition from dry food and less water. Proper feeding of a dog or puppy is a major factor in keeping the animal healthy from puppyhood into adulthood. A 2-pound dog will eat 1.3 ounces of dry dog food each day. A 10-pound dog will eat 4.3 ounces of dry dog food each day. A 75-pound dog will eat 20.4 ounces of dry dog food each day.

Commercial dog food has been tested and is a complete ration. The bag of food will indicate how much the dog should be fed; it should be what the dog will consume in 15 to 20 minutes. The dog should be fed once a day. The exception is a very large dog or a hard-working dog; then, twice a day is acceptable. The dog should be fed the same food on the same schedule each day. Table scraps are not recommended.

When buying dog food, read the label. Compare and contrast nutrient content. Determine the age of dog the food was manufactured to feed. Puppies need high protein content. Also, compare and contrast dry food with canned food. Dry food is often preferred because it is said to promote tooth and gum health. Commercial foods are often of three types based on nutrients and age of the dog to be fed: puppy, adult, and senior. Choose food based on the dog's stage of a life. Failing to feed a dog properly may lead to health problems.

HEALTH CARE

Dog health care begins with good nutrition, sanitation, and the environment in which they are kept. Dirty, cluttered pens and houses harbor disease. Failing to feed dogs properly results in inadequate nutrition.

PARASITES

Dogs and puppies are subject to a number of internal and external parasites. It is best to use practices that prevent parasites. Puppies may acquire parasites from their mothers or other dogs they are around. Keeping pens and other facilities clean helps reduce the chance of parasites spreading from one dog or puppy to another. Removing feces from pens helps prevent the spread of certain internal parasites.

Internal Parasites

The major internal parasites of dogs are heartworms, hookworms, roundworms (ascarids), tapeworms, and whipworms. Various treatments are available. Some are effective for only one parasite; others supposedly help prevent several parasites. The assistance of a veterinarian is desirable.

Heartworms infect the hearts of dogs, causing shortness of breath, coughing, lack of stamina, easy tiring, and death. Heartworms are spread by mosquito bites. Treating a dog can also cause death. If too many heartworms are killed at one time, the dead worms can block blood vessels.

Hookworms, roundworms, tapeworms, and whipworms are found in the digestive systems of dogs. These worms get into the body in various ways. Hookworms may enter by going through the skin and traveling through the bloodstream to digestive system. Fleas can transport tapeworms. Contact with feces from infected dogs is a major source of parasites. Keeping dog pens and other areas clean helps prevent the spread of internal parasites. Puppies can be born with roundworms!

Fecal exams are often used to determine the presence of internal parasites. Worms, eggs, or other evidence may be in the feces. Dewormers are used to control worms. The kind of dewormer depends on the kind of worm parasite. Use the assistance of a veterinarian or other qualified dog health person in choosing and administering dewormers.

16–21. Medications are available for treating external parasites of dogs, including ringworm, fleas, mange (mites), and ticks. (Courtesy, Education Images)

External Parasites

The major external parasites of dogs are fleas, lice, ticks, mange (mites), and ringworm. Some of these can also infest humans. Signs of external parasites include scratching, loss of hair, discoloration of skin, and sores. Some external parasites are large enough to be easily seen, such as ticks.

Sanitation is an important part of external parasite control. Dips and baths are often used to treat external parasites and skin diseases. Pens, beds, and other facilities where dogs are kept may need to be treated to eliminate

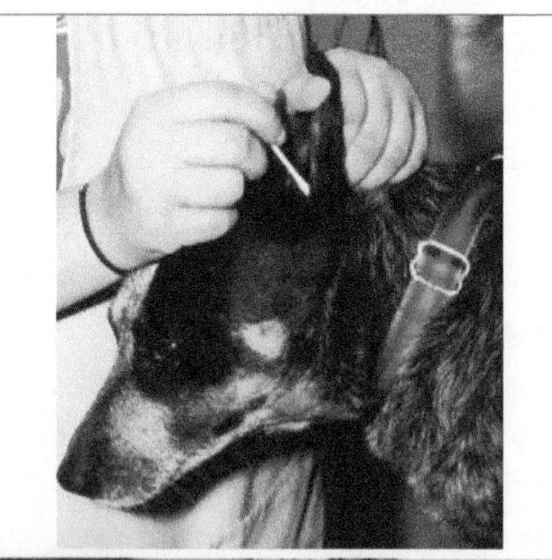
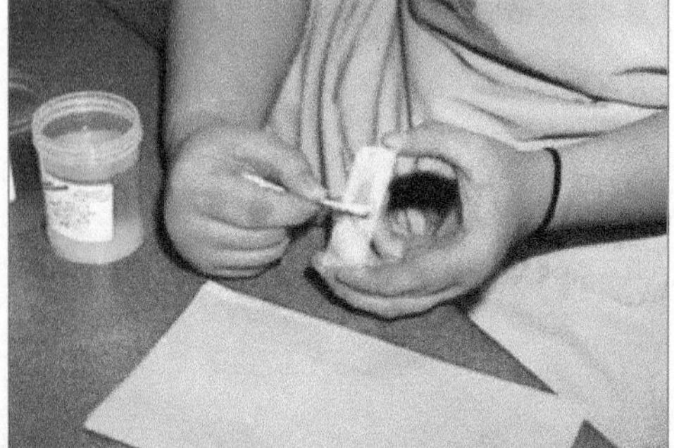
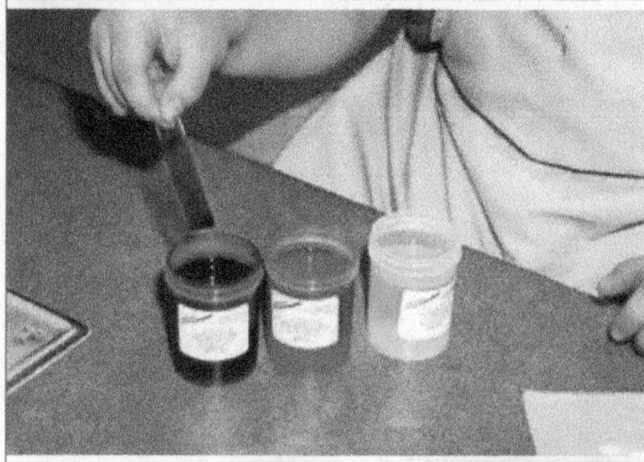
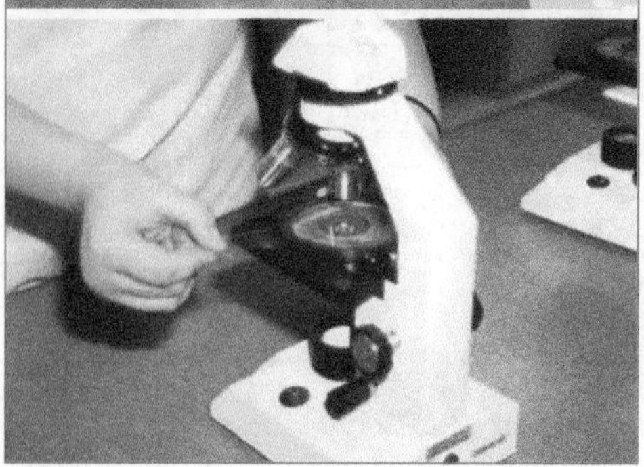

16–22. How to check a dog's ear for mites: Use a cotton swab to collect a specimen of ear wax/material, rub the swab on a microscope slide, dip the slide in fixative and stain solutions, and view the slide with a microscope. (Courtesy, Education Images)

parasites. Special collars can be put on some dogs to help control external parasites.

INFECTIOUS DISEASES

Dogs are subject to a number of infectious diseases. These can be caused by bacteria, viruses, fungi, or protozoa.

Examples of bacterial diseases are brucellosis, leptospirosis, tetanus (lockjaw), tuberculosis, salmonellosis, and campylobacteriosis. Lyme disease, transmitted to dogs by ticks, is also a bacterial disease.

Viral diseases include distemper, herpes, infectious canine hepatitis, rabies, canine parvovirus, and canine coronavirus. Some of these can be effectively controlled with vaccinations.

Fungal diseases include histoplasmosis, coccidioidomycosis, and blastomycosis. These diseases may be difficult to diagnose. Lab tests, X-rays, biopsies, and other means of gaining fungal cultures may be needed.

Protozoan diseases include coccidiosis, toxoplasmosis, trichomoniasis, giardiasis, and Nantucket disease. Some of these cause diarrhea. Medications to stop diarrhea should be used. Other drugs may also be needed to control the diseases.

Other infectious diseases include Rocky Mountain spotted fever and canine ehrlichiosis. Ticks transfer Rocky Mountain spotted fever. Controlling ticks helps prevent the disease. Veterinarians will help identify and treat these diseases.

OTHER HEALTH CONCERNS

Dogs are subject to a wide range of diseases and other health problems. Some have allergies that cause considerable difficulty. Mechanical blockages in the digestive system can result from ingesting foreign bodies, hairballs, sticks of wood, and other materials. Dogs may have system failures, such as kidney failure or heart disease. Digestive system problems can result in vomiting, diarrhea, loss of appetite, and emaciation.

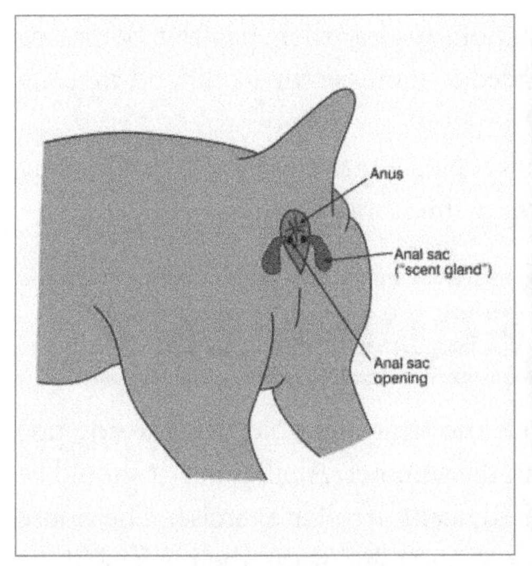

16–23. Location of anal sacs (scent glands).

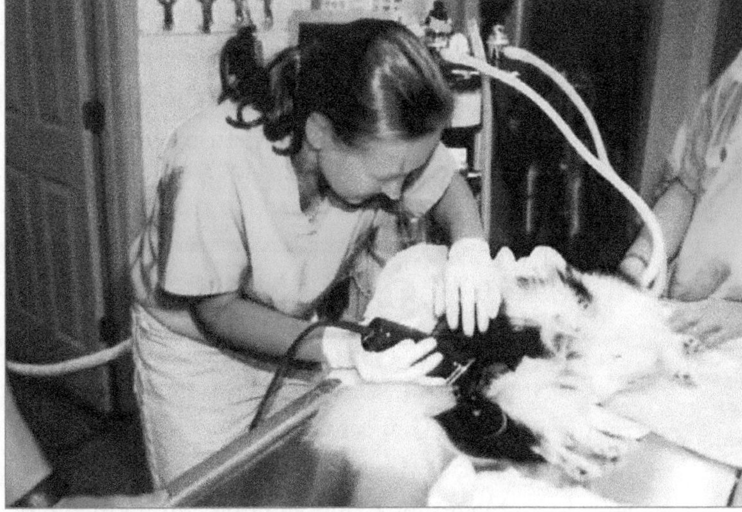

16–24. Using clippers on a long-haired dog to gain access to the anal sacs for expressing. (Courtesy, Education Images)

An annoying health problem is anal sac infections and odor. Anal sacs (scent glands) are on each side of a dog's rectum. These sacs are typically emptied during a bowel movement. Sometimes the sacs become infected and produce a foul odor. Dogs that drag or scoot their hind ends along grass or carpet are experiencing anal sac problems. Experienced individuals know how to "express" (squeeze out) these glands and remove the foul odor and irritation. Sometimes treatment by a veterinarian is needed. Expressing these sacs may be a routine part of bathing a dog. This helps assure a clean odor after the bath.

VACCINATIONS

Good health care includes a vaccination program. Vaccinations help a puppy develop immunity to certain diseases. An adult dog may be vaccinated to assure continuing immunity.

Vaccinations are given to prevent the following diseases: distemper, infectious hepatitis, leptospirosis, coronavirus, rabies, and canine parainfluenza (kennel cough). Good records of vaccinations are needed. A veterinarian will provide information on a vaccination program and provide the vaccinations. A possible vaccination schedule for a puppy is presented in Table 16–1.

Table 16–1. Possible Vaccination Schedule for Puppies and Dogs

Age	Vaccination
6 to 8 weeks	distemper-measles-parainfluenza (CPI), parvovirus
8 to 12 weeks	DHLPP (distemper, hepatitis, leptospirosis, parainfluenza, parvovirus)
12 weeks	rabies
16 weeks	DHLPP
12 months	rabies
Boosters and other vaccinations may be needed.	

Adapted from: Delbert G. Carlson, and James M. Giffin. *Dog Owner's Home Veterinary Handbook*. New York: Howell Book House.

Dogs and puppies should routinely receive vaccinations to keep them healthy. Some dog keepers will re-vaccinate a bitch after she has been bred to increase the likelihood that the puppies will have antibodies when they are born.

Puppies often acquire disease from their mothers. Keeping females in good condition will help prevent puppy disease problems, including parasite infestations.

FACILITIES AND EQUIPMENT

Dogs and puppies kept outside need shelter from the heat and cold. An outside dog requires a house that is warm in the winter and cool in the summer. At all times, it should be draft free, dry, and easy to clean. It should have an adjacent area for exercise. The house should be big enough that the dog can lie in the back at night and warm the house with its body heat and lie in the front during the day with its head out to protect the area. Some owners even put a partition between the two areas.

The biggest problem is that a house is often too large and the animal's body heat cannot warm the area. A dog the size of a beagle needs about 2 × 2 feet (0.7 × 0.7 m) of floor space and a roof height of no more than 18 inches. A medium-sized dog of about 40 to 75 pounds (18 to 34 kg) requires floor space of 4 × 4 feet (1.4 × 1.4 m).

16–25. Brushing is important in caring for the well-being of a dog. (Courtesy, Education Images)

Fences are needed to keep dogs that are outside from straying. Wire, wood, and other materials may be used to build a "dog-proof" fence. Invisible sonic and radio containment systems can also be used. With such systems, the dog has a collar with a receiver that emits an audible signal followed by a mild electrical shock. With training, the dog soon learns the boundaries of its yard.

An owner should also provide a collar, warm bedding, water and feed dishes, and a brush for the animal. An identification tag should be attached to the collar. This is useful if the dog runs away or is lost. The tag should have the owner's name, address, and telephone number.

TRAINING AND GROOMING

Training and grooming help make dogs manageable and appealing.

TRAINING

Training is the process of teaching a dog desired habits or qualities. With a puppy, it may be not to urinate or defecate on the floor. This is known as housebreaking.

Training includes obedience to verbal commands, such as "Heel," "Come," and "Stay." Training often involves a collar with a leash. Walking the dog around and giving commands associated with movements of the leash are basic in training.

Some puppies attend obedience school. This is an organized program to train young dogs and their owners in a relationship of trust and following commands. Kindness and reward are fundamental in all training efforts. It is best to learn from an experienced trainer.

GROOMING

Grooming is washing, combing, brushing, trimming, and otherwise caring for the external appearance of a dog. Grooming may involve cleaning the hair, removing trash and parasites, cutting the nails and hair, and treating minor wounds. Show animals must be groomed according to the breed and exhibit rules.

16–26. A dog being trained to obey a radio containment system. (The small white flags mark the boundaries of the yard for the dog. The flags are beneficial to the trainer. Once the dog is trained, the flags are no longer needed.) (Courtesy, Education Images)

16–27. Dogs used in hunting receive special training.

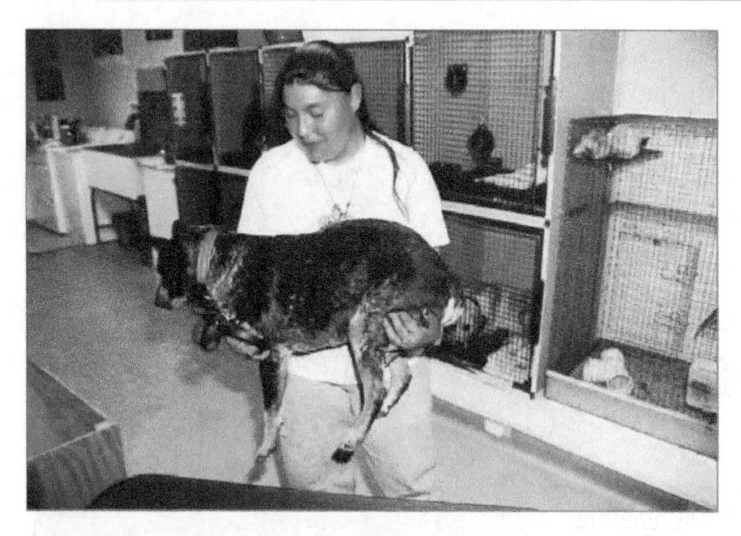

16–28. How to lift and carry a large dog without injuring the dog or yourself. (Courtesy, Education Images)

Grooming can highlight the strong features of an appealing animal. It is best to learn grooming from an experienced dog groomer.

Hair Coat

The hair coat of a dog should be brushed regularly. Bathing is not a routine part of maintaining the hair coat of a dog. Bathing is done when the dog is especially dirty. Of course, a dog that has been running loose outside may need a bath before it can be brought inside. Too much bathing will remove oils from the coat and can cause skin irritation.

Long-haired dogs will require special care to remove plant burrs, excess hair, and hair mats (masses of hair usually found under the legs and behind the ears). Hair mats can often be removed by teasing them out but may occasionally need to be cut out. Care should be taken not to cut the skin when removing hair mats.

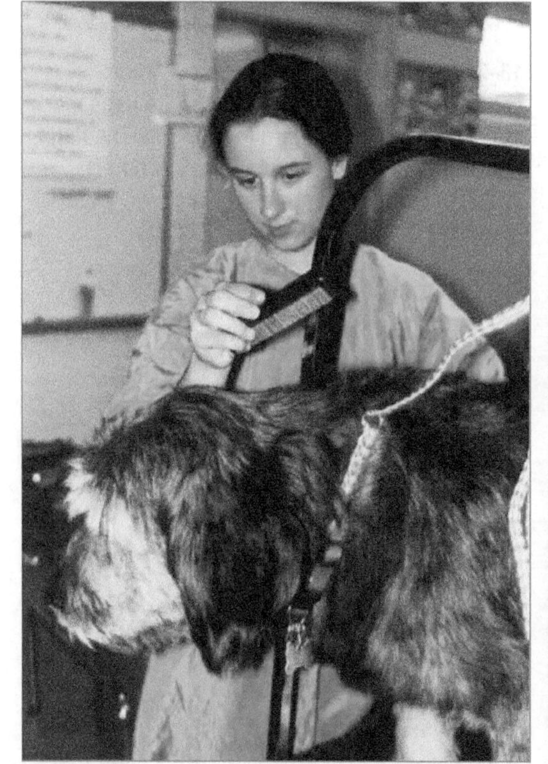

16–29. Using a dressing comb to remove dirt from a long-haired dog. (Courtesy, Education Images)

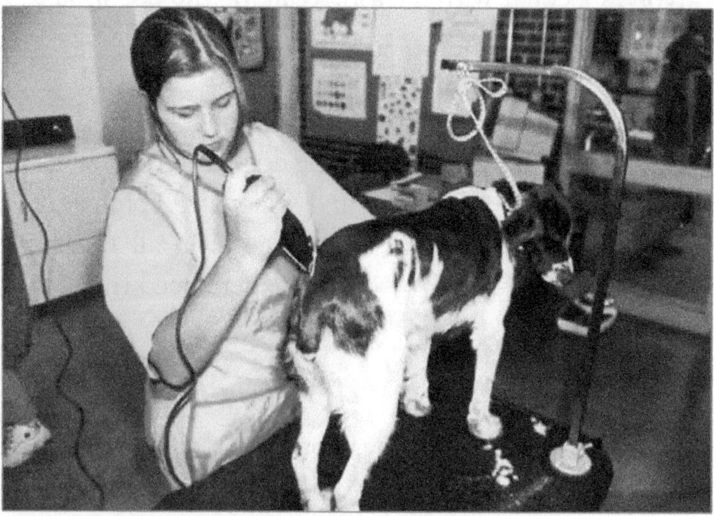

16–30. Using clippers to trim and shape the hair of a dog. (Courtesy, Education Images)

Nails

The need for nail trimming depends on the surface the dog is kept on. Inside dogs will need trimming more often than outside dogs. Be careful to trim each nail above the ***nail bed*** (the growing point of the nail). The nail bed is easy to spot on animals with light-colored nails; it is the darker area close to the paw. Use sharp clippers that will cut the nail, not crush it. A crushed nail can be painful and can cause bleeding.

Ears

A dog's ears should be cleaned monthly with a cotton swab and mineral oil or alcohol. Clean only the surface you can see, and never put sharp objects in the dog's ear. Always check for parasites, such as ticks, around the ears. Check your dog's ears for mites. Dark, grainy deposits inside the ear are an indication of mites.

Eyes

Dogs may get irritants in their eyes. If this happens to your dog, flush the affected eye with an eyewash, and watch the eye to see that it heals. Serious injuries and irritations that do not heal on their own should be called to the attention of your veterinarian.

Teeth

The dog's teeth are generally resistant to decay. You may want to clean them regularly to prevent plaque and tartar. Use a small toothbrush with toothpaste, salt water, or a salt water–baking soda solution. Kits are available for cleaning the teeth of dogs. Clean the teeth from the gum line toward the tips. Providing crunchy food can help maintain healthy teeth. "Kibbled" food and dog biscuits that are hard help clean the teeth.

Note: Always be careful in grooming and performing other care of a dog. The dog may be surprised and become aggressive. It may growl, bite, or claw.

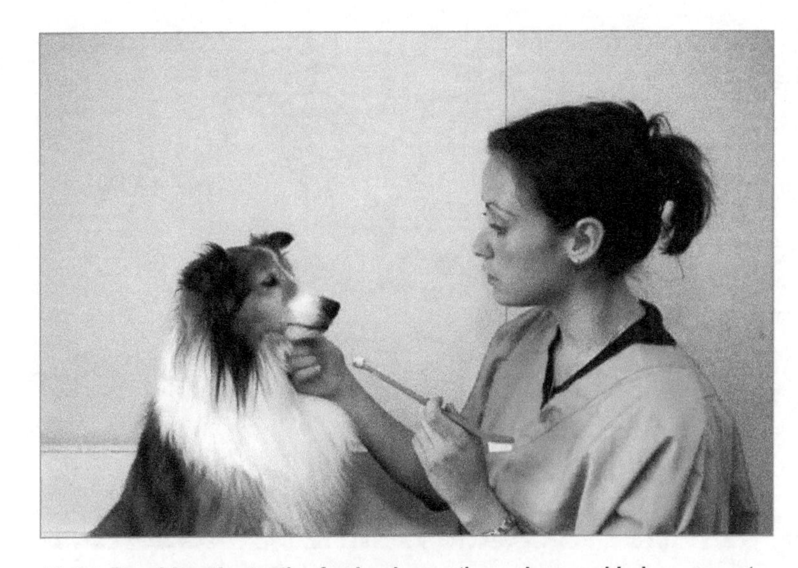

16–31. Brushing the teeth of a dog keeps them clean and helps prevent decay. (Courtesy, Education Images)

SANITATION AND WELL-BEING PRACTICES

All animals deserve a clean and comfortable environment in which to live. A dog is no exception.

Animals that are in captivity (kept in cages or pens or within a fenced area) have special needs when it comes to sanitation and cleaning. Animals in the wild distribute their wastes away from areas in which they choose to live. Since much space cannot normally be provided, steps must be taken to ensure sanitary and clean conditions.

- Remove body wastes—Body wastes (feces and urine) must be removed from the animal's quarters daily. This practice will keep the animal clean, keep the cage fresh, and prevent diseases and parasites.
- Provide clean bedding—Bedding materials should be changed on a regular schedule. Bedding provides warmth and comfort for the animal. Dirty bedding can promote parasites, make the animal dirty, and cause illnesses, like respiratory infections.
- Clean the living area—Clean animal pens and cages regularly with a good soap and disinfectant. The animal should be removed beforehand and placed in a temporary holding area. Remove all old bedding and scrub the cage/pen with a brush. Remove all material and stains from the animal's living area.

16–32. A clean kennel cage is important to a young puppy. (Courtesy, Education Images)

Look after the dog's well-being. Sometimes, people fail to realize that what they do may not be in the best interest of an animal. Here are a few examples to overcome potential injury to a dog:

- Never leave a dog in a hot car in the summer.
- Always have water available for a dog.

- Provide secure facilities that offer protection from a harsh environment.
- Never tie a dog in the hot sun, cold, or rain.
- Provide appropriate food on a regular basis.
- Neuter animals to prevent unwanted baby animals.
- Provide proper care for newborn animals.
- Have routine vaccinations, wormings, and other health care provided by a qualified individual.
- Groom, trim, and otherwise help an animal maintain an attractive appearance.
- Follow all local regulations with dogs and other companion animals. This may include paying tax assessments.

16–33. Proper housing is needed. This plastic doghouse is for a large dog. (Courtesy, Education Images)

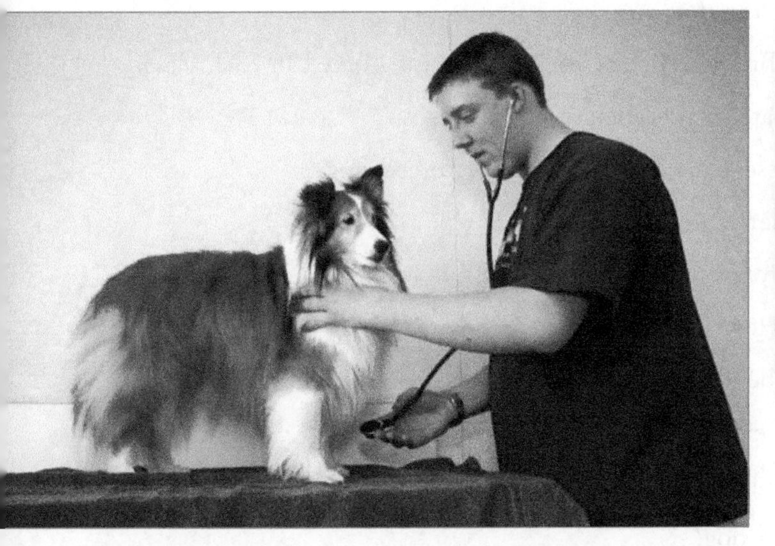

6–35. A stethoscope may be used to listen to the heartbeat of a dog–one f its vital signs. (Courtesy, Education Images)

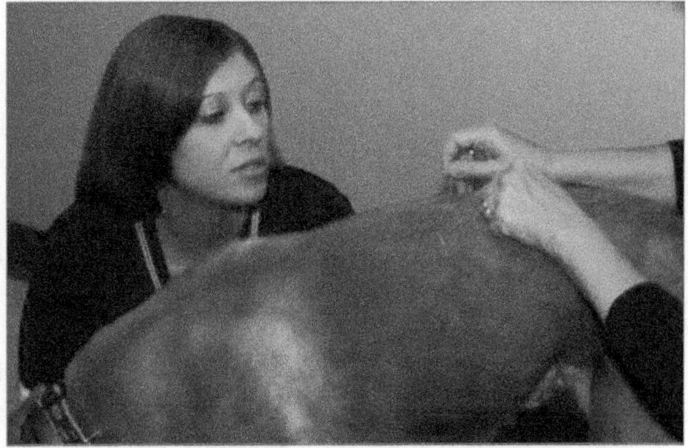

16–34. Acupuncture is being administered to a dog as a treatment for lameness and hip dysplasia. (Acupuncture involves the placement of needles in locations that will promote healing. Sometimes an electrical charge is applied to the needles.) (Courtesy, Education Images)

REVIEWING

MAIN IDEAS

Dogs are popular. They are versatile animals and may be used for many purposes—companionship, security, hunting, retrieving, herding, tracking, and guiding. The selection of a companion dog should be based on good reasoning. Getting a dog is a long-term responsibility. A dog needs feed, water, and exercise daily. Its living quarters must be cleaned on a daily basis, and the animal's health must be routinely checked. All companion animals, including dogs, require a commitment of time and money. A pet could be a part of the family for 10 years or more, so make the decision a family decision.

If you want an animal to do certain tasks beyond that of being a companion, study the types of animals that fall into the proper category. Talk to people who own these breeds of dogs and find out the animals' strengths and weaknesses.

Dogs should be groomed regularly. Combing and brushing the hair of long-haired dogs is very important. Many animals are self-groomers and can indicate when they are not feeling well by their ungroomed appearance.

QUESTIONS

Answer the following questions using correct spelling and complete sentences.

1. What is the scientific name of the dog?
2. What are the classes of dogs? Briefly describe each class. List a breed in each class.
3. How does the age of a dog related to its dietary needs?
4. What is a mixed-breed dog? Assess the merits of a mixed-breed dog.
5. What factors should be considered in selecting a dog?
6. What is the general reproductive process of dogs?
7. What is an orphaned puppy? How is an orphaned puppy fed?
8. What is weaning? What should be the stage of development when puppies are weaned?
9. What health problems may be experienced by dogs?
10. What vaccination program is followed with puppies and dogs?
11. What is involved in grooming a dog?
12. What are sanitation and cleaning? Why are they important?
13. What is involved in training a dog?
14. What should be considered in looking after the well-being of a dog?
15. Why is dry dog food usually preferred over canned dog food?

EVALUATING

Match the term with the correct definition. Write the letter by the term in the blank provided.

a. training
b. whelp
c. hound
d. toy breeds
e. nail bed
f. bitch
g. stud dog
h. herding dogs
i. sporting dogs
j. tracking dogs

______1. A class of dog based on the use of the breeds for tracking.
______2. A class of dog known for herding sheep and cattle.
______3. A class of small dogs weighing between 4 and 16 pounds.
______4. To give birth to puppies.
______5. A female dog.
______6. The growing point of a nail in a dog's foot.
______7. A male dog used for breeding.
______8. Getting a dog to have the desired traits or habits.
______9. A group of dogs used for sporting purposes.
_____10. A type of dog used for following scents.

EXPLORING

1. Visit a nearby store that sells dogs. Make a list of the dog breeds in the store. For each breed, give your opinion of the advantages and disadvantages of the breed.

2. Volunteer to dog-sit or care for someone else's dog. (You can establish a profit-making enterprise by looking after pets while their owners are away.) Be sure to visit with the owner about the needs of the dog and how it is managed. Develop a schedule of activities for the time you are responsible for the animal.

3. Tour a veterinary medical facility that treats small animals. Interview the veterinarian about the work. Determine the kinds of animals treated and the common veterinary medical procedures performed. Study how animals are restrained and moved. Draw a floor plan of the facility, including the animal boarding area. Note how the pens or cages are constructed and maintained.

4. Observe labels on different dog foods sold in a store. Compare and contrast nutrient sources and content as related to the stage of life of the dog to be fed. Identify the food you think best for a puppy, adult dog, and senior dog. Compare the daily feed cost of several different foods. Discuss your findings in class.

5. Devise a health maintenance program for a puppy to prevent and control infectious and parasitic disease. Extend the program for an adult and senior dog. Include pregnant and lactating females. Compare and contrast approaches to the health of dogs. Give an oral report in class.

CHAPTER 17

Cats

OBJECTIVES

This chapter provides basic information on cats as companion animals. It has the following objectives:

1. **Describe the biology of a cat.**
2. **Name and explain the classes of cats.**
3. **Assess important factors in selecting a cat.**
4. **Describe management practices with cats.**
5. **Identify training and grooming practices with cats.**
6. **Explain the nutrition and feeding of cats.**
7. **Describe cat reproduction.**
8. **Describe important health practices with cats.**
9. **Identify equipment and facility needs of cats.**

TERMS

declaw
ELISA
hairball
immunoassay
kitten
kittening
litter
litter pan
long-haired cat
maternity box
pads
pet carrier
queen
scratching post
short-haired cat
tomcat

17–1. The family cat will likely find a way to be a part of many activities. (Courtesy, Dmitriy Shironosov/Shutterstock)

CUTE, playful kittens and cats quickly grab our attention. People enjoy holding them and listening to their purrs. A tiny straw or string can get a kitten moving, slapping, and jumping. In short, cats make loving and wonderful pets once they are accustomed to their homes. Of course, cat owners must enjoy and appreciate them.

Today in the United States, the number of cats is right at 85.8 million. This population figure means that there are nearly 10 million more cats than dogs. Cat populations have been increasing since the year 2000. Keeping a cat varies somewhat from a dog. Part of the difference relates to their personalities and needs as they are kept. Cats are smaller than dogs making them better suited to apartments and small homes. Cats use litter boxes whereas dogs must be let or walked outside a few times each day.

Cats are sometimes referred to as "felines." This name is derived from their scientific name, *Felis domesticus*. These felines are small carnivorous animals that are also commonly known as the domestic cat or house cat.

CAT BIOLOGY

17–2. Cats have a wide range of physical appearances.

The life processes and biology of house cats are similar to those of other animals. Knowing these processes helps in meeting their needs. Most cats are kept as companions, or pets. A few are kept for services they provide in keeping down rodents around barns and warehouses. In this case, the may be referred to as "rat cats."

Cats may live 12 to 18 years, with some living as long as 30 years. A female cat is known as a ***queen***, while a male cat is known as a ***tomcat*** or tom. A baby cat is a ***kitten***. Gestation is 60 to 65 days. Queens often have litters of three or four kittens, though sometimes they have larger litters. The eyes and ears of newborn kittens are sealed; therefore, they cannot see or hear. Both eyes and ears are open at about two weeks of age. The kittens may nurse their mother for about six weeks.

Though cats and dogs are known for not getting along, they often do and live together in the same house. Biologically, they both descended from the same ancient mammal, Miamis. Cats are carnivores and will often catch birds, mice, and other small animals as prey. Many breeds of cats with unique features are used as companion animals. Going to a local cat show is a good way to see a number of breeds.

17–3. A Sphynx being judged at a cat show. (The Sphynx, or hairless cat, is a rare breed that originated in Canada. The breed is the result of a mutation from a cat born without hair in 1966. Though it is called "hairless," most of these cats have a covering of fine down on their skin. In 2002, the Sphynx was accepted for competition by the Cat Fanciers' Association.) (Courtesy, Education Images)

SIZE AND SKELETON

Typical domestic cats weigh 6 to 15 pounds (2.7 to 7 kg), though some are larger and some are smaller. Cats with longer hair appear larger, but their weight may be no more than that of cats that appear smaller.

Cats have about 250 bones. The exact number depends on the size of a cat and the length of its tail. Their muscles, bones, and structure are designed for speed and quickness. Cats have about 500 muscles, with the largest muscles in the hind legs.

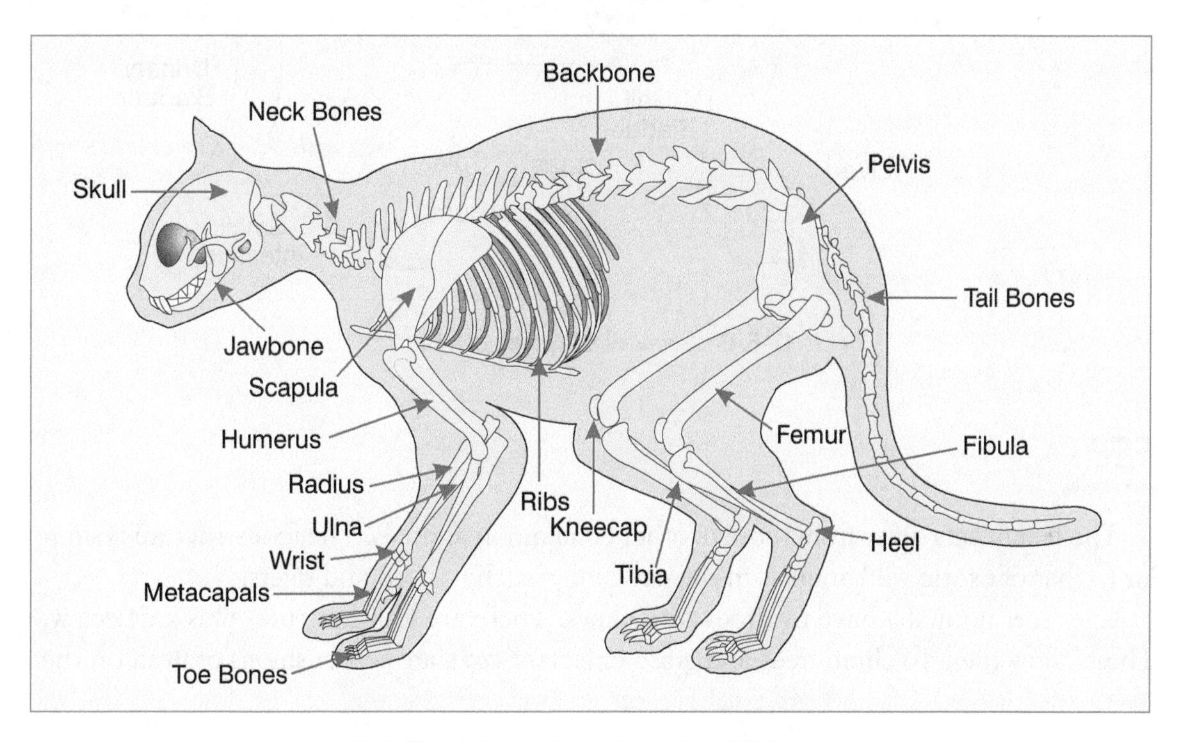

17–4. The skeleton of a cat has about 250 bones.

BODY SYSTEMS

Cats have body systems similar to those of other animals. They are carnivorous animals that require high-protein diets.

The sensory systems of cats are well developed. The nose and paws are especially sensitive. Cats have extra smell and taste organs. These allow them to distinguish between foods.

The ears of cats are quite sensitive; they hear the slightest movements. Cats have three-dimensional vision as well as interesting eye-color combinations. Their whiskers are tactile hairs; they are connected to sensitive nerves that help cats protect themselves and find their way in the darkness.

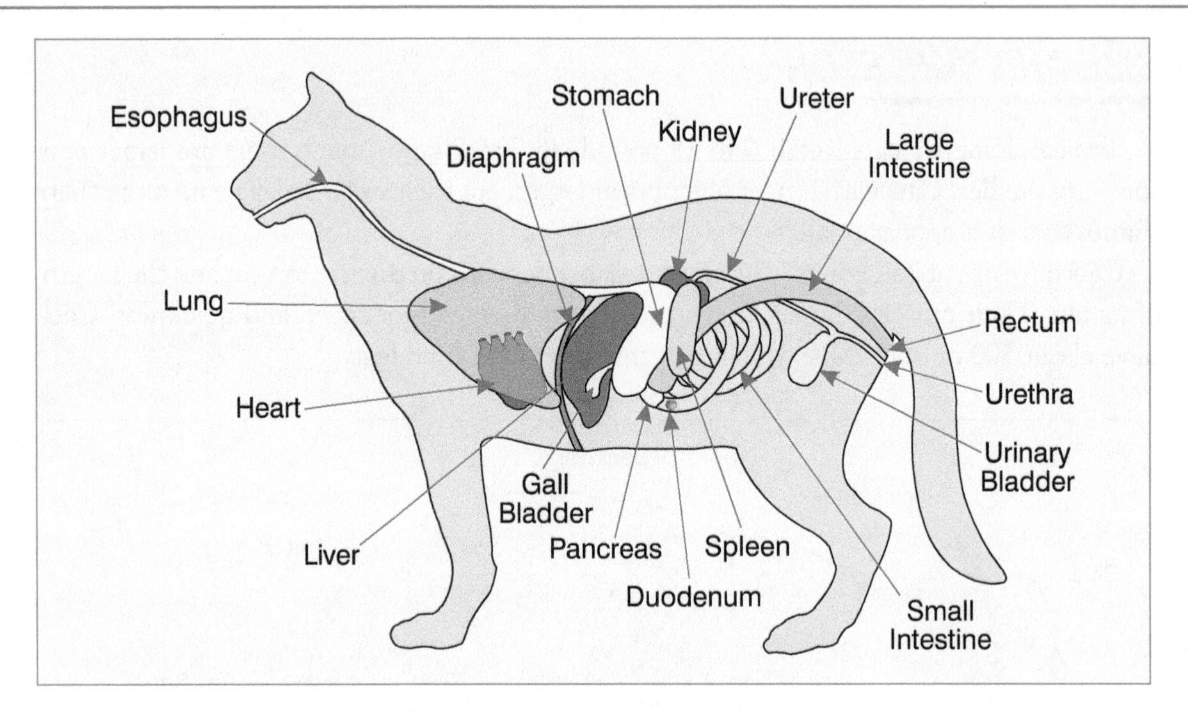

17–5. Gastrointestinal system of a cat.

FEET

The feet of cats differ from those of other companion animals. The foot structure is similar to that of exotic wild animals in the cat family, such as lions and tigers.

Cats' feet normally have five toes with claws. There are four long toes plus a dewclaw. These allow them to climb trees and grasp objects. ***Pads*** are soft cushions of flesh on the

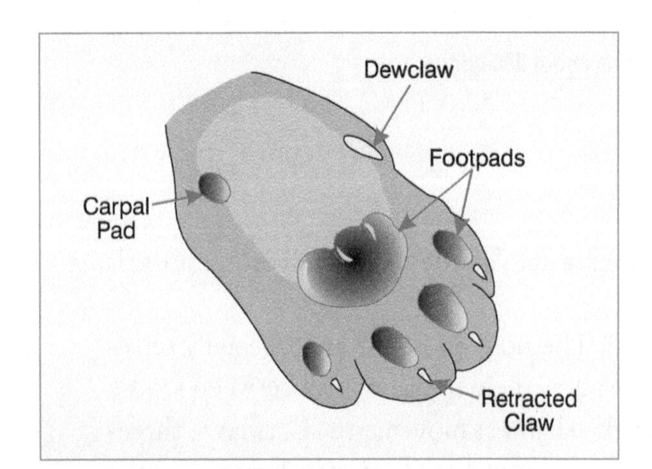

17–6. A bottom view of the foot of a cat shows the spring pads, which allow a cat to walk very quietly. The claws are retracted.

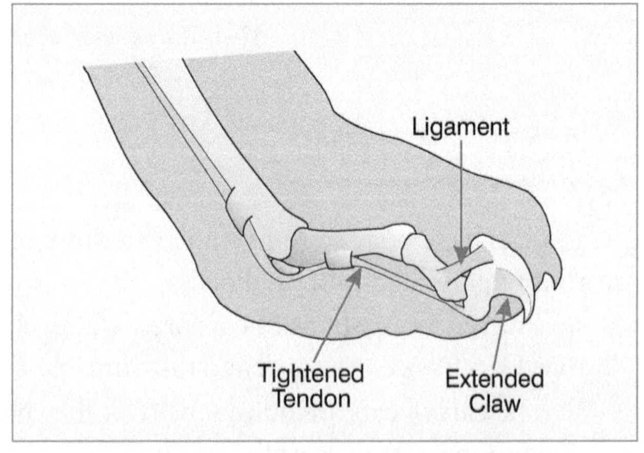

17–7. A side drawing of the foot of a cat shows the claws extended. Declawing is removing the claws at their base.

bottoms of cats' feet that help them approach quietly when stalking prey or move about a house without making noise.

The claws are dangerous to people and can damage furniture. Some cats are declawed. To ***declaw*** a cat is to surgically remove the claws from the cat's feet. The procedure is done by a veterinarian. A time of healing is needed for the cat to recover. A cat that has been declawed should never be turned outside; it is less able to protect itself.

17–8. There is a reason this cat appears to have large feet. (The cat has seven toes! This photo was made at the former home and now museum of famous writer Ernest Hemingway in Key West, Florida. Hemingway kept several of these mutant cats in his home. A large number of the cats continue to roam about today as part of the museum.) (Courtesy, Education Images)

CLASSES AND BREEDS

The type of cat to select is a matter of personal choice. Basically, there are two classes from which to choose: the common house cat and the purebred cat.

The common house cat is not a specific breed, but a mix of various breeds. It is produced by chance and not by planning for a specific breed.

The purebred cat is bred by design. Careful planning and selection of bloodlines are used with purebred cats. There are two categories of purebred cats: short-haired and long-haired. Thirty-six breeds of cats are included in these categories in the United States. The American Cat Fanciers Association and the Cat Fanciers Association are the most recognized cat registries.

- Short-haired breeds—A ***short-haired cat*** is low maintenance—it requires little or no brushing compared with a long-haired cat. Some of the most common short-haired breeds are the Abyssinian, American Shorthair, Burmese, Colorpoint Shorthair, Egyptian Mau, Exotic (Shorthair and Longhair), Havana Brown, Japanese Bobtail, Korat, Manx, Ocicat, Rex, Russian Blue, and Siamese.

- Long-haired breeds—A ***long-haired cat*** requires more attention to look its best, but it is very pretty when groomed. Long-haired cats, however, tend to get hairballs, which they must have removed. A ***hairball*** is a wad of hair that collects in an animal's digestive tract. It can clog the movement of food in digestion and cause serious

17–9. Two examples of short-haired cat breeds are the Abyssinian (left) and the Manx. (Courtesy, Education Images)

17–10. Two examples of long-haired cats are the Persian (left) and Maine Coon. (Courtesy, Education Images)

problems for the cat. There are products available that will prevent hairballs from forming or will cause their elimination. Some of the most popular long-haired breeds are the Persian, Balinese, Birman, Himalayan, Maine Coon, and Turkish Angora.

SELECTING A CAT

Because cats require less care, they make better pets than dogs for some people. Cats are independent and self-sufficient. They are also more economical than dogs and some other companion animals. In selecting a cat, there are several considerations:

- Age—Kittens are usually cute and adorable, but they are also more work than older cats. They require special attention, care, feeding, training, and time to learn. Kittens are more active and playful than grown cats. Kittens adapt to new homes and other animals much quicker and with less stress than adult cats. It usually takes less than three weeks for a kitten to adapt—it does not have a very long memory of living in another place. Kittens change as they mature. The adult may have little resemblance to a tiny kitten.
- Gender—Male and female cats are much alike in temperament as kittens, but major behavioral differences develop at maturity. If the owner wants to raise kittens, then the obvious choice is a female. If the owner does not want kittens, a cat of either sex can be chosen. Having the cat spayed or neutered at a young age should be considered. Female cats are spayed to eliminate the period of heat and unwanted pregnancy. Female cats may have strange behavior during heat. Male cats that are not neutered urinate frequently to mark their territory. This is

17–11. The eyes of a 7-day-old kitten are sealed, as are the eyes of all newborn kittens for about 10 days. (Kittens can be selected and taken from their mothers at four to five weeks of age if they have learned to eat.) (Courtesy, Education Images)

CONNECTION

SAFETY IN RESTRAINING A CAT

A cat may be nervous about being handled. It generally doesn't like new things being done to it. It may hiss, bite, and scratch. A cat's claws can make severe skin wounds on a person's arm or hand. The cat can also run, jump, and try to escape. Sometimes, a cat needs to be restrained, even though it doesn't like it.

This shows one method of restraining a cat. The entire cat is placed in a specially designed zippered feline bag. Its head is out. Using the different arrangements of zippers, one foot at a time can be extended from the bag. The well-being of the cat is considered. (Courtesy, Education Images)

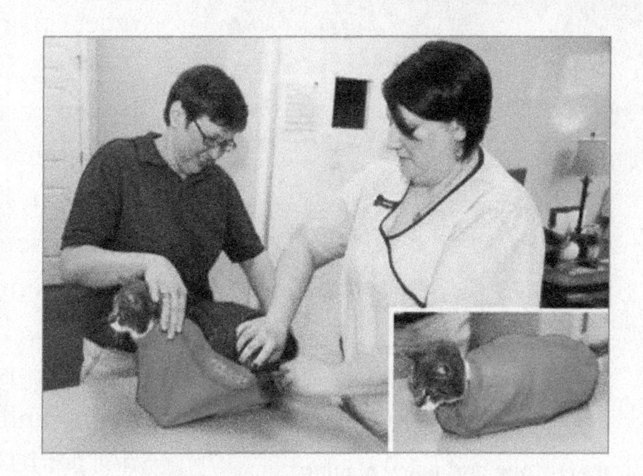

17–12. Cats are interesting animals. (Courtesy, Education Images)

17–13. Cats are curious and enjoy playing. (Courtesy, Education Images)

annoying in a home. Urine stains can ruin carpet and furnishings. When let outside, tomcats may stray from home searching for females to breed or male cats to fight.

- Purebred or mixed breed—Most people enjoy mixed-breed cats. Some prefer purebred or pedigreed cats. If the owner plans to raise kittens or show the cat, he or she may want to consider a pedigreed cat. A pedigree will allow the owner to trace the animal's family history. However, buying a pedigreed cat is more expensive.

MANAGEMENT

Exercise care when handling a cat. To pick up a cat, slip one hand under the cat's chest and hold its front legs gently but firmly with your fingers. Put the other hand under the cat's hindquarters. Treat your cat carefully, and make sure it feels secure. Kittens are sometimes picked up by the nape of the neck—this is how a queen moves her babies!

Owners should remember that every cat is different. Give a cat time to get used to its new home and owner. It is wise to keep the cat confined until it is familiar with its new environment and owner. Do not force attention on the cat. While working with the cat, repeat its name often. This will reenforce the animal's learning. If the cat is frightened, it will fight and try to run away. Patience and time allow the cat to feel loved and comfortable in its new surroundings.

A cat, if kept inside, requires a bed, a carrier, a litter pan, a scratching post, and possibly some toys. The equipment does not have to be expensive, but for your cat to be happy and healthy, the items are needed. If your cat stays outside, it needs protection from the elements and a safe place to eat where other animals do not get its food and water.

TRAINING AND GROOMING

Training and grooming are needed to ensure that a cat fits in well with its owners.

TRAINING

Most training focuses on toilet training. It is easier with cats than dogs. Queens usually help train their kittens. A kitten can be trained with patience (and a few messy accidents). Put the kitten in the litter box. Use your hands to help it scratch with its front paws in the litter. Repeat this every time the kitten appears to be looking for a place to urinate or defecate.

Some furniture and other places are off limits. Providing a comfortable place for scratching and sleeping will help a cat adjust. Rewards for desired behavior work well. Firm voice commands are recognized by a cat when its behavior is undesirable.

17–14. Management includes providing for the needs of a cat, including personal attention. (Courtesy, Education Images)

GROOMING AND CARE

Grooming a cat includes combing and brushing its hair, trimming its nails, and cleaning its ears, eyes, and teeth and gums. Most cats spend hours each day self-grooming.

Hair Coat

With few exceptions, a cat's hair coat should be brushed daily. Use a medium-bristle brush, along with a flea comb, on a long-haired cat and a flea comb on a short-haired breed. A cat may occasionally need to be bathed. Do so carefully in a tub or sink with about 4 inches of warm water. Use a mild soap.

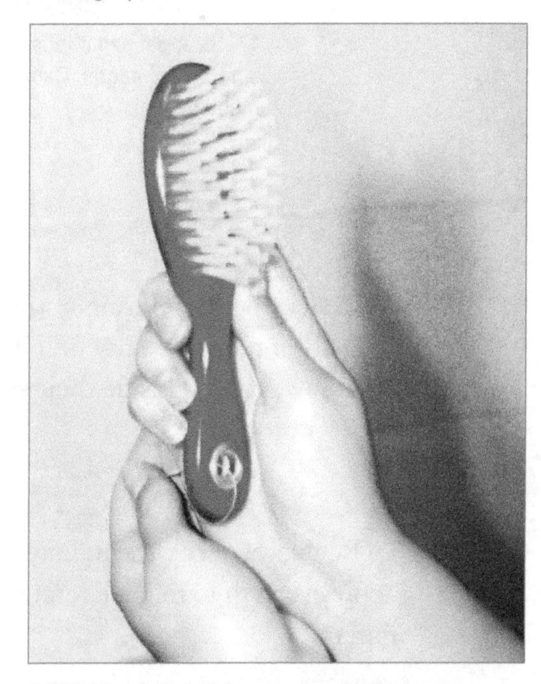

17–15. A small, soft brush should be used in brushing a cat. (Courtesy, Education Images)

Nails

A cat that has a scratching post or spends time outside will probably not need to have its claws trimmed. If claws do require trimming, be careful not to trim below the nail bed. Use sharp clippers or a special cat clipper and file. Do not use scissors.

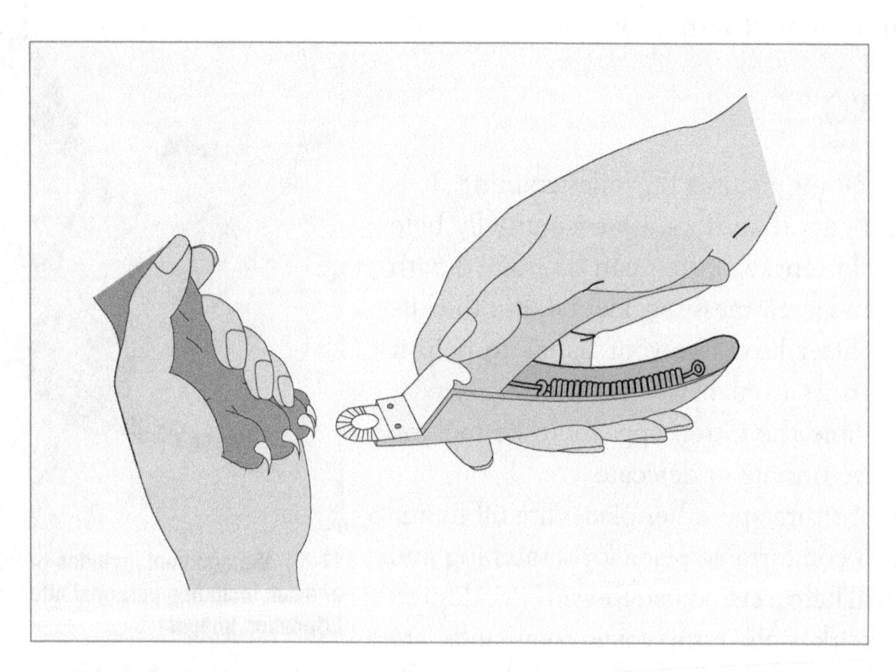

17–16. A pet nail clipper can be used to nip the tips of the nails, and a nail file to smooth the edges. Caution: Make small cuts to avoid cutting into the skin or a blood vessel.

Ears

Check the ears during normal grooming. A discharge or stopped-up ear canal requires help. If ear mites are present, buy ear drops from the pet store and apply as directed. Outside cats may get ticks inside their ears. These should be carefully removed.

Eyes

Regularly check the animal's eyes for any discharge, which indicates infection. You can apply eyewash to remove foreign materials. Serious problems need the attention of a veterinarian.

Teeth and Gums

Teeth and gums should be checked regularly to detect any soreness. Use hard foods occasionally to promote clean teeth. Tartar and plaque should be removed by your veterinarian. In some cases, an owner can use a toothbrush kit designed for cats.

NUTRITION AND FEEDING

The proper commercial food will promote cat health and growth. Most cat foods are manufactured for kittens, adults, and senior cats. Compare and contrast information on labels to get the right food for your cat. A cat should not be fed dog food, as it does not contain enough protein. Always have water available.

A cat requires a high-protein, high-fat diet. Provide a high-quality commercial cat food. Food is available in several forms: dry, semi-dry, canned specialty, and canned maintenance. Labels list the information to be used in the selecting the proper food for a cat. Feeding dry feed promotes tooth and gum health.

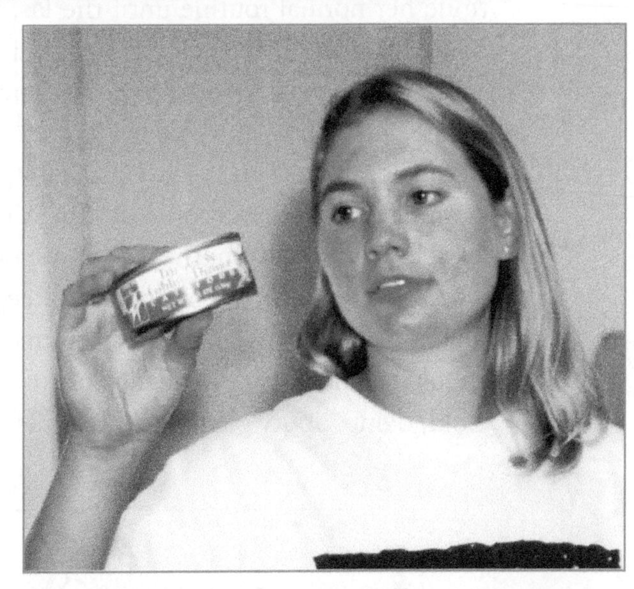

17–17. Read the nutrition information on cat food. This shows a can of turkey-based meat products. (Courtesy, Education Images)

If a cat is pregnant, she should continue to receive her same daily food allowance until the last 20 days of her pregnancy. Then she should be allowed to eat all she wants. After the kittens arrive, the mother may not eat the first few days. Make sure she has both food and water available. During the time the mother is nursing the kittens, feed her all the food she wants. Producing milk requires more nutrients.

17–18. Two separate bowls are used to provide a continual supply of food and water. (Courtesy, Education Images)

On the day the kittens are weaned, take all food from the mother. On the second day, feed one-fourth of what the mother used to eat. On the third day, increase this to one-half the amount she used to eat. On the fourth day, feed three-fourths of the normal amount. Finally, on the fifth day, feed the normal amount. Limiting the mother's intake of food will help decrease her milk production.

REPRODUCTION

The female cat's heat period is usually easy to detect. She will growl, chew, maul, and call for the male. Gestation is 60 to 65 days. Giving birth is ***kittening***. The female may continue her normal routine until the last week of pregnancy, when she should no longer be running up and down stairs and jumping on furniture.

During the last week of her pregnancy, the female will be restless as she looks for a place to have her young. A ***maternity box*** should be prepared for kittening. A maternity box is a quiet, draft-free, and secure place for parturition. Place the maternity box in a dark, quiet, warm, traffic-free location. All that is needed for the maternity box is a cardboard carton large enough for the mother to stretch out at full length on her side and have room left over. The top should be left attached to keep the area dark. A door should be cut, with 3 to 5 inches of cardboard left at the bottom of the door so the kittens do not escape. The bottom should be lined with shredded paper and a towel. A temperature of 85°F is best for kittening. Cold, damp, drafty places are the chief cause of kitten deaths.

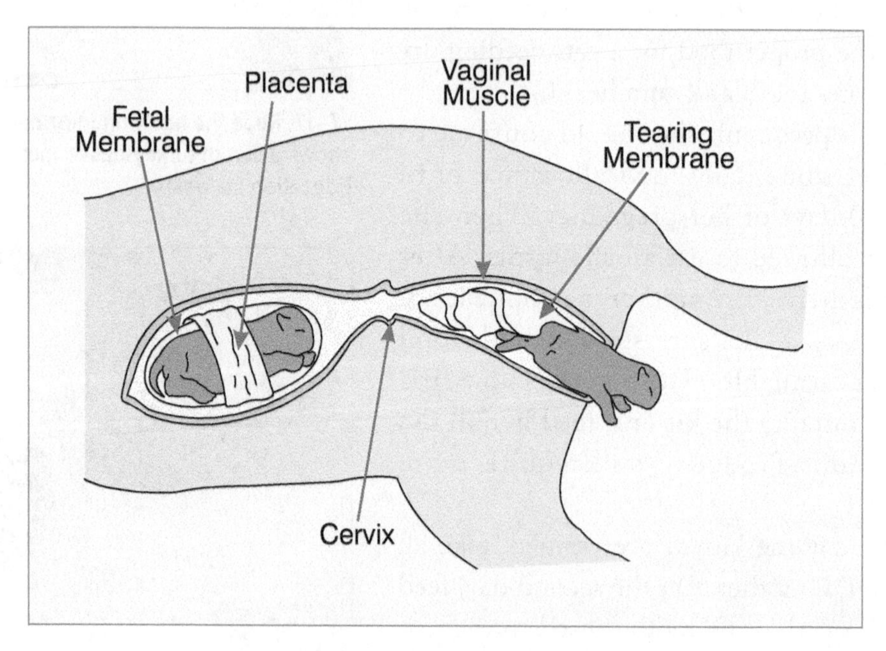

17–19. The birth process of kittens.

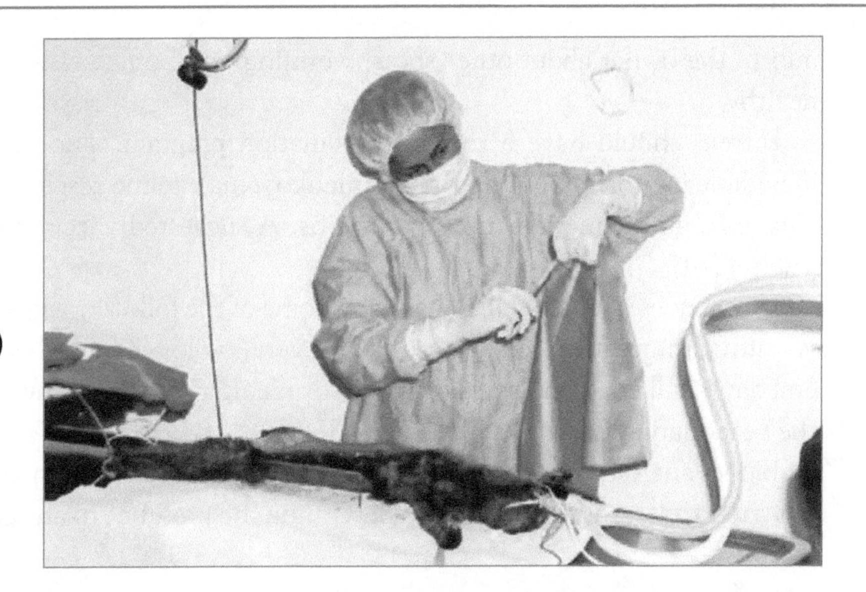

17–20. A veterinarian is preparing to spay a cat. (Courtesy, Education Images)

The queen will give birth to a litter, which is usually three or four kittens. After the mother delivers the kittens, she should be left alone. No strangers should be allowed to see the mother or her kittens for at least three days. Most births are normal, with the front feet and head first through the birth canal. Occasionally a kitten will be presented wrong. Assistance of a qualified person is needed. The queen will lick the kittens to dry and clean them following birth. Kittens should nurse within an hour. Queens need more calories during lactation.

Kittens are ready to begin eating food by three weeks of age. Feed the kittens the same food as the mother. If it is dry food, moisten it with water to make it easier for the kittens to eat. Have water available. Some kittens are weaned at 25 days of age. Queens will often continue nursing for six to seven weeks. At weaning, kittens should be eating adequate food for their nutritional needs. After weaning the kittens, do not overfeed the queen because it will cause her to gain weight and overproduce milk. This can lead to problems in the mammary system.

Cats may be neutered to limit reproduction. Male cats are castrated. Female cats are spayed. Castration removes the testicles of male cats, and spaying removes the ovaries of female cats, making them incapable of reproduction.

HEALTH CARE

Cats are generally healthy animals, especially if proper precautions are taken. A new owner of a cat should take it to a veterinarian to have it completely checked out. Vaccinations may be needed to prevent the animal from getting some diseases. The veterinarian will inform the owner about other shots, wormings, and office visits the animal needs to stay healthy.

Kittens should have a routine vaccination program. The diseases to protect against include feline infectious enteritis (panleukopenia), feline respiratory disease, rabies, feline leukemia, and feline infectious peritonitis. A suggested vaccination schedule is shown in Table 17–1.

If a cat does not feel well, it may show some of the following signs: coughing and sneezing; vomiting; runny nose; red, watery eyes; severe, prolonged diarrhea; loss of appetite for several days; dull coat; and a non-caring attitude. If a cat shows any of these signs, a visit with the veterinarian is in order.

Blood analysis may be used to check for disease. Immunoassay methods are used. ***Immunoassay*** is a procedure to determine the presence of antigens, antibodies, and pro-

Table 17–1. Possible Vaccination Schedule for Kittens and Cats

Age	Vaccination
6 to 8 weeks	panleukopenia (FPV), phinotracheitis (FVR), calicivirus (FCV)
12 weeks	second FPV, FVR, and FCV; give first feline leukemia (FeLV) shot if negative to ELISA test
16 weeks	first rabies; second FeLV, and third FPV, FVR, and FCV
16 months and each year	FPV, FVR, FCV, FeLV, rabies

Note: Get the assistance of a veterinarian in designing a vaccination program.

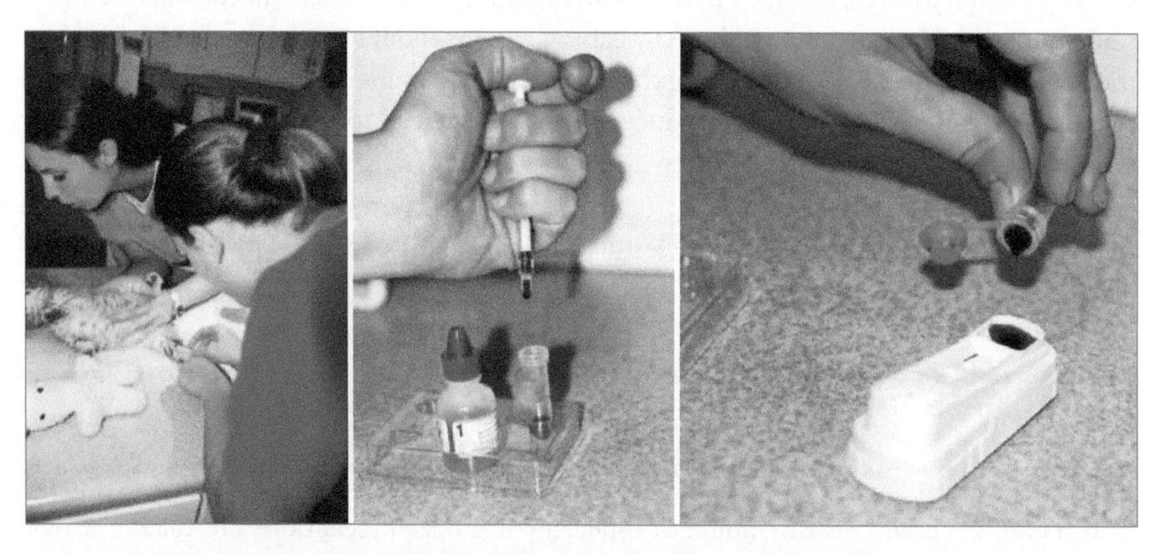

17–21. A SNAP® test is being used to check a blood sample for feline heartworms. (Blood is drawn from a cat, placed in a tube with anti-FHW solution, centrifuged, and visually observed after 10 minutes in a special device. Training is needed to use the kit accurately.) (Courtesy, Education Images)

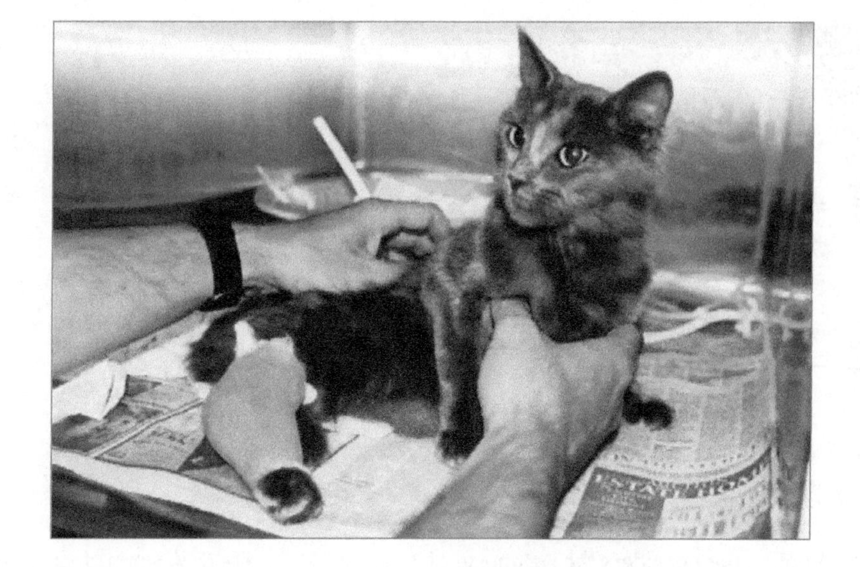

17–22. Veterinary hospitals have special facilities for animals to recuperate following surgery. (Courtesy, Education Images)

teins. New kinds of immunoassay tests use ELISA technology. ***ELISA*** is the acronym for Enzyme-Linked Immunosorbent Assay. These tests are used with blood, urine, and other solutions to check for antibodies. Kits are available to make diagnosis easier. An example is the SNAP® FHW (feline heartworm) test that takes about 10 minutes. It uses a sample of blood that has been treated with an anti-FHW solution and centrifuged. Blood color is observed in a special device.

17–23. Condominium poles provide a place for cats to play and exercise. (Courtesy, Education Images)

FACILITIES AND EQUIPMENT

The equipment for a cat does not have to be elaborate. A cat needs something to occupy its time and give it exercise. An outside cat needs protection from the weather and a safe place where other animals do not bother it.

A ***pet carrier*** is a carrying case for a cat or other animal. Since all cats must be taken to the veterinarian, they will, at some time, travel. The case should be large enough for the animal to turn around in, well ventilated, easy to clean, and lined with a towel in the bottom.

17–24. A pet carrier considers the well-being of a cat. (Courtesy, Education Images)

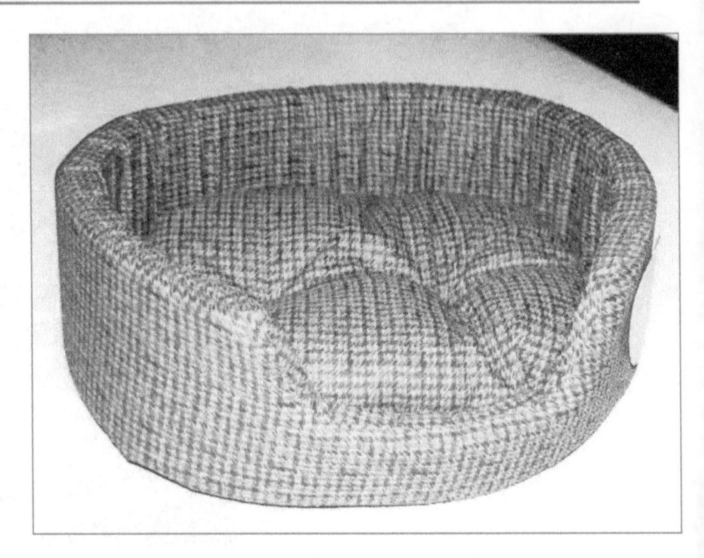

17–25. A warm, secure bed for a large cat. (Courtesy, Education Images)

The bed for a cat does not have to be a fancy bed purchased from a pet store. It may be something as simple as a cardboard box with fairly high sides and the bottom lined with a cushion. Cats prefer quiet places to sleep that are secure, warm, and draft-free. To avoid drafts, it is a good idea to place the bed several inches off the floor.

17–26. A good litter box makes cleaning easier and serves the cat better. (Courtesy, Education Images)

A ***litter pan*** is a container designed for the cat to use in urinating and defecating. Most litter pans are made of plastic or of metal with baked enamel, which will not rust. The pan should be washed with soapy water when the litter is changed.

Litter is an absorptive material that helps prevent odor. Many types of litter are available, such as prepackaged sawdust, shredded newspaper, sand, clean soil, peat moss, or wood shavings, which are often used with a small amount of baking soda to help eliminate odor. If droppings are removed daily, changing the litter once a week should be sufficient. Use a slatted scoop to remove the soiled litter, and place it in a plastic bag that is then sealed. Soiled litter in a bag is typically disposed of with the household garbage.

Cats are curious and energetic animals. They need scratching posts for exercise—not to sharpen their claws, as is commonly believed. A ***scratching post*** is a fabric-covered cylinder mounted in a vertical position on a wide base. Carpet or cloth may be used to cover the cylinder, and the wide base will ensure stability. A scratching post can be made easily. Since it may be hard for a scratching cat to know the difference between carpet on the post and carpet on the floor, set the post on a bare floor.

Cats also enjoy toys. However, avoid toys that are rubber, furry, or woolen and those that can be torn apart and eaten. Watch the cat to see the types of toys it enjoys. Many toys can be made from things already around the house, such as a paper bag, a shoe box, or a spool attached to a strong string and hung in a doorway.

CONNECTION

KITTENS WHEN YOU WANT KITTENS

Kittens are cute and cuddly. They must be fed, watered, trained, and otherwise cared for. Kittens grow up to be cats! Their upkeep requires effort. Unwanted kittens may be abused and not receive proper care. Responsible owners prevent unwanted kittens. Cat birth control is used.

Neutering is the most effective method. However, a queen that has been spayed or a tom that has been castrated can never be used for breeding. Tubal ligations with queens and vasectomies with toms are options that may be reversible.

Cat birth control drugs may become available in the United States. Two examples are megestrol acetate and delmadinone acetate. These products are for females. They alter the estrous cycle and delay heat. Once removed from the drugs, a queen can cycle into heat and be bred. Biotechnology will help make good animal birth control possible.

REVIEWING

MAIN IDEAS

Cats make good companions for some people. They are less trouble and more economical than dogs. Providing proper care is essential for the well-being of a cat.

Mixed-breed and purebred cats are chosen as pets. Purebred cats are classified as long-haired and short-haired. Several breeds are in each class. Most people keep mixed-breed cats. Purebred cats are needed for showing and raising kittens for sale.

Selecting a cat that is right for you is important. Consider its age and gender. Female cats are typically better in homes. Toms (male adult cats) urinate on the floor and walls to mark their territory. Neutering can help prevent some problems.

Provide a good environment for a cat. Cats need scratching posts. They need to be trained in how to use a litter box and in appropriate behavior in a home. Grooming is needed to keep them healthy and attractive and to reduce cat hairs over the house and furniture. Particularly, cats should not be around food preparation areas and infants.

Cats need a proper diet. They need higher-protein feed than dogs. Select a quality cat food designed for the age and condition of the cat. Read the label. Cats like meat, such as animal organs, fowl, and fish. The food can be dry or in cans. Some variety is good. Table scraps should usually not be fed. Cat food label information can be compared and contrasted based on stage of life: young (kitten), adult, and senior adult.

A pregnant queen will kitten 60 to 65 days following conception. A kittening box provides a safe, appropriate environment. Most queens have litters of three or four kittens.

All cats need regular health care. A vaccination program is recommended for kittens, beginning at six weeks of age. This is continued as adult cats.

QUESTIONS

Answer the following questions using correct spelling and complete sentences.

1. How widely kept are cats as companion animals? How long do cats live?
2. What parts of the sensory system are especially well developed in cats?
3. What are the important features of the feet of cats?
4. What are the two classes of purebred cats? Name two breeds in each class.
5. What should be considered in selecting a cat?
6. What are some considerations in managing a cat?
7. What training is needed by cats kept in a home?
8. What are the grooming needs of cats?
9. What are the general nutrition needs of cats?
10. What is the reproductive process of cats?
11. What should be established to have good health with kittens?
12. What is a litter pan? How is a litter pan managed?
13. What three stages of life of a cat are important in food selection? How is this important in comparing and contrasting foods?
14. Which form of feed is usually best for a cat: dry or canned (moist)? Why?

EVALUATING

Match the term with the correct definition. Write the letter of the term in the blank provided.

a. queen
b. kittening
c. hairball
d. declaw
e. maternity box
f. litter
g. pet carrier
h. scratching post
i. immunoassay
j. pads

______1. The process of a queen giving birth.
______2. A female cat.
______3. A carpet- or cloth-covered exercise structure for the cat.
______4. A container used to transport a cat.
______5. The absorptive material placed in a litter box.
______6. To remove the claws of the cat surgically.
______7. A wad of hair that may form in the digestive system of a cat.
______8. A box for kittening.
______9. Soft, fleshy areas on the bottoms of cat feet.
_____10. A procedure to determine the presence of antigens and other substances.

EXPLORING

1. Arrange to cat-sit for a neighbor or family member. Learn all the details so you can take good care of the cat. Keep an activity log each day you are responsible for cat-sitting.

2. For one day, job shadow a person in a pet store, grooming business, or veterinary medical clinic where cats are served. Write a report on your experience.

3. Prepare a poster or electronic presentation on some aspect of cats. You may choose to do a presentation on the care, breeds, nutrition, health, or other feature of cats. You may wish to refer to The Cat Fanciers' Association, Inc., Web site: **www.fanciers.com**. Be sure to include details of the subject you choose. Give an oral report to the class.

4. Compare and contrast commercially available cat foods. Review labels on products in stores or as otherwise available. What ingredients are listed? How does protein content compare? Chose the food that is best for the cat based on its stage. Give a brief oral report in class on your findings.

5. Develop a health maintenance program for a kitten. Include veterinary care as well as daily routine care to prevent diseases and parasites. Expand the program to include its adult and senior cat stages. Be sure to include the control of parasites and diseases. Discuss your planned program in class.

CHAPTER 18

Birds, Rodents, and Reptiles

OBJECTIVES

This chapter provides background information on several species of companion animals, including birds, rodents, and reptiles. It has the following objectives:

1. Describe the kinds of companion birds and their management.
2. Describe the kinds of companion rodents and their management.
3. Describe the kinds of companion reptiles and their management.
4. Describe the kinds of rabbits and their management.
5. Identify other companion animals and describe their keeping.

TERMS

amphibian
aviary
bird
bird cage
birdkeeping
diurnal
fledgling
greenfood
grit
hutch
nocturnal
quill
reptile
rodent
solitary

18–1. Replacing the shavings in a guinea pig habitat. (Courtesy, Education Images)

BIRDS, rodents, and reptiles–just how many different species are kept as companions? We can add ferrets, hedgehogs, rabbits, chinchillas, tarantulas, snakes, and several others to the list. Each species is unique though most all animals have the same basic needs: food, water, air, and space or shelter. Regardless of the species that is kept always remember to provide for its well-being.

If you are thinking about getting one of the species in this chapter, you need to learn about its upkeep. You need to know its habitat needs, feed and nutrition, health care, reproduction, and general personality. Some of the species may pose hazards such as bites by a snake, claw scratches by a rabbit, or the elusive behavior of a gerbil.

Habitat is the environment in which an animal is kept to promote its well-being. Most animals have unique needs based on their natural life needs. Cages, pens, and other equipment may be used to create an artificial habitat.

COMPANION BIRDS

18–2. A keel-billed toucan (*Ramphastos sulfuratus*) has been trained to perch on its owner's arm. (Courtesy, Meaning/Shutterstock)

Several kinds of birds make excellent pets in homes and apartments. Some 4.5 million households in the U.S. have one or more birds as companion animals. Study the needs of each species and provide for its well-being. Special equipment and facilities will be needed to care for a bird. Birds must have food, water, and a good environment.

A ***bird*** is an animal with feathers. It is the only animal with feathers. Birds have four appendages: two wings and two legs. All birds have wings. Some birds can fly quite well; others cannot fly. Birds stand upright on two feet. Birds have bills instead of chewing mouths. Birds reproduce by laying eggs. All birds are in the class Aves. Scientists have further divided birds into 28 orders. For example, the parrot is in the order Psittaciformes.

Birdkeeping is raising birds in captivity. All areas of bird care are included—feeding, breeding, health care, and cleaning. Pet stores, discount stores, and other places sell equipment needed for birdkeeping. Many of these same stores sell birds.

GROUPS OF COMPANION BIRDS

Seven groups of birds are most popular with birdkeepers. Each group has several species and varieties. These have unique features that make them appealing to birdkeepers. Varieties represent different colors and other distinct features. The groups of companion birds are:

- Cockatiels and Parakeets—Cockatiels and parakeets are in the same order as the parrot. Cockatiels are easy to train and are lovable birds. Colorful yellows, oranges, and greys are present on the adult birds. They are good whistlers. Incubation is 25 to 26 days with fledgling occurring 55 to 56 days later. Cockatiels like cereal grains, fruit, and greenfood. Some are fed commercial parrot food. Length is 10 to 18 inches (20 to 45 cm) or more.

- Finches—Finches are 4 to 5 inches long and come in several varieties and colors. Incubation period for eggs is 12 days. The fledgling stage begins at 18 to 21 days after hatching. A ***fledgling*** is a young bird that has left the nest but is not feeding independently. The males of some varieties have unique songs. Standard bird seed can be used as feed, especially small cereal grains, such as milo. Adults are typically 4 to 5 inches (10 to 12.5 cm) long.
- Lovebirds—Lovebirds are short-tail parrots from Ethiopia. They are excellent cage birds and are quite colorful. Young lovebirds are easy to tame and show affection. Older birds are difficult to tame. Since lovebirds are in the parrot order, they often receive the same care as parrots. Lovebirds do not talk. Lovebirds like cereal grains, fruit, and greenfood. Adults are 5½ inches (14 cm) long.

18–3. Parakeets in a comfortable cage with food, water, and natural limb materials. (Courtesy, Education Images)

- Canaries—The canaries were introduced to the United States from the Canary Islands. Noted for bright colors, canaries are popular with birdkeepers. Some canaries are better singers than others, especially the males. Canaries generally prefer larger cages. Some varieties prefer outdoor aviaries year round. Incubation is 14 days. The fledgling stage is 14 days later. Canaries prefer a seed diet along with some greenfood. Adults are typically 5 to 7 inches (12 to 17 cm) long.
- Budgerigars—Sometimes called budgies, budgerigars are native to Australia. They are friendly and talkative, with some knowing 600 words. Budgies often have health problems. They are colorful and appealing. Use a commercial budgerigar feed mix. Adults are typically 7 inches (18 cm) long.

18–4. Colorful parrots are appealing to bird keepers. (Courtesy, Ronen/Shutterstock)

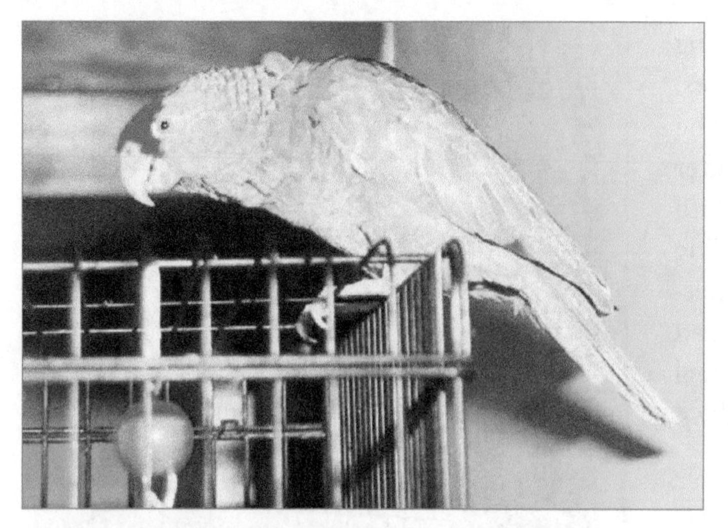
18–5. This Mexican redhead Amazon parrot is an interesting bird. (Courtesy, Education Images)

- Amazon Parrots—Amazon parrots are quite talkative and entertaining as adults. They are typically cuddly. Many colors and interesting plumage make them appealing to watch. Feed a commercial parrot food, fruit, and greenfood each day. Parrots are often kept in large cages. The length of an adult may be 15 inches (37.5 cm).
- Macaws—Macaws are types of parrots. They are large, colorful, playful, and intelligent birds. They need more room than most birds and care is important. Macaws are hardy, but require protection from bad weather. Incubation is 25 days, with fledgling occurring 55 days later. They like seeds, fruit, and greenfoods. Some macaws prefer a little meat occasionally in their diets. The length of an adult is 13 to 36 inches (32.5 to 90 cm).

Table 18–1. Features of Selected Birds

Kind	Weight (grams)	Normal Life Span (years)	Considerations/Features
FInches	10–15	5–10	Easy to care for
Canaries	15–30	10–20	Easy to care for; males are singers
Budgerigars	30–35	10–20	Easy to care for
Cockatiels	80–90	10–20	Easy to care for and train
Lovebirds	40–50	10–20	Taming and training are more difficult
African greys	400–500	60–70	Can be trained to talk; enjoy talking
Macaws	1,000–1,400	60–70	Some can be trained to talk; interesting; colorful

MANAGEMENT OF COMPANION BIRDS

As living organisms, birds have similar needs for life as other animals.

Reproduction

In general, birds reproduce by the females laying fertile eggs. The male has mated with the female several days prior to the laying of the fertile eggs. Eggs are incubated naturally or can be removed for artificial incubation. Natural incubation is preferred. Nesting facilities must be in the cage or aviary. Incubation periods vary with the species and variety of bird. Incubation typically ranges from two to four weeks.

18–6. Commercial feeds are prepared for the needs of various kinds of birds, such as this parrot food. (Courtesy, Education Images)

Nutrition and Feeding

Use commercial feeds designed especially for the kind of bird. Read nutrition and species information on the label of a commercial food container to compare and contrast feeds. Some birds need supplemental fruit and greenfood. Apples are commonly fed fruits. ***Greenfood*** is non-dried vegetables with juice still present. It includes green peas, fresh corn, carrots, cabbage, and bean sprouts.

Grit is sometimes needed to promote action of the gizzard. Grit is finely ground abrasive material similar to sand. Ground oyster shell or similar shell materials may be used. Small amounts of the ground shell are placed in the cage. Calcium and other minerals may be obtained from the grit.

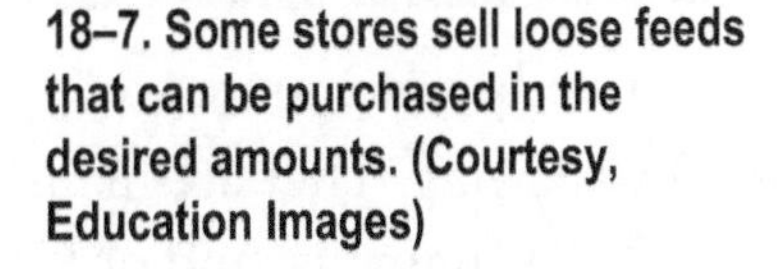

18–7. Some stores sell loose feeds that can be purchased in the desired amounts. (Courtesy, Education Images)

Health Care

Birds are subject to a number of diseases. Good nutrition and protection from inclement weather help prevent disease. All new birds should be quarantined from existing birds for a period to determine if the new birds have any diseases. Sick birds lose their interest in food and water. They are inactive. Feathers become tattered. The services of a veterinarian may be needed as soon as a disease is suspected. Birds can get sick and die rather quickly from disease.

18–8. This cockatiel has a cage that promotes its well-being. (Courtesy, Education Images)

Facilities and Equipment

Birdkeeping is typically in one of two kinds of facilities: cage or aviary. The goal is to have a good habitat.

A ***bird cage*** is a box-like enclosure that keeps birds captive. Cages are made of wire; metal, wood, or plastic bars; and other materials. Some cages are made to be hung from stands. Other cages are boxes that sit on tables or the floor. Cages must have feeders and waterers as well as perches.

An ***aviary*** is a house-like enclosure for birdkeeping. Aviaries are large enough for people to walk inside. They allow plenty of room for the birds to move around and get exercise. Aviaries have roosts, nesting areas, feeders, and other equipment.

The size of the cage or aviary depends on the number and needs of the birds. Some birds need more room for flight than others. Obviously, many birds need more space than one or two birds.

RODENTS

Rodents have increased in popularity as pets. Several species have become more widely kept. A ***rodent*** is an animal with two large front teeth designed for gnawing. They are mammals. The scientific classification of rodents is phylum Chordata, class Mammalia, and order Rodentia. Several species of rodents, such as gerbils, hamsters, mice, and rats, are kept as pets. They require a clean cage, a well-balanced diet, clean water always available, and attention from their owner. The owner is repaid with an attentive, loving pet.

In the wild, gerbils, hamsters, guinea pigs, and mice and rats damage crops since they are rodents. It is important to make sure the animals do not escape. If they escape and multiply, they will cause serious crop damage. Gerbils are illegal to have as pets in California. The U.S. Department of Agriculture has warned about taking them into desert regions of the west.

GERBILS

Gerbils make fascinating pets because they are very quick and curious. The gerbil came to America in the 1950s for medical research. Today, there are some 430,000 kept as companion animals in the U.S. Unlike most rodents, gerbils sleep during the night and are awake during the day, they are ***diurnal***.

18–9. Gerbils in a comfortable cage with litter and a waterer. (Courtesy, Education Images)

The adult gerbil's body is 4 inches long. Its tail is another 4 inches long. A typical adult weighs about 3 ounces. Gerbils are dark brown on their backs, light brown on their sides, and light gray underneath. They have black and white whiskers.

Gerbils are shy animals. They must be handled gently. Scratching the ears and back of the gerbil will relax it. When holding a gerbil, keep your hand firmly around its entire body to make sure it does not fall or wriggle free.

Reproduction

Gerbils are ready to mate at three months of age. A gerbil has only one mate. Potential mates should be kept separated from each other until they grow accustomed to each other's scent. The gestation period is 24 to 25 days. Most females have litters of about five babies. Both parents stay with the litter until weaning at six weeks of age. People should avoid contact with the young until their eyes are open.

Nutrition and Feeding

Gerbils eat a tablespoon of food each day. They only need to be fed once per day. Water should always be available. Gerbils eat grain, seeds, roots, grasses, sunflower seeds, corn, oats, wheat, watermelon seeds, bits of apple and lettuce, and may enjoy fresh grass as a treat. Using a commercial gerbil food is okay. A commercial gerbil food can be used. Gerbils like variety. Compare and contrast ingredient variety as listed on the label.

Health Care

Gerbils are fairly disease free. Gerbils will use one corner of their cage as their toilet, so it is important to clean it every one or two days. Gerbils do not like cold temperatures or drafty areas. Keep the temperature 65 to 80°F (18.3 to 26.6°C) and, if the gerbil is in the sunlight provide it with an area for shade. Gerbils also need hard, dry food or a block of wood to chew on. This will keep their teeth filed down, so they can chew properly.

Facilities and Equipment

Gerbils, because of their curiosity, prefer interesting homes, which may be purchased or made by the owner. A cage with two floors is best so the animal exercises by running up and down and all around. To create fun and excitement for the animal, the owner may want to create stairs, holes, and other play things for the gerbil. Make sure the cage has a top and keep it covered so the gerbil does not escape.

Gerbils work to make their cage a comfortable place. They need bedding in their cage. The bedding should be absorbent and clean. Cedar chips, sawdust, and small-animal litter work well when put into the cage about an inch thick. Gerbils build nests. Put a small piece of burlap, paper, or cardboard in the cage for use in nest building.

HAMSTERS

Hamsters are fun animals. The hamster came to America in 1938. Today, there are 1.2 million companion hamsters in the U.S. It is ***nocturnal***, which means that it sleeps during the day and is awake at night.

CONNECTION

NEVER AROUND YOUR NECK

Snakes like to wrap around things. They have strong bodies. Many obtain their food by wrapping around it and strangling it to death. Their strength can crush a small animal. Even small snakes are quite strong; larger snakes are even stronger. Some snakes can kill a pig or deer by tightening–known as constriction.

A snake draped over your shoulders or around your neck could suddenly tighten. If it is around your neck, you could quickly suffocate. The damage would be done before you could get it off or summon help. You would be found by someone but not before you were dead!

Make it a rule: Never drape a snake around your body! (Courtesy, Education Images)

Three types of hamsters are used as pets. The Syrian or golden hamster has a golden body with white patches on its cheeks, front legs, and hind feet. The angora hamster has long, fuzzy hair, which comes in many colors, but is most common in tan, gray, or white. The Chinese hamster is solid black or gray and has short hair, which distinguishes it from the angora hamster.

18–10. A mature hamster fits comfortably in a hand. (Courtesy, Education Images)

Hamsters must be handled frequently to remain tame. It is best to stroke their backs until they get used to your touch. Pick up a hamster by placing your hand underneath for support. After a hamster trusts a person, it will crawl over him or her and ride in a shirt pocket. Always protect a hamster from dogs, cats, and anything else that may scare it, or it may try to escape. A good way to train a hamster is to give it a treat for doing a thing you like. Peanuts, cashews, and other nuts are good treats.

Reproduction

Male and female hamsters kept together in a cage will breed. If you do not want to raise hamsters, keep only one since they do not get lonely nor require the attention of another hamster.

Hamsters should not be bred until they are eight weeks old. Many keepers do not recommend breeding until three months old. After the female is one year old, it is very rare for her to produce offspring. Only well-developed, healthy animals should be used for breeding. Gestation is 16 days. You can have too many hamsters if you do not manage them carefully.

Start out the breeding process with hamsters by putting the cages of the male and female close to each other so they get used to one another. Place the female in the male's cage. If the male is put into the female's cage, she may kill him because she feels threatened. Wear gloves when handling the hamsters at this time. If the female is not in heat, they may fight and she will need to be removed. Put the female into the male's cage at night about 9 o'clock because most breeding takes place between 9:00 and 10:00 p.m.

Leave the female alone for nine days after she gives birth. Do not clean her cage for those days. A frightened mother may kill and eat her young. When the young reach two weeks of age, they are old enough to be weaned.

18–11. An exercise wheel in the cage of a hamster provides health-promoting exercise. (Courtesy, AlexKalashnikov/Shutterstock)

Nutrition and Feeding

Hamsters eat about ½ounce of food each day. They should be fed at night when they are active. The natural food of hamsters is grains, seeds, and vegetables. Corn, oats, or wheat may be mixed with a prepared dog food. Commercial hamster food is likely the best way to feed hamsters.

As they eat, hamsters stuff their cheek pouches, hide the food, and return to do the same. In the wild, hamsters burrow into the ground and make one main tunnel with many side chambers in which they store food. As you clean the cage, you may find pockets of food. It is okay to put the food back where you found it if it is not spoiled. It is easy to compare and contrast feeding behavior and nutrition with day-feeding species.

Health Care

Hamsters that are healthy have the following signs: soft, silken fur; a plump body; and prominent, bright eyes. A hamster that is long or skinny, has watery eyes, a runny nose, or a wet tail may be ill. If clumps of fur are missing, it may have a disease called mange.

Hamsters and guinea pigs may get diseases or parasites from wild mice or birds. If they do get fleas, lice, or other insects, it is okay to treat them with an insecticide for use on cats. Never use an insecticide for dogs on a hamster.

Facilities and Equipment

Many hamster owners buy bird cages or use old aquariums to house their animals. If a wooden cage is used, be sure to cover the wood on the inside so the hamster does not gnaw its way out.

Hamsters need a clean, dry cage away from drafts—55 to 80°F (12.8 to 26.7°C). If the temperature falls below 45°F (7.2°C), the hamster will go into hibernation. It may appear as though the animal is dead, but it is hibernating. Hold the animal to slowly warm it and feed it warm milk, one drop at a time, until it revives. When returning the animal to its cage, cover the cage with a heavy cloth. Hamsters also need a hard, dry food or even a block of wood. Chewing on a hard material will keep their teeth filed down so they can chew properly. Give the female scraps of cloth, tissue, paper, or cotton to line her nest. Hamsters use one corner of their cages for a toilet, much like gerbils.

MICE AND RATS

Mice and rats can be good company because they are friendly and curious. They can be trained to climb ladders, beg for food, do tricks, and ride on people's shoulders and in their pockets. They are used in psychological, biological, medical, and nutritional studies. Tamed mice and rats have gotten a bad reputation because of their wild relatives, which cause damage and are not loving and fun.

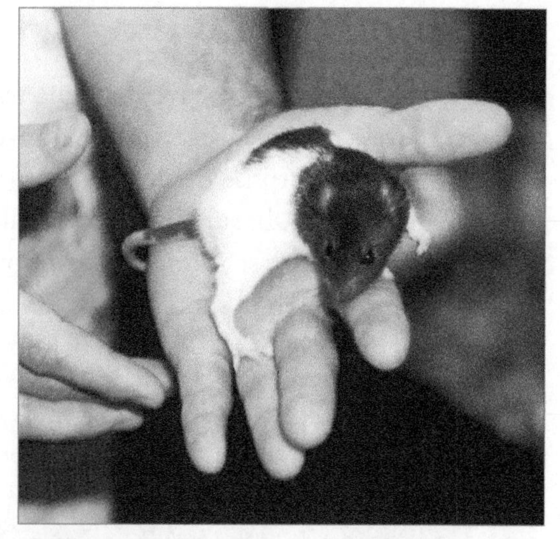

18–12. A hooded rat. (Courtesy, Education Images)

Mice and rats come in various colors. Rats may be all white, black, brown, or hooded. Hooded coloring is white with brown or black around the head and shoulders. White rats are easier to tame and make nicer pets. Mice come in a variety of colors, such as white, black, tan, brown, or spotted. There is no difference in their personality based on their color.

Mice and rats are usually very tame animals, especially when the animal is acquired at a very young age. The younger the animal, the easier it is to tame, and the fastest way to tame it is with hand feeding. When picking up a mouse or rat, grab it by the tail near its body, place it in your other hand and then gently stroke its head and back. After the animal gets to know its owner, it will be at the door waiting to get out of its cage to crawl around on its owner. Treats may be used to encourage the animals. Since the animals are nocturnal, it is best to play with them in the early evening.

18–13. Pick up a white mouse by its tail. (Courtesy, Education Images)

Reproduction

A female mouse may be bred at 8 to 10 weeks of age and a female rat at three months. It takes both animals about 21 days to give birth after they are bred. The male and female may be left together until the female gives birth and then the male must be removed until the babies are weaned at about three weeks old. Baby mice have a habit of jumping, so anyone holding the animals should be aware. Also, since the babies reach maturity so quickly, it is important to separate the males and females or more babies will be born. Since a litter consists of six to eight babies, it does not take long to be overwhelmed with young.

Nutrition and Feeding

Mice and rats have the same water requirements as all other animals. So, make sure they always have plenty of clean, fresh water. They eat dry dog food and that is enough to keep them healthy. Surprise them with new foods, such as seeds, nuts, rabbit food pellets, pieces of hard-boiled eggs, bread, breakfast cereals, rice, leafy foods, or raw potatoes. Feed dishes should be small enough so the animals cannot sleep in them and attached to the side of the cage above the floor. The food should stay clean and not get the animal's droppings in it. Nutrition requirements of mice and rats can be compared and contrasted with other species.

18–14. White mice can be kept in an old aquarium with litter, waterers, and feeders. (Courtesy, Education Images)

Health Care

Mice and rats will live a healthy life if they are kept in good health. They need a block of wood to chew on to keep their teeth in good condition and a clean, dry cage. Like hamsters, it is okay to dust them with cat flea powder and disinfect their cage. However, never use a flea powder designed for dogs.

Facilities and Equipment

Mice and rats must be kept in wire or metal cages because they will gnaw through wooden cages. They need a hard, dry food or even a block of wood to chew on to keep their teeth in top condition. A solid upper platform should be provided for the animal to rest on and platforms and ramps should connect the floor and platform.

Since mice and rats are curious animals, toys, such as swings, perches, and exercise wheels should be included. Sawdust, cedar shavings, or cat litter should be used as bedding to keep the cage clean and dry. For nesting, cotton or shredded paper should be supplied.

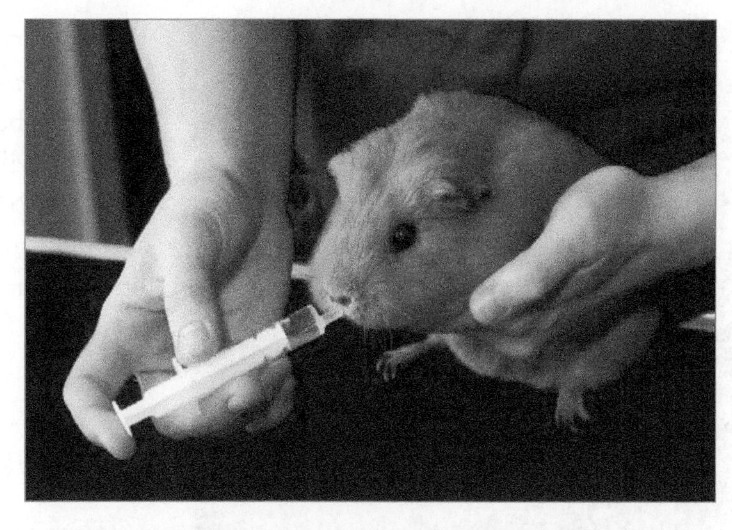

18–15. Administering liquid to a guinea pig with a syringe with needle removed. (Courtesy, Mark William Penny/Shutterstock)

GUINEA PIGS

Guinea pigs make friends and greet their owners with whistles. Slightly over 1 million guinea pigs are kept as companion animals in the U.S. The guinea pig has a short, heavy body, short legs, and no tail. Three strains of guinea pigs are: the English, Peruvian, and Abyssinian. The English and Peruvian strains of guinea pigs both have short hair and are found in a wide variety of colors, but the Peruvian has rough hair that stands up in all directions, which distinguishes it from the English strain. The Abyssinian strain has long hair and has either solid or mixed colors.

18–16. A silky guinea pig. (Courtesy, Education Images)

Guinea pigs require attention and enjoy being talked to, especially when eating. When handling the guinea pig, even when it is small enough to fit into your hand, hold the animal with your thumb and forefinger just behind the animal's head and in front of its front legs. As the animal is lifted, use your other hand to support its back end. If the animal remains comfortable, it will remain calm, but if it feels uncomfortable, it may scratch you.

Reproduction

Female guinea pigs are called sows and male guinea pigs are called boars. Sows should not be exposed to the boar until they are three to five months old. The boar is always put into

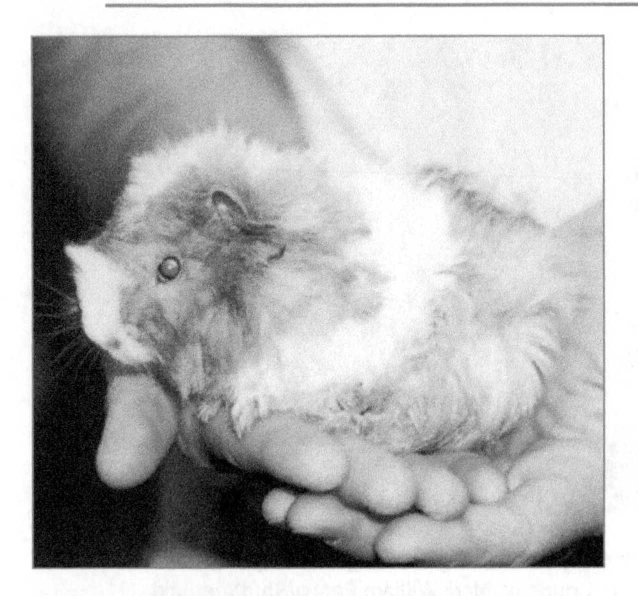

18–17. An abbey guinea pig does not have the smooth coat of a silky guinea pig. (Courtesy, Education Images)

the sow's cage and she is never put into his cage. They are left together for three weeks. The sow will give birth between 63 and 72 days after she is bred. The sow will breed again within a few hours of giving birth to her litter. If she is not bred at that time, she will wait until the litter has been weaned. The litter should be weaned at three weeks old. The males and females in the litter should be separated to prevent breeding.

Nutrition and Feeding

The nutrition and feeding of guinea pigs can be compared to and contrasted with other small animals. Contrasted to other species, they drink very little water because the eat so many high-moisture greens. However, make sure they always have some clean water. Feed should be available at all times. At the end of the day, old food should be removed because the animal can get sick from rotten food. Favorite foods are alfalfa, apples, carrots, corn, dandelions, lettuce, cauliflower, clover, celery, lawn clippings, spinach, and tomatoes. Dry foods, such as rabbit food pellets, and a salt spool must also be available.

Guinea pigs, compared to other species such as mice, like different foods, so experiment to see what the animal prefers. Guinea pigs, monkeys, and humans are the only three animals that cannot make their own vitamin C, so make sure the guinea pig gets some of the foods mentioned above because they all contain some vitamin C.

Health Care

Guinea pigs are healthy animals. If they are not feeling well, the owner will be able to tell because the animal will sit perfectly still and hunched up. Its coat may be ruffled and messy. It will not eat and will lose weight. The droppings may be loose and watery. If any of these symptoms appear, the animal should be separated from other animals and taken to the veterinarian. If the animal picks up parasites, it may be treated the same way as the hamster.

Facilities and Equipment

Guinea pigs are not fighters and they do not jump or climb, so their cages may be small. Their cages should have some type of cover, even if it is just wire, to keep other animals away. A guinea pig needs at least 1 square foot of floor space, but many animals may be housed in one cage, as long as the cage is large enough. For example, a cage large enough for one male and three females would be 36 by 24 inches (0.9 by 0.6 m).

Guinea pigs need a dry floor of which at least half of it is solid and not made of wire. To keep the cage dry, a thin layer of shavings, sawdust, or straw may be put down. Do not use dusty material. Dust can make the animal sick. Guinea pigs will not grow if the temperature gets below 65°F (18.3°C) in the winter or if it is very hot in the summer.

REPTILES AND AMPHIBIANS

Reptiles and amphibians are similar in many regards. Both are kept by some people as companion animals.

REPTILES

A ***reptile*** is an ectothermic animal with dry, scaly skin and lungs for breathing. Ectothermic means that the animal's body temperature adjusts to its environment. Some people refer to reptiles as cold-blooded animals because they feel cold to the touch.

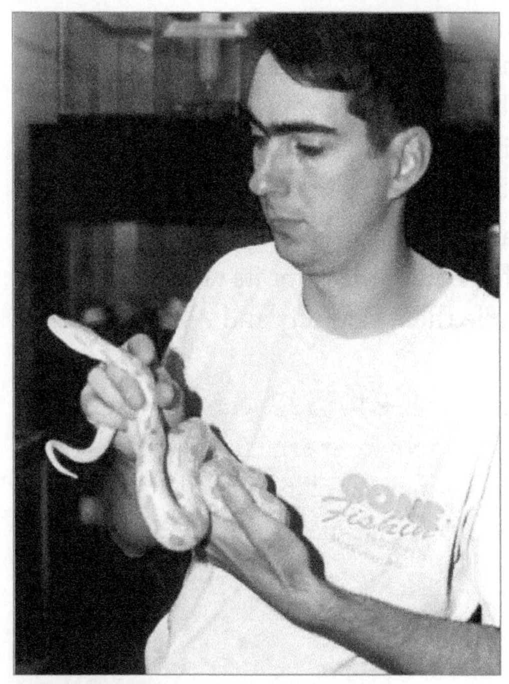

18–18. An Albino Burmese Python is a snake that may be kept. (Courtesy, Education Images)

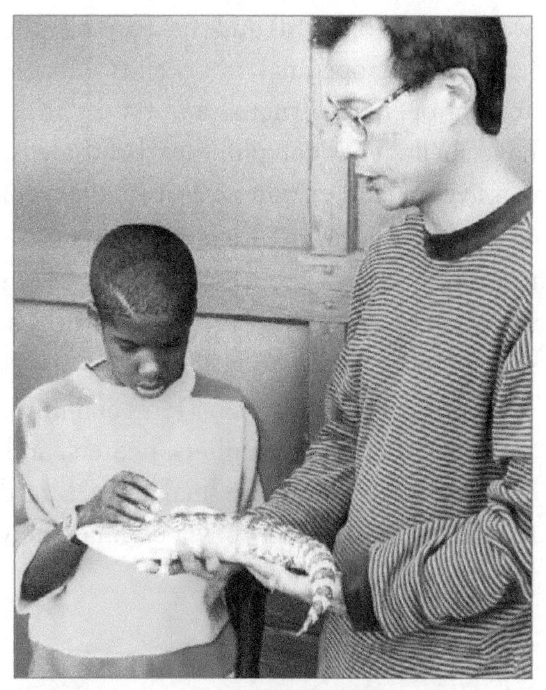

18–19. Keeping a blue-tongued skink requires special knowledge of its needs. (Courtesy, Education Images)

18–20. An iguana is a popular reptile pet. (Courtesy, Thomas Zobl/Shutterstock)

18–21. Snakes prefer gravel and stones in their containers. This shows two red tail boa snakes in a securely covered aquarium. (Courtesy, Education Images)

Two groups of reptiles are sometimes kept as pets: 1-lizards and snakes and 2-turtles. More species of lizards and snakes are kept than of turtles. Snakes have long tails but do not have legs, ear openings, nor eyelids. Most lizards have long tails, four legs, movable eyelids, and ear openings. Iguanas and skinks are types of lizards that are sometimes kept. Turtles have shells. They can pull their heads, legs, and tail inside the shell for protection. Turtles can live on land or water. Most reptiles reproduce by laying eggs.

People planning to use reptiles as pets need to carefully study the animal they propose to use. They need to know how to safely keep it and the requirements for its well-being. Since many people are afraid of reptiles, owners must handle them carefully and keep the animal secure so it does not escape.

Keeping a reptile is a long-term commitment. They can live 10 to 20 years or more. Many are kept in aquaria with filters, heaters, and lights. They require special foods to meet their dietary needs. Snakes like small mice. Lizards and turtles may eat plants, small animals, and insects. Keepers usually buy a commercially prepared feed that meets the dietary needs of the species they are raising.

AMPHIBIANS

An ***amphibian*** is an animal whose body is covered with skin and lives part of its life in water and part on land. Common amphibians are frogs, toads, and salamanders. Frogs and toads are quite similar. As adults, they have four legs and no tail. Salamanders typically have long tails and four legs, though a few species have only two legs.

Amphibians are often kept in an aquarium. Part of the area is covered with water and part is above the water line. The dry area is often of pebbles, stones, and wood. Food must be provided. Salamanders eat algae and insect larvae. Frogs and toads eat insects. Amphibians reproduce by laying eggs. The amphibians kept as companion animals are often captured from the wild or are from farms that specialize in the species.

18–22. A salamander. (Courtesy, Education Images)

RABBITS

Rabbits are widely kept as companions and for commercial food and pelt production. Information is presented on keeping rabbits as companions here. Detailed information on rabbit production is available in Rabbit Production, a book available from Prentice Hall Interstate.

KINDS OF RABBITS

Rabbits are sometimes mistakenly referred to as rodents. Rabbits have four large, upper, front teeth, and this alone distinguishes them from rodents, which have two large teeth. Rabbits are classified in the order Lagomorpha and family Leporidae. This family includes rabbits and hares. Domestic rabbits descended from the European wild rabbit, Oryctolagus cunniculus. Hares are larger than rabbits and have long ears with black tips. Colors of rabbits range from solid white to solid black and includes browns, grays, and various spotted color patterns.

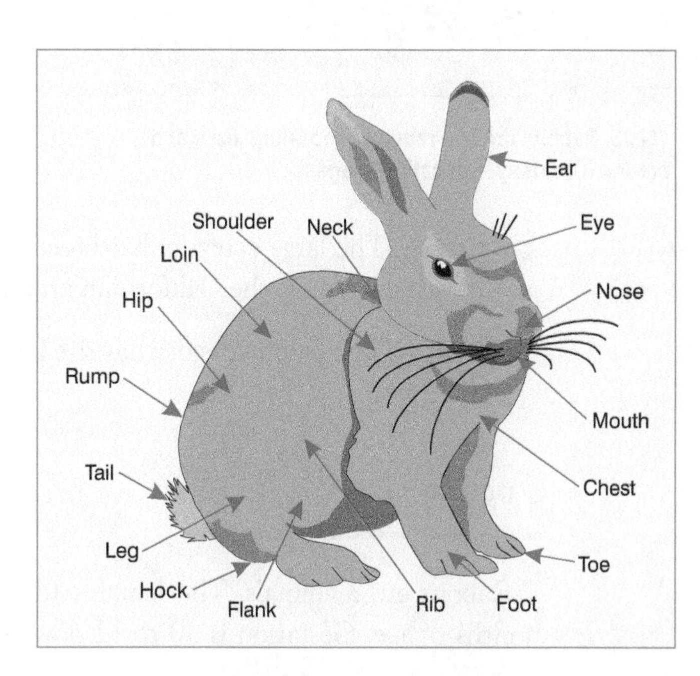

18–23. Major parts of a rabbit.

18–24. Dressed rabbit in a supermarket. (The rabbit is also known by its French name of lapin.) (Courtesy, Education Images)

18–25. Rabbits are in a range of appealing sizes and colors. (Courtesy, Education Images)

The 45 breeds of rabbits are placed in categories by weight. The weight categories and examples of breeds:

- Miniature—The miniature category is the smallest. It is sometimes known as the dwarf category. The Jersey Wooly and Holland Lop are two miniature breeds.
- Small—The breeds in the small category are slightly larger than those in the miniature category. Examples of breeds include Dutch and Mini Lop.
- Medium—The medium category is larger than the small category. Examples of breeds include English Angora and Sable.
- Large—The large category has breeds that are next to the largest category. Examples of breeds include the Californian and Satin.
- Giant—The giant category has the largest breeds. Examples are French Lops and Giant Angora.

REPRODUCTION

Rabbits are mammals. The females (does) may be bred for the first time at about six months of age. Gestation is 30 to 32 days. Females need to have materials (straw, paper,

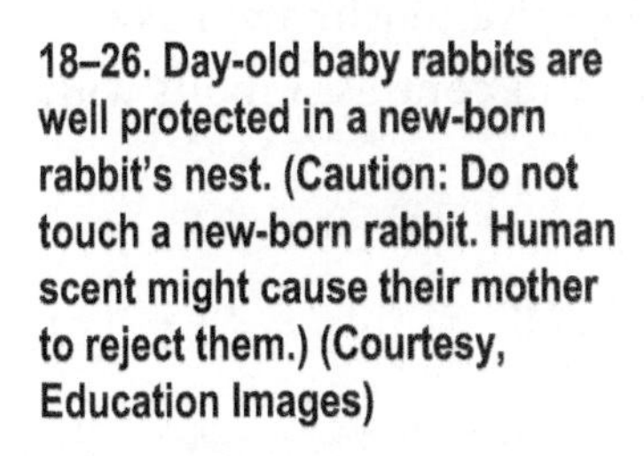

18–26. Day-old baby rabbits are well protected in a new-born rabbit's nest. (Caution: Do not touch a new-born rabbit. Human scent might cause their mother to reject them.) (Courtesy, Education Images)

etc.) for nest building just before parturition. Litters average six babies. Baby rabbits are hairless and have closed eyes. They are weaned at six to eight weeks of age.

NUTRITION AND FEEDING

Use a special pelleted commercial rabbit food. Select a feed based on the age and condition of the rabbit being fed. Changes in feed sometimes upset the digestive systems of rabbits.

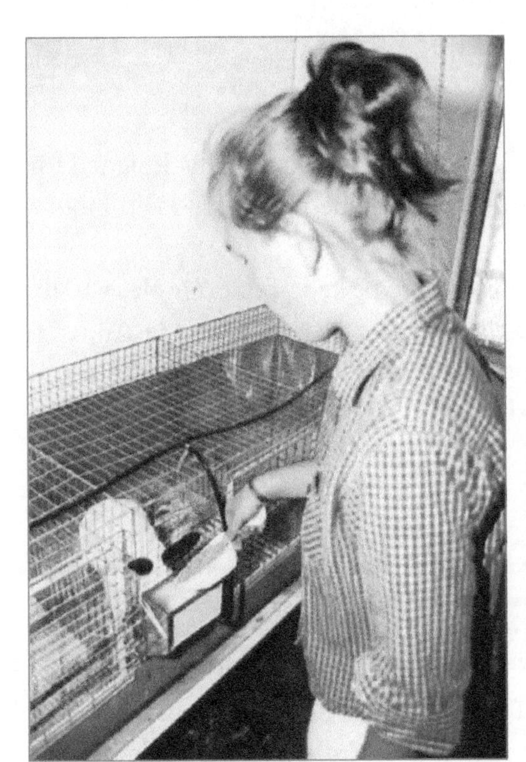

18–27. A rabbit is receiving a commercial feed. (Courtesy, Education Images)

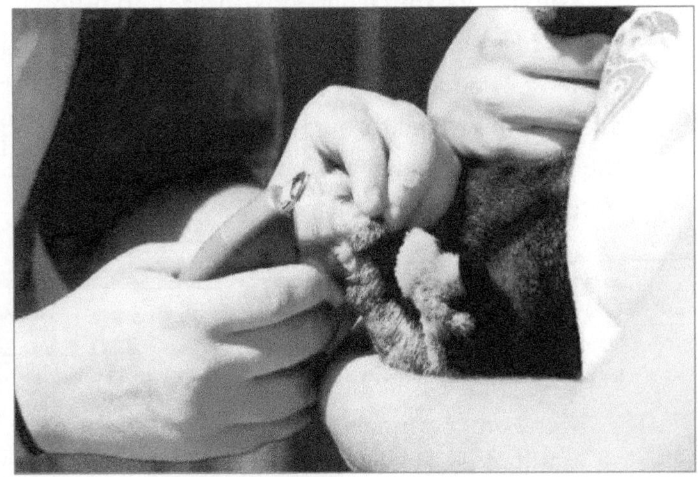

18–28. Clipping the claws on a rabbit. (Courtesy, Education Images)

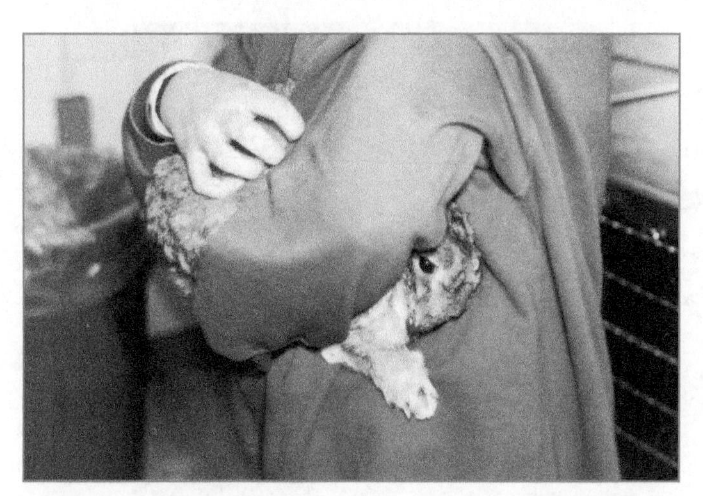

18–29. How to hold and carry a rabbit. (Courtesy, Education Images)

Fruit, vegetables, and greenfoods may be fed sparingly. Too much of these can also upset the digestive system. Do not overfeed rabbits. Digestive sensitivity makes it easy to compare and contrast rabbits with other species.

HEALTH CARE

A good environment helps prevent disease problems. Avoid cool, drafty hutches. Health problems include colds, diarrhea, constipation, pot belly, eye infections, and sore hocks (feet). A change in the normal behavior of a rabbit is a sign of disease. Get the assistance of a veterinarian or other person qualified in rabbit diseases.

FACILITIES AND EQUIPMENT

Rabbits are often kept in cages known as hutches. A **hutch** is a cage specially designed for rabbits. Hutches should be covered to keep out rain. Walls may be needed to keep rabbits warm in cold weather.

The size of a hutch depends on size and number of rabbits to be kept. For example, a 10-lb. rabbit needs a hutch that is 18 inches high by 36 inches long and 24 inches wide. Wire attached to wooden frames is often used. Wire bottoms of hardware cloth are easy to clean and allow wastes to fall through and out of the cage. With large rabbits, wire can lead to sores on the feet; therefore, a wooden floor may be needed. Waterers and feeders should be in each hutch. Keep hutches clean.

OTHERS

Many other animals are sometimes used as pets. These include insects, ferrets, hedgehogs, chinchillas, deer, raccoons, and farm animals, such as horses, goats, pigs, and cattle. People who have wild animals near their homes often put out food and become friendly with them. Squirrels commonly live in trees in urban areas and become friendly to people. Some of these are wild animals that have not been domesticated. They can suddenly change behavior and attack people.

FERRETS

18–30. A ferret should be carefully handled. (Courtesy, Education Images)

Ferrets (*Mustela putorius furo*) are mammals. They are members of the weasel family and descended from the European polecat. Nearly 1.1 million ferrets are kept as pets in about a half million households in the U.S.

Ferrets have long, slender bodies with long tails. A male is known as a hob and a female as a jill. They are primarily nocturnal. When well cared for, ferrets can live 11 years or so.

Ferrets are kept in cages. Most cages should be 24 inches wide by 24 inches long by 14 inches high. Use special ferret or cat food. Water should be available. Keep cages clean. Ferrets sometimes appear to create more filth than other species.

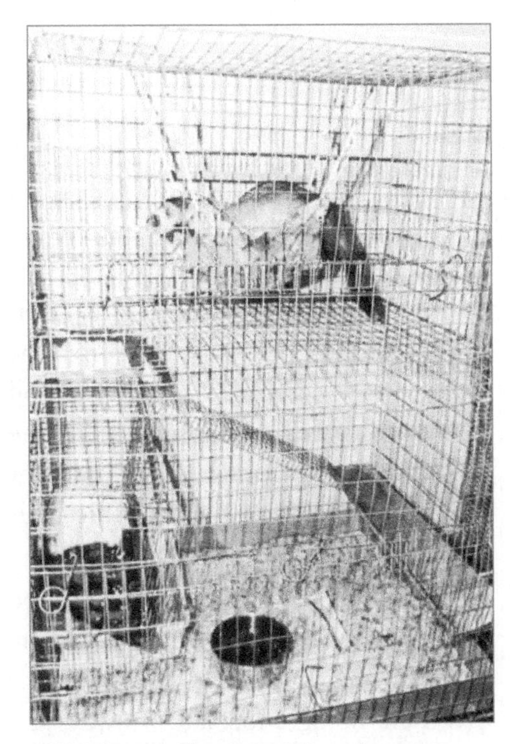

18–31. Ferrets like cages where they can exercise and play. (Courtesy, Education Images)

18–32. Water container attached to the outside of a cage. (Courtesy, Education Images)

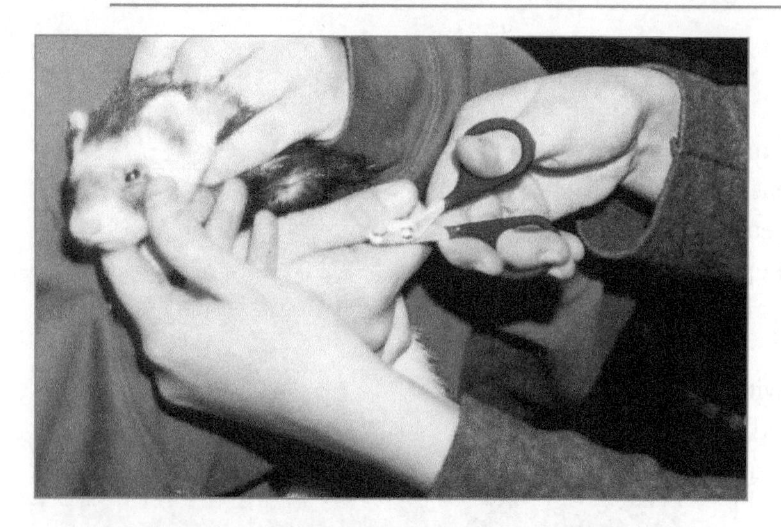

18–33. Trimming the nails of a ferret. (Courtesy, Education Images)

Female ferrets breed in the spring after they are 10 months old. Gestation is about 42 days. Litter size is six to eight babies. Eyes open in three to four weeks. Babies begin eating bread or water-softened dry cat food when their eyes open. Young can be weaned after six weeks. They reach nearly full size in 14 weeks.

Some health care and grooming are needed. Ferrets are susceptible to several diseases, with canine distemper usually fatal. It is best to establish a health care program with a veterinarian when the ferret is young. Nails need to be trimmed regularly. Use care and only remove a small amount of the tip. The trimmers used with dogs and cats work well with ferrets.

Ferrets are sometimes removed from their cages. Always keep them under careful watch and never outside where they can escape. Some incidents of ferrets attacking infants have been reported.

HEDGEHOGS

Hedgehogs have increased in popularity as companion animals in recent years. A hedgehog somewhat resembles a miniature porcupine because of the quills that cover its body. A ***quill*** is a modified hollow hair that is stiff and has a point. Quills on hedgehogs are ½ to 1 inch (1.3 to 2.5 cm) long. The quills on a hedgehog are not particularly dangerous when the hedgehog is handled properly. When frightened, hedgehogs roll into a ball. Their muscles cause the quills to stick outward and prevent attack. Hedgehogs also roll into balls for sleeping.

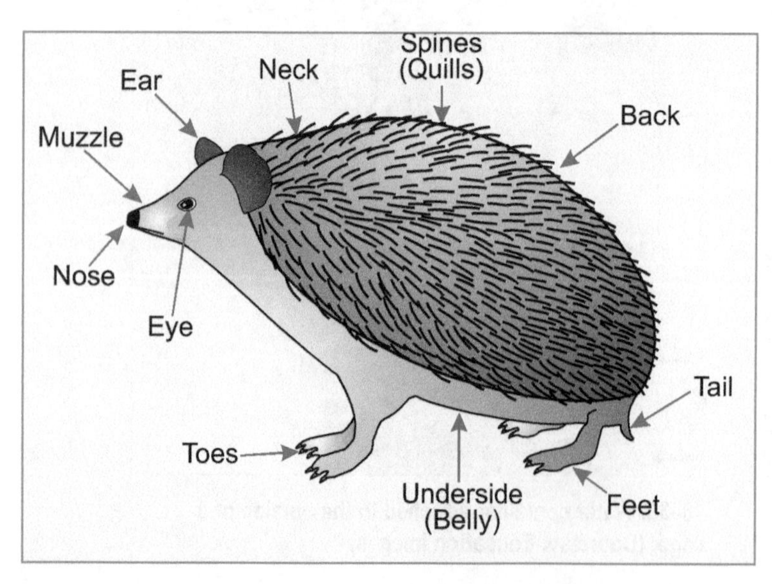

18–34. Major external parts of a hedgehog.

Hedgehogs are native to Europe and Africa. Colors vary but many resemble "salt and pepper," with others being chocolate brown. The scientific name of the hedgehog is *Erinaceus europaeus*. An adult European hedgehog is about 12 inches

(30 cm) long. They may weigh as much as 3 pounds. An African hedgehog is smaller and shorter. African hedgehogs are most popular in the United States.

In the wild, hedgehogs move quite slowly except for quick darts. Many eat dead animals, worms, and other easy-prey foods. Hedgehogs are kept in cages or unfilled aquaria. A small plastic "swimming" pool as used with children can be arranged nicely. Hedgehogs will not get over the sides of the pool and escape unless equipment placed in the pool allows them to do so. Exercise equipment similar to that used with hamsters should be in their home. Use wood shavings or other litter on the bottom of the cage or container.

18–35. An African hedgehog. (Courtesy, Education Images)

Hedgehog health is evident by clear, bright eyes. Their nose should be dry and free of discharge. Ears should be erect and clean. Their body covering should be soft and appear in good condition. Areas around the anus and genitals should be clean and free of feces.

Clean water should always be available. Hedgehogs are meat-eaters and are often fed live mealworms. A small amount of vegetable or fruit may be fed, such as two peas or kernels of corn. Commercial catfood may also be used with hedgehogs.

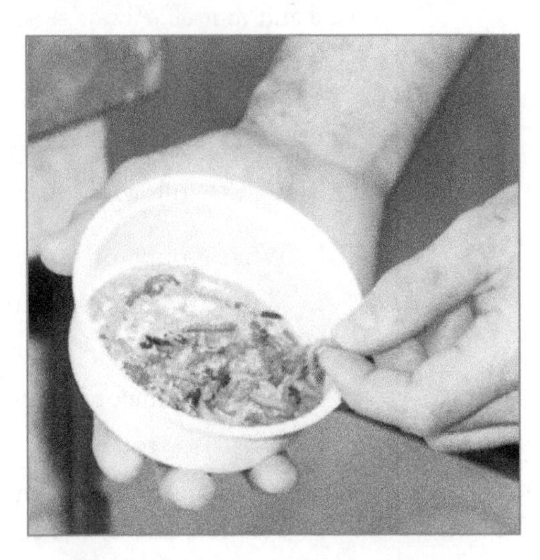

18–36. Mealworms for feeding an African hedgehog. (Courtesy, Education Images)

Hedgehogs of either sex may be kept. Normal social behavior is solitary. ***Solitary*** means that hedgehogs do not need companions. They typically get as far away from each other as possible. Males in particular move far apart as adults. Two hedgehogs are not needed. Of course, a male and female are needed if breeding is to be done. Most hedgehogs prefer separate cages unless the male and female are to be bred. Hedgehogs can breed after about seven weeks of age. After a short gestation, the female gives birth to a small litter. The male will eat the babies if present in the cage. Females nurse the young for about five weeks, at which time they can be weaned.

REVIEWING

MAIN IDEAS

Birds, rodents, reptiles, and other species may be successfully kept as companion animals. Study and learn all that you can before attempting to keep any companion animal.

Birdkeepers often have favorite birds. Companion birds are in seven groups: finches, canaries, budgerigars, cockatiels and parakeets, lovebirds, Amazon parrots, and macaws. Good management is essential for the well-being of the bird. This includes a cage or aviary, food and water, and providing a good environment.

Several species of rodents are used for companion animals. Gerbils, hamsters, mice and rats, and guinea pigs are most widely kept. Proper equipment is needed to assure their well-being.

Reptiles and amphibians are kept by some people. Other people are frightened by them. Snakes, lizards, and turtles are the reptiles that are kept. Amphibians include salamanders and frogs. Amphibians must have an environment of water and land.

Rabbits are widely kept as companions as well as animals for food and fur. Hutches are needed to provide for their well-being. Rabbits are typically provided a pelleted commercial food.

Ferrets, hedgehogs, and chinchillas are other species that may be kept. Both are mammals. A good environment must be provided. Proper food, water, and nesting materials are needed.

Care must be appropriate for each animal species. This includes health, food, handling, and other practices that promote well-being. It is essential to compare and contrast nutritional requirements which vary by species and stage of life (young, adult, and senior) as well as use. Properly lift, move, and restrain an animal. Know each species as compared and contrasted with others. Use safe practices to prevent injuring the animal and yourself. Always seek the assistance of a trained individual in learning animal care.

QUESTIONS

Answer the following questions using correct spelling and complete sentences.

1. What groups of birds are kept? Briefly describe each group.
2. What are the distinguishing features of birds as organisms?
3. What is the role of grit in bird nutrition?
4. What two major kinds of facilities are used with birds? Distinguish between the two.
5. What is a rodent?
6. What species of rodents are kept as companion animals? Select one species and briefly describe how it is kept.
7. What is a reptile? Amphibian? Name one example of each that is kept.
8. How are rabbits kept? What housing is needed? Equipment?
9. How are ferrets kept?
10. What social behavior of hedgehogs is important in keeping them?
11. What is habitat? Name habitat needs for any three species of animal in this chapter.
12. What is housing? How does it relate to habitat?
13. What housing and equipment are needed with your favorite animal species covered in this chapter? List.
14. What are the nutrition and feeding needs of small animals? Select two species and compare and contrast their needs.
15. How is the health of small animals promoted? Diseases and parasites controlled? Select any two species and compare and contrast health care needs.
16. How is a mouse lifted? Rabbit? Compare and contrast lifting and carrying the two species.

EVALUATING

Match the term with the correct definition. Write the letter by the term in the blank provided.

a. birdkeeping
b. fledgling
c. greenfood
d. aviary
e. diurnal
f. hutch
g. solitary
h. nocturnal

_______1. An animal that is active at night.
_______2. A type of facility used for keeping rabbits.
_______3. Raising birds in captivity.
_______4. Cabbage, peas, and similar foods.
_______5. A young bird that has left the nest but is not capable of providing food for itself.
_______6. An animal that is active in the daylight.
_______7. Behavior of an animal that lives alone.
_______8. A facility for keeping birds.

EXPLORING

1. Make a tour to a birdkeeping facility. Observe the cages and aviaries that are in use. Interview the manager about the kinds of birds that are kept, how they are managed, and the problems that are encountered. Prepare a written report on your observations.Be sure to include how cages and aviaries create appropriate habitat for birds.

2. Job shadow a worker in a companion animal store for a day. Observe the species of animals and how they are managed. Prepare a report that includes detailed information on the care provided different species. Be sure to include the kinds of feed, cages, and other equipment that are sold.

3. Observe the safe handling of animal species in the school lab, a pet shop, kennel/cattery, or other facility. This includes lifting, moving, and restraining animals. Practice these skills yourself under direction of a capable individual. Be sure to follow all safety practices to prevent injury to the animal and yourself. Describe your experiences in an oral report. Be sure to include your thoughts from practice in handling each of the animal species.

4. Practice performing appropriate care for small animals under the direction of a capable individual experienced in small animal care. Perform various procedures using the laboratory equipment that is available. Ready/clean a cage and properly dispose of wastes. Provide feeders and waterers with feed and water. Prepare bedding. Use devices such as nail/claw clippers, medical devices/application equipment, and others as may be appropriate. Be sure to follow appropriate safety and sanitation practices. Give an oral report on your experiences. Describe the housing and equipment that are used. Also describe your thoughts on what it was like to perform the procedures using the laboratory equipment.

CHAPTER 19

Ornamental Fish

OBJECTIVES

This chapter is an introduction to keeping ornamental fish. It has the following objectives:

1. Explain ornamental species and list examples.
2. Explain the equipment and facilities needed for ornamental species.
3. Describe the water environments for ornamental species.
4. Describe how to care for ornamental species.

TERMS

aeration
aging
aquarium
aquarium maintenance schedule (AMS)
biological filtration
biological oxygenation
chemical filtration
companion fish
egg-laying fish
filtration
freshwater ornamental fish
habitat
livebearing fish
mechanical filtration
nitrogen cycle
ornamental fish
oxygenation
salinity
saltwater ornamental fish
shoal
synthetic seawater
tapwater
thermostat
tropical fish
vat

19–1. Koi are popular ornamental fish in outdoor pools, ponds, and tanks. (Courtesy, Johnny Lye/Shutterstock)

ORNAMENTAL fish are popular for their beauty, grace, and entertainment. Ornamentals form a large industry in the United States. They are more valuable than food fish on a weight basis. Over 9 million households in the U.S. keep such fish ranging from a goldfish in a bowl to an aquarium with several species to an outside fish pond.

Aquariums are often found in homes. These are typically glass tanks in which the owner tries to replicate a natural environment for fish and other species. The environment in an aquarium is made attractive with crushed gravel, growing plants, decorative materials, and filtered water. The latter—filtered water—is critical in keeping the aquarium clean and bright. Lighting is often used to enhance the appearance of an aquarium.

Just as with terrestrial animals, aquatic ornamentals require care. The care provided should be based on the needs of the species being kept. You will need to invest in equipment, fish and other species, and feed for the animals kept.

ORNAMENTAL SPECIES

Ornamental fish are fish kept for their appearance and personal appeal to people. Some ornamental fish have bright colors and large, fancy fins. These fish are not typically used for food.

Tropical fish form a subgroup of the ornamental species. A ***tropical fish*** is a small, brightly colored fish that is popular in home and office aquaria. These fish thrive in warm water, breed rapidly, and tolerate conditions in home aquaria. Tropical fish range from 1 to 12 inches in length.

Companion fish are kept in homes for human companionship or as pets. An aquarium with fish may provide an important use of leisure time. Companion fish amuse, entertain, and provide enjoyment. Companion fish may be tropical fish or they may be other species of fish, as well as other aquatic animals, such as snails or turtles, and aquatic plants.

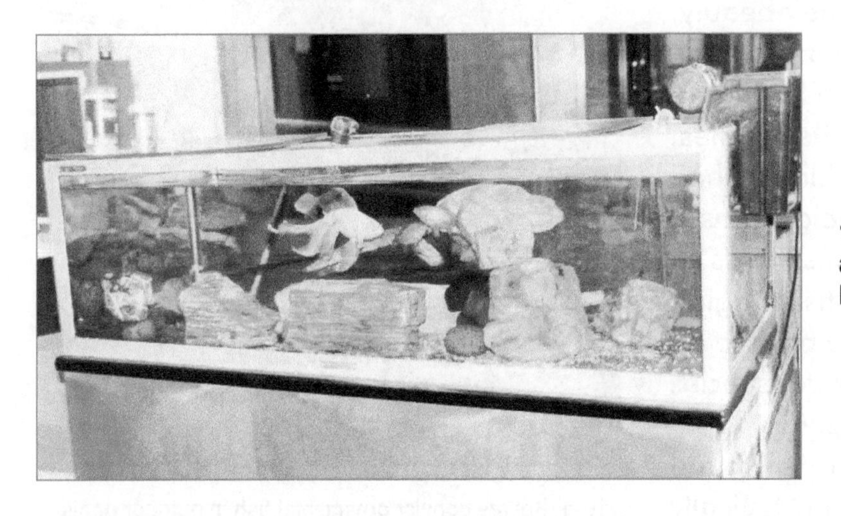

19–2. An attractive freshwater aquarium. (Courtesy, Education Images)

SOURCES OF FISH

Most ornamental fish are not native species. They have been brought into North America from another place. Releasing these fish into streams is often illegal.

Ornamental fish are captured from the wild or raised in confinement. In the past, most ornamental fish were caught from streams and lakes. People have found cultured ornamental fish to survive better in aquaria. A few people raise ornamental fish for stores and pet shops. Florida is the leading state in the production of ornamental fish.

Wild ornamental fish are typically caught in the tropical waters of the Pacific Islands, Asia, and South America. Proper treatment of captured fish can rid them of disease and help them survive in captivity.

FRESHWATER ORNAMENTAL FISH

Freshwater ornamental fish are most popular. These are species that grow in freshwater. They will tolerate only a very small amount of salt in the water.

The freshwater ornamental fish reproduce in two ways: laying eggs and giving birth to live young.

19–3. A volume producer of ornamental fish in Maryland uses large tanks. (Courtesy, Education Images)

Egg-Laying Fish

Egg-laying fish reproduce by the female fish releasing eggs that are fertilized by sperm from the male fish. Incubation usually lasts only a few days.

Popular egg-laying ornamental fish include koi, goldfish, gouramis, tetras, barbs, and catfish.

- Koi—The koi is a variety of common carp (Cyprinus carpio) that was developed by selecting fish with distinctive coloring. The colors may be bright red, black, gold, or white. Koi, a freshwater fish covered with scales, may reach a length of 37.5 inches (95 cm) and weigh 3 to 10 pounds or more. Most koi grow in pools or large tanks. Koi clubs are found in some places. The members raise and enter koi in various competitive events.

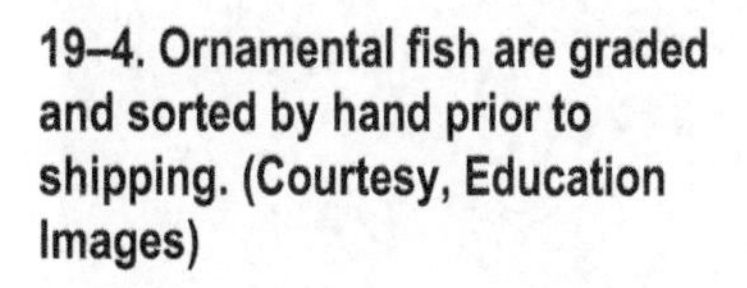

19–4. Ornamental fish are graded and sorted by hand prior to shipping. (Courtesy, Education Images)

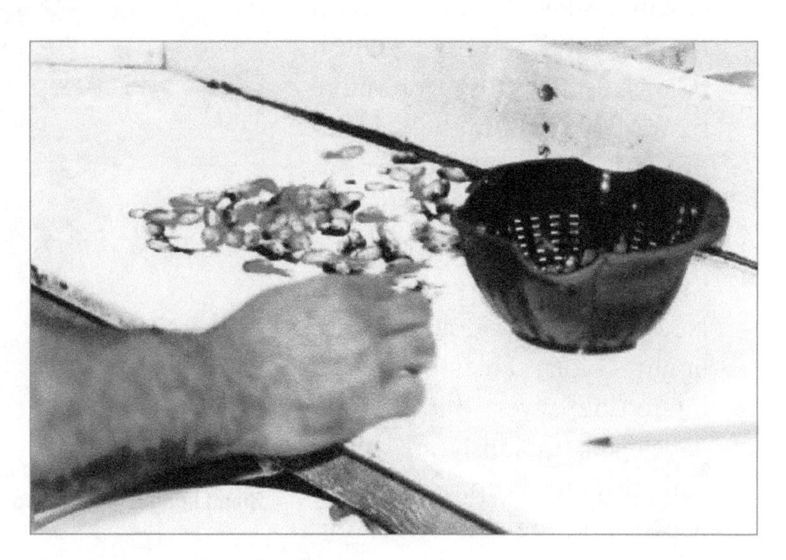

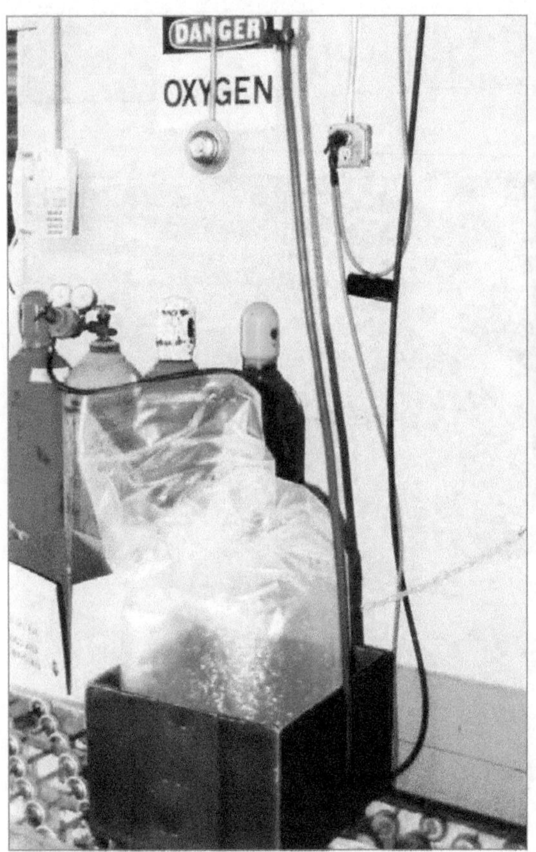

19–5. Ornamental fish may be shipped in plastic bags with water and oxygen. For protection, the bags are placed in boxes. (Courtesy, Education Images)

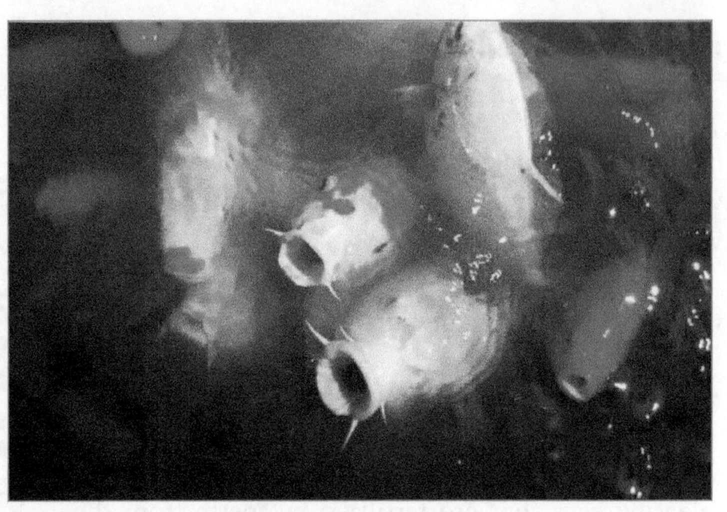

19–6. Koi come to the water surface to eat food thrown into the water. (Courtesy, B. Speckart/Shutterstock)

19–7. Goldfish are often kept in small bowls. (Courtesy, Blend Images/Shutterstock)

- Goldfish—People have been keeping and raising goldfish (Carassius auratus) for more than 2,000 years. In the koi family and sometimes known as golden carp, goldfish are smaller than koi. They are adapted to a wide range of environments. New varieties of goldfish with bright colors, fancy fins, and interesting eyes have been developed. Goldfish are hardy and easy to keep. Their size depends on the size of the con-

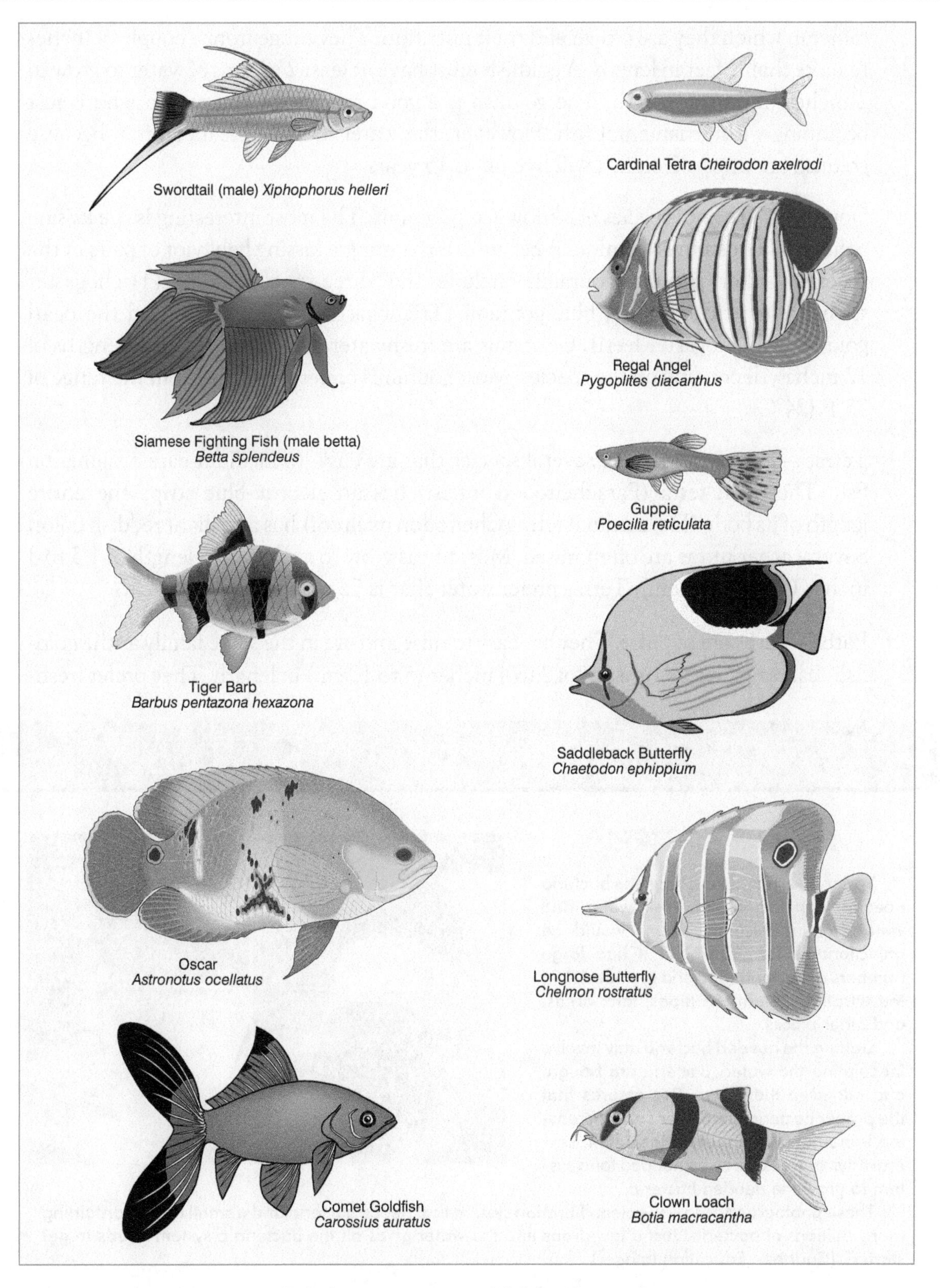

19–8. Common ornamental fish kept in aquaria.

tainer in which they are grown and their nutrition. They range from a couple of inches to more than 2 feet in length. A goldfish must have at least 2 gallons of water to grow to 2 inches (5 cm) in length. The goldfish is a good species for the person who is just beginning with ornamental fish. However, the water must be cleaned often because goldfish are messy. Goldfish will live up to 15 years.

- Gouramis—Several species are known as gouramis. The most interesting is the kissing gourami (Helostoma temmincki) because of the unique kissing behavior of pairs of the species. Other popular gouramis include the three-spot gourami (Trichogaster trichopterus trichopterus), blue gourami (Trichogaster trichopterus), and the pearl gourami (Trichogaster leeri). Gouramis are freshwater fish that may reach lengths of 12 inches, depending on the species. Most gouramis prefer water that is in the range of 75°F (24°C).
- Tetras—The tetras include several species that are easy- to medium-care ornamental fish. The neon tetra (Paracheirodon innesi) has an electric-blue stripe the entire length of its body. The cardinal tetra (Cheirodon axelrodi) has a brilliant reddish color. Several other tetras are often raised. Most tetras grow to a maximum length of 1.5 to 3 inches (3.75 to 7.5 cm). Tetras prefer water that is 72 to 85°F (22 to 28°C).
- Barbs—Barbs are popular. They are easy to raise and are in the same family as the goldfish. Barbs reach a mature size of 2 to 4 inches (5 to 10 cm) in length. They prefer fresh-

CONNECTION

GETTING BACTERIA

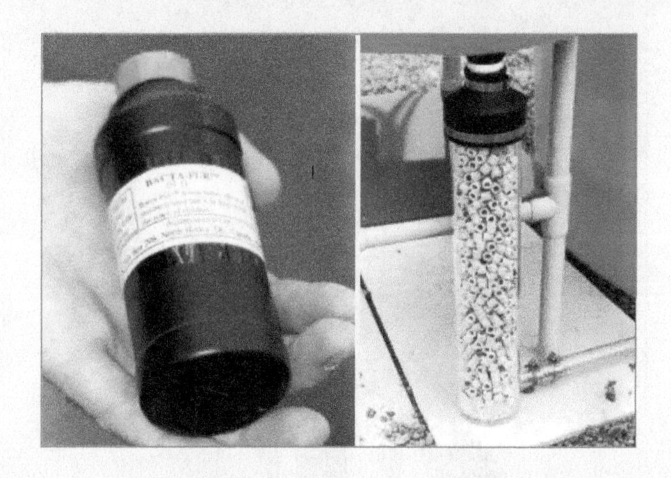

Water does not always have the bacteria needed with fish tanks. Well water and water from municipal systems would be objectionable for drinking if it had large numbers of bacteria. Bacteria are needed in the filtration system, on pipes, tank areas, and other places.

Getting the needed bacteria may involve inoculating the water. Bacteria are bought and added to the water. This assures that the proper bacteria are present and increasing in numbers. Without an inoculant, several days or weeks may be needed for a system to grow the needed bacteria.

These photographs show the clean filtration system that needs bacteria and a small bottle containing many millions of bacteria. Just a few drops into the water gives all the bacteria a system needs to get started. (Courtesy, Education Images)

water that is 70 to 80°F (21 to 27°C). Most barbs like aquaria with plenty of light. Three common barbs are the spotted barb (Barbus binotatus), rosy barb (Barbus conchonius), and tinfoil barb (Barbus schwanenfeldi). Some barbs eat their eggs after spawning.

- Catfish—Some species of America's most popular foodfish are grown as ornamentals. The upside-down catfish (Synodontis angelicus) is interesting to watch in an aquarium because of its behavior in swimming upside-down. The glass catfish (Kryptoptereus bicirrhis) and electric catfish (Malapterurus electricus) are also interesting species. The glass catfish has a clear-like body that may appear as a rainbow. The electric catfish can give off an electric shock that will kill smaller fish. Catfish prefer water that is 70 to 80°F (21 to 27°C). Most catfish prefer the darker areas of an aquarium.

Livebearing Fish

Livebearing fish give birth to live young. They like to live in a group of five or more, known as a ***shoal***. Rather than releasing sperm over the eggs, as with egg-laying fish, the males deposit sperm inside the female fish. The fertilized eggs develop into nearly fully formed fish. Females of some species store sperm from the male until it is needed for reproduction. Four common livebearers are guppies, swordtails, mollies, and platys.

- Guppies—The guppy (Poecilia reticulata) is the most popular livebearer. Many varieties have been developed, with the primary difference between them being the shapes of their fins and tails. Guppies like plenty of food and will eat different foods. Most producers feed a dried commercial food. Guppies may reach lengths of 2.5 inches (6.35

19–9. Ornamental fish in an aquarium need regular feeding but no more than will be eaten in a few minutes. (Courtesy, Education Images)

cm). Females may give birth to 200 baby guppies at a time, though the average is about 50 babies. They prefer water that is 68 to 75°F (20 to 24°C). Guppies reproduce profusely, but the adults may eat the young.

- Swordtails—Similar to guppies, a major distinction of the swordtail (Xiphophorus helleri) is the long, sword-like appearance of the caudal fin. Many different colorings are found on swordtails. They may reach lengths of 3 to 4.75 inches (8 to 12 cm) and are easily kept with other livebearers. Swordtails often eat their young when they are born, so producers use special traps through which the young can escape. Swordtails, like guppies, eat live and dried food. Their environmental requirements are similar to the guppy.
- Mollies—Most mollies are of two species: small fin (Poecilia sphenops) and large fin (Poecilia latipinna). With some mollies, the fins get so large that swimming is nearly impossible. Most mollies are black. Other colors are increasingly in demand, such as the white molly. Mollies prefer water that is 72 to 82°F (22 to 28° C). Mollies like to eat vegetable foods, such as algae and cooked spinach. They like to form large schools in tanks with other mollies. It is best to keep several mollies together in an aquarium.
- Platys—Platys are popular in aquaria and have been bred to achieve certain desired characteristics. The variatus platy (Xiphophororus variatus hybrid) is especially popular. It has a yellow color with a distinctive orange-red caudal fin and dark-lined vertical markings. Some platys may be brownish-yellow in color. They grow to about 2.5 inches long (6 cm) and prefer water that is 68 to 77°F (20 to 25°C). They are hearty eaters, being especially fond of live and dried food. Platys are easy to keep with other livebearers.

SALTWATER ORNAMENTAL FISH

Saltwater ornamental fish are those that live in saltwater. The water is a complex mixture of various minerals, with salt (sodium chloride) being one of the substances that is found in the largest amount. Other minerals in seawater include magnesium chloride, magnesium sulfate, and calcium carbonate. Very small amounts of several other minerals are found in seawater, such as zinc and molybdenum. Most saltwater ornamental fish require natural or synthetic seawater. A saltwater aquarium is a marine aquarium. All of the species included here are egg-layers.

- Angelfish—There are many different species of angelfish; some can live in freshwater. Most angelfish appear delicate, but they are hardy and capable of living a long time in a well-managed aquarium. A popular angelfish is the French angelfish (Pomacanthus paru), which has a black body and vertical yellow stripes. The coral beauty angelfish

(Centropyge bispinosus) has a wide range of colors, including orange and black. Most angelfish prefer a water temperature of 77 to 86°F (25 to 30°C). Angelfish may be 6 inches (15 cm) long and 10 inches (26 cm) high (tip of fin to tip of fin). Angelfish eat a wide range of live food in natural environments, but will eat commercial food in captivity. Their eggs hatch in about 36 hours. As part of the incubation process, the eggs are carried in their parents' mouths and placed on the leaves of water plants and later in the sand.

- Butterfly fish—Butterfly fish are beautiful and popular marine aquarium fish. Most of the butterfly fish need to have ample space in an aquarium. The long-nosed butterfly fish (Forcipiger flavissimus) has beautiful bright yellow and black colors, with the snout and throat being green. This fish will grow to lengths of 6 inches (16 cm). Butterfly fish prefer temperatures of 75 to 82°F (24 to 28°C). Butterfly fish are territorial and are preferably isolated from other butterfly species. Some people keep only one butterfly fish in a tank.
- Basslets—Basslets are small colorful fish that are popular in marine aquaria. The royal gramma (Gramma loreto) is among the best-known basslet. The royal gramma is a hardy fish and particularly good for the person who is just beginning a marine aquarium. Royal gramma are very aggressive toward others of their species. They are, therefore, usually isolated from other royal gramma. They can be mixed with other species of fish. Another popular basslet is the bicolor basslet (Pseudochromis paccagnellae), which has a yellow and violet color. The bicolor basslet is also an aggressive fish. Basslets prefer tiny, living food, such as brine shrimp. They like water that is 79 to 82°F (26 to 28°C).

EQUIPMENT AND FACILITIES

Ornamental fish need a good environment to survive and grow. This environment is artificially created by the equipment and facilities that are used. Needs vary with the species being grown. For example, goldfish can exist in simple systems, but other fish, such as angelfish, require more elaborate equipment.

The basic equipment begins with a water container. Other equipment makes the environment better for fish. Water containers are often decorated with colorful gravel, aquatic plants, and other items.

WATER CONTAINERS

Several different kinds of water containers are used. Some are used to display fish; others are used to reproduce the fish and keep them in good health.

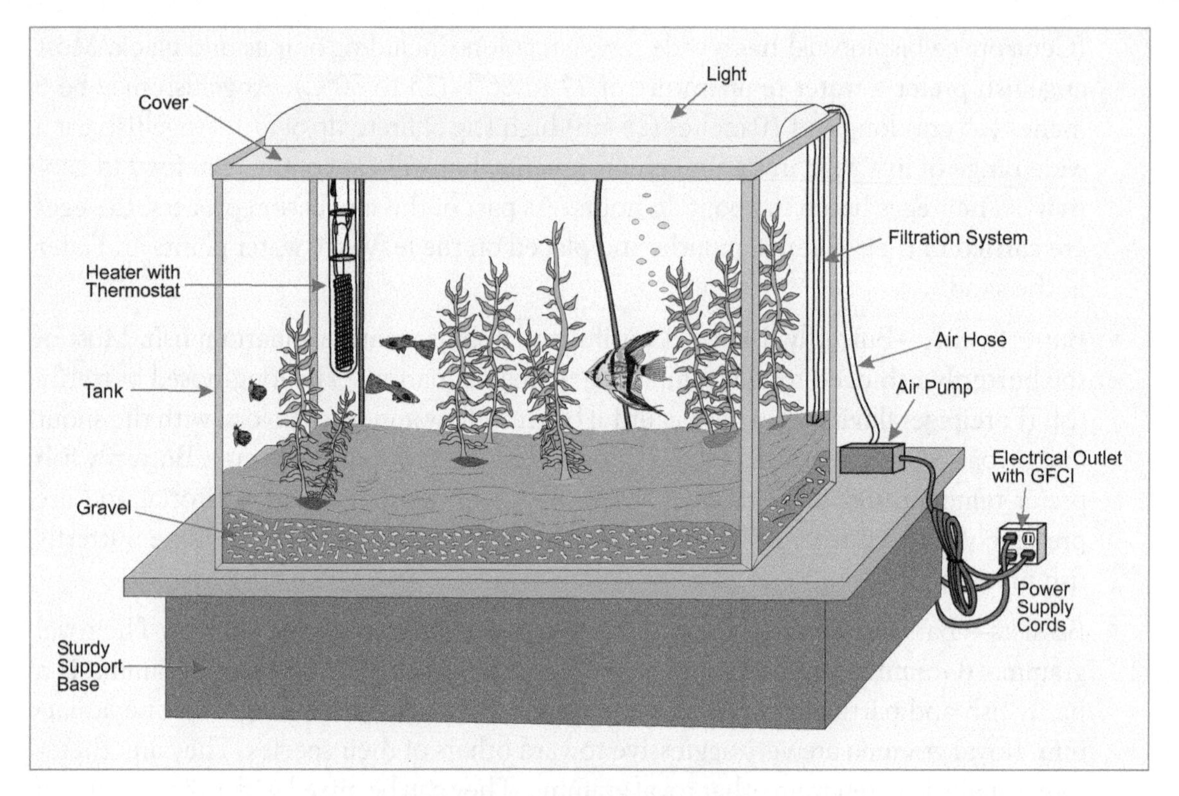

19–10. The basic equipment for an aquarium.

Tanks (Aquaria)

Tanks vary widely. Some are for production; others for enjoyment and often have clear walls for good viewing.

An ***aquarium*** is a container used to hold water. Sometimes the containers are known as "fish tanks" or "tanks." An aquarium should be watertight. Begin with small aquaria and expand into larger sizes. Several smaller aquaria also have advantages over one larger aquarium, especially in disease control and water management.

A simple aquarium is the goldfish bowl. This is a rounded glass bowl with a water capacity of about 1 gallon (4 liters), but it may be smaller or larger. It is used with species that are hardy and can survive in water that may get fouled and is not regularly aerated.

The most common aquaria are rectangular, though some are square or spherical. Specially built aquaria are available, but are often expensive. Aquaria range in size from 10 gallons (42 l) to 30 gallons (127 l), 50 gallons (212 l), 100 gallons (424 l), or larger.

Aquaria of all-glass materials are preferred, especially with saltwater. Aquaria that have metal corners or other parts will corrode and are more likely to leak. All-glass aquaria are easier to clean. The glass in an aquarium should be at least $^{1}/_{4}$ inch (64 mm) thick for small aquaria and $^{3}/_{8}$ inch (96 mm) thick for larger aquaria. A disadvantage of all-glass construction is that the larger aquaria are very heavy—thicker glass is used in making them. Only high-strength glass should be used in building an aquarium.

Aquaria need to be on stands that will support their weight. The stands should be level and located where they will not need to be moved often. Moving an aquarium is a big job.

Pools and Fountains

Decorative pools and fountains may be used for goldfish, koi, and similar species that are hardy. These structures are made of plastic, fiberglass, concrete, or other materials. The construction should be watertight and easy to clean. The arrangement of the facilities should make it easy to manage the water and empty the container, should the need arise.

19–11. Koi in a decorative interior water garden. (Courtesy, Education Images)

Vats

A ***vat*** is a large tank made of concrete, fiberglass, or a similar material. Vats are in a fixed location and cannot be moved about. These structures are used to raise and reproduce fish and not usually for displaying fish. Vats are rather large and may hold hundreds or thousands of gallons of water. The structures should be designed for ease of management. Water circulation, aeration, drainage, and other features should be designed into vats. Vats should be watertight and designed to meet the needs of the species that will be produced in them.

Earthen Ponds

Adapted ornamental species may be produced in small ponds made of earth, known as earthen ponds. These are frequently used with koi, goldfish, sunfish, and other species that

19–12. Students have cultured a sturgeon in a school tank. (The sturgeon is an ancient fish, with fossils dating 200 million years ago. Their bodies are covered with bony plates called scutes rather than scales or scaleless-skin.) (Courtesy, Education Images)

are adapted to the local climate. The pond's construction should prevent seepage or overflow from nearby streams. Select a location that keeps predators and undesirable species out of the ponds. Grassy areas are needed around the edges to keep the water from getting muddy and to reduce erosion.

WATER QUALITY EQUIPMENT

Equipment is needed to keep the water in a fish tank appropriate for the growth of the species.

Oxygenation

Oxygenation is the process of keeping adequate dissolved oxygen (DO) in the water. Oxygen that has been dissolved in the water is removed by the gills of the fish and used for life processes. Water that has little movement in a tank takes on little new oxygen. Fish can quickly use up the DO and die.

Several methods are available to keep dissolved oxygen in the water. Some are much more practical than others.

Aeration is the process of bubbling air into water or splashing water in the air. This uses mechanical methods of moving the water and air. Aeration promotes the gas transfer between the water and the air. The supply of oxygen in the water is replenished by aeration.

With aquaria, air or oxygen may be injected into the water. An air pump may be used to force air through a plastic tube into the water at the bottom of an aquarium. With vats, aerators may be suspended on the surface of the water and fans used to splash the water in the air. Splashing increases the rate of diffusion of oxygen in the water.

Biological oxygenation is using phytoplankton (tiny plants) and other aquatic plants to replenish oxygen. These plants carry on photosynthesis to produce food. In the process, oxygen is released into the water.

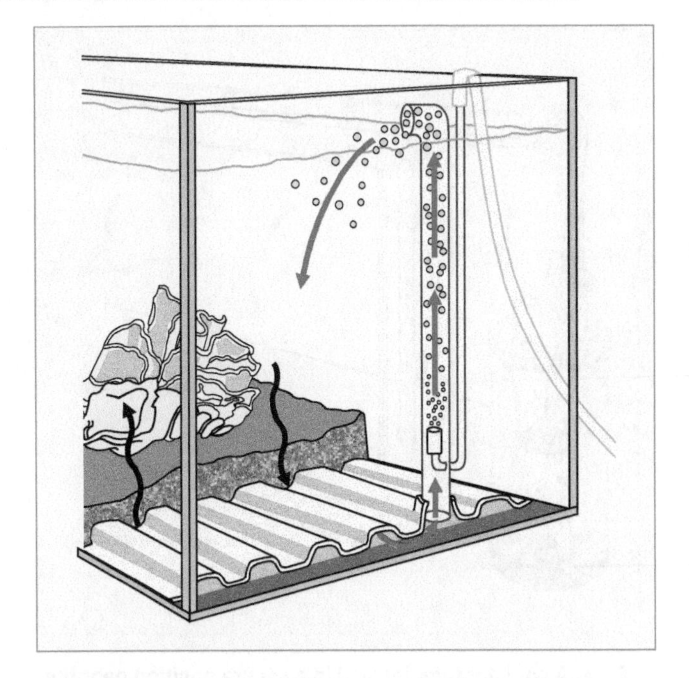

19–13. Underwater gravel filters are sometimes used in aquaria.

Filtration

Fish create solid wastes in the water. These wastes foul the water and make it unfit for fish. ***Filtration*** is the process of removing solid particles and gases from the water to keep a good environment for the fish. Three kinds of filtration are used: biological, mechanical, and chemical.

Biological filtration is using bacteria and other living organisms to convert harmful materials into forms that are less harmful. The bacteria feed on the uneaten feed, feces, and gases, such as nitrogen, in the water. Snails are sometimes used in aquaria to remove organic matter from the water. Crawfish may be used to scavenge food particles at the bottom of the aquarium.

Bacterial action changes nitrogen compounds into less harmful forms and replenishes the DO in the water. Wastes may form ammonia, a compound containing nitrogen. This material undergoes denitrification by bacteria in the nitrogen cycle. The bacteria are primarily found in the filter bed of the aquarium.

Mechanical filtration is using various kinds of filtration devices to remove particles from the water and keep the water clear. It involves flowing water over and through filters made of gravel, charcoal, and fibrous materials (known as floss). Filtering materials must be cleaned often; otherwise, they will become clogged with wastes and will not work very well. Many mechanical filtration systems are tied in with oxygenation systems.

Chemical filtration is using chemical processes to filter water. Special kinds of chemical filters are used. Ozone and ultraviolet irradiation may be used. Chemical filtration often uses activated charcoal to help keep the water clear and prevent water yellowing.

Aquaria often use combinations of biological, mechanical, and chemical filtration. Devices are added to a fish tank so filtration can take place.

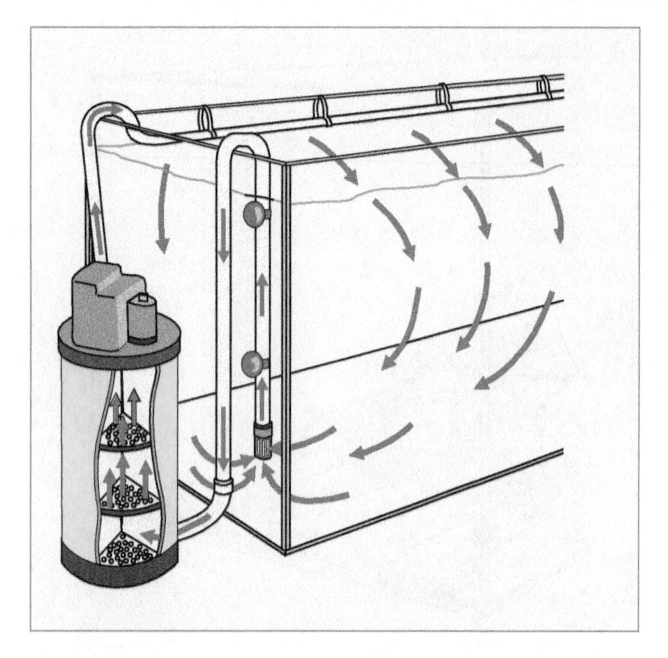

19–14. A canister filter located in a secure position near the aquarium.

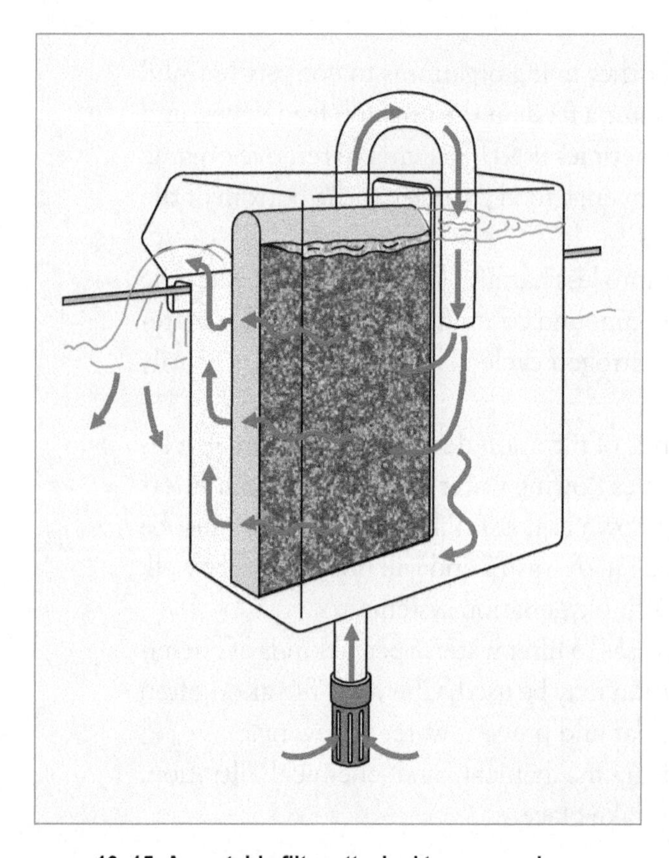

19–15. An outside filter attached to an aquarium.

- Undergravel filters—Undergravel filters are placed at the bottom of the fish tank. The water is pulled through the bed of gravel. The gravel bed is both a mechanical and biological filtration system. The gravel particles screen solid materials out of the water and provide a place for beneficial bacteria to grow.
- Canister filters—Canister filters are filtration systems outside the aquarium. Pumps and a system of tubes move the water through several layers of filtering material, such as activated charcoal and filter floss. The material removes solid wastes and provides a location for the bacteria to act. The material must be cleaned or replaced every few weeks, depending on the extent of fouling of the water.
- Outside filters—Outside power filters hang on the back of the aquarium. Water is moved through the filtration material much as a canister filter. These are common on small aquaria.
- Nitrifier/denitrifier filters—Nitrifier/denitrifier filters supplement the action of bacteria in the nitrogen cycle. These filters are expensive and used only in special situations in aquaria.

Thermometers and Heaters

Thermometers measure the temperature of the water. They often have thermostats attached to them. A ***thermostat*** is a control device that turns on the heater when the water temperature gets below a certain level.

Heaters help keep the water warm enough for fish. Most heaters are glass-enclosed, electric, heating elements that extend into the water. The thermostat in the heater turns the heating element on when the water is cool and off when the water has been warmed sufficiently.

Thermometers may be independent of thermostats and heaters. Each species of fish has its own temperature requirements. Most of the ornamental fish in aquaria prefer temperatures in the 70 to 80°F (21 to 27°C) range.

COVERS AND LIGHTING EQUIPMENT

Most fish tanks should be kept covered and properly lighted. Covers keep the fish from jumping out of the container and prevent predators or objects from falling into the fish tank. Covers are designed for easy removal or opening to feed the fish and remove or add fish.

Lighting may be built into the cover. Lighting is needed so people can see the fish as well as to encourage biological processes. Water plants need light to live and grow. Algae and other organisms may need light to grow. Lights often have reflective hoods that allow better lighting. Caution: Be very careful when using electricity around the water in a fish tank.

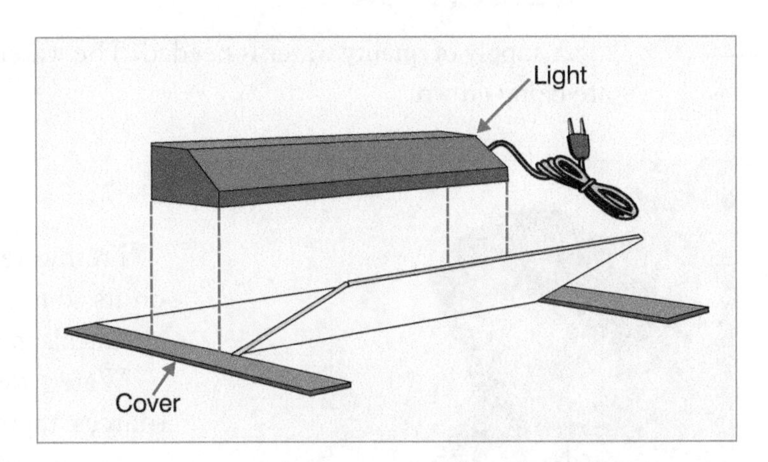

19–16. A combination cover and light for an aquarium.

OTHER EQUIPMENT

A wide range of other equipment may be useful. Examples include dip nets, transfer containers, containers for aging water, decorative items for a fish tank, and hydrometers to determine the salt content of the water in a marine aquarium.

THE WATER ENVIRONMENT

Keeping ornamental fish requires a good water environment. Freshwater and saltwater species usually cannot be mixed. Some species also prefer brackish water, which is a mixture of freshwater and saltwater. Managing the systems varies with freshwater and saltwater; however, many of the principles are the same.

ARTIFICIAL HABITAT

Setting up an aquarium is establishing a desired habitat. A ***habitat*** is a place where a plant or animal naturally lives. The aquarium is set up to duplicate the environment that nature provides. Knowing what to include requires careful study of natural habitats. In some cases, aquaria are designed for a particular climate, such as the Amazon River.

Various plants, animals, minerals, rocks, and other materials may be used in trying to get balance in the artificial habitat. The items are arranged to provide a pleasing appearance as well as a good place for the fish to live and grow.

WATER SOURCE

A supply of quality water is needed. The water must be appropriate for the species that are being grown.

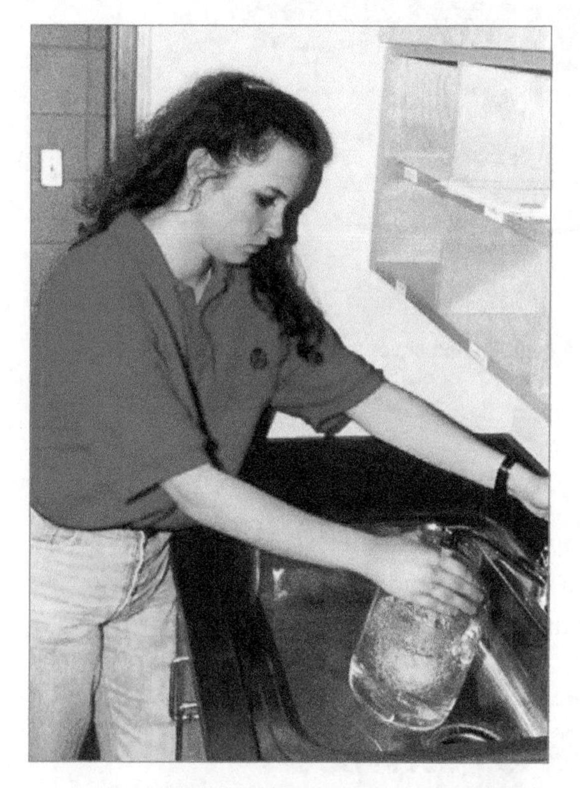

19–17. Water for a small aquarium can be collected in a jar for aging. (Courtesy, Education Images)

Freshwater

Freshwater varies in quality. Water quality depends on its source and the substances in it. The dissolved chemicals in water can make it unfit for fish.

Water for fish tanks comes from three major sources: tapwater, rainwater, and wells.

Tapwater is the water at the faucets in homes, offices, and businesses. Tapwater is readily available but usually cannot be used as it comes from the tap. Municipal water systems often add chlorine or other chemicals to water. These substances keep the water safe for humans to use, but make it dangerous to fish. Tapwater can often be readied by "aging." ***Aging*** is collecting the water in an open container and allowing it to stand for several days. Chlorine and other substances are released into the atmosphere. Of course, a large aquarium will require a lot of "aged" water. Containers in which water is collected and stored must be clean. Dirty containers pollute the water. Tablets that prepare water for use can be purchased at a pet supply store and put into the tapwater.

Rainwater can be used in aquaria, but must be collected so it is not polluted. Acid rain and other substances in rainwater may make it unfit for fish tanks.

Large-scale producers use wells. The water should be tested to determine its mineral content and other substances. In some cases, well water may be unfit for fish tanks. Gases in water can be released by storing the water in an open container for a few days.

Saltwater

Saltwater may be from seawater or made using a synthetic salt mix. Natural seawater is often used near oceans or lakes with saltwater. Seawater is a complex solution of many chemical compounds. Natural seawater must be collected away from sources of pollution. This may involve going out into the ocean, away from inflowing streams, and collecting the seawater in a large tank that is brought back to shore.

Synthetic seawater is made by mixing freshwater with a saltwater mix. These mixes can be bought at pet supply or aquaculture stores. Follow the instructions that are included with the mix for preparing the seawater. In most cases, it is easier and safer to make synthetic seawater than to use natural seawater.

19–18. Synthetic seawater can be made using a specially prepared mixture of salt and minerals. (Courtesy, Education Images)

WATER QUALITY

Water quality is the suitability of the water for use in a fish tank. Water has to be prepared before placing fish in it. This section is about maintaining water quality.

Temperature

Fish thrive best if the temperature of the water is within an appropriate range. The requirements for each species should be determined.

In some cases, the water will need to be warmed. In moving fish, the water into which they are placed should vary only a few degrees from the water where they have been. Sudden changes in water temperature cause stress.

Most ornamental species prefer water that is 70 to 80°F (21 to 26°C). Saltwater species often prefer water that is slightly warmer than the freshwater species. Heaters may be needed to keep the water at a uniform temperature.

pH

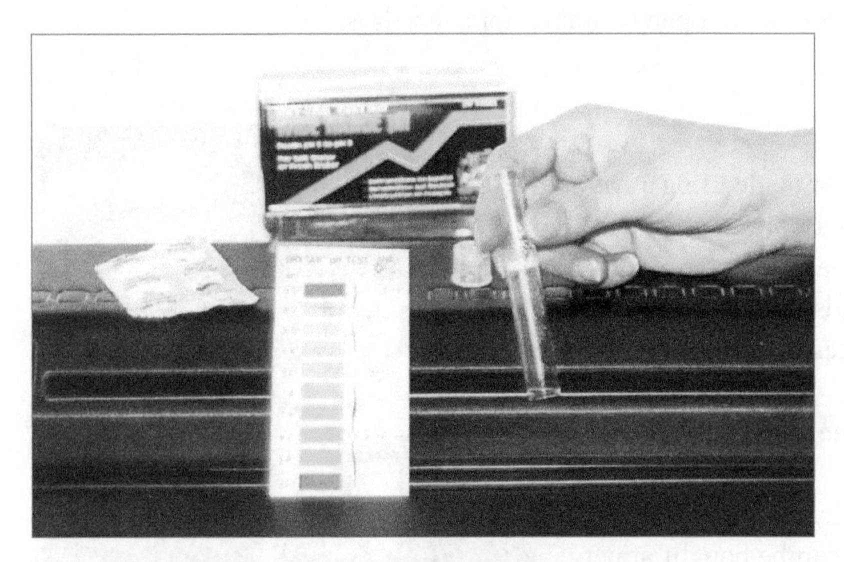

19–19. Using a kit to test pH. (Courtesy, Education Images)

pH is the acidity or alkalinity of the water. It is measured on a scale of 0 to 14, with 7.0 being neutral. pH readings below 7.0 are acidic; the lower the number, the higher the acidity. pH readings are increasingly alkaline as the number goes above 7.0. Ornamental species prefer water with a pH of 7.8 to 8.3.

Changes in pH result from reactions in the water. Many things in an aquarium influence pH. For example, the materials used in a filter are continually being dissolved into the water. Those made from calcium-based materials can result in higher pH readings. The nitrification process and carbon dioxide in the water can lower the pH. Buffering solutions and powders can be bought at pet supply and aquaculture stores to help maintain water within the right pH range.

Nitrogen Cycle

The ***nitrogen cycle*** is the process of converting animal wastes into ammonia, ammonia into nitrites, and nitrites into nitrates. It is a natural process that occurs in water.

Ammonia is toxic to fish. Ammonia can sometimes be noticed as an odor above the water. It is more likely to be a problem when fish are fed too much or when solid wastes are not filtered out of the water. Commercial test kits can be used to test for ammonia. They must be used quickly and properly because, as a gas, ammonia can quickly go into the air. Overfeeding and overstocking the fish tank contribute to ammonia problems.

Dissolved Oxygen

Dissolved oxygen (DO) is gaseous oxygen in the water. Fish and other aquatic life remove the DO from the water. Most fish need at least 5.0 ppm of DO. Below 3.0 ppm DO level, some species show stress and die. Most fish die in a few minutes when DO is as low as 1.0 ppm. Oxygen meters and test kits can be used to measure DO.

19–20. Using a hydrometer to check salinity. (Water in a marine aquarium should have a specific gravity of 1.020 to 1.025.) (Courtesy, Education Images)

Salinity

Salinity is the salt content of water. Freshwater species do not want and cannot survive in water with much salt. Saltwater species require sufficient salt in the water. Salinity is measured as specific gravity using a hydrometer. Water has a specific gravity of 1.0. Saltwater has a slightly higher specific gravity. Most saltwater species want water with a specific gravity of 1.020 to 1.024 at a water temperature of 77 to 80°F (25 to 26°C). Salinity can be increased by using commercially prepared sea salts.

CHANGING THE WATER

Water in fish tanks should be changed periodically. Since this is a big job with large tanks, careful management should be followed to prevent water fouling. Water should be

19–21. Use a sponge brush to clean the inside walls of an aquarium.

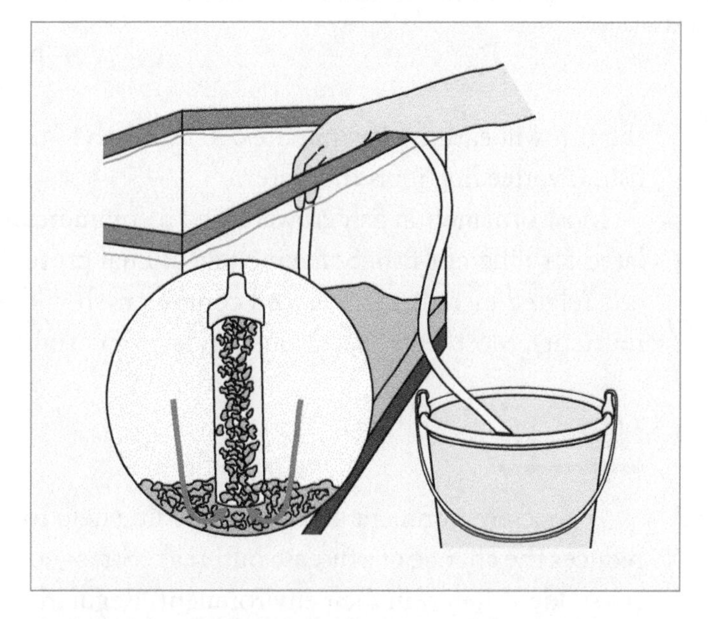

19–22. Using a siphoning tube to remove solid materials from the bottom of an aquarium.

obtained and aged, as with a new aquarium, before it is used. Fish will need to be temporarily moved to another water tank. Care should be used to avoid stressing the fish at this time.

CARING FOR ORNAMENTAL FISH

19–23. Properly feed and care for fish each day. (Courtesy, Education Images)

Ornamental fish have needs that must be met. How long fish live and their rate of reproduction is often related to the care they receive.

FEEDING

Fish need the right amounts as well as the proper nutrients in their feed. The amount a fish is fed each day is its ration. The ration should provide a diet that meets the nutrient needs of the fish. The amount of feed that fish eat is related to the temperature of the water as well as the kind of fish and their stage of life. As the water temperature rises, the use of food by the fish increases; therefore, they should be fed more.

The amount to feed is no more than the fish will eat in a few minutes. This involves careful observation of the behavior of the fish. Overfeeding fouls the water.

Most ornamental fish growers feed a commercial food. These feeds are specially formulated for different fish. Some include animal protein, such as shrimp. Others include grain and related ingredients. All feed containers should have labels that list the ingredients and nutrients. Most feeds are recommended for certain species.

HEALTH

A good environment and proper feeding help to keep fish healthy. Minimizing stress also reduces the chance of a disease outbreak. Stress occurs when fish are handled or when there is a sudden change in their environment. Regularly and carefully observing fish helps people know the behavior of healthy fish. A disease should be suspected when one or more of the following are observed:

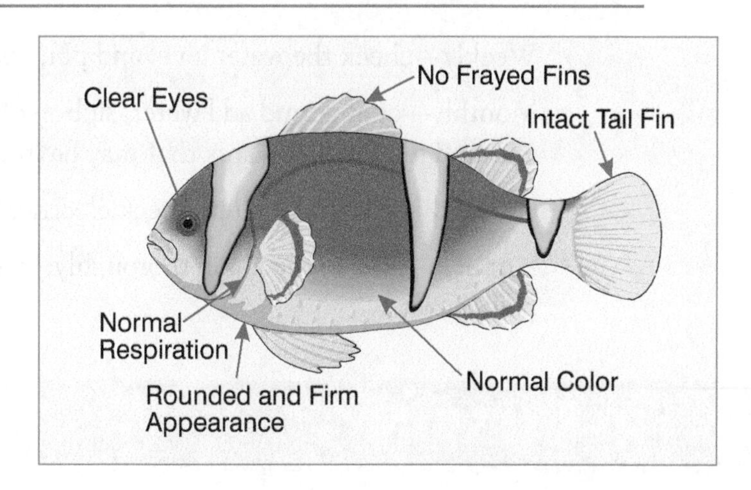

19–24. Signs of good health are important in selecting fish.

- Fish scratching on objects
- Fish failing to eat
- Increased respiration (gill movement) and gasping at the surface of the water
- Clamped or folded fins
- Cloudy eyes
- Bloody spots or fuzzy patches on the body
- White spots on the body
- Frayed fins
- Color changes
- Protruding eyes or abdomen
- Material hanging from the anus of the fish (could be a worm parasite)

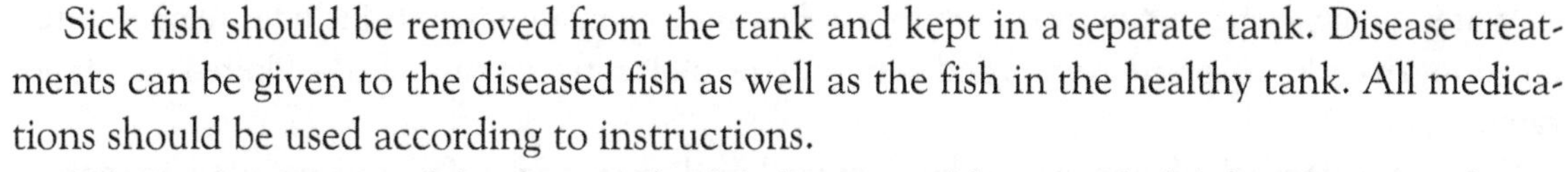

Sick fish should be removed from the tank and kept in a separate tank. Disease treatments can be given to the diseased fish as well as the fish in the healthy tank. All medications should be used according to instructions.

When selecting new fish, observe the fish for signs of disease. If a fish has the signs, do not get the fish. Also, do not get healthy fish from an aquarium where other fish are sick. You will likely be bringing disease to your aquarium!

MAINTENANCE SCHEDULE

Following a regular maintenance schedule can be valuable in keeping an aquarium in good condition. An ***aquarium maintenance schedule (AMS)*** is a listing of important activities and the frequency with which they should be performed in keeping a good aquarium environment. Some activities must be done each day, while others can be done weekly, monthly, quarterly, or annually.

- Daily—provide feed; check the heater, temperature, aeration, and filtration; remove any dead fish; and observe the fish for unusual behavior.

- Weekly—check the water level and pH; add water and chemicals as appropriate.
- Monthly—change and add water; siphon off dead material on the bottom; remove algae; and tend to any plants that may be in the aquarium.
- Quarterly—clean the filter; check electrical connections; check hoses and pump.
- Annually—clean the tank thoroughly; rinse the bottom gravel; and replace light bulbs/tubes.

REVIEWING

MAIN IDEAS

Ornamental fish are kept for beauty and personal appeal. Some ornamental fish are captured from the wild; others are produced in fish tanks. Many different species of ornamental fish are kept. Some are egg-layers; others are livebearers. Some live in freshwater; others live in saltwater.

Producing ornamental fish involves establishing and maintaining a good environment. The needed facilities include water containers, water-quality equipment, and tank covers and lighting.

Water source is important to the success of ornamental fish production. Freshwater can be obtained from drilled wells, municipal water systems, and rain. Saltwater can be obtained from the sea or synthetically made using commercially available salts. In nearly all cases, the water must be conditioned or aged before it is used.

Water quality—the suitability of water for fish—must be maintained. It includes temperature, pH, nitrogen cycle, dissolved oxygen, and salinity. The solid materials must also be filtered out of the water.

Care of ornamental fish includes feeding, controlling disease, and performing maintenance activities on daily, weekly, monthly, quarterly, and annual bases.

QUESTIONS

Answer the following questions using correct spelling and complete sentences.

1. What are the sources of ornamental fish?
2. What are examples of ornamental fish? Name two examples each of freshwater egg-laying fish, freshwater livebearing fish, and saltwater fish. Briefly describe each species.
3. What equipment is needed to produce ornamental fish?
4. What is oxygenation? Why is it important?
5. What kinds of filtration are used? Distinguish between the three common kinds.
6. What should be considered in selecting a source of water?

7. What are the important areas of water quality with ornamental fish?
8. What is the purpose of feeding? What guidelines should be followed in feeding?
9. Name three fish behaviors that indicate a possible disease problem.
10. Describe a maintenance schedule for ornamental fish production.

EVALUATING

Match the term with the correct definition. Write the letter by the term in the space provided.

a. egg-laying fish
b. aquarium
c. oxygenation
d. filtration
e. nitrogen cycle
f. shoal
g. guppy
h. habitat
i. salinity
j. dissolved oxygen

______1. A container used to hold water for ornamental fish.
______2. Fish that reproduce by the female laying eggs.
______3. The process of removing solids and gases from water.
______4. Gaseous oxygen in water.
______5. Salt content of water.
______6. The process of keeping adequate DO in the water.
______7. The process of converting wastes into ammonia, nitrites, and nitrates.
______8. Place where a plant or animal naturally lives.
______9. Group of fish living together.
_____10. Popular ornamental livebearing fish.

EXPLORING

1. Set up two fish tanks in the classroom. Establish a saltwater environment in one tank and freshwater in the other. Select the appropriate equipment and install the oxygenation, filtration, and lighting systems. Select the appropriate species and establish the tank. Establish an artificial habitat that is pleasing to view as well as appropriate for the species selected. Set up a routine aquarium maintenance schedule (AMS) for the tanks. (Additional resources may be needed. A visit to a local pet store will provide access to materials and information on the species to use.)

2. Take a tour of a large aquarium and study the different aquatic environments that have been installed. Ask the manager to let you observe the water filtration systems and other aspects of managing the aquarium.

CHAPTER 20

Wildlife Animals

OBJECTIVES

This chapter is an introduction to animal wildlife and management. It has the following objectives:

1. Explain the importance of wildlife animals.
2. Classify the major game animals.
3. Explain species endangerment and practices to prevent endangerment.
4. Explain important practices in wildlife management.
5. List important considerations for sports enthusiasts.
6. Describe ways to enjoy wildlife.

TERMS

animal wildlife
biodiversity
birding
carrying capacity
endangered species
game
hunting
limiting factor
niche
non-game animal
pelt
territory
trapping
urbanization

20–1. An American black bear has captured its next meal from a rapidly flowing stream. (Courtesy, Sorin Colac/Shutterstock)

PEOPLE enjoy wildlife! Watching it is a favorite pastime for some people. Others enjoy fishing, hunting, trapping, and other ways of harvesting wild animals. People differ in how they view the use of wildlife. Some view legal hunting as a good sport; others disagree. Regardless, reasonable uses are needed to assure that a species doesn't become extinct.

The Endangered Species Act helps protect animals from extinction and promotes the recovery of endangered species. One example is the gray wolf (*Canis lupis*). This animal was hunted to near extinction in some locations. With efforts to promote its repopulation in the Big Rivers Region of the Great Lakes, the gray wolf was taken off the endangered species list in 2006.

People have come to realize that a healthy wildlife population requires planned management. Protective laws may be needed. Habitat for a particular species may need to be established. Regardless, wildlife are essential to ensure biodiversity on Earth.

THE IMPORTANCE OF WILDLIFE

Animal wildlife is any animal that has not been domesticated. This includes insects, birds, fish, rodents, and many others, such as deer, elk, and bear. These animals benefit humans in many ways. Those hunted for food and other uses are known as **game**.

Why should we be concerned with wildlife populations? Our society has evolved to a point where we can go to the grocery store and buy any kind of meat or fish we could possibly want. There is no longer a need to hunt or fish for our food.

This argument would be true if the only reason we wanted to have fish and wildlife was to provide food. There are many other reasons to be concerned with wildlife and game fish populations. In this chapter, economics, beauty, recreation, and biodiversity are covered.

20–2. Once endangered animals, bald eagles are among favorite birds to watch in some places. (Courtesy, Education Images)

ECONOMIC VALUE OF WILDLIFE

Large sums of money are spent on wildlife in North America each year. Millions of dollars are made every year in department stores that sell hunting and fishing supplies. The communities

20–3. Hunters spend a great deal of money on guns, travel, and other things, such as camping.

around popular hunting, fishing, and wildlife areas and parks depend on the income from nature lovers, hunters, and fishers to operate the services of the community. Sports enthusiasts in many states pay an additional tax for hunting and fishing supplies. In most cases, the funds raised from this tax is returned to support wildlife in the state.

Hunting is taking animals with guns or other weapons. The animals are used for meat, hides, or other products and to make trophies for home or office decorations.

Trapping is capturing animals for their products. The animals are usually caught in traps and remain alive until someone checks the traps. Most of the animals are valued for their hides. Some trapped animals have other products. For example, such as people in local areas may use the meat of the raccoon for food. Some animals are trapped to move them to other places. People who trap animals are trappers.

A ***pelt*** is the whole hide with fur attached. Pelts are used to make valuable fur coats and other products. Some animals used for pelts include fox, rabbit, squirrel, raccoon, and mink.

20–4. Pelts in a Sitka, Alaska, store were taken by trappers. (Courtesy, Education Images)

BEAUTY AND AESTHETICS

People stop to look at a deer grazing in a field or a young family of mallard ducks swimming on a pond. We enjoy the

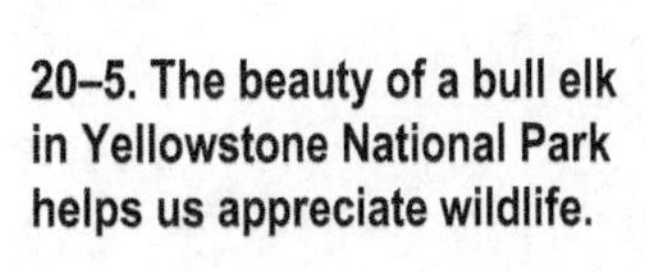

20–5. The beauty of a bull elk in Yellowstone National Park helps us appreciate wildlife.

show wildlife animals provide. Many people want to vacation in the country or at a park in hopes of seeing wildlife animals. The beauty of wildlife captures our attention and imagination.

Chances are that you have visited a zoo or been a part of a wildlife officer's program that let you get close to a wild animal. You may even be in a club formed to enjoy wildlife. Bird watching clubs, conservation clubs, and organizations at local parks have become a popular way to enjoy wildlife.

20–6. Sport fishing is popular with people of all ages. (Courtesy, Rick Rudd, Virginia Tech)

RECREATION

Every year, hundreds of thousands of people enjoy fish and wildlife through hunting, fishing, photography, scouting, bird watching, and other recreational activities. Adults and children can enjoy nature through recreational activities.

Although hunting for food is no longer required for us to survive, it gives hunters a chance to relive what it must have been like for our ancestors. Most hunters are true sports-people who eat what they kill. Hunting is one of the oldest sports in the world.

CONNECTION

BAG LIMITS

Bag limit is the maximum number of game that a hunter can take. The number may be on a daily, seasonal, or other basis. The limits may apply to any game–fish, fowl, deer, or others that are covered by the regulations. Non-game species are not protected by bag limits.

The purpose of bag limits is to protect wildlife and have a harvest that maintains the population. If one hunter takes more than his or her share, wildlife is threatened and other hunters may not have animals to take.

The number in a bag limit is based on game populations. Changes in wildlife populations may result in bag limits being raised or lowered depending on what is needed to protect the wildlife. In some cases, the bag limit is only one per season for a hunter, such as this hunter has with the turkey.

Fishing with a hook and line has been called relaxing, tense, laid back, and exciting—all are true! Many people who fish today still fish for food. Many others fish for sport, releasing most of their catch. We are fortunate to live where we are never far from a place to fish.

Photography, scouting for wildlife, bird watching, and other recreation activities are enjoyed by people who live in the country, suburbs, and in the city.

20–7. A jackrabbit represents biodiversity. (These rabbits are readily identified by their large ears.) (Courtesy, Fremme/Shutterstock)

BIODIVERSITY

North America has a variety of wildlife animals. From the desert southwest to the maple-aspen forests of the northeast, many unique environments and many unique species of wildlife are found.

Having a variety of wildlife species adds to the ***biodiversity*** of our planet. Biologically diverse populations are less likely to be harmed by disease or illness. One concern is the growing number of wildlife animals that are ***endangered species***. Endangered species are animals that are close to becoming extinct. Every time an animal reaches extinction, the biodiversity of our planet decreases.

MAJOR TYPES OF WILDLIFE

It is impossible to list all of the wildlife in North America. A few of the major types of fish and wildlife animals follow.

GAME ANIMALS

Animals hunted for sport, meat, and other products are game animals. The kinds of animals used for game varies with where you live. Game animals can be large, like the white-tailed deer and the antelope. They can also be small animals, like the rabbit or squirrel.

In most places, game animals can be harvested by hunters during a regulated hunting or trapping season. Personnel in state departments of game and fisheries decide when the

20–8. Examples of mammal wildlife.

season occurs and how many animals can be harvested. A license is usually needed to hunt. Hunter education is expected before a person begins hunting with a gun.

Popular species of large game animals include the white-tailed deer, mule deer, elk, antelope, bighorn sheep, and alligator. Small game species include the cottontail rabbit, squirrel, fox, mink, raccoon, and muskrat.

NON-GAME ANIMALS

More non-game animal species than game animal species exist. These animals can be found deep in the forest or in urban areas, or just about any place in between! The diversity of non-game species is amazing.

A ***non-game animal*** is one that does not provide a consumable product like meat or furs. It provides beauty and adds to the biodiversity of our planet. Some non-game animals serve as a food source for predators. Others are scavengers and help nature recycle by eating dead animals. Non-game animals are an important part of nature.

Most people do not think of wildlife in their neighborhood. Many non-game animals can be found in urban and rural areas across the country. Mice, rats, opossum, and others are comfortable in an urban setting.

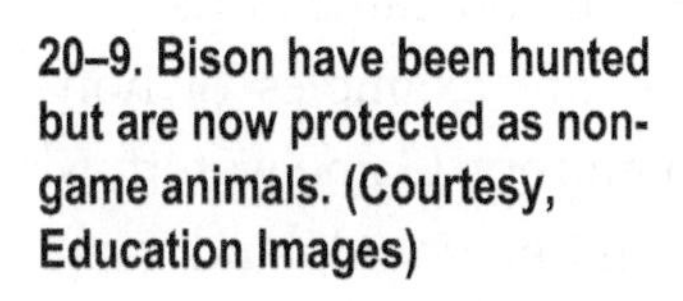

20–9. Bison have been hunted but are now protected as non-game animals. (Courtesy, Education Images)

GAME FISH

Game fish are sought to provide food or sport to the fisher. Game fish exist in both fresh and salt water and come in a variety of sizes. Most game fish species are caught with a hook and line.

20–10. A sport fisher with a 14-pound coho salmon.

Freshwater game fish can be caught in every state. Again, the types of game fish will differ from state to state. Major types of freshwater game fish include black bass (largemouth, smallmouth, and spotted), sunfish, walleye, trout, pike, muskellunge, salmon, striped bass, crappie, and others. Popular saltwater species include groupers, sea trout, cobia, snapper, flounder, redfish, mackerel, tuna, sailfish, dolphin, and many more.

NON-GAME FISH

Just as non-game animals outnumber game animals, non-game fish outnumber game fish. Non-game species can be found in small freshwater creeks, ponds, lakes, and in saltwater habitats. Many species of non-game fish are too small to be caught on a hook and line, but some non-game species are very large. For example, the sturgeon can grow to over 2,000 pounds!

GAME BIRDS

Game birds are hunted for food and sport. They can be divided into three basic categories that include migratory (non-waterfowl), waterfowl, and upland game birds.

The woodcock and the crow are examples of non-waterfowl game birds that are migratory. Game waterfowl include many species of ducks and geese. Upland game birds include the pheasant, grouse, and quail.

Harvest of game birds is regulated by state and federal law. Since migratory birds can travel to other countries, laws governing their harvest often have international input.

20–11. A red tail hawk is a bird of prey.

BIRDS OF PREY

Eagles, osprey, owls, and hawks are all examples of birds of prey. Most birds of prey are protected by laws that make shooting them or attempting to catch or harm them illegal. These birds are hunters. They eat small rodents, fish, snakes, reptiles, amphibians, and small mammals.

NON-GAME BIRDS

Non-game birds include all other birds. The sparrow, blue jay, robin, and pelican are all examples of non-game wild birds. Some are known as song birds and are found in areas where people have homes. Non-game birds live in a variety of environments. You can find these animals in wetlands, plains, uplands, and anywhere in between.

20–12. Binoculars aid in bird watching. (Courtesy, Education Images)

ENDANGERED SPECIES

Unfortunately, people have had a negative impact on many wild animals. Through ***urbanization*** (building cities, suburbs, and all of the support services needed by people in these areas), agricultural growth, pollution, and neglect, many species of wildlife are extinct or endangered.

With work, the fate of some endangered species can be changed. The American alligator was once close to extinction. Today, the alligator is doing very well. In fact, the population is in such good shape that alligators can be legally harvested in some places in the southeastern United States! Other examples include bald eagles and bison.

WILDLIFE MANAGEMENT PRACTICES

Think for a moment about where you live. How many people live in your home? Does everyone have a bed and a chair at the table? People need space in which to live. You would not be very comfortable if 10 people shared your bedroom, or if 50 of your classmates came to your house for dinner. There are limits to how many people your home can accommodate. Wildlife species are limited by the same factors.

20–13. A grizzly bear has unique needs for survival.

HABITAT

Habitat is the place where animal wildlife live. It has four components: food, space, water, and cover (shelter). Think of your home as a habitat. It is a place where your basic survival needs are met. You have cover in a house or apartment. You also have food, water, and space to live.

Wildlife must have the proper habitat to meet their needs. All four components must be in place to support a wildlife species. If one component is missing or lacking, it is called a ***limiting factor***. A deer needs to have browse (woody, broadleaf plants used for forage). If there is no browse, the deer cannot survive in that habitat—food is the limiting factor. Pheasants need tall grass to nest. If there is no tall grass in the habitat, there will be no pheasants—cover is the limiting factor.

Even if a habitat is perfect for a particular species, there are limits to how many animals the habitat can sustain. The habitat's ***carrying capacity*** is the number of animals it can support. Carrying capacity is usually low in the winter and high in the summer. The population in a given habitat is directly related to the carrying capacity.

INCREASING WILDLIFE POPULATIONS

The best way to increase wildlife populations is to meet their basic needs in the habitat. While this is true, there are other considerations.

Niches

Every animal has a ***niche*** or a special place in a habitat where it belongs. The animal fills an important role in consuming and producing for the overall habitat. For example, white-tail deer and squirrels both live in the same habitat. The deer eats small branches and likes to bed in fields of high grass—this is its niche. The squirrel lives in the trees and eats nuts, berries, etc. The squirrel fills a different niche. Some animals may share similar niches, but they are seldom the same.

20–14. A coyote enjoys a special niche in its territory.

Interspersion

When things are interspersed, they are mixed. For example, in an 8-acre plot of land, 2 acres could be dedicated to food, 2 acres to water, and the remaining 4 acres to cover. If these areas are arranged in large blocks, there would be very few places where animals could have easy access to all three areas. Placing these resources in smaller blocks would allow quicker access to all three and increase the number of animals in the habitat.

Territory

Some animals claim an area, or ***territory***, which they protect and keep other animals from entering. Bears will mark their territory to warn others from invading. Space can quickly become a limiting factor for territorial animals.

CONSIDERATIONS FOR WILDLIFE SPORTS ENTHUSIASTS

Wildlife sports are important traditions in some communities. Sports include traditional hunting and trapping, as well as watching (observation), attracting wildlife, and photography. Sports enthusiasts need to keep some simple rules in mind when enjoying wildlife.

EXERCISE CAUTION

20–15. Hunter orange should be worn when hunting to make it easy for other hunters to see you. It is also the law in most states. (Courtesy, Education Images)

Hunting and trapping are sports that involve deadly force. Guns can kill or injure people as well as animals. Hunter and trapper safety courses are required in many states before a license can be obtained. These courses will teach you how to use traps and firearms and to respect their power.

Another potential area of danger involves human/animal interaction. Wild animals are wild! They are not pets or zoo animals. They are driven by instinct and will try to get away from you and out of danger any way they can. Often, this includes biting, scratching, goring (puncturing with horns), or attacking, using any means available to them. When hunting, trapping, observing, or photographing, remember to stay at a safe distance.

TAKE ONLY WHAT YOU NEED

Wild animals are a renewable resource. Game harvests are closely regulated by state and federal agencies. These agencies set limits on the number of animals that can be taken. To maintain wildlife populations, it is necessary for sports enthusiasts to obey those regulations.

There are times when a fisher or hunter may be having an excellent season and can harvest limits of fish or game on a daily basis. The sports enthusiast needs to ask, “How much will I use?" If the harvest will be wasted, should you take the animal even though you are within the legal limits? A conscientious sportsperson would answer, "No."

RESPECT THE HOME OF ANIMALS

How would you like to have visitors in your home leave their trash on the floor? Would you be upset if someone came to your house and broke all of the furniture? You probably would not be happy with either situation.

People are often in the homes of animals. While we are in their habitat, we need to respect where they live. Leaving paper, cans, and bottles, or damaging trees and animal homes, is not the way a responsible sports enthusiast uses the resource.

WAYS TO ENJOY WILDLIFE

Most people enjoy wildlife in one way or another. Someone who lives in the city can enjoy many kinds of birds and small animals that have adapted to city life. Some ways to enjoy wildlife without hunting follow.

BIRD WATCHING

Many people enjoy watching birds in the wild or even in their own backyard! Watching and studying wild birds is ***birding***. We can attract birds with feeders or with plants that

20–16. Wildlife photography is a favorite activity with some people. (Courtesy, Education Images)

20–17. Birding provides wonderful insights into the lives of birds. This shows a flycatcher with four young. (Courtesy, Education Images)

20–18. The short-eared owl has an interesting appearance. (Courtesy, Education Images)

they need for food. It is easy to attract birds to your home with a simple birdfeeder. You can purchase a feeder or make one that will hold seeds and grains the birds eat. Hanging flowering plants like fuchsia will even attract hummingbirds! If you start "birding" be aware that it can be very addicting! Some people will keep logs of the kinds of birds they spot. Others will even take trips to faraway places in hopes of adding new birds to their list!

PHOTOGRAPHY

Taking pictures of wildlife animals can be very exciting! Capturing an animal on film is not as easy as it sounds. Often the photographer must plan carefully and wait for long periods of time to get their shot! In many ways, the photographer uses the same skills that a hunter uses, but the animals are left in the wild and the trophy is the photograph.

ANIMAL WATCHING

Some people simply enjoy watching wildlife animals. It is special to see a doe and her fawn feeding in a field or even watching a family of skunks scurry around a park. There are probably several places in your community where you can go to watch animals. From bat houses to public parks there are usually many options for you to choose from. Some people will set up animal feeders or plant crops that will provide the wildlife with food in areas where they can watch them.

REVIEWING

MAIN IDEAS

Animal wildlife exists in large numbers across the country. These animals are important to us economically. They add beauty to our world. They provide recreation. In addition, they add to the biodiversity of our planet. Major types of wildlife species include game and non-game animals, game and non-game fish, game and non-game birds, birds of prey, and endangered species.

Managers of wildlife must be concerned with providing food, water, cover, and space for the animals. Wildlife populations can be increased by providing niches for animals, practicing interspersion, and meeting territorial needs of animals.

Sports enthusiasts should exercise caution when dealing with wild animals. Those that are conscientious take wildlife that they will use. All people who use wildlife need to respect the animal's habitat and home.

There are many ways to enjoy wildlife animals including birding, photography, and animal watching. For more information on wildlife, refer to Wildlife Management: Science and Technology available from Prentice Hall Interstate.

QUESTIONS

Answer the following questions using correct spelling and complete sentences.

1. What are the four reasons that animal wildlife is important? Briefly explain each reason.
2. Distinguish between animal wildlife and game animals.
3. What is an endangered species? Give an example of an animal that was once endangered, but is no longer.
4. What is habitat? What are the four components of a habitat?
5. What can be done to increase wildlife populations?
6. How should sports enthusiasts respond to animal wildlife?
7. What is your favorite way to enjoy wildlife?

EVALUATING

Match the term with the correct definition. Write the letter by the term in the blank provided.

a. hunting
b. niche
c. endangered species
d. game
e. animal wildlife
f. territory
g. carrying capacity
h. pelt
i. birding

______1. The natural skin covering an animal used in making clothing and other products.
______2. Animal close to becoming extinct.
______3. The number of animals a habitat can support.
______4. The place and role of an animal in its habitat.
______5. Taking animals with guns and other weapons.
______6. An animal harvested for food or other products.
______7. The area an animal will defend as its own.
______8. Animals that have not been domesticated.
______9. Watching and studying wild birds.

EXPLORING

1. Develop a personal library of information on game animals in your area. Collect brochures and other materials in your library. These are available from a local conservation officer, game and fish commission, or office of the Cooperative Extension Service.

2. Tour a wildlife refuge. Learn the kinds of animals in the refuge and the practices followed in improving habitat for the animals. Prepare a written or oral report on your findings.

3. Select a species of wildlife found in your area. Prepare a report that describes the animal, its habitat requirements, and other conditions that provide for its well-being.

UNIT FIVE

(Courtesy, Boris Djuranovic/Shutterstock)

Service, Laboratory, and Exotic Animal Technology

CHAPTER 21

Service and Safety Animals

OBJECTIVES

This chapter has information on using animals for service and safety. It has the following objectives:

1. Explain service and safety animals and give examples of each.
2. Describe practices with service and safety dogs.
3. Describe practices with service and safety donkeys.
4. Describe practices with service and safety llamas.
5. Describe practices with service and safety geese.

TERMS

cria
guard dog
handler
jack
jenny
K-9
multipurpose dog
safety animal
service animal
single-purpose dog

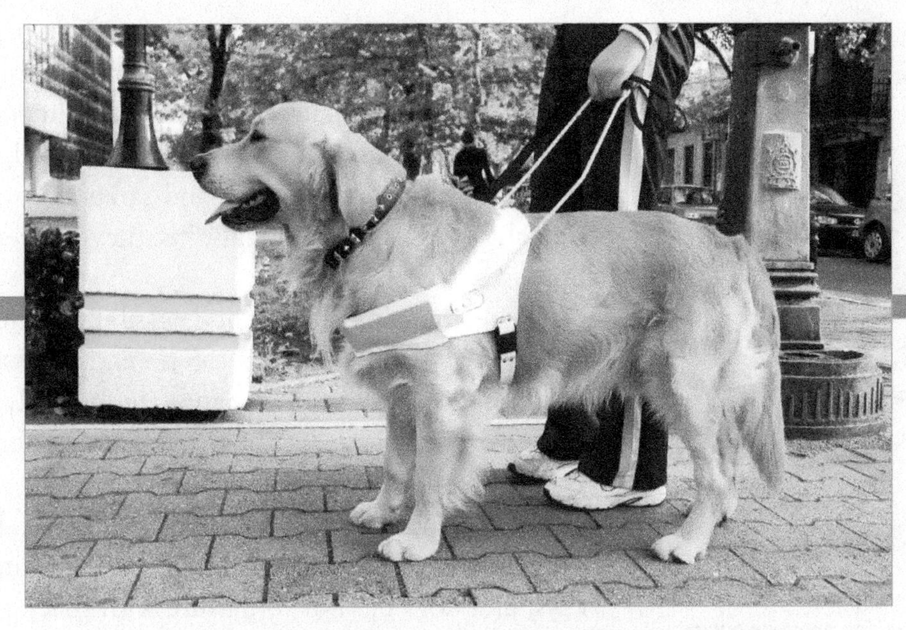

21–1. A trained seeing-eye dog. (Courtesy, Boris Djuranovic/Shutterstock)

ANIMALS help people in many ways! Think about the people you know who have animals. What kinds of animals do these people have? How do their animals help them?

Some animals are used in ways that do not provide food or pleasure. These animals help people live better lives. In some cases, they help people live independently. Close your eyes for a moment and think about what it would be like to be unable to see. Can you find your way to school with your eyes closed? A trained leader dog for the blind helps visually impaired people live independently.

Many different animal species are used for service and safety. Training is usually required so the animals will know what to do and how to do it. The training usually includes close bonding between the animal and its human owner.

KINDS OF SERVICE AND SAFETY ANIMALS

21–2. A specially trained dog helps its owner to be active and independent. (Courtesy, Micimakin/Shutterstock)

Service and safety animals help people in many ways. A ***service animal*** is an animal that gives assistance in some way. These animals often provide an important service. A ***safety animal*** is an animal that helps protect people and property. Service and safety are not widely thought of as uses for animals, but they are very important uses.

There are many kinds of and possibilities with service and safety animals. Animals can do various tasks if they have the correct instincts and are properly trained. Examples of service and safety animals are dogs, donkeys, llamas, and geese.

Service and safety animals help and protect. Dogs and geese in a person's yard warn of intruders. Llamas, donkeys, and dogs put out to pasture with sheep and goats protect them from predators. Today, many ranchers are using other animals to control the aggressive predators and protect their animals.

DOGS

21–3. Trained border collie dogs are herding sheep. (Courtesy, Rickshu/Shutterstock)

Dogs are common service and safety animals. Many homes have dogs. Dogs are used not only as companions, but to warn the occupants of the home when someone arrives. Dogs help with law enforcement, assist people who have visual or mobility problems, guard property, and herd livestock.

An example of how dogs are helpful is the police dog. The ***K-9*** is a dog trained to assist with law enforcement work. The K-9 is a highly versatile, dedicated member of a police department. When properly trained, the police K-9 can perform a variety of services.

K-9 unit dogs are commonly the German Shepherd breed because they are easily trained, strong, aggressive, and good tempered. An aggressive dog is needed for police work; however, a mild-tempered dog is best with the public. Some law enforcement agencies use standard poodles and Rottweilers.

21–4. A German Shepherd is being trained for security work. (Courtesy, Marcel Jancovic/Shutterstock)

Farmers, ranchers, and business owners use trained dogs as guards. A ***guard dog*** is one that is trained to protect property. It may bark, bite, and attack potential danger. With sheep, a guard dog stays with a flock of sheep and protects them. With businesses, it stays inside a building or fenced area. Its presence deters thefts and vandalism.

The breeds most common for guard dogs in the United States are large dogs weighing 80 to 120 pounds (36 to 54 kg). They stand at least 25 inches (63 cm) at the shoulder and have flat ears covered with hair, short and blunt muzzles, and long tails. They are usually fawn, gray, or all white with dark muzzles. The most common breeds used for guard dogs are Great Pyrenees from France, Komondor and Kuvasz from Hungary, Maremma Sheepdog from Italy, Cão de Castro Laboreiro from Portugal, Shar Planinetz from Yugoslavia, Tibetan Mastiff from Tibet, and Akbash Dog and Anatolian Shepherd Dog, both from Turkey. The most successful breeds are the Akbash and Pyrenees.

Some ranchers use dogs for herding. A trained herding dog can move sheep, cattle, and other animals from one pasture to another or into a corral. Border Collies are well known as herding dogs.

21–5. A trained beagle has identified contraband in a bag belonging to an air traveler. (Courtesy, USDA)

TRAINING

K-9 units require special management practices. First, a handler is selected. A ***handler*** is the person who works solely with a dog. They go through training together and will work together later.

Dogs are trained at one to three years of age. They go through 12 weeks of training with their handlers. The dogs are trained to respond to commands and to bite and hold rather than to maul. Dogs trained in specialty areas, such as explosives and narcotics, require at least another 8 to 10 weeks of training. After that, the dog still goes through two days of training per month. K-9 units may be used for searching buildings for burglary suspects, tracking suspects of crimes, locating lost persons, locating explosive or incendiary devices, searching out hidden narcotics, and protecting their officer partner.

A dog is used in law enforcement until it is eight or nine years old. This depends largely on the area of work. The harder the job, the earlier the dog is retired. The handler usually has the opportunity to adopt the retiring dog.

Dogs may be trained as single-purpose or multipurpose animals. A **single-purpose dog** is trained and used for just one activity, such as patrols, tracking, narcotics, or explosives. A **multipurpose dog** is trained and used for more than one activity. The choice of training depends on how the dog is to be used in law enforcement. An agency that is small and has only one dog may want a multipurpose dog. Large agencies may have several dogs with specialized training.

Training and care are needed. Guard dogs are not like donkeys or llamas who need very little training and care. Training a guard dog takes time. First, the dog must be acquired when it is young, about six to eight weeks old. It should be intelligent, alert, and confident. There is no difference between the performance of male and female pups. However, if a female pup is selected, it is recommended that she be neutered.

21–6. A K-9 in active training for law enforcement work. (Courtesy, Jody Pollock, Michigan)

Puppies that will work sheep must be reared with them. They are given little human contact from six to eight weeks of age. The pup must be monitored and bad behaviors must be corrected immediately. The training for the pup will take up to a year. Patience is essential. Give the pup the opportunity to learn with the flock. Start the pup out in small areas and keep working with it.

When a pup is weaned from the litter and ready for training, the socialization with the sheep begins. The pup is placed in an area of about 150 square feet (14 m^2) with three to six lambs. Orphan lambs work well because all of the animals are lonely and they are more likely to establish bonds. The pup should have an area it can escape to that the lambs cannot get to, which should contain the pup's food. The water dish for the pup and lambs should be shared to force bonding between the two. The pen should be checked several times the first few days to make sure all animals are healthy and adjusting to each other.

Gentle play between the pup and the sheep is okay. Be alert if the play gets too rough. A lamb that dominates the pup should be removed and another lamb put in its place. The reason to put the pup alone with the lambs is to build confidence and to bond with the sheep. The pup may be petted by the owner each day when it is checked. Excessive handling must be avoided.

The animals may have to be moved to bigger pens periodically if the pup grows too big for the pen. After the pup is 16 weeks of age, it and the group it was raised with can all be put out with the flock. The pup should not be put into an area where there are predators until it is ready to take on the guarding role. The age varies with dogs, but it is sometimes as young as $4^1/_2$ months. The dog will begin patrolling and taking on other activities of guarding. It should be large enough to defend itself from predators.

It takes pups up to two years or more to mature. Puppy foolishness and guard dog genetics make them independent. Also, keep the goal in mind. This pup is not a herding dog, but a guard dog. Its job is to stay with or near the sheep and protect them from predators. Herding dogs, such as Border Collies, Australian Shepherds, and Australian Kelpies are used to move sheep from area to area by nipping, chasing, and barking at the sheep.

21–7. A Border Collie pup lives next to lambs as part of his training. (Courtesy, Jody Pollock, Michigan)

One-fourth of the dogs trained as guards will not be successful. Failure may be due to starting training of the pup when it was too old, or improper rearing by the owner. Some dogs also will harass, injure, or kill stock, and roam away from the flock. Always post warning signs to protect unknowing people from a guard dog. Never allow a guard dog to roam where it can attack innocent people.

The number of guard dogs needed varies. It depends on how old the dog is and how much area the dog can handle. The sheep may graze close together or scatter. The rule is to start with one dog and let it get established before adding others. A first-time guard dog user should start with one pup.

Three behaviors for an effective guard dog are:

- be trustworthy to not injure the stock;
- be attentive and stay close to the flock; and
- be protective whenever a predator shows up.

A guard dog that is not on duty (with the sheep) is not where it is supposed to be!

FEEDING AND NUTRITION

A dog used for service and safety work must receive proper food if it is to perform effectively. Each breed varies somewhat in requirements. In general, a dog should receive needed energy to do its work and other needed nutrients for it to stay healthy. Commercial feeds selected for the age, condition, and work activity are likely the best to use.

Throughout the work day, dogs should receive plenty of good water. Never let them drink from a puddle in the street or another dog's bowl if the dog is a disease suspect. Information on the feeding of K-9 unit dogs and guard dogs can be found in Chapter 16.

CONNECTION

BEWARE OF DOG

Some dogs are dangerous. They will attack and bite people. Bites can cause severe injuries and deaths, especially with children. This can happen quickly when the owner is not around.

The owner of a dog is responsible for its behavior. Dogs trained for law enforcement and retired may retain their aggressive instinct. These dogs must be controlled at all times. They must be penned and on leashes. A dog that may attack should never be allowed to run loose. In some cases, euthanasia is appropriate. Posting a sign is informative but not adequate if a dog runs free.

A dog should never run free in residential areas. All dogs cause damage to property. Some can cause human injury, pain, and death. (Courtesy, Education Images)

HEALTH CARE

21–8. Maremma pups are eating highly nutritious goat's milk. (Courtesy, Luck-E-G Nubians, Idaho)

A service and safety dog should be properly vaccinated, wormed, and otherwise receive health care. Those that work around people will also need grooming to maintain an attractive appearance. K-9s are usually under routine care of a veterinarian.

Guard dogs do not usually live to a very old age. It is easy for a rancher to forget to worm and vaccinate the dog since it is out with the sheep. Besides disease, some dogs just disappear, others are poisoned, shot, or hit by vehicles. Guard dogs may only live to be about three years old, so it is important to start training a new pup. Ideally, the training will occur while the older dog is working well with the sheep.

FACILITIES AND EQUIPMENT

Proper facilities and equipment are needed to care for the well-being of service and safety dogs.

Using a K-9 unit is a 24-hour-per-day, 365-day-per-year decision. The K-9 unit dogs have varying facilities. Some law enforcement departments mandate the dog go home with

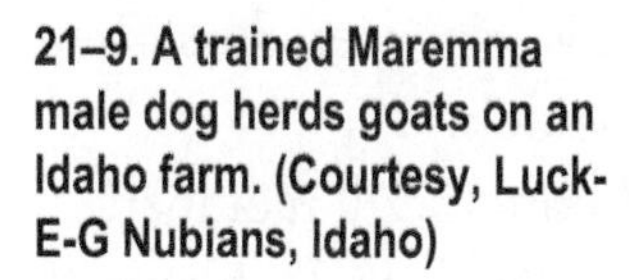

21–9. A trained Maremma male dog herds goats on an Idaho farm. (Courtesy, Luck-E-G Nubians, Idaho)

21–10. The skills of a German Shepherd are demonstrated at a dog show. (Courtesy, U.S. Department of Agriculture)

its handler. The dog is with the handler 24 hours per day. The dog rides with the handler, does duty with the handler, and lives in the home with the handler and his or her family. Some officers say they spend more time with their dog than their spouse or children. Other K-9 unit dogs are kept in kennels at the police department or in a kennel at the handler's home.

Sheep guard dogs spend all of their time with the sheep. Special facilities, separate from the sheep, are not needed, except, possibly, in the winter if the sheep are penned.

Leashes, food and water bowls, grooming equipment, and other items may be needed. These vary with the nature of the dog and its work.

DONKEYS

Some farms and ranches use donkeys for service and safety. Ranchers may use them to protect sheep and goats. Donkeys may be used to teach a show calf to be halter broken. Donkeys were first used to guard sheep in Montana and Utah.

Donkeys are classified as either a ***jenny***, which is a female, or a ***jack***, which is a breeding male. A gelding is a castrated male. When selecting a donkey, know that jacks and geldings may be too aggressive with the sheep and kick or bite them. Geldings are less likely to be aggressive. They are selected over jennies because they are stronger. Jennies work well with the sheep because they are less aggressive, unless they have newborn babies of their own. This may be remedied by removing the jenny and her newborn for a week or ten days, then return the two to the sheep flock.

21–11. Young donkeys.

Donkeys have a natural hate for all members of the dog family such as fox and wolves. Donkeys do not kill predators, but they chase them with their teeth bared, ears back, and striking at their feet. Guarding is a natural part of the donkey's temperament. Donkeys are replacing dogs in some places because they are more economical to buy and feed than dogs.

21–12. The head of a mature donkey. (Courtesy, Martina I. Meyer/Shutterstock)

TRAINING AND MANAGEMENT

The key to managing donkeys as security animals with sheep is to make sure they bond with the sheep. Donkeys are gregarious, so they want to be around other animals. If there are no donkeys to be around, they will assume the role of a sheep and fit in with the crowd. Bonding takes time. The donkey needs to get familiar with the flock. One donkey is needed for every 150 sheep.

Donkeys make good guards with no training. Not every donkey bonds with sheep. It is important that a donkey is tested with sheep before it is released with them. Often, it is best for the producer is to buy a young female donkey raised on a sheep farm. If more than one donkey is with the sheep, they will bond and spend time with each other rather than with the sheep.

Donkeys have limitations; though, they can kill dogs, coyotes, and foxes. A donkey can be killed by bears, mountain lions, and wolves. Donkeys know this and are afraid of these animals. It is important to know what sheep need protection from to ensure that donkeys are the correct animals for the job.

Besides knowing the limitations of the donkey, it is important to understand the techniques the donkey uses to

21–13. A guard donkey has bonded with this flock of sheep.

protect the sheep. The donkey will bray loudly and stamp its feet at night if it hears or sees something unusual. They will chase dogs and coyotes that get too near.

FEEDING AND NUTRITION

Donkeys usually graze and feed along with the sheep flock. If additional feed is provided, the donkey should not be fed in a separate area. This will lessen the bond between the donkey and the sheep. The donkey may even drive the sheep away so it can eat, especially early in the relationship. For a strong bond to exist, the animals must graze, sleep, and live together and establish the same routine.

HEALTH CARE

Donkeys are healthy animals. They do not require very much veterinary care, because they are hardy. Donkeys may be susceptible to most of the diseases of horses. Injuries may also occur when fighting off attacks by predators. Chapters 4 and 7 have more information on health care with donkeys.

FACILITIES AND EQUIPMENT

Donkeys do not need special facilities and equipment. They should be kept with the sheep. They should sleep with the sheep since this is when the predators attack. If the donkey is not there during the hours it is needed, it cannot protect the sheep. It is important that the donkey is kept with the sheep so it bonds with the sheep. Donkeys can survive in the same climate as sheep, so no special housing is required.

LLAMAS

Llamas are members of the camel family. They are from South America. Llamas have been used as pack animals for many years. Their usefulness as guard animals has come about in recent years.

Llamas are used on some farms as protection for other animals. Llamas were first brought to the United States as novelty animals. Initially, llamas were put with sheep because of convenience. Producers quickly realized that they were losing fewer sheep to predators when the llamas were with the sheep.

Llamas that encounter predators get very aggressive and protective. They are proud animals that remain calm and do not get scared and run from intruders. They take an aggressive role. When they encounter coyotes, dogs, foxes, or bears, they give an alarm call, walk or run toward the intruder, chase it, kick it, paw it, and sometimes kill it.

The female lays down to be bred. Gestation is 350 days. Most give birth without assistance. The young are called **cria** and are usually born in the morning. The female, if approached by the male after she has been bred, will spit at him.

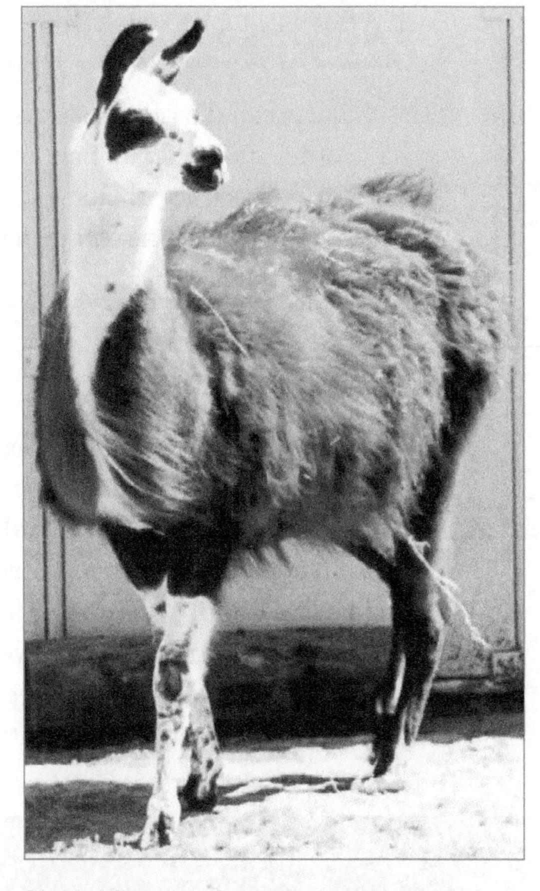

21–14. A llama can become aggressive when danger approaches. (Courtesy, Education Images)

TRAINING AND MANAGEMENT

Unlike the donkey, llamas quickly bond to sheep. The two species get along well. The llama is also very instinctive about what to do with the sheep. No training is needed for a llama. They take over and protect sheep as their companions.

FEEDING AND NUTRITION

Similar to donkeys, llamas work well as guard animals with sheep because they eat the same food as sheep and sleep with them at night. This helps them know when threats occur to the flock's safety. Llamas have the same lifestyle as sheep, so they fit naturally alongside them.

Llamas may be fed with the sheep because they will stand aside and let the sheep eat while they look on. They eat the same foods as sheep and are very satisfied with a bale of hay and water.

HEALTH CARE

Llamas, much like donkeys, do not have many health threats. Also, they may last as long as 20 years as a guard animal with the sheep. In the end, they cost less money than a guard dog, which initially costs less, but requires more in care. A dog lasts only three years, of which, the first is only training. The llama is automatically trained because it is all instinct.

FACILITIES AND EQUIPMENT

Llamas can die of heat stress and require shade in hot weather. However, they do not have other special requirements. They can live happily on $^1/_8$ of an acre with only a three-sided windbreak for shelter. The "toilet habits" of the llama are clean and disease-free because they select one corner of their pen for defecation and do not deviate from that area.

GEESE

Geese are used to protect homes and their owners from intruders. A goose can be a very mean and ruthless animal. When an intruder arrives, the goose begins honking and then chases the intruder. If the intruder is caught, the goose will hold onto the person with its beak and beat the person repeatedly with its strong wings. This may not seem too threatening, but wait until a goose comes to attack you!

Another use of geese is in weeding crops, especially cotton. This use has declined in importance. Large gaggles of geese could be placed in a field to remove the weeds and grass. Geese do not readily eat the cotton plants.

21–15. A Gray Lag goose.

TRAINING AND MANAGEMENT

Geese used as protection should be obtained as goslings or baby geese. They should arrive in the late spring or early summer. Caring for the birds is easier when the weather conditions are more favorable.

Geese normally do not need training. They need to know the boundaries of the property where they live. Fences usually help in this regard.

FEEDING AND NUTRITION

Geese need to receive feed that meets their needs. Grains and other supplements are used.

Goslings (young geese) should be fed all they can eat and drink. A 20- to 24-percent protein commercial feed should be used. A chick starter or a duck or waterfowl formula will be fine if the goslings are allowed outside to forage for food on their own. Goslings grow faster than chicks, up to 10 pounds in ten weeks. They need food in addition to the starter. Goslings will eat grass and insects to get the necessary food to balance their ration. Goslings need water at all times. If weather permits, goslings should be outside during the day and inside at night.

Adult geese can be fed whole corn for feed and then left to forage on their own. Many times, geese are used to weed crops, such as strawberries, asparagus, sugar beets, and orchards. The geese will eat the weeds and not harm the plants. However, the geese should be removed from the strawberries when they start to ripen so they do not eat or ruin the berries.

HEALTH CARE

Goslings and geese are healthy animals. They need ample water to drink and to use to stay cool in the warm weather. Goslings, when they are very young, should not be outside in the rain or they become waterlogged—their down soaks up all the moisture it will hold. Goslings have no feathers to repel the water. Their tiny legs cannot support the weight of the water. Geese are known for being disease-resistant, so little has to be done to maintain their health.

FACILITIES AND EQUIPMENT

Adult geese may need shelter from inclement weather. A shed where they can go to escape predators and roost is beneficial. The feeder and waterer may be in this area.

Goslings require a warm, dry place for the first four weeks of their lives. If, in late May or June the weather is still cold, the goslings must be kept dry, warm, and away from drafts. A 250-watt heat lamp may be installed 15 to 18 inches above the floor to keep the goslings warm. If the goslings are piling up on each other, it is too cold and the weaker birds may get smothered. If the goslings are comfortable, they will sit evenly distributed on the floor under the heat lamp.

Goslings require litter on the floor of their pen. Litter, such as straw, sawdust, shavings, or crushed corn cobs, will work. The material must be dry and absorbent since goslings are very messy.

REVIEWING

MAIN IDEAS

Several species of animals are used for service and safety. Dogs, donkeys, llamas, and geese are three examples.

The K-9 unit has become important in law enforcement. When properly used, the trained police dog is an asset that cannot be overlooked. It increases the chances of capture and the safety of the officer.

Guard dogs are used effectively as protectors for sheep. They require more work than either llamas or donkeys. The guard dog is a highly trained animal that works effectively with the sheep if it is reared correctly.

Donkeys make good protection providers for sheep. Not all donkeys perform as they should. A donkey should be tested for how well it bonds with sheep.

Llamas are proving to be successful protection animals because of their instinct to bond with and protect sheep. The viciousness of a llama can be provoked by a predator. Their popularity as companion animals is also growing.

Geese are animals that may not be used by everyone, but for those that choose the birds, they work out well. The birds are vocal and very aggressive, but also easy to care for and disease resistant.

When selecting and considering types of animal protection, it is important to note the time commitment to train the animal, the efficiency of the animal, the length of time the animal can perform the job, the type of protection needed, and which animals are best suited for that type of protection.

QUESTIONS

Answer the following questions using correct spelling and complete sentences.

1. What animals are commonly used for service and safety?
2. What are the major considerations in using dogs as guard animals?
3. What are the major considerations in using donkeys as guard animals?
4. What are the uses of llamas as guard animals?
5. What are the uses of geese as service and safety animals?
6. Why are donkeys and dogs not used together?

EVALUATING

Match the term with the correct definition. Write the letter by the term in the blank provided.

a. handler	d. K-9	g. jack
b. service animal	e. jenny	h. single-purpose dog
c. safety animal	f. cria	

______1. A male donkey.

______2. A baby llama.

______3. A female donkey.

______4. A dog used for one safety or service activity.

______5. The person who directs a K-9.

______6. Animals that give assistance.

______7. A trained police dog.

______8. Animals used to protect people and property.

EXPLORING

1. Contact local law enforcement officials and learn the use they make of K-9 dogs. Interview the handler and determine how the dog is trained, what it does, and how it is managed. Prepare a written or oral report on your findings.

2. Prepare a report on using llamas, geese, or donkeys as service and safety animals. Interview owners of such animals and use library resources for information. Give an oral report in class on your findings.

3. Investigate the services provided by small animals in your local community. Use the information in Chapter 21 as a guide. Interview a local resident who has one or more small service animals. Consider at least two species or breeds. Prepare a short oral report on your findings and discuss the services you found.

CHAPTER 22

Scientific and Laboratory Animals

OBJECTIVES

The purpose of this chapter is to introduce the use of laboratory research animals. It has the following objectives:

1. **Describe types and methods of animal research.**
2. **Identify the kinds of animals used in laboratory research.**
3. **Discuss animal models used in modern research.**
4. **Explain important management practices used in animal research.**

TERMS

animal by-product tissue
animal tissue culture
applied research
basic research
cavy
clinical research
laboratory animal
living research animal
mathematical and computer model
models for animal research
nonhuman primate
nonliving research system
primate
research

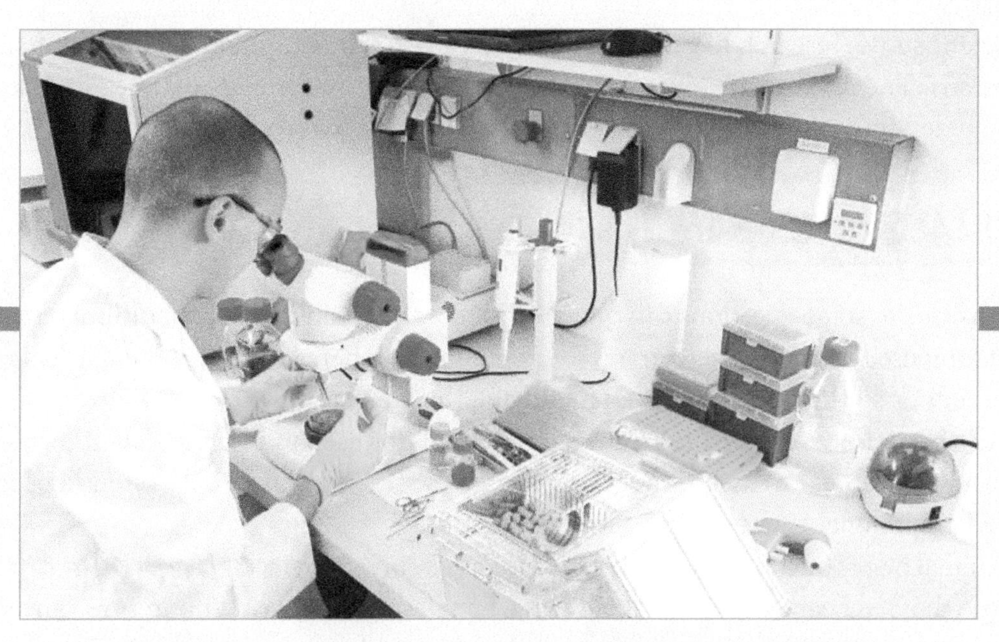

22–1. A scientist is investigating the effects of a protocol on animal tissue. (Courtesy, anyaivanova/Shutterstock)

USE of animals in research helps humans in many ways. Polio was once a dreaded disease. A similar vaccine was developed with the help of laboratory animals. The same methods are used in other areas.

Animals had a major role in the development of antibiotics, insulin for diabetics, anesthetics, radiation and chemotherapy for cancer, open-heart surgery, joint replacements, and hundreds of other medical advancements. In addition to helping humans, research animals have also contributed greatly to animal health. People need to understand just how important animals are in making scientific advances.

Opinions differ on the use of animals. What do you think about using animals in research? Regardless, animal well-being is always important. People and animals live better and longer thanks to animals.

LABORATORY RESEARCH METHODS

Scientists use research to gain information. ***Research*** is a systematic attempt to answer questions. The research involves using procedures that are carefully planned and designed to provide accurate findings. Animals are sometimes used in the research.

ANIMALS IN RESEARCH

Answers to some questions can only be found by using animals. Agricultural researchers have long used animals in research. How would researchers know how animals respond to new feeds or methods of care without studying the animals?

A ***laboratory animal*** is an animal raised and/or used in research laboratories. They are carefully tended to assure their well-being. Laboratory animals are often kept in controlled environments to assure accurate research.

Animal by-product tissues are also used in research. An ***animal by-product tissue*** is a tissue that has been taken from an animal when it was used for other purposes. The tissues, organs, or other parts of animals slaughtered for human food may be used in research. These materials are collected in slaughter plants and carefully preserved.

Animal products are also used in research and to produce medicines and similar products. A good example is the chicken egg. Eggs contain albumin, which is rich in nutrition and an excellent medium for the production of many vaccines.

22–2. A white mouse is being injected with an experimental medicine being developed for use with humans. (Courtesy, Vit Kovalcik/Shutterstock)

TYPES OF RESEARCH

22–3. Cattle by-product pituitary glands will be used to synthesize and artificially culture enzymes in a laboratory at Georgia Tech. (This package contains 100 pituitary glands.) (Courtesy, Corinne Mounier-Lee, Georgia Tech)

Animals are used in three types of research: basic, applied, and clinical. Each type of research helps answer important questions.

Basic research is carried out in a laboratory, with a computer, or in nature. According to the American Medical Association, the goal of basic research is to increase the knowledge and understanding of life processes and disease. Basic research does not have a predetermined focus. It uses observing, describing, measuring, and experimenting to provide data that are often used in the other types of research.

Scientists may have specific purposes in mind for their research. A good example is developing a new vaccine or a medical procedure that builds upon existing knowledge. This is known as ***applied research***. Animals are used in applied research, but this type of research could involve computers, nonanimal substitutes, or even people.

Research that takes place in a medical facility or clinical setting is called ***clinical research***. Clinical research focuses on a specific human or animal problem. It often involves the animal species that the research is intended to help. Clinical research is supported by knowledge gained in basic and applied research.

Scientific and laboratory animals are used in basic and applied research for human- and animal-related research objectives. They are used in clinical research when their species is the subject being studied. For example, research on chickens involves using chickens in carefully controlled settings.

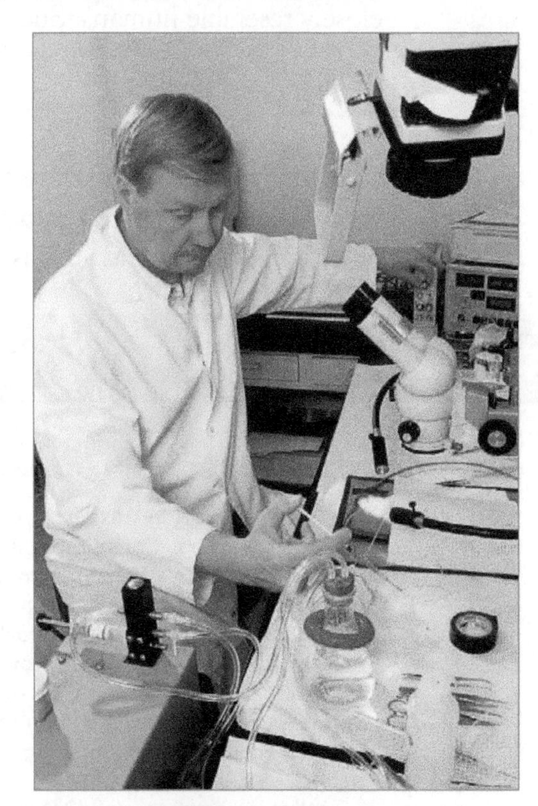

22–4. A physiologist measures heart performance and blood pressure in a copper-deficient laboratory rat. (Courtesy, Agricultural Research Service, USDA)

ANIMALS USED IN RESEARCH

Some animals are better than others for particular kinds of research. A few selected species of animals are used for basic and applied research. Clinical research animals involve any species that is being studied.

Most basic and applied animal research takes place in a laboratory setting. Scientists need animals with certain traits that make a particular species useful for research purposes. Most research facilities have a limited amount of space. Animals that require small amounts of space are most desirable. Larger animals also eat more food and create more waste. It is not surprising to learn that small animals are the animals of choice! In fact, more than 90 percent of the animals used in biomedical research are mice, rats, and other rodents. Recent human research with tissue has focused on pigs. Important similarities exist between pigs and humans.

Another consideration is the similarity between the animal and the benefactor of the research. For example, a strain of mice has been developed that have immune systems that closely resemble humans. Such mice are used in HIV and AIDS research.

MICE

Mice (Mus musculus) are the most popular animals for scientific and laboratory use. They are small, relatively inexpensive animals that need little space and food. Mice have a mild temperament and are easy to handle and work with.

CONNECTION

SENSING INFORMATION

Animals provide research information about themselves. Sometimes, devices are attached to animals to help in collecting the information.

This shows a sensor being attached to the neck of the steer. The sensor provides information about the grazing habits of the steer. This lets scientists know the amount of time a steer spends each day grazing and other details.

Is this information useful? Does collecting the information injure the steer? The well-being of the animal is a top priority in this research. (Courtesy, Agricultural Research Service, USDA)

Because they are efficient breeders, they are especially suited to research that requires large numbers of animals. Mice are most often used in genetic, immunology, and infectious disease research.

There are more than 100 documented bloodlines of mice available to researchers. The bloodlines are designated by a system of letters and numbers. Examples are C3H, C57BL/6, and CBA.

Mice, like most other rodents, are nocturnal. They spend much time grooming themselves. This is useful in medical research because an early sign of disease is an ungroomed appearance. Like all rodents, mice's teeth grow constantly. This is not usually a problem in laboratory settings, as the food provided is hard and keeps the teeth trimmed.

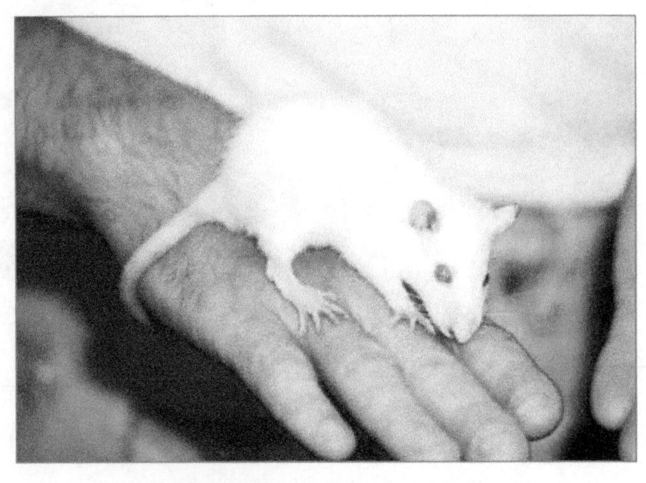

22–5. White mice are often used in laboratory research. (Courtesy, Education Images)

RATS

A larger cousin of the mouse is the laboratory rat (Rattus norvegicus). It was developed from the wild brown Norwegian rat. Rats are commonly used in behavioral and nutritional research.

Five strains of rats account for most of the animals used in modern research. Examples of the strains are Wistar (WI), Long Evans (LE), F344, and LEW.

Rats are also nocturnal rodents. They are usually docile and easily handled. They are larger and stronger than mice and greater effort is required to restrain them. Since they seldom fight, male rats can be kept in the same cage together.

HAMSTERS

The golden Syrian hamster (Mesocricetus auratus) is the most commonly used hamster for research. It has a stout body, short tail, and reddish-gold fur on the head, back, and sides, with grey-white fur on the underside. All hamsters have cheek pouches, which they use to transport and temporarily store food.

Hamsters are aggressive and will bite when startled. They should be handled carefully and gently. When they are handled properly, they will become easier to work with. Females usually dominate the males and larger animals dominate smaller ones. Animals kept in the same cage develop a "pecking order" and can live peacefully.

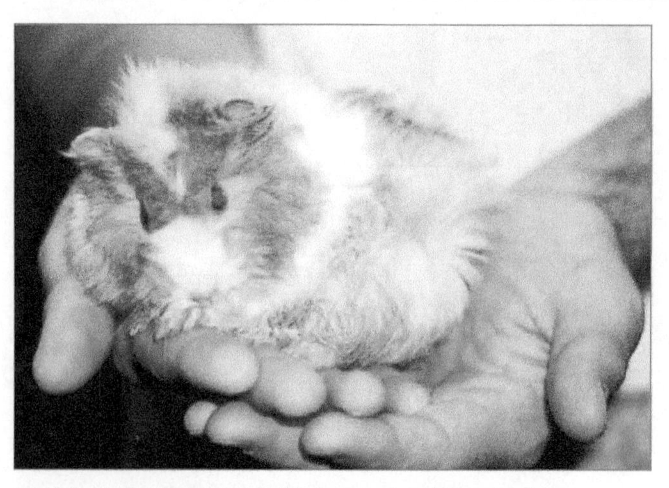

22–6. Guinea pigs are used in human nutrition studies. (Courtesy, Education Images)

GUINEA PIGS

A Guinea pig (Cavia porcellus) is also known as a **cavy**. Guinea pigs differ socially and anatomically in several ways from other rodents. Important research strains include strain 13, strain 2, Dunkin Harlet (DH), and Shorthair (SH).

The cavy has a stocky body and short legs. It cannot climb as well as most other rodents and lacks agility. These characteristics, along with their docile nature, make cavies easier to handle than most laboratory rodents.

One nutritional difference between guinea pigs and other rodents is their inability to produce vitamin C. Like humans, guinea pigs must supplement their diet with this vitamin. Vitamin C deficiency causes a fatal disease known as scurvy.

RABBITS

Rabbits (Oryctolagus cuniculus) were first domesticated in the 16th century. They serve as models for human and animal diseases, produce serum antibodies, and are used for drug testing. The New Zealand white rabbit is the most commonly used research animal.

Rabbits are used in areas of human research where rabbits and humans suffer from similar diseases and conditions. For example, rabbits have atherosclerosis (hardening of the arteries), emphysema, spina bifida, and cleft pallet. Laboratory rabbit research helps scientists understand these conditions in humans.

Rabbits are curious animals that enjoy exploring their surroundings. They are gentle, usually docile, and easy to handle if not excited. Rabbits can be frightened by loud or sudden noise and often show fear by stomping their hind feet. A frightened rabbit may try to bite or scratch.

22–7. Rabbits are used in antibody research and drug testing. (Courtesy, Education Images)

CATS

The cat (Felis catus) has been living with humans for more than 5,000 years. The domestic cat is of particular importance in brain research. The brain of a cat seems to be at a development stage between primates and lower mammals. Much of the present human and animal brain research is with cats. The American Shorthair is the dominant research breed.

Healthy cats are curious and aware of their surroundings. Their ears are erect and eyes are bright and alert. Most cats enjoy being petted and purr when handled. Some cats are raised specifically for laboratory use. Other lab cats are caught as abandoned cats or taken from local animal shelters.

22–8. Kittens being well cared for in a clean facility. (Courtesy, Education Images)

DOGS

Dogs (Canis familiaris) have been used for experiments for more than 300 years. The most common research with dogs includes physiology, pharmacology, and surgical studies. Beagles are the breed of choice for most nonsurgical research.

Daily contact improves the dog's physical and psychological well-being. A healthy dog greets its handler with bright eyes, alert ears, and a wagging tail.

Good facilities for dogs include a large cage with waterer, feeder, and rest area. Indoor and outdoor runs are used to ensure that the animal is able to exercise.

22–9. A beagle represents the breed that is used in lab research involving dogs. (Courtesy, Education Images)

NONHUMAN PRIMATES

A ***primate*** is an animal with an opposable thumb used for grasping things. The human being is a primate. Look at your thumb. How important is it to you?

Several species of nonhuman primates are used in a broad range of scientific and laboratory settings. A ***nonhuman primate*** is an animal very similar to humans, with thumb-forefinger opposition being a big similarity. Examples include chimpanzees and monkeys.

22–10. Chimpanzees are nonhuman, higher-order primates that are used in researching areas of human behavior. (Courtesy, Education Images)

Monkeys can be divided into two basic categories: old world and new world. Old world monkeys come primarily from Africa and Asia. They have close-set nostrils that open downward and cheek pouches capable of carrying food. Old world monkeys also have pads on their buttocks.

New world monkeys have nostrils that open to the side, and do not have buttocks' pads or cheek pouches. New world monkeys also have long tails used for grasping and climbing. A unique characteristic of nonhuman primates that separates them from other scientific and laboratory animals is their social behavior. Communication through verbal and nonverbal means is of special interest to researchers.

Most research is with conditioned animals that are taken from the wild. Very few nonhuman primates are bred domestically for scientific and laboratory use.

MODELS FOR SCIENTIFIC AND LABORATORY INQUIRY

The use of animals in research is strictly controlled and monitored. Scientists have developed a set of acceptable ***models for animal research***. These ensure that scientific and laboratory animals will contribute to knowledge and not be misused.

Four basic models exist for the use of animals in research:

1. Living animals
2. Living animal systems (tissue culture)
3. Nonliving systems
4. Mathematical or computer models

Using these models, scientists can better understand the human body, human behavior, diseases, treatments, and the effects of drugs on the body.

Living research animals (including humans) provide the best vehicles for psychological and human research. A ***living research animal*** is an animal that is alive and responds to treatments. They can react to their environment and are readily examined. The trend is away from using living animals and toward other models.

Animal tissue cultures can tell scientists a great deal about how body systems will react to stimuli. An ***animal tissue culture*** is the use of living tissues cultured in a lab. The tissues are taken from animals, including humans, and grown in a laboratory. A great deal of the research in cancer treatment, through radiation and chemotherapy, is done using tissue cultures.

22–11. This llama, a large laboratory animal, is being used in veterinary medical research. (The shaved area on its neck does not represent anything that would threaten its well-being.) (Courtesy, Education Images)

Another model is with nonliving systems. A ***nonliving research system*** is using mechanical models developed by scientists that mirror animal activity. They are used in research that is related to movement or effects of injury on living animal systems. An example related to humans is the development of the artificial hip. Many models had to be developed in the laboratory before a working design could be developed.

Computers have opened new approaches. The ***mathematical and computer model*** is a model that mimics animal behaviors and systems. It was developed after many hours of basic research. The models do not involve animal subjects in research. Scientists can manipulate these models on the computer and see results of their manipulation in seconds; conducting the experiment with live animals in the laboratory would take much longer. Computer and mathematical models are limited to what we know about a particular animal or situation and, therefore, are not perfect.

MANAGEMENT IN ANIMAL RESEARCH

Research with animals must meet acceptable standards. These standards focus on working with living beings. The well-being of animals is a top priority.

REGULATION OF ANIMAL RESEARCH

Animal research is heavily regulated. Researchers themselves are much involved. Animal research also has other regulations.

Universities doing animal research must follow strict regulations. Research institutions are members of the Institutional Animal Care and Use Committee (IACUC). All animal research

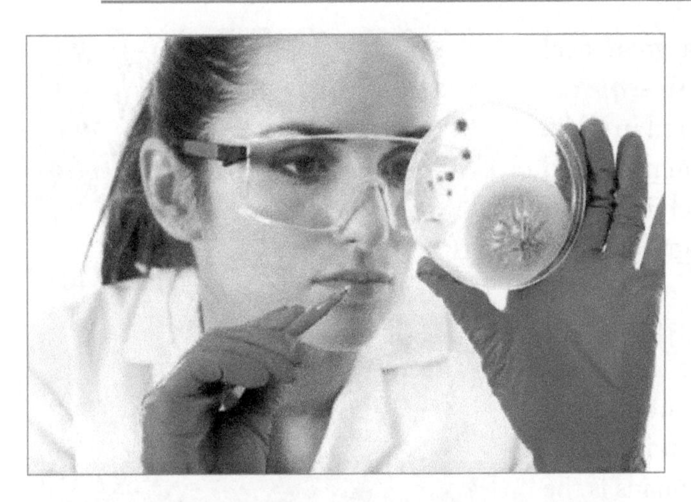

22–12. A bacterial culture in a petri dish with agar medium is being observed for growth. (Courtesy, Michal Kowalski/Shutterstock)

conducted must be approved by this committee. This committee looks at the research proposal to make sure that protocols of animal treatment are followed. For example, the committee will look at medical care provided for the animals, the qualifications of the personnel conducting the research, and the use of anesthetics.

External review of the use of animals in research is based on the Animal Welfare Act of 1985. This federal legislation helps assure the proper use of animals in research. The U.S. Department of Agriculture is the federal agency responsible for ensuring animal welfare in research. This agency regulates and inspects animal research facilities through the Animal and Plant Health Inspection Service (APHIS).

In addition, the Public Health Service (PHS), the Environmental Protection Agency, and the Food and Drug Administration have animal welfare regulations. Most states also have regulations that are intended to protect the welfare of animals.

LAB ANIMAL NUTRITION

It is important that the animals used in research are fed nutritious, balanced diets. If an animal is malnourished, the results of the research may be clouded by the effect of nutrition and not by the effect of the experimental treatment! Care is needed on the part of researchers and laboratory technicians to give animals involved in research a proper diet.

Animal diets will vary greatly depending on the type of animal, the animal's size, maturity level, and activity level, and the environment. The animal's diet includes water, protein, carbohydrates, fats, vitamins, and minerals. These elements provide energy and nutrients that help the animal grow and maintain itself.

LAB ANIMAL ENVIRONMENT

How would you feel if you were reading this book in a room with poor lighting, loud noise in the background, and a temperature below freezing? You probably would not be concentrating on what is written here! Just as humans have comfort levels in their environment, animals also have ranges in which they are comfortable. The only life experience most scientific and research animals have is in the laboratory environment. Making their existence comfortable is a major concern of animal researchers.

Five areas related to the environment are included: space, temperature, light, ventilation, and noise.

22–13. Scientists are examining an image of the results of an experiment comparing genes of 1,700 strains of *campylobacter jejuni*. (These bacteria, found in animal feces, are responsible for common gastroenteritis in humans. Keeping food clean is essential.) (Courtesy, Agricultural Research Service, USDA)

- Space—Probably the biggest concern for animal welfare in scientific and laboratory settings is the animal's living space. An animal must be given enough space to eat, sleep, and live out its life in relative comfort. Space requirements will vary with the type and size of the animal. Suggested minimums range from 6 square inches (15 cm^2) for a small mouse to 144 square feet (4.4 m^2) for horses. Space requirements are available from the National Institutes of Health in publication 86-23.
- Temperature—Animals are usually most comfortable at a temperature between 65 and 85EF (16.6E to 29.4E C). Relative humidity of the room should fall in the 30 to 70 percent range for most animals.
- Light—Lighting in the facility should be uniformly distributed and be bright enough for researchers to see and effectively work with the animals. Animals require a daily period of darkness to help them rest. The light and dark periods should remain on a set schedule to avoid stressing the animals.
- Ventilation—Fresh air provided through adequate ventilation eliminates odors and reduces the number of airborne diseases. Fresh air is a necessity in maintaining a healthy environment.
- Noise—Although it is impossible to eliminate noise from the environment, the animals should not be exposed to loud noises or areas of high activity (loading docks, building entrances). Noise can cause stress in animals and even lead to health and behavioral problems.

LAB ANIMAL HYGIENE

Hygiene is a primary concern for researchers in a laboratory setting. Diseases result from poor sanitation. Diseases can ruin experiments by causing animal sickness and even death.

22–14. Guinea pigs are widely known for their roles in research. (This guinea pig has a clean cage and is receiving good care.) (Courtesy, Education Images)

Sterilization of equipment and materials that come in contact with animals is a fact of life in the laboratory. Bedding that is used to line the cages, instruments used for examination, even the cages used to house the animals must be sterilized to avoid spreading disease.

Disinfectants are used to kill or inhibit the growth of bacteria and viruses. These chemicals are usually too strong to use directly on animals and are applied to cages and equipment used by animals.

Sanitation is the practice of controlling bacteria and other organisms. This may be accomplished by removing contaminated objects with which animals come in contact. For example, removing animal waste, washing the cages, and cleaning examination tables are common sanitation practices. Sanitation includes changing the animal's bedding or cleaning the cage floor.

REVIEWING

MAIN IDEAS

Animals contribute to the welfare of people and of other animals. Scientific and laboratory animals are used in basic and applied research and to a lesser extent in clinical research.

There are many types of scientific and laboratory animals, but the most popular animals are the rodents. More than 90 percent of the animals in biomedical research are rodents. Other popular research animals include rabbits, cats, dogs, and nonhuman primates.

The advancements made in animal research have allowed scientists to develop four basic models that utilize animals in research. Only one model, the living animal model, actually conducts research with live animals.

Management in the laboratory or research facility is a top priority for scientists. Animal welfare is the goal of animal research scientists. Scientific and laboratory animal research is strictly regulated, internally and externally.

The living environment is regulated to provide a healthy, comfortable life for the animals. Space requirements, temperature, humidity, noise, and lighting are all components of the animals' environment. Keeping the animals healthy through hygiene is accomplished through sterilization, disinfection, and sanitation.

QUESTIONS

Answer the following questions using correct spelling and complete sentences.

1. Distinguish between using animals and using animal by-products in research.
2. What are three kinds of research? Briefly define each.
3. What animals are used in laboratory research? Which is most popular?
4. What animal models are used in scientific and laboratory research?
5. How is laboratory animal research regulated?
6. How is the well-being of animals provided for in a laboratory environment?

EVALUATING

Match the term with the correct definition. Write the letter by the term in the blank provided.

a. cavy
b. laboratory animal
c. basic research
d. animal by-product tissue
e. nonhuman primate
f. living animal model
g. Animal Welfare Act
h. clinical research

______1. A federal law that assures proper use of animals in research.
______2. A guinea pig.
______3. Research animals that are kept alive for study.
______4. Animals similar to humans.
______5. Research in a medical facility.
______6. An animal raised for and/or used in laboratory research.
______7. Tissues from an animal used primarily for purposes other than research.
______8. Research that increases knowledge to help understand animals.

EXPLORING

1. Prepare a report on the use of laboratory animals to alleviate a human disease. Be sure to investigate the disease and its effect on humans. Do you feel that animals should have been used for this research?

2. Organize a class debate or discussion to explore the use of animals in research. What is the conclusion drawn from the information that is shared? Do you agree with the positions taken by different individuals in the debate or discussion? Why?

CHAPTER 23

Exotic Animals

OBJECTIVES

This chapter introduces exotic animals. It has the following objectives:

1. **Identify and describe major species of exotic animals.**
2. **Explain how to assess opportunities in exotic animal production.**
3. **Explain important management practices with exotic animals.**

TERMS

animal refuge
biologist
circus
exotic animal
exotic animal dealer
exotic animal investment species
feral animal
performing animal
rare breed
zoo
zoo curator

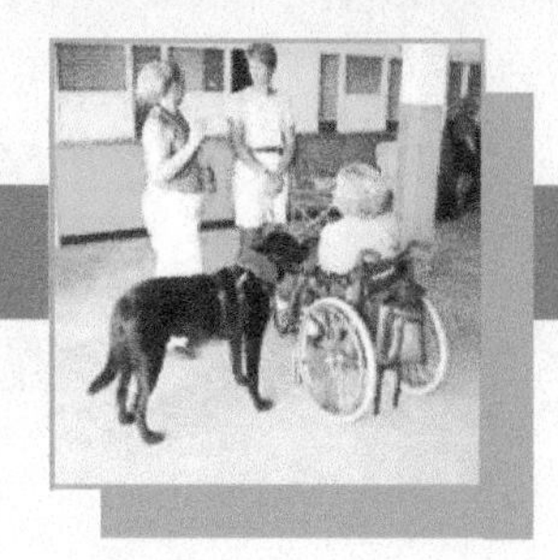

23–1. A Dama Gazelle is a sophisticated exotic animal. (Courtesy, Education Images)

ANIMALS are special in many ways! What would a circus be without elephants, or a zoo without a zebra? We look forward to the entertainment animals can provide. Just watching them go about life is fascinating.

People enjoy working with animals. Some people have jobs caring for them; others own and raise them. New and interesting kinds of animals are occasionally grown. The people who grow them are pioneers. They learn by trial and error. They do not have the benefit of research and written information to help them.

Special animals often need a little help from humans to survive. The animals may need to be cared for in protected areas or receive a little extra attention. There may be at least one of these that is highly interesting to you.

MAJOR KINDS OF EXOTIC ANIMALS

An ***exotic animal*** is an animal that is not native to where it is being raised or is an animal that is rarely seen. Exotic animals are often new or are used in new ways. In some cases, they need extra care to survive. Exotic animals are used for a wide range of purposes. Often, formal marketing structures are not in place for exotic animals or their products.

Exotic animals are relatively newly domesticated animals or animals that are beginning to be used for food and fiber production. They may also be used as investment animals and zoo and performing animals.

EXOTIC FOOD AND FIBER ANIMALS

Some exotic animals are used for human food and other products. In a broad sense, all current domesticated livestock started as an exotic species. Exotic animals used for food are new agricultural animals or animals being used for new purposes. A few species in the United States that are now considered exotic are the llama, alligator, elk, and bison. Many other animal species are raised to produce agricultural commodities. Some exotic animals are harvested from the wild.

23–2. Camels proudly hold their heads high. (Courtesy, ErickN/Shutterstock)

Llama

Llamas are mammals that have the potential of being raised commercially for their coat. Llama meat has a low fat content and is delicious. At this time, few llamas are eaten. Breeding animals can sell for more than $10,000 and to eat one would make for a very expensive meal. Some llamas are used as guard animals with sheep and goats. In some parts of the world, llamas are used as pack animals.

Alligator

The American alligator, once on the brink of extinction, is now considered an exotic food animal. Alligators are raised on farms in many southern states. Their meat is light in color, has a firm texture, and a distinct flavor. Their skin is used for making expensive leather products. Alligators grow and reproduce well in captivity.

23–3. An American alligator. (Courtesy, Ekaterina Pokrovsky/ Shutterstock)

Bison

The American bison (sometimes called buffalo) is being used to produce an alternative to beef. Although few purebred bison are grown for meat, the genetics of the bison are used in cross-breeding programs with domesticated beef animals. This cross, often called a beefalo, yields meat similar to beef with a lower fat content and less cholesterol.

23–4. A bison that has been raised from a day-old calf for research in developing a brucellosis vaccine for bison. (Courtesy, Agricultural Research Service, USDA)

Elk

Elk are typically thought of as wild animals. Several individuals are now producing elk much as other livestock are raised. The producers are learning the requirements for successfully reproducing and caring for elk. Little information has been available on how to raise elk. Practical experience is important.

Rare Breeds

Breeds of livestock and poultry that were common just a few years ago are now rare. A ***rare breed*** is a breed of domesticated animal that was once popular and is now disappearing. Producers often take a special interest in a breed and strive to continue its existence. Producing these breeds is an alternative similar to keeping other exotic animals. Efforts are

23–5. An elk undergoing domestication is tame. (Courtesy, North American Elk Breeders Association)

made to keep the bloodline of the rare breeds pure for future breeding and research. This includes breeds of hogs, chickens, cattle, and other farm animals that were popular 50 or more years ago.

23–6. The Barred Plymouth Rock is becoming a rare poultry breed. (Courtesy, Education Images)

Other Exotic Food and Fiber Animals

Several species of animals are raised for food, fur, and fiber that are considered exotic. Eel, turtles, mink, and upland game birds, deer, and game animals are a few examples.

EXOTIC SPECIES HARVESTED IN THE WILD

Some businesses have started to market products from exotic wild animals. Meat from the antelope, deer, elk, and small game is sold throughout the country. Other wild products include snake meat, shark steaks, and skins, feathers, hides, and hair from a variety of birds and animals.

EXOTIC INVESTMENT ANIMALS

Since exotic animals are new animals, or animals being used for a new purpose, most people are afraid to invest the money needed to start an exotic animal business. However, there are some entrepreneurs who take the risk. Investing in exotic animals can pay off well or it could leave you broke.

An ***exotic animal investment species*** is an animal raised in hopes of making a large sum of money. If successful, they are more profitable than cattle. There is a high degree of risk and many people fail to earn big profits.

Early investors in the pot-bellied pig market paid thousands of dollars for the first breeding animals that came to the United States. They hoped to make their money back selling these animals as pets. Although there was interest in the animals, the American public did not buy many pot-bellied pigs. Some investors were stuck with expensive pets.

Other investors in exotic animals have fared better. A trend for sports enthusiasts has been hunting in animal preserves. With the areas available to hunters decreasing, animal preserves have become a good alternative. Investors are realizing a boom in business, while sports enthusiasts are provided with a place to practice their sport.

CONNECTION

ELEPHANTS ARE LARGE IN MANY WAYS

Elephants are favorite exotic animals. As the largest of all terrestrial animals, their size makes them impressive. They can weigh 12,000 pounds (5,400 kg) and stand over 12 feet (4 m) tall. They never stop growing as long as they live!

An animal the size of a mature elephant takes a lot of food. Elephants eat up to 300 pounds (136 kg) of food a day when on the range. They like grasses, tree fruits, leaves, bark, small branches, and roots. Large elephants can push over 30-foot (9 m) tall trees to get food. An elephant can drink up to 40 gallons (150 L) of water each day.

When kept in captivity, elephants need open areas enclosed with strong fences. They can live 60 or more years if they receive good care. This shows an elephant breaking the ice over its watering facility in a Colorado Springs zoo in the winter. The elephant ingested the ice. (Courtesy, Education Images)

Investing in Agriculture

One area of investment is in agriculture. Several species are available for an entrepreneur to choose to invest in today. Good business people keep their eyes open for new investment opportunities. No one knows for sure what the next big exotic species opportunity will be!

Investing in Pets

There will always be people who want to own unique pets. New companion animals could be a good investment for the investor. Roadblocks for the new companion animals include state and federal laws restricting new animals, international law, and the compatibility of the animal with the environment and other animals.

23–7. In the 1980s, there was high interest in the pot-bellied pig. (Courtesy, Education Images)

Investing in Research

Many animals, especially strains of mice and rats, are used in research. A potential investor could fund development of a new strain or improvement of an existing strain. New research animals could also be added to assist with research.

EXOTIC ZOO AND PERFORMING ANIMALS

Zoo and performing animals are raised in captivity or captured from the wild. Training often begins at a young age to prepare animals that will be performers.

Zoo

A **zoo** is a zoological garden. It is a parklike area where animals are kept for viewing. Many of the species are exotic animals that are not native to the area where the zoo is located. Sometimes, the animals will breed in captivity. Environments are constructed that resemble the wild environments for the animals. Probably the best-known zoo in North America is the San Diego Zoo in California, which has a total of 3,200 animals representing about 800 species. The Smithsonian Institute Zoo in Washington, D.C., is well known for its panda bears. Many cities have one or more zoological parks.

Zoos have four roles: education, entertainment, research, and animal conservation. Many people with special training in zoology work in a zoo. A **zoo curator** is a person trained in exotic animal zoology who looks after the animals.

Zoos get animals in several ways. Some get animals by breeding and raising them. In some cases, zoos exchange animals. For example, one that has

23–8. Zoological parks fill important educational roles. (This display at the San Diego Zoo explains why female polar bears now have single cubs rather than twins: Loss of the Arctic ice cap through global warming has resulted in inadequate food for bears, resulting in decreased fertility.) (Courtesy, Education Images)

23–9. Four animals commonly found in a zoo are (clockwise from top left) sable antelope, tiger, grizzly bear, and zebra. (Note the long nails on the grizzly!)

raised a young giraffe might exchange it with another zoo for an animal of equal value. Many zoos get their animals from animal dealers. An ***exotic animal dealer*** buys and sells exotic animals. They help meet the needs of zoos for certain species of animals.

Animal well-being is a high priority. Each animal is fed according to its needs. Health care is provided to keep the animals disease-free. Occasionally, the animals must be protected from people who are diseased. Some species get diseases from humans. An example is the closure of a popular and attractive monkey exhibit at a major zoo because the animals contracted tuberculosis. They got the disease from eating food scraps thrown into their area that had been infected by humans!

Some zoos have trained animals and provide performances. Others limit their work to caring for the well-being of the animals.

Performing Animals

A ***performing animal*** is one taught to go through a routine of unusual activity. Performing animals may be domesticated or exotic species. Dogs and horses are often taught to perform tricks. Exotic animals, such as elephants, tigers, and monkeys, are also taught similar activities.

Many circuses have trained animals as part of the show. A ***circus*** is a variety show with animal and human performances. Humans are closely involved with the animals.

Who do you think breeds and raises circus animals? The people who do so have skills in agriscience! It is true that some animals come from the wild, but many are bred and raised in captivity.

23–10. An African lion in its native habitat. (Courtesy, Keith Levit/Shutterstock)

ANIMAL REFUGES

An ***animal refuge*** is a protected place where animals are cared for to assure their well-being. Some animal refuges are for wildlife; others include exotic animals. A bird sanctuary is a kind of refuge.

There are about 475 National Wildlife Refuges in the United States. These range from small areas to protect one species of bird to large land areas for the protection of several species. Private individuals also have refuges that cater to exotic animals.

Some animal refuges are homes to feral species. A ***feral animal*** is one that was once domesticated but now lives in the wild. Horses, dogs, hogs, and others live as feral animals in various areas of the United States.

Animal refuges promote animals and develop new ways of meeting their needs. They often try to maintain and reestablish a threatened species. Biologists often work with the animals. A ***biologist*** is a person trained in areas dealing with living organisms. Wildlife biologists have specific training in the biology of wildlife.

23–11. People can view animals, such as this river otter that is sunning on a limb, at the Northwest Trek.

23–12. Visitors to the Chincoteague National Wildlife Refuge can often see the feral ponies. (Courtesy, U.S. Department of Agriculture)

OPPORTUNITIES IN EXOTIC ANIMAL PRODUCTION

Many opportunities exist for the entrepreneur to get into the exotic animal business. But, before you get started, you should examine the requirements to start an exotic animal production business.

PROFITABILITY

Making a profit is the goal of all businesses. The exotic animal business is no different. If you decide to enter this part of livestock and poultry production, you need to know the costs, advantages, and regulations that will affect your business.

Start-up Costs

Start-up cost depends greatly on the species you choose to raise. As mentioned earlier, llama production would be a very expensive enterprise. Investing in a species of tropical fish, on the other hand, would be much less expensive. One must consider start-up costs of breeding stock, facilities, and equipment.

Operating Costs

Operating costs include feed, medicine, labor, marketing, utilities, permits, insurance, and any other cost of being in business. You can get information to estimate these costs from Extension agents, feed salespeople, or others in the same business.

CHOOSING A SPECIES TO RAISE

Suppose you have decided to raise an exotic animal. How will you decide which species to raise? Here are areas to consider.

Location Limitations

Many exotic species are environment-specific. For example, alligators would not grow well in Ohio. They need to be in a subtropical region to grow and reproduce. Fortunately, there are other species that are not as environmentally sensitive. Also, some species (tropical fish for example) can be kept indoors. You need to be aware of limitations you have because of your location.

Comparative Advantages

You should also compare advantages and disadvantages between species. If you are considering several alternatives, look at ease of production, costs, marketing, and ease of leaving the industry.

Regulations

Ask the proper agencies in your state and community about regulations for exotic animals. You must find out if you are allowed to possess the species or if you need a license to produce the animals.

Regulations on some exotic animals are strict. States like Florida, Louisiana, and Mississippi have experienced many problems with exotic animals invading the wild. One example is the nutria. This large rodent is not native to the United States. It was accidentally released and has now become a major problem. The animal is a vegetarian with a big appetite! These animals are destroying thousands of acres of wetland vegetation every year.

Regulations are also in place to prevent the spread of disease, protect people and property, and ensure that animals are treated humanely. Housing, zoning, shipping, and purchasing regulations must be explored before starting your business.

23–13. A veterinarian is checking the vital signs of a whitetail deer. (Courtesy, Education Images)

Market

You could be an excellent producer and go out of business if there is no market for your product. Make sure there is a market before making an investment. You should decide if you are going to sell directly to consumers or sell to a dealer or processor. You will likely need to develop a market before beginning production.

EXOTIC ANIMAL MANAGEMENT PRACTICES

Although there are many similarities between exotic and other agricultural animal species, there are important differences. To raise an animal is to be responsible for its well-being. Responsible producers will know the needs of animals.

SIMILARITIES WITH OTHER ANIMALS

The basic rules for animal production apply to producing exotic animals. They must have a balanced ration. They must be given adequate space to live. Their living area should be kept sanitary and safe. The environment must be comfortable. They must receive preventive care and timely treatment for illness and injury. In these ways, raising exotic animals is very much like raising hogs or sheep.

DIFFERENCES WITH OTHER ANIMALS

While there are many similarities, the differences are key to being a successful exotic animal producer. Each species has its own characteristics that make it a unique animal. Major areas of difference between traditional domestic livestock and exotic species are nutritional requirements, health needs, housing and living space, and reproduction.

Nutritional Requirements

Although every animal needs a balanced ration, every animal differs in its nutritional requirements. Domestic livestock rations have been carefully studied. We know with a great degree of certainty what they require in their diet. Information on most exotic animals is sketchy at best.

What would you feed alligators? How can you find out what they need? Fortunately, alligator diets have been researched by universities and alligator producers. This information would be easy to find from these sources. Most other exotic animals are the subject of research and you can learn about their requirements from universities or the Cooperative Extension Service.

23–14. A trained panda performs in a circus.

Health Needs

Health care may be needed for exotic animals. Some veterinarians and others in animal health care are familiar with the health needs of domestic animals. Unfortunately, they may not know about particular needs of exotic animals. The veterinarians trained to work with exotic, zoo, circus, and performing animals are usually at colleges of veterinary medicine or work for agencies that have exotic animals.

Housing and Space

Housing needs will vary with the species. You need to find out whether animals can be kept together, how much space they need, what temperature they prefer, and the type of housing they need.

The exotic animal owner will also need to know how to restrain their animals. For example, keeping an ostrich in confinement requires strong fences!

Reproduction

Animal production requires reproduction. Some species will reproduce well in captivity, while others need to be given great care to encourage reproduction.

REVIEWING

MAIN IDEAS

An exotic animal is any new species of animal that is being used for agricultural purposes, or an existing animal that is being used for a new purpose. The major types of exotic animals include food and fiber animals, investment animals, and zoo and performing animals.

Opportunities exist for entrepreneurs in the exotic animal industry. One must consider the profitability of the animal, the costs of production, and choose a species to raise.

Exotic animals and domestic animals have many characteristics in common. They all need proper nutrition, health care, sanitary living conditions, and space to live. They also have differences that must be attended to. To find out about these differences, you should seek help from universities, the Cooperative Extension Service, or veterinarians that specialize in exotic animal care.

QUESTIONS

Answer the following questions using correct spelling and complete sentences.

1. What is an exotic animal? How are they different from other domesticated animals?
2. What are three species of exotic animals used for food and fiber?
3. Why did investors often lose money with the pot-bellied pig?
4. How does a zoo get its animals?

5. Give an example where zoo animals contracted disease from humans.
6. What are the major questions to answer in trying to determine if an exotic animal business will be profitable?
7. What should you consider in deciding on the species of exotic animal to raise?
8. What are the major practices in managing exotic animals?

EVALUATING

Match the term with the correct definition. Write the letter by the term in the blank provided.

a. zoo curator	d. animal refuge	g. biologist
b. performing animal	e. exotic animal	h. exotic animal investment species
c. zoo	f. exotic animal dealer	

______1. An animal that is raised in hopes of having a large profit.
______2. Person who buys and sells exotic animals.
______3. Animal that is not native to the area where it is raised and is being used for new purposes.
______4. Parklike area where animals are kept for viewing.
______5. Animal taught to perform a routine of unusual activities.
______6. Person trained in understanding living organisms.
______7. Person trained in exotic animal zoology.
______8. Place where animals are protected to assure their well-being.

EXPLORING

1. Visit a zoo or animal refuge. Develop a list of the animals that you see. Study the kind of environment provided. Interview a zoo curator about feeding and other management practices with the animals. Prepare a report on what you observed.

2. Select an exotic animal. Use various reference materials and prepare a report on the origin of the animal and the kind of environment it needs to survive. Report in class.

UNIT SIX

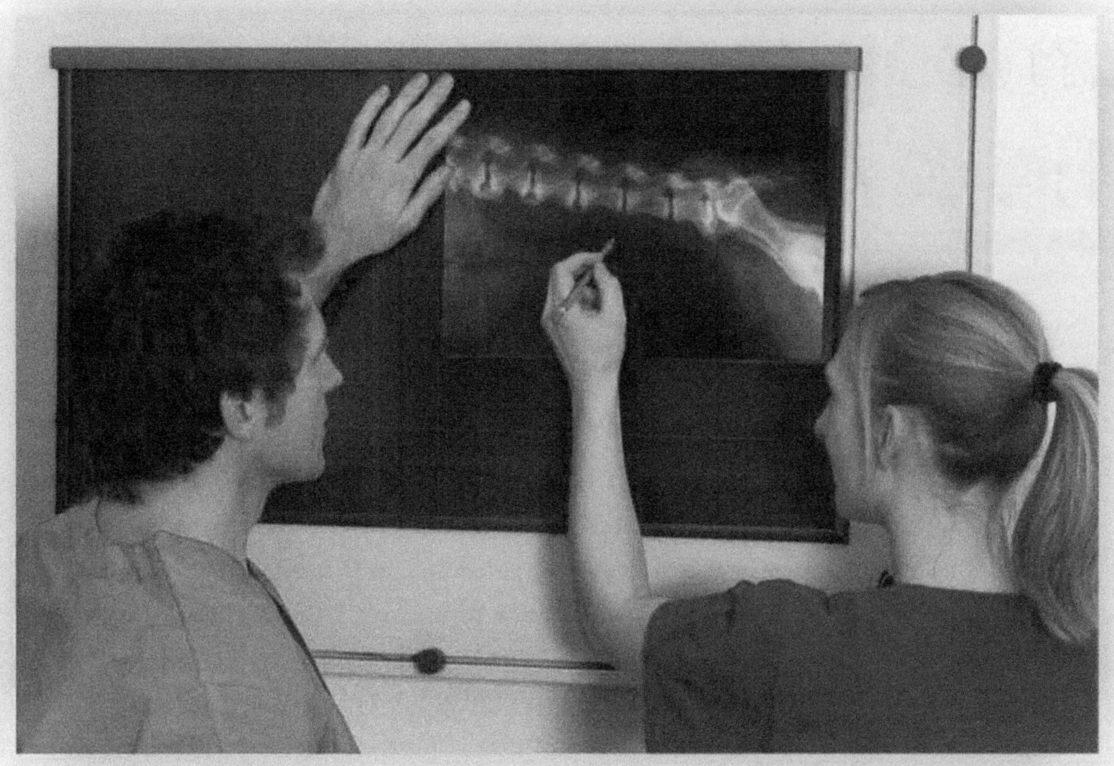

(Courtesy, Fotokostic/Shutterstock)

The Animal Industry

CHAPTER 24

Entrepreneurship in the Animal Industry

OBJECTIVES

This chapter introduces the opportunities for entrepreneurs in the animal industry. It has the following objectives:

1. Explain consumers and the role of consumption in animal production.
2. Describe the scope of animal production in the United States.
3. Explain entrepreneurship in livestock and poultry production, and list available opportunities.
4. Identify important areas of management in successful entrepreneurship.

TERMS

animal producer
beef
business plan
consumer
consumption
dairy product
decision making
entrepreneur
management
part-time animal producer
per capita consumption
risk
system

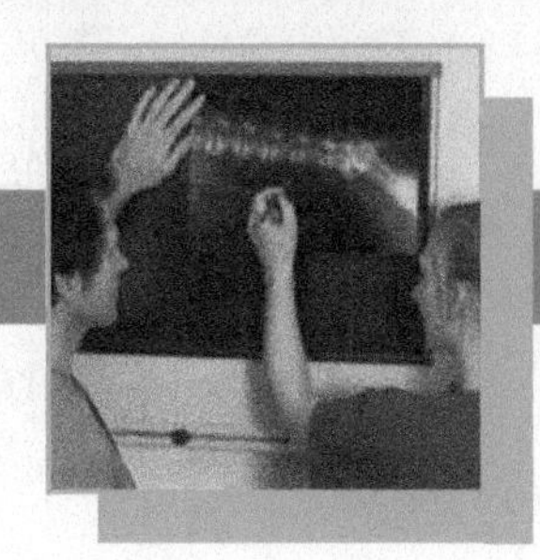

24–1. A beginning entrepreneur beef producer with her show heifer. (Courtesy, Education Images)

MANY opportunities are available in the animal industry. Some people start their own ranches, dairies, or other animal businesses. Others work at jobs in the animal industry. People who like to work with animals often go into animal production. What are your preferences?

People can work for themselves and reap the benefits of their hard work. The animal industry is large. It accounts for nearly one-half the money from agricultural sales. People need to get the right education and experience to be successful.

Begin planning for a future in the animal industry while in high school. Take courses in agriculture, science, and business. Get on-the-job experience by working for other people. This will help you understand what is involved in having an animal enterprise.

THE ROLE OF CONSUMERS

The animal industry is set up to meet the needs of people. It provides a supply of wholesome products in the forms that people want. And people want many products!

24–2. A consumer is choosing from among a wide variety of cheese products. (Courtesy, Education Images)

A ***consumer*** is a person who uses goods and services. All people are consumers! Consumers want products that satisfy needs.

What consumers want is important. They have strong ideas about what they will buy. They want good value for their money. Every time people go to the store to buy food or other items, they are consumers. They make decisions based on price, the money they have, and the desirability of the products.

POPULATION

The world's population is more than 6 billion people. Extraordinary amounts of food and clothing are needed. The number of people will increase sharply in the future. Some forecasters say that the number may reach 10 billion by the year 2030. Of course, animal products are more important in some parts of the world than others.

In North America, people like animal products. Canada, Mexico, and the United States have a combined population of about 450 million people. The United States has about 275 million people, with California, New York, Florida, and Texas being the states with the largest populations.

CONSUMPTION

Consumption is the use of animal products and other goods and services that have value. Every time you drink milk, consumption occurs. Overall, Americans have a high consumption of animal products.

Information is compiled on how much people consume. This is often reported as ***per capita consumption***, which is the average amount consumed by a person. It may be reported in weight or another measure and usually on an annual basis.

A typical American eats more than 220 pounds of meat per year. This includes 66.4 pounds of beef and veal, 49.7 pounds of pork, 76.9 pounds of chicken, and more than 30 pounds of other meats, like turkey, lamb, and fish and shellfish. It does not include game, such as deer and rabbit.

Annual per capita egg consumption is about 235 eggs in the United States. Some of these are served as egg foods, and others are used in cooking.

Dairies have about 9.2 million cows that produce 18,201 pounds of milk (about 2,275 gallons) each, for a total of more than 20.9 billion gallons of milk. That is more than 76 gallons of milk for every man, woman, and child in the United States.

These production numbers are impressive! Animal production for food is big business. When the number of animals used in research, for service, and as companions is added to the number of animals in production, the result is a huge industry!

SCOPE OF ANIMAL PRODUCTION

The scope of the five main areas of food animal production is briefly described here.

DAIRY

Nationwide, milk production has been profitable over a number of years. Milk is sold by the hundredweight (100 pounds, or about 12.5 gallons). Many dairy products are made from milk. A ***dairy product*** is milk or any product in which milk is the major ingredient, such as cheese or ice cream. Fluid milk is the beverage product we all know. It is sold as whole milk, reduced-fat milk, low-fat milk, or fat-free milk. Other forms sold include evaporated milk, condensed milk, and dry milk.

The annual per capita consumption of dairy products is 597.9 pounds. In the last 25 years, dairy product consumption has varied. The peak consumption was 601.2 pounds per capita, and the low was 535 pounds per capita.

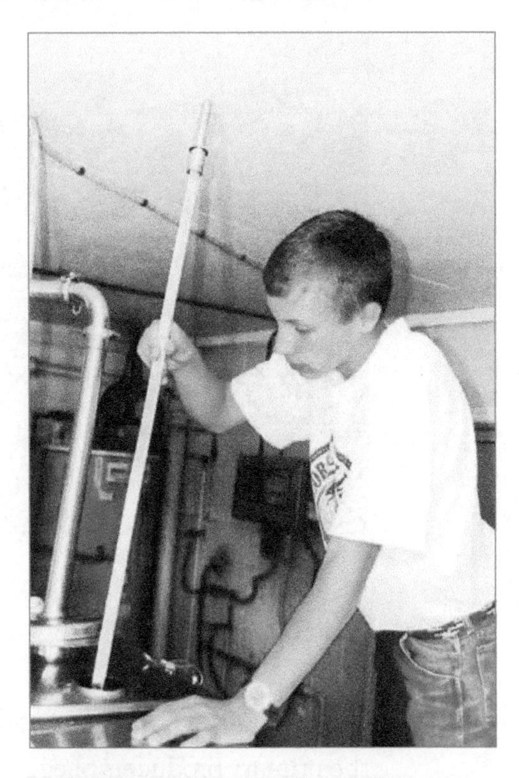

24–3. Checking the amount in a milk tank on a dairy farm. (Courtesy, Education Images)

There were 9.2 million dairy cattle in the United States in 2002. California had 1.53 million, followed by Wisconsin, which was the top dairy production state until the early 1990s. Other important dairy states are New York, Pennsylvania, and Minnesota. The nature of dairy farming varies. Some places have small family-owned dairies. Other places have large corporate-owned dairies. The trend has been toward larger dairy farm operations and fewer farms.

BEEF

Beef is the meat produced by cattle. For many years, more beef has been eaten in the United States than any other meat. Per capita beef consumption has declined in recent years. In 1994, there were more than 100 million beef animals in the United States. Today, there are about 96.7 million beef animals. As with most food animals, beef cattle have varied a few million in number in the last 10 years.

There are more than 900,000 beef producers in the United States. These producers sell nearly 54 billion pounds of beef per year. Texas has been the number one beef cattle state for many years. Texas had 13.7 million beef animals in 2001. Kansas ranked second, with 6.7 million beef animals. Rounding out the top five states were Nebraska, with 6.6 million; Oklahoma, with 5.1 million; and Missouri, with 4.25 million beef cattle.

24–4. A beef cattle producer is checking his herd of crossbred cattle. (Courtesy, U.S. Department of Agriculture)

POULTRY

The poultry industry is composed of two distinct segments: meat and eggs. Chickens are the primary producers of eggs. Chickens, turkeys, ducks, and a few other species are used for meat.

The poultry meat industry has expanded in the last decade. Production is now more than 35 billion pounds per year. Many chickens are marketed as broilers. Most broilers grow fast and are marketed at seven weeks of age. People feel that eating broilers has dietary health benefits. More than 1.2 billion broilers and chickens were produced in a recent year.

24–5. A poultry technician examines embryos in fertile chicken eggs to determine if the embryos are virus-free. (Courtesy, Agricultural Research Service, USDA)

Egg production has had only a slight increase in recent years. Current annual production is about 71 billion eggs. Concern about cholesterol in the human diet has resulted in some drop in egg consumption. More than 280 million hens are used to produce eggs in the United States.

California is the leading state in egg production. Pennsylvania is second, followed by Indiana, Ohio, and Georgia.

In addition to chickens, turkeys are produced in the United States. Nearly 300 million turkeys are raised each year.

SWINE

The swine industry has increased steadily over the last decade. Though the number of breeding animals has decreased in the last 10 years, the number of market animals has increased. Market hogs are young hogs grown for processing. They typically weigh about 240 pounds. In 2000, there were about 60 million hogs on farms in the United States. About 25 billion pounds of pork are annually marketed in the United States.

Iowa leads in hog production, with nearly 15.4 million head. North Carolina, Missouri, Illinois, Indiana, Minnesota, and Nebraska follow.

SHEEP

The United States has about 7.03 million sheep and lambs. About 6 million lambs are marketed each year, and 43 million pounds of wool valued at $15.3 million were produced in 2001. Sixty-five thousand farms have sheep and lambs.

Texas is the top state in sheep production, with nearly 20 percent of the total U.S. production. Other states in the top five are California, Wyoming, Colorado, and South Dakota.

ENTREPRENEURSHIP IN LIVESTOCK AND POULTRY PRODUCTION

24–6. A student raising a lamb is receiving important instruction from her teacher. (Courtesy, Education Images)

An ***entrepreneur*** is a person who takes a risk by trying to meet the demands of people for a product. Creativity is needed. For the entrepreneur to be successful, the animal or product must meet a unique demand. The entrepreneur typically owns and operates a business, such as a farm or ranch.

A key part of being an entrepreneur is risk. ***Risk*** is the possibility of losing what has been invested. People must have money—often large amounts—to start some kinds of animal entrepreneurships. If their businesses fail, they will lose their money. Often, the money is borrowed from banks and other sources.

Entrepreneurs are thought of as their own bosses. They have a big responsibility. The success or failure of their businesses depends on how they perform. Being an entrepreneur offers many advantages. You can set your own work schedule, have the opportunity to be creative, reap the rewards of your success, and get a great deal of self-satisfaction. There are, however, no guarantees of success.

There are many opportunities in animal production for entrepreneurs. These relate to raising animals and meeting the needs of producers for supplies, services, and marketing.

ANIMAL PRODUCER

An ***animal producer*** is an entrepreneur who raises livestock, poultry, companion, or other animals. Animal producers may be known as farmers or ranchers. Often, they are further known by the kind of animal they raise—for example, trout farmers or horse ranchers. Producers of companion animals may be known as kennel operators if they produce dogs or as aviary operators if they produce birds.

Livestock and poultry farmers engage in the breeding, feeding, and general business of raising commercial livestock and poultry. They can be classified as beef cattle farmers, dairy

24–7. A poultry producer is checking the operation of a canopy heater in a house with new chicks. (Courtesy, U.S. Department of Agriculture)

farmers, sheep farmers, swine farmers, small-animal farmers, and fish and aquatic-animal producers.

Entering the livestock and poultry production industry as a farmer requires a great deal of capital. Most types of farming require land, equipment, and specialized facilities. Many young or new entrepreneurs begin by working at home or developing partnerships with established farmers. Some start as tenant farmers and eventually are able to save the start-up capital required.

CONNECTION

ADVERTISING WITH A BIG COW

Successful animal businesses need customers. They need people to buy and use their products. This is true in all areas, from pet supplies stores to ranches with fine breeding cattle to processors of food products.

Advertising is promoting the ideas, goods, or services available. Newspapers, radio, television, billboards, and other means are used. The goal is to get the attention of potential customers.

Big cows get attention! This photo shows a big cow being pulled through city streets promoting a brand of milk and ice cream products. The big cow just barely goes under traffic lights and low overpasses. The cow will be stopped in parking lots of stores that sell dairy products. It gets attention! (Courtesy, Education Images)

It is best to select an area of interest and concentrate on one or two species of livestock. For example, an entrepreneur could choose to specialize in dairy production. Raising dairy cattle takes much time and effort. The producer must be familiar with approved practices for production and make good business decisions. Although it is possible to have a diversified operation with many types of livestock, a farmer who concentrates on one species can operate more efficiently.

A drawback to focusing on one species is that the producer is tied to one product in the market. When the market is favorable, profits are good. When the market falls, profits also fall.

If you are considering becoming an entrepreneur, collect as much information as possible. You should know what kind of facility is required, how much it will cost to get started, what the market for your product is like, and how to produce your livestock successfully.

PART-TIME ANIMAL PRODUCER

Another option for entrepreneurship lies in becoming a part-time animal producer. A **part-time animal producer** is one who raises animals only part of the time. The individual may do other work on a full-time basis.

Many farms in America today are operated part time. Some people like to be around nature and animals. They like the thought of making a little extra money from their animals. Some families that have traditionally lived in the cities or suburbs are also starting to enter the part-time farm business. Part-time farms can provide some income for the families involved, but the major source of income is usually their full-time jobs.

Part-time farms have marked differences from full-time operations. The part-time farms are usually smaller, with fewer acres and fewer numbers of livestock. These operations are commonly more diversified than full-time operations. Several species of livestock, fruit and vegetable production, and some crop land are the norm.

These operations are less expensive to start and can provide some additional income. Many small operations attempt to fill a marketing niche, like Angora goats or Siamese potbellied pigs. If you decide to enter such an enterprise, be very cautious. Be certain that a market exists for your product and that you can raise the animals. Many entrepreneurs have lost money because of bad decisions.

NONPRODUCTION ANIMAL ENTREPRENEUR

Producers rely on many supplies, services, and marketing activities to get products to consumers. People are needed to start and operate businesses that provide these. Without such businesses, today's animal producers would not have a high level of production.

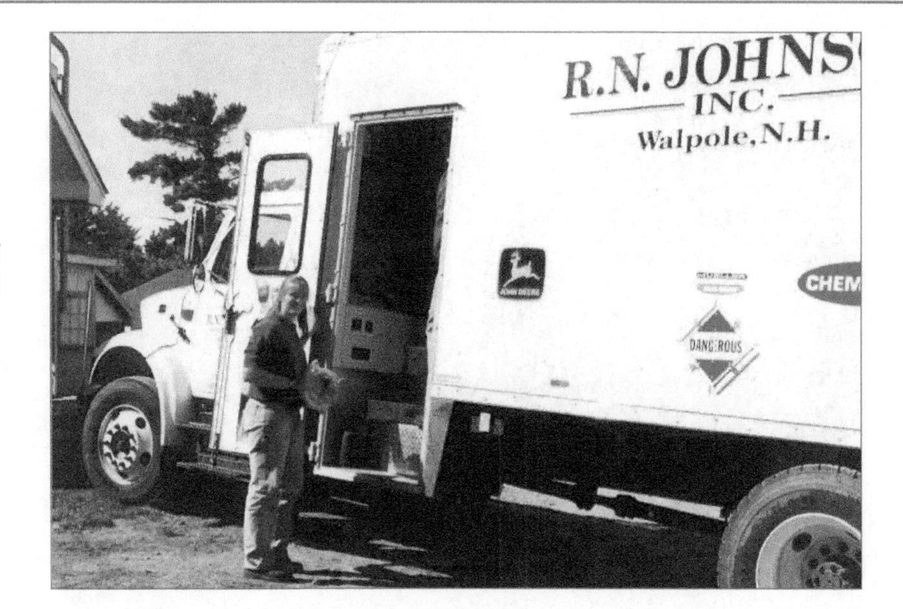

24–8. On-farm delivery of supplies for dairying and other enterprises fills a need. (Courtesy, Education Images)

Animal Supplies

Providing supplies to animal producers can result in a good opportunity for some individuals. Animal supplies entrepreneurs set up businesses to meet the needs of animal producers for supplies. Many kinds of animal supplies businesses are possible. The kinds needed depend on the animals and products produced. The following are a few examples of animal supplies businesses.

- Feed mill—manufactures feed for animal producers
- Feed store—sells feed to animal producers
- Animal medicine store—sells animal medicines and other supplies
- Animal equipment store—sells equipment used in animal production, such as feeders, waterers, milkers, gates, fencing, chutes, and pens
- Pet supplies store—sells animals, feed, medications, equipment, and other items needed with small animals

Animal Services

Animal producers often need specialized services to help with animal production. Opportunities may exist for people to be entrepreneurs in these areas. The kinds of services depend on the animals and products produced. A few examples of animal services follow.

- Groomer—trims, bathes, and otherwise cares for companion animals

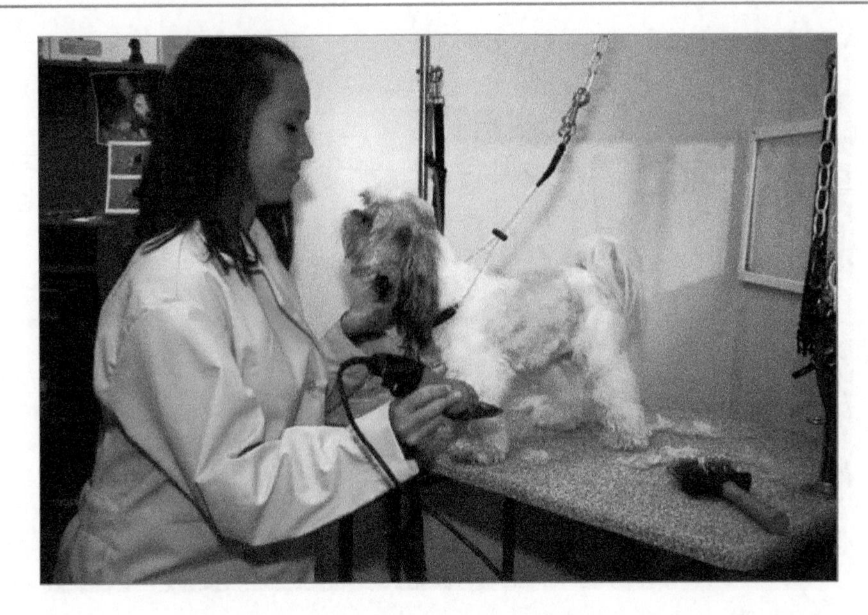

24–9. Skills are needed to provide pet grooming services. (Courtesy, Education Images)

- Animal sitter—cares for an animal when the owner is away, such as for a dog when its owner is traveling
- Veterinarian—cares for the health of animals
- Farrier—puts shoes on the feet of horses
- Custom shearer—shears wool from sheep
- Hauler—has specialized equipment to haul live animals (e.g., cattle trucks and trailers) or their products (e.g., refrigerated milk tanks on trucks)
- Construction contractor—constructs buildings, fences, and other facilities on farms and ranches that produce animals

Animal Marketing

Entrepreneurs in animal marketing may fill a wide range of roles. They link the animal producer and the consumer of the animal product. They must have good skills in working with producers and be informed on market information. The following are a few examples of animal marketers.

- Cattle buyer—buys cattle from a producer for a processor or other user
- Sale barn owner—operates a selling facility where producers bring animals to sell to buyers
- Packing house operator—operates a facility that packs and/or processes animal products

24–10. Auctions have been used in marketing livestock. (Courtesy, Mississippi State University)

- Food processor—takes raw animal products and prepares other products in forms that consumers want, such as cheese from milk

THE ROLE OF MANAGEMENT

Working hard and producing good animals is important, but the success of an animal business requires more—good management. ***Management*** is the use of resources to achieve the objectives or goals of the business. The resources are used efficiently and directed toward the objectives.

BUSINESS PLAN

A business plan is essential. A ***business plan*** is a written plan on how the animal enterprise will be operated so it is successful. Many people try to operate without written plans. With good effort, they may be quite successful on a small scale. Banks and other lending agencies often require business plans before granting loans. Business plans are usually revised each year.

Business plans help in decision making, which is an important part of management. ***Decision making*** is choosing among alternatives that are available. These relate to the kind of animal, the facilities to use, the feed and care provided, and the time to market the animal. People need good information for decision making. Bad information may result in bad decisions. And no one wants to make bad decisions!

Using information or data is an integral part of planning and decision-making with an animal enterprise. Developing business plans may involve using data to forecast productivity or assess what has actually happened. Income and profit are very important data that may be organized into charts, tables, or graphs. Numbers are often involved and known as quantitative information. The data used must be accurate. Good records and careful analysis help a manager make good choices. This is just as true with a small pet shop as it is with a large cattle ranch.

MANAGEMENT FUNCTIONS

Management involves many important duties. These are often grouped into what are called the five functions of management.

- Planning—Planning is deciding on a course of action. It is important in preparing a business plan as well as in the daily operation of an animal enterprise.
- Organizing—Organizing is identifying and arranging important activities and resources. This is needed to assure good use of time and effort in meeting objectives.
- Staffing—Staffing is getting people to help operate an animal enterprise. In some cases, one person can do the work. Larger animal enterprises require more people.
- Leading—Leading involves getting people to do what needs to be done. It is the influence a manager has over other people.
- Controlling—Controlling is assessing how well the enterprise has done. It involves keeping records and following appropriate financial procedures. The information gained is used in revising the goals and objectives.

KNOWING AND USING SYSTEMS OF OPERATION

A ***system*** is a combination of parts or processes that form one complete function or collection. Managers may deal with several kinds of systems. You have probably heard of delivery systems, marketing systems, digestive systems, mechanical systems, social systems, and many more. A simple way to grasp the meaning of system is to think of a plumbing system. A water source with pressure moves water through pipes to water heaters and faucets for our use.

System dynamics is often an area of interest to managers. System dynamics is a way of studying the world around us as a group of independent parts that make up a whole. We

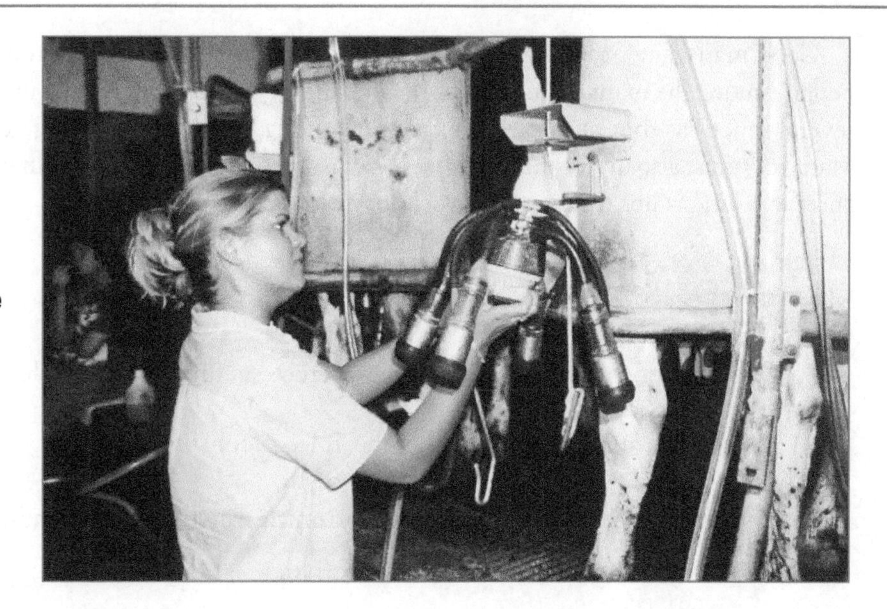

24–11. A dairy producer is readying equipment for the next cow. (Courtesy, Education Images)

often speak of economic systems, such as capitalism in the United States. System dynamics helps people understand relationships between the various parts of a system. If we understand each part, we better understand the system. We often organize systems to understand and achieve goals in the world around us.

A common element in most systems is the division of tasks or labor. This helps in understanding who does what in an entrepreneurship and when, how, and where they do it. If several people work for a manager, their responsibilities vary so that collectively they achieve the desired goals.

REVIEWING

MAIN IDEAS

The animal industry is large and diverse. Annual receipts from livestock and poultry account for nearly one-half the income generated from agriculture. Companion and other animals add greatly to the income.

Consumers are people who use animal products. They have preferences about what they will buy. Producers must produce animals that meet the needs of consumers. If they do not, the consumers will not buy the products, and the prices the producers receive will be low when the animals are sold.

The needs of consumers are not always met. This gives the opportunity for entrepreneurs to provide new products or services. Careful study and preparation are needed to be successful as an entrepreneur.

There are many opportunities for entrepreneurs in the animal production industry. Most opportunities are in animal farming, either in a full- or part-time operation.

Good management is essential for success. Management is an important function of entrepreneurs. Among the management tasks, developing a business plan is most important. Managers have several functions that must be carried out: planning, organizing, staffing, leading, and controlling. Managers must also understand and use systems. System dynamics helps in understanding the world and the relationships that exist between its various components.

QUESTIONS

Answer the following questions using correct spelling and complete sentences.

1. What is a consumer? What is the role of consumption?
2. Define entrepreneur. Discuss opportunities in the animal industry for entrepreneurship.
3. Select one area of animal production and summarize its scope in the United States.
4. What is the role of an animal producer?
5. What is the importance of management in the animal industry?

EVALUATING

Match the term with the correct definition. Write the letter by the term in the blank provided.

a. animal producer
b. consumer
c. per capita consumption
d. entrepreneur
e. management
f. risk
g. decision making
h. system

_____1. The possibility of losing what has been invested.

_____2. A person who uses goods and services.

_____3. The average amount consumed by a person.

_____4. The use of resources to achieve the objectives or goals of an animal enterprise.

_____5. One who assumes risk in a business activity.

_____6. An individual who raises animals.

_____7. A collection of individual parts that work together to achieve a goal.

_____8. Choosing among various alternatives.

EXPLORING

1. Prepare a bar graph that shows amounts of consumption for animal products. Include beef, pork, poultry, eggs, and lamb and mutton.

2. Visit an animal producer. Determine the kinds of animals produced and the general nature of the work. Discuss the job and lifestyle of the entrepreneur in animal agriculture.

3. Investigate the management of an animal enterprise. Select an enterprise, determine the role of management, and develop a list of areas where good decision making is essential. Use references found in the library or on the Internet. Prepare a report on your findings.

4. Develop a list of the systems of operation used in the small or large animal industry. You may wish to make a field trip to a modern animal production facility. Choose one system of operation and identify its components. (Think about how feeding, watering, marketing, and other functions take place at the facility.) Identify how the system functions to achieve the desired outcome. Give an oral report in class.

5. Good management of an animal enterprise (livestock or other) often involves dynamic changes in the business operation. The business plan may be modified to more appropriately serve the needs of the enterprise. To do so often requires research of problems. Develop your skills by designing, conducting, and completing research to identify and solve livestock management problems. Interview a local livestock producer or other person involved in the livestock or small animal industry about problems. Use publications, the internet, and other sources of information. Prepare a report on your findings.

24–12. Entrepreneurship might include forming a small business that provides for the hoof care of horses, such as a farrier. This shows cleaning of a hoof with shoe. (Courtesy, Fotokostic/Shutterstock)

CHAPTER 25

Career Development in the Animal Industry

OBJECTIVES

This chapter introduces careers in the animal industry and presents important information related to career and personal success, safety, and citizenship. It has the following objectives:

1. Explain the importance of career choices and development opportunities in animal science.
2. Describe the role of supervised experience in animal industry careers.
3. Relate resources and information to animal science.
4. Describe career areas in the animal industry.
5. Demonstrate personal and occupational safety with animals.
6. Identify appropriate first aid practices.
7. Demonstrate interpersonal skills in career success.
8. List and explain important employer expectations and work habits.
9. Identify good citizenship skills.
10. Describe the role of the FFA in achieving success in animal industry careers.

TERMS

career
citizenship
employability skills
employment
entry-level job
first aid
gainful employment
goal
goal setting
information
job
MSDS
occupation
patriotism
personal protective equipment
record keeping
resource
safety
supervised experience
training plan
training station
work ethic
zoonotic disease

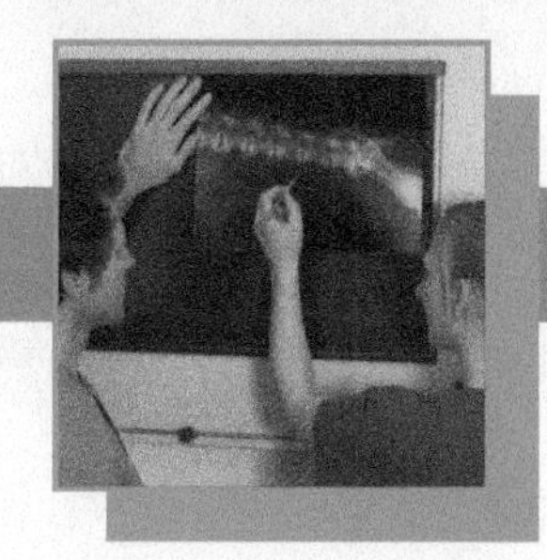

25–1. Scientists observe the behavior of a sow and litter from a distance. (Courtesy, Agricultural Research Service, USDA)

ANIMALS are fun! Some people own cattle or hogs; others have a dog or a cat. A few people don't own any animals but, even so, they enjoy seeing animals. Some people make their living in the animal industry. Every one knows that animals are very important to human beings.

The study of animal science is a part of the Agriculture, Food and Natural Resources Career Cluster. (This is sometimes abbreviated as AFNR.) Within AFNR, there are eight career pathways. One of the pathways is Animal Systems. This pathway has been carefully studied and designed by authorities in agricultural education and the animal industry. Content standards have been developed to guide the delivery of education, including the class in which you are using this book.

The standards help assure that you learn the most important knowledge and skills in school. This book, ***Introduction to Livestock and Companion Animals***, integrates most all of the standards. Some of the advanced standards require education beyond high school. The standards you will achieve with this book are built on a base of science, particularly animal biology. This chapter helps you understand important areas for success in the animal industry. Enjoy yourself!

CAREER DEVELOPMENT IN ANIMAL SCIENCE

People want to do useful things. They want to feel that they are contributing to society and have a good income. Choosing a career is an important decision. People work thousands of hours during the 30 or more years of their employment. Get good information when making choices.

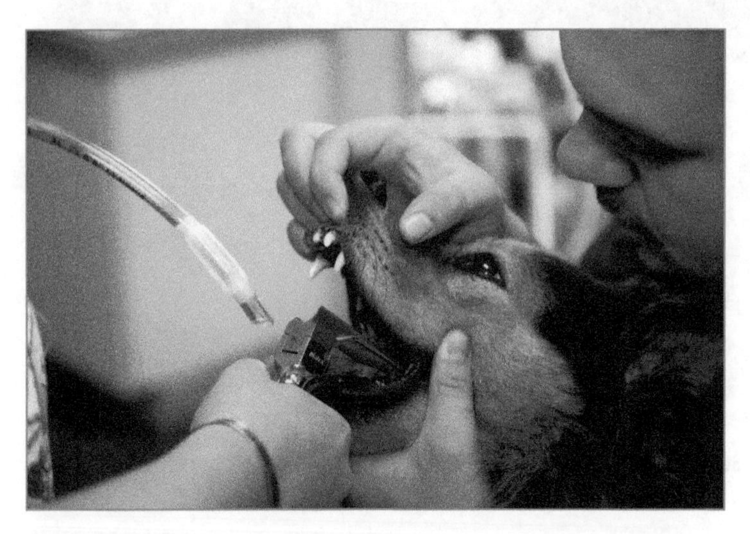

25–2. A veterinary assistant is learning the use of a laryngoscope with a dog. (The instrument is used to see the vocal folds and glottis of a mammal. Its use facilitates the placement of a tracheal tube for administering anesthesia.) (Courtesy, Education Images)

Many off-farm careers related to livestock, as well as careers with companion animals, support animal production and enjoyment. Less than 2 percent of the U.S. work force is in farming. However, nearly 20 percent of all workers are in careers related to agriculture. Although some careers have limited contact with animals, about 535,000 people work in meat and dairy production. In addition, thousands of people are in companion animal, poultry, aquaculture, and equine jobs. Others work in the exotic and service animal fields.

25–3. A counselor is discussing career planning with students. (Courtesy, Education Images)

EMPLOYMENT

People use their time in different ways. ***Employment*** is the primary way people use time and earn a living; employment is often called work.

Sometimes, people serve as volunteers and are not paid. Other times, they work for what they gain to help cover the costs of living.

Gainful employment is having a job that provides benefits, such as salary or wages. Some jobs provide additional benefits. For example, a ranch manager may be given free housing on the ranch. Such benefits are a part of the total returns that a person gets for working.

Entrepreneurs have jobs. They are gainfully employed, though they may not be paid a salary. They receive benefits if their animal production enterprise is a success.

GOALS

People often think about what they want to have and do in their lives. Much planning and work are needed to reach goals.

A ***goal*** is what an individual wants to achieve. Goals have a powerful influence in our lives. We all want to achieve what we set out to do. No one wants to fail. People need to be careful in setting goals to be sure that the goals reflect what is important to them.

Goal setting is describing what we want to accomplish. It includes how we will achieve the accomplishments. People do it all the time. Sometimes, our goals may not be very important in the long run. Goals people set about their careers are definitely important! Begin by understanding careers and the needed education. Match this information with what you want.

25–4. Gaining leadership skills through FFA helps with success in careers. (Courtesy, Education Images)

CAREERS, OCCUPATIONS, AND JOBS

A ***career*** is the general direction that a person takes with work. Most careers involve a sequence. People begin at a lower level and advance to higher levels. Moving up is climbing the career ladder, or advancing.

25–5. Hands-on experience develops career skills, such as how to collect an ear swab sample. (Courtesy, Education Images)

The lowest-level job is the entry level. An ***entry-level job*** is one that requires little or no previous experience. It is the first job for many people. Acquiring education and following instructions will help people advance. Careers may have a series of occupations and jobs.

An ***occupation*** is an area of work with specific duties. The same occupation has similar duties regardless of location. For example, the occupation of a large-animal veterinarian has similar duties regardless of where the veterinarian works.

A ***job*** is specific work that has definite duties. Jobs are carried out at certain sites and with employers. People who have jobs are employed. People can have occupations and not have jobs. For example, if the large-animal veterinarian had no job, the individual would still have an occupation, even though he or she were unemployed.

Sorting through the meanings of career, occupation, and job helps people make good decisions. The next time you hear about employment, check to see if it is an occupation or a job and how it is a part of a career.

DEVELOPMENT OPPORTUNITIES

Education is needed for success in animal production. Practical experience with animals and practical experience in the workplace are also essential.

Preparation for animal science careers can involve a number of approaches.

- Take agriculture courses in high school. Also, take supportive courses in science, mathematics, and communications.
- Pursue education in agriculture beyond high school at a technical school, college, or university.

- Continue education throughout your career with young adult and adult education. Field days, workshops, and other means can be used to stay informed. Employers may provide on-the-job training. Overall, staying up-to-date is the responsibility of each individual, especially if he or she wishes to be able to pursue career development opportunities.

PLANNING AND MANAGING SUPERVISED EXPERIENCE

Supervised experience (SE) is planned activities related to classroom and laboratory learning that are carried out outside normal class time. A variety of "learning by doing" activities are available in areas of animal science. All are carried out under the direction of your teacher. (Supervised experience is also known as supervised agricultural experience, or SAE.)

The object of SE is to provide planned and practical activities that help develop skills needed to be successful in the workplace or in life. Supervised experiences are usually planned around the goals and career plans of each student.

Teachers try to make supervised experiences significant and related to classroom instruction. Teachers usually cooperate with parents and other adults to provide the most meaningful experiences for students. Older students may work in animal-related businesses, where they are supervised by their employers in cooperation with their teacher.

Students are involved in deciding and planning their SE. This helps assure that the experiences are interesting and important to them. Students should list their career goals before beginning to plan their SE.

25–6. A student uses dog care and walking for her supervised experience. (Courtesy, Education Images)

STRATEGIC PLANNING

Supervised experience is often planned as a program. Strategic planning processes may be used to create a series of individualized learning activities. The series is planned by you, your

25–7. Practical experience with animals can be gained in many ways, such as in a school laboratory. (Courtesy, Education Images)

25–8. A Texas student carries out her supervised experience with a lamb in a school facility. (Courtesy, Education Images)

teacher, your parents or guardians, and, if applicable, an employer. SE should be based on your interests and career goals.

Planning usually involves a sequence of activities that, in this case, are related to animal careers. We begin with simple or introductory activities and move to more advanced activities. The activities are also tied to the level of instruction in the classroom and laboratory. For example, a student may begin with one animal and gradually increase the number.

Kinds of SE

Four kinds of supervised experience are used in animal production. Planning and conducting often involve choosing the kind you will have. The kinds are:

- Exploratory—Exploratory SE is supervised experience that helps students explore and develop their interests. It provides the opportunity to look at a number of subjects. Exploratory SE is often the first kind of supervised experience a student has upon enrolling in agriculture. The purpose of exploratory SE is to help you understand and appreciate animals. Students in the earliest stages of instruction in agricultural education benefit the most from exploratory SE.

- Placement—Placement SE is supervised experience that involves carrying out activities on a farm or ranch or in an agribusiness related to animal careers. School labs, government agency offices, and community facilities are also used for placement. In this experience, you are typically working for another individual. The purpose is to provide practical experiences needed to enter and advance in a particular occupation. The employer provides the facilities and other resources needed to provide the training.

25–9. The Internet has sites with good information to guide decision making. (A good site to begin your investigation is www.ffa.org.) (Courtesy, Education Images)

The location where a student is placed is called a ***training station***. Students working at training stations are often paid for their work. Younger students may start with non-paid supervised experiences. Students who are placed in school labs are said to have directed lab experience. This is because they are not away from the school under the direction of an agribusiness employer.

- Entrepreneurship (or ownership)—Entrepreneurship SE (or ownership SE) is supervised experience in which the student owns an enterprise. Students may own calves, hogs, lambs, rabbits, dogs, or other animal enterprises. Outside of production agriculture, students may own small businesses, such as pet walking services or animal grooming businesses. Students may carry out entrepreneurship SE from their homes or other locations. Entrepreneurship SE helps students develop skills needed to own and manage production agriculture or agribusiness enter-

25–10. Entrepreneurship supervised experience may involve a beef animal. (Courtesy, Education Images)

prises. Students with entrepreneurship supervised experience programs own materials and other resources for their enterprises.

- Research/experimentation—Research/experimentation SE is supervised experience that involves identifying a particular science-based problem, searching for information about it, and conducting a scientific experiment to arrive at a conclusion. The student makes recommendations about how to solve a problem related to animals. Always remember to practice animal well-being and never abuse animals.

  Students with research/experimentation SE often get the assistance of science teachers in their school or local community scientists. They may have displays at FFA agriscience fairs. The activities are often closely related to instruction in areas of science.

Training Plans

A ***training plan*** is a written statement of your SE intentions. Information in the plan relates what you learn in class to your SE. A first step in planning is to propose a tentative career or education goal. A goal is where you want to be in a few years. Activities in the plan are intended to lead you toward the goal. These activities involve the job skills you will need. From these activities, you will develop the ability to do the work of an occupation.

Another part of planning is to assess FFA activities that relate to your interests. Integrate SE plans that will lead to the achievement of your FFA goals.

CONNECTION

PARTICIPATION DEVELOPS SKILLS

People often have many opportunities to develop job skills. Some people, however, are reluctant to participate. Not participating can later be a handicap in getting a job and in succeeding on the job.

Students in agriculture classes are in FFA. Many FFA activities develop skills for job success. There is one catch. You must be active. Get involved. Participate.

Leadership, personal, and career skills are developed through FFA. Those who are most active gain the most from participation. The Montana student in this photo is developing valuable skills in relating to other people. She is presenting a project she has developed to an adult leader. (Courtesy, Education Images)

Training plans may include improvement activities and supplementary practices. These are carried out in addition to the supervised experience skills. Often, they support student achievement.

MANAGING AND CONDUCTING

Managing and conducting supervised experience begins after it is underway. You will be able to use the selected record system to document your experiences.

Record Keeping

Record keeping is recording SE activities. The information may be written on paper or prepared using a special computer program. Your state or school may have approved SE record-keeping systems. You will be able to use what has been selected.

25–11. A student and her teacher are reviewing her supervised experience plans and records. (Courtesy, Education Images)

Many states have online record-keeping systems. These often allow students to design and structure records specific to their situations. Some students may use the Agricultural Experience Tracker (AET). The AET allows students to customize their supervised experience and keep appropriate records. These include educational and financial records associated with supervised experience and details of their FFA activities. The AET was created by a team of agricultural educators from high schools and universities in Texas.

The AET system will guide you through creating your own record book. The system will help you capture resources to promote learning in agricultural education. Two major resources are involved: time and money. Making wise use of these two resources is essential to success. The entries you make should result in a well-designed and customized system that meets your unique needs. To some extent, the record book is like a management information system that is related to agricultural education and the FFA.

To use the AET system (except for a time of exploration), you will need to login with information provided by your teacher. You can also temporarily go the Web site to explore what the AET has to offer. For additional information, including viewing a video on AET and how it is individually tailored, go to: www.theaet.com.

Several kinds of records are kept. Most important is a list of the experiences you have gained. This list will be valuable as you apply for FFA awards. Records are also needed of the income and expenses related to your SE. Financial records are necessary in determining the success of an enterprise.

Good records are also important in making FFA advancements. Filling out awards forms is easier when you have good records. The degree of accuracy is also improved.

Evaluating and Expanding

As you develop skills, you will want to expand into new or more advanced areas. Expanding involves continual planning, assessing (evaluating), and adjusting your direction.

Several factors involved in expanding SE are:

- Follow good work practices.
- Seek to develop new skills and learn new things.
- Expand the scope of your SE. Add more animals, or take on additional responsibility.
- Strive for success! You want to learn and achieve just as much as you possibly can.
- Gain recognition for accomplishments through FFA. The records you keep are important. Fill out application forms in detail to explain what you have done.

RESOURCES AND INFORMATION

25–12. Feed is a resource used in animal production. (Courtesy, Education Images)

Career success often depends on getting and using resources and information to assure that an individual has employability skills.

RESOURCES

A ***resource*** is a supply, a skill, money or property, or an item of information needed to achieve your goals. We often speak of natural resources, raw materials resources, human resources, financial resources, or education/knowledge resources.

In the animal industry, the resources may include all the inputs we need to produce animals successfully. The nature of these resources varies with the kinds of animals and the production systems used. For example, beef cattle cow-and-calf production requires, among other resources, good pasture. On the other hand, a rabbit production enterprise requires hutches, feed, waterers, feeders, and other items.

Managing resources wisely is essential. Good management helps assure the success of an animal enterprise. It also promotes sustainability, or the ability to produce on a long-term basis. Further, we must meet financial responsibilities and legal obligations with our resources, such as repaying borrowed money in a timely and agreed-upon manner.

INFORMATION

Information is knowledge or news that is communicated. Good information is a valuable resource in animal care and production. People need good information in order to make the best possible decisions. Locating, assessing, and using information is important in most plant and animal careers today. We must try to stay up-to-date in order to be successful. Some of the information deals with interpersonal skills and systems operation. Other information deals with technical subjects.

25–13. Books may contain detailed and useful information about a subject. (Courtesy, Education Images)

Locating Information

When looking for information, consider the forms it is in and the sources it is from. Also, determine whether the information is up-to-date and accurate.

Sources are the businesses or agencies that produce information. Examples of sources are employers, publishers, agencies, universities, industries, and associations. Agencies include the U.S. Department of Agriculture, state departments of agriculture, and other agencies that relate to agriculture and resources. Frequently, a search on the Internet will turn up information. Always carefully assess the information to be sure it is accurate and up-to-date.

The most common forms of information are:

- Books—Books provide basic information about a subject. Most books have been carefully written and reviewed to assure accuracy.
- Bulletins—A bulletin is a small publication that often reports the findings of research or development work. Some bulletins are very scientific in nature. Readers must often draw their own conclusions from the research.
- Magazines—Magazines are published weekly, monthly, or on some other schedule. Some magazines address plant and animal production.
- Newspapers—Newspapers are published daily or weekly. Many have a wide range of news stories and report prices of agricultural commodities. Some trade papers relate specifically to plants and animals.
- Manuals—Manuals are publications that accompany products to provide details about the products. Manuals may also include assembly and use instructions.
- Electronic forms—Electronic forms of information include the Internet and disc-stored information. The Internet has many sites that offer agricultural information. DVDs and

CDs are often used to produce published information. These are used on a personal computer, with Adobe Acrobat Reader being a popular software program.

Assessing Information

Assess information to see whether it is accurate and appropriate for your needs. Using incorrect or out-of-date information may result in poor decisions and relationships. Consider the following:

- Source—Is the source of the information reliable? Does the source use reviewers or other procedures to assure technical accuracy?
- Legal accuracy—Is the information accurate in terms of laws and policies?
- Operation—Does the material show systems and relationships? Does the material promote teamwork and harmony in the workplace?
- Date—When was the information prepared? Up-to-date information is essential.
- Illustrations—Is the material illustrated with photographs and line drawings? Are the illustrations attractive and accurate?
- Safety—Does the material promote safe practices?
- Availability—Is the information readily available at a reasonable cost?

CAREERS IN FOOD ANIMAL AREAS

Owning an animal production business is one way to be involved in the production of livestock. Many people, however, want to be involved with animals, but not as entrepreneurs. They want employment with other people.

FARM OR RANCH WORKER

Many people who eventually want to own and operate their own farms start out as farm or ranch workers. Opportunities for this type of employment will vary with the type of agriculture in your community.

Farm and ranch work varies based on the knowledge and ability of the person employed. Duties range from general farm labor in doing routine jobs, such as feeding, maintaining sanitation, and moving animals, to more complicated tasks, such as treating sick animals. A small farm may employ a person to do a wide range of tasks. A worker in a large operation may specialize in a particular area, such as milking cows.

Many entry-level positions are as farm or ranch workers. New employees are expected to have a general knowledge of livestock and to work competently with animals. They need to learn their jobs quickly. High school agricultural education is helpful.

FARM MANAGER

The farm manager carries out the everyday operation of the farm. Responsibilities may also include doing long-range planning and setting production goals. Farm managers are usually hired by corporate farms, by institutional farms, or by individuals who own farms but do not want to be involved in the daily operation.

Becoming a farm manager requires extensive experience or a combination of experience and education. Employers are often looking for people with education beyond high school.

Tasks a farm manager might do include selecting and marketing animals, formulating feed, maintaining herd health, and supervising employees. Helpful preparation includes high school agricultural education, farm work, and postsecondary education in agriculture.

CAREERS IN SMALL ANIMAL CARE AND MANAGEMENT

A wide range of career opportunities are found in small animal care and management. Some of these overlap with other areas of careers in the animals industry. In some cases, an individual might begin at an entry-level job in small animals and, with good work ethic, advance to other higher-level jobs.

Success will require having the needed education and training and being productive at work. Good personal skills are as important with these careers as with other areas of work. Knowing how to safely manage animals is essential.

A few small animal care and management occupations are listed here and briefly described. You will find other information about occupations elsewhere in this book. Information should be gathered locally related to job openings, rate of pay, and requirements. This information will allow you to compare and contrast occupations.

- Pet Groomer—Pet groomers mostly work with dogs but cats and a few other species may be included. A pet groomer bathes/shampoos, cuts hair, clips nails/claws, inspects for signs of disease, greets customers, restrains animals, and places animals in crates or cages. Groomers also give customers information about the care of the animal and offers suggestions about health. Education and training may begin in a high school small animal care class and include supervised experience, an apprenticeship (on-the-job) under the direction of a skilled groomer, or otherwise. A local business license or certificate may be needed. Jobs may be at grooming and boarding facilities, veterinary clinics/hospitals, and pet stores. Income varies widely depending on volume of business and the like but is in the range of $20,000 a year.

- Pet Shop Employee—Pet shops or retail stores usually employ several people with varying skill needs and responsibilities. Individuals greet customers, advise customers on animal selection and care, demonstrate animal management, answer questions about animal care and products, refer owners for animal health needs, stock shelves, care for animals in the store, and receive payments from customers. Most learn on-the-job though many employees take courses in high school in animal management. Rate of pay is minimum wage upward to $24,000 or so annually for store managers.

- Kennel/Cattery Worker—Kennels and catteries have employees who care for animals, greet customers, manage animals, clean facilities and install bedding, and provide feed and water. Jobs are usually at kennels or catteries, with many being part-time and some full time. Most are small businesses and have few employees. Rate of compensation is minimum wage upward to $20,000 or so annually.

- Veterinary Assistant—A veterinary assistant assists a veterinarian in providing health care for animals. Duties of a veterinary assistant include greeting customers, receiving animals and delivering animals to owners, setting up veterinary facilities for the veterinarian's work, restraining animals, observing animals during anesthesia, cleaning examining tables and other facilities, using autoclaves to sterilize instruments, stocking supplies, and assisting veterinarians with a wide range of duties. Most work in veterinary clinics or hospitals and complete a veterinary assistant program of education. Licensing may be required. Issuance of a license may be based on achieving a satisfactory test score. Rate of pay varies with level of education, amount of experience, size and location of clinic or hospital, and other conditions. Rate of pay varies from entry-level in the low $20,000s to $30,000 or over.

- Animal Control Officer—Animal control officers typically work for local government agencies to assist with removal of nuisance animals, capture loose animals, place animals in shelters, relocate homeless animals with new homes, care for animals, observe animals for health problems, and interpret animal control regulations to citizens. The work may involve operating vehicles, snaring animals, restraining animals, transporting animals, and placing animals in crates. Sometimes the animals can exhibit aggressive behavior; caution is needed to avoid attack and injury. On-the-job education is often needed. Rate of pay is minimum wage and above for a few thousand dollars.

CAREERS IN PLEASURE AND SERVICE ANIMAL AREAS

Careers in pleasure and service animal areas are numerous. You could have a career raising animals for research or being a pet store owner. Horses, game and fish, service animals, and laboratory animals are a few areas.

HORSE INDUSTRY WORKER

25–14. Some people have occupations training animals, such as this mule on a Colorado ranch. (Courtesy, Lucky Three Ranch, Inc.)

Many people enjoy riding and racing horses. Stables provide places to keep and ride horses. In addition, there are plenty of places to enjoy horseback riding.

Increased interest in light horses for pleasure and racing has opened new jobs. These range from stable attendant to racehorse trainer. As with many other careers, you begin by doing hand labor.

Stable hands spend most of their time doing essential work, such as cleaning the barn and feeding the horses. As you get more experience, you may become a horse trainer, riding instructor, stable manager, or barn boss. Job titles in the race industry differ from those in the pleasure horse industry.

GAME AND FISHERIES WORKER

25–15. A fish hatchery technician tends tanks of fish. (Courtesy, Ed Grateau, U.S. Fish and Wildlife Service)

Do you like wildlife? Do you like being outdoors? Are you concerned about fish and wildlife? Then you may be interested in a career in game and fisheries.

A wildlife conservation officer works to educate the public about fish and wildlife and enforces game and fish laws. The duties may include conducting workshops on wildlife and the environment. Conservation officers are usually hired by state or federal agencies.

Game farm workers raise and manage different species of wildlife. Some of these are returned to the wild. Others are kept on game farms. Hunting clubs, conservation groups, and government agencies need wildlife to enhance populations. Although no specific qualifications exist, a high school diploma and an interest in working with game birds and animals are desirable.

Fish hatchery technicians raise and manage fish. The fish may be for stocking in public and private ponds, lakes, and waterways. People employed in this field care for brood fish, tend eggs, maintain water quality, feed fish, provide a healthy environment for the fish, and manage the facilities. Hatchery workers are generally under the direction of a fisheries biologist. High school courses in mathematics, science, chemistry, biology, and aquaculture are helpful.

People in game and fish positions should have an interest in wildlife, fisheries, and conservation. Usually a wildlife conservation officer must pass a test and have needed training. Experience and a college degree in a wildlife area are beneficial.

SERVICE ANIMAL WORKER

Raising animals that serve humans is important. Service animals, such as leader dogs for the blind, companion animals, and draft animals, fill many roles.

Draft animals are used for work, primarily providing the power to pull machinery. Most machines today, however, are powered by engines. Some draft animals are used as hobbies and, in a few cases, to do specialized work. In a few countries, animal power is used to grow crops.

Although there are large producers, many service animals are raised by small producers or individuals.

25–16. Grooming a steer for show develops skills in working with animals. (Courtesy, U.S. Department of Agriculture)

LABORATORY ANIMAL WORKER

A laboratory animal is used in research to improve human and animal life. Laboratory animals are raised and kept under the best possible conditions. Without these animals, progress in human medicine and related areas would nearly stop.

Many laboratory animals are delicate. They are raised to meet specific qualifications for research. For example, animals are bred with immune systems that are almost identical to human immune systems for AIDS research. Laboratory animal production is a complex field. It requires an in-depth knowledge of science, chemistry, and biology, as well as an interest in animal research.

CAREERS IN SCIENCE-ORIENTED AREAS

The science of animals is expanding. Animal scientists study animals and use laboratory equipment to solve problems related to animals.

ANIMAL SCIENTIST

Animal scientists study animals and develop improved methods of production. They provide services to the livestock, poultry, and companion animal industries. Positions related to animal science are animal nutrition specialists, animal reproduction specialists, geneticists, chemists, and biotechnologists.

An animal research scientist concentrates on a specific area of interest in animal production. Research scientists work in laboratories with experimental animals and animal production inputs and with other scientists. Most people who choose this career have an advanced college degree in the area. Personal needs are a high interest in scientific inquiry and a knowledge of biology, chemistry, and microbiology.

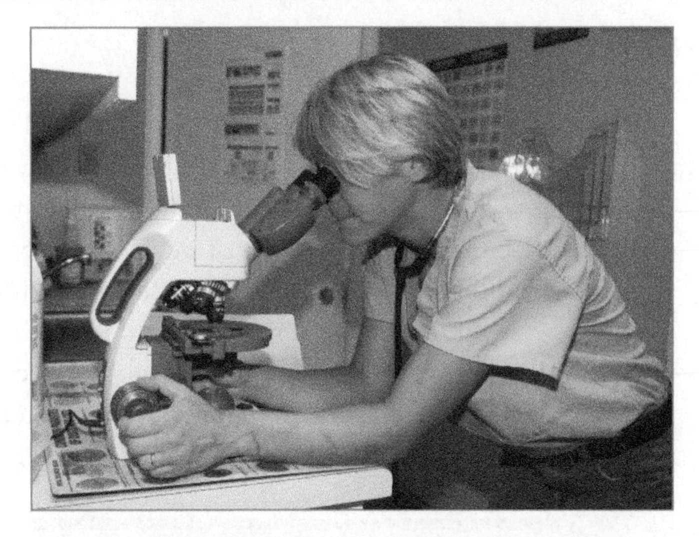

25–17. A slide prepared with a blood sample from a dog is being viewed for evidence of heartworms with a microscope. (Courtesy, Education Images)

LABORATORY ASSISTANT

Assisting with laboratory animal production or experimentation can be a rewarding career. Animal laboratory assistants help with experiments, care for laboratory animals, and raise laboratory animals. Attention to detail is required. A high school diploma with an emphasis in animal agriculture is a minimum. Most laboratory assistants have some college education. Some have a master's degree in biology, chemistry, animal science, or a related discipline.

AGRICULTURAL ENGINEER

Agricultural engineers often work in the animal industry. They work with animal facilities, waste management, environmental impact, and animal product improvement. Engineers must have a bachelor's degree and the ability to assess a problem and prepare a solution to it. Knowledge of agriculture and the animal industry is helpful.

25–18. An owl is being studied in terms of health and well-being. (Courtesy, College of Veterinary Medicine, Texas A&M University)

WILDLIFE BIOLOGIST

A wildlife biologist does many of the tasks of a general biologist. The focus is on wildlife animals and habitat. A wildlife biologist may study an endangered animal species or concentrate on improving the environment for many species. Work can include genetics, studying the environmental impact of industry, or repopulating an animal where it no longer exists. Wildlife biologists help with the living conditions of wild animals and fish.

A wildlife biologist must be committed to studying animals and science. A college education is required for entry, and higher degrees are needed for advancement. If you want a career in wildlife biology, you need a high interest in biology, chemistry, animal science, mathematics, and research.

SERVICE CAREER AREAS

People who provide for the needs of animals and help with marketing are important in the animal industry. Producers often solicit help from animal service professionals.

25–19. A veterinarian is floating the teeth of a horse. (Courtesy, Education Images)

VETERINARIAN

Veterinarians promote animal health. They provide care for sick and injured animals, as well as prevent illness. A veterinarian may specialize with specific livestock or generalize in all animals. A few veterinarians now specialize in fish health.

Becoming a veterinarian takes time and dedication. Preveterinary students often earn a bachelor's degree in animal science or biology. Then, they apply for admission to a school of veterinary medicine. If they are accepted, they complete several more years of education and training. Although much is required to become a veterinarian, it is a very rewarding career.

SALES AND SERVICE

The animal industry has many professionals in supplies sales and service. These people sell feed, ensure quality products, sell retail products, service equipment, operate stockyards, and perform many other services. Agricultural education in high school and experience in the animal industry are beneficial.

Many sales positions related to livestock and poultry production require a college education. The education should be in animal science or a related area.

Although not a requirement for many service positions, some postsecondary education is desirable. Examples of positions are machinery serviceperson, stockyard worker, veterinary assistant, and feedmill worker.

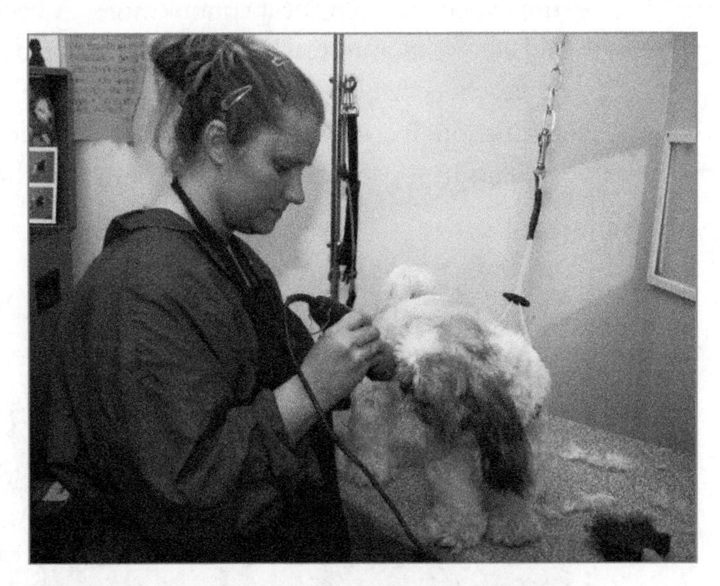

25–20. Pet grooming is a service occupation with good opportunities in some cities. (Courtesy, Education Images)

LIVESTOCK BUYER/SELLER

Knowing what the industry and consumers demand in a meat animal leads to success in livestock buying and selling. A good eye for livestock quality on the hoof is essential. Buyers and sellers may work for themselves or have clients in the production or packing industry.

Training in animal selection is a requirement for this career. High school education in agriculture, biology, and mathematics is helpful. Additional training in animal science, business management, and public relations at the postsecondary level will be beneficial.

EDUCATION, EXTENSION, AND COMMUNICATION AREAS

Careers available in education, extension, and communication emphasize the "people" side of the animal industry. All levels are involved. Teaching and communicating information are necessary.

AGRISCIENCE TEACHER

Agriscience teachers work in the public schools. They teach basic agriculture and specialized classes. They work with students of all ages—kindergarten through adult. They are agriculture experts in their local communities. A bachelor's degree in agriculture or a related area and certification to teach agriculture are needed.

Teachers involve students in FFA. FFA is a student organization that enhances agriculture instruction. It also provides personal and leadership skill development. Teachers also involve students in supervised experience programs that include areas of the animal industry.

25–21. An agriscience teacher is instructing a class in a classroom with state-of-the-art technology. (Courtesy, Education Images)

AGRICULTURAL EXTENSION AGENT

Agricultural extension agents work with local producers to help them improve production or solve livestock and other agricultural problems. Extension agents who specialize in livestock provide help to producers.

A bachelor's degree in agriculture or a related area is needed to be an extension agent. Many states require a master's degree within a few years. Some extension agents specialize in animals. This requires a strong background in animal production.

AGRICULTURAL COMMUNICATION SPECIALIST

Agricultural communication (agricommunication) specialists share information about animals. They may work with newspapers, in public relations, and with broadcast media (radio and television). An agricultural communicator may host a television show, write for a livestock magazine, or produce videos and computer-based materials.

A career in agricultural communication will require a university degree in agriculture or communication. Sometimes, people combine the study of agriculture and communication. People employed in this field need to have a broad knowledge of animals and have a people-oriented personality.

25–22. Agricommunication specialists are important in keeping the public informed. This shows an agriculture television editor. (Courtesy, Education Images)

SAFE PRACTICES IN WORKING WITH ANIMALS

Working with animals can be dangerous. Some animals are large, and others have sharp teeth and claws. Animals are often frightened when people work with them. People can be injured when they try to protect themselves.

Safety is preventing injury or loss. It is important to prevent injury to yourself, other people, and animals.Safety is often presented in two areas: occupational and personal.

OCCUPATIONAL HEALTH AND SAFETY PRACTICES

Occupational safety is being safe in our work. It involves safety at the workplace. Sometimes distinguishing between occupational and personal safety is difficult. In the work environment, the two tend to be associated in promoting our well-being and that of our co-workers. Since humans may contract some diseases from animals, health safety in animal work is closely associated with personal safety.

Here are several practices that promote occupational safety with animals:

- Avoid zoonotic disease—A ***zoonotic disease*** is an illness that is passed from animals to humans. A few zoonotic diseases are ringworm, cat scratch fever, salmonella, rabies, brucellosis, and tuberculosis. The best way to prevent zoonoses from spreading is to keep healthy animals! Make sure the animals have clean quarters and are vaccinated

25–23. Knowing animals and how to approach them improves safety. This shows a farrier repairing a horse's shoe. (Courtesy, American Paint Horse Association)

against diseases. Avoid contact with infected animals, and take precautions to avoid bites and scratches. Always wash your hands after handling animals. Wearing protective gloves and clothing is an important health safety practice.

- Never work alone with an unfamiliar animal—Learn how to handle animals before attempting to work with them. You may not be familiar with specific behaviors or dangers. Never work alone with an animal without someone knowing where you are and what you are doing. Never work with an animal without permission from your instructor, supervisor, or its owner.

25–24. Properly restrain an animal that you are working with. (Courtesy, Education Images)

- Work in teams or groups–When possible, you should have a team or group to work with. At the very least, you should have a partner when working with animals in a laboratory or in a production facility. Accidents happen in the best facilities with the best workers. Always have another person with you just in case!
- Close doors and gates you open—Make sure entrances and exits are secured. Having escaped dogs or an animal escape can create big problems, especially if it is a large animal. There are many stories of escaped dogs or cattle running through the streets of towns! Take precautions to keep animals restrained.
- Handle one kind of animal at a time—This keeps you focused on your work, and you will be better able to handle the animal. Remember that animals have good senses and can detect the smell of other species easily. Working with mice and then with snakes could end in a bite to your hand! Remember to always use health safety practices with all animals.
- Remain calm—Animals can sense if you are nervous or excited. Try to remain calm at all times.

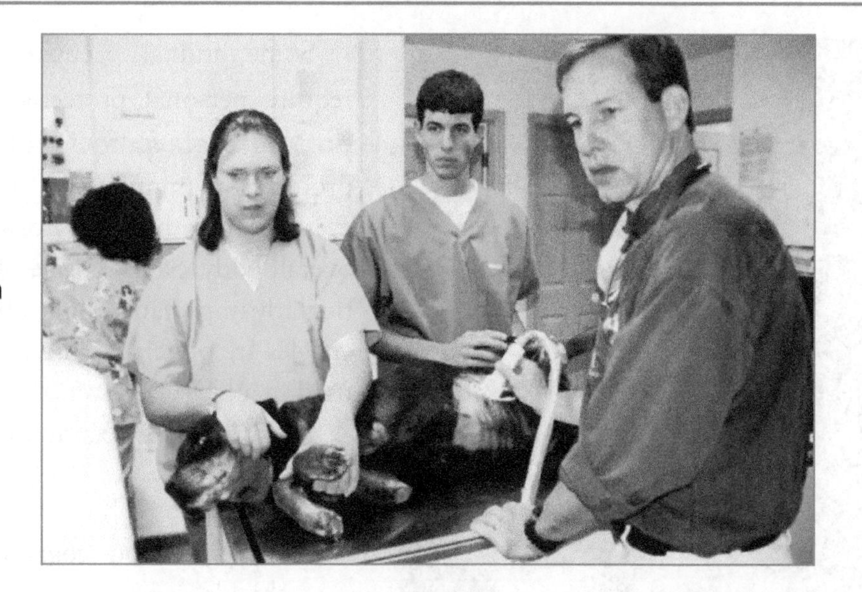

25–25. People often work together as a team, such as in performing an ultra sound examination of a dog. (Courtesy, Education Images)

Do not get frustrated or angry when animals are not doing what you want. By remaining calm, you will help the animals stay at ease. This will make them easier to work with and reduce the possibilities of injuries to them and you.

- Be gentle with animals—Avoid using excessive force to get animals to cooperate. Most of the time, gentle nudging, encouragement, and patience will far outperform brute force. Using force increases the possibility of injuring the animal and yourself. Do not abuse animals.
- Get away from an aggressive animal as soon as possible—Getting away from an aggressive animal is the top priority. Then, properly report the behavior. There is a reason the animal became aggressive. Work with someone to find the reason and correct the behavior.
- Properly restrain animals—Make sure the animals that you are treating with medicine or that you are grooming are properly restrained. If you are working with large animals, they should not be able to injure themselves or others by sudden movements. Small animals should also be handled correctly and restrained when being given medical treatment or being groomed.
- Use mechanical devices safely —Animal work often involves using equipment and tools. Examples include exam tables, head squeezes, x-ray equipment, clippers, brushes and combs, and syringes. Always know how to safely use such devices. Know and follow safe practices. Never take risk by violating mechanical safety practices, as with hazardous devices used to restrain large animals.

PERSONAL SAFETY

The practices that we follow in being safe vary with the nature of the work. Employers are expected to inform workers of hazards and provide for their safety. This, however, does not relieve employees of their safety responsibilities.

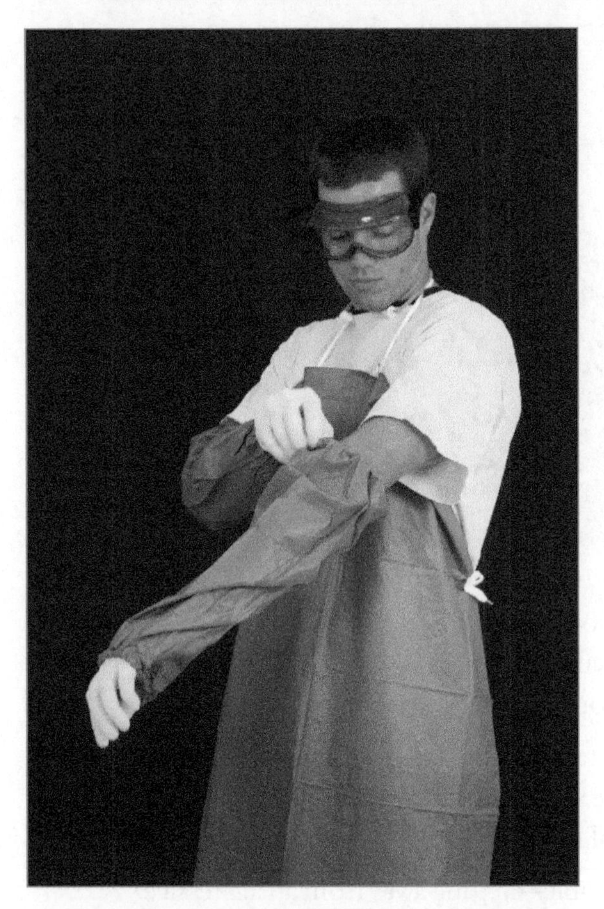

25–26. Safety goggles and vinyl gloves, apron, and sleeves are being worn. (Courtesy, Education Images)

Some animal occupations have hazards that require personal protective equipment. ***Personal protective equipment*** (PPE) is equipment used to protect from injury. It may be designed to protect the eyes, hearing, the respiratory system, the skin, or the overall body. Safety goggles or safety glasses are particularly important in any lab activity. Properly fitted goggles are preferred. PPE should always be properly used in a lab.

Following are examples of PPE that may be needed:

- Eye protection–goggles, safety glasses, and face shields
- Hearing protection–ear muffs and ear plugs
- Skin protection–rubber or plastic gloves, aprons, and other protective clothing specific to the work
- Foot protection–rubber boots and steel-toed shoes
- Head protection–hard hats
- Respiratory protection–particle masks and respirators

SAFETY INFORMATION

Safety information on chemicals, compounds, and mixtures is provided by MSDS, which is the abbreviation for Material Safety Data Sheet. ***MSDS*** is a widely used filing and retrieval system of information about substances that may pose hazards. This information includes instructions for safe use and hazards associated with a particular product. Of particular use is the action to take in case of an accident. In the animal industry, many of the substances used to treat and prevent disease are included in MSDS.

MSDS may be available and kept in a paper notebook or retrieved via the Internet. Some people prefer to have a paper copy readily available and use the Internet to gain information on new or rarely used products. A good Web site for products used in animal health care is this MSDS retrieval service: **www.midwestvet.net**.

25–27. Contents of an MSDS notebook are being reviewed. (Courtesy, Education Images)

FIRST AID

Emergencies don't wait! When an accident happens, be ready to respond appropriately. Don't waste time and fear to react. Remaining calm and using common sense is essential with an emergency situation. An animal bite or kick, a fall, a cut, or other injury event to a person may require first aid. If an accident happens at school, tell the teacher and follow school procedures.

First aid is providing limited care for an illness or injury. This care is provided by a lay person rather than an experienced emergency responder. In general, first aid consists of life-saving techniques a person can perform with or without first aid training using minimal equipment. The techniques are used only until qualified medical help is available. In most locations of the United States, emergency medical help is available by calling 911.

Three aims of first aid are to preserve life, prevent further injury, and promote recovery. Some people take first aid training so that they know the best techniques to use. Most people, however, have not had such training (but should get it). Regardless, a person should take the necessary first aid steps. Always have a first aid kit and step-by-step information handy. The local Red Cross can provide training and recommendations on first aid supplies that should be readily available. In administering first aid, be sure to protect yourself from body fluids of the injured individual.

In any situation, the key first aid steps are sometimes referred to as the ABCDs: airway, breathing, circulation, and deadly bleeding or defibrillation. Remember to keep the accident victim's airways open and promote breathing. Cardiopulmonary resuscitation (CPR) may be needed if a person is not breathing. Broken bones, bleeding and wounds, amputation, burns and scalds, chemical exposure and poisoning, choking, sunburn, heat exhaustion and sunstroke, cold exposure and frostbite, splinters, stings, sprains, and shock all require care specific to the problem. Defibrillation is using a device to deliver a shock to restore rhythm of the heart in case of heart failure. Only individuals qualified to do so should use a defibrillator.

Moving an injured person may be critical to his or her recovery. How a person is moved is based on an assessment of the injury. If the injury is a small cut on the hand, use the contents of a first aid kit to stop bleeding, clean the injury, apply antiseptic, and bandage the site. If the injury is severe, such as a broken bone, including a broken neck or back, seek the immediate assistance of a qualified emergency responder before moving the person. Some injuries must have immediate care of a physician and hospital medical facilities. Occasionally a site poses the potential of additional injury if a person is not moved. An example would be a nearby fire.

25–28. The contents of a first aid kit are being inventoried to be sure that all needed items are present. (Courtesy, Education Images)

If an individual is bleeding, take steps to control the loss of blood. Use pressure on the wound and on the arteries. In severe cases, get qualified help. Get training in how to take the necessary steps. Generally with minor wounds, apply pressure to stop the bleeding, clean the area with an appropriate antiseptic, and apply a sterile protective bandage. First aid kits typically have the supplies to take these actions.

Recommendation: Get first aid training and get help if a person is injured!

PERSONAL AND INTERPERSONAL SKILLS FOR CAREER SUCCESS

Personal skills help people to be successful on the job. These include skills in relating to other people. Eight key personal attributes are included here.

Employability skills are the abilities an individual needs to gain and maintain a job. They include specific job skills as well as work ethic and human relations abilities. Keeping current with information is also necessary. This assures that a person is up-to-date in his or her career field. Employability skills help a person to be a productive worker. These are skills employers are seeking when they hire employees.

Personal skills are those that help us do our work. Interpersonal skills are those that help us relate to other people. It is sometimes difficult to distinguish between the two.

Here are a few examples of personal and interpersonal skills:

25–29. The abilities to speak, maintain eye contact, and respect others promote relationships with people. (Courtesy, Education Images)

- Have a positive lifestyle—Live a life that speaks well of you. Get plenty of rest. Avoid substances that impair your ability to work. Obey laws, and be courteous to others.
- Get along with co-workers—No one likes to work with troublesome people. They are unpleasant to be around. Make it a point to get along with others. Avoid creating trouble between people.

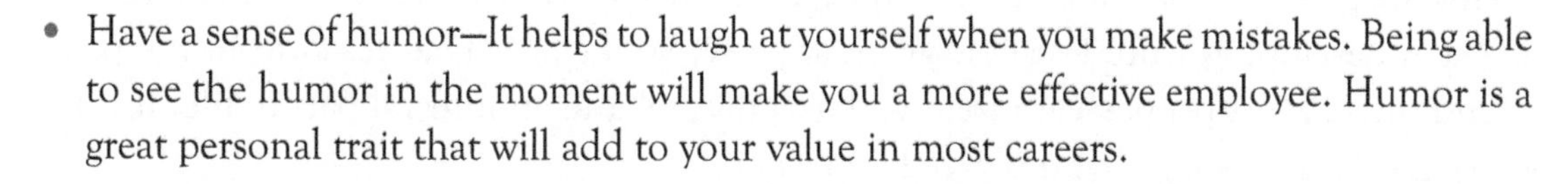

- Have a sense of humor—It helps to laugh at yourself when you make mistakes. Being able to see the humor in the moment will make you a more effective employee. Humor is a great personal trait that will add to your value in most careers.
- Be honest—Be honest in all you do every day. Honesty is a virtue that will carry you far in all walks of life. It is especially important in your career. Honesty is measured in what you say, in what you do, and in the promises you keep. It takes much time to gain the trust of others, but you can lose trust in a matter of seconds through dishonest acts.
- Be willing to learn—One of the most important personal traits is the willingness to learn new things. Knowledge keeps your mind alive and keeps your curiosity healthy. Being unwilling to learn new things leads to closed-mindedness, stale ideas, stubbornness, and a decline in knowledge and skill.
- Eliminate biased opinions—Eliminating bias of others is a positive step toward including all people in the decisions we make. The more minds working on a problem, the better the solution!
- Love your work—You can often hear people complaining about their work. Work is easy for people who love their jobs. They look forward to going to work. They like the people and customers they work with, and they have a purpose for doing their work. People who love their work are in great demand. They are the best employees. Find a career you can love.

- Assume your share–Always perform the duties of your job. Do a little extra, and lend a hand to others. Since our success increasingly depends on working with other people, considering the needs of others in the work environment is highly beneficial.

WORK ETHIC AND EMPLOYER EXPECTATIONS

Employers expect many things from their employees. Simple tasks like coming to work on time or working for a full day are easily identified. These relate to how we view work.

WORK ETHIC

People view work differently. The ***work ethic*** is how we feel about work. Do we think it is essential? Do we think it is the most important thing in our lives?

How we go about choosing a career and the one we choose are part of our work ethic. People begin developing attitudes about work at a very young age. They learn from the people around them. Most people learn that work is a natural and an integral part of life. It is how people go about having what they need and want.

The way people go about work is also a part of their work ethic. People are expected to be good, dependable, honest workers. They are expected to work hard and do a job well. These attributes help people to be successful.

EMPLOYER EXPECTATIONS

Seven employer expectations that can make you more successful at work are listed here:

- Communicate with others–When employers are asked to list expectations of employees, they often identify communication skills as a top priority. The ability to speak and write–to present your ideas clearly–is necessary in most jobs today. Practice writing and speaking at every opportunity. Become a good communicator in preparation for entering the work force.
- Look professional–Looking the part is important in all careers. If you went to the dentist's office and were met by a woman who claimed to be a dentist but was wearing cutoff shorts and a tank top and was missing two front teeth, you would probably want to find another dentist! The woman may be a fine dentist, but your first impression would make you

doubt that fact. You should be committed to looking the part for your job. Employers also expect good personal hygiene habits with your appearance, clothes, and manners.

- Acquire knowledge and skills—Employers would like to hire people who have the skills they need to do the job. In reality, employers will not be able to find individuals who possess all the needed skills. Employers hire employees with the expectation that they will acquire the knowledge and skills to perform the job. Modern animal production is changing at a rapid pace. Even if employers could hire people with all the skills needed today, next year those skills will be outdated. Prepare yourself with current skills, but be willing and able to learn new skills to keep you on top of your profession.
- Be a team player—Employers expect their workers to be able to get along with each other and be able to operate as a team. Being a team player means that you are able to work with others to accomplish a common goal. You are not a team player if you are concerned only with your personal success and recognition. Team players are happy to be a part of the group and work to support the group's efforts and goals.
- Be responsible—Being responsible for your own actions is an expectation of most employers. If you make a mistake, admit it. There are few things that will make an employer angrier than an employee who cannot accept responsibility for his or her actions.
- Practice good work ethic and ethical conduct—Employers expect employees to have and practice a strong work ethic and demonstrate ethic conduct. A strong work ethic is shown by having pride in work, working a full day for a day's pay, and doing the best work you are capable of doing. Never miss work except for an emergency or a scheduled vacation. Always be at work early, and stay on the job. Do not leave work early. A strong work ethic will make your employer happy and will serve customers well. A strong work

25–30. Employers expect employees to be responsible in operating equipment. (Courtesy, Education Images)

ethic makes you a valuable employee. Ethical conduct refers to honesty and respect for others in the workplace and includes fairness, equality, and respecting the rights of others.

25–31. Signs in a poultry processing plant instruct employees in proper safety and sanitation. (Courtesy, Education Images)

- Legal responsibilities—Employees are expected to know and follow the law. They should comply with Federal, state, and local regulations, including those on safety. Read and follow workplace safety and health posters. Report hazards and worker injuries to the employer. Employees should follow laws related to their work such as animal well-being, use of animal medicines, and animal care.

GOOD CITIZENSHIP SKILLS

Citizenship is participation as a member of the population of the United States. It also means that a person is a legal resident of the country and is entitled to certain rights. Being a citizen has certain duties and responsibilities. These form the attributes of a good citizen.

Laws of the United States provide conditions to participate as a citizen. Among the rights of citizenship are the right to vote, the ability to own property, and freedom of speech, religion, and the press. Each of these has certain conditions that protect people and the overall well-being of the nation.

People who are good citizens have desired traits. Examples of traits of good citizens are:

- Abiding by the law—Laws are followed in all regards. This includes paying taxes, not violating the rights of others, and safely operating motor vehicles.
- Supporting charity—Needy people benefit from the help of others. Good citizens support legitimate charities.
- Voting—Voting is a privilege. Citizens can vote at age 18 provided they have registered and not lost the right to vote because of criminal convictions. Before casting a vote, always get information about the issues or candidates so you will make good choices.
- Supporting the well-being of others—Citizens participate in decision making for their communities. They look at the overall well-being of their communities rather than at their personal interests.
- Being patriotic—***Patriotism*** is admiration for and loyalty to one's country. Citizens take pride in their country. Good citizens pledge allegiance to the flag, ascribe to ser-

vice, and demonstrate enthusiasm for their nation. Patriotism is also shown by service in the armed forces.

- Taking pride–Good citizens take pride in their communities, jobs, and homes. They appreciate the role of a quality environment in promoting well-being.
- Contributing–Being a productive worker, good student in school, or other socially useful person contributes to the overall betterment of society.

25–32. FFA members are saying the Pledge of Allegiance to the Flag as part of their patriotism development. (Courtesy, Education Images)

LEARNING THROUGH THE FFA

The FFA is the organization for students enrolled in agriculture classes. Chapters are found in most schools with such classes. The FFA offers many opportunities and benefits to students. Details about the FFA are found in the National FFA Manual and its Web site at: **www.ffa.org**.

Active membership in the FFA is open to students between the ages of 12 and 21. Enrollment in an agriculture class is required. Annual dues are collected by most chapters though in some schools or states the dues are covered by education funding. FFA membership involves degrees through which a member can advance. Some students enroll in middle or junior high school and pursue the Discovery FFA Degree. The Greenhand FFA Degree is the first for students in high school. It is followed by the Chapter FFA Degree. Beyond the chapter, there is the State FFA Degree and the American FFA Degree. Moving up in degrees of membership is based on participation and accomplishment. Full requirements are available from an FFA advisor, who is also known as an agriculture teacher.

The FFA is organized into local, state, and national organizations. The local FFA is known as a chapter. It is associated with a school's agriculture classes. Approximately 7,350 schools in the United States have FFA chapters. At the state level, the FFA is known as an association. This is because it is associated with the National FFA Organization. The major headquarters for FFA activities is in Indianapolis, Indiana. The National FFA Con-

25–33. An FFA member is presiding over a meeting. (Courtesy, Education Images)

vention is the major gathering of students from throughout the United States. It is currently held in Indianapolis, Indiana.

Benefits of Membership

The FFA offers many benefits to members. Benefits are best realized by students who get actively involved. Some of the major benefits are:

- Being involved with activities that promote school achievement.
- Gaining recognition for your accomplishments.
- Developing a positive self-concept.
- Becoming self-confident.
- Gaining career and work experience. (This is in conjunction with supervised experience, which was covered earlier in the chapter.)
- Being part of a team.
- Getting to travel to interesting places and meet students from other schools.
- Learning how to solve problems and make good decisions.
- Developing skills in money management.
- Developing leadership and social skills.

Local Chapters

Local chapters are associated with the agriculture classes and program at schools. These chapters have members who develop programs of activities, serve as officers and committees, and carry out important activities. Chapters use a team of officers and several committees to run the organization. These individuals gain valuable skills in speaking, leading, organizing, and getting along with people.

The team of officers in a chapter typically consists of the following: president, vice president, secretary, treasurer, reporter, sentinel, and advisor (teacher). Some chapters also have historians, parliamentarians, and chaplains.

FFA meetings follow a fairly set order of business. Meetings are run by the officers. An opening ceremony is used to initiate a meeting. Minutes of the previous meeting are presented. Old and new business may be considered. An informative program may be presented. A closing ceremony is used to adjourn a meeting. In some cases, refreshments are a part of an FFA chapter meeting.

Committees are very important in a local chapter. Because there are several committees, each student can be involved. The committee names and duties often relate to the program of activities that is carried.

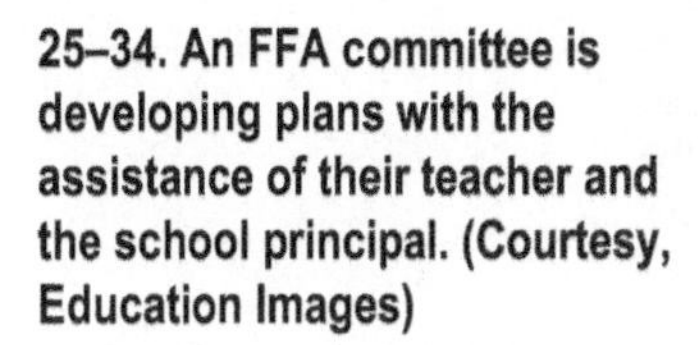

25–34. An FFA committee is developing plans with the assistance of their teacher and the school principal. (Courtesy, Education Images)

Awards and Events

The FFA has proficiency awards, chapter awards, career development events (CDEs), and other awards such as Agri-Entrepreneurship and the Agriscience Award. Students who excel gain recognition. Many times students choose to do projects involving animals and animal science. They are careful to use animals only in ways that do not cause harm and promote their well-being.

Proficiency awards are in a number of areas, including several in animal agriculture. These awards are used to recognize students who have excelled in their supervised experience. Many awards are presented locally, at the state level, and by the National FFA Organization. Awards are based on achievement as an owner/entrepreneur or for placement supervised experience.

Examples of proficiency awards in animal systems include: equine science; beef, dairy, swine, sheep, small animal, and wildlife production; aquaculture; and diversified livestock. These sometimes change. You will want to get the latest information on the opportunities for you from the FFA organization.

25–35. A student group is practicing swine evaluation in as part of the livestock evaluation CDE. (Courtesy, Education Images)

CDEs are activities which students compete with other students. These activities are based on the skills a person develops related to entering and advancing in a career. Examples include those in dairy cattle, livestock evaluation, poultry evaluation, horse evaluation, meats evaluation and technology, and dairy foods.

The FFA also has leadership events. Participation in these can focus on animal agriculture. Examples of leadership involvement include public speaking, creed speaking, job interview's, and parliamentary procedure. The creed speaking event is for members holding the Greenhand Degree.

25–36. An FFA member is practicing for the prepared public speaking FFA event. (Courtesy, Education Images)

REVIEWING

MAIN IDEAS

Exciting careers are available in the animal industry. There are opportunities in many areas. The meat animal industry provides wholesome food. The pleasure and service animal areas meet important needs. Careers are also found in science-based fields, in animal management and service, and in agricultural communication.

Careers in the meat animal industry include becoming an entrepreneur in animal or poultry production, working for someone on a farm or ranch, or managing a farm or ranch. The pleasure and service animal industries also offer many opportunities that include working with horses, game and fisheries, draft animals, and laboratory animals.

Science-related careers include working in animal laboratories, conducting animal research, discovering new medical treatments, and designing animal facilities. Animal management and service careers include working in veterinary science, providing inputs for animal production, and marketing animal products. Many workers are involved in the people side of the animal industry through education, extension, and agricultural communications.

Safety is important when working with animals. This includes both occupational and personal safety. You should respect animals and remain calm when working with them. Avoid zoonotic diseases by maintaining animal health and using preventive measures. You may sometimes need to use personal protective equipment (PPE).

The wise use of resources and information promotes career success. Acquiring needed education, pursuing development opportunities, and planning and managing supervised experience are part of preparation for animal science career success.

There are several employer expectations and personal skills you should develop in yourself to make you a more valuable employee. A good work ethic will carry you toward success. Citizenship and patriotism are expectations in many occupations.

QUESTIONS

Answer the following questions using correct spelling and complete sentences.

1. List and compare several careers in small animal care and management.
2. What attributes of a person indicate a good work ethic?
3. How is goal setting a part of career planning?
4. What are the responsibilities of a farm manager?
5. Describe one career area in pleasure and service animals.
6. What is the nature of the work of an animal scientist?
7. Why is safety important in working with animals?
8. What personal safety practices should be considered in working with animals? In using mechanical devices with animals?

9. What interpersonal skills are important to job success? Name eight. Select two of the skills and describe them in some detail. Include ways to apply the skills (competencies) in the animal industry.
10. What expectations do employers have of workers? Name six, and describe one in detail. Be sure to include resources, interpersonal skills and systems of operation.
11. What is supervised experience? How is a training plan proposed and managed? How is supervised experience conducted, evaluated, and documented?
12. What resources may be needed in animal careers? Indicate how these resources may be applied in such careers.
13. What is information? Why is information important? Indicate how the information is applied to the competencies needed in specialty areas of the animal industry.
14. What is citizenship? Patriotism? How does a person demonstrate good citizenship characteristics? (Name at least two ways.)
15. What opportunities are available through the FFA for students to participate in leadership and career development?

EVALUATING

Match the term with the correct definition. Write the letter by the term in the blank provided.

a. work ethic
b. goal
c. career
d. job
e. entry-level job
f. safety
g. record keeping
h. personal protective equipment
i. resource
j. information

______1. Prevention of injury or loss.
______2. How a person feels about work.
______3. Specific work carried out at a site.
______4. The level of achievement people set for themselves.
______5. The general direction that a person takes with work.
______6. The first job for many people.
______7. Recording SE activities (and other activities associated with work and personal lives).
______8. Knowledge or news that is communicated.
______9. Something needed to do work and achieve goals.
_____10. Equipment used to protect from injury.

EXPLORING

1. Interview professionals in the small animal care and management industry. Discuss their job descriptions, daily activities, career preparation, and opportunities in this field. Also, find out what the salaries for beginning professionals would be and educational requirements. Compare the information that you find. Prepare a written report that summarizes what you learned.

2. Discuss post–high school education for career opportunities with your guidance counselor and your agriculture teacher. Set long-term goals. Support these with short-term goals. Write the ways and means required to obtain your goals.

3. Construct a poster or a computer-based presentation about a specialty career in animal agricultural enterprises that interests you. Use a computer technology and the Internet to research Web sites with animal careers. Examples of Web sites to get you started are: Animal Careers – https://asas.org; http:animalcareers.about.com and http://animalcareerguide.com. Include in your search career development opportunities. Review information on the nature of the work, the education needed, and the likely compensation upon entering the career. Also, think about why your choice is best for you when compared with other jobs.

4. Prepare a wall chart or poster that lists rules for personal safety when handling animals. Post the chart in your school lab or classroom. Give an oral report on your chart and demonstrate one of the personal safety rules such as wearing gloves, safety glasses or goggles, or boots and using ear plugs. Be sure to include mechanical safety when equipment is used.

5. Create a record keeping system that will serve your unique needs and interests in agricultural education and in conducting supervised experience. Records document and help in evaluating your supervised experience program as experiential learning activity. Ask your teacher, parents, and individuals involved with your supervised experience about the kinds of records you might need. Give careful thought to your situation and needs as well as your supervised experience. You may use the AET system or other record keeping system that is available in your school. In some cases, you may use a record book printed on paper. Regardless, enter accurate information, prepare detail summaries, and analyze your progress. Determine the profitability of projects, such as raising pigs or calves. Assess if you are making desired advancement toward education, FFA, and career goals. Prepare a brief summary to discuss with your teacher or a parent.

6. Prepare a career development and entrepreneurship plan for opportunities in specialty fields of the animal (agricultural) industry. List opportunities in the area surrounding your home community. Identify needed preparation/education/job skills for career entry into these opportunities. Investigate how gain the needed preparation and list the schools, colleges, or internship experiences that provide the preparation.

7. Prepare a poster that demonstrates the use of health practices in the workplace. Be sure to include the use of personal protective equipment as well as the role of practices that minimize or prevent zoonosis. Orally review your poster with the class.

8. Investigate ethical conduct in the workplace. Relate how this is a part of employer expectations. Demonstrate your understanding of ethical conduct by writing a one-page paper on employer expectations as related to ethical conduct.

9. Prepare a short presentation that summarizes employer expectations of worker legal responsibilities. Demonstrate that you know the role of OSHA and following legal safety practices, as well as other legal aspects of a particular job.

10. Strategic planning processes are often used with supervised experience development. The process involves defining a direction and allocating resources to help assure success. Investigate the process of strategic planning. Determine how it applies to your supervised experience. Plan, conduct, and evaluate your supervised experience accordingly.

11. Working with animals sometimes involves the use of hazardous materials such as chemicals, sanitation solutions, vaccines, antibiotics, and topical medicines. Develop a list of materials used in the local animal industry or in the school lab or on a local farm/ranch. Investigate the safety hazards these products may pose to humans. Determine the guidelines to follow when handling or using the product. Give a short oral report in class.

12. Choose an animal species and identify the equipment and tools that may be used when working with it in the laboratory, on a farm/ranch, or in placement supervised experience. Investigate safety needs and associated hazards with the equipment. Study its safe use. Demonstrate the proper use of the laboratory equipment and tools. Examples with small animals include pet carriers, litter boxes, kennel gates, waterers, feeders, scales, examining tables, and restraints. With larger animals, chutes, headsqueezes, restraints, scales, fences, and gates. Read and follow instructions. Get the assistance of authorities on the use of the equipment, including your teacher or placement station supervisor.

CHAPTER 26

Issues in the Animal Industry

OBJECTIVES

This chapter provides readers with an awareness of emerging issues in the animal industry. It has the following objectives:

1. **Discuss the importance of issues in animal science.**
2. **Identify and explain issues associated with companion animals.**
3. **Identify and explain animal issues associated with global warming.**
4. **Identify and discuss issues associated with animal biotechnology.**
5. **Explain animal identification as an issue.**
6. **Discuss changes in grain consumption as an animal issue.**
7. **Discuss two meat animal controversies.**
8. **Relate nutrient management as an animal issue.**

TERMS

agrichar
anaerobic
biogas
carbon credit
carbon sequestration
chevaline
downer cow
effluent
ethanol
greenhouse effect
hippotherapy
issue
malnutrition
methane
sink

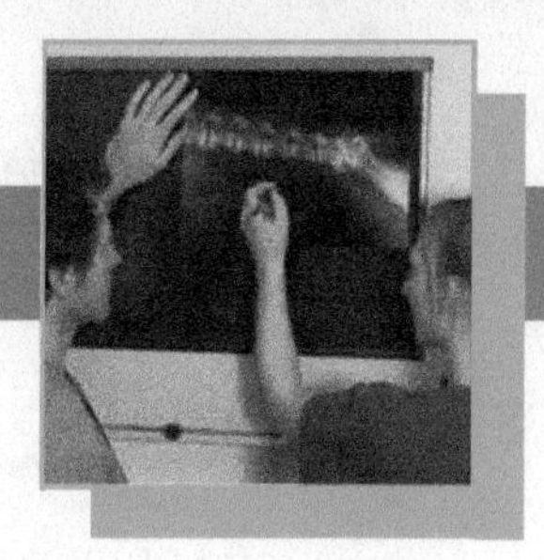

26–1. People often develop special bonds with animals. (Dealing with issues surrounding them becomes a challenge.) (Courtesy, Phil Date/ Shutterstock)

ANIMAL-RELATED issues exist today that weren't envisioned just a few years ago. Why has this happened? It has happened because changes have occurred that may not appear related to animals.

Disasters and weather-related catastrophes impact animals. Hurricane Katrina left thousands of animals homeless and dead. The human population of this world continues to grow, putting pressures on some parts of our environment. When the earth's environment changes, it has an effect on animals that live in that environment. Some change is rapid, like a major weather tragedy. Other change is much slower, such as global warming. And not everyone agrees that global warming is occurring!

Food supply is important to all people. Certain animal issues have resulted because they threaten or benefit food supply. Closely related is food safety. The national animal ID system was started to improve food safety and reduce threats to humans resulting from contamination or sabotage of animal food products.

ISSUES AND ANIMALS

As animal use changes, opinions about their use become issues. Controversy sometimes emerges. But, changes in animal use are often associated with changes in society. Some are very gradual; others more rapid.

An ***issue*** is a problem or subject that has more than one point of view. You may view an issue one way and those around you may hold another view. Sometimes issues are stated as questions. Identifying the best answer is not easy. Often an issue must be examined from a larger societal perspective.

Today, animals are used for different kinds of work than they were a century ago. Draft horses, donkeys, camels, cutting horses, and water buffalo have all been used for physical work throughout history. Now companion animals are used for treatment in elderly housing or special education classes in our schools. Horses are used for therapy in handicapped populations. Seeing Eye dogs give our blind population more mobility and a semblance of normal life.

26–2. Bonding with an animal may at first appear challenging but bonds typically form rather quickly. (Courtesy, Education Images)

Steps to consider with animal-related issues are:

- Identify the right issue. Without the right identification, wrong issues or issues that do not exist may be pursued.
- Get good information. Having the facts requires staying informed and avoiding hype. Become active and get involved.
- Identify possible solutions. In doing so, also identify the consequences of the various solutions.
- Make a decision about an issue. Weigh the facts and remove any emotion. In some cases, one can only recommend actions to government officials. It is important to let them know what we think. Call or write these officials. Go to public meetings to gain information and express your opinions.
- Implement the decision that has been made.

ANIMAL MISUSE

Animal misuse frequently comes under the title of inhumane treatment. This treatment is often dealt with by the Humane Society. One would think that proper feeding, care, housing, and cleanliness are common sense and should be done by anyone who owns or cares for an animal. That is far from true. Malnutrition, poor housing and filthy conditions are the most common problems that lead to companion animal mistreatment. In addition, some people simply treat their animals with little respect or even abuse them.

PUPPY MILLS

Breeding facilities that breed and produce dogs in large numbers are known as puppy mills. The goal of these breeders is to make quick financial returns by selling the puppies via the Internet, to animal brokers, or to pet shops. Problems begin to happen when puppies are overcrowded due to over-breeding. This can result in poor veterinary care or poor food and shelter. It can also result in excess animals that cannot be sold. These surplus animals may be killed in an economical, though sometimes inhumane way.

26–3. A malnourished two-year-old cow has just given birth to a calf. (Most likely, this cow has not been fed properly. Neither she nor the calf may survive. She may be unable to produce sufficient milk for the calf. Low weight and a poor hair coat are signs of a deficiency.) (Courtesy, Agricultural Research Service, USDA)

Continued use of related females and males sometimes happens in puppy mills that are promoting "purebred animals." A common problem associated with inbreeding is the increased frequency of genetic defects. These animals are simply destroyed or if they are sold, a new owner may be faced with an animal with increased health problems.

Other problems that are sometimes not associated with puppy mills have their roots in these facilities. If pets are unhealthy or more costly to their owners, they are frequently abandoned or killed. Abandoned animals put increased strain on animal shelters and law enforcement that may be called to catch roaming animals. Housing these animals may also overcrowd humane societies.

The federal Animal Welfare Act (AWA) is administered by the USDA and makes every effort to oversee this issue along with other animal facilities such as zoos and laboratories. Although there are nearly 100 inspectors in the United States, the increasing scope of the pet industry is challenging this animal safety net.

MALNUTRITION

Malnutrition is a condition that results when animals are fed diets that do not meet their daily needs over a sustained period of time. This includes maintenance, growth, reproduction, lactation, and work (see Chapter 3, Animal Nutrition and Feeding).

Malnutrition and, later, starvation often go unnoticed to the general public. It can be a slow process and a very cruel death. Sometimes people simply move out and leave their animals behind locked doors without food. However malnutrition might happen, it requires the goodwill of neighbors or friends to let someone know of animals that may not be properly fed. In our society today, people sometimes are hesitant to become involved in someone else's affairs; instead letting unusual signs go unreported.

ANIMAL FIGHTING

Dog fighting is getting two dogs to fight until one can no longer fight or has died. Often the owner of the losing dog will kill the dog or abandon it. These dogs are initially bred and trained to fight. Although dog fighting is illegal in all states and considered a felony in 48 states, it continues. High-profile individuals in the United States have actually been charged with this crime.

Hog-dog fighting is featured as more acceptable than dog fighting. It allows some protection to a dog as it fights with a wild hog that no longer has tusks. The dog is timed to see how fast it can pin down a hog. Unfortunately, both animals can sustain serious injuries. Hog-dog fighting violates cruelty laws in some states.

26–4. A cockfight is underway. (You will see human feet at the top–they are watching the fight-to-death event.) (Courtesy, Dmitry Savinov/Shutterstock)

Cockfighting is an even older sport than dog fighting and takes place in many countries, including the United States. Specially bred birds are placed in a small area to fight, usually until one is dead. Sometimes birds are fitted with artificial spurs which are even more dangerous than their normal spurs. Laws against cockfighting make it illegal in most states.

Animal fighting brings several questionable practices with it. Spectators (though prohibited) often gamble on one of the animals. Drug enforcement agencies frequently determine that illegal drugs are used on animals and sold to other spectators. Grave concern about this questionable sport lies in part on that it may be conditioning children and other spectators to become accepting of animal and even human abuse.

DISPLACED AND UNCLAIMED PETS

Sometimes animals are abandoned. They may be purposefully dumped out without a home. Natural disasters can also create major problems for animals and their owners. An animal may be lost or separated from its owner.

Hurricane Katrina is a well-known natural catastrophe that claimed human lives and caused massive amounts of property damage in the Gulf Coast area, with New Orleans being the largest affected city. Less publicized are the devastating results Hurricane Katrina had on pets. "Ali" was one of those pets. Ali was a Chesapeake Bay Retriever that was less than a year of age. Ali survived, but not without challenges.

Ali was rescued and moved to Atlanta along with many other dogs to be restored to health. She had acquired heartworms that would take many months in which to recover. Ali was listed on the pet adoption site for Hurricane Katrina pet victims. A young couple in Chicago noticed the advertisement and adopted the beautiful Chesapeake Bay Retriever. It was a happy ending for Ali, but many other abandoned pets go without being adopted.

Dealing with abandoned pets is a growing problem. People need to take full responsibility for the pets they own. Education is the key to managing pets properly and reducing the deaths or mistreatment given to unwanted pets. Small animal classes in agricultural education programs across the United States are educating the next generation of pet owners to take better care of their pets.

26–5. "Ali" has adapted comfortably to her Chicago home after being rescued following Hurricane Katrina. (Courtesy, Ryan and Sarah Westrom, Chicago)

ANIMAL ISSUES AND THE ENVIRONMENT

Several issues related to the environment are increasingly sources of conversation. It is becoming more evident that global warming is occurring. The rate of global warming is debatable, but shrinking glaciers are one of several measures that indicate it is increasing at a rapid rate; a rate that some consider alarming. Some people call this the ***greenhouse effect*** because it results from more gases in the atmosphere. These gases absorb more infrared radiation resulting in an increase in atmospheric temperature. This rising temperature causes global changes to occur over time.

Carbon dioxide is the largest contributor to the world's greenhouse gases. One of the largest contributors to greenhouse gas is automobile emissions. To a lesser degree, methane produced by animals has been suggested as a contributor to greenhouse gases. Realistic calculations suggest that all methane emissions by the world's beef cattle are equivalent to 0.1% of the total greenhouse gases.

Animal owners and producers must be conscientious about doing their part in reducing this problem. Additionally, the solution may actually contribute to the profitability of livestock operations. The following section will address the issue and speak to methane production from manure being an alternative energy option available to the animal industry.

METHANE FROM MANURE

Methane (CH_4) is the most abundant gas in the atmosphere. It is the principal component of natural gas. Because methane amounts in the atmosphere have been increasing each year, pressure is mounting to convert this methane into useful energy. This conversion would also reduce the amount of methane in a gaseous state in the atmosphere.

Methane is produced in large quantities by ruminant animals. Grazing increases emission of methane by animals. Cattle being grazed convert nearly eight percent of their gross energy to methane. Scientists have determined that 16 percent of the earth's annual methane emissions are from the belching of cattle. Chickens, hogs, and cattle combined produce 37 percent of the methane released into the atmosphere each year. These amounts can be compared to the remaining 63 percent from natural and industrial sources.

Manure stored in a dry form or deposited on pastures greatly reduces methane production. However, larger operations today are more apt to store manure in cost-effective liquid manure and disposal facilities. These facilities use anaerobic systems that significantly increase methane emissions. (***Anaerobic*** refers to activity that can be carried out by microorganisms without the presence of oxygen.)

Livestock manure is composed mainly of organic matter and water. Dairy and swine farms use more liquid and slurry systems; these are more adaptable to methane production. Liquid manure management (lagoons, tanks, ponds, etc.) is a more cost- effective option and serves to reduce emissions more than solid systems.

The key to methane capture from manure is to use one of three methods:

- Manure can be mixed with water and pumped into covered lagoons which provide an anaerobic state that produces and recovers methane.
- Manure can be heated and mixed (8% solids) to produce biogas and stable effluent that allows methane to be caught in a covered digester. (**Effluent** is water that remains from a process or activity. Typically, effluent is associated with water that flows from a waste treatment facility. It is considered to be wastewater.)
- Dairy manure can be scraped into a long tank with a tight cover in which biogas is collected at a rate of about 40 cubic feet per cow. (**Biogas** is the combination of gases produced by the biological breakdown of organic matter in the absence of oxygen. It is comprised primarily of methane and carbon dioxide.)

26–6. A lagoon collects wastes at an Illinois hog farm. (Courtesy, Education Images)

CARBON SEQUESTRATION

How does carbon enter the atmosphere? The primary source is through the use of fossil fuels such as coal or oil. When automobiles, furnaces, or manufacturing plants use fossil fuels, carbon is a byproduct that ends up in the atmosphere. Deforestation contributes to carbon dioxide buildup in the atmosphere. When trees are harvested and not replaced, carbon must find a new home–the trees are no longer present to use it in photosynthesis!

Where does carbon go if it is released? Forests and other plant growth can use some. A **sink** is a place that buffers or holds carbon. The ocean is a huge sink, but it appears to be losing some of its buffering capacity. Scientists don't know why. Land-based sinks also hold much carbon. The soil itself, prairie pot holes, peat bogs, and reforestation all sequester or grab much carbon. Schools that plant a tree for Arbor Day are contributing to the important process of adding to the land sink. The remainder of carbon is emitted into the air. The air always contains some carbon, but excessive amounts contribute to smog, air pollution and airborne allergies.

Tropical deforestation is decreasing land sink capacity and has scientists concerned. This concern combined with the oceans losing some of their sink capacity has caused many of our nation's leaders and scientists to consider ways the trend can be reversed.

A recent example of a way to reverse this trend is by using biodiesel in school buses or in government vehicles. Biodiesel reduces by 82% the amount of carbon released into the atmosphere. Hybrid cars also reduce the amount of air emissions.

CARBON CREDITS

Private industry and agricultural organizations are offering carbon credits as a means of rewarding producers who use good conservation practices. A **carbon credit** is a measure of carbon release. It is an approach to controlling emissions of carbon materials into the air. A monetary value is assigned to carbon release. Aggregate credits achieved through large blocks can be sold. Farmers and ranchers can also sell carbon credits on the Chicago Climate Exchange at a minimal cost and risk.

The Natural Resources Conservation Service (NRCS) in the USDA offers many conservation programs, some of which target the storing of carbon in the soil. Private practice can also reap dividends by selling credits to companies who buy these credits to compensate for company pollution. Practices that offer promise include reforestation or new forest land, manure methane digesters, switch grass and agrichar.

The notion of carbon credits grew out of the Kyoto Protocol, which is an agreement among 170 countries. Carbon emissions were viewed as a worldwide problem. The United States government did not embrace the Kyoto Protocol approach which created widespread disagreement throughout the country.

26–7. Plowing and digging in the soil can result in carbon being released into the air.

AGRICHAR

Soil is dark and extremely fertile for a reason: agrichar. Char is a black, porous, finely granular, carbon material. The carbon in char is stable for thousands of years. It is used in agriculture and known as **agrichar**. Native prairie is adept at capturing carbon from the air and sinking it into the earth. Plowing the soil promotes the breaking away of the carbon that has been stored. This exposure of soil particles to the air increases the rate at which carbon leaves the soil.

The capture and storage of carbon in the soil is **carbon sequestration**. New pasture and field crop practices may enhance the increase of carbon sequestration. Perhaps the reintroduction of mixed prairie grasses may hold the key to returning more char into the soil and improving the high fertility that was so common when the prairie was first plowed.

When something holds more carbon through either char or massive roots holding it, it is said to be carbon negative. Practices that produce carbon negative situations will be viewed more and more positively as global warming becomes a threat to the well-being of the earth.

REPRODUCTIVE FAILURES

The reproductive capacity of some species, if not all, is tied to the environment in which they live. Some factors that influence environmental quality are naturally occurring events, such as the pollution from a volcano. Other factors are human-made.

Any human-made product that enters the air, soil, or water has the potential of causing harm to animals. Thorough testing is essential before products are released. Some of these substances are for industrial use; others are used in agriculture.

An example is the relationship between a widely used herbicide and reproductive problems in animals. A University of Illinois reports indicates that exposure of animals to atrazine causes reproductive dysfunction in amphibians, fish, reptiles, and mammals. Atrazine is the second most widely used herbicide in the U.S. Crop producers may need to find alternative and safe herbicides. (A source for more information is: **http://news.illinois.edu/news/11/1128atrazine_ValBeasley.html**.)

TRANSGENIC AND CLONED ANIMALS

Transgenics and cloning are cutting-edge technologies. They pose the potential of huge gains in some areas of productivity. These technologies pose major issues for human investigation, understanding and resolution. (You may wish to refer to Chapter 6, Animal Biotechnology.)

TRANSGENICS AND BIOENGINEERING

Bioengineering is able to gain the novel traits not normally found in a species. Gaining such traits can yield advantages to a particular animal species–and to the producers of the animals. Bioengineering is done through scientific methods rather than selective breeding. Research efforts are underway and show positive gains such as the elimination of certain animal diseases.

Three categories of bioengineering are most likely to raise serious issues:

- Plant-animal- human combinations
- Animal-animal combinations
- Animal-human combinations

A few examples: Genes from fireflies can be placed into tobacco plants to make them lightly glow at night. Swine organs are so similar to those of humans that they can be grown to alleviate the shortage of human organs. Milk from transgenic animals can be used to produce treatments for human disorders.

A transgenic animal is one whose genetic material has been artificially altered during the formative stages of embryo development. With transgenic animal production comes related issues such as patented animals. Should a company that produced a transgenic animal be allowed to patent the animal and make money off any sales of those animals or animal products obtained from patented animals? On the other hand, should future scientists be allowed to use those patented animals to further scientific discoveries? Ethical issues quickly arise along with possible risks associated with unknown diseases or future crosses of transgenic animals.

26–8. Choosing the gender of offspring is an issue to some people. (This shows a scientist using high-speed equipment to sort X and Y sperm. Sorting allows the sperm to be used to fertilize an egg to choose the sex of the offspring.) (Courtesy, Agricultural Research Service, USDA)

Would it be ethical to genetically engineer a cow using human genes to produce milk similar to human breast milk? Scientists in Argentina have all ready genetically engineered cows using human genes to give milk with very low lactose. This allows individuals with lactose intolerance to drink it. The milk is said to be similar to human milk. Should an animal with human genes be given human rights?

CLONING

Cloning is the process of gaining a new organism without sexual reproduction. Cells or tissues from an existing animal are used. The new organism has genetic material identical to its single parent–source of the cells/tissues. This makes possible the production of large numbers of identical animals. There are clearly multiple issues that surface with cloning; some are positive and others are negative.

Cloning will allow an enhanced ability to spread superior genetics throughout the animal world. Ultimately this will result in lower food costs. High milk producing animals will produce more milk and rapidly growing animals will produce meat more quickly. Both will likely produce more pounds of product per nutrient eaten and hence increase efficiency.

Opponents of cloning point out numerous issues. Monocultures in plants have resulted in disease outbreaks that can damage large acreages. Likewise animals with common genes run an increased risk of diseases and genetic defects that could cause large numbers of animals to become sick or even die. Perhaps an even more passionate argument surfaces around the ethics of cloning. Most people feel the cloning of humans would be unethical. Many view cloning of animals as a slippery slope that could lead to discussion of cloning humans.

Cloned animals have not been allowed to enter the food chain. In 2008, agencies that oversee that issue reversed the decision and now allow cloned animals to enter the human food chain. The unknown in this decision is consumer reaction. Americans are likely to accept it just as they accept genetically modified crops and many other food technologies. Foreign markets are another issue. Many countries may refuse meat from cloned animals; forcing separate slaughter facilities for overseas markets.

NATIONAL ANIMAL ID SYSTEM

The events of September 11, 2001, (known as 9/11) have become household symbols reflecting a need for homeland security. One homeland security issue deals with animal safety. A national data base for livestock would allow for quick identification of animals that had been exposed to foreign or domestic animal disease. This is known as source verification. The goal would be identification of these animals within 48 hours of exposure to a disease. Currently this program is voluntary.

Animals included in the identification system are dairy, beef, goats, sheep, horses, hogs, poultry, llamas, and alpaca. There are three phases to the system. These include premise identification, individual identification, and enhanced tracking in the market place. An animal identification number is assigned to an individual animal unless exceptions are given. A group or premise identification number is used in for some species in certain situations (sheep, pigs, or poultry). Herd owners can use a portable handheld reader while feedlots or packers may use a stationary panel.

26–9. The tag in the ear of a goat can be used to gain background information. (Courtesy, Education Images)

The management possibilities with a national ID system are great. Animals could quickly be identified for management purposes such as vaccinating, treatments, and all aspects of working cattle. A fringe benefit would be reduced cattle rustling on western ranches. Data storage would allow quick retrieval throughout an animal's lifetime. Lost animals that result in lost data important in herd analysis would be virtually eliminated.

ANIMALS IN THE WORK FORCE

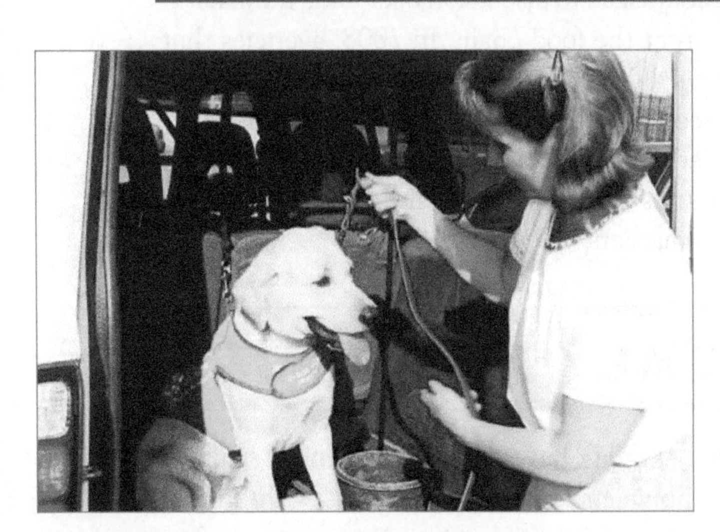

26–10. Guide dogs provide useful services to blind people. (This young dog is in guide dog training.) (Courtesy, Education Images)

Animals are often used for human benefit without their being harvested. Some are known as "working animals."

Dogs make wonderful pets, but they are also used for work. K-9 dogs work for a police force and track drugs. Leader dogs serve as guides to the blind. Dogs used in therapy may be located in a children's hospital or in an elderly care unit. Some dogs go to school, not to learn, but to promote learning in special education classrooms or elementary schools.

Horses have multiple work-uses. Some are used in handicapped camps and hippotherapy units. ***Hippotherapy*** is a human treatment that uses the movement of a horse as therapy. It is used with people who have movement dysfunctions. In addition, horses are used to cut out beef cattle from a herd when they need treatment. Horses carry people and cargo across rugged terrain. Draft horses are still used for farm work. They are used by the Amish in the United States and in many other countries throughout the world.

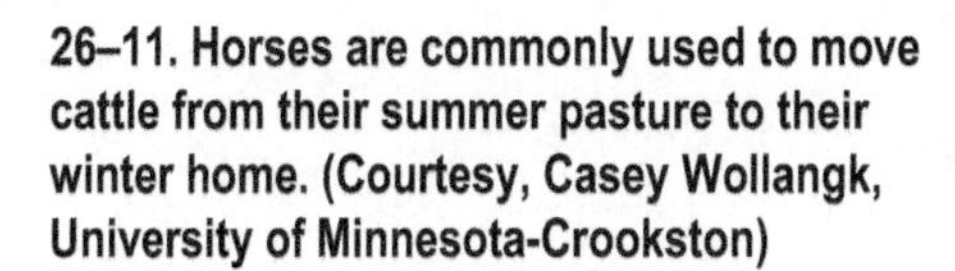
26–11. Horses are commonly used to move cattle from their summer pasture to their winter home. (Courtesy, Casey Wollangk, University of Minnesota-Crookston)

GRAIN USE

Animal production is impacted by the use of grain to produce energy rather than it being used for feed. The impact of this change is emerging. Already, we know that the cost of grain is increasing even though more acres are being planted to grain crops.

It is not uncommon today to see a house being heated with corn rather than natural gas or electricity. Automobiles today burn from 10-85% ethanol rather than gasoline made from oil products. ***Ethanol*** is a flammable, colorless compound that is commonly known as ethyl alcohol (C_2H_6O). It can be made from corn, other grains or switch grass. Biodiesel made from soybeans is fueling many school buses and government trucks. A byproduct of the poultry industry is chicken fat. Some trucks and other diesel-equipped vehicles are now being operated with a mixture of diesel and chicken fat.

The use of renewable energy sources is a rapidly expanding trend today. While these energy sources have many supporters, they are causing a change that may forever impact worldwide animal production and even human food consumption in developing nations. Rising oil prices are making it economical to use grains and other renewable energy sources as fuels. In fact, as oil prices go even higher, it allows renewable energy companies to pay

26–12. Some of the gain at this facility will go for ethanol production. (Courtesy, Education Images)

even more for corn, soybeans, and other raw products. But this is also driving up the cost of feeds derived from corn and soybeans that are fed to livestock. Will that leave enough profit to raise livestock that are fed corn? Perhaps it will force livestock producers to utilize more forages to grow their animals.

While we can envision the reduced profits this may have for livestock grown in the United States, it may be an even larger challenge for countries that depend on imports of grains for either animal or human food. Market prices of grains were formerly a result of the demand for food or feed. Now, the demand for fuel production is also a factor. It is not known if this will be a permanent market change. Time will tell, but perhaps this is a historical change of great magnitude.

FOOD CONTROVERSIES

People in the United States are divided on whether or not horse harvest should be allowed. For many years horses that were old or had no function were harvested, with some for animal food and others for human food.

HORSES FOR FOOD

Horse meat (known as ***chevaline***) is popular in some parts of the world, including France and Italy. Chevaline has been exported to countries that desired horse meat. More recently, most slaughter plants have been closed in the United States due in part to pressure from animal rights groups such as PETA and people who view horses as pets. The price of horses dropped when slaughter plants closed. Other people feel that horses are much like cattle and hogs from the perspective of harvest.

Leaders in the horse industry are themselves debating this issue. Many point out that it is more inhumane to not have slaughter plants. Reasons include illegal transport of horses to countries where slaughter laws are much less stringent, hence leading to more abuse of horses than if slaughter was allowed in the United States. Other supporters of horse slaughter argue that allowing a horse to live in severe pain from arthritis or very old age is a painful life of low quality that should not be promoted by the absence of slaughter plants.

Opponents of horse slaughter argue that horses deserve to have rights just as humans do. It is not up to humans to decide when a life for a horse no longer meets some threshhold of quality that should allow it to live or be slaughtered. Others say that in a civilized nation, we should value lives of horses (whom some view as pets and others don't) at a higher level than meat animals and hence should eliminate horse slaughter. Some say that all species deserve equal consideration and no animal should be eaten or given any priority over another.

This issue is reaching a boiling point on both sides. Many concerned with both sides of this issue argue that their position is more humane to horses. Meanwhile, discussions are being held to reach a compromise on this issue. Because both Canada and Mexico allow horse slaughter, added pressure from both sides is creating a desire to come to a middle ground on this issue. Young adults will likely reach a conclusion or at least a compromise on this issue in the years to come. You may be part of that discussion.

CONNECTION

HORSES FOR MEAT

Horse meat is known as chevaline. It is not available in most U.S. supermarkets, but chevaline is popular in France, Italy, and other European countries. Many horses are used for meat products.

The United States has a Horse Meat Act. This law covers inspecting and labeling horse meat for human consumption. Inspection stamps use green ink rather than purple as on carcasses of other animals. In past years, most of the horses slaughtered in the United States for human consumption are exported to other nations. Some years, horse meat has been among the largest of the exported meat products.

Meat markets may offer attractively packaged horse meat cuts. Chevaline is wholesome meat that is high in protein. (Courtesy, Education Images)

DOWNER COWS

We want to be sure that our food system has wholesome food products. This includes having meat foods from animals that are healthy and have not been abused. These animals have been produced to promote their well-being. Major meat recalls have been made to remove the possibility of a food product that would be unhealthy.

One animal condition that has caused controversy is the downer cow. A ***downer cow*** is a cow that cannot walk. This condition may be due to injury or disease. A cow can become a downer following giving birth, stepping in a hole, poor long-term nutrition, or acquiring a neurological disease. In recent years, major concern has arisen over possible mad cow disease as related to downer cows.

ANIMAL RIGHTS AND ANIMAL WELFARE

Animal rights proponents believe that animals deserve rights similar to those given to human beings. Peter Singer and Tom Regan are authors that wrote baseline books and articles on this issue. Singer coined the term "speciesism" which is a term that refers to giving one animal species preference over another. It is considered similar to racism or sexism in some people's minds. Regan believes that humans can only kill animals if it is in self-defense. Vegetarians and vegans are often proponents of animal rights.

Animal welfare refers to the well-being of animals. Animal rights proponents argue that animal welfare should not be mixed with animal rights. Others argue that improving animal welfare is a logical middle ground on this controversial issue. Improving the air flow in barns or optimizing the nutritional level of animals are examples of issues that would improve the welfare of animals, but would not give them the added rights or freedom that animal rights proponents favor.

26–13. An injured dog is receiving expert care at a veterinary clinic. (Courtesy, Education Images)

WELL-BEING AND CONFINEMENT

Confinement of animals has become a big issue in some areas of the animal industry. Some restaurants now have guidelines about space requirements for meat animals. Producers are changing some of their production practices based on these concerns and protests. Grocery stores are now selling "free range" chicken and pork from chickens and hogs not raised in confinement. People sometimes are willing to pay 10-15% more for these food products. This does not go far enough for staunch animal rights proponents. People on limited budgets wouldn't consider paying 10-15% more for this type of meat when they are simply trying to find enough money to feed their family. Hence, the issue of animal rights and animal welfare is continuing to be discussed.

26–14. Free-range chickens are allowed to run about in pasture areas that are fenced. (Courtesy, Weldon Schloneger/Shutterstock)

This issue goes far beyond food issues. Extreme animal rights proponents argue that animals should never be used for research. They have even broken into scientific laboratories to release animals. A portion of research on Alzheimer's disease was lost when animal rights extremists released laboratory animals from a renowned Alzheimer's disease laboratory at a major university. Horse lovers who ride horses may find that staunch animal rights proponents view horseback riding as a form of slavery to horses.

26–15. Scientists are now researching the development of a virtual fence system for cattle. (This shows a cow with a prototype neck saddle. Efforts are underway to reduce the size of the device to no more than an ear tag. The system uses electronically generated cues to control movement of the cow.) (Courtesy, Agricultural Research Service, USDA)

Some researchers, such as Stanley Curtis, a well-known scientist at the University of Illinois, have spent their lifetimes improving animal welfare. Others who prefer working with wild animals have spent their lives trying to improve the habitats of wild animals. Meanwhile, animal rights proponents spend most of their money on promoting the animal rights cause. Which do you think contributes more to an animal's life?

BIOSECURITY

Animal producers are increasingly concerned with biosecurity, which is the protection of animals from harm intentionally or unintentionally inflicted on animals. It includes preventing diseases, pests, and bioterrorism acts that threaten animals and their products. Bioterrorism refers to the intentional release of an agent or substance that causes harm to animals. Taking needed actions to protect animals sometimes causes concern among animal producers and citizens of the community.

Some of the actions that may be taken include:

- Keeping visitors to animal facilities to a minimum—Visitors can bring harmful substances on their shoes or clothing. This is particularly a concern if the individuals have visited other animal facilities. Some producers require visitors to change shoes and clothing before going into animal facilities. Of course, visitors with poor intentions can purposefully bring and release agents or substances that are harmful. Locked gates, good fences, restrictive signs, and other steps can be used to prevent human access.

- Using good animal care practices—Providing for the well-being of animals helps prevent problems from arising. Good nutrition along with proper vaccination and other health practices help animals develop resistance. Such animals are less susceptible to disease.

- Cleaning equipment—Equipment that goes from one animal facility to another may not be allowed to enter without being sanitized. Trucks and trailers that haul animals are particular sources of harmful agents. For example, hauling a sick animal in a trailer contaminates the trailer with potential disease organisms.

- Restricting stray or wild animals—Wild animals harbor disease and may carry it to a herd or flock of domesticated animals. There are many ways this can happen. One example is allowing wild birds in a poultry house. Another example is feral pigs getting into hog facilities. Almost every species can be afflicted with a disease by its wild counterparts. Access by wild animals can be minimized with good fences, buildings that are tight and do not allow other animals to enter, and taking steps to reduce wild animal populations in the vicinity.

- Introducing only healthy animals—Only bring animals that are disease-free into a flock or herd. The origin of animals should be known and free of disease. Additions to an animal herd or flock should be properly vaccinated and free of disease, including parasites. It is always a good idea to quarantine animals that are being brought to a farm from another source such as market or another producer. Veterinary testing may be needed to assess the presence of disease.

- Using feed and other inputs that are safe—Feed, water, and other supplies can transfer harmful substances to animals. Use feed from a proven and reliable source. Contaminated feedstuffs can introduce diseases and harmful chemical substances. Water may need to be tested before use with animals if it is from an unknown source.
- Making daily observations—Animals should be observed at least one time a day. In some cases, more frequent observation is beneficial. If changes in behavior, the presence of wild animals, or situations that pose hazards are evident, take necessary actions to solve the matter. Being a conscientious animal producer is a valuable personal quality.

NUTRIENT MANAGEMENT

Nutrients that animals eat either eventually become food for humans or animal waste. Nutrient (manure) management is becoming a large issue today due in part to increasing farm size. Feeding animals a well-balanced and digestible diet is the first step in nutrient management. The second step in this process includes having adequate land to spread manure to limit nutrient application to levels that the soil will hold with minimum runoff or permeability into underground or surface water systems. Many animal producers have established nutrient management plans.

One might argue that large numbers of animals existed in earlier periods in American history. In fact, some species have dropped in total numbers. However, earlier farms were small and each owned a few animals. Manure was spread evenly over the acres each farmer operated. Diversified farms meant that acres and animals were balanced and could handle the nutrients from animal waste. Today, animal operations are much larger. Many own little crop land and must actually contract with crop farming neighbors to spread manure. Environmental agencies monitor this process to ensure that excess nutrients are not applied to each acre of land. Nitrogen, phosphorus, and potassium levels are monitored in neighboring above ground and underground aquifers.

Environmental concern by citizen groups is raising the level of intervention on this issue. Today, many jobs for environmental feedlot specialists or pollution control agents are advertised. Ten years ago, most of these jobs didn't exist. Monitoring our environment by limiting nutrient amounts to match the land holding capacity will be important. The American culture is changing as it values natural environmental resources at a greater level. This concern must be balanced with producing food in an economical way. The balance between these two pressures will be another issue that gets ultimately decided by future generations.

REVIEWING

MAIN IDEAS

Change causes new issues to arise as societal concerns. The animal science world may be a dramatic example as new issues continue to arise at a rapid pace.

Animals are susceptible to a wide variety of issues because different people view animals in varied ways. Some people view animals as pets. When storms displace animals from their homes or force them to be adopted as was the case in the aftermath of Hurricane Katrina, it becomes an important issue.

Other pet owners are concerned with animal abuse or neglect. Animal fighting, malnutrition and puppy mills are examples of issues currently being discussed across the United States.

Animals are playing new roles in society. Many are now part of the work force. Traditional animals such as dogs are used in elderly care units or schools where animals have been shown to improve learning. Dogs are also used in law enforcement and as "leader" dogs for the blind. Horses are used in hippotherapy for mentally handicapped children.

Waste products of animals such as methane and solid waste (manure) are an environmental issue. Global warming may be increasing in part due to methane released from animals. Proper handling of manure reduces pollution and can reduce or increase carbon release into the atmosphere.

Animals can now be altered with genetic engineering by moving genes from one animal to another. Some of these animals can be of great medical assistance to humans and can be patented. However, some people are asking if altering the genetic makeup of animals is ethical. What do you think?

Transgenic animals are only one ethical issue related to animals. Several others exist including animal rights, animal welfare, cloning, and the horse slaughter industry. The threat of disease crossing international boundaries and food safely concerns following 9-11 have given rise to a national ID system for domesticated animals. It is currently a voluntary system. Will it stay voluntary?

QUESTIONS

Answer the following questions using complete sentences and correct spelling.

1. Pets can be abused or placed in challenging life situations. List two ways pets can be abused and one way they can accidentally become homeless.
2. Select one issue in animal science and write a pro and a con for the issue.
3. How are animals being impacted by an energy-driven marketplace (as compared to a food or feed market)?
4. Is it ethical to slaughter horses? Offer at least one pro and one con on horse slaughter. What impact does lack of horse slaughter facilities have on the horse industry?
5. List three methods of reducing global warming that have been discussed in this chapter.

6. Discuss two animal rights issues and two animal welfare issues.
7. Name three jobs that animals perform.
8. What is carbon sequestering?
9. How can a farmer manage waste (manure) to allow the benefits to be greater than the potential negative aspects of manure?
10. What national practice has been put in place to allow for closer supervision of animal food products safety?
11. List one advantage and one disadvantage of animal cloning.
12. How does tree planting or prairie grass restoration improve carbon sequestering?

EVALUATING

Match the term with the correct definition. Write the letter of the term on the line provided.

a. issue
b. sinks
c. animal welfare
d. carbon sequestering
e. transgenic animals
f. carbon credits
g. puppy mills
h. nutrient management
i. malnutrition
j. agrichar
h. animal rights
i. renewable energy

______1. Feeding animals correctly and handling/applying manure properly.
______2. Place that buffers or holds carbon.
______3. A topic or problem that has more than one side.
______4. Human-like rights given to animals.
______5. Animal that has a gene or genes transferred to it from another animal or plant.
______6. Breeders that breed large quantities of an animal, discarding unsold animals.
______7. Energy produced from plants or biomass that can be grown.
______8. Deficiency of nutrients in animals or humans, most often being protein.
______9. Can be bought or sold on the Chicago Climate Exchange.
______10. Concern for an animal's well-being.
______11. A black, porous, finely granular, carbon material.
______12. Methods of holding carbon rather than releasing it into the atmosphere.

EXPLORING

1. Divide the class into small groups. Debate issues such as horse harvest for food or animal fighting.

2. Invite an NRCS specialist into class to discuss conservation practices that reduce carbon escape into the atmosphere.

3. Visit a local Humane Society facility or an animal shelter. Discuss how animals reach the shelter facilities and how animals can be adopted from them.

4. Arrange to tour a renewable energy processing plant. Prepare a report on your observations.

5. Research an important person in animal rights. Here are a few celebrities to begin with: Rachel Ray, Bob Barker, Jane Goodall, Casey Affleck, Hilary Swank, and Howard Lyman. Identify other individuals prominent in animal well-being. In you study, determine the role of the individual(s) in animal rights. Choose one individual and make a brief oral report in class on the individual's work with animals. What is your personal opinion of their work?

6. Design, conduct, and complete research to identify and solve animal management problems and issues. The animals may be livestock, small animals, or other species. Choose an issue presented in this chapter or another issue that you identify. Research background information on the issue using publications, the internet, and other resources. Interview an animal producer about the issue you are researching. Use experimentation on the problem, if appropriate. Prepare a short written report on your findings. Give an oral report in class.

26–16. A transgenic cow appears no different from other cows of a breed. (Scientists are now attempting to improve cattle and their products through genetic engineering. Do you suppose this will matter to consumers? What if a human gene were transferred to the cow?) (Courtesy, Agricultural Research Service, USDA)

Appendixes

APPENDIX A. Animal Mineral Needs, Functions, and Deficiency Symptoms

Macro-minerals	Function	Deficiency Symptoms
Calcium (Ca)	structural component of the skeleton; controls the excitability of nerves and muscles; required for coagulation of the blood	rickets or osteomalacia develop—the bones become soft and deformed; milk fever occurs in cows; blood clotting time increases
Phosphorus (P)	component of the skeleton providing structural support; aids in lipid transport and metabolism; component of AMP, ADP, and ATP	rickets; depressed appetite; may chew on wood or other objects
Magnesium (Mg)	needed for normal skeletal growth; needed in chemical reactions in muscles; activates enzymes during the Kreb cycle	anorexia; reduced weight gain; reduced magnesium in the blood; hyperemia of the ears and other extremities
Potassium (K)	required to move sodium; helps to maintain acid-base balance in the body; helps to uptake glucose and carbohydrates	slowed growth; unsteady walk; overall muscle weakness; Mg deficiency causes potassium deficiency
Sodium (Na)	maintains osmotic pressure; maintains acid-base balance; required in transmission of nerve impulses	reduced rate of growth; reduced feed efficiency; decreased milk production; weight loss in adults
Chlorine (Cl)	regulates extracellular osmotic pressure; maintains acid-base balance in the body	slowed growth; fall forward with legs extended backward when startled by a sudden noise
Sulfur (S)	required for protein synthesis; used in cartilage; in birds sulfur is used in feathers, the lining of the gizzard, and in muscle	reduced growth; reduced weight gain

(Continued)

APPENDIX A (Continued)

Macro-minerals	Function	Deficiency Symptoms
Cobalt (Co)	makes up vitamin B_{12}	loss of appetite; reduced growth; anemia; if untreated, death will result
Iodine (I)	component of thyroxin—controls the oxidation of cells	Goiter—enlargement of the thyroid gland; dry skin; brittle hair; young born without hair; reproductive problems
Zinc (Zn)	activates enzymes; used in DNA and RNA	slowed growth; anorexia; scaling and cracking of paws; poor feathering; rough hair
Iron (Fe)	found in hemoglobin in the blood; found in myoglobin in muscle	anemia—common problem among newborns; pale color; shallow breathing; rough hair; slow growth; iron deficiency affects about half of the world's human population
Copper (Cu)	enzyme activity to uptake iron; needed to maintain the central nervous system	anemia; uncoordinated
Manganese (Mn)	needed in bone structure	skeletal deformation—enlarged joints, lameness, shortening of legs, and bowing of the legs
Selenium (Se)	component of numerous enzymes	nutritional muscular dystrophy

APPENDIX B. Functions and Deficiency Symptoms of Vitamins in Animals

Fat Soluble Vitamins	Function	Deficiency Signs
Vitamin A	required in retinol for night vision; needed in epithelial cells that cover body surfaces; needed for bone growth	night blindness; dry and irritated eyes; respiratory infection; reproductive problems
Vitamin D	enhanced calcium and potassium levels allowing bone mineralization; prevents tetany	abnormal skeletal development—lameness, bowed, and crooked legs; slowed growth
Vitamin E	promotes health	failure of the reproductive system; changed cell permeability; muscular lesions
Vitamin K	required for blood clotting	long blood clot time; hemorrhages; in severe cases, death
Thiamin (B1)	promotes health	anorexia, beriberi in humans—numbness, weakness, and stiffness in thighs, unsteady walk, edema in feet and legs, and pain along the spine
Riboflavin (B2)	functions in coenzymes	reduced growth rate; skin lesions; hair loss
Niacin	used by cells in energy metabolism	retarded growth; decreased appetite; diarrhea; vomiting; dermatitis
Pantothenic Acid	needed in energy metabolism	slowed growth; dermatitis; graying of the hair; fetal death; skin lesions
Vitamin B6	help with protein and nitrogen metabolism; involved in formation of red blood cells	convulsions; lesions around feet, face, and ears
Vitamin B12	needed in several enzyme systems	anemia*; retardation; skin pigmentation
Folacin	used in a variety of metabolic reactions	slow growth rate; anemia
Biotin	needed for several enzyme systems	scaly skin; abnormalities of the circulatory system
Choline	aids in transmission of nerve impulses	fatty liver; hemorrhaging kidney
Ascorbic Acid (Vitamin C)	prevents scurvy; causes several metabolic reactions to occur	scurvy—edema, weight loss, and diarrhea

*This can occur in monogastrics fed entirely from plant material. Babies nursing from vegetarian mothers may also develop a vitamin B12 deficiency.

APPENDIX C. Scientific Classifications of Agricultural Animals (all are in the Domain Eukaryota)

DAIRY AND BEEF

a. Kingdom Animalia—animals collectively; the animal kingdom
b. Phylum Chordata—one of the approximately 21 phyla of the animal kingdom in which there is either a backbone (in the vertebrates) or the rudiment of a backbone, the chorda
c. Class Mammalia—mammals or warm-blooded, hairy animals that produce their young alive and suckle them for a variable period on a secretion from the mammary glands
d. Order Artiodactyla—even-toed, hoofed mammals
e. Family Bovidae—ruminants having polycotyledonary placenta; hollow, nondeciduous, up-branched horns; and nearly universal presence of a gall bladder
f. Genus *Bos*—ruminants quadruped, including wild and domestic cattle, distinguished by a stout body and hollow, curved horns standing out laterally from the skull
g. Species *primigenius*—cattle (Two subspecies: *Bos taurus* and *Bos indicus—Bos taurus* includes ancestors of European cattle and of the majority of cattle found in the United States; *Bos indicus* is represented by the humped cattle (Zebu) of India and Africa and the Brahman breed of America)

SHEEP

a. Kingdom Animalia
b. Phylum Chordata
c. Class Mammalia
d. Order Artiodactyla
e. Family Bovidae
f. Genus *Ovis*—the genus consisting of the domestic sheep and the majority of wild sheep; horns form a lateral spiral
g. Species *Ovis aries*—domesticated sheep

SWINE

a. Kingdom Animalia
b. Phylum Chordata
c. Class Mammalia
d. Order Artiodactyla
e. Family Suidae—the family of nonruminant, ariodactyl ungulates, consisting of wild and domestic swine

f. Genus *Sus*—the typical genus of swine, restricted to the European wild boar and its allies, with the domestic breed derived from them

g. Species *Sus scrofa* and *Sus vittatus*—*Sus scrofa* is a wild hog of continental Europe from which most domestic swine have been derived; *Sus vittatus* was the chief, if not the only, race or species of East Indian pig that contributed to present day domestic swine

HORSE

a. Kingdom Animalia

b. Phylum Chordata

c. Class Mammalia

d. Order Perissodactyla—nonruminant, hoofed mammals, usually with an odd number of toes, the third digit the largest, and in line with the axis of the limb; includes the horse, tapir, and rhinoceros

e. Family Equidae—members of the horse family may be distinguished from the other existing perissodactyla (rhinoceros and tapir) by their comparatively more slender and agile build

f. Genus *Equus*—includes horses, asses, and zebras

g. Species *Equus caballus*—the horse is distinguished from asses and zebras by the longer hair on the mane and tail; the presence of the "chestnut" on the inside of the hind leg, and by other less-constant characteristics, such as larger size, larger hooves, more arched neck, smaller head, and shorter ears

POULTRY

a. Kingdom Animalia

b. Phylum Chordata

c. Class Aves—those animals having feathers

d. Subclass Neornithes—modern birds (as opposed to fossil birds)

e. Suborder Carinatae—those birds with keel-like breast bones; this superorder has 2,810 genera and 8,616 species; in addition, there are over 25,000 subspecies; the genus-species of domesticated birds are:

    1. chicken—*Gallus domesticus*
    2. duck—*Anas domestica*
    3. goose—*Anser domestica*
    4. guinea fowl—*Numida meleagris*
    5. pigeon—*Columba domestica*
    6. turkey—*Mealeagris gallapavo*

Glossary

A

abomasum—the compartment in a ruminant's stomach that secretes digestive juices, which kill microbes that have acted on food in preparation for digestion.

accelerated lambing—the use of out-of-season lambing techniques for the improvement of reproductive efficiency in sheep.

additive—a substance added to feed for a particular purpose, such as to increase vitamin content of the feed.

aeration—the physical process of bubbling air into water or splashing the water.

aging—collecting water in an open container and allowing it to stand for volatilization of substances in the water that are harmful to fish.

agrichar—carbon in the black, porous, finely granular soil.

albumen—the white part of a poultry egg that surrounds the yolk.

allele—an alternate form of a gene.

alveoli—tiny structures in a mammary gland that remove nutrients from the blood and convert the nutrients into milk.

amino acids—the building blocks of protein.

amphibian—a class of animal that spends part of its life in water and part of its life on land.

anaerobic—referring to the activity carried out by microorganisms in the absence of oxygen.

anaplasmosis—a parasitic disease, primarily of cattle, caused by a protozoan that attacks the red blood corpuscles.

anatomy—the study of the form, shape, and appearance of animals.

anestrous—the time when a female does not cycle through estrous; the nonbreeding season.

animal biotechnology—the application of biotechnology methods to improve animals.

animal by-product tissues—tissues taken from animals for secondary purposes, such as tissues taken from slaughtered cattle for uses other than meat.

animal domestication—taking animals from nature and raising them in a controlled environment.

animal health—the condition in which an animal is free of disease.

animal industry—all the activities in raising animals and meeting the needs of people for animal products.

animal marketing—the movement of animals from the farm or ranch to the consumer; includes the preparation of the products for consumption.

animal model—the genetic evaluation for dairy cattle production using pedigree data and production-related factors.

animal processing—the preparation of animals or their products for the consumer; includes slaughtering, grading, fabricating, and other activities.

animal producer—a person who raises livestock, poultry, or companion or other animals.

animal refuge—a protected place where animals are cared for to assure their well-being.

animal rights—the notion that animals have rights, much as humans have rights.

animal selection—the choosing of animals to achieve desired goals, such as the improvement of herd quality.

animal services—types of assistance provided in animal production from off-farm sources.

animal shelter—a place that receives and provides care for abandoned, lost, and stray animals and for those that have been given up by their owners (sometimes known as pounds).

animal supplies—materials needed in animal production that are from off-farm sources.

animal tissue cultures—tissues taken from living animals for laboratory study.

animal welfare—protection of animals' well-being; *see also* animal well-being.

animal well-being—the result of providing of care to animals so their needs are met and they do not suffer.

animal wildlife—any animal that has not been domesticated.

anthelmintic—a chemical compound used for deworming animals.

anthrax—an acute infectious disease that can affect most endothermic animals.

antibiotic—a substance produced by one organism that will inhibit or kill another organism when used in medicine.

antibody—an immune substance produced in the body of an animal.

applied research—research carried out to achieve a specific purpose.

aquacrop—a commercially produced aquatic plant, animal, or other organism.

aquaculture—the production of aquatic plants, animals, and other organisms.

aquarium—a container used to hold water for ornamental fish.

aquarium maintenance schedule—a listing of important activities in keeping a good aquarium.

Arthropoda phylum—the phylum made up of animals with exoskeletons, segmented bodies, and jointed legs.

artificial insemination—the placing of semen collected from a male animal in the reproductive tract of a female of the same species.

asepsis—a sterile condition in which no living microscopic organisms are present.

aseptic technique—the action taken to assure asepsis.

atom—the smallest unit of an element.

automated feeding—mechanical animal feeding.

aviary—a house-like enclosure for birdkeeping.

AVMA—American Veterinary Medical Association; the professional organization for veterinary medicine.

B

backgrounding—the production system that takes a calf from the time of weaning to the feedlot phase; weight is added before the animal goes to a feedlot.

bacteria—one-celled organisms that may cause disease; disease examples are tuberculosis, brucellosis, and mastitis.

balanced ration—a ration that contains all the nutrients an animal needs in the correct proportions.

balling gun—a tool or instrument used for orally administering boluses or capsules to large animals.

barrow—a male hog castrated before sexual maturity.

basic research—research carried out to increase the knowledge and understanding of an area.

beef—the meat of cattle.

behavior—the reaction of an organism to a stimulus; how an organism responds to its environment.

billy—a male goat; also known as a buck.

biodiversity—inclusion of a wide variety of members in a population, especially a wildlife population.

biogas—gases produced by the biological breakdown of organic matter in the absence of oxygen.

biological—a type of medicine used to prevent disease.

biological filtration—the use of bacteria and other organisms to convert harmful materials in aquaculture water into forms that are less harmful.

biological oxygenation—the use of phytoplankton (tiny plants) to replenish oxygen in water.

biologist—a person with specialized training in living organisms.

biotechnology—the management of biological systems for the benefit of humans; biology is used to produce new products.

bird cage—a box-like enclosure for birdkeeping.

birding—watching and studying birds, especially song birds.

birdkeeping—raising birds in captivity, especially ornamental birds.

bitch—a female dog.

bivalve mollusk—any mollusk species with a hinged shell.

blackleg—an acute, highly infectious disease that primarily affects young cattle; usually results in death.

boar—a sexually mature male hog.

Bos indicus—cattle of Indian origin; descendants of the zebu and have humps on their necks, large droopy ears, and loose skin; includes cattle with Brahman blood.

Bos taurus—cattle of European origin; includes common breeds, such as Hereford and Angus.

botany—the study of plants.

brackish water—water that is a mixture of saltwater and freshwater.

brand—a permanent method of cattle identification involving burning a scar on the hide of cattle.

breed—a group of animals of the same species that share common traits.

breeding—helping animals reproduce and grow so the desired offspring result.

breeding period—the time of the year when the female has estrus periods and estrous cycle.

broiler—a young meat-type chicken between 5 and 12 weeks of age.

broodfish—a sexually mature fish kept for reproduction.

browse—woody and broadleaf plants; forage for goats, deer, and certain other animals.

brucellosis—a disease of the reproductive tract of cattle, sheep, goats, and hogs that causes abortion; also known as Bang's disease.

buck—a male goat; also known as a billy.

business plan—a written plan on how an enterprise will be operated to help assure success.

by-product—a product made from a part of an animal that is not used for food.

by-product feeds—feeds made from by-products.

C

cage—a structure for confining aquacrops in water; a cage floats on the water.

calorie—a measure in nutrition of the amount of heat required to raise the temperature of water 1°C.

candling—shining light through an egg to see its features.

capon—a castrated male chicken.

carbohydrate—food component that provides energy; includes sugars, starches, and cellulose.

carbon credit—a measure of carbon release that is assigned a monetary value.

carbon sequestration—the capture and storage of carbon in the soil.

care—the appropriate maintenance of an animal, including food, water, shelter, and health upkeep.

career—the general direction that a person takes with work.

carnivore—a flesh-eating animal.

carrying capacity—with wildlife, the number of animals a given habitat area can support.

castration—removal or destruction of the testicles of a male so it does not breed.

cavy—a guinea pig.

cell—the basic unit of life; the building block for organisms.

cell division—the process of a cell splitting into two cells.

cell specialization—differences in cells that allow them to perform unique activities.

chammy—soft, pliable leather made from the skin of sheep and goats.

chemical filtration—the use of chemical processes to filter water; activated charcoal is often used for chemical filtration in an aquarium.

chevaline—the meat of a horse.

chromosome—one of the thread-like parts of a cell nucleus containing the genetic material.

circulatory system—the body system that moves nutrients, oxygen, and metabolic wastes.

circus—a variety show with animal and human performers.

citizenship—participation as a member of the population of a nation, such as the United States, and assumption of the responsibilities and rights associated with legal residency.

classification system—a scientific approach used to distinguish between animals of different species.

clinical research—research that takes place in a medical clinic or other laboratory.

cloning—the production of one or more exact genetic copies of an animal.

coccidiosis—a parasitic disease affecting poultry; prevented and treated with anticoccidials in feed and water.

cock—a male chicken one year of age or older.

cockerel—a male chicken under one year of age.

cold housing—the use of unheated buildings to house animals.

colostrum—the first milk given by an animal after parturition; baby animals need colostrum to help develop disease immunity.

common name—the name used for an animal species in everyday conversation.

companion animal—a domesticated animal kept by humans for enjoyment in a long-term relationship.

companion fish—ornamental fish kept for human companionship or as a pet.

concentrate—a feed high in energy, such as corn, oats, or wheat.

conception—the time when an egg is fertilized to create new life.

confinement method—the raising of sheep, hogs, and other animals in confined areas; complete environmental control may be used.

conformation—the type, shape, and form of an animal.

connective tissue—tissue that holds and supports body parts.

consumer—a person who uses goods and services.

consumption—the use of products and services that have value.

contagious disease—a disease spread by direct or indirect contact.

contract production—production under an agreement made between a producer and a buyer before the hogs or other animals are raised.

copulation—the sexual union of a male animal and a female animal.

cow-calf system—a cattle production system that involves raising calves for sale to other growers.

creep feeding—feeding calves supplementally so they grow faster while still nursing.

cria—a young llama.

crossbreeding—the mating of animals of the same species but of a different breed.

crustacean—an animal with an exoskeleton, such as a shrimp or lobster.

culling—removing dairy cows from the herd based on established criteria.

curry comb—an oval-shaped plastic or metal device used to loosen sweat, manure, and other foreign materials in animal hair.

cutability—the amount of saleable retail cuts that may be obtained from a carcass.

D

dairy cattle—cattle raised and kept for milk production.

Dairy Herd Improvement Program—a national dairy production testing and record-keeping program.

dairy product—milk or any product in which milk is the major ingredient.

debeaking—removing the tips of beaks so birds cannot peck each other.

decision making—choosing from available alternatives, especially as related to a business enterprise.

declawing—removing the claws from the feet of a cat.

dehorning—removing the horns from horned breeds of cattle.

deoxyribonucleic acid (DNA)—a protein-like nucleic acid on genes that controls inheritance.

diagnostic ultrasound—a noninvasive way of imaging soft tissues in the body; also known as ultrasonography.

diet—the type and amount of feed an animal receives in its ration.

digestive system—the system that breaks food into smaller parts for use by the body.

disease—a disturbance of body functions or structures.

disinfectant—a cleaning substance that destroys the microbial causes of disease.

disposal pit—a pit in the earth for disposing of dead birds; also known as a compost pit; microorganisms decompose the dead birds.

dissolved oxygen—gaseous oxygen suspended in water; used by aquatic organisms; measured as DO.

diurnal—in relation to animals, being awake in the daytime and sleeping at night.

DNA sequencing—determining the order of nucleotides in a DNA fragment.

docking—cutting of all or part of the tail; with sheep, docking keeps manure from soiling the wool.

doe—a female goat; also known as a nanny.

dominant trait—a trait that covers up or masks the allele for a recessive trait.

dose syringe—a tool or instrument for orally administering liquid medicine to animals.

down—the soft, feathery covering on young poultry.

downer cow—a cow that cannot walk because of physical injury, illness, or calving.

draft—an animal's pulling to move an object.

draft animal—an animal trained and used for pulling heavy loads.

draft horse—a horse standing 14.2 to 17.2 hands high and weighing more than 1,400 pounds (635 kg).

drake—a mature male duck.

drenching—giving liquid medications orally.

driving horses—a group of horses used for showing and entertaining.

drug—any substance used to promote life processes in maintaining health and treating disease.

dry cow—a cow that has stopped producing milk; most cows in a herd produce milk 305 days for each lactation.

dual-purpose breed—a breed of cattle raised for both meat and milk.

dubbing—removing the comb and wattles on day-old chicks.

duckling—a young duck that has not grown feathers.

E

earthen pond—a water impoundment dug into the earth; small ponds are used for ornamental fish.

ectotherm—an animal that adjusts its body temperature to its environment.

effluent—water remaining from a process or activity, such as wastewater treatment.

egg—the sex cell of a female animal.

egg injection—a method of vaccinating chicken embryos in the egg on the eighteenth day of incubation.

egg-laying fish—fish that reproduce by laying eggs.

elastrator—a process of castration and docking used primarily in beef cattle and sheep; it involves placing a rubber-band type of device around the scrotum or tail, which cuts off circulation so the scrotum or tail dries up and falls off; a bloodless method of castration and docking.

electromagnetic radiation—the process of using photons of energy traveling at the speed of light to produce an image on film; also known as X-ray.

embryo—a young animal developing in the reproductive tract of a female from fertilization to the completion of differentiation.

embryo transfer—moving an embryo from the reproductive tract of one female animal to that of another female animal of the same species.

employability skills—the abilities an individual needs to gain and maintain employment.

employment—the primary way people use time and earn a living.

endangered species—animal species that are close to becoming extinct.

endocrine system—the body system that secretes hormones for regulating metabolism, growth, and reproduction.

endotherm—an animal that maintains a constant body temperature; sometimes called a warm-blooded animal.

entrepreneur—a person who takes risk by trying to meet the demands of people for a product.

entry-level job—a job that requires little or no previous experience.

environment—the surroundings of an organism.

epithelial tissue—tissue that covers a body surface.

equine sleeping sickness—a viral disease of horses, mules, and wild rodents; transmitted by biting insects.

equitation—the riding and managing of horses; horsemanship.

estrous cycle—the time between periods of estrus.

estrous synchronization—the act of bringing a group of female animals into heat at the same time.

estrus—the time when the female is receptive to the male and will stand for mating; also known as heat.

ethanol—a flammable, colorless compound commonly known as ethyl alcohol.

euthanasia—the act of providing an easy and painless death, thus relieving an animal of suffering.

ewe—a female sheep.

excretory system—the system that rids the body of wastes.

exotic animal—an animal that is not native or is being given a new use; exotic animals often are not domesticated.

exotic animal dealer—a buyer and seller of exotic animals.

exotic animal investment species—an exotic animal raised in hopes of making much money and giving a good return on investment.

expected progeny differences (EPD)—data on an animal that provides estimates of genetic value; identifies traits that could be passed to offspring.

expressing—manually squeezing the eggs from a female fish; sperm may be squeezed from a male using a similar procedure; also applies to veterinary medical procedures to empty glands and other body structures.

external parasite—a parasite that lives on the external parts of its host.

F

factory farm—a large-scale facility that efficiently produces animals in a controlled environment.

farm flock method—the keeping of a flock of sheep on a farm to produce market lambs.

farrier—a person who cares for the feet of horses, especially by designing and fitting shoes.

farrowing—the process of a sow giving birth to pigs.

fat—food component that provides energy; contains 2.25 times more energy than carbohydrates; the form in which excess energy is stored by an animal.

fat-soluble vitamin—a vitamin stored in body fat and released by the fat as the body needs it.

feather—long hair on the legs of horses.

feed—what an animal eats to get nutrients.

feed analysis—the process of determining the nutrients in a feedstuff or manufactured feed.

feeder pig—a pig weighing 30 to 60 pounds sold to another farm to feed to market weight.

feedstuff—an ingredient used in the feed for animals.

feral animal—an animal that was once domesticated but now lives in the wild.

fertilization—the union of sperm and egg.

fetus—a young animal developing in the reproductive tract of a female after the organs have begun to form; the fetus matures for birth.

fiber—the material left after food has been digested.

filly—a female horse under three years of age (four years of age for Thoroughbreds).

filtration—the process of removing solid materials from water.

fin—a structure on a fish that helps with locomotion and balance in the water.

fingerling—an immature fish larger than a fry.

finishing system—a beef cattle production system that feeds growing calves for slaughter.

first aid—providing limited care for an injury or illness by a lay person to preserve life, prevent further injury, and promote recovery.

fish—vertebrate aquatic animals.

fledgling—a young bird that has left the nest but cannot feed on its own.

floating—using a file to remove the sharp edges on the teeth of horses.

foal—a young horse of either sex that has not been weaned.

foaling—the process of a female horse giving birth.

food fish production—the raising of fish for use as human food.

foot and mouth disease—a viral disease affecting animals with cloven hoofs; there is no known treatment.

forage—a feed made mostly of leaves and tender stems of plants.

forager—a bovine that eats grasses and plants and makes good use of these foods in meeting its nutritional needs.

free access—the situation in which animals eat all the feed they want; feed is always available.

freshwater—water with little or no salt content.

freshwater ornamental fish—ornamental fish that live in freshwater.

frog—the V-shaped elastic pad in the middle of the sole of a horse's hoof.

fry—young, newly hatched fish.

functional type—the traits of a cow that are good enough to allow her to complete a useful life in a dairy herd.

fungi—one of five kingdoms of living organisms; fungi are unicellular (one-celled) organisms; some fungi may cause disease.

G

gaggle—a group of geese.

gainful employment—employment that provides benefits, such as salary or wages.

gait—the way a horse moves (walks, runs, etc.).

gallop—a four-beat gait in both Western and English riding.

game—animals hunted for sport and food.

gamete—a sex cell that can unite with another gamete to form a zygote.

gander—a mature male goose.

gelding—a male horse castrated before reaching sexual maturity.

gene—a segment of a chromosome that contains a hereditary trait of an organism.

genetic code—the sequence of nitrogen bases in a DNA molecule.

genetic engineering—a form of biotechnology in which genetic information is changed to make a new product.

genetics—the laws and processes of inheritance by offspring from parents.

gene transfer—the moving of a gene from one organism to another.

genome—the genetic material for an organism.

genotype—the genetic makeup of an organism.

gestation—the period of pregnancy.

gilt—a young female hog that has not farrowed.

gizzard—a muscular organ inside a bird that grinds food.

glucose—a simple sugar in food that is the source of energy for most cells.

goal—what an individual wants to achieve.

goal setting—establishing goals to be achieved and ways and means of reaching them.

gosling—a baby goose of either sex.

grade cattle—cattle that are not registered; may be purebred or of mixed breed.

grass-fed beef —animals whose diet is primarily forage (pasture).

greenfood—nondried vegetables or plant materials with sap still present; examples are cabbage and green peas.

greenhouse effect—an increase in atmospheric gases resulting in rising global temperatures and overall changes.

gregarious behavior—the instinct of animals to form groups, such as flocks or herds.

grit—finely ground abrasive material fed to birds to promote grinding in the gizzard.

grooming—washing, combing, brushing, trimming, and otherwise caring for a dog or other animal.

growth—increase in size of muscles, bones, and organs of the body of an animal.

grubs—internal parasites caused by heel flies; they move through the body, forming bumps under the skin on the backs of cattle; also known as warbles.

guard dog—a dog trained to protect property.

H

habitat—a place where animal wildlife and plant wildlife live.

hairball—a wad of hair that collects in the digestive tracts of cats; hair is from licking their coats while self-grooming.

halter broken—the state in which an animal is comfortable being led with a halter and responding to commands of its handler.

hand—a measurement for horses equal to 4 inches (10.2 cm).

handler—a person who trains and works with K-9 dogs or other animals.

harness—the assemblage of attachments placed on an animal to provide control and allow it to pull.

harvesting—the preparation of animals and animal products for human use and may include one or more functions such as capturing, grading, killing, fabricating, processing, packing, sorting, and transporting.

hatchery—a place where eggs are incubated for hatching.

health—the condition of an animal; how well the functions of life are performed.

heart murmur—an irregular sound caused by disruption of normal blood flow within the heart.

herbivore—an animal that eats foods from plant sources.

herding dog—a dog trained to assist in herding sheep and other animals.

heredity—the passing of traits from parents to offspring.

heterozygous—having different alleles for a particular trait.

hippotherapy—human disability treatment that uses the movement of a horse as therapy.

hog cholera—a viral disease of swine; nearly eradicated in the United States.

homeostasis—a characteristic of animals with organ systems in which a relatively constant internal environment is maintained.

homogenization—the breaking of fat droplets in milk into very small particles so they stay in suspension.

homozygous—having similar alleles for a trait.

horsemanship—equitation; the riding and management of horses.

hound—a dog that tracks; i.e., it follows scent left by animals or people.

hunting—catching or killing wild animals for sport or food; guns or other weapons are often used.

hunting and jumping horse—a horse specially bred and trained for use in fox hunting.

hutch—a cage or similar structure specially designed for keeping rabbits.

hyperthermia—higher than normal body temperature.

hypothermia—lower than normal body temperature.

I

immunity—resistance of an animal to disease.

immunoassay—a procedure to determine the presence of antigens, antibodies, and proteins.

immunoglobulin—an antibody passed from a cow to her calf in colostrum; provides passive immunity to disease.

implant—small pellet-type material placed underneath the skin for the slow release of medicine, growth regulators, or other substances.

incinerator—a device for burning dead chickens or other animals.

incubation—the time required for the embryo to develop in a fertile egg; with fish it is the time between spawning and hatching.

information—knowledge or news that has been communicated.

injection—a shot; the administering of medications directly into the bloodstream, tissues, muscles, or body cavity; normally performed with a hypodermic needle and syringe.

insemination—the act of placing sperm in the reproductive tract of a female.

integumentary system—the skin and other outer covering that protects an animal and helps regulate temperature and body processes.

internal parasite—a parasite that lives inside its host.

invertebrate—an animal that does not have a backbone.

isolation—the separation of diseased and nondiseased animals.

issue—a problem or subject that has more than one point of view.

J

jack—a male donkey of breeding age.

jenny—a female donkey.

job—specific work; work carried out at a site.

jog—a two-beat horse gait.

K

kid—a young goat under a year of age.

kidding—the process of a doe (nanny) giving birth.

Kingdom Animalia—the major division of living organisms that contains all animals.

kitten—a baby cat.

kittening—the process of a female cat giving birth.

K-9—a trained police dog.

L

laboratory animal—an animal raised and/or used in a research laboratory.

laboratory testing—the observing of samples of body fluids or tissues for abnormalities.

lactation—the secretion of milk by the mammary glands; the production of milk by a female mammal.

lactation ration—a ration high in nutrients needed by lactating animals.

lamb—a young sheep less than one year old; the meat from a young sheep.

lamb feeding—specialized lamb production in which the lambs go to feedlots after weaning.

lambing—the process of a ewe giving birth.

layer—a mature female chicken that is producing eggs; a chicken kept for egg production.

leptospirosis—a bacterial disease of cattle, sheep, and other animals.

lice—species of external biting and sucking parasites of cattle, hogs, and other animals.

life science—the science dealing with living things.

light horse—a horse standing 14.2 to 17 hands and weighing 900 to 1,400 pounds (408 to 635 kg).

limiting factor—the missing component of a wildlife habitat without which wildlife cannot survive.

linear evaluation—the coding of 15 primary traits of dairy cattle for use in corrective mating.

lipid—a nutrient that can be dissolved with ether (ether is a solvent used in nutrition research).

litter—the floor covering in a poultry house; also the floor coverings in other animal shelters; a group of young animals born at a single birth, such as pigs or kittens.

litter pan—a container with absorbent material for pets to use in urinating and defecating.

livebearing fish—fish that reproduce by giving birth to live young.

livestock—animals produced on farms and ranches for food and other products.

living research animal—a living animal used in research.

long-haired cats—a class of cat breeds having long hair.

lope—a three-beat horse gait.

lymphatic system—the body system that circulates a fluid known as lymph; lymph protects the body from disease.

M

maintenance—no loss or gain in weight by the body.

malnutrition—a condition that results when an animal has a diet that does not meet its daily needs over a sustained period of time.

mammal—a class of animals whose offspring are fed with milk secreted by the mammary glands of the female.

management—the use of resources to achieve the goals or objectives of an enterprise.

management-intensive grazing—a pasture system that concentrates large numbers of dairy cows on small pasture areas to harvest forage by grazing plant leaves and stems; the cattle are moved to another pasture after 12 to 48 hours of grazing.

marbled—having intramuscular fat; makes beef more palatable to humans.

mare—a mature female horse four years of age or over (five years of age or over for Thoroughbreds).

mariculture—the production of aquacrops in saltwater.

market hog—a young hog 5-8 months of age and weighing 220-250 pounds harvested for human food products.

mastitis—a bacterial disease of the mammary glands; the udder of cattle may be warm and hard to the touch; results in lost dairy production.

maternity box—a box prepared for giving birth, typically for cat parturition.

mathematical and computer model—a model for carrying out research related to animals without using animals; the model mimics animal behavior.

meat animal—an animal raised for meat.

meat animal by-product—a product made from the parts of an animal that is not used as food.

meat-type hog—a hog whose carcass gives the greatest amount of lean meat in the areas of high-value cuts.

mechanical filtration—the use of devices to remove particles from the water and keep the water clear.

mechanical oxygenation—the use of devices that bubble or splash water to add dissolved oxygen.

medication—a substance used to prevent, control, or treat disease.

meiosis—cell division for sexual reproduction.

metabolic disorder—a disease related to nutrition.

methane—an abundant gas found in the atmosphere and a component of natural gas.

microinjection—a process of injecting DNA into a cell using a fine-diameter glass needle and a microscope; used to produce transgenic animals.

milk fever—a metabolic disorder characterized by low blood calcium and paralysis of cows (primarily dairy cows).

milking system—how milking is accomplished on a dairy farm, such as the equipment used and how the cows are positioned for milking.

mineral—an inorganic element needed for a healthy body.

mitosis—cell division for growth and repair.

mixed-breed dog—a dog of unknown ancestry and likely with parents of different breeds.

models for animal research—controls over the use of laboratory animals to assure that the use will contribute to knowledge and not abuse the animals.

mohair—a product from Angora goats used in making clothing.

molecular biotechnology—the science of changing the structure and parts of cells to change an organism.

molecule—the smallest unit of a substance.

Mollusca phylum—the phylum made up of animals with hard shells.

mollusk—an aquatic animal with a thick, hard shell.

molting—the process of poultry shedding and renewing feathers.

MSDS— an abbreviation for Material Safety Data Sheets; a filing and retrieval system for safety information on substances that may pose hazards.

mule—a hybrid offspring of a male ass (jack) and a female horse (mare).

multi-purpose dog—a K-9 dog used for more than one activity.

muscular system—the body system that creates bodily movements; acquires materials and energy.

muscular tissue—tissue that creates movement of body parts.

mutation—a change that naturally occurs in the genetic material of an organism.

mutton—meat from a sheep more than one year old.

N

nail bed—the growing point of a nail, such as the nail on a dog's foot.

nanny—a female goat; also known as a doe.

natural insemination—the act of a male animal depositing semen in the reproductive tract of a female during copulation.

natural selection—the situation in which animals breed without control by humans.

needle teeth—eight sharp teeth in pigs at birth; usually clipped off to prevent injury to the sow.

nervous system—the body system that coordinates body functions and activities.

nervous tissue—tissue that responds to stimuli and transmits nerve impulses.

neutering—the process of sexually altering animals.

niche—the special place in a habitat where every animal fits.

nitrogen cycle—the process of converting animal wastes into ammonia, ammonia into nitrites, and nitrites into nitrates; naturally occurs in water.

nocturnal—in relation to animals, being awake at night and sleeping in the daytime.

noncontagious disease—a disease that is not spread by animal contact; includes nutritional, physiological, and morphological diseases.

non-game animal—an animal that does not provide products, such as meat and fur.

non-human primate—a primate that closely resembling the human; often used in research; examples are monkeys and chimpanzees.

non-living research system—the use of mechanical models that resemble animal activity.

non-sporting dog—a dog developed for a specific purpose; primarily used as a pet.

nutrient—a substance that is necessary in order for an organism to live and grow.

nutrient dense—a food or feed with large amounts of nutrients relative to calories.

nutrition—food needs and how food is used by an animal; the process by which an animal eats and uses food.

O

occupation—specific work that can be described and that has similar duties in different locations.

omasum—the third compartment in the stomach of a ruminant; filters food materials as they move into the next compartment.

omnivore—an animal that eats foods from both plant and animal sources.

oocyte—an immature egg cell in the ovary of a female.

oocyte transfer—the process of transferring an oocyte from a donor to a recipient in which fertilization will occur.

operculum—the structure on a fish that covers the gills.

oral medication—a medication given through the mouth.

organ—a group of tissues with a similar function.

organism—any living thing.

organismic biotechnology—the science of improving intact or complete organisms without artificially changing their genetic makeup.

organ system—a system formed when two or more organs work together to perform an activity.

ornamental fish—fish kept for their appearance and personal appeal.

orphaned lamb—a lamb whose mother has died.

ovary—the primary reproductive organ of the female.

ovulation—the release of a mature ovum by the ovary.

ox—a bovine draft animal.

oxygenation—the process of keeping adequate dissolved oxygen in water.

P

paddock—a small fenced pasture or dry lot area for horses to exercise and, possibly, graze.

palatability—how well an animal likes food; how food feels in the mouth and on the tongue.

parasite—a multicellular animal organism that lives in or on another animal.

particle injection—using a microprojectile unit to shoot tiny particles coated with DNA into cells; a genetic engineering technique.

part-time animal producer—a person who raises animals on a part-time basis; the person may have income from other sources.

parturition—the process of giving birth.

pasteurization—the heating of milk to destroy bacteria.

patriotism—admiration for and loyalty to one's country.

peacock—a male peafowl.

peahen—a female peafowl.

Pearson square method—a simple way to hand calculate a ration for an animal.

pedigree—the record of an individual's heredity.

pelt—the whole hide of an animal with the fur attached.

pen—a structure that confines aquacrops in water; pens are attached to the earth at the bottom of the water.

penicillin—a well-known antibiotic.

per capita consumption—the average amount consumed by a person.

performing animal—an animal taught to go through a routine of unusual activity.

permanent pasture—a pasture comprising perennial grasses and other plants; a pasture grows from year to year without planting.

personal protective equipment (PPE)—equipment individuals use to protect themselves from injury, including goggles and safety glasses, ear plugs and muffs, gloves, shoes and boots, hard hats, and clothing.

pet carrier—a carrying case to carry a pet safely and securely.

pharmaceutical—a medicine used to treat disease.

pharmacology—the study of drugs.

phenotype—an organism's physical or outward appearance.

physiology—the study of the functions of cells, tissues, organs, and systems of an organism.

piglet—a baby pig.

plankton—tiny plants and animals that float in water; serves as food to some aquatic animals.

pleasure animal—an animal kept as a pet or to use in other ways for the enjoyment of people.

plug—a horse with poor conformation and common breeding.

polled—naturally without horns.

polo mount—a horse especially bred for use in a polo match.

pond—a water impoundment made by building earthen dams or levees.

pony—a horse standing under 14.2 hands high and weighing 500 to 900 pounds (227 to 408 kg); a small horse.

porcine somatotropin (pST)—a growth hormone for swine.

porcine stress syndrome—a nonpathological disorder in heavily muscled hogs that results in sudden death.

poult—a young turkey not grown to the point where its sex is readily identifiable.

poultry—fowl raised on farms for use as food.

poultry science—the study and use of science in raising poultry.

power—the combination of pulling capacity and speed; measured in horsepower.

practice—the routine operations and procedures carried out by veterinarians.

preconditioning—preparing animals for stress associated with hauling.

predator—a animal that lives off other animals; kills, mauls, or preys on them.

predicted transmitting ability—an estimate of the traits an animal will transmit to its offspring.

prefix—a syllable, group of syllables, or word placed at the beginning of a root word.

pregnant—the condition in which a female carries unborn young.

primate—an animal with an opposable thumb used for grasping things.

probability—the likelihood or chance that a trait will occur.

probe—a tool for measuring backfat on a hog.

production cycle—the complete cycle in the production of a crop or growth of an animal.

production intensity—the number or weight of aquatic species in a volume of water; biomass.

progeny—the offspring of animals.

prolific—producing large numbers of young or offspring.

protein—a food substance important for animal growth, maintenance, reproduction, and other functions.

protoplasm—the mixture of water and proteins in cells that carry out life processes.

protozoa—unicellular organisms; simplest form of animal life; some protozoa may cause disease.

puberty—the time when sexual maturity is reached; the animal is capable of reproduction.

pullet—a young female chicken.

pulse—the heartbeat as felt through the walls of the arteries under the skin of an animal.

Punnett square—in genetics, a technique for predicting phenotype.

purebred—an animal eligible for registry in a breed association; its parents must meet standards of the association.

purebred flock—a flock of sheep of purebred breeding.

Q

queen—a female cat.

quill—a modified hollow hair that is stiff and has a point; found on hedgehogs and porcupines.

R

rabies—a viral disease in endothermic animals accompanied by frenzied behavior, fever, vomiting, diarrhea, drawn lips, and often foaming at the mouth; an autopsy of the brain is needed for accurate diagnosis.

racehorse—running or harness horses used in racing.

raceway—a water impoundment that uses flowing water.

ram—a male sheep kept for breeding purposes.

range band method—a production system used with sheep in which a herder moves the flock over a large area of land.

rare breed—a formerly popular breed of domesticated animal that is now disappearing.

ration—the food an animal consumes each day; the total amount of feed an animal is given in a 24-hour period.

ratite—a group of large, flightless birds, such as the ostrich and emu.

recessive trait—a genetic trait masked by a dominant trait.

recombinant DNA—gene splicing; genes are cut and moved to a cell to be altered.

record keeping—the recording of activities for safe keeping and for use in assessing accomplishment of goals; includes personal, professional, financial, and other records.

regurgitate—the process by which ruminants return eaten but unchewed food to their mouths for chewing.

reproduction—the process by which offspring are produced.

reproduction ration—a ration high in nutrients needed by breeding animals.

reproductive efficiency—timely and prolific replacement of a species.

reproductive system—the system by which new individuals of the same species are created.

reptile—an ectothermic animal with dry, scaly skin and with lungs for breathing; examples are lizards and snakes.

research—a systematic effort to answer questions; carefully planned and designed procedures are followed.

resource—a supply, a skill, money or property, or information needed or used to achieve goals.

respiratory system—the body system that regulates gas exchange of an organism.

restraint—control of an animal for examining, treating, grooming, or performing other practices.

reticulum—the second compartment of a ruminant's stomach; its many layers trap nonfood materials.

riding horse—a horse ridden for pleasure.

risk—the possibility of losing; in entrepreneurship, the possibility of losing financial investment.

roan—a mixture of white and colored hairs on cattle or horses.

roaster—a young chicken somewhat older and larger than a broiler.

rodent—an animal with two large upper front teeth for gnawing.

root word—a word or part of a word that is the basis for the meaning of a word; used in understanding veterinary medical terminology.

roughage—a feedstuff of stems and leaves; has high fiber content.

rumen—the first and largest section of a ruminant's stomach; receives ingested food materials.

ruminate—the act of a ruminant chewing its cud.

S

sac fry—newly hatched fish with the yolk sac still attached.

safety—preventing injury or loss by following practices that are wise and prudent.

safety animal—an animal that helps protect people and property.

salinity—the amount of salt in water; measured as ppt (parts per thousand).

saltwater—water that has a high salt content; typically 33 to 37 ppt salt.

saltwater ornamental fish—ornamental fish that live and grow in saltwater.

sanitation—using practices that prevent development of disease-causing organisms.

scheduled feeding—providing feed at certain times of the day.

science—knowledge about the world we live in.

scientific name—the name of an organism based on its taxonomy; usually two parts to name though some have three parts; written in italics or underlined.

scratching post—a device for cats to use in scratching; exercise post for cats.

scrotum—two-lobed sac in mammals that holds the testicles and helps regulate testicle temperature.

selection—choosing or picking the animals for a herd, especially with dairy cattle.

self-feeding—when animals have feed available all of the time.

service animal—an animal that assists people in living and work.

sexed semen—semen that has been prepared in a lab to produce all male or all female offspring.

sexual reproduction—union of sperm and egg to form a new individual.

shipping fever—an environmental disease of cattle and sheep caused by stress; more likely to affect underfed animals.

shoal—a group of five or more fish living together.

short-haired cats—a class of cat breeds having short hair.

single-purpose dog—K-9 dog used for just one activity.

sink—a place that holds or stores carbon.

skeletal system—bony structure that gives a framework for the body.

sleeping sickness—a viral disease of horses transmitted by insect and rodent bites; animals appear sleepy, cannot swallow, and die within two to four days.

smolt—immature salmon larger than a fry.

social ranking—the order that an animal falls within a group of animals of the same species.

solitary—living along; not needing companions.

sow—sexually mature female hog.

spawning—release of eggs by a female fish and subsequent fertilization by a male of the same species.

spawning nest—a place where fish spawn; artificial spawning nests are used in aquaculture.

spaying—surgical procedure that removes the ovaries and uterus of a bitch to prevent breeding; also used with other animals.

specific pathogen free—animals free of disease at birth and raised in an aseptic environment; most often associated with hogs but used with other species, including shrimp.

spent hen—hen no longer used for egg production; processed into cooked poultry products.

sperm—sex cell of the male animal.

sporting dog—dog trained to help hunters find and retrieve game.

spring water—water from natural openings in the earth; similar to well water.

stallion—a mature male horse four years of age or over (five years of age or over for Thoroughbreds).

steer—a male bovine castrated while young and before secondary sexual characteristics had developed.

stem cell—a biological cell of a multicellular organism that can divide and differentiate into specialized cells.

stethoscope—an instrument used to hear and amplify the sounds of the heart, lungs, and other internal organs.

stock horse—popular mixed-breed horses often used with cattle.

stomach—muscular organ that stores and prepares food for digestion in the small intestine.

stud dog—a male dog used for breeding.

stud horse—male horse kept specifically for breeding.

suffix—a syllable, group of syllables, or word added at the end of a root word.

superovulation—getting a female to release more than the usual number of eggs during a single estrous cycle.

supervised experience (SE)—planned activities related to classroom and lab learning that are carried out outside of normal class time; also known as supervised agricultural experience (SAE).

supplement—a feed material high in specific nutrients; used to assure an animal has sufficient nutrients.

surface runoff—excess water from precipitation.

swim bladder—a small, balloon-like organ that helps a fish maintain buoyancy in the water.

synthetic biology—using chemical substances to create systems with some of the characteristics of living organisms.

synthetic seawater—made for saltwater aquaria by using a commercially available mix.

system—a combination of parts or processes that form one complete function or collection; includes social systems, biological systems, digestive systems, mechanical systems, and others.

T

tail docking—clipping the tail from baby pigs to prevent tail biting.

tapwater—the water from faucets in homes and offices; usually must be treated before using with fish.

target tissue—the site in an organism where drug action is needed to treat disease.

taxonomy—science of classifying organisms.

temporary pasture—forages used for grazing one season and replanted.

terrier class—a class of 25 breeds of terriers.

territory—the area an animal defends and protects.

testicle—primary reproductive organ of the male.

thermostat—a control device for maintaining a constant environmental temperature in air or water.

tissue—a cluster of cells that are alike in structure and activity.

tom—male turkey.

tomcat—male cat.

topical medication—a medication placed on the skin or surface of an animal.

total mixed ration—a ration that provides all needed feed ingredients in each mouthful of feed a cow eats.

toy breed—small dogs used for companionship.

training—the process of helping a dog or other animal acquire desired habits or qualities.

training plan—a written statement of supervised experience intentions, including skills to be developed and other details.

training station—the location where placement supervised experience is carried out.

transgenic animal—an animal that has stably incorporated a foreign gene into its cells.

trapping—capturing animals for their products using traps that keep the animals alive until detected by the trapper.

triage—a system for making decisions about how an injury or injuries will be handled in an emergency situation.

tri-purpose animal—an animal used for work, milk, and meat; oxen is an example.

tropical fish—small, brightly-colored fish that are popular in home and office aquaria.

type production index—a system for ranking dairy cows based on overall performance.

U

udder—mammary glands, teats, and associated structures on female mammals.

ultrasonics—equipment that uses bursts of high-frequency sound in assessing the conditions of animals, including amount of lean meat and reproductive condition.

urbanization—building cities, suburbs, and other developments for the convenience of people; often destroys wildlife habitat.

V

vat—large tank made of concrete, fiberglass, or similar material for growing fish.

vector—a carrier of disease; an intermediary that transfers disease from one animal to another.

vertebrae—bones that form a segmented spinal column in animals.

vertebrate—animal with a backbone.

vertical integration—when an agribusiness is involved in more than one step in providing agricultural products; chicken industry is a good example.

veterinarian—an individual with a doctor of veterinary medicine degree who provides veterinary medical care.

veterinary assistant—a staff member in a veterinary clinic or hospital at the level of a clinical aide with skill in many areas of animal care; less skill and education than a veterinary technician or veterinary technologist.

veterinary medicine—the discipline that deals with the health and well-being of animals; a branch of medicine dealing with animals.

veterinary technician—a technical-level support person in a veterinary medical clinic or hospital; should complete a two- or three-year program accredited by the AVMA.

veterinary technologist—a support professional in veterinary medicine with a four-year degree from an accredited veterinary technology program.

veterinary technology—the science and art of providing professional support service to veterinarians in their practice; training required as per the American Veterinary Medical Association guidelines.

virus—a tiny particle too small to be seen with an ordinary microscope that often causes disease.

vitamin—organic food substances needed in small amounts for good health; important in body functions.

W

walk—a four-beat horse gait.

warm housing—heating buildings in which animals are kept in the winter.

water facility—structures in which aquacrops are grown.

water quality—suitability of water for a particular use.

water-soluble vitamins—vitamins dissolved by water and need to be consumed each day.

weaning—withdrawing the need or opportunity for a puppy or other young animal to nurse its mother.

well water—water pumped from aquifers deep in the earth.

wether—male sheep castrated before sexual maturity.

whelp—process of a bitch giving birth.

whelping box—a specially made box for dog birthing.

withdrawal time—period before slaughter during which an animal should not receive a medication or other restricted substance.

wool—soft coat on sheep used as a fiber for human clothing and other products.

work ethic—how people feel about work.

working dog—dog used by people to get work done.

X

X-ray—see "electromagnetic radiation".

Y

yoke—(1) wooden bar in a harness that holds two animals together so they work together as a team; primarily used with oxen; also a device placed on the necks of animals to keep them in a pasture or pen. (2) the center, yellow part of a chicken egg; provides nourishment for a developing embryo.

Z

zoo—zoological garden; park-like area where animals are kept for viewing.

zoo curator—person trained in exotic animal zoology who looks after the animals in a zoo.

zoology—the study of animals.

zoonosis—any infectious disease that can be transmitted from animals to humans and, in some cases, vice versa (see zoonotic disease).

zoonotic disease—any disease that humans can contract from other animals.

zygote—a fertilized ovum.

Bibliography

American Kennel Club, The. *The Complete Dog Book*, 20th ed. New York: Wiley Publishers, Inc., 2006.

Animal Health. Animal and Plant Health Inspection Service USDA. accessed on February 24, 2008, at http://www.aphis.usda.gov/animal_health/animal_dis_spec/swine/

Blood, D. C., and V. P. Studdert. *Comprehensive Veterinary Dictionary,* 2nd ed. New York: W. B. Saunders, 2000.

Boleman, Larry L., Dennis B. Herd, and Chris T. Boleman. *Managing Beef Cattle for Show.* College Station: Texas Cooperative Extension Service, Publication AS 1-2, 2001.

Campbell, Karen L., James E. Corbin and John R. Campbell. *Companion Animals: Their Biology, Care, Health, and Management*. Upper Saddle River, NJ: Pearson Prentice Hall, 2005.

Carlson, Delbert G., and James M. Giffin. *Cat Owner's Home Veterinary Handbook.* New York: Howell Book House, 1995.

Cheeke, Peter R. *Applied Animal Nutrition,* 2nd ed. Upper Saddle River, New Jersey: Prentice Hall, 1999.

Cheeke, Peter R. *Contemporary Issues in Animal Agriculture,* 2nd ed. Danville, IL: Interstate Publishers, Inc., 1999.

Cochrane, Phillip E. *Comparative Veterinary Anatomy and Physiology*. Clifton Park, NY: Delmar Learning, 2004.

Colville, Thomas, and Joanna M. Bassert. *Clinical Anatomy and Physiology for Veterinary Technicians.* St. Louis, MO: Mosby, Inc., 2002.

Cunningham, Merle, Mickey A. Latour, and Duane Acker. *Animal Science and Industry*, 7th ed. Upper Saddle River, NJ: Pearson Prentice Hall, 2005.

Duno, Steve. *Guide to Cat Care.* New York: Dorling Kindersley Publishers, 2001.

Ensminger, M. E., and C. J. Hammer. *Ensminger's Equine Science,* 8th ed. Upper Saddle River, NJ: Prentice Hall Interstate, 2004.

Floyd, James G. Vaccinations for the Swine Herd, ANR-0902 publication, Alabama Cooperative Extension System. Accessed on February 24, 2008, at http://www.aces.edu/pubs/docs/A/ANR-0902/

Gibson, Jerry D. *Equine Science and Management.* Danville, IL: Interstate Publishers, Inc., 1999.

Gillespie, James R. *Modern Livestock and Poultry Production,* 7th ed. Albany, NY: Delmar Learning, 2004.

Han, Connie M, and Cheryl D. Hurd. *Practical Diagnostic Imaging for the Veterinary Technician,* 2nd ed. St. Louis, MO: Mosby, Inc., 2000.

Harrington, Rodney B. *Animal Breeding: An Introduction.* Danville, IL: Interstate Publishers, Inc., 1995.

Haynes, N. Bruce. *Keeping Livestock Healthy,* 4th ed. North Adams, MA: Storey Books, 2001.

Henkel, Keri, ed. *Occupational Guidance for Agriculture.* Minneapolis: Finney Company, 2002.

Heren, Ray. *The Science of Animal Agriculture*, 3rd ed. Clifton Park, NY: Thomson Delmar Learning, 2007.

Jackson, Peter, and Peter Cockcroft. *Clinical Examination of Farm Animals.* Oxford, England: Blackwell Publishing Company, 2002.

Kahn, Cynthia M., ed., *The Merck Veterinary Manual,* 9th ed. Whitehouse Station, NJ: Merck & Co., Inc., 2005.

Latimer, Kenneth S., Edward A. Mahaffey and Keith w. Prasse. *Veterinary Laboratory Medicine Clinical Pathology*, 4th ed. Ames, IA: Iowa State Press, 2003.

Lee, Jasper S., Ronald J. Biondo, Lyle E. Westrom, Jim Hutter, and Amanda R. Patrick. *Plants and Animals: Biology and Production.* Upper Saddle River, NJ: Prentice Hall Interstate, 2004.

Lee, Jasper S., Gary J. Burtle and Michael E. Newman. *Aquaculture: An Introduction,* 3rd ed. Upper Saddle River, NJ: Pearson Prentice Hall, 2005.

Lee, Jasper S., and Diana L. Turner. *AgriScience,* 5th ed. Upper Saddle River, NJ: Pearson Prentice Hall Interstate, 2010.

Lee, Stephen J., Christy Mecey-Smith, Elizabeth M. Morgan, Ray E. Chelewski, Randi Hunewill, and Jasper S. Lee. *Biotechnology.* Danville, IL: Interstate Publishers, Inc., 2001.

Marinelli, Deborah A. *Careers in Animal Care and Veterinary Science*. New York, NY: The Rosen Publishing Group, Inc., 2001.

McCrimmon, Dennis M., and Joanna M. Bassert. *Clinical Textbook for Veterinary Technicians,* 7th ed. St. Louis, MO: Saunders Elservier, 2010.

Palika, Liz. *Reptiles and Amphibians.* New York: Alpha Books, 1998.

Parker, Rick. *Equine Science*, 3rd ed. Clifton Park, NY: Thomson Delmar Learning, 2008.

Pasquini, Chris, Tom Spurgeon, and Susan Pasquini. *Anatomy of Domestic Animals,* 9th ed. Pilot Point, TX: SUDZ Publishing, 1995.

Pinney, Chris C. *The Illustrated Veterinary Guide,* 2nd ed. New York: McGraw-Hill, 2000.

Pratt, Paul W. *Principles and Practices of Veterinary Technology.* St. Louis, MO: Mosby, Inc., 1998.

Romans, John R., William J. Costello, C. Wendell Carlson, Marion L. Greaser, and Kevin W. Jones. *The Meat We Eat,* 14th ed. Danville, IL: Interstate Publishers, Inc., 2001.

Romish, Janet A. *An Illustrated Guide to Veterinary Medical Terminology*, 2nd ed. Clifton Park, NY: Thomson Delmar Learning, 2006.

Scanes, Colin G., George Brant, and M. E. Ensminger. *Poultry Science*, 4th ed. Upper Saddle River, NJ: Pearson Prentice Hall, 2004.

Sirois, Margi. *Principles and Practice of Veterinary Technology*, 3rd ed. St. Louis, MO: Mosby Elservier, 2011.

Stutzenbaker, Charles D., Brenda J. Scheil, Michael K. Swan, Jasper S. Lee, and Jeri Mattics Omernick. *Wildlife Management: Science and Technology,* 2nd ed. Upper Saddle River, NJ: Prentice Hall Interstate, 2003.

Swine Care Handbook. Des Moines, IA: National Pork Board, 2002.

Taylor, Robert E. and Tom G. Field. *Scientific Farm Animal Production,* 10th ed. Upper Saddle River, NJ: Pearson Prentice Hall, 2012.

Warren, Dean M. *Small Animal Care and Management,* 2nd ed. Clifton Park, NY: Delmar Learning, 2002.

(Source of data in this section: www.americanpetproducts.org/press_industrytrends.asp, accessed on October 22, 2015.)

Index

A

B

Q

R

S